ROGER O. P—

ROGER O. P—

Switched Reluctance Motor Drives

A Collection of Papers on Switched Reluctance Drives

Including: Motor Design, Finite-Element Analysis,
Variations on the Classical SR Motor,
Early Milestone Papers, Converters and Controls,
and Applications

PCIM REFERENCE SERIES IN POWER ELECTRONICS & INTELLIGENT MOTION

Advisory Board: Dr. S. Meshkat, Motion Research Inc./Robert Martinelli, Analytic Artistry/K. Kit Sum, Summit Electronics/Dr. T.J.E. Miller, University of Glasgow/Marvin W. Smith, Adree Technical Services.

Directed toward graduate and mid-career engineers, the PCIM Reference Series provides original information and significant papers covering the design of power conversion and intelligent motion systems. In addition, the Series includes similar information from English translations of outstanding works in other languages.

Advanced Motion Control Systems	Dr. S. Meshkat, Motion Research, Inc.
Elements & Applications of Servo Sensors	Y. Akiyama, Ph. D. and Y. Oshima, Ph. D.
Recent Developments In Resonant Power Conversion	K. Kit Sum, Summit Electronics
Smart Design for Power Conversion	R.M. Martinelli, Analytic Artistry
Smart Power Economics, Technology & Applications	Marvin W. Smith, Adree Technical Services
Switched Reluctance Motor Drives	Prof. T.J.E. Miller

OFFICIAL CONFERENCE PROCEEDINGS PUBLISHED BY OR ASSOCIATED WITH INTERTEC COMMUNICATIONS INC.

PCI '88 Dearborn	SATECH '88 Conference, October 3-7, 1988, Dearborn, MI, USA
PCI '88 Munich	PCI/MOTOR-CON '88 Conference, June 6-9, 1988, Munich, West Germany
PCI '87 Long Beach	SATECH '87 Conference, September 14-18, 1987, Long Beach, CA, USA
PCI '87 Munich	PCI '87 Conference, May 11-13, 1987, Munich, West Germany
PCI '86 Boston	SATECH '86 Conference, October 27-31, 1986, Boston, MA, USA
PCI '86 Munich	PCI '86 Conference, June 17-19, 1986, Munich, West Germany
PCI '85 Chicago	SATECH '85 Conference, October 21-25, 1985, Chicago, IL, USA
PCI '84 Paris	PCI '84 Conference, October 29-31, 1984, Paris, France
PCI '84 Atlantic City	PCI/MOTOR-CON '84 Conference, April 2-5, 1984, Atlantic City, NJ, USA
PCI '83 Orlando	PCI/MOTOR-CON '83 Conference, April 19-21, 1983, Orlando, FL, USA
PCI '82 Geneva	PCI/MOTOR-CON '82 Conference, September 28-30, 1982, Geneva, Switzerland
PCI '82 San Francisco	PCI/MOTOR-CON Conference, March 29-31, 1982, San Francisco, CA, USA
PCI '81 Munich	PCI '81 Conference, October 14-17, 1981, Munich, West Germany
PCI '80 Munich	PCI '80 Conference, September 3-5, 1980, Munich, West Germany
PCI/MOTOR-CON '83	PCI/MOTOR-CON '83 Conference, October 13-15, 1983, Geneva, Switzerland
MOTOR-CON '88 Dearborn	SATECH '88 Conference, October 3-7, 1988, Dearborn, MI, USA
MOTOR-CON '88 Munich	PCI/MOTOR-CON '88 Conference, June 6-9, 1988, Munich, West Germany
MOTOR-CON '87 Long Beach	SATECH '87 Conference, September 14-18, 1987, Long Beach, CA, USA
MOTOR-CON '87 Hannover	MOTOR-CON '87 Conference, April 1-3, 1987, Hannover, West Germany
MOTOR-CON '86 Boston	SATECH '86 Conference, October 27-31, 1986, Boston, MA, USA
MOTOR-CON '85 Hannover	MOTOR-CON '85 Conference, October 22-24, 1985, Hannover, West Germany
MOTOR-CON '85 Chicago	SATECH '85 Conference, October 21-25, 1985, Chicago, IL, USA
MOTOR-CON '84 Atlantic City	PCI/MOTOR-CON '84 Conference, April 2-5, 1984, Atlantic City, NJ, USA
MOTOR-CON '83 Orlando	PCI/MOTOR-CON '83 Conference, April 19-21, 1983, Orlando, FL, USA
MOTOR-CON '82 Geneva	PCI/MOTOR-CON '82 Conference, September 28-30, 1982, Geneva, Switzerland
MOTOR-CON '82 San Francisco	PCI/MOTOR-CON '82 Conference, March 29-31, 1982, San Francisco, CA, USA
MOTOR-CON '81 Chicago	MOTOR-CON '81 Conference, June 10-13, 1981, Chicago, IL, USA
IMS '87 Long Beach	SATECH '87 Conference, September 14-18, 1987, Long Beach, CA, USA
IMS '86 Boston	SATECH '86 Conference, October 27-31, 1986, Boston, MA, USA
IMS '85 Chicago	SATECH '85 Conference, October 21-25, 1985, Chicago, IL, USA
HFPC '88	High Frequency Power Conversion '88 Conference, May 1-5, 1988, San Diego, CA, USA
HFPC '87	High Frequency Power Conversion '87 Conference, April 21-23, 1987, Washington, D.C., USA
HFPC '86	High Frequency Power Conversion '86 Conference, May 28-30, 1986, Virginia Beach, VA, USA
Factory Electronics/Manufacturers	Factory Electronics Conference, November 14-17, 1983
Factory Electronics/End Users	Factory Electronics Conference, November 14-17, 1983

OTHER PUBLICATIONS BY INTERTEC COMMUNICATIONS INC.

PCIM Magazine	Established in 1975 for Power Electronics & Intelligent Motion Engineers
The Art and Practice of Step Motor Control	Albert C. Leenhouts, Litchfield Engineering Co.

Switched Reluctance Motor Drives

A Reference Book of Collected Papers

Edited by Prof. T. J. E. Miller

GEC Titular Professor in Power Electronics,
Scottish Power Electronics and Electric Drives (SPEED)
University of Glasgow
October 1988

INTERTEC COMMUNICATIONS INC.
Publishers - Ventura, CA (USA)

First published in 1988 by
INTERTEC COMMUNICATIONS INC.
2472 Eastman Avenue, Buildings 33-34
Ventura, California 93003-5774 (USA)

Intertec Communications Inc. is the exclusive distributor

©1988 Intertec Communications Inc.
Library of Congress #88-82364
ISBN #0-931033-17-9
Printed in the United States of America

Foreword

Even though the precise implications and complete range of applications of switched reluctance drives cannot yet be fully discerned, within the last year or two it has become apparent that they have the potential to significantly reorientate, technically and commercially, many industries which manufacture and use electrical machines.

Real knowledge about both theory and practice remains patchy and mainly thin; and by the same token considerable misinformation and, indeed, prejudice against them persists. It is therefore particularly timely that this bibliography with its informed commentary should now be made available.

It is possible to point out that the most obvious principle involved in the operation of switched reluctance motors was first demonstrated by Wheatstone in the early 1830s, and some have chosen to describe SR motors by reference to the stepping motor, but neither idea is really pertinent or helpful. Switched reluctance motors are, in the vogue phrase, something else again. The features which make them so: firstly, their unique combination of excellent performance characteristics, secondly, the distinctive (and subtle) design and control strategies which they require — have only been uncovered and developed in the last 10-15 years. And only during the last five years has something of the full significance of the performance capabilities and attractive cost considerations begun to be at all generally appreciated outside the small circle of pioneers.

Two things were fundamental to the emergence of switched reluctance motors. First was a demonstration that the physical laws governing electromechanical devices could provide for greater energy conversion to be obtained through interaction between electric currents and suitably shaped soft permeable material than through interaction between two sets of currents. Second was the timely availability of power semiconductor devices which could be used reliably and economically to implement the switching and control strategies necessary for the realisation of practical machine systems.

Machines depending upon the forces exerted on suitably configured iron were, of course, already familiar (though never very widely used) under the names of "variable-reluctance" or just "reluctance" machines. These were perceived as being much inferior in performance to machines using currents on both rotor and stator. This perception was, for a considerable period, wrongly presumed to reflect some fundamental law about singly as opposed to doubly excited machines and it had a serious influence on the early acceptance of SR technology. However it overlooked an important point of distinction in that the familiar reluctance machines were essentially "synchronous," whereas switched reluctance machines were "variable speed" (in the same sense as a DC machine). More importantly, it overlooked the fact that SR machines use magnetic structures and control strategies specially designed to maximize the continuously usable, space-rate-of-change of energy.

The term "switched reluctance" was created in 1980 in an effort to point up this distinction whilst at the same time avoiding an excessively cumbersome string of words (or their initial letters). Whilst a purist in the English language could object that "switched" refers not to "reluctance" but to the currents in the excited member of the machine, the term has found wide favour (and has removed the need that would otherwise have been felt by purist electrical engineers, to debate questions such as "what is a synchronous machine?" if some of the cumbersome alternatives had been used).

With regard to the nature of the switched reluctance drive, the most illuminating statement is that the drive is the *complete equivalent of a fully controlled (both armature current and field flux, under closed-loop conditions) separately-excited, DC machine.* This statement will be found to answer almost all questions and arguments about SR drive capability and quality. Insofar as the machine is fed from a switched DC supply and has no commutator or brushes, it could also be described as a brushless DC machine. However, this description does it less than justice because "brushless DC" is commonly used for machines employing permanent magnets and so falling significantly short of the degree of control inherently available in the SR drive.

To summarize the current situation with regard to SR drive technology, inevitably over simply, comments need to be made on three aspects. Firstly, *acceptance* of the technology has increased enormously during the last two years, and particularly the last year, amongst both users and suppliers. Secondly, *expertise* continues to reside principally with the one or two specialist companies or groups which have been developing the technology (and their experience in it) for as long as 15 years. More recently a small number of the top international companies (mainly over the last 5 years or less) nave invested very heavily indeed in research and development programs; and, very recently, one or two smaller companies have announced application specific products — for example for machine tool and servo drives — and numerous teaching institutions have perceived a rich vein for academic research. Thirdly, the *outstanding operating qualities* of SR drives have been established beyond dispute. These include specific power and torque outputs, efficiencies, torque/inertia ratios, controllability, programmability, very high or low speeds, wide speed range, "cold" rotors, robust mechanical and electronic designs...all available at highly competitive costs.

This situation, particularly so far as design and performance are concerned, is filled out in a full and balanced way by the present bibliography with its commentary. It will be immensely valuable to anyone seeking either an introduction or a comprehensive survey and the author is to be thanked and congratulated for this important contribution.

Two concluding comments are pertinent. The selection of papers is arguably the best possible and is as complete as it could reasonably be. However, a truly full exposition of the actual state-of-the-art in SR drives is not possible simply because commercial considerations put, and will continue to put, significant restrictions on what information is disclosed in the public domain.

Finally, SRD technology and its exploitation are, relative to future potential, in their infancy. Even those who have specialized for 15 years are in the lower reaches of a considerable learning curve. The coming few years will see an enormous build up of activity in design and in application made particularly intense by the uniquely wide range of circuit, machine and control possibilities, the particularly sophisticated and subtle design techniques which have to be used, and the exceptionally wide ranges in size, speed, and applications to which SR drives are fitted. We can expect to see, and benefit from, a new verison of this bibliography by the end of the decade!

Peter Lawrenson
Switched Reluctance Drives Limited, Leeds, UK
4 July 1988

PETER JOHN LAWRENSON - Biography

Currently:	Chairman and Chief Executive, Switched Reluctance Drives Limited, Leeds, UK Vice-President, Institution of Electrical Engineers Director, Dale Electric International plc and Simplex Electrical Limited Professor of Electrical Engineering, University of Leeds, UK
Professional Distinctions:	Fellow IEE (1974); Fellow IEEE (1975); Fellow, Royal Society (1982).
Prizes:	Numerous, including James Alfred Ewing Gold Medal (1983) and Esso Energy Gold Medal (1985).
Research and publications:	In excess of 100 papers, 18 patents and 2 books in the areas of electromagnetism, turbo generators, synchronous reluctance machines, superconducting AC generators, stepping motors and systems, switched reluctance drives.

Contents

Foreword, Professor P.J. Lawrenson IV

Review of the Collected Papers on Switched Reluctance Drives, T.J.E. Miller XI

Introduction: Switched Reluctance Drives, T.J.E. Miller 1

Section 1
Switched Reluctance Motors 71

General

Switched Reluctance Motor Drives, P.J. Lawrenson
Electronics & Power, February 1983 72

Switched Reluctance Drives — A Fast Growing Technology, P.J. Lawrenson
Electric Drives and Controls, April/May 1985 77

Brushless Reluctance Motor Drives, Tim Miller
IEE Power Engineering Journal, Vol. 1, November 1987 83

Variable-Speed Switched Reluctance Motors,
P.J. Lawrenson, J.M. Stephenson, P.T. Blenkinsop, J. Corda, and N.N. Fulton
Proceedings IEE, Vol. 127, July 1980 93

Discussion on "Variable-Speed Switched Reluctance Motors," M.R. Harris, H.R. Bolton,
P.A. Ward, J.V. Byrne, G.B. Smith, J. Merrett, F. Devitt, R.J.A. Paul, K.K. Schwartz,
M.F. Mangan, A.F. Anderson, R. Bourne, P.J. Lawrenson, J.M. Stephenson, and N.N. Fulton
Proceedings IEE, Vol. 128, September 1981 107

Motor Design 119

A Review of Switched Reluctance Machine Design, N.N. Fulton and J.M. Stephenson
International Conference on Electric Machines, 1988 120

The Application of CAD to Switched Reluctance Drives, N.N. Fulton
Electric Machines and Drives Conference, IEE Publication No. 282 127

PC CAD for Switched Reluctance Drives, T.J.E. Miller and M. McGilp
Electric Machines and Drives Conference, December 1987; IEE Publication No. 282 132

Computation of Torque and Current in Doubly-Salient Reluctance Motors From Nonlinear
Magnetization Data, J.M. Stephenson and J. Corda
Proceedings IEE, Vol. 126, May 1979 140

Analytical Estimation of the Minimum and Maximum Inductances of a Double-Salient Motor,
J. Corda and J.M. Stephenson
*Proceedings of the International Conference on Stepping Motors and Systems,
University of Leeds, September 1979* 144

Saturable Variable Reluctance Machine Simulation Using Exponential Functions,
J.V. Byrne and J.B. O'Dwyer
*Proceedings of the International Conference on Stepping Motors and Systems,
University of Leeds, July 1976*   154

Limitation of Reluctance Torque in Doubly-Salient Structures,
M.R. Harris, V. Andjargholi, A. Hughes, P.J. Lawrenson, and B. Ertan
*Proceedings of the International Conference on Stepping Motors and Systems,
University of Leeds, July 1974*   160

Switched Reluctance Motor Excitation Current: Scope for Improvement,
J.W. Finch, H.M.B. Metwally and M.R. Harris
Power Electronics and Variable-Speed Drives Conference, 1986; IEE Publication No. 264   171

Estimation of Switched Reluctance Motor Losses, P. Materu and R. Krishnan
Virginia Polytechnic Institute and State University; IEEE-IAS Proceedings, October 1988   176

Sensitivity of Pole Arc/Pole Pitch Ratio on Switched Reluctance Motor Performance, R. Arumugam and
J.F. Lindsay, Concordia University; R. Krishnan, Virginia Polytechnic Institute and State University;
IEEE 1988   188

Finite-Element Analysis   193

Analysis of Variable Reluctance Motor Parameters Through Magnetic Field Simulations, Karl Konecny
MOTOR-CON Proceedings, 1981   194

Switched Reluctance Motor Torque Characteristics: Finite Element Analysis and Test Results
G.E. Dawson, A.R. Eastham and J. Mizia
IEEE Transactions, Vol. IA-23, May/June 1987   205

Finite-Element Analysis Characterization of a Switched Reluctance Motor with
Multi-Tooth per Stator Pole, J.F. Lindsay, R. Arumugam and R. Krishnan
Proceedings IEE, Vol. 133, November 1986   211

Variations on the Classical SR Motor   219

Variable-Speed Drives Using Multi-Tooth per Pole Switched Reluctance Motors
J.W. Finch, M.R. Harris, A. Musoke and H.M.B. Metwally
13th Incremental Motion Control Systems Symposium, 1984   220

Analysis and Optimization of the 2-Phase Self-Starting Switched Reluctance Motor
M.A. El-Khazendar and J.M. Stephenson
International Conference on Electrical Machines, 1986   229

Microprocessor-Controlled Single-Phase Reluctance Motor, J.C. Compter
Drives/Motors/Controls, 1984; IEE 1984   233

Performances of a Multi-Disk Variable Reluctance Machine
J.P. Bastos, R. Goyet, J. Lucidarme, C. Quichaud and F. Rioux-Damidau
International Conference on Electrical Machines, 1982   238

Experiment, As a Generator, of a Multi-Disk Reluctance Machine Having a 200kW Nominal
Power at 500RPM, R. Goyet, C. Rioux, J. Lucidarme, R. Guillet, D. Griffault, and C. Bleijs
International Conference on Electrical Machines   242

Electrical Control of a Linear Reluctance Motor Prototype
R. Goyet, R. Gheysens, J. Lucidarme, D. Matt, and C. Rioux
European Power Electronics Conference, September 1987   245

Early Milestone Papers   251

Tangential Forces in Overlapped Pole Geometries Incorporating Ideally Saturable Material, J.V. Byrne
IEEE Transactions, Vol. MAG-8, March 1972   252

Compatible Brushless Reluctance Motors and Controlled Switch Circuits, B.D. Bedford
U.S. Patent No. 3,679,953, July 1972   260

Compatible Permanent Magnet or Reluctance Brushless Motors and Controlled Switch Circuits
B.D. Bedford
U.S. Patent No. 3,678,352, July 1972 268

An Axial Air-Gap Reluctance Motor for Variable Speed Applications, L.E. Unnewehr and W.H. Koch
IEEE Transactions, Vol. PAS-93, 1974 281

Characteristics of Saturable Stepper and Reluctance Motors, J.V. Byrne and J.G. Lacy
IEE Publication No. 136, March 1976 291

Section 2
Converters and Controls

Converters 295

Inverter Drive for Doubly Salient Reluctance Motor:
Its Fundamental Behaviour, Linear Analysis and Cost Implications, W.F. Ray and R.M. Davis
IEE Electric Power Applications, Vol. 2, December 1979 296

Inverter Drive for Switched Reluctance Motor:
Circuits and Component Ratings, R.M. Davis, W.F. Ray and R.J. Blake
Proceedings IEE, Vol. 128, March 1981 306

Development of a Unipolar Converter for Variable Reluctance Motor Drives
J.T. Bass, M. Ehsani, T.J.E. Miller and R.L. Steigerwald
IEEE Transactions, Vol. IA-23, 1987 319

Converter Volt-Ampere Requirements of the Switched Reluctance Motor Drive, T.J.E. Miller
IEEE Transactions, Vol. IA-21, 1985 328

Controls 337

The Control of SR Motors, W.F. Ray, R.M. Davis and R.J. Blake
Conference on Applied Motion Control, June 1986 338

Speed Control of Switched Reluctance Motors, J. Corda and J.M. Stephenson
International Conference on Electrical Machines, 1982 347

Microprocessor Control of a Variable Reluctance Motor, P.H. Chappell, W.F. Ray and R.J. Blake
Proceedings IEE, Vol. 131, March 1984 352

Four-Quadrant Brushless Reluctance Motor Drive
T.J.E. Miller, P.G. Bower, R. Becerra and M. Ehsani
IEE Conference on Power Electronics and Variable Speed Drives, July 1988; IEE 1987 363

Robust Torque Control of Switched-Reluctance Motors Without a Shaft Position Sensor
J.T. Bass, M. Ehsani and T.J.E. Miller
IEEE Transactions, Vol. IE-33, August 1986 368

Simplified Electronics for Torque Control of Sensorless Switched Reluctance Motor
J.T. Bass, M. Ehsani and T.J.E. Miller
IEEE Transactions, Vol. IE-34, May 1987 373

Detection of Rotor Position in Stepping and Switched Reluctance Motors by
Monitoring of Current Waveforms, P.P. Acarnley, R.J. Hill and C.W. Hooper
IEEE Transactions, Vol. IE-32, August 1985 379

A Simple Motion Estimator for VR Motors, W.D. Harris and J.H. Lang
IEEE Industry Applications Society Annual Meeting, October 1988 387

A State Observer for Variable Reluctance Motors, A.H. Lumsdaine, J.H. Lang and M.J. Balas
15th Incremental Motion Control Systems Symposium, June 1986 394

Section 3
Applications   401

A High Performance Variable Reluctance Drive: A New Brushless Servo
J.V. Byrne, J.B. O'Dwyer and M.F. McMullin
PowerConversion International magazine, February 1986   402

Design of a Reluctance Motor As a 10kW Spindle Drive, J.V. Byrne and M.F. McMullin
MOTOR-CON Proceedings, September 1982   407

A Low-Cost, Efficient 1kW Motor Driver, Bruce Powell
MOTOR-CON Proceedings, April 1984   422

Ultra-High Torque Motor System for Direct Drive Robotics, Ross Welburn
MOTOR-CON Proceedings, April 1984   431

High Performance Switched Reluctance Brushless Drives
W.F. Ray, P.J. Lawrenson, R.M. Davis, J.M. Stephenson, N.N. Fulton and R.J. Blake
IEEE Transactions, Vol. IA-22, July/August 1986   439

High Performance MOSFET Switched Reluctance Drives
D.M. Sugden, R.J. Blake, S.P. Randall, J.M. Stephenson and P.J. Lawrenson
IEEE Industry Applications Society Annual Meeting, October 1987   448

A Current-Controlled Switched-Reluctance Drive for FHP Applications, T.J.E. Miller and T.M. Jahns
Conference on Applied Motion Control, June 1986   454

A Review of the Integral-Horsepower Switched Reluctance Motor Drive
M.R. Harris, J.W. Finch, J.A. Mallick and T.J.E. Miller
IEEE Transactions, Vol. IA-22, July/August 1986   463

Step Motors That Perform Like Servos, K.A. Regas and S.D. Kendig
Machine Design, Penton Publishing Inc., December 1987   469

Switched Reluctance Motor Drives for Rail Traction: Relative Assessment, P.S.R. French
Proceedings IEE, Vol. 131, September 1984   474

Switched Reluctance Motor Drives for Rail Traction: A Second View
W.F. Ray, R.M. Davis, P.J. Lawrenson, J.M. Stephenson, N.N. Fulton and R.J. Blake
Proceedings IEE, Vol. 131, September 1984   486

Additional References   495

Review of the Collected Papers On Switched Reluctance Drives

T. J. E. Miller

The switched reluctance motor drive now has the maturity to take its place in the catalogue of drive technologies as an efficient brushless drive with cost advantages, a wide speed range, and inherent simplicity and ruggedness. These characteristics have been painstakingly wrought from an old concept that dates back early into the nineteenth century, but which could not flourish until modern power electronics, computer-aided design, and controls provided the key to unlock the extraordinary electromagnetic potential of these machines.

The SR drive has had to fight for its place in a technology that is already rich in alternatives, and it is to the credit of its developers and champions that it has made such progress. Most of the more recent contributors are represented by their own works in this collected volume, and it is intended that the inclusion of these works should mark a general recognition of their significance. Regrettably, the task of selection leaves out many valuable contributions and it is hoped that the list of References includes most of these. The selection has been made with the objective of stitching together a story which, to the editor at least, gives a coherent but economical account of the technology without excessive digression into specialist areas. One or two papers have been left out for copyright reasons, but any other omissions are the responsibility of the editor and his apologies are offered.

Although the history of the SR drive is a long one, the most important chapter of it begins in the 1960s and 1970s with the work of French, Unnewehr and Koch, Bausch, Bedford, Byrne, Lawrenson and Stephenson, the brothers Jarret; and of course the co-workers of all these well-known names.

We begin with an introductory chapter that covers the basic principles of the SR drive from a tutorial technical viewpoint.

Section 1.1 - Switched Reluctance Motors - General

Following the introductory chapter, there is a general section that begins with a number of discussion or review papers including the landmark paper by Professor Lawrenson [1980], 'Variable-speed switched reluctance motors'. It is here that the term 'switched reluctance' was used for the first time in relation to the radial-gap motor that is the focus of attention today. Many equally acceptable alternative terms have been used. The switched reluctance motor is a very pure form of reluctance motor; and the most important aspect of its control is that its current is switched on and off at particular rotor positions. Perhaps it would be more accurate to use the term 'commutated' rather than 'switched', giving rise to 'commutated reluctance motor' or 'electronically commutated reluctance motor'; but the SR term has been perhaps the most widely used in recent times. The term 'current-regulated stepping motor' proposed by Prof. Lipo is incorrect on two counts: first, it is not a stepping motor, and secondly, it is not always current-regulated. To call it a 'variable-reluctance motor' introduces an air of variability which is perhaps not the right image in the tightly controlled world of modern motion control systems.

Section 1.2 - Motor Design

The review section is followed by the serious business of designing the motor. This begins with an authoritative review of design approaches by Fulton and Stephenson. Their own formidable contributions to this subject are also reflected in this section. The 1987 paper by Fulton represents a state-of-the-art account of design techniques, emphasizing the integrated approach in which the motor and its control are designed simultaneously by advanced CAD techniques. A less powerful technique based on the IBM PC is described by Miller and McGilp, providing a convenient way to 'size' a switched reluctance drive using a fast menu-driven package that fits on a single PC diskette.

The two papers by Corda and Stephenson are milestone papers that establish the basic foundations of the computer-aided design procedure. Together with Byrne's 1976 paper on the representation of the magnetization curves, these two papers tackle the difficulties of electromagnetic design in a direct but elegant way, dealing effectively with the double saliency, the sharply nonlinear magnetic characteristics, and the time-variation of parameters that preclude the use of conventional motor design theory.

The fundamental question of the maximum torque per unit volume is elegantly analyzed by Harris et al [1974] in a way which makes the underlying physics abundantly clear. In this era, it would be easy for these principles to have been lost in a mass of computer-aided design data, but Harris' analysis is well worth careful study even though the most

powerful finite-element tools are available. The basis of
the design objectives can be found in Harris' treatment of
the energy-conversion areas in the flux-linkage/current
plane: it is here that the limits of performance, and the
comparison with other motor concepts, can be understood at
the most fundamental level.

Harris' work on the limitations of reluctance torque
extends to the linear reluctance motor (not covered in this
work), and it also led to the development of the SR motor
with multiple teeth per pole, reported in his 1984 paper
with Finch, Musoke, and Metwally in Section 1.4. But it also
develops another line of approach in the 1986 paper with
Finch and Metwally, namely the analysis of the most
effective current and voltage waveforms. Again they produce
a practical analysis that is as important for the power
electronics as their earlier work was for the motor.

Section 1.3 - Finite-element analysis

This short section contains a sample of papers on
finite-element analysis of the SR motor. Numerical
field-analysis methods such as the finite-element method,
the boundary-element method, and indeed finite-difference
methods are important because the geometry of the SR motor
is not amenable to simple calculation. This is particularly
important in the calculation of magnetization curves at
different rotor orientations -- data which is essential for
accurate design. For very accurate results three-dimensional
methods are likely to be essential, but two-dimensional
field solutions are helpful in the search for the best
lamination geometry, and also for more advanced studies such
as the distribution of core losses and the effects of local
saturation.

Section 1.4 - Variations on the classical SR motor

Many are the variations on the doubly-salient
'commutated' reluctance motor, and the selection here
includes single-phase and 2-phase drives; axial-gap disk
motors, and linear motors, as well as the
multiple-tooth-per-pole motor mentioned earlier.

Section 1.5 - Early milestone papers

While the emphasis in this collection is on recent developments, a selection of classic papers is included to represent the fine work of some of the earlier attempts to perfect the switched reluctance motor. The 1972 paper by Byrne on the effect of saturation on overlapping poles was the first in a series which established the full understanding of these effects. Many will remember the arguments in the 1970s about whether saturation gives 'something for nothing'; Byrne's analysis makes fascinating reading and will do so for a long time. Most of the papers in this section show how all the key concepts of motor design and control, without exception, were well understood and established more than 15 years ago. It is interesting to observe that power electronics and associated control IC's were at that time not quite ready to be easily applied to exploit the electromagnetic concepts, and there was a delay before the 'critical mass' was achieved some years later, particularly with the work of Ray and Davis at the University of Nottingham, collaborating with Lawrenson and Stephenson at the University of Leeds in the U.K. and resulting in the first commercial application of the modern SR drive.

Section 2.1 - Converters

The papers by Ray and Davis [1979 and 1981] gave the most detailed account of thyristor circuits for SRM control up to that time. In particular the 1979 paper includes an elegant mathematical analysis of the current waveforms and the torque production in the magnetically linear case with the assumption of negligible fringing around the corners of partially overlapping poles. These papers were concerned with the problems of thyristor commutation and the control of current at low speeds by chopping.

By 1986 the GTO thyristor was well established and the converter reported by Torrey and Lang was the first in which these devices were applied to a 'large' SR drive of 60kW, designed for an experimental electric vehicle under a Department of Energy program in the United States.

More recently the focus of attention at the leading edge of SR technology has been on transistor-controlled drives, as evidenced in the 1987 paper by Sugden and colleagues of Switched Reluctance Drives Ltd. The development of transistor and MOSFET converters has had important consequences in the control of small SR drives, as we shall see in Section 3.

Because the SR motor requires unidirectional phase currents rather than alternating current, there is a theoretical possibility of using only one main switching device per phase, instead of the two that are necessary in a.c. and PM brushless drives. The 1987 paper by Bass et al reviews several attempts to achieve this objective and describes a capacitive suppression circuit with energy recovery to the d.c. supply. This particular circuit has been made obsolete by the n+1 switch circuit described in the introductory chapter, but it is included because the theory of its operation throws light on the basic problem of reactive energy in the SR motor. A more thorough theoretical analysis of this problem is presented in the 1985 paper by Miller on the converter volt-ampere requirements. This paper extends the earlier work of Byrne in describing the effects on saturation, and shows why magnetic saturation has a beneficial effect on the converter volt-ampere requirement even though it limits the torque available from a given motor volume.

Section 2.2 - Controls

The 1986 paper by Ray, Davis, and Blake is one of the more recent accounts of the basic control principles of the SR motor. Following this, the 1982 paper by Corda and Stephenson describes a speed control algorithm that generates appropriate gating and commutation angles for any combination of torque and speed. This paper also derives the functional relationship between torque and speed for different forms of control, and is one of very few to address this important area. With voltage control and switching angle modulation the SR motor is by no means confined to its natural 'series' characteristic: in fact its torque/speed characteristic can be 'programmed' more flexibly than that of many other motors without resort to oversizing the power electronic converter.

The SR motor is a natural candidate for microprocessor control, as reflected in the 1984 paper by Chappell and colleagues.

The 1988 paper by Miller, Becerra, Bower and Ehsani introduces a much simpler scheme called 'multimode commutation' in which a small number of discrete combinations of switching angles are used, with p.w.m.

control to modulate the torque between mode changes. This technique, suitable for implementation in an integrated circuit, provides virtually all of the performance available from much more complex and expensive schemes, including high-speed operation at constant power which is one of the main advantages of the SR motor over the PM brushless d.c. motor. The multimode commutation scheme is, in each mode, exactly equivalent to the commutation and control of the PM brushless d.c. motor: thus this paper firmly establishes the equivalence of the two machines for all practical control purposes, with the important difference that the SR motor has by far the larger speed range for the same control cost and the same converter volt-amperes.

In all the papers mentioned so far, there has been very little mention of the shaft position transducer. The most expensive and sophisticated controls use resolvers, while the least expensive use optical interrupters, Hall effect sensors, or other simple transducers. Ideally the controller should be able to operate without any shaft position sensor, not only to reduce cost but to improve reliability and avoid the problems that often arise with additional auxiliary devices. The remaining papers in this section describe various attempts to control SR drives with no shaft position sensor. The simplest schemes follow the principle of stepper motor control, that is, open-loop control, with some form of stabilization and special arrangements for starting. Conduction angle modulation is an important aspect of this type of scheme, a technique not normally applied to stepper motors. The method described by Acarnley, Hill and Hooper [1985] is a relatively inexpensive method based on the fundamental properties of the current waveform in an inductance that depends on rotor position. Finally, Lumsdaine, Lang and Balas [1985] describe an extremely sophisticated but complex scheme in which a microcomputer performs a parallel simulation of the motor operation, while feedback from the terminal voltages and currents is used to cause the results of the simulation to converge, in real time, on the operation of the motor. The simulation includes a parameter representing the rotor position and this is taken as the actual rotor position once the simulation has converged on the actual operation.

An important difference between the SR motor and the brushless d.c. PM motor is that in the SR motor at standstill it is possible to determine which phase should be fired first in order to start rotation in either direction. In the PM d.c. motor with magnets on the rotor surface this is not possible: until the magnets are rotating, there is no way of determining the rotor position from measurements at the terminals. This means that in theory, control schemes can be devised for SR motors without shaft position sensors, which can operate right down to zero speed. For the surface-magnet PM brushless motor this is not possible.

Section 3 - Applications

The applications for which the SR motor has been tried divide naturally into 'large drives', where the main competition is from a.c. induction motor drives; and 'small drives', where the main competition is from brushless d.c. drives. There is no hard-and-fast power level defining the boundary but it is probably somewhere around 5-10kW. The large drives typically use thyristors or GTO's and include the general-purpose drives marketed by Tasc Drives Ltd. for a very wide range of industrial applications mainly in the range from 4-22 kW. While these drives have been in production for some years, the electric vehicle drives reported by Vallese and Lang, and the rail traction drives described in the paper by French and the reply by Ray and his colleagues appear to have proceeded no further than the prototype or study stage.

The review paper by Harris et al [1986] includes a comparison between the SR motor and the induction motor using quoted test data, with the emphasis in the 5-10hp range; and the 1986 paper by Miller and Jahns includes a similar comparison at the fractional-horsepower level between switched reluctance and d.c. drives.

The 1986 paper by Byrne and colleagues from Inland Motor describes a servo drive and this paper also provides valuable comparison data between he SR drive and, this time, the brushless d.c. as well as the d.c. equivalent drives. They demonstrated the naturally high torque/inertia ratio of the SR motor and concluded that its performance was roughly on a par with that of the d.c. drive, but not as good as that of the rare-earth PM brushless d.c. drive.

Claims of much higher performance are found in the paper by Regas and Sendig [1987] who describe a small brushless reluctance servo drive with an advanced magnetic design and very sophisticated control using a special-purpose integrated circuit. A computer plotter servo drive is given as an application example.

In the future...

    What is immediately clear from the papers that discuss
applications is the wide range of applications for which the
SR motor has been or is a candidate. The motor drives
business is composed of a myriad of niche products and niche
applications, and is a mature business with a heavy
investment in existing technology. It is therefore not
surprising that the SR motor has been slow to realise its
full commercial potential, despite its technical
characteristics.

    It can be argued that in spite of the excellent
development work already done, the SR motor is only just now
becoming mature enough to take on some of these
applications, and with the steady development of power
electronics and controls, this is likely to accelerate, as
many of the papers in this section demonstrate.

    What technical improvements are still needed to
maintain this progress? First, the motor itself: this is
perhaps where the least amount of development is needed,
although there is undoubtedly a need for reduced noise. In
the power electronics, the introduction of power devices
packaged in a way that is suitable for the SR motor, rather
than purely for a.c. drives, is much to be desired; and
indeed it is welcomed from those device manufacturers who
have already done so. It is to be hoped that they will
benefit themselves. From the point of view of the power
switches, the SR motor as a load has several advantages over
a.c. and brushless d.c. drives: as we have seen, it is
essentially immune from shoot-through faults, does not
require lock-out protection, has zero short-circuit current,
and cannot generate any overvoltage under open-circuit
conditions.

    The development of integrated-circuit controls
specifically for SR motors will make a big contribution to
their acceptability by relieving design and application
engineers of the burden of designing controls which,
although simple and inexpensive in the end, require
sophisticated techniques and understanding to achieve. These
IC controls may be doubly useful if they can commutate both
SR and brushless PM d.c. motors. For the future, we can look
forward to IC controls that operate without the shaft
position sensor.

    Taking all these factors into account, it would seem
likely that within a few years the SR motor will be common
not only in industry, which has already found many uses for
it, but also in kitchens and automobiles.

Introduction

T. J. E. Miller

Introduction
Switched Reluctance Drives
T.J.E. Miller

1. The switched reluctance motor

The concept of the switched reluctance motor was established by 1838 (see Byrne et al, [1986]), but the motor could not realise its full potential until the modern era of power electronics and computer-aided electromagnetic design. Since the mid-1960s these developments have given the SR motor a fresh start and have raised its performance to levels competitive with d.c. and a.c. drives and brushless d.c. drives.

It is difficult to be certain about the origin of the term 'switched reluctance', but one of the earliest occurrences is in Nasar [1969] in relation to a rudimentary disk motor employing switched d.c. Professor Lawrenson [1980] was perhaps the first to adopt the term in relation to the radial-airgap motor which is the focus of attention today, but the terms 'brushless reluctance motor', 'variable reluctance motor', and 'commutated reluctance motor' are among several equally acceptable alternatives that were in use long before this time. It could perhaps be most accurately described as a 'statically commutated doubly-salient vernier reluctance motor'.

Apart from the well-known work by Lawrenson and his colleagues at the University of Leeds, and subsequently at Switched Reluctance Drives Ltd., there have been many other substantial contributions to the technology since the mid-1950s. Among the most notable are the works of French [1967]; GE (Bedford, [1972], and several other authors subsequently); Ford (Unnewehr and Koch, [1974]); Professor J.V. Byrne of University College, Dublin ([1972], [1976], [1982], [1986]); the Jarret Company in France; Inland-Kollmorgen (Ireland); Professor M.R. Harris of the University of Newcastle upon Tyne; and Professor J.H. Lang of the Massachusetts Institute of Technology.

By the time of Unnewehr and Koch's paper in 1974, most of the basic design principles of both the motor and the control were well understood, but modern technology has facilitated much refinement since that time.

With the exception of large-diameter 'direct drive' robot motors, the only commercially produced SR drives at the time of writing are the Oulton drive (Tasc Drives Ltd.) in the U.K. and a computer plotter servo (Hewlett-Packard/Warner Electric) in the U.S. However, this situation is likely to change rapidly as the technology matures and becomes more widely understood.

The switched reluctance motor is a doubly-salient, singly-excited motor. This means that it has salient poles on both the rotor and the stator, but only one member (usually the stator) carries windings. The rotor has no windings, magnets, or cage winding, but is built up from a stack of salient-pole laminations, Fig. 1.

There are two essentials that distinguish the SR motor from the variable-reluctance stepper [Kenjo 1985]. One is that the conduction angle for phase currents is controlled and synchronized with the rotor position, usually by means of a shaft position sensor. In this respect the SR motor is exactly like the PM brushless d.c. motor, but unlike the stepper motor, which is usually fed with a squarewave of phase current without rotor position feedback. The second distinction between SR and stepper motors is that the SR motor is designed for efficient power conversion at high speeds comparable with those of the PM brushless d.c. motor; the stepper, on the other hand, is usually designed as a torque motor with a limited speed range. Although this may seem a fine distinction, it leads to fundamental differences in the geometry, power electronics, control, and design technique.

The SR motor is more than a high-speed stepper motor. It combines many of the desirable qualities of both induction-motor drives and d.c. commutator motor drives, as well as PM brushless d.c. systems. Its performance and inherently low manufacturing cost make it a vigorous challenger to these drives. Its particular advantages may be summarized as follows:

1. The rotor is simple and requires relatively few manufacturing steps; it also tends to have a low inertia.

2. The stator is simple to wind; the end-turns are short and robust and have no phase-phase crossovers.

3. In most applications the bulk of the losses appear on the stator, which is relatively easy to cool.

4

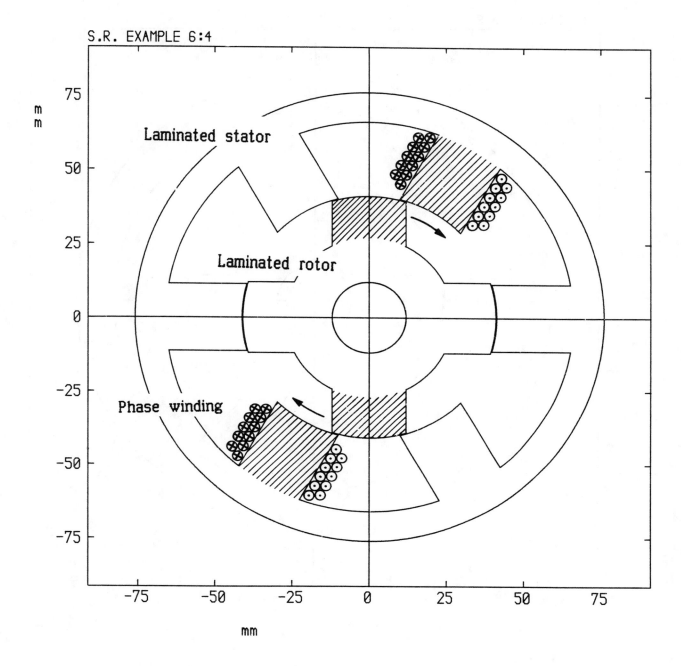

S.R. EXAMPLE 6:4

Fig. 1. Cross-sections of switched reluctance motors
generated by CAD program. One phase comprises windings
on opposite poles.

(a) 3-phase 6:4

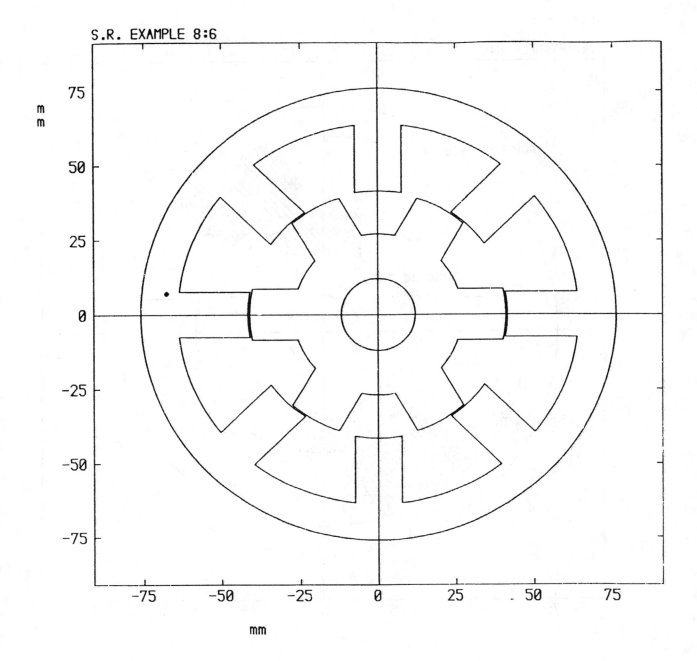

S.R. EXAMPLE 8:6

(b) 4-phase 8:6

6

4. Because there are no magnets the maximum permissible rotor temperature may be higher than in PM motors.

5. The torque is independent of the polarity of phase current; for certain applications this permits a reduction in the number of power semiconductor switches needed in the controller.

6. Under fault conditions the open circuit voltage and short-circuit current are zero or very small.

7. Most converter circuits used with SR motors are immune from shoot-through faults, unlike the inverters used with a.c. and brushless d.c. drives.

8. Starting torque can be very high, without the problem of excessive inrush currents, as for example the large starting current of induction motors at high slip.

9. Extremely high speeds are possible

10. The torque/speed characteristic can be 'tailored' to the application requirements more easily than in the case of induction motors or PM motors.

These are clear advantages that require little or no qualification. Other advantages must be seen in the light of the trade-offs that go with them. Because there is no fixed magnet flux the maximum speed at constant power is not as restricted by controller voltage as it is in PM motors. However, the absence of 'free' PM excitation imposes the burden of excitation on the stator windings and the controller, and increases the per-unit copper losses. Particularly in small motors this is a disadvantage that limits the efficiency.

The SR motor also has some clear disadvantages. The most important is the pulsed, or at least non-uniform, nature of the torque production, which leads to torque ripple and may contribute to acoustic noise. Over a narrow speed range, it is possible to reduce the torque ripple to less than 10% r.m.s., which is comparable with the levels attainable in induction motors and other brushless d.c. drives; but it is practically impossible to maintain this level of smoothness over a wide speed range. Fortunately it is easier to achieve smooth torque at low speeds, where many loads are most sensitive to torque ripple effects.

The acoustic noise can be severe in large machines where ultrasonic chopping frequencies are not practical. But even in small ones, when all steps have been taken to

Fig. 2   Stator of SR motor

Note the essential simplicity of construction
and windings. This motor, designed for a highly noise-sensitive
application, has compression fingers bearing on the stator
poles. The short, robust endwindings are also evident,
with no phase crossovers. Cooling air has excellent access
all along the stator slots. The slot fill in this motor is
less than 30%.

(Scottish Power Electronics and Electric Drives, Glasgow)

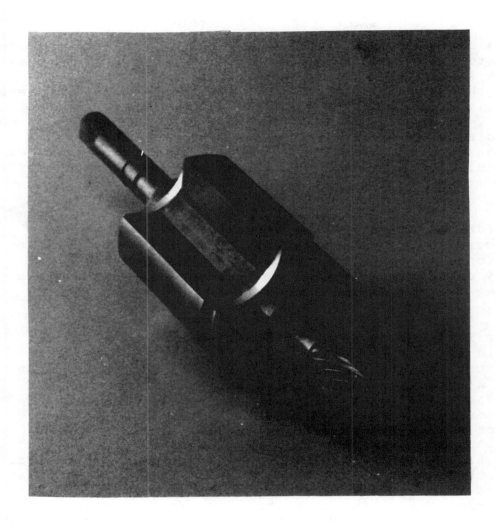

Fig. 3 Rotor of SR motor

Comprising only a stack of laminations, the SR rotor
is the simplest of all electric motor rotors. The
ground surface of the rotor poles helps to achieve a
uniform airgap, as in induction motors. Windage loss is
not as great as the salient-pole construction might suggest;
but the air turbulence helps to cool both the rotor and the
stator.

(Scottish Power Electronics and Electric Drives, Glasgow)

minimize chopper noise, there remains a characteristic sound
similar to 'tickover' noise in internal combustion engines
at light load; under heavy load this tends to become a
'growl' that may be difficult to eliminate. The noise level
is sensitive to the size, being much less severe in small
machines. It also depends on the mechanical construction and
the precision of the firing angles. The torque ripple is
also sensitive to these factors. Although the construction
is simple, electrical and mechanical precision are essential
to keep it quiet, and this tends to increase the cost.

A further aspect of the torque ripple is that the
ripple current in the d.c. supply tends to be quite large,
making for a large filter capacitance requirement. This in
turn may cause significant a.c. line harmonics in systems
operating from rectified a.c.

The SR motor makes use of the 'vernier' principle
common in stepper motors, in which an internal torque
multiplication is achieved with a rotor speed slower than
that of a rotating-field machine with the same number of
phases and rotor poles. Without this multiplication the
torque per unit volume would be much less than that of
induction motors and PM motors, but the price paid is a
substantial increase in commutation frequency, which may
lead to higher core losses and converter switching losses.
The effect is compensated by a smaller volume of iron than
that of a comparable a.c. motor, and also by the fact that
in some sections of the core the flux excursions are
unipolar, which helps to limit the hysteresis losses. The
internal torque multiplication or 'vernier' effect
compensates for the relatively poor utilization of converter
volt-amperes and restores the effective power factor to a
'competitive' level at the expense of switching frequency
and magnetic losses. This mechanism is not applicable to the
synchronous reluctance motor, and this is the main reason
why its weak performance relative to its PM stablemates does
not carry over to the switched reluctance motor.

For optimum performance the airgap needs to be about
the same as that of an induction motor of comparable
diameter, or perhaps slightly larger; PM brushless motors,
however, can operate with larger airgaps and therefore
slightly larger manufacturing tolerances.

The pole shape of an SR motor cannot be made 'square'
as is the normal tendency in a.c. and d.c. machines. This is
true of virtually all known designs of SR motor, yet very
little analysis has appeared in the literature, even though

10

it undoubtedly has a profound effect on its characteristics relative to those of more conventional machines. Long narrow poles tend to produce the best designs by reducing the effects of end-winding inductance and resistance; but this also has the effect of reducing the flux and inductance, and the SR motor typically requires more turns of thinner wire than an a.c. motor wound for the same voltage. In small drives with a wide speed range, this tends to require a lower minimum duty-cycle in the chopping of the supply voltage, and if the chopping frequency is high, special high-frequency pulse techniques and very fast power switches and diodes may be necessary.

Much effort has been expended in attempts to compare the power output and efficiency of SR motors with those of competing drive technologies. For any such comparison to be meaningful it is necessary to restrict it to a narrow set of specifications. It is impossible to make completely general statements about relative performance; far too many variable parameters are involved. There are surprisingly few detailed comparisons in the literature. In larger sizes it is likely to be found that when all aspects of the performance are put on an equal basis, the SR motor is no smaller than an induction motor designed to the same specification. In small sizes the power density, or equivalently the efficiency, of both these motors falls off and neither of them can attain the performance of the brushless PM motor.

The SR motor cannot start or run from an a.c. voltage source, and it is not normally possible to operate more than one motor from one inverter. It is normally necessary to use a shaft position sensor for commutation and speed feedback. Serious attempts have been made to operate without the sensor but inevitably there is a price to be paid either in performance or in control complexity.

The cabling for SR motors is typically more complex than for induction motor drives: a minimum of four wires, and more usually six, are required for a three-phase motor, in addition to the sensor cabling.

This somewhat lengthy review of the disadvantages of SR motors is included in the interests of making a balanced appraisal of it as it stands today. The weighting attached to each advantage and each disadvantage is different for every application, and the weighted sum can only be evaluated relative to a detailed specification. In view of the distinct characteristics of the SR drive, its likely application is where a brushless drive is required with a

wide speed range, with a cost saving over the conventional
PM brushless d.c. motor drive. The control is simpler than
the field-oriented induction-motor drive but in larger
machines this does not necessarily mean that it will be less
expensive, because control costs for a given level of
functionality are tending to decrease. The noise and torque
ripple are likely to remain worse than brushless d.c. PM
motors, but in small drives this may not be a particular
problem. By comparison with small commutator motors (a.c. or
d.c.), the SR motor can fairly be claimed to be
significantly quieter.

2. Poles, phases, and windings

The 'classical' forms of switched reluctance motor are those in Fig. 1, with stator:rotor pole numbers of 6:4 and 8:6. Others are possible, including 4:2, 6:2, 10:4, 12:8, and variants with more than one tooth per pole such as 12:10, [Harris 1985]. Only the two shown in Fig. 1 are considered here.

Many of the basic rules constraining the choice of pole numbers, pole arcs, and phase number were expounded by Lawrenson [1980]. The relationship between speed and fundamental switching frequency follows from the fact that if the poles are wound oppositely in pairs to form the phases, then each phase produces a pulse of torque on each passing rotor pole; the fundamental switching frequency in one phase is therefore

$$f_1 = nN_r = \frac{rpm}{60} \cdot N_r \qquad Hz$$

where n is the speed in rev/s and N_r is the number of rotor poles. If there are q phases there are qN_r steps per revolution and the 'step angle' or 'stroke' is

$$\epsilon = \frac{2\pi}{qN_r} \qquad rad$$

The number of stator poles usually exceeds the number of rotor poles.

The pole arcs are determined by the essential torque-production mechanism, which is the tendency of the poles to align. If fringing is neglected there must be overlap between a pair of rotor poles and the poles of the excited stator phase; in this case torque can be produced through an angle β, which is the smaller of the stator and rotor pole arcs. To produce unidirectional torque through 360 degrees of rotation it is obvious that β must not be smaller than the step angle, otherwise there will be 'gaps' where no torque is produced: thus

$$\beta > \epsilon.$$

In order to get the largest possible variation of phase inductance with rotor position, the interpolar arc of the rotor must exceed the stator pole arc. This leads to the condition

13

$$\frac{2\pi}{N_r} - \beta_r > \beta_s$$

which ensures that when the rotor is in the 'unaligned' position relative to the stator poles of one phase, there will be no overlap and therefore a very low inductance. The unaligned position is defined as the conjunction of any rotor interpolar axis with the axis of the stator poles of the phase in question. In Fig. 1b the phase on the vertical axis is 'unaligned' while the phase on the horizontal axis is 'aligned'.

A further constraint on the pole arcs is that usually the stator pole arc is made slightly smaller than the rotor pole arc. This permits slight increases in the slot area, the copper winding cross-section, and the aligned/unaligned inductance ratio.

The constraints on pole arcs can be expressed graphically as in Fig. 4, in which the 'feasible triangles' [Lawrenson, 1980] define the range of combinations normally permissible. As might be expected, the variation in performance of machines defined by different points in these triangles is considerable. Fig. 5 shows, for a three-phase motor, the cross-sections corresponding to the vertices A, B and C on Fig. 4a. Design C is likely to have too high an unaligned inductance and too little winding area. Design B has more copper area but still the unaligned inductance will be high because of fringing. Design A has a large winding area and a high inductance ratio, leading to a high efficiency and power density, but its torque ripple is higher than in the others.

The 'optimum' tooth width/tooth pitch ratio used in stepper motor design is not applicable to the SR motor. It is of course possible to determine a combination of pole arcs that gives the highest inductance ratio and therefore the highest 'static torque per ampere'. But too many other factors have to be considered to make this the universal choice. Among them are the torque ripple, the starting torque, and the effects of saturation. Curvature effects are also more pronounced than in steppers because of the small number of poles. As in steppers, pole taper is likely to be of benefit in reducing core losses, the m.m.f. drop in the rotor and stator steel, and the adverse effects of saturation. Stator pole taper also reduces the unaligned inductance, but it slightly decreases the winding area.

Several other detailed modifications to the simple geometry of Fig. 5 are permissible and advantageous, such as the use of a hexagonal stator blank which increases the winding area and can produce a mechanically stiffer core, at

14

Fig. 4. Pole-arc constraints — 'feasible triangles' [Lawrenson 1980]

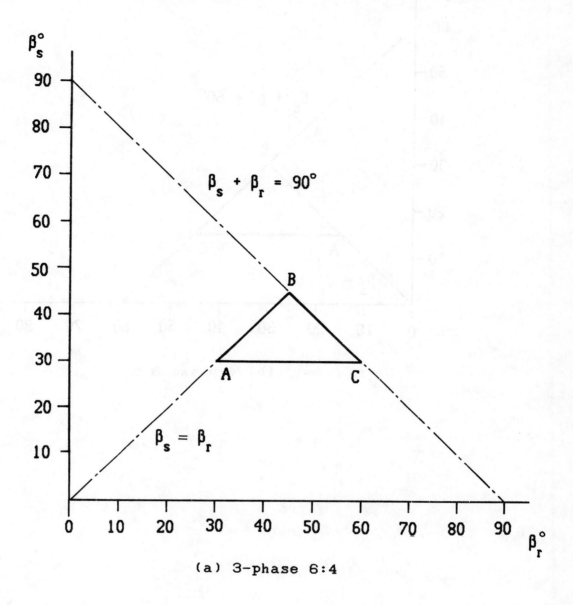

(a) 3-phase 6:4

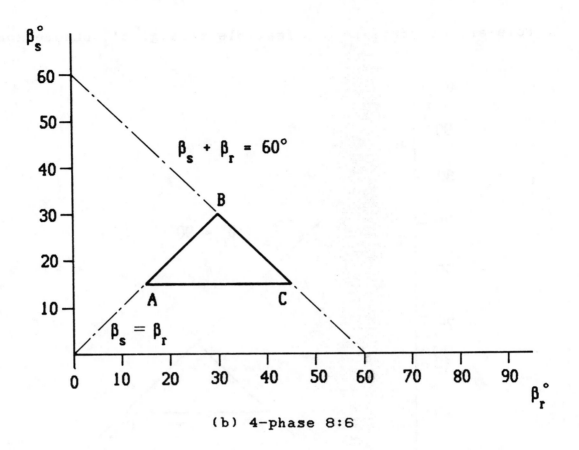

(b) 4-phase 8:6

16

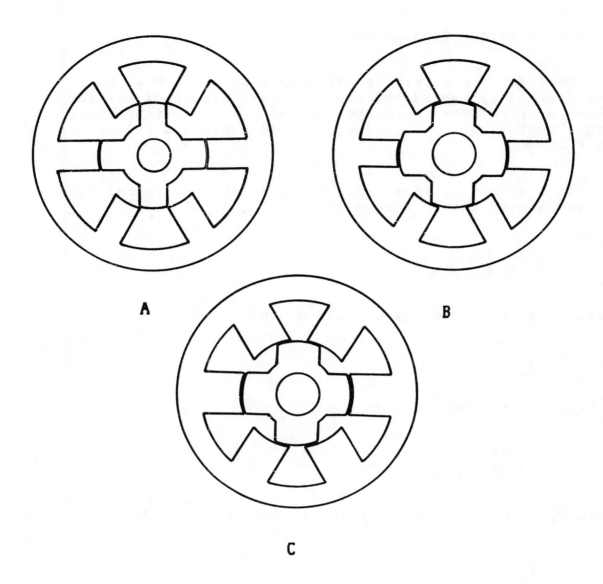

Fig. 5. Cross-sections obtained from the vertices of the feasible
triangles of Fig. 4a. The relative proportions determine
the inductance ratio and the copper and core losses.

the same time reducing the scrap from the punching process.
Pole overhangs can be used to control the local saturation
during the initial overlap period. Welding the outside of
the stator stack is permissible as in a.c. motors, but this
cannot be used on the rotor, which requires mechanical means
for compressing the laminations together. The rotor may be
skewed slightly to reduce noise.

3. Static torque production

Consider the primitive reluctance motor in Fig. 6a. When current is passed through the phase winding the rotor tends to align with the stator poles; that is, it produces a torque that tends to move the rotor to a minimum-reluctance position.

In a device of this type, the most general expression for the instantaneous torque is

$$T = \left[\frac{\partial W'}{\partial \theta}\right]_{i\text{-const.}}$$

where W' is the coenergy defined as in Fig. 6b:

$$W' = \int_0^i \psi \, di$$

An equivalent expression is

$$T = -\left[\frac{\partial W_f}{\partial \theta}\right]_{\psi\text{-const.}}$$

where W_f is the stored field energy defined as in Fig. 6b:

$$W_f = \int_0^\psi i \, d\psi$$

When evaluating the partial derivatives it is essential to keep the indicated variables constant. If the differentiation is done analytically, then W_f must first be expressed as a function of flux (or flux-linkage) and rotor position only, with current i absent from the expression. Likewise W' must first be expressed as a function of current (or m.m.f.) and rotor position only, with flux (or flux-linkage) absent from the expression. If the differentiation is performed by taking differences or interpolated differences from a look-up table, then the same principle must be observed [Stephenson and Corda 1979].

If magnetic saturation is negligible, then the relationship between flux-linkage and current at the instantaneous rotor position θ is a straight line whose slope is the instantaneous inductance L. Thus

18

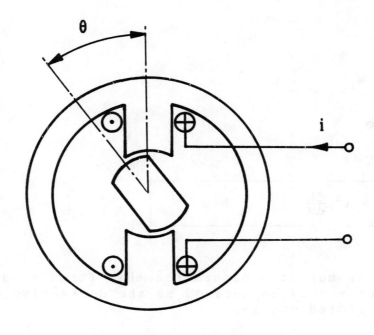

(a) primitive motor

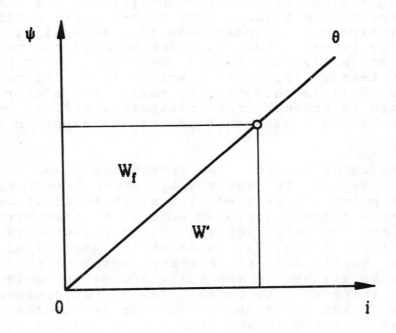

(b) field energy and coenergy

Fig. 6. Elementary reluctance motor showing principle of torque
production

19

$$\psi = Li$$

and

$$W' = W_f = \frac{1}{2} Li^2 \qquad\qquad J$$

Therefore

$$\boxed{T = \frac{1}{2} i^2 \frac{dL}{d\theta} \qquad\qquad N\text{-}m}$$

If there is magnetic saturation this formula is invalid and the torque should be derived as the derivative of coenergy or field stored energy.

Although saturation plays an important role in determining the characteristics and performance limits of switched reluctance motors, most of the basic control characteristics can be understood from an analysis of the magnetically linear motor, considering only one phase in isolation as in Fig. 6a. Mutual coupling between phases is ignored in this analysis. In practice it is desirable to keep it as small as possible, by making the stator yoke thick enough to prevent cross-saturation effects between phases that share common sections of the magnetic circuit.

As the rotor rotates, the inductance L varies between two extreme values. The maximum L_a occurs when the rotor and stator poles are aligned. The minimum inductance L_u occurs when a rotor interpolar axis is aligned with the stator poles. The variation with rotor position is shown in idealised form in Fig. 7a, in which the neglect of fringing results in sharply defined 'corners' which coincide with particular positions. If the rotor and stator pole arcs are different, there will be a small 'dwell' at maximum inductance. Likewise if the interpolar arc of the rotor exceeds the stator pole arc, there is a 'dwell' at minimum inductance. The upper and lower 'corners' occur when rotor and stator pole corners are in conjunction, and between these positions the inductance varies more or less linearly as the overlap area varies. If the steel is assumed to be infinitely permeable and fringing is neglected, the inductance can be estimated roughly as

$$L(\theta) = 2N_p^2 P_g + L_u = 2N_p^2 \frac{\mu_0 r_1 \ell \alpha}{g} + L_u$$

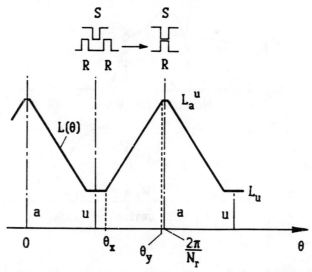

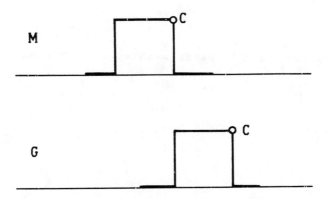

(a) Idealised inductance variation, ignoring fringing and saturation

(b) Ideal current waveforms for motoring and generating at low and medium speeds. (C = commutation)

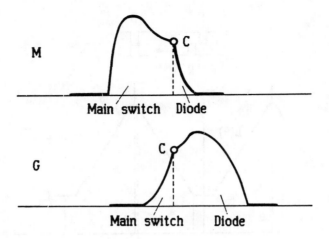

(c) High-speed current waveforms depart from
    the ideal, owing to the self-e.m.f. of the
    motor. Note the phase advance in the turn-on
    angles relative to those in (b).

Fig. 7. Variation of inductance with rotor position.

where α is the overlap angle between the rotor and stator poles. If the origin of θ for a particular phase is taken to be the alignment of the previous pair of rotor poles, then α and θ are related by the following expression throughout the rising-inductance interval:

$$\alpha \ = \ \theta - \theta_x \ ; \ \ \theta_x < \theta < \theta_y$$

where

$$\theta_x \ = \ \frac{2\pi}{N_r} - \frac{\beta_r + \beta_s}{2} \ ; \ \ \theta_y \ = \ \frac{2\pi}{N_r} - \frac{\beta_r - \beta_s}{2}$$

The form of Fig. 7a reflects the variation of the overlap angle α as the rotor rotates. The unaligned inductance includes the end-turn inductance and a contribution due to leakage flux passing across the stator slots.

The torque is independent of the direction of the current. Its direction depends only on the sign of $dL/d\theta$. When the rotor poles are approaching the aligned position this is positive, and positive (i.e. motoring) torque is produced, regardless of the direction of the current. When the rotor poles are leaving the aligned position and approaching the unaligned position, the torque is negative (i.e. braking or regenerating), regardless of the direction of the current. Therefore the ideal motoring current waveform is a rectangular pulse that coincides with the rising inductance. Similarly, the ideal braking current waveform is a rectangular pulse that coincides with the falling inductance. The implication is that the current must be switched on and off in synchronism with the rotor position; in other words, the SR motor is a shaft-position-switched machine just like the squarewave brushless d.c. motor.

To produce torque at all rotor positions the entire 360 degrees must be 'covered' by segments of rising inductance from different phases, as shown in Fig. 8, and the phase currents must be commutated and sequenced to coincide with the appropriate segments as shown. There is no fundamental reason why the conduction period on each phase should not exceed the step angle, and indeed a small amount of overlap is desirable to minimize torque ripple in the form of notches in the instantaneous torque waveform at the commutation instant. Too much overlap can lead to positive impulses of torque at the commutation angles. While these

23

Fig. 8. Commutation sequence produces continuous torque

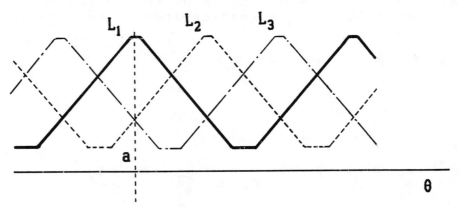

(a) inductance profiles of three phases

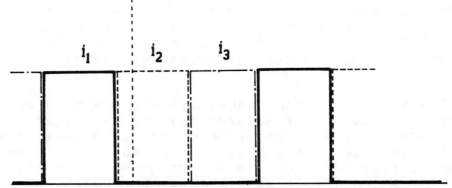

(b) ideal current waveforms of the three phases at low
and medium speeds. With suitable pole geometry these
waveforms produce a relatively smooth torque.

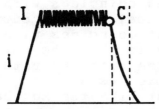

(c) practical phase-current waveform at low speed, controlled
by chopping or p.w.m.

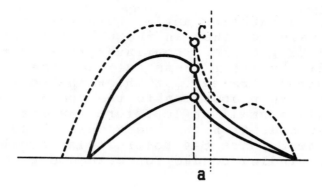

(d) practical current waveforms at very high speed, limited
    by the self-e.m.f. of the motor. These waveforms produce
    significant torque ripple. The lower solid curve is at
    a higher speed than the upper solid curve, but has the
    same switching angles and the same supply voltage. This
    effect causes the torque to fall off at high speed.
    Advancing the turn-on angle, as shown by the dotted curve,
    increases the current and the torque, up to the point where
    the total conduction angle equals the rotor pole pitch.

add to the average torque, they impose transient or
vibratory stresses on the shaft, coupling, and load.
'Commutation underlap', on the other hand, is permissible
only at speeds high enough so that the rotor inertia
(including any load inertia) can maintain rotation through
the torque notches. Any underlap at zero speed may result
in failure to start if the rotor position happens to fall
between the turn-off angle of one phase and the turn-on
angle of the next. In the 6:4 motor the minimum conduction
angle is 30deg, and in the 8:6 motor it is 15deg.

     Not all SR motors admit the same degree of overlap
between phases, because of the geometrical constraints of
the pole geometry. With 6:4 3-phase motors the maximum
conduction angle is 45deg., i.e. 1.5 times the step angle.
In the 8:6 4-phase motor, the maximum conduction angle is
30deg., i.e. 2.0 times the step angle. The 4-phase motor can
therefore have more conduction overlap between phases, but
it will only be useful if the stator poles can be made
sufficiently wide to provide a correspondingly wider angle
of rising inductance, without reducing the slot area to the
point where copper losses become too high. In other words,
the 4-phase motor might have less torque ripple but only at
the expense of efficiency.

Note that the 'conduction angle' here refers to the conduction of the main switches in the phaselegs of the controller. The winding current continues to flow through the freewheeling diodes after these switches are turned off. With the usual control strategy, in which the voltage is regulated by p.w.m. in proportion to the speed, the rotor rotates through only a small angle before the freewheeling current is extinguished. At high speeds this is no longer the case, and in the 3-phase 6:4 motor it is possible to approach 90deg. of winding conduction with 45deg. of main switch conduction.

Energy conversion loop

The average torque can be estimated from the 'energy conversion loop' which is the locus described on the flux-linkage/current diagram by the point whose coordinates are (i, ψ) during each step or working stroke. This is shown in Figs. 9, 12 and 15. Note the saturation in the aligned magnetization curve in each of these figures. Rectangular current pulses can be realized at low speeds if the phase is switched on at the unaligned position and off at the aligned position, and the current is maintained constant by some external means such as p.w.m. regulation or a large external reactor. There is also a 'natural' speed at which the current waveform is flat-topped; this speed and the corresponding current are related by the equation

$$V - Ri = i\omega \frac{dL}{d\theta}$$

(section 4). Again, this equation is valid only if saturation is negligible.

The electromagnetic energy that is available to be converted into mechanical work is equal to the area W. In one revolution each of the q phases conducts as many strokes as there are rotor poles, so that there are qN_r strokes or steps per revolution. The average torque is therefore given by

Average torque = Work/stroke X $\dfrac{\text{No. of strokes/rev}}{2\pi}$

or

$$T_a = W \, \frac{qN_r}{2\pi} \qquad\qquad \text{N-m}$$

The average electromagnetic power converted is

$$P_e = \omega T_a$$

where ω is the speed in rad/s. From this must be subtracted the friction and windage and rotor core losses.

The area W shown within the dotted line in Fig. 9 represents the maximum energy available for conversion with the current and flux-linkage limited to the values shown. Obviously the torque per ampere will be maximised if the

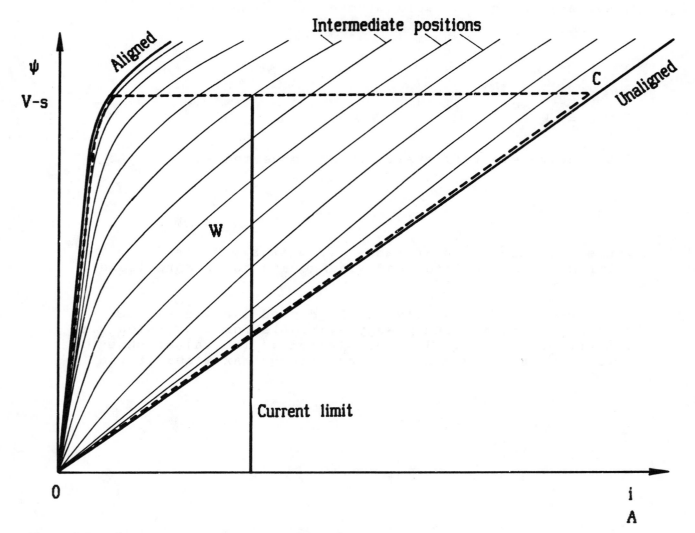

Fig. 9. Energy conversion loop showing
maximum available energy conversion (dotted trajectory).
In practice the current is limited, either thermally or
by the converter, as shown by the vertical line. Similarly
the maximum flux-linkage is limited by core losses and the
tendency towards a spiky current waveform and noisy operation.
In small motors only a small fraction of the
available conversion energy can be utilized.

area between the aligned and unaligned curves is maximized. Ideally this requires

1. the largest possible unsaturated aligned inductance, implying a small airgap with wide poles;

2. the smallest possible unaligned inductance, implying a large interpolar arc on the rotor, narrow stator poles, and correspondingly deep slotting on both the stator and the rotor;

3. the highest possible saturation flux-density.

While the geometry is simple, it is by no means easy to achieve these objectives in design calculations and computer methods are essential to get a good result. The pole geometry that maximises the 'static' torque per ampere (as derived from Fig. 9) can be uniquely determined, following procedures that have long been established for stepper motors. The problem is more complicated for the SR motor because of the unequal rotor and stator slotting, and when the dynamics are taken into account it becomes impossible to establish a single lamination geometry that is optimum for all applications. The main reason for this is that the SR motor is usually applied as a high-speed machine with a wide range of variable speed, and therefore the current waveform is almost never a pure rectangular wave. The current waveform may vary from the one shown in Fig. 8c at low speeds to one of those shown in Fig. 8d at high speeds. At low speeds the current is forced to have an approximately rectangular shape by chopping or p.w.m. in the controller. At high speeds there is no chopping and the current waveform takes up a natural shape determined by the speed, the turn-on and turn-off angles, the applied voltage, and the rate of change of inductance.

The static energy-conversion diagram is a useful first step in design. It requires the calculation of only the aligned and unaligned magnetization curves, which are both amenable to finite-element analysis as shown in Fig. 10. It provides an upper bound on the torque capability, which can be approached at low speeds.

If the vertical line representing the maximum permissible current is swept from left to right, the area W initially increases with the square of the current, but as saturation sets in it becomes more nearly linear. The torque per ampere thus becomes more nearly constant as the current increases, but it is never constant in the way that it is in d.c. and surface-magnet brushless d.c. PM motors.

In small motors only a small fraction of the available energy can be converted because of the thermal current limit. An increase in scale naturally permits more of the

29

Fig. 10. Finite-element flux-plot for SR motor

(Courtesy Lucas Engineering & Systems Ltd.)

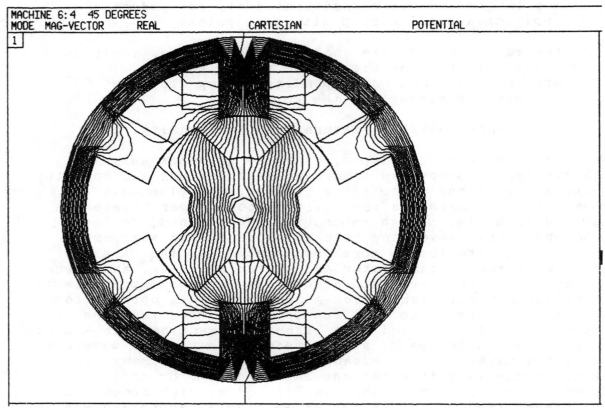

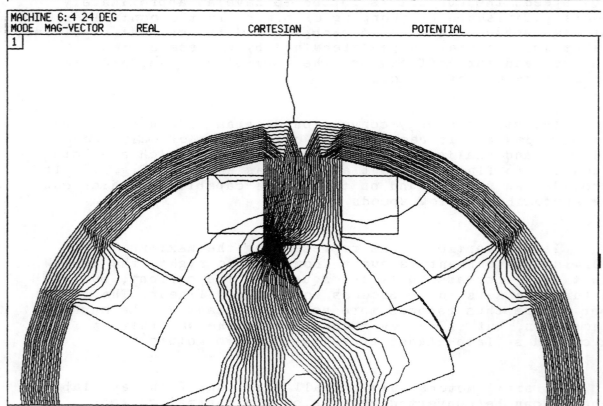

available energy to be converted. The same increase could be obtained with more intense cooling; or, alternatively, during intermittent operation. However, operation with an extreme conversion loop may be inefficient and noisy, with a poor power factor and peaky currents.

A rough estimate of the maximum attainable torque per unit rotor volume can be derived from an idealised triangular area approximating the dotted trajectory in Fig. 9. Following methods of Harris [1975] the result is

$$\frac{T}{V} = \frac{N_r B_s^2 (\lambda - 1) q}{2\pi^2 \mu_0} \cdot \frac{\beta g}{R} \quad N\text{-}m/m^3$$

where B_s is the flux-density in the stator poles at the maximum flux-linkage ψ_s in the aligned position; λ is the aligned/unaligned unsaturated inductance ratio; β is the pole arc (assumed equal for stator and rotor); and g is the airgap. For a three-phase 6:4 motor having a pole arc of 30 deg. and an airgap of 0.25 mm, at a rotor radius R of 25mm, it should be possible to achieve $\lambda = 10$ and B_s = 1.6 T, giving a specific torque of 60 kN-m/m^3 from the extreme trajectory. With very small motors (say, less than 100W) the fraction of this theoretically available torque that can be achieved continuously with quiet operation and acceptable losses may be only of the order of 5%. This figure improves rapidly with scale so that at 5kW it should be possible to achieve up to 25% or 15 kN-m/m^3, and in highly-rated machines with special cooling perhaps double this; transient ratings may be still higher.

4. Partition of energy and the effects of saturation

The shape of the energy conversion loop, and its area W, depend on the variation of current with rotor angle, and this in turn depends on the control parameters and the speed, as well as on the motor design. The current waveform may vary widely over the operating range. To determine both the current and the torque at speed it is necessary to simulate the operation of the motor (and converter) for at least one stroke. This means solving the terminal voltage equation as a function of time, and since this is a differential equation a time-stepping method such as Euler's method or the Runge-Kutta procedure is required. In the case of the ideal machine with no fringing or magnetic saturation, represented by the inductance variation shown in Fig. 7, an analytical solution is possible, as elegantly presented by Ray and Davis [1979], in which many of the basic characteristics of the SR drive are identified and illustrated. Here, however, the solution by numerical techniques will be outlined: such methods are necessary in practice to deal with saturation and fringing effects.

The terminal voltage equation for one phase is

$$v = Ri + \frac{d\psi}{dt}$$

Suppose that the flux-linkage ψ is a function of both current i and rotor angle θ:

$$\psi = \psi(i, \theta)$$

Then

$$\frac{d\psi}{dt} = \frac{\delta\psi}{\delta i} \cdot \frac{di}{dt} + \omega \frac{\delta\psi}{\delta\theta}$$

$$= L \frac{di}{dt} + e$$

where L is the incremental inductance (the slope of the magnetization curve) and e is a 'back-e.m.f.' This equation is quite general and shows that from the terminals the SR motor appears to have an equivalent circuit that comprises, in each phase, a resistance, an incremental inductance, and a back-e.m.f., e. In a general way this is similar to other motors. But the back-e.m.f. is different in that it depends on the phase flux-linkage and therefore on the phase

32

current. e also varies with rotor position. It cannot be regarded as the only term in the voltage equation that contributes to the torque production, as it can in the d.c. motor. The term 'self-e.m.f.' will be used instead, as a reminder that the product ei includes an energy-storage component as well as an energy-conversion component. During the interval of rising inductance, e varies strongly with current and weakly with rotor position, but L varies strongly with rotor position and weakly with current.

Apart from losses, the electrical energy supplied at the terminals during a small rotation is partitioned between stored magnetic field energy and mechanical work. Ideally all the energy would be converted to mechanical work, but this cannot be achieved in practice. The proportion in which the energy divides between magnetic field energy and mechanical work depends on the shape of the magnetization (ψ–i) curves in the neighborhood of the particular rotor position.

If there is no saturation the incremental inductance is the total inductance at the particular rotor angle, and this is equal to the ratio of flux-linkage to current. In this case

$$\frac{d\psi}{dt} = L\frac{di}{dt} + i\omega\frac{dL}{d\theta}$$

The first term has the appearance of inductive voltage drop across a fixed inductance, while the second term is the self-e.m.f. proportional to current, speed, and rate of change of inductance with rotor angle. If the current is flat-topped, then during the flat-top period the self-e.m.f. is constant and the first term is zero. This defines the 'natural' speed at which a flat-topped current waveform is achieved without chopping, as discussed in section 3. On the other hand, if the inductance is constant (as for example around the unaligned position), the self-e.m.f. is zero and the first term absorbs all the applied voltage. The equivalent circuit can change from being mainly an inductance to mainly an e.m.f., depending on the rotor angle and the current waveform.

If the above equation is multiplied by i, the left-hand side represents the electrical power supplied (after resistive losses have been subtracted). The rate of change of stored magnetic energy is

$$\frac{d}{dt}\left[\frac{1}{2}Li^2\right] = iL\frac{di}{d\theta} + \frac{1}{2}i^2\frac{dL}{d\theta}$$

Subtracting this equation from the previous one multiplied by i, the electromechanical energy conversion is

$$P_m = T\omega_m = \frac{1}{2} i^2 \frac{dL}{d\theta} \cdot \omega_m$$

which gives the 'linear' expression for instantaneous torque as before. Of the electrical power supplied, however, this represents less than half. The remainder is the rate of magnetic energy storage, which exceeds the electromechanical conversion by the term

$$iL \frac{di}{dt}.$$

The most 'effective' use of the energy supplied is when the current is maintained constant (during a period of rising inductance), and even then the highest level of 'effectiveness' is only 50%. This is illustrated in Fig. 11a, which shows the energy exchanges over a small rotation $\Delta\theta$. The triangular area ΔW_m representing the mechanical energy conversion is one-half the rectangular area ΔW_e representing the electrical energy supplied, since it has the same base ($\Delta\psi$) and the same height i, the current i being constant through the small rotation.

Note that the energy stored in the magnetic field is not necessarily dissipated. With the appropriate converter circuit it can be recovered to the supply at the end of the period of rising inductance. This is why the term 'effectiveness' is used, not 'efficiency'. The consequence of a low 'effectiveness' is to increase the volt-ampere rating of the converter for a given power conversion in the motor. 'Effectiveness' is therefore akin to power factor in a.c. machines, and it can be defined more precisely in terms of the energy ratio (see below).

The partition of input energy into mechanical work and stored field energy is improved if the motor saturates. This has been discussed very clearly by Byrne [1972]. If, at a given rotor position, the magnetization curve is saturated as in Fig. 11b, then the area representing the mechanical work can exceed half the area of the 'supply rectangle'. An extreme case is shown in Fig. 11c, with very sharp saturation occurring at a low current level, and extremely high inductance below this level. In this case practically all the energy supplied is converted to mechanical work and very little is stored in the magnetic field. The effectiveness theoretically approaches 100%. But such curves could not be realised with practical electrical steels, and they would require very small or zero airgap clearances.

34

Fig. 11. Partition of input electrical energy

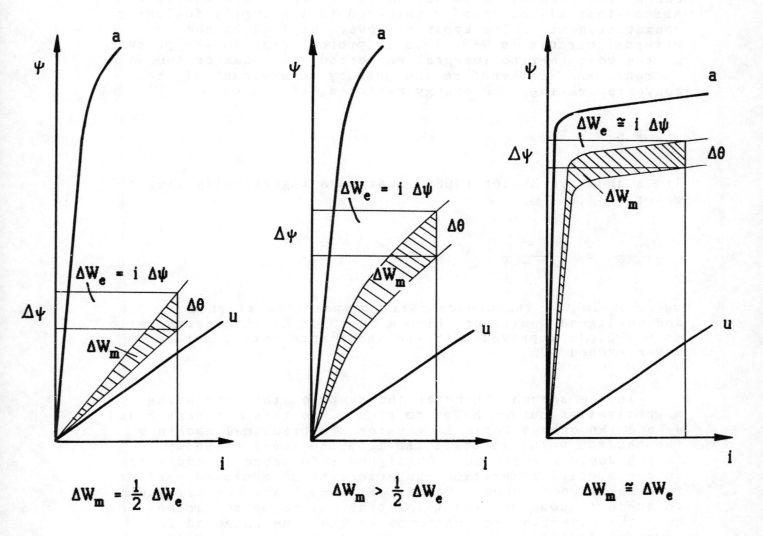

(a) Linear (no saturation) (b) Typical practical case (c) Idealised case with extreme saturation

This argument has been developed for incremental exchanges of energy taking place over very small rotations. Similar principles apply to the total exchanges taking place in a complete working stroke, Fig. 12. The total electromechanical energy converted is W, as before. The energy returned by the motor to the external circuit is R; assume that all of this is returned to the supply following commutation at C. The apparent power supplied by the external circuit is W+R. This is proportional to the product of the voltage-time integral and either the peak or the mean current, and is therefore the primary determinant of the converter rating. The energy ratio is defined as

$$\frac{W}{W + R}.$$

It is shown in Miller [1985] that in a magnetically linear motor

$$\frac{W}{W + R} = \frac{\lambda_u - 1}{2\lambda_u - 1}$$

where λ_u is the inductance ratio between the aligned and unaligned positions. With a ratio of 6, the energy ratio is 0.455. It improves with the inductance ratio, but can never exceed 0.5.

In the saturating motor the shape of the saturating magnetization curves helps to reduce the ratio between R and W, and the energy ratio is greater. In practical machines the maximum value is still far from the ideal of unity (which would require zero unaligned inductance in addition to the special saturation charactersitics mentioned earlier for the aligned curve). Values of 0.6-0.7 are not difficult to achieve, however, and this turns out to be sufficient to keep the converter volt-amperes in the same range as for inverter-fed induction motors [Harris et al 1986; Miller 1985]. However, this is only possible because of the increased switching frequency (the vernier effect). For the same number of rotor poles the SR motor has roughly half the electromagnetic 'effectiveness' of a good a.c. motor, but it traverses the energy conversion loop twice as often, restoring the converter volt-amperes to about the same level as in a.c. drives. Of course the penalty is a higher switching (commutation) frequency.

The concept of energy ratio can be applied also to a.c. machines (on a per-phase basis); when this is done it is found that even for quite low values of energy ratio the terminal power factor remains high and the converter volt-amperes remain low.

36

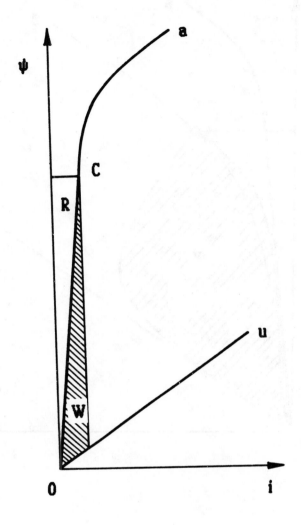

Fig. 12. Energy ratio and energy exchanges during one complete working
stroke

    (a) Linear case

   W = energy converted into mechanical work
   R = energy returned to the d.c. supply via freewheel diodes

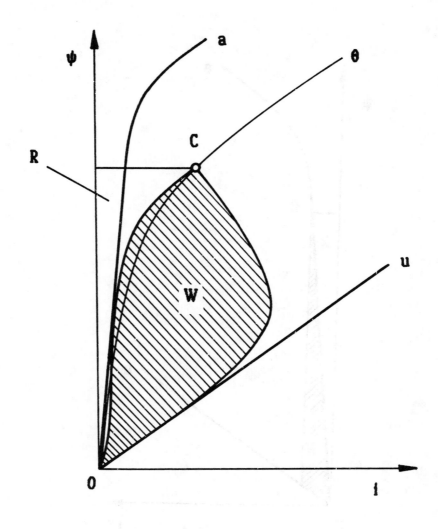

(b) Typical practical case

In the literature there has been much discussion of the 'torque doubling' effect of saturation. While saturation improves the energy ratio, it decreases the conversion area W per unit of motor volume. The decrease in converter kVA/kW outweighs the decrease in power per unit volume, and for this reason saturation is desirable up to the point where the increased core losses associated with it limit the power density. It is not the case that high-speed reluctance machines can produce 'double the torque' and it is equally untrue that they require 'double the kVA'. When all the complexities of design are taken fully into account, their performance in terms of these parameters is very roughly on a par with that of induction motors.

5. Dynamic torque production

Under normal operating conditions at speed, the energy exchanges, both incremental and total, can be determined by integrating the voltage equation and developing the conversion loop in the ψ-i diagram. The necessary time-stepping procedure was developed by Stephenson and Corda [1979] and only an outline of their method is described here.

The voltage equation is integrated in the form

$$\psi := \int (v - R.i) \, dt$$

through one time-step, giving a new value of ψ. If the speed is assumed constant, the integration can be done with respect to rotor angle θ. Otherwise the rotor angle must be determined by a simultaneous integration of the mechanical equations of motion, as is normal in such simulations. At the end of the time-step θ and ψ are both known, and the current i can be determined from the magnetization curve for that rotor angle. To minimize this computation Stephenson and Corda used a set of polynomials to represent the magnetization curves at a number of rotor angles between the aligned and unaligned positions, and then applied an interpolation procedure at the end of each time-step to determine the current from the flux-linkage at the particular rotor position. The instantaneous torque can be determined from a difference-approximation to the partial derivative of coenergy at constant current, by a second interpolation procedure that uses stored field energies precalculated at discrete current levels at each of a number of rotor angles. By this method the waveforms of instantaneous phase current and torque can be developed from the integration. An example is shown in Figs. 13-15.

Phase Current vs. Angle

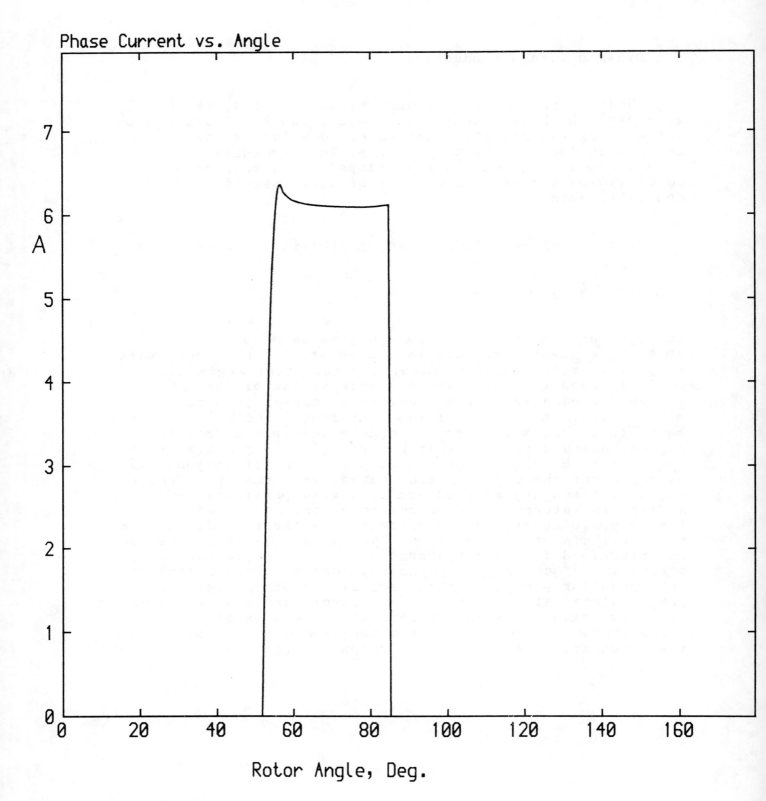

Rotor Angle, Deg.

Fig. 13. Current waveform produced by Stephenson and Corda's method

(a) Low and medium speeds. The natural flat-topped
waveform arises when the self-e.m.f. equals the
applied voltage.

40

Phase Current vs. Angle

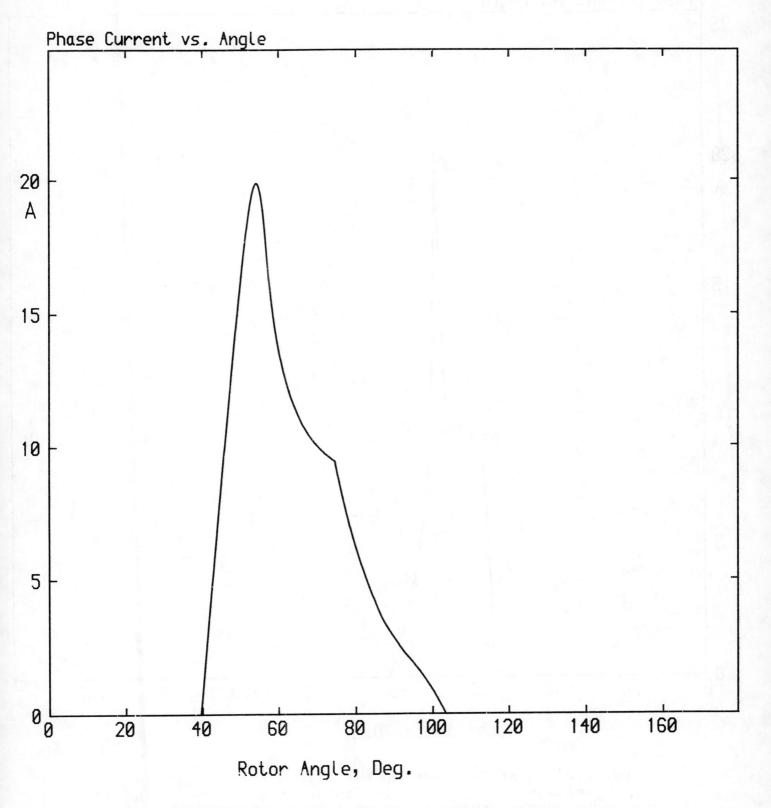

Rotor Angle, Deg.

(b) Very high speed. The self-e.m.f. exceeds the applied
    voltage once the poles begin to overlap. Note the
    phase advance; the turn-on angle is at 40deg., i.e.
    5deg. before the unaligned position. Commutation is
    also advanced relative to that in (a).

41

Phase Current vs. Angle

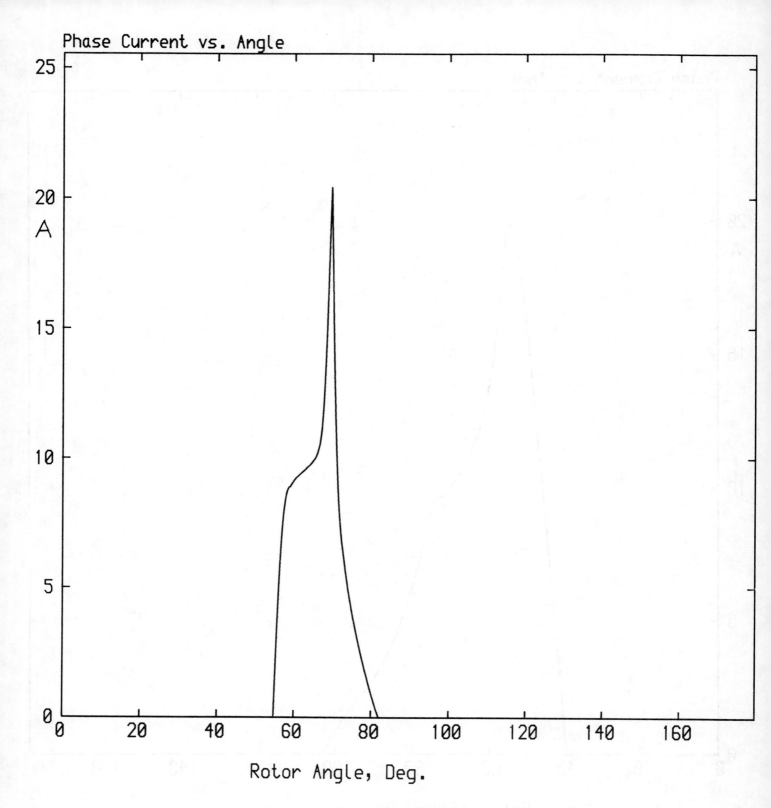

Rotor Angle, Deg.

(c) Low and medium speeds, but with a higher applied
    voltage than in (a). The motor is being driven harder
    to convert more of the available energy than in (a).
    The firing anlges are the same as in (a) but because
    of saturation a peak appears in the current waveform.
    To avoid this, the voltage must be reduced, or
    the commutation angle advanced, or the stator yoke
    must be made thicker.

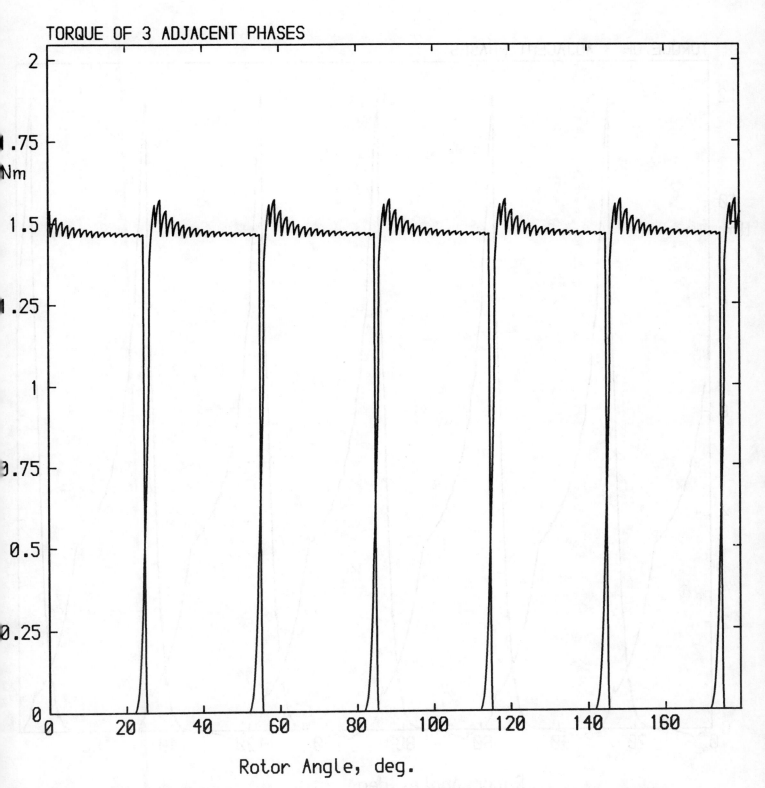

Fig. 14. Torque waveforms of the three phases corresponding to Fig. 13
The high-frequency ripple effect, especially in (a), is due
to numerical approximation of the magnetization curves.

(a) Low and medium speeds

43

TORQUE OF 3 ADJACENT PHASES

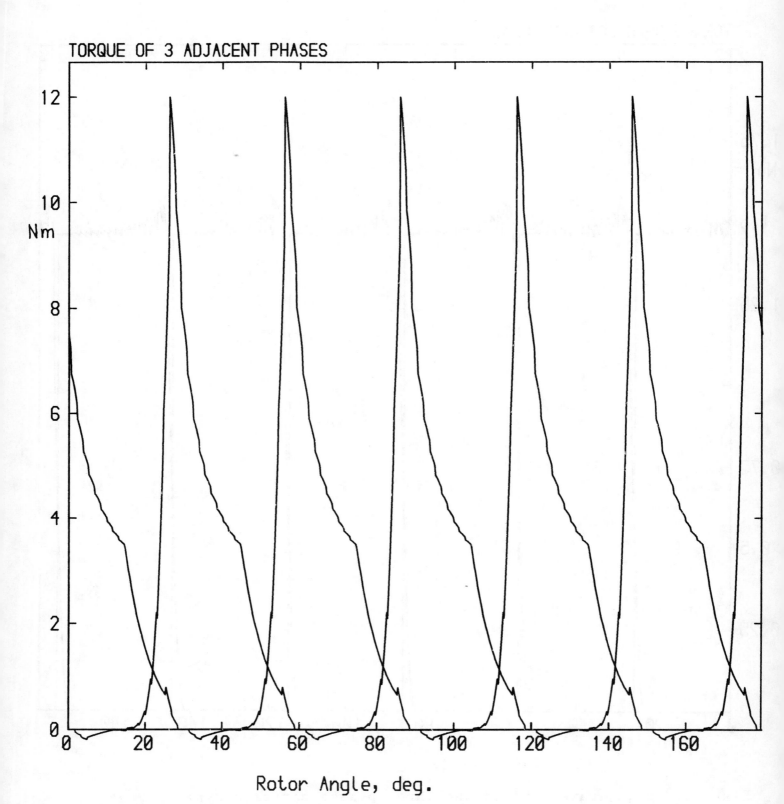

Rotor Angle, deg.

(b) High speed

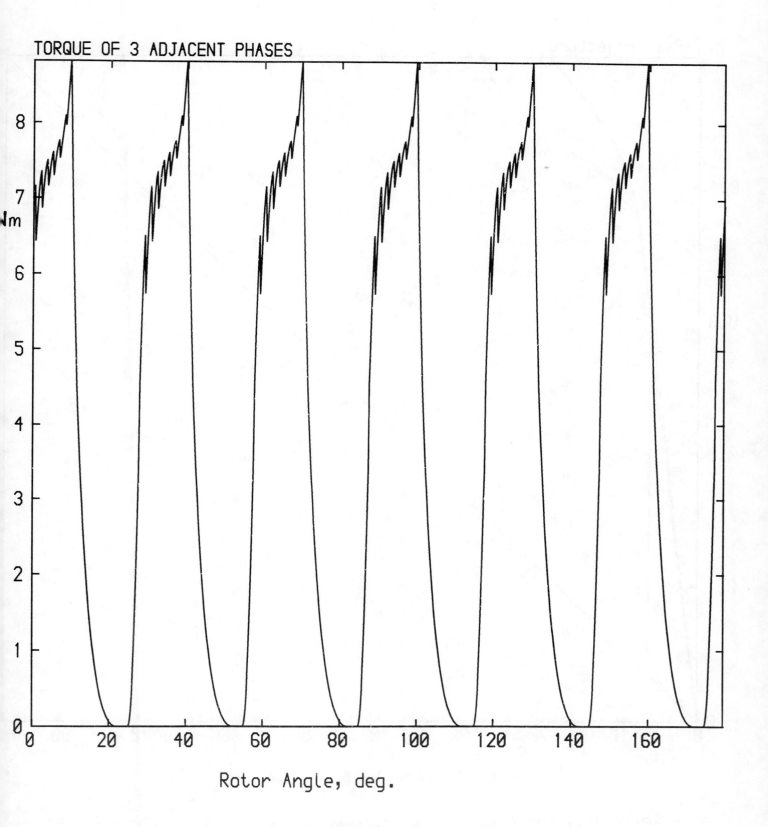

(c) Low and medium speed with increased drive voltage

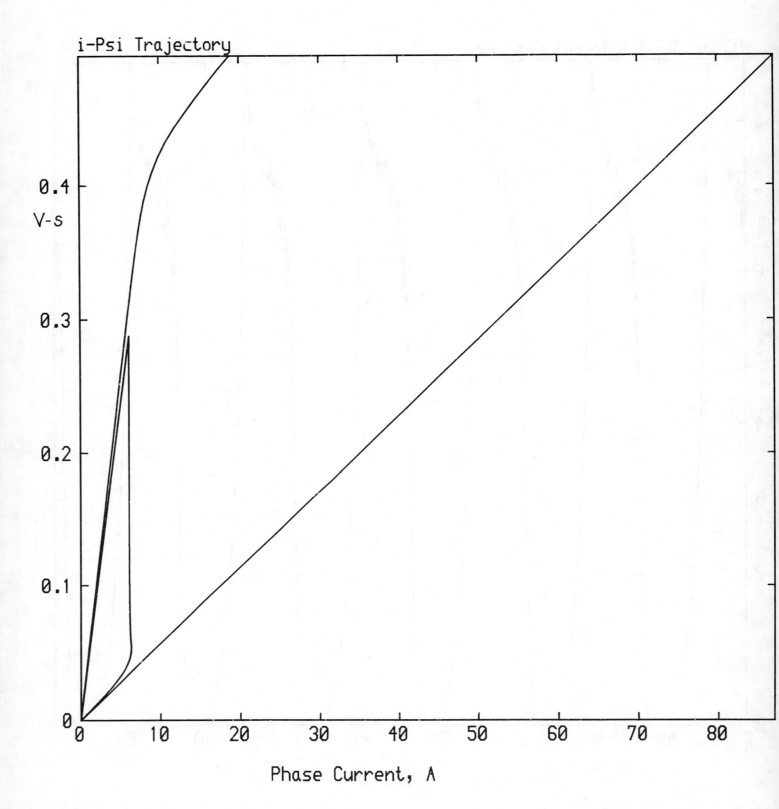

i-Psi Trajectory

Fig. 15. Energy conversion loops corresponding to Figs. 13 and 14.

(a) Low and medium speeds

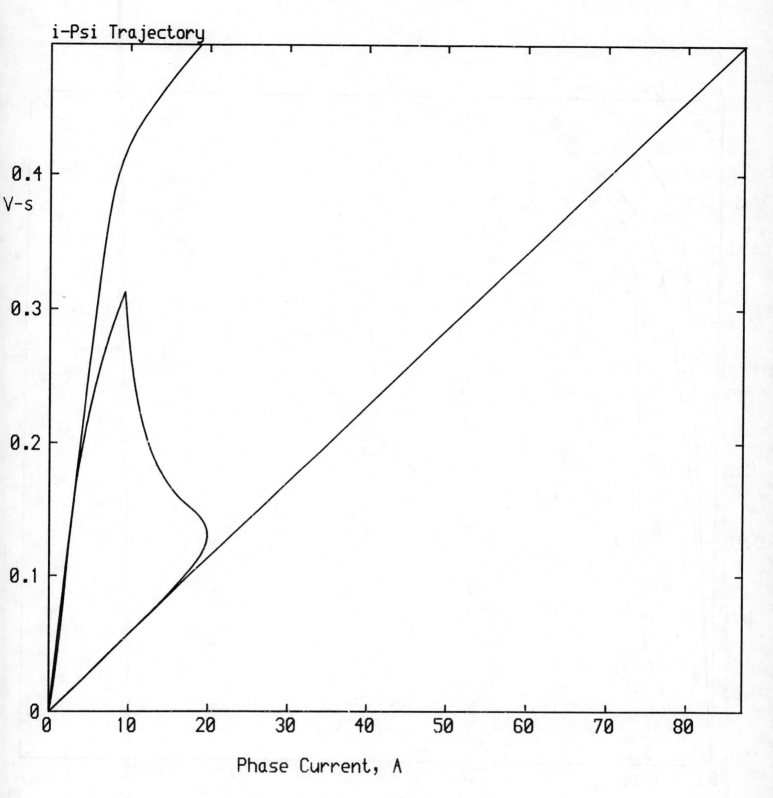

i-Psi Trajectory

0.4
V-s

0.3

0.2

0.1

0

0   10   20   30   40   50   60   70   80

Phase Current, A

(b) High speed

The converted energy (and average torque) is much higher in
(b), showing how the increase in conduction angle overcomes
the effects of increasing self-e.m.f. at high speeds. This is
an example of the 'programmability' of the speed/torque
characteristics [Byrne 1976]

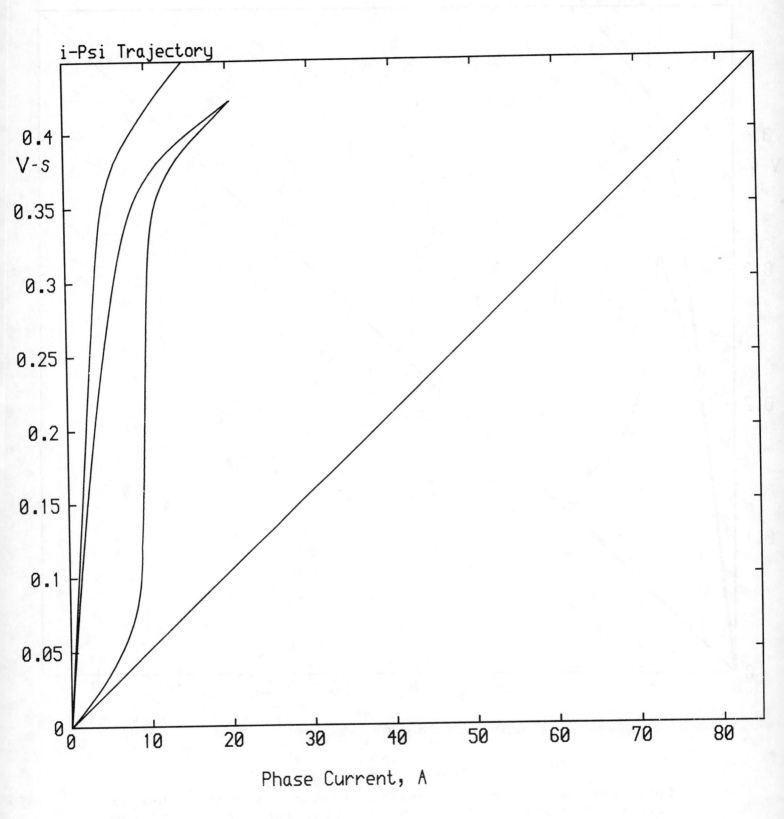

i-Psi Trajectory

Phase Current, A

(c) Low and medium speed with increased drive voltage

drives (say, above 20 kW) it becomes progressively more difficult to chop at such a high frequency because of the limited switching speed and losses of larger devices (such as GTO's). This makes it more difficult to achieve quiet operation in larger motors, particularly at low speeds.

In small drives it is often acceptable to use p.w.m. control over the entire speed range. This is also the usual approach with surface-magnet PM brushless motors. In such cases the SR controller circuit can profitably be reduced to the circuit of Fig. 16b, in which the chopping is performed by one transistor in common for all the phases. The lower transistors commutate the chopped voltage to the phases in proper sequence, under the control of the shaft position sensor and gating logic. This circuit requires only q+1 transistors and q+1 diodes for a motor with q phases. A three-phase motor thus requires only 4 transistors and 4 diodes. There is practically no loss of functionality with this circuit relative to the full circuit having 2n transistors in Fig. 16a, and indeed at low speeds it tends to run more smoothly. Its main limitation is that at very high speeds the phases cannot be 'de-fluxed' or de-energised fast enough through the diodes, because the control transistor keeps switching on, with a long duty cycle. If there is still a significant freewheeling current by the time the rotor reaches the aligned position, it may start to increase as the self-e.m.f. becomes negative, and braking torque is produced. As the chopping duty cycle and/or the speed increase further, the net torque decreases rapidly and the losses increase. These problems only arise when the speed range is very wide, typically more than 20:1.

Many other circuits have been developed in attempts to reduce the number of switches all the way down to q (i.e., one per phase) and take full advantage of unipolar operation. (See Bass et al [1987] for a review). The split-link circuit in Fig. 16d has been successfully used by Tasc Drives Ltd. with GTO thyristors in a range of highly efficient drives from 4-80kW. In other cases it seems that when the device count is reduced to one per phase, there is a penalty in the form of extra passive components or control limitations. The bifilar winding in Fig. 16c suffers from double the number of connections, a poor utilization of copper, and voltage spikes due to imperfect coupling between the bifilar windings. In Fig. 16e the device count is reduced to q plus one additional device to bleed the stored energy from the dump capacitor C back to the supply via the step-down chopper circuit. The mean capacitor voltage is maintained well above the supply rail to permit rapid de-fluxing after commutation. A control failure in the energy-recovery circuit would result in the rapid build-up of charge on the dump capacitor, and if protective measures were not taken the entire converter could fail from overvoltage.

6. Converter circuits

The torque is independent of the direction of the phase current, which can therefore be unidirectional. This permits the use of unipolar controller circuits with a number of advantages over the corresponding circuits for a.c. or PM brushless motors, which require alternating current. Although the SR motor could be operated with alternating (but nonsinusoidal) current, unidirectional (d.c.) current has the added advantage of reducing hysteresis losses.

In Fig. 16a is shown a circuit well suited for use with transistors (bipolar, field-effect, or insulated-gate). The phases are independent, and in this respect the SR controller differs from the a.c. inverter, in which the motor windings are connected between the midpoints of adjacent inverter phaselegs. The winding is in series with both switches, providing valuable protection against faults. In the a.c. inverter the upper and lower phaseleg switches must be prevented from switching on simultaneously and shorting the d.c. supply; this is possible only by means of additional control circuitry, which is unnecessary in the SR controller.

The upper and lower phaseleg switches are switched on together at the start of each conduction period or working stroke. At the commutation point (C in Fig. 7 and 8) they are both switched off. During the conduction period either or both of them may be chopped according to some control strategy, such as maintaining the current within a prescribed 'hysteresis band'. This mode of operation is necessary at low speeds when the self-e.m.f. of the motor is much smaller than the supply voltage. At high speeds both transistors remain on throughout the conduction period and the current waveform adopts a 'natural' shape depending on the speed and torque. It is convenient in the logic design to use one transistor primarily for 'commutation' and the other for regulation or chopping. At the end of the conduction period when both switches are turned off, any stored magnetic energy that has not been converted to mechanical work is returned to the supply by the current freewheeling through the diodes. Note that when they become forward-biased, the diodes connect the negative of the supply voltage across the winding to reduce its flux-linkage quickly to zero.

The inductance varies with rotor position. Therefore if fixed-frequency chopping is used, the current ripple varies. If hysteresis-type current regulation is used, the chopping frequency varies as the poles approach alignment. Figs. 17-19 show operation in the chopping mode. Chopping frequencies above 10kHz are usually desirable, as in other types of drive, to minimize acoustic noise. In larger

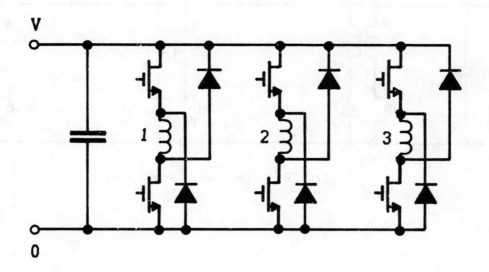

Fig. 16.　Converter circuits for 3-phase SR motor

(a) 2-transistor/phase circuit

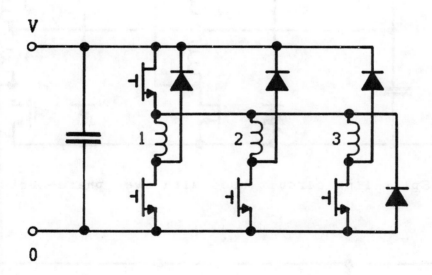

(b) n+1 transistors for n-phase motor

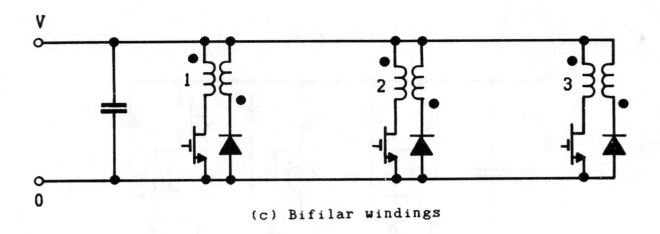

(c) Bifilar windings

(d) Split-link circuit used with even phase-number

(e) C-dump circuit

7. Control: current regulation, commutation

For motoring operation the pulses of phase current must coincide with a period of increasing inductance, i.e. when a pair of rotor poles is approaching alignment with the stator poles of the excited phase. The timing and dwell of the current pulse determine the torque, the efficiency, and other parameters. In d.c. and brushless d.c. motors the torque per ampere is more or less constant, but in the SR motor no such simple relationship emerges naturally. With fixed firing angles, there is a monotonic relationship between average torque and r.m.s. phase current, but in general it is not very linear. This may present some complications in feedback-controlled systems although it does not prevent the SR motor from achieving 'near-servo-quality' dynamic performance, particularly in respect of speed range, torque/inertia, and reversing capability.

It is characteristic of good operating conditions that the conversion loop fits snugly in the space between the unaligned and aligned magnetization curves, as in Figs. 15 and 19. This principle was recognized and clearly explained by Byrne [1976], following ideas that had previously been published by Melcher at M.I.T. Fig. 15b corresponds to high-speed operation where the peak current is limited by the self-e.m.f. of the phase winding. A smooth current waveform is obtained with a peak/r.m.s ratio similar to that of a half sinewave.

At low speeds the self-e.m.f. of the winding is small and the current must be limited by chopping or p.w.m. of the applied voltage. The situation here is exactly similar to that in the brushless d.c. PM motor. The regulating strategy employed has a marked effect on the performance and the operating characteristics. Fig. 17 shows a current waveform controlled by a 'hysteresis-type' current-regulator that maintains a more or less constant current throughout the conduction period in each phase. Fig. 20a shows schematically the method of control. As the current reference increases, the torque increases. At low currents the torque is roughly proportional to current squared, but at higher currents it becomes more nearly linear. At very high currents saturation decreases the torque per ampere again. This type of control produces a constant-torque type of characteristic as indicated in Fig. 21. With loads whose torque increases monotonically with speed, such as fans and blowers, speed adjustment is possible without tachometer feedback, but in general feedback is needed to provide accurate speed control. In some cases the pulse train from the shaft position sensor may be used for speed feedback, but only at relatively high speeds. At low speeds a larger number of pulses per revolution is necessary, and this can

Phase Current vs. Angle

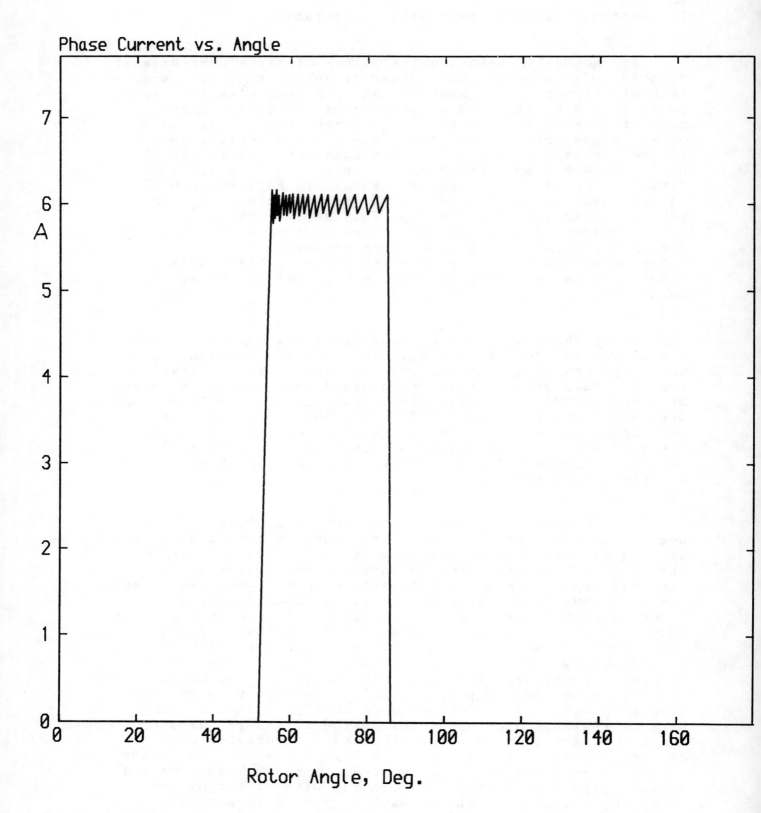

Fig. 17. Current waveform in chopping mode at low speed. Note the
reduction in chopping frequency as the phase inductance
increases with increasing overlap.

54

TORQUE OF 3 ADJACENT PHASES

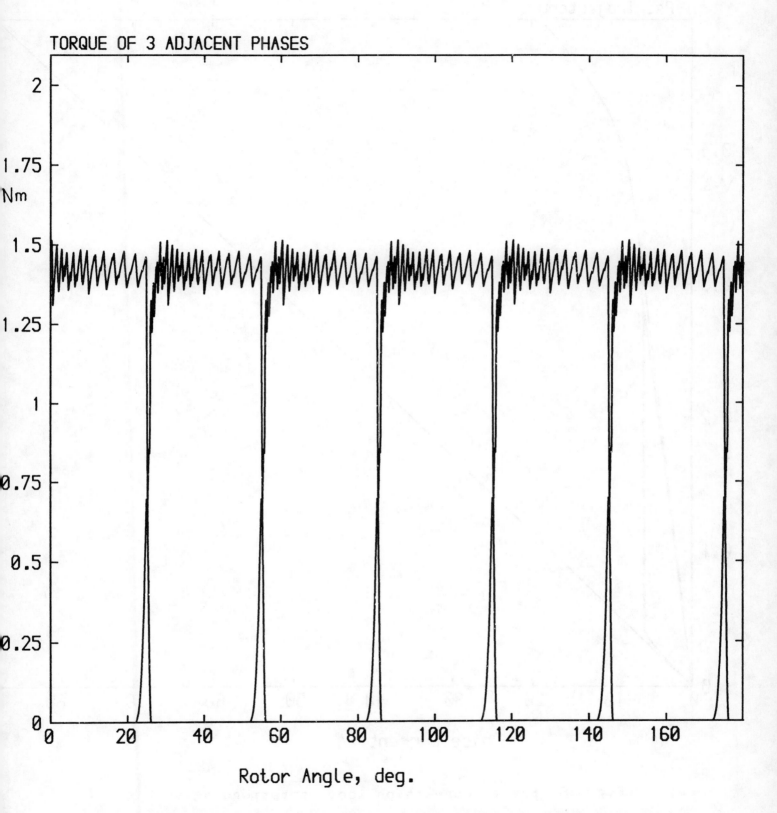

Rotor Angle, deg.

Fig. 18. Torque waveforms of the three phases, corresponding to
Fig. 17.

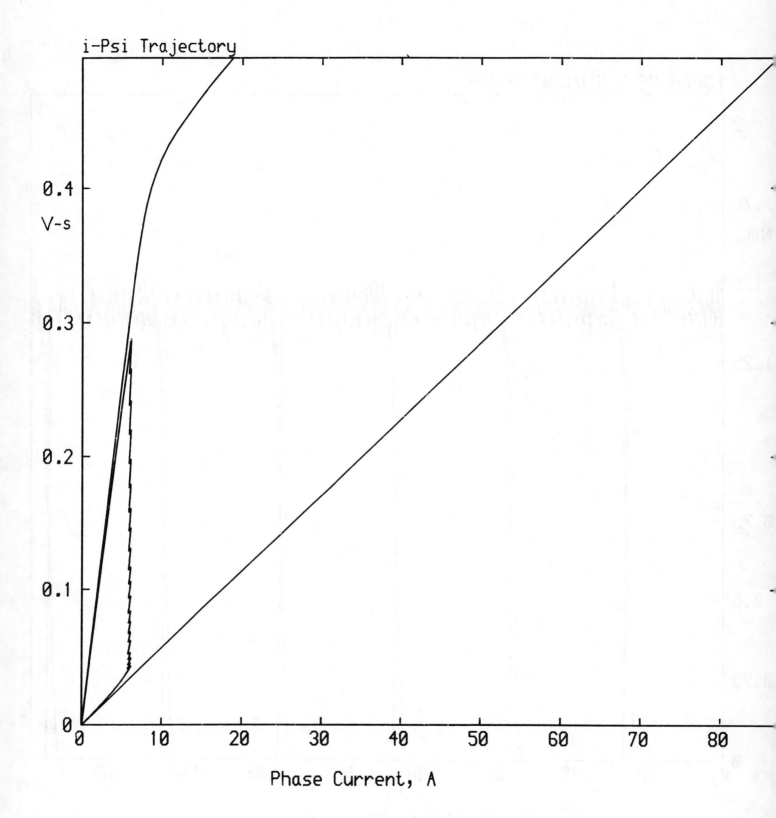

i-Psi Trajectory

Fig. 19. Energy conversion loop corresponding to Fig. 17.

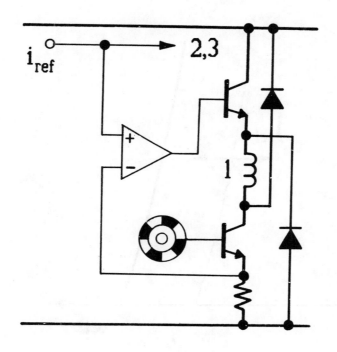

(a) hysteresis-type

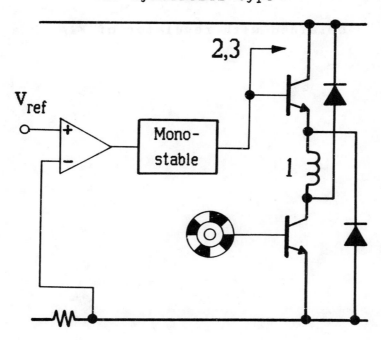

(b) voltage-p.w.m. type (duty-cycle control)

Fig. 20. Schematic of current-regulator for one phase

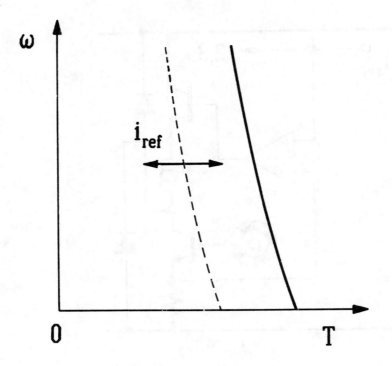

Fig. 21. Constant-torque characteristic

obtained with regulator of Fig. 20a

58

be generated by an optical encoder or resolver, or alternatively by phase-locking a high-frequency oscillator to the pulses of the commutation sensor [Bose 1986]. Systems with resolver-feedback or high-resolution optical encoders can work right down to zero speed. The 'hysteresis-type' current regulator may require current transducers of wide bandwidth, but the SR drive has the advantage that they can be grounded at one end, with the other end connected to the negative terminal of the lower phaseleg switch. Shunts or Hall-effect sensors can be used, or alternatively, 'Sensefets' with in-built current sensing. Much of the published literature on SR drives describes this form of control.

Fig. 20b shows an alternative regulator using fixed-frequency p.w.m. of the voltage with variable duty-cycle. Again the situation is exactly similar to that in a brushless d.c. PM motor, and indeed the control circuits used for the SR motor may be adapted directly from those developed for the PM motor. The current waveform is similar to that shown in Fig. 13, except that after commutation the current decays through the diodes somewhat more rapidly because the reverse voltage applied is effectively d times the forward voltage applied before commutation. (d = duty cycle). The torque and energy-conversion loop are similar to Figs. 14 and 15. The duty-cycle (or "off-time") of the p.w.m. can be varied by a simple monostable circuit. This form of control is similar to armature-voltage control in a d.c. motor. A purely theoretical analysis of the relationship of speed and torque given by Corda and Stephenson [1982] explains why this type of control produces a characteristic that is closer to constant-speed than that of the current-regulator control just described.

Current feedback can be added to the circuit of Fig. 20b to provide a signal which, when subtracted from the voltage reference, modulates the duty cycle of the p.w.m. and 'compounds' the torque-speed characteristic. It is possible in this way to achieve under-compounding, over-compounding, or flat compounding just as in a d.c. motor with a wound field. For many applications the speed regulation obtained by this simple scheme will be adequate. For precision speed control, normal speed feedback can be added. The current feedback can also be used for thermal overcurrent sensing.

A desirable feature of both the 'hysteresis-type' current-regulator and the voltage p.w.m. regulator is that the current waveform tends to retain much the same shape over a wide speed range.

Phase advance

When the p.w.m. duty cycle reaches 100% the motor speed can be increased by increasing the dwell (the conduction period) or the advance of the current-pulse relative to the rotor position; or both. These increases eventually reach maximum practical values, after which the torque becomes inversely proportional to speed squared, but they can typically double the speed range at constant torque. The speed range over which constant power can be maintained is also quite wide, and very high maximum speeds can be obtained, as in the synchronous reluctance motor and induction motor, because there is not the limitation imposed by fixed excitation as in PM motors.

This principle of phase advance was recognized in the earliest work in variable-reluctance motor drives. The reasons for its effectiveness were eventually clarified by Byrne in his 1976 paper, which discusses phase advance angles that may be large enough to start conduction well before the unaligned position. Byrne worked with a 4:2 motor with wide poles and was able to maintain an essentially rectangular current waveform even with large values of phase advance, although his waveforms show a tendency to become peaky with extreme values. Later workers, notably Ray and Davis [1979], applied this principle to motors of higher pole number, probably with much narrower poles, and as a result they obtained much more peaky waveforms, in which the peak current exceeds the commutated current by a larger margin than in Byrne's motors. Fig. 13b is an example of such a waveform. It has the undesirable characteristic of a large peak/r.m.s. ratio (undesirable, that is, when using transistors; with SCR's it is more acceptable). Also, it is inherent with this waveform that until the peak is reached, current is flowing without producing positive torque, and therefore the efficiency is compromised. Nevertheless, in narrow-pole motors these penalties may have to be paid in order to achieve a wide speed range after the p.w.m. duty cycle has reached 100%.

The same principle of phase advance has, of course, been widely used in other motor drives including the brushless d.c. PM motor, but in the case of the surface-magnet motor the benefits are not very pronounced. It can be shown that this is directly associated with the fixed flux of this motor, and that there is a trade-off between torque per ampere and the speed range at constant power. The surface-magnet PM motor has a high torque per ampere and a limited speed range, whereas the SR motor has a wide speed range and a limited torque per ampere.

A simplified SR control system that provides a limited choice of fixed firing angles together with p.w.m. voltage control is decribed by Miller et al [1988]. This control effectively establishes the equivalence of the SR motor and

60

the brushless d.c. PM motor from a control point of view, and is extremely simple to implement. A schematic diagram of this control is shown in Fig. 22.

Torque/speed characteristic

The generic form of the torque/speed capability curve is shown in Fig. 23. For speeds below ω_b the torque is limited by the motor current (or the controller current, whichever is less). Up to the 'base speed' ω_b it is possible, by means of the regulators in Fig. 20, to get any value of current into the motor, up to the maximum. The precise value of current at a given operating point depends on the load characteristics, the speed, and the regulator and control strategy. In the speed range below ω_b the firing angles can be chosen to optimize efficiency or minimize torque ripple. If the load never needs to operate at high speeds above ω_b, it will usually be possible to design the pole geometry to optimize these parameters without regard to the efficiency at high speeds, and this provides considerable design freedom to obtain smooth torque and simplify the control.

The 'corner point' or base speed ω_b is the highest speed at which maximum current can be supplied at rated voltage, with fixed firing angles. If these angles are still kept fixed, the maximum torque at rated voltage decreases with speed squared. However, if the conduction angle is increased and/or advanced, as discussed earlier, there is a considerable speed range over which maximum current can still be forced into the motor, and this sustains the torque at a level high enough to maintain a constant-power characteristic, even though the core losses and windage losses increase quite rapidly with speed. This is shown in Fig. 23 between points B and P. The angle θ_D is the 'dwell' or conduction angle of the main switching device in each phase. It should generally be possible to maintain constant power up to 2-3 times base speed.

The increase in conduction angle may be limited by the need to avoid peaky currents, or to avoid continuous conduction, which occurs when the conduction angle exceeds half the rotor pole-pitch. It may have to be limited to lower values because of core loss or other factors. At P the increase in θ_D is halted and higher speeds can now only be achieved with the natural characteristic, i.e. torque decreasing with speed squared.

At very low speeds the torque/speed capability curve may deviate from the flat-torque characteristic. If the chopping frequency is limited (as with GTO thyristors, for

61

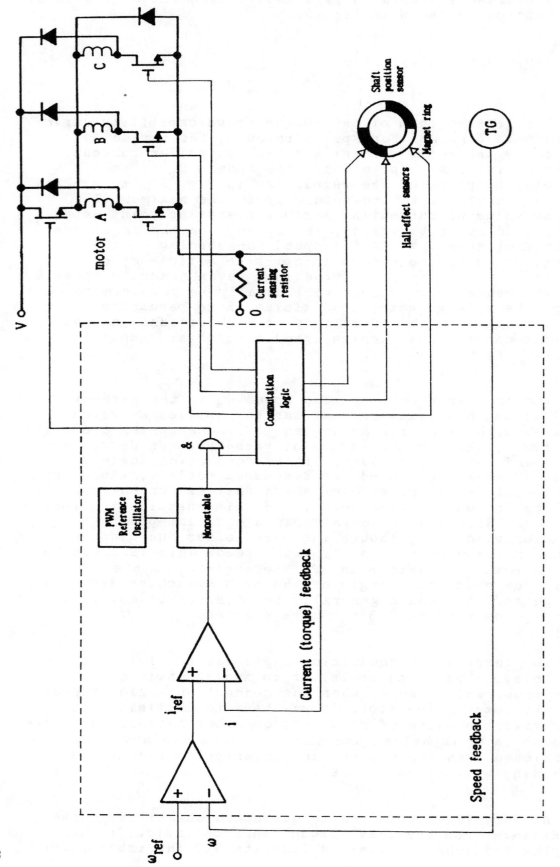

Fig. 22. General form of controller for operating the SR motor as a brushless d.c. motor.

example), or if the bandwidth of the current regulator is limited, it may be difficult to limit the peak current without the help of the self-e.m.f. of the motor, and the current reference may have to be reduced. This is shown in curve (i) in Fig. 23. On the other hand, if this is not a problem, the very low windage and core losses may permit the copper losses to be increased, so that with higher current a higher torque is obtained, as shown in curve (ii). Under intermittent conditions, of course, very much higher torques can be obtained in any part of the speed range up to base speed. In Fig. 24 this can be seen by extrapolating the constant-duty-cycle curves above the maximum current locus.

It is important to note that the current which limits the torque below base speed is the motor current (or converter output current). The d.c. supply current increases from a small value near zero speed to a maximum value at base speed. Basically this is because the power increases in proportion to the speed as long as the torque is constant. With fixed d.c. supply voltage at the input to the converter, the d.c. supply current is approximately proportional to the product of speed and torque.

Fig. 24 shows the computed torque/speed characteristics of a small SR motor. The computation was performed using PC-SRD, a commercially licenced CAD package for SR drives, based on the IBM PC and developed at Glasgow University. The essentially constant-torque characteristic is maintained up to point B at 2250rpm, limited by maximum motor current which corresponds to a winding current-density of 4.8A/mm^2. At speeds above the base speed the natural characteristic is shown at rated voltage, with torque decreasing roughly as speed squared. In the computation, the windage loss was neglected. In practice the effect of windage loss is to increase the torque roll-off particularly above 4,000rpm.

The natural characteristics for different fixed values of the chopping duty-cycle d are shown in Fig. 24. This parameter has much the same effect as varying the d.c. supply voltage. Also shown in Fig. 24 is the effect of conduction angle control. With the chopper saturated, i.e. d=1, the applied voltage remains at its rated value and as the speed is increased, the maximum torque is sustained by advancing the turn-on angle with fixed commutation angle. The decrease in the torque is mainly due to the fact that more of the current is being conducted when the rotor is in a position of low dL/dθ, but core losses also contribute to the decrease. If windage losses are included, the net characteristic is still constant-power up to more than 5,000rpm.

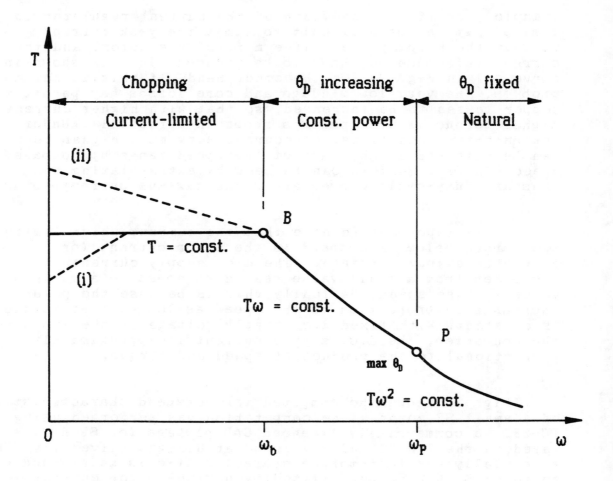

Fig. 23. General torque/speed characteristic of switched reluctance motor

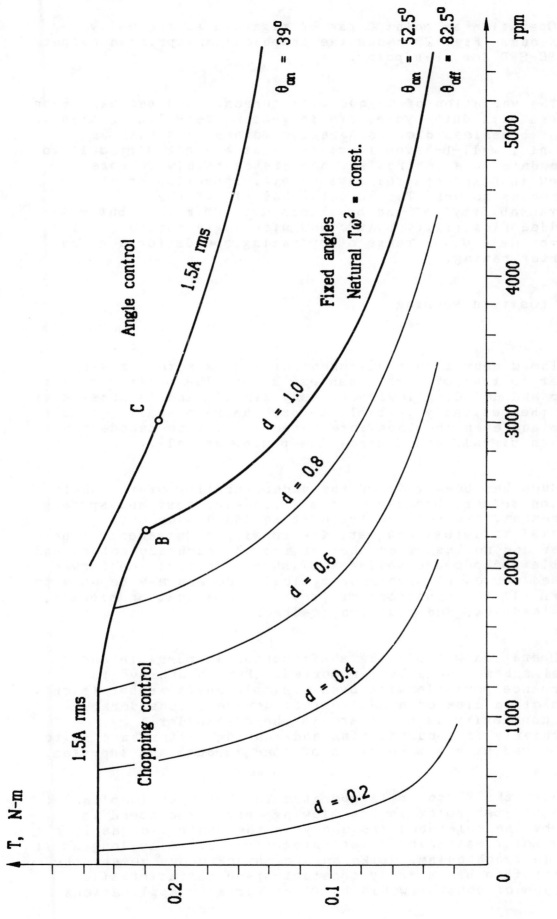

Fig. 24. Torque/speed characteristic computed by PC-SRD for small SR motor

Operation at point C can be regarded as thermally continuous. Fig. 25 shows the graphical and printed output from PC-SRD for this point.

The variation of torque with current, and the variation of speed with duty-cycle, are in general less linear than in d.c. or brushless d.c. squarewave motors, but they are monotonic, well-behaved functions that are not difficult to accommodate in a controller, and are certainly no more complex than the control laws of a.c. induction or PM synchronous motors. The SR motor has all of the 'programmability' of the brushless d.c. PM motor, but with the added flexibility that comes with angle control, which can provide a wider range of operating speeds for a given converter rating.

Shaft position sensing

The commutation requirement of the SR motor is very similar to that of a PM brushless motor. The shaft position sensor and decoding logic are very similar and in some cases it is theoretically possible to use the same shaft position sensor and even the same integrated circuit to decode the position signals and control the p.w.m. as well.

Much has been made of the undesirability of the shaft position sensor, because of the associated cost and space requirement, and because there is an added source of potential failures. However, the sensing requirement is no greater and no less than that of the PM brushless motor, and reliable methods are well established. In position servos or speed servos resolvers or optical encoders may be used to perform all the functions of providing commutation signals, speed feedback, and position feedback.

Operation without the shaft sensor is possible and several schemes have been reported. But to achieve the performance possible with even a simple shaft sensor (such as a slotted disk or a Hall-effect device), considerable extra complexity is necessary in the controller, particularly if good starting and running performance is to be achieved with a wide range of load torques and inertias.

When the SR motor is operated in the 'open-loop' mode, like a stepper motor in its slewing range, the speed is fixed by the reference frequency in the controller as long as the motor maintains 'step integrity', i.e., as long as it stays in synchronism. Like an a.c. synchronous motor, the SR motor then has a truly constant-speed characteristic. This type of control would be ideal for many applications

66

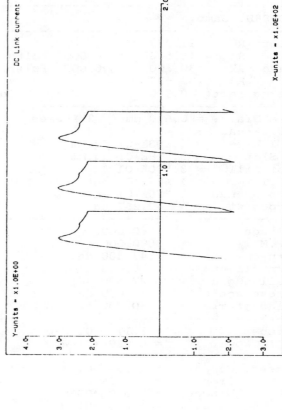

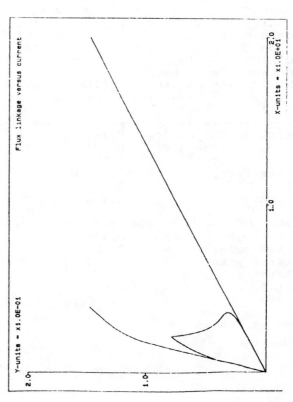

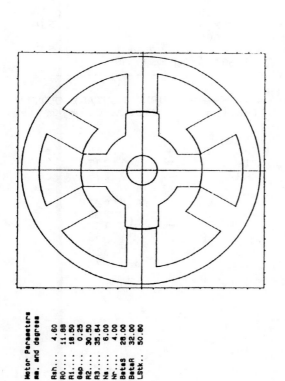

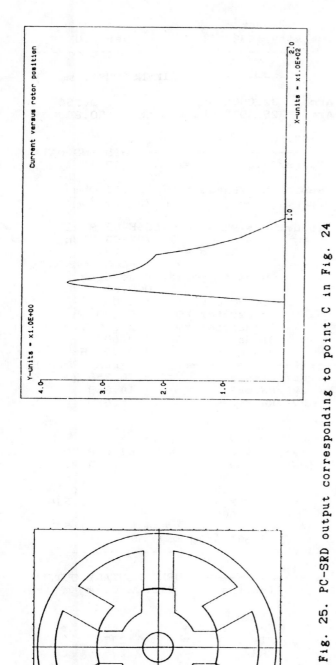

Fig. 25. PC-SRD output corresponding to point C in Fig. 24

67

```
--------------------------------------------------------------------
  PC-SRD. Sample
-------------------------------------------------------DIMENSIONS ( mm  )
Rotor  RO     11.879
       R1     18.500    4.000  Poles  Arc   32.000° Gap          0.250
Stator R2     30.500    6.000  Poles  Arc   28.000° Stack Lgth   50.800
       R3     35.637
Stckng fact.  0.970         Steel : Losil 500/50
-----------------------------------------------------------WINDING DATA
Wire Dia.  = 0.629 mm   3 Phases     Turns/Pole      =       110
C/S. Areas  :
   1 Wire   = 3.108E-01 Sq.mm        Resistance/Phase =     2.163 Ohm
   Slot Cu  = 3.418E+01 Sq.mm        Temp.                 75.000 °C
Slot fill    3.500E-01
M.L.T      = 1.467E+02 mm            Inductance/Phase = 5.556E-02 H Al
Lgth o/ends= 7.370E+01 mm                            = 7.379E-03 H Un
Copper wt. =    0.267 kg                   Ratio =       7.529
-------------------------------------------------------------CONTROL DATA
Voltage            60.000            Current Regulator setting
R.P.M            3000.000                          1000.000 A
Turn-on angle      47.400 deg.       Duty Cycle       1.000
Turn-off angle     75.000 deg.       Transistor RQ    0.000 Ohm
Dwell angle    =   27.600 deg.       Transistor VQ    2.000 V
Stroke angle   =   30.000 deg.       Diode      VD    0.600 V
O/lap starts   =   30.000 deg. BTC   Phase freq. =  200.000 Hz
-------------------------------------------------------------PERFORMANCE
Torque         = 2.113E-01 N-m       Efficiency      =    76.358 %
Shaft Power    = 6.639E+01 W         kVA/kW(pk)      =    14.868
                                     kVA/kW(rms)     =     5.850

Losses: Copper = 1.464E+01 W
        Iron   = 5.912E+00 W
        Windage= 0.000E+00 W         Deg. C / W      =     3.000
        Total  = 2.056E+01 W         Temp. rise      =    61.668 °C
CURRENTS  =                  PEAK           MEAN               R.M.S
Winding                      3.591          0.954              1.502
Transistor                   3.591          0.745              1.413
Diode                        2.143          0.209              0.510
DC Link (Supply)                            1.609
DC Link (Capacitor)                                            1.517
RMS Current Density  =  3118.934 A/SQ.in. =    4.834 A/SQ.mm.

-----------------------------------------------------SUPPLEMENTARY OUTPUT
WEIGHTS: Copper =    0.267 kg        Inertia   = 2.589E-05 kg-m²
         Iron   =    0.896
         Total  =    1.163
Resistivity     = 8.197E-07 Ohm-m    Temp. fact =       1.216
CPU             =    3.741           ETF        =       1.277
PSlot           =    1.496           PRS        =       2.245

IRON LOSSES      Eddy current                   Hysteresis
Rotor yoke      =    0.444 W                =       0.243 W
Rotor poles     =    0.329                  =       0.158
Stator yoke     =    2.530                  =       1.224
Stator poles    =    0.671                  =       0.313
Total           =    3.974                  =       1.938
Sigma           =    0.280 psi

End of design
```

but it suffers from two difficulties: one is to ensure that synchronism is maintained even though the load torque (and inertia) may vary; the other is to ensure reliable starting. Because of the large step angle and a lower torque/inertia ratio, the SR motor usually does not have the reliable 'starting rate' of the stepper motor, and some form of 'inductance sensing' or controlled current modulation (such as sinewave modulation) may be necessary in the control at low speeds.

Section 1
Switched Reluctance Motors

1.1 - General

Switched Reluctance Motor Drives

P.J. Lawrenson

Electronics & Power, February 1983, pp. 144-147

The pressures to improve product quality, manufacturing flexibility and speed of production and, more recently, to improve the efficiency of energy utilisation, have generated worldwide efforts during the last 10-15 years to exploit semiconductor technology to develop improved forms of controlled electrical drive. Of these new developments, the drive based on the salient-pole switched-reluctance motor, although the most recent and the most unexpected, must now be regarded as being of particular potential importance for a very wide range of applications.

Drives for the control of speed (most commonly), but also for the control of torque, power or acceleration, have long been essential elements in a huge range of industrial applications, in transportation systems and to a lesser degree in medical and domestic equipment. Examples which come to mind include: machinery throughout the wide range of the process industries; the diverse applications of pumps and fans; shaping and positioning by machine tools; servosystems generally; lifts and cranes; traction and propulsion on road, rail and water; mixing and blending; printing; production lines; packaging; sewing; washing; audio systems; mining; and many others.

Until recently these drives used almost exclusively DC motors with a few significant applications which favoured AC commutator motors. Even now the DC motor dominates controlled-drive technology. Increasingly, however, applications are being found for 'brushless' motors — in which the function of the commutator is performed by transistors or thyristors, and frequently the field flux is provided by a permanent magnet — and for inverter-fed AC machines, particularly induction motors. For some 10 or 15 years there have been many who have predicted, because of increased motor robustness and the falling costs of power-electronics, that AC or brushless-motor based systems would largely eliminate conventional DC machines. So far this

1 7·5kW, 1500 rev/min TEFC switched-reluctance motor for industrial use (Courtesy TASC Drives Ltd.)

has not happened, both because conventional DC machine systems have been improved significantly and because power electronics costs have not fallen sufficiently for general price competitiveness with the DC drives to be achieved.

However, the point has now been reached at which, for a number of applications, the performance/cost combination for inverter-fed induction motors makes them commercially attractive, for example in the very large pump and fan market. There is now every reason to expect the balance to swing progressively away from the DC machine system, not only as the part of the costs attributable to electronic components decreases as compared with that of the motors, but as reliability improves and users gain confidence and come to appreciate the improved capabilities achievable with electronically controlled commutatorless drives.

Switched-reluctance motor

In the competition with conventional DC-motor based systems (whether or not supplied by electronic means), the inverter-fed induction motor drive has generally been regarded as the one most likely to succeed (in industrial applications at least), and it is on this system that the greatest effort has been expended around the world. The main reasons for

this have been the low cost, ruggedness and freedom from maintenance of the induction motor which has been accepted as setting the standards in these respects. However, resulting from intensive academic and, more recently, industrial research and development over the past 15 years, the leading position of the induction motor is now open to question, and there is a growing appreciation that switched-reluctance (SR) motor drives are able to offer significant advantages: they can achieve unusually good combinations of high specific power output and, particularly, high system efficiency; their costs are clearly below those of AC-motor based systems (and even now of many DC ones also); they offer a range and quality of control usually only associated with the best DC-motor systems; and they provide a range of significant operational advantages in terms of robustness and reliability, both mechanical and electronic.

These qualities remain to be fully demonstrated in the field, but large-scale production of motors applicable to a wide range of industrial applications is now established, and hard evidence from the field is accumulating rapidly. Figs. 1 and 2 show a motor and controller from a range designed for general industrial applications. The rate of acceptance in the market place will, of course, continue to be influenced by normal caution in the face of a new product, particularly one incorporating semiconductor electronics but, as illustrated by the recent significant penetration by inverter-fed systems, the time really does seem at last to be ripe for the full benefits of semiconductor drives to be exploited in the market place.

Reluctance machines have been familiar as synchronous motors with conventional cylindrical stators, distributed 3-phase windings and 'salient'-pole rotors with cage windings. Until recently they were thought (a) to offer only very poor specific power outputs and efficiencies and (b) to be practicable only in relatively small sizes, and it is for these reasons that, to many people, reluctance motors have appeared as unexpected elements in modern drive systems. However, it has been known for a little while that even conventional reluctance motors, when properly designed, could surpass induction motors in specific power output and efficiency.

When developed in a fundamentally different way with salient-pole stators as well as salient-pole rotors (but without a cage winding), and when the currents in the armature are controlled in an optimum way, it has now been established that the advantages (of the resulting SR motors) in output and efficiency over the induction motor, particularly over wide speed and power ranges, can be very significant indeed. Moreover, because the operation of reluctance motor is inherently independent of the direction of current flow in the windings, very important advantages follow as it is possible to supply the windings with unidirectional currents, yielding particularly economical and reliable power conditioning circuits having fewer power semiconductors and providing low cost overall.

Fig. 3 illustrates the rudiments of a switched-reluctance motor and one of its driving circuits. The diagram illustrates eight stator poles and six rotor poles, although many different pole numbers and rotor/stator combinations are possible. Also, while in the case illustrated there would be four separate circuits or 'phases', SR motors may be designed, depending on the application, with one, two, three, four or even more phases, and there is no fixed relationship between the numbers of poles and phases. The supply to the motor is effectively DC, and there is no parallel with conventional polyphase AC machines, whether induction or synchronous. As will become increasingly clear from what follows, the machine is more properly described as a brushless (although unexcited) DC machine.

The salient poles on the stator carry concentrated windings of particularly simple form (rather like the field coils of a conventional DC machine), but the salient poles of the rotor carry no windings of any kind. The extreme simplicity and robustness of a typical rotor is shown in Fig.4. Both stator and rotor cores are constructed, to reduce iron loss and for manufacturing convenience, from laminated material. As seen in Fig.3, diametrically opposite stator poles are excited simultaneously, and excitation of one pair of poles causes a pair of rotor poles to be attracted magnetically into

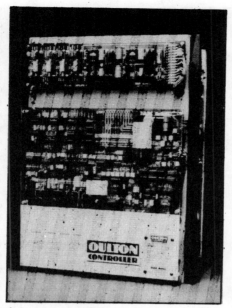

2 Control electronics and power convertor (Courtesy TASC Drives Ltd.)

alignment producing the basic torque of the device. The Figure implies excitation of poles AA', and if subsequently poles BB' are excited then the rotor poles bb' would move into alignment with them (with clockwise rotation). The switching sequence of the stator circuits is determined by the rotor position using some suitable transducer which can be of optical, magnetic or capacitance type.

The switching of the motor winding current may be effected by transistors or thyristors of varying types, as appropriate to the duty and, in conjunction with the rotor-position sensor, they perform a role similar to that of the commutator in the conventional DC machine. In the light of this and the basic torque producing mechanism, it might be expected that the machine will have operating characteristics like those of a conventional DC machine and, indeed, this proves to be the case. Operating the machine from a constant voltage source with current pulses switched on and off at fixed rotor positions (independent of speed), the motor produces a DC series-motor torque/speed characteristic. If, however, control is excercised over the rotor angles at which switching is effected and also over magnitude of the current, then very great flexibility of the operating characteristic is available — providing, for example, constant torque, constant power or series, or combinations of these in both motoring and regenerating quadrants. Moreover, this control is

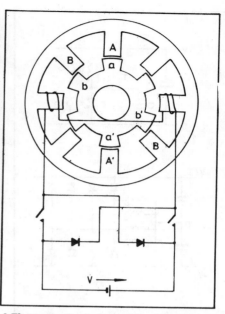

3 Elements of a 4-phase SR motor showing one circuit

effected at the logic level, and so the possibility of convenient (and continuous in principle) adjustment of characteristics to meet operational or market requirements is opened up.

Drive circuit and control

While the minimum requirement of the power-electronic circuits is to supply unidirectional current pulses, it is also necessary for efficient and fully controlled operation (including 4-quadrant operation with regeneration and dynamic braking) for the circuits to be able to handle negative as well as positive power flows, and also to be able to 'position' the current pulses in a continuously variable way in relation to rotor position. There is, not surprisingly, a considerable variety of circuits to achieve this, and these have to be chosen in the most suitable and cost-effective ways to suit different applications.

A principal advantage of systems based on the reluctance motor as compared with the induction motor is, as noted, that only one power switch per phase is necessary. Two sample circuits which demonstrate this advantage and achieve the necessary flexibility and control are illustrated in Fig. 5. Fig. 5a, which shows (for simplicity) one phase only, relates to a motor winding with a primary P and closely-coupled (bifilar) secondary S. Energy is injected into the motor by way of the current supplied through the thyristor T and energy is returned to the supply through the secondary circuit

73

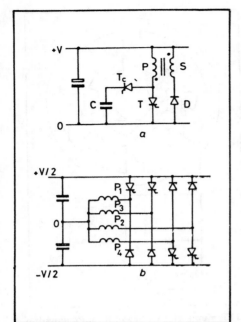

5 Elements of power convertor circuits with one switch per phase: (*a*) **Bifilar system (one phase only);** (*b*) **Split supply system for four-phase motors**

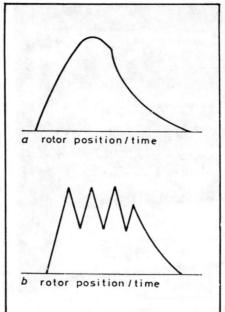

a rotor position / time

b rotor position / time

6 Winding current wave forms: (*a*) **High speed;** (*b*) **Low speed**

Table 1 Comparison of performance and cost of drives 7·5 kW, 1500 rev/min constant torque

criterion / drive type	slip coupling	DC	AC PWM	SRD
cost	0·8	1·0	1·5	1·0
efficiency, %				
FLT & FS	75	76	77	83
FLT & ½FS	38	65	65	80
power per frame size	0·8*	1·0*	0·9†	>1·0†
controllability	0·3	1·0	0·5	0·9
control complexity	0·2	1·0	1·8	1·2
reliability/serviceability	1·3	1·0	0·9	1·1
noise, dB	69	65	74	74

* vent † TEFC

and the diode D (when T is switched off). Fig. 5*b* shows all phases ($P_1...P_4$) of a 4-phase machine in a configuration which avoids a bifilar winding on the motor but retains only one switch per phase. Both circuits, depending on application and voltage of operation, provide attractive practical arrangements which are plainly much more economical than induction-motor inverter circuits.

Thyristors have been shown in the above circuits, and they are being used in various SR drives but, of course, transistors can be used with advantage in many cases and, indeed, are the best choice at lower power levels. When thyristors are used, commutation circuits have also be incorporated (see T_c and C

in Fig. 5) and, as is the case with inverters, the design of these commutating circuits is one of the major elements in the overall design. However, in this respect also, reluctance-motor systems have important advantages over induction-motor/inverter systems and important economies can be achieved since it is possible, for example, to commutate all the main power devices from a single commutating circuit in many cases.

Typical motor winding current waveforms in the SR drive are shown in Figs. 6*a* and *b*, applicable to high- and low-speed operation, respectively. The complete divergence from the sinusoidal shape associated with conventional machines is apparent. Superficially this may be thought disadvantageous, but it

is in fact consonant with the achievement of the high powers and particularly the high efficiencies which characterise SR drives over wide ranges of speed and load. The information-level electronics has continuous control over the switch-on instant of the winding current, the switch-off instant and the effective amplitude of the pulse, and these variables are programmed to provide the optimum operating conditions for given load and speed demands.

Performance characteristics

Fig. 7 is included to illustrate, on the one hand, something of the flexibility of the operating characteristics which are readily achievable and, on the other, the high level of efficiency and particularly the wide operating ranges over which these can be maintained. Fig. 7*a* is, in fact, the operating characteristic typical of a traction drive with a constant torque region up to base speed, followed by a constant power region up to three times base speed and a series region up to top speed. By way of guidance with regard to relative performance, the power characteristic was achieved within a frame equivalent to an 8-pole induction motor C 180M frame, class-B insulated, rated at 11kW at 720 rev/min. Fig. 7*b* shows contours of constant efficiency for the whole system (including losses in the convertor). They are for the same motor as Fig. 7*a* and, for comparison purposes, the full-load efficiency of the induction motor operating on a pure sineware supply is 86%. The considerable superiority of the SR system is apparent, particularly remembering the loss of system efficiency which would occur when using the induction motor in a variable-speed mode with an inverter, and particularly bearing in mind the relatively rapid falloff of its efficiency with speed and load.

Current status and future potential

SR drives have so far been built in a range of types and sizes from 10W at 10 000 rev/min to 50kW (peak) at 750 rev/min. Detailed design studies have been made for machines rated at 0·25 MW, and other designs have been projected to 1 MW. These confirm the general properties and qualities of the SR system as outlined above. Indeed there is evidence, at least up to the sizes studied, and contrary to previously propagated opinion, that performance of the machine improves further with increase in size relative to that of its more conventional rivals.

Cost estimates for the system are perhaps most easily made in comparison with those of induction-motor systems. For the motor, the opinion of many manufacturers is that manufacturing costs will be below those of even the squirrel-cage induction motor — the saving coming from the complete absence of a rotor winding and the simple form of the stator winding. Convertor costs also are projected to lie well below those of the equivalent inverter for an induction motor — because of the reduction in the number of main devices and the special economy in commutation circuits which is possible in the SR system. Moreover, the control is also expected to be economical.

Reliability of the convertor should be good because of the reduced component count compared with the inverter, and it is particularly helped by the complete absence of any 'shoot-through' path. The reliability of the motor should be essentially the same as that of the cage induction motor but with the expectation of some advantages following from the absence of any rotor winding, the extreme simplicity and robustness of the stator winding, and the excellent thermal properties of the motor.

Mention should be made of two possible weaknesses of the system, namely acoustic noise and the possibly large number of interconnections between motor and convertor. The basic mode of operation of the motor generates pulsations in torque (24 per revolution in the simplest 4-phase machine of Fig. 3) and can be a source of noise, particularly through the excitation of structural resonances. Nevertheless, experience has shown, at least for a substantial number of applications, that this problem is not critical (being of similar magnitude to that with PWM inverters) and further reductions in noise level can be confidently expected. So far as the number of interconnections is concerned, this is strongly influenced by whether or not a bifilar winding is chosen. Bifilar system can involve many connections but in a singly wound motor the number can be as low as five for a 4-phase machine.

A constructional arrangement which has advantages in a variety of situations, and particularly when the number of leads is high, is illustrated in Fig. 8, which shows an integrated configuration of motor and power electronics. The system illustrated is a drive rated at 50kW peak at 750 rev/min for operation in a road-going vehicle. It is worth

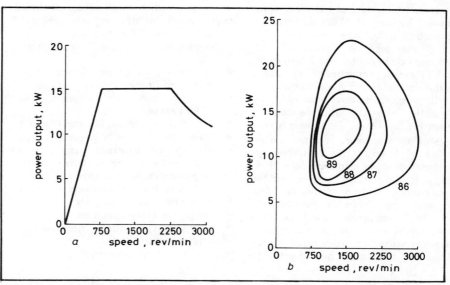

7 Illustrative performance characteristics for 180M frame SR motor: (a) Power-output/speed; (b) Constant efficiency contours

8 Integrated motor/power-convertor design, 50kW (peak) at 750 rev/min (Courtesy Chloride Technical Ltd.)

noting the compactness, not only of the motor, but also of the power electronics in comparison with the motor (the scale graduations are of 1cm).

Because of the possible variations in the magnetic circuit configuration and the number of motor phases, and because of numerous power convertor designs, there is a very large family of SR drive systems, and so simple all-embracing statements about their qualities are very difficult to make. However, Table 1 is included to give some general overall guidance. The data reflect a commercial manufacturer's overall assessment of the relevant properties of SR drives in comparison with three more conventional systems —

PWM inverter-fed induction motor, DC motor and eddy-current slip-coupling. They are all 7·5kW, 1500 rev/min, constant-torque industrial systems. Several of the parameters are normalised to unit value for the DC machine, as providing the current standard.

The values ascribed to the qualities of 'controllability', 'control complexity' and 'reliability/serviceability' are not precisely formulated. So controllability reflects an overall weighting of dynamic response, speed range, stability, accuracy, regulation, smoothness of torque etc.; control complexity mainly reflects the nature and extent of the power and control electronics; and reliability/serviceability reflects again the nature and extent of the electronics (not least the level of inherent protection and security), but also the necessary level of maintenance (taking into account any commutation and brush gear, and the thermal properties of both motor and the control), and the suitability for hazardous and dirty environments.

From the Table, the SR drive emerges against all the others as having a considerable attractiveness and, particularly, very favourable efficiency. Moreover, its costs also emerge as low, being potentially much cheaper than the inverter-fed induction motor. No data are given in the Table for brushless PM machines. These machines may perhaps be developed with superior performance characteristics to induction-motor systems but they must be expected to cost more, and so would not appear to be capable of achieving any general advantage over SR systems.

Overall, therefore, the SR drive offers:

- a combination of high specific output and high system efficiency effective over wide ranges of both speed and power
- low cost — both capital and running
- valuable and easily provided flexibility of control, including 4-quadrant operation with regeneration and dynamic braking, soft start, torque limitation etc.
- robustness and reliability
- simplicity of manufacture
- suitability for hazardous environments
- freedom from maintenance.

All these factors make it potentially attractive for a wide range of applications throughout industry, in transportation systems and in domestic appliances. In all these areas the strong compatibility of SR systems with microprocessor control is a further significant attraction with exciting possibilities for the ready adjustment of operating characteristics and for the provision of 'intelligence'. It remains to be seen just how great the exploitation of SR systems will be in various fields, in this country and elsewhere.

Acknowledgment

The author wishes to acknowledge not only his indebtedness for assistance with this article but also the outstanding contributions to the whole development of SR systems theory and technology which has been made by Michael Stephenson and Norman Fulton and others at Leeds University, Rex Davis, Bill Ray and Roy Blake at Nottingham University and associates at Chloride Technical Ltd. and TASC Drives Ltd.

Switched Reluctance Drives—A Fast-Growing Technology

P.J. Lawrenson

Possibly the most striking feature of switched reluctance drive technology is the recognition which has suddenly built up concerning its value for a very wide range of sizes and applications. This is particularly remarkable in view of the short time which has elapsed since the reported performance and qualities of SR drives were greeted with disbelief.

Independently proven

One reason for this change is the confirmation of the advantageous properties of SR drives by a variety of independent groups and companies, bringing with it appreciation of the engineering and market opportunities which are opened up. As a consequence, it is now possible to point to advantages which are no longer in dispute. So, for example, SR drives offer unequalled levels of efficiency, particularly when considered over wide ranges of power and speed. The motors are exceptionally robust, particularly well suited to hazardous environments and they are small in size—being never larger, and sometimes significantly smaller, than the more conventional d.c. or induction motors.

Fewer power switches

The controllers involve fewer main power devices than can ever be achieved with inverter fed induction motors (for comparable motor utilization) and, even though the total area of semiconductor may be similar, the overall converter costs are smaller (having regard to assembly costs), and they have features which make them in principle more reliable than inverters.

DC motor capability

From the point of view of control, the drives are inherently four-quadrant without any added complexity, they offer high torque to inertia ratios and very good dynamic performance. They are readily adapted to operate under any form of closed-loop control and, within themselves, are capable of being adapted to provide any required torque speed characteristic. Of course, certain conventional drives can equal or even improve upon the SR drive in relation to one or more of the above characteristics and to particular applications. However, it is the capacity of the SR

system to combine most or all of the above qualities as the norm which is its real attraction and fits it for such a wide variety of applications over very wide ranges of sizes and speeds.

Doubts overturned

That this should be possible comes about as a result of (a) the recent demonstration (contrary to previous beliefs) that reluctance motors are capable of at least equalling the capacity of other machine types to convert electrical into mechanical energy and (b) the combination of such motors with modern power electronic technology—not only in a form which uses the minimum possible number of electronic components but which, by proper design of control strategies, provides for the optimum utilization of the electrical energy input.

Operating principles

Reluctance motors all operate on the principle that a piece of soft, magnetic iron will align itself with a magnetic field imposed upon it and, in the long-established forms of synchronous reluctance motor, salient poles on the rotor synchronize with the magnetic field produced by electric currents in windings on the stator. Such machines, however, also carried cage windings like induction motors on the rotor to enable the starting process to take place.

Simple motor

Switched reluctance motors differ from these synchronous reluctance motors in two ways: firstly, they completely eliminate the need for any winding on the rotating member; and secondly, they employ salient poles on both the rotor and stator—with the latter being the only ones to carry electrical windings. Fig. 1 shows, the cross-section of the magnetic circuit of an SR motor with eight stator poles and six rotor poles. This is a combination which has found considerable general purpose use but there are very wide ranges of stator or rotor pole numbers, and different combinations between stator and rotor which are advantageous for different applications.

To some readers this structure will

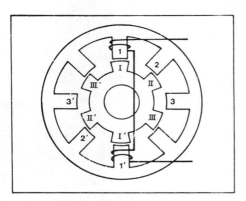

Fig 1. Elements of the magnetic circuit of a switched reluctance motor with eight stator and six rotor poles.

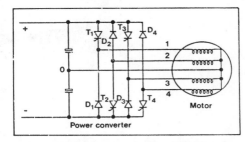

Fig 2. The elements of one example of a power switching circuit for a 4-phase SR motor.

appear virtually identical to that of a VR stepping motor and indeed this is the case although design proportions and principles differ in various significant ways. Fig. 1 shows the stator poles 1 and 1' on which stator coils are illustrated and these establish a magnetic field along the vertical axis of the machine so that the rotor lines up as shown with its poles I and I' aligned with the stator. If now the currents on poles 1, 1' are switched off and the currents on poles 2 and 2' are switched on, then it is apparent that the rotor will move so that its poles II and II': will align themselves with the stator poles 2 and 2'. Again, if the excitation is shifted to the stator poles 3, 3', then the rotor will move so its poles III, III' come into line with 3 and 3'. There is thus the situation where progressive switching of the stator coils giving the effect of a magnetic field rotating clockwise produces a steady motion of

Reprinted with permission from Electric Drives and Controls, April/May 1985.

Fig 3. A 7.5kW switched reluctance motor with its g.t.o. converter and control (courtesy of Tasc Drives Ltd).

the rotor moving anti-clockwise. The instant at which the stator currents are switched on and off is controlled basically from observation of the rotor position, and some appropriate rotor position sensor is incorporated in the machine for this purpose. By superimposing appropriate additional control on to the basic signal provided by the rotor position, it is possible to achieve simultaneously high torque output, high electrical efficiency and smooth control over rotor velocity and position.

Simple power converter

This assumes the availability of a d.c. supply and also a means of switching it to the different stator circuits in turn. Notice that there is no need to distinguish between the possible directions of current flow in the stator coils. The behaviour of the rotor in generating torque and power is independent of whether the currents flow positively or negatively in the circuits. It is this feature which leads to the particularly simple form of power circuits used to control the motor.

Fig. 2 shows the elements of the switching power circuit—which is assumed to be supplied either from a battery or from rectified mains, as would be the case with an inverter-fed induction motor or a d.c. motor. It can be seen that the motor circuits 1, 2, 3 and 4 can accept energy via a power switch, successively T1 to T4, from one half of the d.c. supply and return surplus energy via diodes, D1 to D4, to the other half, when the phase winding ceases to be advantageously torque productive. The switches are shown as being thyristors but, in practice, they could be g.t.o.s, transistors or power field-effect devices, depending on size, cost and performance requirements. Thyristors of course require commutation circuits in addition, but a further

advantage of SR technology is that it provides for more economical commutation circuits than induction motor/inverter circuitry.

Simple regeneration

All of the above discussion has been in terms of motoring operation. However, in order for the system to act as a generator, or as a brake, or to provide for full 4-quadrant operation, it is only necessary to adjust the timing of the switching operations. Of course, for regenerative or 4-quadrant operation it is necessary for the converter to be connected to a battery or to a controlled bridge as with d.c. or induction motors.

Sensing rotor position

The signals for the switching of the devices come from some suitable sensor coupled to the rotor, and both optical and magnetic devices have been used with considerable success. The same kind of sensors are used of course in

78

Fig 4. The construction of a switched reluctance motor stator.

that this converter is less than one half of the volume of the first converter produced commercially two years ago and made available for evaluation at a number of installations in the UK, Europe and North America.

The size and complexity of the control and the drive depends on the technology employed, including the nature of the cooling, and developments will take place progressively as the available technology and customer requirements take shape.

Microprocessor-based controllers, or others based on special purpose chips, will lead to further great increases not only in simplicity but also in reliability and testability with associated cost advantages.

Figs. 4 and 5 illustrate the nature of the motor itself. Fig. 4 shows a typical stator construction—note the simple form of the coils, especially the short overhang, and the advantages which come from this including ease of cooling. The thermal characteristics of the SR motor are advantageous in a

number of ways. Firstly, the rotor is effectively 'cold' with only a small amount of iron loss being generated in it. All of the heat in the machine is produced in the stator (particularly in connection with any kind of brushless machine and there are many examples in both industrial and domestic applications. It is, however, the control strategy which is superimposed upon the basic sequence set by rotor position which is at the heart of the success of the SR system and, in this matter, and indeed in the choice of basic motor configuration, a great deal of know-how is involved.

Constructional Advantages

The external appearance of the SR motor is essentially identical to an induction motor. Similarly, the converter and control looks like the inverter for an induction motor. Fig. 3 shows a composite picture of a 7.5kW SR motor alongside its g.t.o.-based converter. The motor is built in a 132 frame and so

the compactness of the control and converter is apparent. It is worth noting the coils), a feature which makes it easy to exploit new cooling methods.

Power ranges

Fig. 5 illustrates three different sizes of rotor and brings out in a striking way both the wide range of sizes which are possible and also the extraordinary simplicity and robustness of the rotor. Due to the latter, there is virtually no limit to speed with the important advantages this can have for many applications. In terms of size the smallest rotor illustrated was designed to produce rather less than 10 watts at 10,000 rev/min whilst the larger one produces 50kW peak at 750 rev/min with a constant power characteristic up to 2,250 rev/min. The range of motors so far built considerably exceeds that illustrated and successful operation has been obtained with motors in excess of 200kW at 1,500 rev/min and designs have been carried out to 1MW ratings.

*Fig 5. Rotors for three reluctance motors
with power ratings from 10W to 50kW.*

Performance and control

Fig. 6 shows the variation of output power with speed for a drive designed to meet a specified traction characteristic —a constant torque up to base speed, a constant power region over a speed range of 4:1 and, above that, a series roll-off characteristic. This particular output characteristic was obtained with a motor built into a 180 M frame which, as an induction motor, would have been rated at a lower level. This characteris- tic is illustrative only and one of the more remarkable features of the SR system is that the output power (or torque) versus speed characteristic can be given virtually any shape, and designed to suit any particular duty.

Intelligence & software control

The form of the characteristic is effectively under software control or equivalent (by way of a pre- programmed chip or similar) and, from the point of view of production economy, this can be an important advantage. In due course, and in line with other developments the motor can be imbued with 'intelligence' and its characteristics can be made to be self- adjusting to suit changing operating circumstances encountered.

Fig. 7 shows for the same motor as Fig. 6, contours of constant efficiency and the extremely wide range, both in output power and in speed, approxi- mately 4 to 1 in each case, over which a very high efficiency can be maintained.

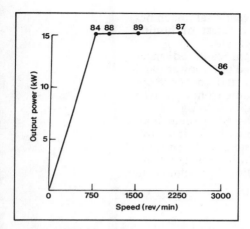

Fig 6. Output power/speed characteristic —showing constant torque, constant power and series sections, for a 180 M-frame motor.

In order to put the figure of 86 per cent into context, this was the maximum efficiency which a typical induction motor could achieve at its operating point when supplied from a pure sine wave. The efficiencies given in Fig. 7 include the whole system, and allow for losses in the converter. When comparing the different efficiencies of different drives it is rarely meaningful to compare the simple maximum achievable as, typically, the usual operating condition is well away from the maximum: it is for this reason that the SR drive is so attractive.

Applications

Turning now to the matter of the application of switched reluctance drives, much the greatest interest is where some form of controlled drive is called for. Traditionally, such drives have been met by d.c. machines operating within appropriate control schemes and, to a great extent, d.c. machines still set the standard by which control drives are assessed.

The demise of the d.c. machine has been anticipated over many years because of commutation and maintenance difficulties, and upper speed limitations. It is only relatively recently, however, that the inverter-fed induction motor has begun to make real impact on the d.c. motor-dominated market, but now inverter-fed machines are making real progress. Some assessments put the a.c. motor's share of the market now as high as some 35 per cent compared with less than 10 per cent only a few years ago.

DC motor control

The switched reluctance motor provides inherently characteristics and control possibilities which are the complete equivalent of the fully controlled d.c. machine. Indeed, the SR machine is strictly a brushless d.c. machine, and it combines these qualities with robustness at least the equivalent of the squirrel cage induction motor. It is therefore not surprising that manufacturers and users are seeing advantageous applications for SR drives in a very wide variety of situations. The first commercial applications of the system have been in connection with **general purpose industrial drives** where the motor is attractive because of its versatility, its low cost, its very high efficiency over the whole working range, and its outstanding robustness and suitability for hazardous environments.

SR systems are, however, being developed rapidly for a variety of other widely varying applications including: mining and **explosion proof machinery** for which the basic construction of the motor is ideally suited for where its 4-quadrant capability is frequently important; **traction applications**, for which low-cost implementation is possible and where the torque/speed characteristics can be tailored to suit both accelerating, steady running and, perhaps particularly important, braking operation; **domestic appliances** where, along with the extreme pressure to achieve low costs, there is the possibility of providing much more attractive levels of control and where the elimination of the commutator would bring various operational advantages including significantly reduced noise; **battery-powered applications** where the very high efficiencies are important in the context of improved operating range and where the controllability and possibility of achieving overall system packages of low cost are valuable. There is also considerable interest in SR systems for servo applications generally, including **robotics**.

Future developments

More generally, important application advantages of the SR system include:

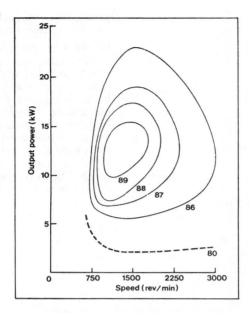

Fig 7. Contours of constant efficiency for the 180 M-frame motor (maximum induction motor efficiency 86%).

☐ running at much higher speeds than conventional machines with advantages in terms of size and weight and possibly gear box considerations;
☐ designs specially adapted to producing very high torques at low speeds;
☐ the basic thermal characteristics in which the rotor is cool and the heat produced in the stator can be readily removed with the possibility of new techniques for doing this advantageously; and
☐ using the motor in a stepping mode to achieve precise and discrete position control.

New markets

Because of the inbuilt feedback of information about rotor position and speed, and because of the essentially 'software' nature of the control, the SR system is inherently suitable for self-adaptive or 'intelligent' operation. In this connection, it may be noted that General Electric recently announced their 'programmable' motor, for application in the first instance to domestic and subsequently low-power hand-tool applications. It was envisaged that this motor would open up completely new markets of great size and it would be right to observe here that the SR system ought to have identical performance but with lower costs because of the elimination of a permanent magnet and because of the fundamental feature that only one switch per phase is required (the permanent magnet system requiring two for efficient operation).

Development potential

The last mentioned development, that associated with self-adaptive control, is only one where the future will bring significant further advances and marketing opportunities. Notwithstanding the levels of performance and the attractiveness of costs which have been indicated, the whole development of SR systems is still at a very early stage. This is in marked contrast to the developments which have gone on for 100 years on conventional machines and where only the finest of honing remains to be carried out.

The further significant advances which are expected in power and control electronics, and the developments which may come from improved magnetic materials, will bring advantages equally to the SR system and, coupled with the potential for improvement within the system itself, promise to give SR drives an increasing advantage over conventional machines in many applications.

Brushless Reluctance Motor Drives

T.J.E. Miller

IEE Power Engineering Journal, Vol. 1, November 1987, pp. 325-331

The variable-reluctance or switched-reluctance motor has some remarkable characteristics that make it attractive for dozens of applications. With CAD software now available, and many prototypes established, there should be few technical barriers between the laboratory and a number of successful niches in the drives market

Reluctance motors have a history of not quite achieving the power density, efficiency or power factor of other established motors such as the induction motor and the various forms of permanent-magnet (PM) motor. But recent advances in electromagnetic design and power electronics have resulted in much improved reluctance motors, with a wider range of possible characteristics; and the ever-increasing variety and volume of the adjustable-speed drives market is creating opportunities for which the modern reluctance motor is very competitive.[1]

This article is mostly about the switched-reluctance (SR) motor and its control, and presents a few new ideas along with a review of some of the well known ones. The synchronous reluctance motor is also discussed as a potentially useful technology for applications where the advantages of the reluctance motor are needed, but where very low torque ripple, or compatibility with AC control techniques, is or are required. Most of the article is relevant to drives in the range from a few watts up to perhaps 50 kW.

Among the advantages of the reluctance motor are the simple construction and the absence of permanent magnets, which eliminates cost in both raw material and manufacturing processes (see Fig. 1). The absence of rotor windings and low rotor losses help to make the machine robust and suitable for high-speed and high-temperature applications; or in other environments where brushless PM motors could be hazardous because of their open-circuit voltage or short-circuit current. Under almost all electrical fault conditions the reluctance motor is inert and completely safe, and it can fairly be described as fault tolerant; the same is true for most inverter faults, and indeed certain of the inverter faults that plague AC drives are not possible with SR drives.

On the negative side, the SR motor tends to have more torque ripple and a higher noise level than other motors. The reputation for noise is almost certainly derived from early models, since quiet SR motors have been developed more recently (early induction motors had the same development problem).

The torque ripple, which can be in the range 10-30%, is indeed a concern; but this has sometimes been used against the SR motor without regard to the torque ripple produced by induction motors or PM brushless motors, which can be just as bad. Many applications are not sensitive to torque ripple even of this magnitude; for those that are, the SR drive should be evaluated with caution, and if reluctance-motor properties are needed, the synchronous version should be considered as an alternative.

It is also true that the SR motor cannot be controlled from a conventional AC inverter. However, to some this is an advantage, because the SR controller is arguably simpler and requires less protection. But the kVA requirement is typically higher than for AC drives: anywhere from 0 to 30% higher in small integral-horsepower sizes. Depending on the torque/speed characteristics required, and the duty cycle, this does not necessarily imply that the AC drive always has the advantage. Over a wide speed range the SR motor may have the advantage.

The SR motor cannot start or run from an alternating-voltage source, and it is not normally possible to operate more than one motor from one inverter.

The interest in reluctance motors is much increased today compared with only five years ago, and there are now innumerable development groups working on it worldwide. But because the design is difficult, particularly in dealing with the magnetic circuit and the calculation of losses, it is likely that many

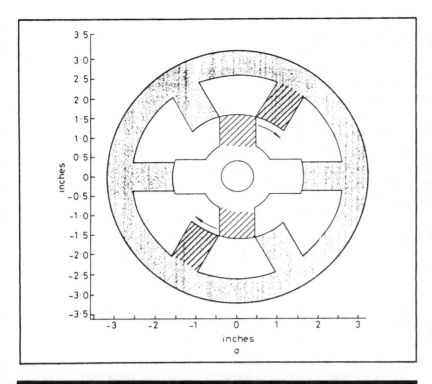

disappointing models will yet be built before the technology is mature enough to be generally viable. Meanwhile, it can be expected that the successes will steadily establish useful niches in the drives market where the special characteristics fit the requirements at the right price.

Torque capability

Because the SR motor is fundamentally a step motor, it produces torque in impulses. During one step the phase current and the corresponding flux linkage follow a closed trajectory as shown in Fig. 2. The trajectory lies between the two extreme magnetisation curves corresponding to the 'unaligned' and 'aligned' positions. Given that a phase winding comprises two opposite poles, the unaligned position is when the stator poles lie midway between two adjacent rotor poles. The aligned position is when the stator poles and rotor poles are in line. For this discussion it suffices to consider just one pair of stator poles and one pair of rotor poles in isolation, such as those highlighted in Fig. 1.

The energy W converted from electrical to mechanical during one step is equal to the area enclosed by the trajectory in Fig. 2. The average electromagnetic torque T is then given by

$$T = qN_r \frac{W}{2\pi} \text{ newton metres}$$

where qN_r, the number of steps per revolution, is given by the product of the phase number q and the rotor pole number[3] N_r. Clearly it is desirable to design the motor to maximise the available conversion area between the unaligned and the aligned curves in order to get the most torque per ampere of phase current. This requires a large aligned inductance, a small unaligned inductance, and a high saturation flux linkage. While the geometry is simple, it is by no means easy to achieve these objectives in design calculations, and computer methods are essential to get a good result.[2] For very detailed design work the finite-element method is helpful, but simpler methods suffice in many cases.

In Fig. 2 only a small fraction of the available energy is converted. This is typical of small motors where the current is thermally limited. An increase in scale naturally permits more of the available energy to be converted. The same increase could be obtained with more intense cooling, or alternatively during intermittent operation. However, operation with an extreme trajectory (such as the dotted curve in Fig. 2)

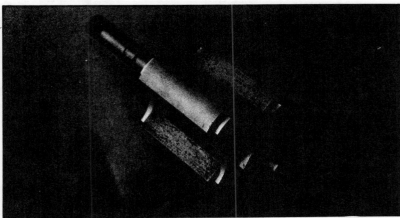

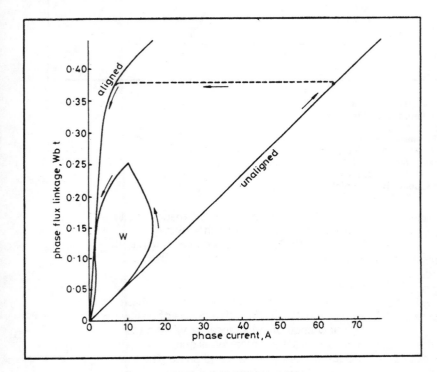

1 *(a)* Schematic cross-section of 1·5 kW switched-reluctance motor with two opposite poles of one phase excited; *(b)* partially wound SR motor stator; *(c)* rotor of SR motor

2 Locus of operating point in the current/flux-linkage plane, showing the energy W converted in each step

may be inefficient and noisy, with a poor power factor and peaky currents.

A rough estimate of the maximum attainable torque per unit *rotor* volume can be derived from an idealised triangular area approximating to the dotted trajectory in Fig. 2. Following the methods of Reference 4, the result is

$$\frac{T}{V} = \frac{N_r B_s^2 (\lambda - 1) q}{2\pi^2 \mu_0} \frac{\beta g}{R} \text{ newton metres/m}^3$$

where B_s is the flux density in the stator poles at the maximum flux linkage ψ_s in the aligned position; λ is the aligned/unaligned unsaturated inductance ratio; β is the pole arc (assumed equal for stator and rotor); and g is the airgap. For a three-phase motor with six stator and four rotor poles having a pole arc of 30° and an air gap of 0·25 mm, at a rotor radius R of 25 mm, it should be possible to achieve $\lambda = 10$ and $B_s = 1·6T$, giving a specific torque of 60 kN m/m³ from the extreme trajectory. With the small trajectory in Fig. 2 the specific torque will be closer to 15 kN m/m³, and this figure is definitely competitive with induction and PM brushless motors.[5]

This simple analysis does not confirm that such values can be obtained with acceptable losses, efficiency and power factor; for this, much more calculation is necessary.[2] For example, the simple torque formula indicates an increase of torque with air gap g, but this can be sustained only if the inductance ratio can be maintained constant and the same flux-density levels can be reached without overheating the windings. The simple formula gives no guidance on these questions. However, it can be said that sufficient SR motors have been built and tested, by many independent engineering groups, to confirm that the values quoted are practicable and can be obtained with high efficiency and quiet operation.

The balance of copper and iron losses is different from that found in AC and PM brushless motors. In the SR motor the flux-density waveform is very nonsinusoidal and differs from one part of the magnetic circuit to another. The largest component of iron loss is often in the stator yoke, simply because this section has the greatest volume. Here the dominant frequency is equal to the step rate, especially at low speeds when the current is limited by PWM of the voltage. For motors with the same number of rotor poles, the step rate of the SR motor exceeds the frequency of the AC (or PM brushless) motor by the factor $2q$; i.e. by 6 times in a three-phase motor. However, the hysteresis losses tend to be reduced by a factor of perhaps 2 or 3, by the unipolar nature of the flux pulsations in most sections of the core. Moreover, the peak flux density is generally lower than in AC motors because of the need to avoid coupling between the phases. These factors bring the iron losses more into line with those of AC motors, and indeed they may be relatively lower because the SR motor has significantly less iron than an AC motor of the same frame size. **85**

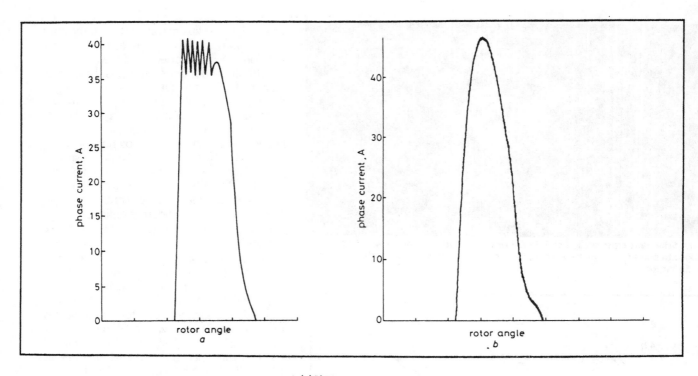

3 (a) Phase-current
waveform obtained with
current-regulated PWM.
(b) Phase-current
waveform obtained with
fixed-frequency voltage
PWM; the chopping
frequency is much higher
than in (a)

At all but the highest speeds the copper losses are usually more significant, particularly in small motors[6] (below, say, 5 kW). To minimise these, a large slot is desirable, which results in an optimum rotor diameter which tends to be a little smaller than the corresponding one for AC or PM brushless motors. Together with the heavily notched shape of the rotor, this usually leads to a very low inertia.

It is impossible to generalise about the relative efficiencies of SR and other types of motor without specifying very tightly the parameters that are kept the same in the comparison. For the same *frame size* the SR motor should be expected to have a lower efficiency than that of the best PM motors, but better than that of a standard induction motor. (It must be added that non-standard induction motors, relieved of the line-start requirement, can be significantly more efficient than standard motors.) However, if the motor *cost* is kept the same, the SR motor should be the most competitive, because it requires fewer manufacturing processes and should have less raw material for the same torque. In 'larger' sizes (say, above 10-20 kW), the induction

motor improves rapidly while the magnet cost puts the PM motor out of contention. Efficiency differences then become marginal, and other factors will determine the choice.

Control

For motoring operation the pulses of phase current must coincide with a period of positive rate of change of inductance, i.e. when a pair of rotor poles is approaching alignment with the stator poles of the excited phase. The timing and dwell of the current pulse are both important in determining the efficiency, the torque per ampere, the smoothness of operation, and other parameters. It is characteristic of good operating conditions that the trajectory of Fig. 2 fits snugly in the space between the unaligned and aligned magnetisation curves.

The trajectory of Fig. 2 corresponds to high-speed operation where the current is limited by the self-EMF of the phase winding. A smooth current waveform is obtained with a peak/RMS ratio very close to that of a half sine wave. This waveform is not stressful to the power semiconductors; indeed the duty cycle is higher

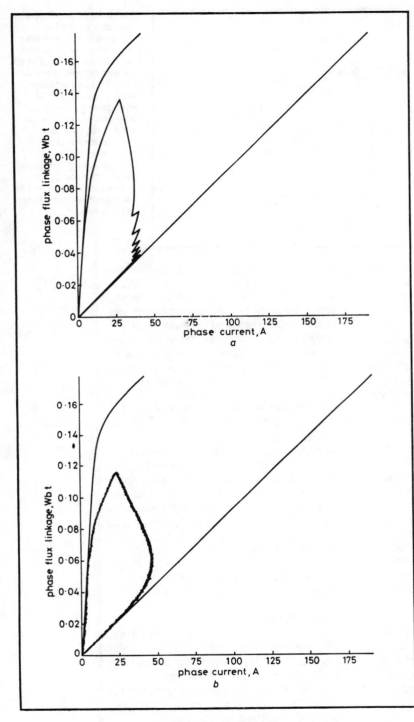

4 (a) Energy-conversion diagram with current regulation; (b) energy conversion diagram with voltage PWM

than that of a comparable AC or PM brushless motor controller, which should benefit the reliability.

At low speeds the self-EMF of the winding is small and the current must be limited by chopping or pulse-width modulation (PWM) of the applied voltage. The strategy employed has a profound effect on the operating characteristics. Fig. 3a shows a current waveform controlled by a current regulator that maintains a more or less constant current throughout the conduction period in each phase. The chopping frequency varies because of the changing motor inductance. Fig. 4a shows the corresponding flux-linkage/current trajectory, and Fig. 5a shows schematically the method of control. As the current reference increases, the torque increases roughly linearly. This type of control produces a constant-torque type of characteristic as indicated in Fig. 6a. To obtain speed control, a speed feedback loop is necessary. This requires a speed sensor in addition to (or integral with) the shaft-position sensor that is already assumed to be present for providing the commutation signals. It can be noted that this form of current regulation requires current transducers of wide bandwidth, in series with the windings. Most of the published literature on SR drives describes this form of control.

Fig. 3b shows, for the same motor, the current waveform obtained with fixed-frequency PWM of the voltage. This can be implemented with one of the transistors in each phase leg; or alternatively as described below. The corresponding energy-conversion trajectory is shown in Fig. 4b, and the control schematic in Fig. 5b. The duty cycle (or 'off time') of the PWM can be varied by a simple monostable circuit. What is found is that the motor now has a constant-*speed* characteristic as shown in Fig. 6b. This is not so surprising if it is realised that this form of control is essentially similar to armature-voltage control in a DC motor. Note that a controlled constant-speed characteristic is now obtained without the expense of current sensors or a speed transducer.

A simple form of current feedback can be added to the circuit of Fig. 6b with interesting benefits. A single inexpensive resistor in the return DC line provides a signal that, suitably filtered, can modulate the duty cycle of the PWM in such a way as to 'compound' the torque/speed characteristic; it is possible in this way to achieve under-compounding, over-compounding, or flat compounding, just as in a DC motor with a wound field. For many applications the speed regulation obtained by

87

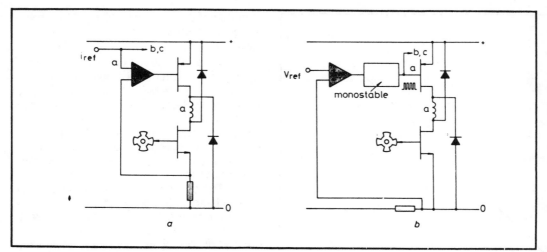

5 (a) Schematic of current-regulator control, showing one phase only; wide-bandwidth current sensors are required in all phases: bottom transistor used for commutation only. (b) Schematic of voltage-PWM control; only one current sensor is used, with very low bandwidth for thermal protection and 'compounding' the speed/torque characteristic: bottom transistor used for commutation only (Patent applied for)

this simple scheme will be perfectly adequate. For precision speed control, of course, a normal speed feedback loop can be added. The single current sensor also serves for thermal-overcurrent sensing.

A further remarkable feature of the voltage PWM scheme is that the current waveform tends to retain much the same shape over the entire operating range of speed and torque. This permits a high degree of utilisation of the available conversion energy over the whole range, ensuring high overall efficiency and power factor.

When the PWM duty cycle reaches 100%, the motor speed can be increased by increasing the dwell (the conduction period) or the advance of the current pulse relative to the rotor position, or both. These increases eventually reach maximum practical values, after which the torque becomes inversely proportional to speed-squared. The speed range over which constant power can be maintained is comparable to that for induction motors, and is markedly better than for PM brushless motors.

It is interesting to compare the two alternative control schemes for the SR motor with the one normally used for PM brushless motors. In the PM motor the flux is fixed by the magnet. Fixed-frequency PWM (with a variable duty cycle) results in a constant-torque characteristic, so that speed feedback is necessary for speed control in many applications.

Shaft-position sensing

The commutation requirement of the SR motor is very similar to that of a PM brushless motor; it is even possible to use the same shaft-position sensor and, in some cases, the same integrated circuit to decode the signals

therefrom and control the PWM. (Several such ICs are now commercially available.) Much has been made of the undesirability of the shaft sensor, because of the associated cost and space requirement, and because there is an added source of potential failures. However, the sensing requirement is no greater and no less than that of the PM brushless motor, and reliable methods are well established.

Operation without the shaft sensor *is* possible and several schemes have been reported.[7] But to achieve the performance possible with even a simple shaft sensor (such as a slotted disc or a Hall-effect device), considerable extra complexity is necessary in the controller, particularly if good starting and running performance is to be achieved with a wide range of load torques and inertias. For rapid acceleration and/or deceleration cycles, or for position control, there is a long way to go before the sensor can be eliminated. Probably the same is true of the PM brushless motor, and even the induction motor is not without its problems in this area.

Controller circuits

The torque is independent of the direction of the phase current, which can therefore be unidirectional. This permits the use of unipolar controller circuits, with a number of advantages over the corresponding circuits for AC or PM brushless motors, which require alternating current.

Fig. 7 shows a circuit well suited for use with transistors (bipolar, field effect, or insulated gate). The phases are independent, and in this respect the SR controller differs from the AC inverter, in which the motor windings are connected between the midpoints of adjacent inverter phase legs. The winding is in series

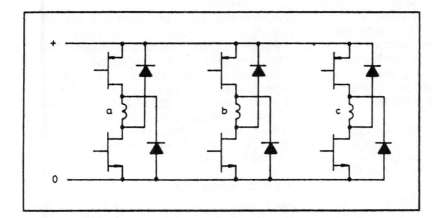

7 Controller circuit for SR motors, suitable for use with transistors

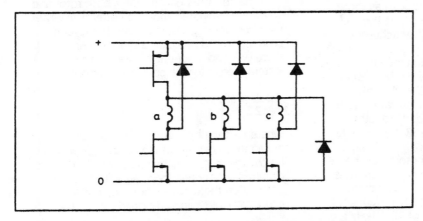

8 Controller circuit with only four transistors and four diodes for a three-phase SR motor. The circuit has nearly the same function as Fig. 7, but only operates in the PWM mode and cannot admit overlap between phases. Bottom transistors are used for commutation only (Patent applied for)

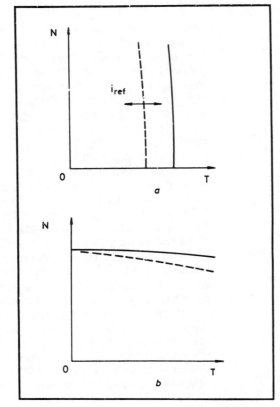

6 (a) Speed/torque characteristic obtained with current regulation. (b) Speed/torque curve obtained with voltage PWM; the dotted curve shows the effect of reduced current-feedback gain ('under-compounding')

with both switches, providing valuable protection against faults. In the AC inverter the upper and lower phase-leg switches must be prevented from switching on simultaneously and short-circuiting the DC supply; this is possible only by means of additional control circuitry, which is unnecessary in the SR controller.

The upper and lower phase-leg switches may be switched on and off together, but a better mode of operation is to use one for commutation and the other for PWM. At the start of the conduction period both switches are turned on. One remains on during the whole conduction period, while the other is 'chopped' by a control strategy such as those described in the previous Section. At the end of the conduction period both switches are turned off and any remaining stored magnetic energy that is not converted to mechanical work is returned to the supply by the current freewheeling through the diodes. At high speeds, if the PWM regulator 'saturates', both switches will be turned on and off together.

The inductance level of the SR motor, while it varies with the rotor position, is generally higher than in PM brushless motors. For the same chopping frequency the ripple current is therefore smaller. This helps to ensure quiet operation, especially if the chopping frequency is high. Chopping frequencies of 10-20 kHz are are desirable, as in other types of drive, to minimise acoustic noise. In large drives (say, above 20 kW) it becomes much more difficult to chop at such a high frequency because of the switching-speed limitations and losses of larger devices (such as GTOs). This makes it more difficult to achieve quiet operation in larger motors, particularly at low speeds.

In small drives it is often acceptable to use PWM control over the entire speed range. This is the normal approach with surface-magnet PM brushless motors. In such cases the SR controller circuit can profitably be reduced to the circuit of Fig. 8, in which the chopping is performed by one transistor in common for all the phases. The lower transistors commutate the chopped voltage to the phases in proper sequence, under the control of the shaft-position sensor. The circuit requires only $n+1$ transistors and $n+1$ diodes for a motor with n phases. A three-phase motor thus requires only four transistors and four diodes. There is practically no loss of functionality with this circuit relative to the full circuit having $2n$ transistors in Fig. 7. Its main limitation is that there can be essentially no overlap between the conduction periods of adjacent phases; but this limitation only becomes a problem at very

low (crawling) speeds and possibly at very high speeds. Otherwise the circuit has only two-thirds of the power devices of a brushless PM motor controller, and it is simpler to control.

Many other circuits have been developed in attempts to reduce the number of switches all the way down to n, and to take full advantage of the unipolar operation of the SR motor. But it seems that in every case there is a penalty in the form of extra passive components or control limitations.[8]

New developments

Solid rotors

In the conventional SR motor both rotor and stator are laminated; induced currents in either member would generally impair the torque production and produce additional losses. If there were no magnetic saturation the instantaneous torque could be expressed as

$$T = \frac{1}{2} i^2 \frac{dL}{d\theta}$$

where i is the phase current; L is the phase inductance; and θ is the rotor angular position. If now the rotor is made solid, or if short-circuited conducting loops are affixed to the poles, then the torque is given by the same expression, but with L replaced by

$$L' = L (1 - k^2)$$

where k is the coupling coefficient between the stator winding and the rotor circuit. L' is the leakage inductance and is completely defined by this equation. It is, of course, meaningful only when the stator flux is changing at a sufficiently rapid rate to ensure that the induced currents in the rotor are 'inductance limited'; this, however, is not difficult to achieve. Fig. 9 shows an experimental rotor of this type.

With suitable pole geometry the minimum value of L' occurs when the stator and rotor poles are aligned, and the maximum value is when they are unaligned. Should it be possible to achieve a higher inductance ratio between these two extreme positions than with the conventional laminated rotor, this machine would be able to achieve a higher torque per ampere. Even without this advantage, the solid rotor has intriguing possibilities for very high speeds because its lateral stiffness is inevitably greater than that of the normal laminated rotor. Rotor losses may seriously limit the viability of this concept, which has yet to be developed beyond a laboratory machine. However, the possibilities for liquid cooling are

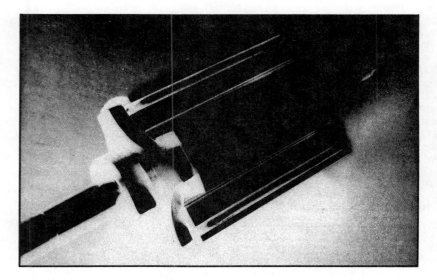

9 Rotor of solid-rotor SR motor with conducting loops to provide reduced secondary resistance; the ideal loop resistance is zero. This motor would be an interesting way to exploit room-temperature superconductors if they ever become available with sufficient current-density capability

much better in a solid rotor than in a laminated one. Interestingly, the current pulses must now be phased to coincide with the separation of the poles, and not with their approach. The torque is produced by repulsion, not by attraction.

Synchronous-reluctance motors

The synchronous-reluctance motor is well known in its line-start version. Both the rotor and the stator present cylindrical surfaces to the air gap. The stator is a conventional polyphase AC stator, while the rotor has internal flux barriers shaped to minimise the ratio of d-axis to q-axis reactance. The rotor has a starting cage to provide across-the-line starting, and operation is normally from a voltage source in an open-loop mode (without shaft-position feedback). It is necessary to operate at a safe fraction of the pull-out torque, and this requirement, together with the need to design for stable operation over a wide speed range, constrains the design in such a way that both power factor and efficiency are poor by comparison with modern AC or PM brushless drives.

At the same time the *switched* reluctance motor has a certain torque ripple, and it would be beneficial to eliminate this by converting to a synchronous-reluctance motor with a PWM sine-wave inverter. The question remains as to whether the synchronous motor could equal the efficiency, power density and power factor of the switched motor; and how would they both compare with induction and PM motors? It is difficult to answer this question in general. But for certain application requirements it may be that the *cageless* synchronous reluctance motor, fed with current-controlled PWM sine

waves oriented to the rotor position, and designed without the stability constraints of the line-start motor, can achieve competitive performance levels. This suggests an interesting candidate for applications with very high ambient temperatures, or others where permanent magnets are undesirable for safety or reliability reasons. There is evidently plenty of fundamental research still to be done.

Conclusion

For applications requiring high-temperature operation or a high degree of fault tolerance, or in cases where a brushless motor is needed but without the cost or the operational problems of permanent-magnet motors, switched- and synchronous-reluctance motors are viable candidates. With CAD to optimise the motor and predict performance, and with the exploitation of modern power-electronics and control techniques, these drives can be expected to find several applications in the ever-expanding market for adjustable-speed motor drives.

Acknowledgments

The author would like to acknowledge many colleagues in UK and US industry and universities, particularly the co-authors mentioned in the references. P. G. Bower of Glasgow University has helped considerably in the development of the control techniques, Alan Hutton with motor design, and Malcolm McGilp with design software. Acknowledgment is also due to the Science & Engineering Research Council for a grant to study the cageless synchronous-reluctance motor; to General Electric (USA); and to the subscribing companies of the Glasgow University SPEED programme.

References

1 MILLER, T. J. E.: 'Small motor drives expand their technology horizons', *Power Engng. J.* 1987, **1**, pp. 283-289

2 MILLER, T. J. E.: 'PC CAD for switched reluctance drives', IEE Conference on Electric Machines and Drives, London, 16th-18th November, 1987

3 MILLER, T. J. E.: 'Converter volt-ampere requirements of the switched reluctance drive', *IEEE Trans.*, 1985, **IA-21**, pp. 1136-1144

4 HARRIS, M. R.: 'Static torque production in saturated doubly-salient machines', *Proc. IEE*, 1975, **122**, (10), pp. 1121-1127

5 HARRIS, M. R., FINCH, J. W., MALLICK, J. A., and MILLER, T. J. E.: 'A review of the integral-horsepower switched reluctance drive', *IEEE Trans.*, 1986, **IA-22**, pp. 716-721

6 JAHNS, T. M., and MILLER, T. J. E.: 'A current-controlled switched reluctance drive for FHP applications', Conference on Applied Motion Control, Minneapolis, USA, 10th-12th June, 1986

7 BASS, J. T., EHSANI, M., and MILLER, T. J. E.: 'Simplified electronics for torque control of sensorless switched reluctance motor', *IEEE Trans.*, 1987, **IE-34**, pp. 234-239

8 BASS, J. T., EHSANI, M., MILLER, T. J. E., and STEIGERWALD, R.: 'Development of a unipolar converter for variable-reluctance motor drives', *IEEE Trans*, 1987, **IA-23**, pp. 545-553

Variable-Speed Switched Reluctance Motors

P.J. Lawrenson, J.M. Stephenson, P.T. Blenkinsop, J. Corda, and N.N. Fulton

Proceedings IEE, Vol. 127, Pt. B, No. 4, July 1980, pp. 253-265

Indexing term: Reluctance motors

Abstract: The paper explores the theory and potential of a family of doubly salient electronically-switched reluctance motors. It is demonstrated that the machine provides the basis for fully-controllable variable-speed systems, which are shown to be superior to conventional systems in many respects. The motor retains all the advantages normally associated with induction motors and brings significant economy in the drive electronics. The basic modes of operation, analysis, design considerations and experimental results from a range of prototype motors up to 15 kW at 750 rev/min are described. The most recent prototype has achieved a continuous rating which is 1·4 times that of the equivalent induction motor.

List of principal symbols

d = rotor diameter
ER = energy ratio (see Section 2.2.3)
f = frequency
g_i = interpolar airgap
i = instantaneous current
k_L = coefficient of inductance overlap
L = self-inductance of stator circuit
N = pole number
P = output power
q = number of phases
R = resistance of stator circuit
r,s = subscripts denoting rotor and stator, respectively
T = electromagnetic torque
v = applied voltage
W^1 = co-energy
β = pole arc
ϕ = rotor pole pitch
ψ = flux linkage
ω = rotor speed
$\hat{}$ = maximum value
$\check{}$ = minimum value

1   Introduction

This paper lays general foundations for the practical design of a family of switched reluctance motors, and, further, it demonstrates that machines in this family are capable of extremely high levels of performance, can be controlled in exceptionally simple and flexible ways, are simple and cheap to manufacture and can offer important operational advantages in both industrial and domestic applications. Because of these qualities, and because of many unfamiliar features, mainly arising from the highly nonlinear nature of most aspects of their operation, an attempt is made here, with due reference to related work, to provide a comprehensive basic treatment of switched reluctance (s.r.) motors.

Reluctance motors are most familiar with conventional, cylindrical stators and distributed, 3-phase windings and, in this form, have been developed over the last two decades to

Paper 795B, received 2nd January and in revised form 29th April 1980

Prof. Lawrenson, Dr. Stephenson and Dr. Fulton are with, and Dr. Blenkinsop and Dr. Corda were formerly with, the Department of Electrical and Electronic Engineering, University of Leeds, Leeds LS2 9JT, England. Dr. Blenkinsop is now with Patscentre International, Melbourn, Royston, Herts. SG8 6DP and Dr. Corda is with the Faculty of Electrical Engineering, University of Sarajevo, 71113 Sarajevo, Yugoslavia

provide high efficiencies and high specific outputs comparable with those achievable from induction motors.[1] They are thought of as orthodox members of the a.c. motor family, of use when true synchronous speed (or position synchronisation) is needed and when minimum first cost is not an overriding consideration, though they are also used with variable-frequency inverters to provide variable-speed drives.

By contrast, the reluctance machines discussed here are of inherently variable-speed type. They may be thought of as 'brushless' machines having parallels with other such machines, either of the d.c. type, but without field excitation, or of the a.c. self-synchronous type, again without field excitation. It will be shown that the absence of field excitation does not lead to inferior performance, as might be imagined, but brings important advantages in cost savings with the additional advantage of the elimination of a commutator.

The motor has salient poles on both stator and rotor (i.e. it is doubly salient), the windings on the stator are of particularly simple form and there are no windings of any kind on the rotor. Currents in the stator circuits are switched on and off in accordance with the rotor position and, with this simplest form of control, the motor inherently develops the torque speed characteristics typical of a series-connected d.c. machine. However, using appropriate strategies, it is both easy and extremely cheap to give the motor a wide range of different characteristics.

The use of reluctance motors with rotor position switching has a long history which has now come full circle Within the last 15 years or so it has appeared frequently in the guise of advanced, closed-loop control schemes for stepping motors and also for various special applications; prior to that it was used with mechanical contactors in low-cost single-phase versions, as in shaver motors, for example; more than 100 years ago, it was employed in the original 'electromagnetic engines' with which modern variable-reluctance stepping motors have much in common. More recently, a number of workers have directed their attention to various particular design developments of stepping motors aimed at considerably increasing the power levels.

Of this work, perhaps the most straightforward extension of stepping-motor strategies has been that by Bausch and Rieke. They have briefly described[2] a 4-phase, double-stack motor using an earlier drive system[3] to give low-speed running in the manner familiar in small-angle stepping motors. They were particularly interested to apply the inherent series characteristic to a vehicle drive. Development closely related to those discussed in this paper have been presented by Unnewehr and Koch,[4] Koch,[5] and Byrne and Lacy.[6] Unnewehr and Koch described a multiple-disc-rotor machine using a separate set of laminations for each stator phase, but the mechanical complexities and cost of this form of construction seem to have held back its development. Subsequently, Koch presented a theoretical treatment of a reversible 3-phase single-stack motor of the kind discussed in the present paper, but his treatment was seriously incomplete in that, not only was it restricted through using a linear model, but it failed to recognise the fundamental effects of the negative torque developed during part of the operating cycle. Furthermore, no consideration was given to the possibility of switching current into a winding before the onset of the period of increasing winding inductance which, as discussed low, is an essential feature of a practical system. The motor described by Byrne and Lacy was also of single-stack type but was restricted to a 4-pole 2-phase stator and 2-pole rotor, a design which is inherently unidirectional. The discussion, as in the underlying patent specification,[7] gave particular attention to the creation of abnormal levels of saturation with the objective of increasing the torque output as proposed by Jarret.[8] By contrast, the present authors do not follow this method of increasing the saturation to 'unnecessarily' high levels (e.g. by reducing the 'packing factor' of the iron) and it will be demonstrated below that very high levels of specific torque are developed well in line with those foreshadowed for stepping motors, albeit of small frame size, in the paper by Harris et al.[9]

The work described here was initiated in rudimentary form at Leeds University more than 10 years ago. The objectives were to build on the existing experience w..h high-performance reluctance motors and to exploit the fact that the reluctance motor (unlike conventional   ction or synchronous motors) can be operated with unidirectional phase currents and, hence, with the minimum number of switches (This economy in switches has, of course, been recognised by several of the authors referred to above and by Ray and Davis.[10]) Transistor or thyristor switches can be used but, in either case, the system costs are minimised. These costs can be expected to be significantly below those of equivalent induction motor systems and, in a number of cases, can already be seen to be below those of equivalent d.c. motor systems.

Following early work on 1-phase motors, detailed studies have been made since 1972 of lower-power motors (from 10W to over 1kW at various speeds up to 10 000 rev/min) using mainly transistor drive circuits. Since 1975 a substantial programme, in close collaboration with colleagues from Nottingham University, has been in progress, directed towards the development of a 50 kW drive (using thyristors) for a battery-powered road vehicle. The part of this programme at Nottingham has been oncerned with electronic circuits for use with s.r. motors and, in their recent paper,[10] Ray and Davis have presented a study, using a linear model, of switching device rating and of control techniques. That paper and this one were prepared essentially as companion papers. Discussion of circuits here is accordingly limited to the minimum of principles necessary to discuss the operation of the motor.

In the following Section the basic principles of operation of the motor (including the patterns of energy flow) are summarised, and a concept appropriate in place of power factor is introduced. Section 3 describes fundamental design considerations raised by the motor. This is followed in Section 4 by a brief discussion of the performance characteristics, both the inherent one and the many achievable ones, and of the methods of controlling and s.e. characteristics. The highly nonlinear nature of the machine and some of the important implications of this for practical realisations of the system are discussed in Section 5. In Section 6, experimental results for different machine sizes and types demonstrate the considerable attractiveness of the s.r. motor system for variable-speed applications. These results give particular attention to the specially demanding characteristics required for an electric-vehicle drive, including a constant-power capability over a wide speed range.

2 Basic principles

Fig. 1 shows the basic elements of a doubly-salient reluctance motor, where it is to be understood that only two of a larger number of stator poles are included. The motor comprises a single stack and both stator and rotor are constructed from laminations. Most simply, diametrically opposite stator poles carry coils connected in series to give a 2-pole field pattern. In this Section, the principles of operation are explained and the parameters fundamental to the operation are i. entified.

2.1 Inductance variation and torque production

Torque is developed by the tendency for the magnetic circuit to adopt a configuration of minimum reluctance, i.e. for the rotor to move into line with the stator poles and to maximise the inductance of the excited coils. Note that the torque is inde endent of the direction of current flow, so that unidirectional currents can be used, permitting a simplification of the electronic driving circuits. In general, because of magnetic nonlinearities, the torque T must be calculated in terms of co-energy W' as

$$T(\theta, i) = \frac{\partial W'(\theta, i)}{\partial \theta} \qquad (1)$$

where θ is the angle describing the rotor position (see Fig.1) and i is the current in the coils. Note that changes in co-energy depend both on the angular position of the rotor and on the instantaneous value of the current. Proper allowance for both of these factors must be made in predicting performance, but it is helpful to consider first a simplified model in which magnetic nonlinearity is neglected and eqn. 1 can be simplified to

$$T(\theta, i) = \frac{i^2}{2} \frac{dL}{d\theta} \qquad (2)$$

94

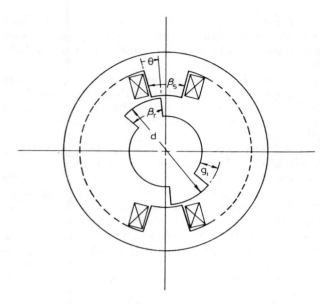

Fig. 1 *Elements of a doubly-salient reluctance motor*

where L is the self inductance of the circuit at any value of θ.

Fig. 2a shows the variation of inductance with rotor position for the pair of stator poles shown in Fig. 1, idealised in that magnetic saturation and the 'rounding' effect of the fringing fields are neglected. The number of cycles of inductance variation per revolution is proportional to the number of rotor pole pairs, and the 'length' of the cycle is equal to the rotor pole pitch ϕ. The physical significance of the different regions R of the variation needs to be

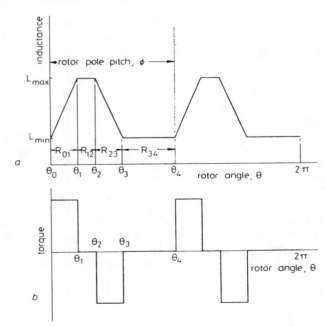

Fig. 2 a *Variation of inductance of one stator circuit as a function of rotor position*
   b *Variation of torque with constant current as a function of rotor position*

recognised for later discussion:

R_{01} : at θ_0 the 'leading' edges of rotor poles meet the edges of stator poles and the inductance starts a linear increase with rotation, continuing until the poles are fully overlapped at θ_1, when the inductance reaches its maximum value L_{max}.

R_{12} : from θ_1 to θ_2 the inductance remains constant at L_{max}, through the region of complete overlap. This region is generally known as the 'dead zone'.

R_{23} : from θ_2 to θ_3 the inductance decreases linearly to the minimum value, L_{min} .

R_{34} : from θ_3 to θ_4 the stator and rotor poles are not overlapped and the inductance remains constant at L_{min}.

The associated variation of torque for a constant coil current follows from eqn. 2 and is shown in Fig. 2b. The torque can be controlled to give a resultant which is positive (i.e. motor action) or is negative (i.e. generator action) simply by switching the current in the coil on and off at appropriate instants during the inductance cycle. There seems to have been an assumption in many earlier publications that maximum operating torque and power are achieved by switching the current on only when the rotor is in the region R_{01}. It is important to recognise that this is not so. For example, to maximise motoring torque, current must be switched on when the rotor is in the region R_{34} or even R_{23}.

2.2 Energy flow and current and flux waveforms

2.2.1 Switching circuits and energy flow: Fig. 3a shows a simple form of switching circuit for controlling the current in the stator coils which, though simple, is sufficiently general to study the possible patterns of energy flow in the motor. At this stage it remains sufficient to consider only a single circuit on one pair of stator poles (along with one pair of rotor poles). When the switch S is closed and the rotor is stationary, current builds up in the winding and energy is taken from the d.c. source 1. When S is opened, the current continues to flow, but now through the diode so that the stored magnetic energy is transferred to d.c. source 2. The general equation governing the flow of stator current may be written

$$v = Ri + \frac{\mathrm{d}\psi}{\mathrm{d}t} \qquad (3)$$

where v is the voltage (of appropriate polarity) applied across the winding and ψ is the flux linking the coil. Again it is helpful to consider first a simplified model which again assumes magnetic linearity and also negligible resistance. On this basis eqn. 3 may be rewritten

$$v = L\frac{\mathrm{d}i}{\mathrm{d}t} + i\frac{\mathrm{d}L}{\mathrm{d}\theta}\,\omega \qquad (4)$$

where ω $(=\mathrm{d}\theta/\mathrm{d}t)$ is the speed of rotation. The rate of flow of energy is given by

$$vi = \frac{\mathrm{d}}{\mathrm{d}t}\left(\frac{1}{2}Li^2\right) + \frac{i^2}{2}\frac{\mathrm{d}L}{\mathrm{d}\theta}\,\omega \qquad (5)$$

95

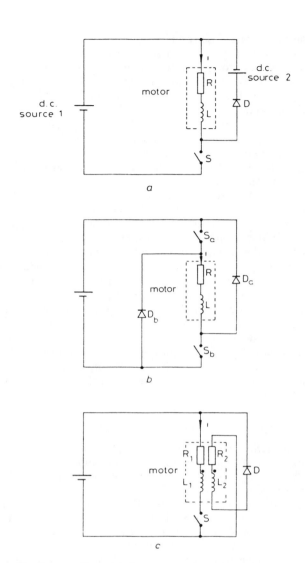

Fig. 3
    a Essential elements of a general switching circuit
    b Circuit with two switches per coil pair
    c Circuit with one switch and bifilar-wound coils

This equation shows that, when running as a motor, the input electrical power goes partly to increase the stored magnetic energy $(\frac{1}{2}Li^2)$ and partly to provide mechanical

output power $\left(\dfrac{i^2}{2}\dfrac{dL}{d\theta}\omega,\right)$ the latter being associated

with the 'motional e.m.f.' in the stator circuit. Thus, with the switch S closed during the region of rising inductance R_{01}, part of the energy from source 1 is converted into mechanical output and part is stored magnetically; but with S open during R_{01}, stored magnetic energy is partly converted to mechanical output and partly transferred to source 2. If current is still flowing during the (constant) maximum inductance period, the energy is simply transferred to source 2. Finally, if current flows during the decreasing inductance period, energy is transferred to the

96

source not only from that stored magnetically but also, through the generation of negative torque, from a mechanical source, which is the condition of regenerative operation.

The possibility of full four-quadrant operation, controlled simply through the instants at which circuits are switched (without the familiar inconveniences of contactors and reversing arrangements), is an important feature of the present system.

Figs. 3b and c show different forms of switching circuit. In the circuit of Fig. 3b, closing S_a and S_b allows energy to flow into the winding and opening both switches causes the current to flow through D_a and D_b, so that the stored energy is returned to the (single) supply. In the circuit of Fig. 3c it has been possible to halve the number of switches and diodes by the introduction of a bifilar winding of the stator (i.e. a primary and secondary circuit wound together on each pole for maximum coupling). When S is closed, current flows into the primary (L_1, R_1), and when S is opened the primary current falls abruptly to zero but a corresponding current is established in the secondary (L_2, R_2) so as to maintain constant flux linkages (stored magnetic energy). This secondary current, flowing through D, returns energy to the supply. This circuit of Fig. 3c is the most economical form, requiring only one electronic switch per coil pair, but a price for this great simplicity has to be paid in that the voltage to be withstood by the switch is twice the supply voltage (assuming an equal number of turns on primary and secondary) and, of course, an extra winding is required on the pole.

Results are presented in Section 5 for machines operated with circuits basically of the type shown in Figs. 3b and c.

2.2.2 Current and flux waveforms: All current and flux waveforms are wholly nonsinusoidal and those of current vary widely with operating conditions. Fig. 4 shows diagrammatically one cycle of the waveforms of current and flux (in accordance with eqn. 4), typical of the important motoring condition in which the winding is switched on at some angle θ_i, *in advance* of the onset of the rising inductance region (Section 2.1). The effective inductance of the circuit is L_{min}, initially, so allowing the current to build up rapidly to its maximum value (and to maximise its torque-producing effect). Subsequently, the rising inductance and the motional e.m.f. cause the current to fall until the switch is opened at some angle θ_x, (typically) before the maximum inductance is reached. Thereafter, the current falls more rapidly because an opposing polarity is applied to the winding by virtue of the current flowing into the supply. Note, in accordance with the introductory comments, that the direction of current during the cycle never reverses. The angle between θ_i and θ_x is referred to as the conduction angle θ_c, and is of considerable importance for the control of the machine (Section 4).

The simplicity of the waveform of flux linking the winding is particularly interesting; recalling that, for the moment, consideration is being restricted to the case of zero resistance, as long as a positive constant voltage is applied, the flux increases at a constant rate (see eqn. 3) and, conversely, when a constant negative voltage is applied the flux decreases also uniformly. The maximum flux always occurs at the instant of switch-off defined by θ_x.

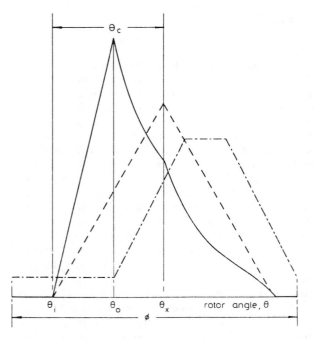

Fig. 4 *Current and flux waveforms derived using a linear variation of inductance*

current
flux
inductance

2.2.3 Phase relationship and 'power factor': The above waveforms clearly indicate fundamental differences between the switched reluctance motor and conventional a.c. machines. Two other distinctive differences should be taken up at this stage: the first is the straightforward one of the relationship between currents in different stator circuits; the second is the more subtle one as to what concept is appropriate in place of power factor.

In connection with the first, it has simply to be noted that the stator circuits are not 'phases' in the conventional sense, there being no automatically fixed relationship between the time when one 'phase' is switched off and another is switched on. The conduction time of a 'phase' winding is chosen in relation to the inductance variation of that winding and to the required operating conditions. The conduction periods of adjacent phases may or may not overlap. The operation of the multiphase motor can be determined by combining the current and torque (and flux) functions as shown in Figs. 2 and 4 for the individual phases. Appropriate allowance must be made for the angular displacement between adjacent pairs of stator poles and, of course, for the magnetic nonlinearity of the machine.

Turning to the matter of the 'quality' of performance which, for a.c. machines is measured by power factor, it seems simplest and most helpful to seek a direct measure in terms of the *useful* energy flow as a proportion of the *total* energy flow (which includes, of course, the circulating magnetic field energy). This leads to the concept of an 'energy ratio' defined, over a cycle, as

$$ER = \frac{\begin{array}{c}\text{total energy supplied to machine} - \\ \text{energy returned to supply}\end{array}}{\text{total energy supplied}} \qquad (6)$$

The significance of this, as with power factor in a.c. systems, is particularly in relation to device ratings and costs. (Note that, of course, device ratings depend on several factors in addition to ER/PF, e.g. voltage rating, current waveform, duty cycle.) The ER values, as will be seen from Section 6, are generally very satisfactory and lead to total device ratings which are less than those of induction-motor-based systems. (The detailed discussion of electronic circuit design will be the subject of a future publication by the authors' colleagues at Nottingham University.)

3 Fundamental design considerations

The discussion so far has been restricted to basic aspects of behaviour in the context of the simplest model of an s.r. motor. It is necessary now to explore the questions of design strategy which immediately arise when a practical system is contemplated. The number of questions is very large because many of the constraints which 'simplify' a.c. machines do not apply: the number of phases to be employed is open to choice between one and many; the ratio of phase number to stator pole number is not fixed; the ratio of rotor pole number to stator pole number is open to a wide variety of choices; the 'best' values of rotor pole arc and stator pole arc have to be considered; attention has to be paid to starting capability; matters concerned with core losses, switching frequencies and varying flux distributions in different parts of the magnetic circuit have to be studied. It is not possible in the context of this paper to cover all these questions adequately, nor can the implications of different applications be explored, but several are dealt with below in broad terms. Attention is given only to self-starting reversible designs.

A basic observation is first made concerning mutual inductance. It is known from stepping motor studies that it is desirable, from the point of view of maximising machine output, to eliminate mutual inductance between phases. Careful consideration will show that mutual inductance due to main (airgap) flux will indeed be zero for the features of the model assumed so far, namely:

(*a*) the iron is effectively infinitely permeable
(*b*) the stator poles are excited in diametrically opposite pairs
(*c*) the rotor has an even number of poles.

3.1 Necessary conditions on pole numbers and pole arcs

Considerable detailed argument is involved in establishing the conditions which must be satisfied by the numbers and the arcs of the stator and rotor poles. However, taking into account:

(i) the above point about mutual inductance

(ii) the possible basic patterns of 'repeatability' between relative stator and rotor pole dispositions

(iii) the need to minimise the permenance associated with L_{min}

(iv) the requirement for self-starting capability in either direction from any rotor position

(v) the desirability of minimising switching frequency, it can be shown, for pole numbers, that

$$\text{LCM }(N_s, N_r) = qN_r \tag{7}$$

and

$$\text{LCM }(N_s, N_r) > N_s > N_r \tag{8}$$

where N_s and N_r are even, the number of phases q is greater than 2 and the symbol LCM denotes the lowest common multiple. It can also be shown, for pole arcs, that

$$\min (\beta_r, \beta_s) > \frac{2\pi}{qN_r} \tag{9}$$

and

$$\beta_s \leqslant \frac{2\pi}{N_r} - \beta_r \tag{10}$$

3.2 Preferred values of pole numbers and pole arcs

Eqns. 7–10 define necessary conditions to be satisfied by N_s, N_r, β_s and β_r, but further consideration is necessary to determine the values of these parameters to give a 'good' design. First, consider β_s and the constraints on its possible values. When $\beta_s > \beta_r$, $\check{\beta}_r = 2\pi/qN_r$, so that

$$\hat{\beta}_s = \frac{2\pi}{N_r} \left(1 - \frac{1}{q}\right) \tag{11}$$

When $\beta_s < \beta_r$

$$\check{\beta}_s = \frac{2\pi}{qN_r} \tag{12}$$

The range of β_s is therefore

$$\hat{\beta}_s - \check{\beta}_s = \frac{2\pi}{N_r} \left(1 - \frac{2}{q}\right) \tag{13}$$

and similarly it can be shown that the range of β_r is

$$\hat{\beta}_r - \check{\beta}_r = \frac{2\pi}{N_r} \left(1 - \frac{2}{q}\right) \tag{14}$$

Thus, both β_r and β_s have the same *range* (increasing with increasing q and decreasing with increasing N_r), although they are at the same time governed by

$$(\beta_r + \beta_s) \leqslant \frac{2\pi}{N_r} \tag{15}$$

These relationships may be shown diagrammatically as in Fig. 5, where the limits of combinations of the values of β_r and β_s are denoted by the sides of the triangle XYZ. At point X (where the machine would contain the least amount of iron) the inductance variation has no dead zone (since $\beta_s = \beta_r$) and is of the form shown in Fig. 6a. As the geometry changes from the point represented by X along XY or XZ to Y or Z, a dead zone appears and the length of the minimum inductance period diminishes accordingly until, at Y or Z, the inductance has the form shown in Fig. 6b. Point Z corresponds to a machine with maximum winding space (as at X) and point Y to a machine with zero winding space (since stator pole arc = stator pole pitch when $\beta_s = \beta_s$). The latter is not physically realisable, but it should be noted that keeping β_r towards $\check{\beta}_r$ implies low rotor inertia.

A special case should be noted at point W, where $\beta_s = \beta_r = \phi/2$. At this point, both the period of minimum inductance of Fig. 6a and the dead zone of Fig. 6b have vanished, as shown in Fig. 6c. Any geometry corresponding to a point inside the triangle has an inductance pattern which exhibits features of all three patterns shown, depending on the position of the point. The line of $\beta_s = \beta_r$ divides the triangle as shown into regions $\beta_s > \beta_r$ and $\beta_s < \beta_r$.

A detailed consideration (not given here, for brevity) of design, manufacturing and performance constraints leads to the conclusion that practical designs have $\beta_r > \beta_s$ and therefore lie in the minor triangle of XWZ. It can be noted in

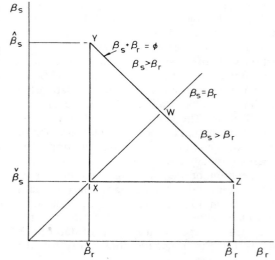

Fig. 5 *Diagrammatic description of possible values for stator and rotor pole arcs*

passing that any such practical design will have a corresponding counterpart in the minor triangle XWY (where it is represented by a point symmetrical about the line XW) and will have the rotor and stator pole arc values interchanged.

Attention is turned now to the number of rotor poles. As the torque at any point is proportional to the slope of the inductance, it follows that, to increase torque, the permeance corresponding to L_{min} should be made as small as possible. It can be shown that if the airgap due to the minor rotor diameter (dimension g_i in Fig. 1) is greater than $d(\phi - \beta_r)/2$, the inductance mainly depends on the rotor interpolar arc $(\phi - \beta_r)$ since fringing effects become predominant in the value of inductance. Hence,

$$(\phi - \beta_r)_{max} = \phi - \check{\beta}_r = \frac{2\pi}{N_r} \left(1 - \frac{1}{q} \right) \tag{16}$$

Hence, keeping N_r as small as possible (subject to the conditions previously derived) will give low values of L_{min}.

A fuller study of the above results leads to the conclusion that, for self-starting reversible drives, useful combinations of pole numbers are:

for 3-phase motors: $N_s = 6, N_r = 4$

for 4-phase motors: $N_s = 8, N_r = 6$

for 5-phase motors: $N_s = 10, N_r = 4$

3.3 Frequency and form of flux variations

The flux variations in the machine are quite different from those familiar in either a.c. or d.c. machines, and lead to a number of difficult questions when detailed motor design and system performance have to be considered. Fig. 7 illustrates, by way of example, the voltage and flux waveforms in various parts of the magnetic circuit for a 3-phase 6/4 pole machine (in accordance with eqn. 3 with $R = 0$). The voltage switching frequency of any phase f_{ph} is (see Fig. 7a):

$$f_{ph} = N_r \omega / 2\pi \tag{17}$$

and of the supply is

$$f_s = q N_r \omega / 2\pi \tag{18}$$

The period and form of the flux variations in the iron vary between the different parts of the magnetic circuit, and may be deduced by superposition of separate variations, of the type shown in Fig. 4, corresponding to the voltages applied to the individual phases. It will be seen that the stator poles experience unidirectional flux pulsations (the sense of which depends on the sense of the winding connection) of frequency f_{ph}.

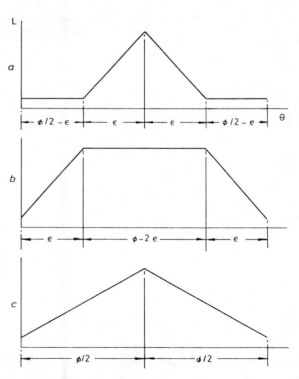

Fig. 6   Inductance profiles corresponding to:
a point X b points Y and Z c point W of Fig. 5
$\epsilon = \phi / q$

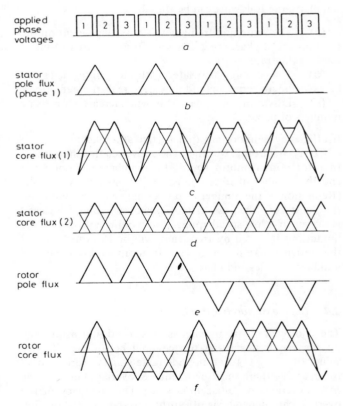

Fig. 7   Voltage and flux waveforms for a 3-phase 6/4 pole s.r. motor

99

The flux patterns in the stator back iron are illustrated by considering the back iron divided into sections bounded by pole centre lines. Assuming that the winding connections are 1, 2, 3 on adjacent poles, followed by $1', 2', 3'$ on opposite poles, then the sections between the centre lines of poles 1 and 2, 2 and 3, $1'$ and $2'$, $2'$ and $3'$ will have flux waveforms of the type shown in Fig. 7c. The waveform is made up of a triangular component of frequency f_{ph} and a d.c. component, the sense of which depends on the particular section. The remaining sections, i.e. between the centre lines of poles 3 and $1'$, $3'$ and 1 have the type of waveform shown in Fig. 7d, which has a d.c. component with a 'ripple' component of frequency f_s.

The poles of the rotor experience bidirectional pulses which have an overall frequency of $\omega/2\pi$ as shown in Fig. 7e, and the core of the rotor experiences bidirectional pulses of flux as shown in Fig. 7f, which again have an overall frequency of $\omega/2\pi$, although the waveform is considerably more complex and contains higher-order frequencies. It is because the rotor sees a minimum frequency $\omega/2\pi$ that it has to be laminated:

Any stator with N_s teeth will have (Int $(N_s/4)+1$) different waveforms occurring in the core (this is derivable from the symmetry of the magnetic circuit) and, therefore, a 4-phase machine has 3 different waveforms, compared to 2 waveforms for the 3-phase case. In the same way as noted for the 3-phase case, those waveforms with d.c. bias occur symmetrically about the core with opposite bias. In summary, the following can be stated:

(a)   The switching frequency is very dependent on the the number of phases (e.g. a change from 3-phase to 4-phase doubles the frequency).

(b)   The core loss associated with these frequencies will be exceedingly complex (and may defy realistic prediction).

(c)   Losses in the driver unit will increase with rising numbers of phases.

If efficiency is important in a particular application, e.g. in a battery traction drive, then the above relationships imply that for normal motor speeds the number of rotor poles and the number of phases must be kept as low as possible. (Reduction of the number of phases may well also minimise the cost of the associated power-switching devices.) However, if a low-speed application is being considered, the limitations imposed by frequency will be less onerous, and the designer has much greater freedom in his choice of numbers of poles and phases.

3.4   Choice of number of phases

The choice of phase number is influenced in a major way by the required starting torque (and hence the effective values of $dL/d\theta$). To ensure adequate starting torque at all rotor angles there must be adequate 'overlap' between the $L(\theta)$ variations of 'adjacent' phases. The adequacy of the overlap can depend significantly on the effects of flux fringing which lead to rounding of the corners of the $L(\theta)$

function and to reduced values of $dL/d\theta$ (and hence torque) at either end of the rising inductance period R_{01} (see Section 5). However, useful guidance about the effects of overlap can be inferred from the ideal $L(\theta)$ variations as follows:

If k_L is defined as the ratio of inductance overlap of two adjacent phases to the angle over which the inductance is changing, then, from Fig. 2a,

$$k_L = \frac{(\theta_1 - \theta_0) - \theta/q}{(\theta_1 - \theta_0)} = \frac{\min(\beta_r, \beta_s) - 2\pi/qN_r}{\min(\beta_s)} \qquad (19)$$

When $\beta_r > \beta_s$

$$k_L = 1 - \frac{2\pi}{qN_r\beta_s} \qquad (20)$$

The variation of k_L is best described by a diagram, since values of N_r and β_s are, to some extent, dependent on q. It is shown in Fig. 8 as a function of β_s, where β_s varies between the maximum and minimum values allowed by a combination of q and N_r. Curve A corresponds to $q = 3$, $N_r = 4$ and curve B corresponds to $q = 4$, $N_r = 6$. It will be seen that higher values of k_L are achieved at relatively low values of β_s for the 4-phase case. For example, M_1 and M_2 mark the midpoints of the two ranges of β_s, and it will be seen that the 4-phase machine has 20% more overlap. Put another way, for a specified amount of overlap, the 4-phase machine can work much closer to $\check{\beta}_s$, a condition which has already been demonstrated to be beneficial. To this extent, the 4-phase machine allows greater flexibility in design and will have the better starting performance.

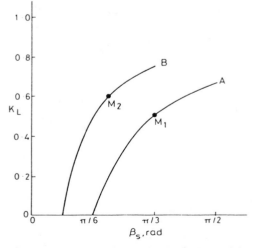

Fig. 8   *Variation of inductance overlap coefficient k_L with stator pole arc*
A $q = 3, N_r = 4$
B $q = 4, N_r = 6$

4 Control and performance characteristics

It is helpful in discussing the control of the s.r. motor to begin by considering its 'natural' or 'inherent' characteristics, i.e. that which occurs under conditions of fixed supply voltage and fixed switching angles. Linear analysis shows that the torque/speed curve is then the same as that of a d.c. series motor.[11,12] This can readily be seen to follow from the fact that, as the motor slows down, the time during which a phase winding is excited increases inversely with the fall in speed and, therefore, so also does the flux. The torque, however, as in all such devices, is proportional to the square of the flux and so the resulting torque/speed curve is defined by

$$T = \frac{k}{\omega^2} \qquad (21)$$

and the power is given by

$$p = \frac{k}{\omega} \qquad (22)$$

The analogy with the d.c. series machine immediately points to the possibility of control through terminal voltage or supply current, and both of these can be readily implemented in the overall system. Additionally, however, there are two further important parameters available to the designer. These are the switch-on angle θ_i and the switch-off angle θ_x (or its equivalent, the conduction angle θ_c). Control of these angles is easily and economically achieved, involving only the appropriate conditioning of timing pulses, and makes available a very wide range of performance characteristics and control possibilities. In practice, of course, control parameters are chosen so as to optimise overall system performance (e.g. to minimise currents or to maximise efficiency), as well as to achieve particular characteristics. It is impossible to set out here the full dependence of performance on the control parameters, but basic modes of operation are explained briefly below.

The first mode is the natural one with fixed supply voltage and fixed switching angles. There is, of course, a family of series characteristics for varying supply voltages (at a given speed the flux is proportional to the voltage V, and the torque varies as V^2), with an upper limiting characteristic set by maximum rated voltage. As the speed falls, the flux rises and 'base speed' ω_b is defined as that which corresponds to maximum flux (and current) at maximum voltage. Base speed is the lowest speed at which maximum power can be obtained and the highest speed for maximum torque.

The second important mode of operation involves the control of speed below ω_b. By analogy with the d.c. series motor, this can be achieved by varying the effective applied voltage either literally, using a variable d.c. link, or by modulation or chopping using the main switching devices. As in the d.c. machine, a constant torque characteristic can be obtained in this mode by 'current limit' current chopping, or the torque can be varied by control of the chopping level. An alternative possible means of controlling speed

below ω_b with the s.r. motor is to reduce the conduction angle θ_c at constant voltage, but this entails an increasing peak flux for constant average torque.

The third important mode of operation is that of controlled speed above ω_b. Although, as explained above, torque falls 'naturally' as ω^{-2}, the natural fall in torque with increasing speed can be offset by increasing the conduction angle proportionately. The conduction time does not then fall as ω^{-1}, nor does the flux, and a variety of characteristics can be obtained. A characteristic of particular interest (especially for traction applications) is that of constant power over a range of speeds. This requires the flux to fall as $\omega^{-\frac{1}{2}}$, so that the torque falls as ω^{-1}. As the conduction angle is increased, there comes a point at which the 'switched on' period for a phase winding is equal to the 'switched off' period ($\theta_c = \phi/2$) and, for an ideal case of zero resistance, if θ_c becomes larger than this, the flux level would tend towards infinity (because a net d.c. voltage would be applied to a purely inductive circuit). This may be taken as defining the upper limit of the third mode of control, but it will be seen in Section 6 that, within this limit, a wide speed range of constant power operation can be achieved with excellent overall performance. At speeds above the highest speed obtained at constant power in the third control mode, the motor reverts to its 'natural' characteristic appropriate to $\theta_c = \phi/2$.

It will be appreciated that changes in θ_c as discussed in the context of constant torque and constant power operation can be achieved by associated changes in θ_i and θ_x, and the proper choice has to be made in relation to application and performance optimisation.

5 Nonlinearities and performance prediction

In predicting the performance of conventional machines it is usually satisfactory to use a model in which the parameters are constant, provided that these constant values are also 'effective' values allowing appropriately for magnetic nonlinearities and other influences, such as skin effect etc. (Transient performance computations continue, of course, to involve nonlinear equations, because of the speed dependence, though they can be handled with varying degrees of simplification depending upon the information required.) With the s.r. motor, however, behaviour depends so much on the nonlinear 'magnetic' parameters that no corresponding simplification using constant 'effective' parameters is acceptable. It becomes essential for all serious design work to recognise the dependence of inductance on instantaneous current as well as position. Consideration is given, first, to the nonlinear nature of the flux-linkage, inductance and static torque characteristics and, secondly, to performance predictions using these characteristics. The basic data presented were measured using a representative 4-phase motor built into a D90 induction-motor frame.

5.1 Static characteristics: flux-linkage inductance and torque

Typical variations of flux-linkage with current, inductance with angle and torque with angle are shown in normalised terms in Fig. 9a, b and c, respectively. The very significant divergence of these characteristics from the ideal ones (as in Fig. 2 for $L(\theta)$ and $T(\theta)$) is immediately apparent. Only at very low values of excitation or for rotor positions near the middle of the minimum inductance region R_{34} do they approximate to the ideal: for currents above $0.15i$ the ψ/i relationship ceases to be linear; the rising section of the inductance curve R_{01} becomes very rounded at both ends; and the torque curve quickly loses its rectangular shape. The differences are owing, of course, mainly to saturation effects in the magnetic circuit, but near the ends of region R_{01} the influence of flux fringing is significant.

Two distinct effects of saturation must be recognised. The first is the 'bulk' effect on the magnetic circuit as a whole when the excitation is raised to practical levels, and this is similar to the effect in other types of machine. The second is the effect of the intense 'local' saturation in the pole tips when rotor and stator poles are only partially overlappped. It is particularly severe when the degree of overlap is small and, as known from experience with stepping motors, requires careful consideration.[13] In general, both local and bulk effects are present and interact, but their effects can be isolated by observations at particular rotor angles in Fig. 9a.

Bulk saturation effects can be observed best for positions with full overlap between stator and rotor poles (when there is no influence from local saturation). From Fig. 9a it can be seen, considering, say, the curve for $\theta = 0.38\phi$, that bulk saturation occurs at all flux levels in excess of $0.5\bar{\psi}$; from Fig. 9c the associated large loss of torque at large overlap and high current is apparent. Clearly, local saturation can only be separated from bulk saturation below this flux level of $0.5\bar{\psi}$, and is best observed at small amounts of overlap. It is most evident in the nonlinearity of the $\psi-i$ curves for $0.05\phi < \theta < -0.12\phi$ at linkage levels below $0.5\bar{\psi}$. (A measure of it is provided by estimating from Fig. 9a, by extrapolation of the linear parts of the curves, the extra current needed: 26% at -0.12ϕ and 40% at 0.05ϕ.)

Turning to the influence of flux fringing, the existence of any torque for negative θ is evidence of its presence; the decrease in slope of the $L(\theta)$ curve (for all excitation levels) at angles above 0.2ϕ is evidence of decreasing fringe flux as local saturation falls. The effect of the strong interaction between local and fringe flux is very plain from the rapid changes in torque with current for negative θ (Fig. 9c). In general, local saturation is encountered first as poles approach overlap, but progressively gives way to bulk saturation as overlap increases (as under running conditions).

An important overall result arising from the combined effects of saturation (particularly local) and fringing is that the effective pole arc on the stator (remembering $\beta_s < \beta_r$) changes with current. If the effective pole arc is taken to be that which is appropriate to a reasonable piecewise linear equivalent of the actual $L(\theta)$ curve between the points $1.05\,L_{min}$ and $0.95\,L_{max}$, then it can be expressed as a percentage of the actual arc: 106% at $0.07i$, 95% at $0.29i$ and 81% at $1.0i$. This is an important effect, and appropriate consideration must therefore be given to the choice of rotor and stator pole arcs when designing a machine which works at high excitation levels.

In concluding this Section it should be noted that the above change in 'effective' pole arc underlines very clearly the dependence of performance, not only on magnetic nonlinearities, but also on the overall geometry and field configuration in the machine. A discussion of methods of predicting inductance values from dimensional and material data lies well outside the scope of this paper, the necessary numerical analysis lying at the limit of capability both of modern techniques and large computers. A comparison of experimental and computed results using a complete 3-dimensional model has been published[14] and, for the limited cases considered, shows good agreement. Various approximate methods suited to routine design have been developed, and two of the present authors recently described a method for the analytical estimation of L_{min}.[15]

5.2 Prediction of performance

The prediction of the dynamic performance requires the simultaneous solution of eqns. 1 and 3 and the mechanical equation, including due allowance for the nature of the flux linkage and inductance characteristics as above. The problem is, in principle, identical with that of the dynamic performance of stepping motors, and various levels at which it can be approached have been summarised by Hughes et al.[16] However, no published method proved sufficiently accurate for the s.r. motor, even at constant speed, and an efficient new technique has been developed.[17]

Further discussion of this subject here is restricted to consideration of the current waveform under steady-state running conditions. This waveform, computed using the new method with proper allowance for the effects of changing saturation levels and inductance values during the cycle, is shown in Fig. 10 for conditions similar to those shown in Fig. 4. The Figure also shows the current waveform calculated using constant parameter models, both models having the same value of L_{min} as the nonlinear model, one having $L_{max}/L_{min} = 8$ and the other $L_{max}/L_{min} = 2$. The inadequacy of the constant parameter models is plain. Measured waveforms are shown in the following Section.

6 Practical experience

The authors have constructed a considerable number of switched reluctance motors, of both single-winding and bifilar-wound types, with power outputs ranging from 10W at 10 000 rev/min to 25kW at 750 rev/min. For ease of construction, and to permit unambiguous comparison with conventional motors and systems, several of these have been built into commercially available induction-motor frames. They have also employed airgaps which are consistent with conventional techniques of induction-motor manufacture. The smaller motors have employed transistor convertors only up to a rating of 1·5kW, although, with the transistors now available, this type of convertor could be used at much higher powers. Alternatively, gate-turn-off thyristors could be used. The larger motors (10kW

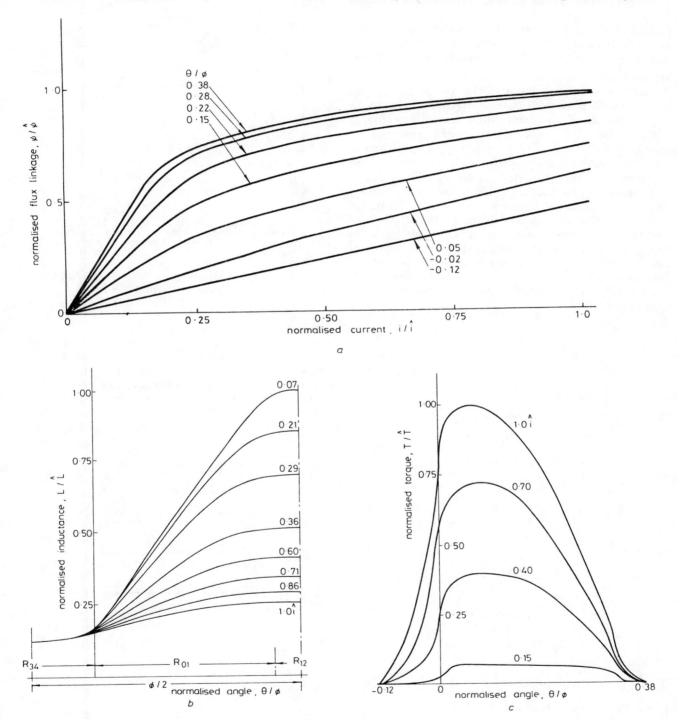

Fig. 9   a   *Flux-linkage/current curves*
      b   *Inductance/rotor-angle curves*
      c   *Torque/rotor-angle curves*

4-phase, 0·75kW s.r. motor

and above) have been developed specifically for battery-powered traction applications, and have employed thyristor convertors developed by colleagues at the University of Nottingham. This Section presents representative results of the authors' experience.

Fig. 11 shows the measured performance of a 3-phase nonbifilar wound motor in a D90 frame, having $N_r = 4$ and $N_s = 6$. The core size was identical with that of the standard induction motor in the frame, and the same airgap diameter was retained (to facilitate comparisons), even though these dimensional choices are unfavourable to the s.r. motor. The Figure shows the measured power/speed characteristics and values of measured efficiency at important load points. The characteristics show regions of constant torque, constant power and series-motor roll-off, together with the associated modes of control (all in accordance with the discussion of Section 4). The efficiency of the standard induction motor at full load (0·75kW, 920 rev/min) is 77% on a sinusoidal supply. The efficiency of the s.r. motor system at the same load point (including losses in the convertor) is 78%, a particularly gratifying result since the design is nonoptimal, and it remains at a high level to the top of the constant power range. Thereafter, owing to the design limitations of this motor, the efficiency falls.

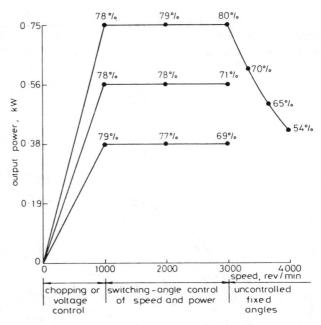

Fig. 11 *Power/speed characteristics of a 3-phase 0·75kW s.r. motor, showing measured system efficiencies*

Fig. 12 gives characteristics similar to the above, illustrative of the performance of a 4-phase bifilar-wound motor. This was built into a conventional D180M frame which would be rated, as an induction motor on a sinusoidal supply, at 9·25kW, 86% efficiency. For operation as a variable-speed motor on an inverter supply, the induction motor would be derated, probably by 10% and possibly by as much as 20% on a p.w.m. supply.[18] The efficiencies of

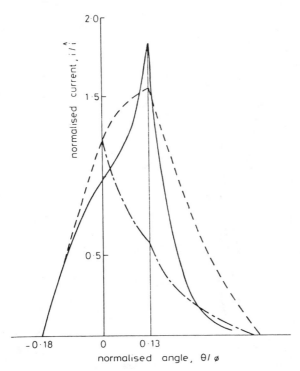

Fig. 10 *Computed linear and nonlinear current waveforms*

Zero winding resistance
nonlinear method
linear method, $L_{max}/L_{min} = 2$
linear method, $L_{max}/L_{min} = 8$

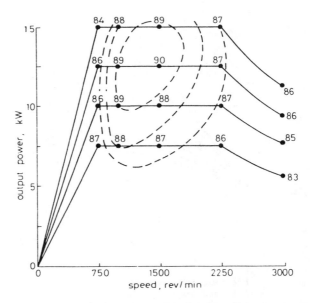

Fig. 12 *Power/speed characteristics of a 4-phase D180M s.r. motor, showing measured system efficiencies*

104

the s.r. motor, including all convertor losses, are seen not only to equal the efficiency of the fully-rated sinusoidally-fed motor at the base speed of 750 rev/min, but to remain very high over a 2:1 range of power output and a 3:1 range of speed. The drive has a peak power rating comparable with that of the induction motor, and can maintain this high power over the 3:1 speed range. The very high levels of efficiency and the wide range of operating conditions over which they are available are emphasised by the loci of constant efficiency shown dotted in Fig. 12.

Fig. 13 shows current waveforms measured for this motor. The correlation of the waveform in part *a* with that predicted in Fig. 10, using the nonlinear analysis should be noted. A detailed study of the accuracy of the computer model has already been presented.[17] The pronounced effect of bulk magnetic saturation is evident in Fig. 13*a*, and this contrasts with the absence of any such effect in the reduced-flux condition at the top of the constant power region shown in Fig. 13*c*. Fig. 14 shows flux waveforms, for this 4-phase motor, which confirm the method for predicting flux patterns given in Section 3. They were obtained using search coils wound round the back iron of the machine at appropriate places and integrating the resulting e.m.f.s.

It is to be noted that, although only motoring characteristics have been presented here for clarity, regeneration has been achieved equally successfully over the whole torque/speed range. These machines used digital control of switching angles, a method which is very convenient for research and development purposes. The method of rotor-position detection is extremely simple; a pair of optical switches and an appropriately slotted disc clamped to the rotor shaft are the only components required on the motor, and from the resulting signals appropriate firing pulses are generated by the electronic controller.

7   Conclusions

A family of variable-speed drives has been described which uses doubly salient rotor-position-switched reluctance motors in conjunction with various forms of electronic power convertor. Particular attention has been paid to various highly nonlinear features, which do not occur in conventional a.c. or d.c. machines, with a view to laying a general foundation for the practical design of s.r. motors.

The motor retains all the advantages of robustness, cheapness, safety and minimal maintenance associated with the cage induction motor, and permits a power-convertor design which uses only half the number of main switching devices needed in the normal inverter for induction motor use. In terms of cost, therefore, the new system is significantly cheaper than equivalent, a.c. motor based, variable-speed systems, and it is expected to retain this advantage as relative costs of machine and electronic components vary. It is also cheaper than many d.c. motor based systems and, with the high proportions of the cost of such systems being in the d.c. motor, the s.r. motor system should gain progressively. One example of this is in the application to

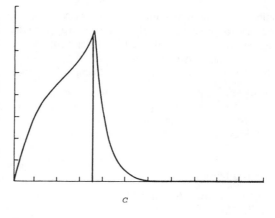

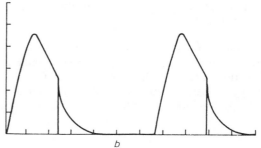

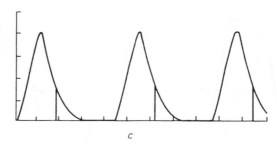

Fig. 13   *Current waveforms for the same motor as Fig. 12, 10kW output:*

a 750 rev/min   *ER* = 0·72
b 1500 rev/min   *ER* = 0·84
c 2250 rev/min   *ER* = 0·87
x-scale: 1 ms/div; *y*-scale: 20A/div

electric-vehicle drives which has received attention in association with colleagues at Nottingham University.

The performances already obtained from s.r. systems shows them to be attractive over a wide range of sizes for many applications. Ratings and efficiencies of *total s.r. systems* have been obtained which better those of a standard induction motor in the same frame, even when fed from a fixed-frequency sinusoidal supply. Indeed, recent results have been obtained in which the continuously-rated specific output of a 15kW s.r. motor is some 1·3 – 1·4 times higher than that of the equivalent induction motor.

It is also appropriate to emphasise the range and convenience of control which s.r. systems permit; they are fully reversible, regenerative and provide very flexible torque/speed characteristics merely through appropriate timing of the switching angles, giving full 4-quadrant operation with no additional convertor components. Additionally, of course, they have the sought-after operational advantages of being 'brushless' designs. It seems entirely reasonable to conclude that, with fuller development either in general or in relation to specific applications, the new system will prove commercially and technically very attractive.

8 Acknowledgments

Acknowledgment is due to the UK Science Research Council and the University of Leeds for the financing of Dr. P.T. Blenkinsop and Dr. J. Corda, respectively, to enable them to work on the theory and design of switched reluctance motors.

The development of the traction system, to which reference has been made in the paper, involved extremely close collaboration between those responsible for the motor. The authors wish to acknowledge the part played in the success of this development by Mr. Davis and Mr. Ray, and more recently Mr. Blake of the University of Nottingham. They also wish to thank Chloride Technical Ltd. for financing the combined project, and in particular Dr. M.F. Mangan for many constructive discussions. The authors are indebted to Brook Crompton Parkinson Ltd. for their assistance, particularly in the provision of data on induction machines.

9 References

1 LAWRENSON, P.J.: 'Synthesis and performance of improved reluctance motors'. Proceedings of the international conference on electrical machines, London, 1974, pp. C3-1 – C3-10
2 BAUSCH, H., and RIEKE, B.: 'Performance of thyristor-fed electric car reluctance machines'. Proceedings of the international conference on electrical machines, Brussels, 1978, pp. E4/2-1 – E4/2-10
3 BAUSCH, H., and RIEKE, B.: 'Speed and torque control of thyristor-fed reluctance motors'. Proceedings of the international conference on electrical machines, Vienna, 1976, Part I, pp. 128-1 – 128-10
4 UNNEWEHR, L.E., and KOCH, W.H.: 'An axial air-gap reluctance motor for variable-speed applications', *IEEE Trans.*, 1974, PAS-93, pp.367 – 376
5 KOCH, W.H.: 'Thyristor controlled pulsating field reluctance motor system', *Electric Machines and Electromechanics*, 1977, 1, pp. 201-215
6 BYRNE, J.V., and LACY, J.G.: 'Characteristics of saturable stepper and reluctance motors', *in* 'Small electrical machines'. IEE Conf. Publ. 136, 1976, pp. 93-96
7 BYRNE, J.V., and LACY, J.G.: UK Patent 1321110, 1973
8 JARRET, J.: 'Machines électriques à réluctance variable et à dents saturées', *Tech. Mod.*, 1976, 2, pp.78-80
9 HARRIS, M.R., ANDJARGHOLI, V., LAWRENSON, P.J., HUGHES, A., and ERTAN, B.: 'Unifying approach to the static torque of stepping-motor structures', *Proc. IEE*, 1977, 124, (12), pp. 1215-1224
10 RAY, W.F., and DAVIS, R.M.: 'Inverter drive for doubly salient reluctance motor', *IEE. J. Electr. Power Appl.*, 1979, 2, (6), pp. 185-193
11 BLENKINSOP, P.T.: 'A novel, self-commutating, singly-excited motor'. Ph.D. thesis, University of Leeds, 1976
12 CORDA, J.: 'Switched reluctance machine as a variable-speed drive'. Ph.D. thesis, University of Leeds, 1979
13 ERTAN, H.B., HUGHES, A., LAWRENSON, P.J., and HARRIS, M.R.: 'A new approach to the prediction of the static torque curve of saturated VR stepping motors'. Proceedings of the 8th incremental motion control systems and devices symposium, University of Illinois, May 1979
14 SIMKIN, J., and TROWBRIDGE, C.W.: 'Three dimensional computer program (TOSCA) for nonlinear electromagnetic fields'. Paper RL-79-097, SRC Rutherford Laboratory, Dec. 1979
15 CORDA, J., and STEPHENSON, J.M.: 'An analytical estimation of the minimum and maximum inductances of a double-salient motor'. Proceedings of the international conference on stepping motors and systems, Leeds, 1979, pp.50-59
16 HUGHES, A., LAWRENSON, P.J., STEELE, M.E., and STEPHENSON, J.M.: 'Prediction of stepping motor performance'. Proceedings of the international conference on stepping motors and systems, Leeds, 1974, pp.67-76
17 STEPHENSON, J.M., and CORDA, J.: 'Computation of torque and current in doubly salient reluctance motors from nonlinear magnetisation data', *Proc. IEE*, 1979, 126, (5), pp.393-396
18 CREIGHTON, G.K., SMITH, I.R., and MERGEN, A.F.: 'Loss minimisation in 3-phase induction motors with p.w.m. inverter supplies', *IEE J. Electr. Power Appl*, 1979, 2, (5), pp. 167-173

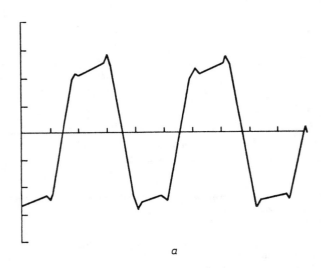

a

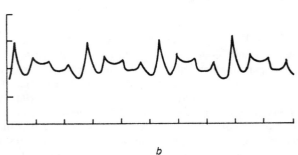

b

Fig. 14 *Measured flux waveforms of a 4-phase s.r. motor*

a Waveform corresponding to the 3-phase waveform of Fig. 7c
b Waveform corresponding to the 3-phase waveform of Fig. 7d

Discussion on "Variable-Speed Switched Reluctance Motors"

M.R. Harris, H.R. Bolton, P.A. Ward, J.V. Byrne, G.B. Smith, J. Merrett, F. Devitt, R.J.A. Paul, K.K. Schwartz, M.F. Mangan, A.F. Anderson, R. Bourne, P.J. Lawrenson, J.M. Stephenson, and N.N. Fulton

Proceedings IEE, Vol. 128, Pt. B, No. 5, September 1981, pp. 260-276

Prof. M.R. Harris (*University of Newcastle-Upon-Tyne*): The authors have described the results of an excellent programme of work, carried through by two universities in collaboration, which shows every sign of being highly important in relation to future electrical variable-speed drives over a wide range of power rating. They have achieved, to date, overall system efficiences which better that of the equivalent polyphase induction motor and maintain this over a 3:1 speed range, with specific outputs in some cases significantly greater than that of the induction motor in the same frame. Construction of the motor is particulary simple and cheap, and the electronic drive circuits are of comparatively moderate complexity and cost.

Several matters of scientific interest, as well as of engineering importance, are raised by the work. I should like to pursue a few of these in some detail, and to invite the authors to comment on my arguments in reply, on the basis of the practical experience that they have accumulated in recent years. In one or two cases, it would be helpful if further technical information could possibly be made available.

The variable-speed system described is unusual and immediately interesting, because it is based on a doubly-salient variable-reluctance motor, and it is now well established [A, B] that this machine has some distinctive electromechanical features. A close comparison with the induction motor is instructive, first on the basis of available torque per unit active volume, and secondly on that of stator electric loading required to achieve optimum performance. Defining D = rotor diameter, λ = rotor circumferential tooth pitch, g = radial gap, and t = tooth width (average of stator and rotor values, where these differ), we can calculate by established theory [B] that, for example, the 3-phase doubly-salient motor with 6:4 slotting should be capable of producing a static mean torque (averaged over a half cycle of the static $T - \theta$ curve) per unit of cylindrical rotor volume of roughly 70 000 Nm/m^3. This calculation assumes that stator excitation MMF per pole F, equal to the critical value [B] F_c, is applied to one pair of poles; that $\lambda/g = 300$ (a possible value for this class of motor) and $t/\lambda = 0.35$; also that gap geometry can be treated as rectilinear (a significant approximation).

Papers 795B, T481P and 8300P by LAWRENSON, P.J., STEPHENSON, J.M., BLENKINSOP, P.T., ČORDA, J., and FULTON, N.N. [see *IEE Proc. B, Electr. Power Appl.*, 1980, 127, (4), pp. 253–265], RAY, W.F., and DAVIS, R.M. [see *IEE J. Electr. Power Appl.*, 1979, 2, (6), pp. 185–193] and STEPHENSON, J.M., and ČORDA, J. [see *Proc. IEE*, 1979, 126, (5), pp. 393–396], respectively

Read before IEE Power Division Professional Groups P1 and P6, 8th December 1980

Assuming now that each phase of the motor is excited with halfcycle square-wave MMF per pole equal to F_c (a highly idealised picture that is nevertheless relevant for order-of-magnitude calculations), then the corresponding calculated value of running torque is about 105 000 Nm/m^3. To achieve a specific torque significantly greater than this value begins to imply a disproportionate increase in excitation MMF that must be unattractive in engineering terms. On the other hand, a lower value of specific torque may well have to be accepted, if insufficient excitation MMF is available in design.

The best commercial value for induction-motor rated torque per unit rotor volume, for a modern 10 kW design, 4-pole, TEFV type, class B temperature rise, is about 36 000 Nm/m^3. We see immediately that the doubly-salient motor can, in principle, produce about treble this specific torque and we therefore expect that it may possibly be a competitive machine at this size, as indeed the authors have demonstrated. However, electric loading in the induction motor (AC = 6 × series turns per phase × RMS phase current ÷ circumference at stator bore) is typically 35 A/mm. We will now establish an equivalent value of electric loading for the 6:4-slotted motor, and investigate what level of torque this equivalent loading might be expected to produce.

Assume, for the sake of argument, that D is the same proportion (about 62%) of stator core OD in both machines. Further, assume that the same total number of stator conductors N, is employed in both machines, and that the copper CSA per conductor is the same, $N/12$ turns being wound around each of the six stator poles of the doubly-salient motor, treating the pole winding as monofilar for simplicity. Neglecting differences in end-winding length in the two machines (the doubly-salient design being favourable in this respect), we may now calculate the permissible halfcycle square-wave current I to give the same stator ohmic loss as in the induction motor. The answer is readily found to be $I = \sqrt{2} \pi D$ AC/N, and the corresponding permissible MMF per pole F is conveniently expressed as

$$F/\lambda = \sqrt{2} \text{ AC}/3 \tag{A}$$

noting that $\lambda = \pi D/4$ for the 6:4-slotted motor. We see that F/λ could readily have a value of $\sqrt{2} \times 35/3 = 16.5$ A/mm. But this estimate is pessimistic, particularly because of the absence of rotor ohmic loss, which permits greater stator ohmic loss. Overall, a very reasonable guide (borne out in design practice) is that F/λ should not exceed about 25 A/mm in this 6:4-slotted, 10 kW motor, interpreting F now as the halfcycle square-wave MMF amplitude that has the same RMS value as the actual wave of excitation MMF per pole. (A somewhat higher value of F/λ is possible if the percentage value of D is substantially reduced, but although this gives improved specific torque, the overall motor torque is only marginally improved.)

It is helpful to express the available MMF in terms of F/λ, because it is readily shown that F_c/λ is independent of machine size if the gap geometry (t/λ and λ/g) is assumed constant, and shows only a moderate variation over the likely practical range of gap geometry. For $\lambda/g = 300$ and $t/\lambda = 0.35$, F_c/λ is found to be 131 A/mm, and this value, or at least something in excess of 100 A/mm, might be regarded as an approximate target value for F/λ, in design, other considerations being equal. Applying essentially the design method described elsewhere [B, C], but modified to calculate running torque in place of static holding torque, the following is a summary of typical results, for a 6:4-slotted motor:

F/λ, A/mm	0	13	26	39	52	66	92	131
Running torque per unit rotor volume, Nm/m^3 × 1000	0	18	41	57	70	82	96	110

The calculation of torque in the above Table allows for F being the RMS-equivalent square-wave of an actual current that is nonsquare, the actual current producing slightly more running torque — about 20% in the highest case — than a true square-wave current would do.

The above calculations are sufficiently accurate to show that, at an F/λ-value which is certainly obtainable in this 10 kW machine, a specific torque comparable with the 4-pole induction motor (and therefore somewhat excelling the 8-pole motor, which is the authors' basis of comparison) is predictably available. The authors' findings are supported. However, the design is nevertheless seen to be short of MMF, compared with what could be effectively utilised for further torque production, if it were available, and that appears to be a basic feature of the design.

It follows that there might be a handsome advantage in overall performance if F/λ could be increased by any means, without increasing ohmic loss. A very distinctive characteristic of doubly-salient motors [A, B] is that this can be achieved simply by increasing the number of teeth per pole, from one in the present case to two or more, so that λ reduces while MMF and ohmic loss are unaffected. (Merely increasing the stator pole number, retaining one tooth per pole, does not offer advantage.) Whether this is an improvement overall depends on whether the consequent increase of core loss is tolerable, owing to the rise in magnetic frequency for a given shaft speed. Specifically, it seems very likely that a motor with two teeth per pole would be attractive. Design studies at Newcastle University have concentrated on a 3-phase motor with 12 stator teeth on six poles and a 10-tooth rotor, and this machine is to be built and tested. In particular cases, notably slow-speed or small machines, an even higher tooth number might be employed. It appears that the 12:10-slotted motor

may show a 60% increase in specific torque compared to 6:4, which is plainly an exciting possibility. Efficiency might also improve, although that depends on core loss and is uncertain.

With all the preceding calculations and discussion, field fringing at the core ends has been neglected. The torque produced will certainly be significantly modified by this effect in the 6:4 motor, but predictably less so in the 12:10 design, because of the increased ratio, core length: tooth pitch.

A remarkable feature of doubly-salient motors, not widely recognised, is that they can still in principle compete with the specific output of induction motors, even at 5 MW size. A typical rated torque per unit active rotor volume (radial ventilating ducts not counted) of a 5 MW, 4-pole, TEFV induction motor is 85 000 Nm/m^3. In the equivalent doubly-salient design, dimensional scale effects permit F/λ to increase from 25 A/mm (appropriate for a 10 kW motor) to perhaps 50 A/mm at 5 MW, 6:4-slotted, and to 125 A/mm in the 12:10 case. The corresponding specific running torques are about 68 000 and 85 000 Nm/m^3, respectively. However, it seems likely that doubly-salient motors will not match induction-motor efficiencies in the largest sizes, because core loss becomes a more important consideration as size increases and induction motors are probably better in this respect. But it does seem clear that these unusual motors should find application in substantial sizes, and it will be most interesting to observe what limit emerges in practice.

Turning to the subject of core loss, we note that a 6:4-slotted motor has a high fundamental magnetic frequency of 200 Hz at 3000 rev/min. This is admittedly no worse than an inverter-driven 8-pole induction motor, but in the doubly-salient motor the harmonic content of flux waveforms is high and this must significantly increase the core loss. The problem is mitigated to some extent by two features of these motors, the first being that, in the stator poles, the flux density excursion is not from $+\hat{B}$ to $-\hat{B}$, but from $\hat{B}$ to 0. Also, one-third of the back-of-core region (which predictably contributes the majority of the core loss) has a low amplitude of flux density, as the authors show. However, there is, additionally, a small but significant core loss in the rotor. The subject is important, as discussed earlier, and it would be helpful if the authors could provide any information on core loss as a function of operating condition, obtained perhaps by subtracting known losses from the measured total loss. It seems likely that the use of a low-loss grade of silicon steel, in 0.35 mm thickness (as opposed to 0.5 mm), would be well justified in these motors, and confirmation of this point would be helpful.

The authors have confronted the awkward problem of defining the equivalent of power factor for switched waveforms in which neither voltage nor current is constrained to be sinusoidal. They propose 'energy ratio' (ER), where, for the arbitrary waveform of power flow shown in Fig. A,

$$ER = \frac{a+b-c-d}{a+b} \tag{B}$$

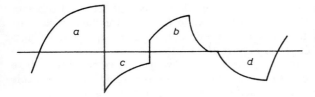

Fig. A *Arbitrary waveform of power flow*

a, *b*, *c*, *d* being four energy areas which comprise this particular complete waveform. This is an interesting proposal, but it appears to me to place too much emphasis on the proportion of the power flow in a complete cycle that is of reversed sign, and none on the *shapes* of the voltage and current waves. Power factor is concerned with how effectively a circuit is utilised for power transfer, and expresses the actual power transfer as a fraction of the greatest transfer possible, subject to some physically meaningful constraints; for sinusoidal working, the constraints are, of course, simply that voltage and current are both sinusoids of constant amplitude, although variable phase. In nonsinusoidal working, one must surely recognise that, if either or both waveforms are narrow and peaky, the circuit is poorly utilised for power flow, even if the flow never reverses in sign.

One conclusion might then be that the proper constraint on current is that its RMS value should be held constant, on the basis that this value is related to the important heating effect, in the load supplied and/or in the supply devices. For want of any better simple way of proceeding, the RMS value of the voltage wave may also be held constant, and this leads to the general definition of power factor (PF) [D], i.e.

$$ PF = \frac{\text{mean power flow}}{v_{rms} i_{rms}} \tag{C} $$

Eqn. C can certainly be criticised, but it does reflect adverse wave shapes and has also the merit of equalling the usual PF value, when circuits are supplied with sinusoidal voltage and current; ER does not equal PF for sinusoidal working, except at the values of zero and unity.

Various definitions are possible for a function of voltage, current and power, that might permit a fair comparison of switched and sinusoidally excited devices, in respect of their power-transfer qualities. All, however, seem to be open to fairly serious objections in one way or another. It is inappropriate to pursue the matter much further here, but the example of Fig. B illustrates the difficulty with energy ratio. If *a* is a voltage-wave, and *b*, *c* and *d* are alternative current-waves that transfer the same mean power, it is readily appreciated that the energy ratios are unity for all cases. But few would agree that the power-transfer 'quality' is as 'good' in *d* as in *b*, and the power-factor values of 1, 0.25 and 0.1, respectively, arguably give a fairer picture, therefore.

It may well be that the most meaningful definition of power factor (or its equivalent) is in relation to the ultimate supply terminals, rather than the motor terminals. According to this view, the motor-phase currents and voltages should be referred back to the terminals of either the supply battery or

(through an appropriate bridge rectifier) the AC supply sytem, as the case may be, and then evaluated at that point by eqn. C.

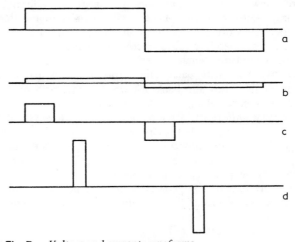

Fig. B *Voltage and current waveforms*

a Voltage waveform
b, *c*, *d* Alternative current waveforms

The authors refer to the proposal by Jarret, to increase torque output by reducing the packing factor of the laminated steel in the pole face, or elsewhere, so creating abnormal levels of saturation. They make it clear that they have not themselves pursued this approach. One quite commonly encounters the view among electrical engineers that the removal of iron from the magnetic circuit in some way offers the prospect of greatly increased torque — a view, to be fair, encouraged by the strong claims that have been made on behalf of this proposal. I should like to state, unequivocally, that this is essentially misconceived; indeed, the static mean torque (as defined earlier) is always reduced by such measures, (assuming, of course, that the t/λ and λ/g values are held constant during the modification.) Production of increased saturation by *reducing the gap* does offer advantage, as is well known [A], the greatest possible increase of torque being a factor of two, but this is a quite separate matter.

However, reducing the packing factor in the pole-face *does* offer the possibility in some cases of altering the shape of the static torque against angle diagram at constant current, with little, if any, significant loss of static mean torque. This altered shape, taken in combination with the practically attainable waveforms of phase current, in dynamic working, may produce some increase in running torque for a given RMS phase current. The running torque will also then be smoother, with less pronounced ripple, which is particularly valuable at low speeds. Although this is definitely possible, I must say that I have yet to see a proposal for iron removal argued clearly in these terms, or experimental results quoted in support. The value of reshaping the torque against angle diagram is most obvious in designs where there is already an unavoidable strong magnetic constriction remote from the gap region, the natural tendency then being for a high peak of torque to occur when stator and rotor teeth first overlap, with a rapid reduction as overlap increases. Iron removal then produces a flattened, more uniform torque against angle diagram.

The interesting claim is made that it is desirable, from the point of view of maximising motor output, to eliminate mutual inductance between phases. It is obvious to me that the absence of mutual inductance in the present machines gives significant benefits of a secondary nature. A cyclic pattern of excitation is made possible that does not vary in identical manner with each step; this reduces magnetic frequency in the rotor and back-of-core, and creates a region of the back-of-core which sees a low flux amplitude, all of which are very important in reducing core loss. However, the authors seem to imply that there is a further, more basic, advantage. This is more difficult to understand, and is perhaps open to some question. It would be very helpful if additional explanation of the point could be supplied.

In connection with gap geometry and slot number combinations, it is first apparent that designs are favoured in which stator and rotor pole arcs (at the gap) are unequal. This certainly has one effect of reducing the ratio of maximum to minimum reluctances (as functions of rotor position) in the magnetic circuit per pole pair, and that must reduce the static mean torque. It would appear (although it is not stated) that this measure is introduced to give some control of the shape of the static torque against angle diagram, the loss of mean torque being tolerated. Also, among the specific slot-number combinations quoted as being practically useful, the ratio 10:4 stands out as unfamiliar and not an obvious extension of known practice in stepping motor design. Indeed, it would seem that although 10:8 is entirely acceptable, giving specific running torque closely comparable with 6:4 and 8:6 at a given level of ohmic loss, 10:4 sacrifices considerable torque, for no obvious advantage. Further explanation of these two matters would again be helpful.

I should like to acknowledge three colleagues, J.W. Finch, A. Musoke and J.E. Brown, whose advice on many matters has contributed to the several points that I have raised. Also, Laurence, Scott & Electromotors Ltd and GEC Machines (Blackheath) Ltd, are thanked for much useful information on machine rating and design.

H.R. Bolton (*Imperial College*): These are three excellent papers of considerable significance which may well come to be regarded as milestones in the history of variable-speed drive systems and the authors are to be congratulated on a remarkable achievement. One may cite the comments in the Report of the Parliamentary Select Committee on battery electric vehicles as evidence that the potential merits of SRM drives for this category of application have been recognised at the highest levels, largely as a result of the work we have heard about this evening. In view of the availability of solid-state switching devices for many years and of an awareness of SRM principles extending back to the 1840s, the most obvious question that arises is similar to the 'why did no-one do it before?' cry, often following announcements of new developments combining effectiveness and apparent simplicity. Here one would ask: why has the development of successful SRM drives been so delayed?

A number of minor points seem worth raising:

(*a*) SRM drives are compared briefly with DC and with autopiloted synchronous machine drives in Section 1 of Paper 795B. Could the authors state the basis for the comparison? Would the SRM drive's advantages still be apparent in high-power drives, say above 100 kW?

(*b*) How does the regenerative performance of the drive compare with the motoring performance, and is there a role for the system specifically as a brushless generator?

(*c*) Is there a case for the use of segmented rotor designs or is the restriction on N_S, N_R and winding combinations that would occur too severe?

(*d*) Do the authors see the advantages of unipolar drive circuits (namely: only one active switching element per phase; no possibility of rail-to-rail 'shoot-throughs') as sufficiently strong to make their use in drives based on other machine types (suitably wound and fed) worthwhile?

(*e*) Some idea of the parameter values (say V_S/V_R, L_{max}/R_1, V_S/L_{max}, ω and ϕ) used for the results of Fig. 6 in Paper 8300P would be useful.

P.A. Ward (*GEC Small Machines Ltd*): The papers are the result of outstanding research work. Although various forms of doubly-salient reluctance machine have been known for many years, there has been little interest in them until quite recently, presumably because suitable electronic controllers have not been available. In relatively few years, the authors have produced a motor whose performance is arguably better than an equivalent size induction motor without the benefit of the years of experience we have had with induction-motor design.

For variable-speed switched reluctance drives to become generally accepted by industry, they must compete economically with phase-controlled DC commutator motors and AC inverter-controlled induction motors. Cost comparisons are particularly difficult for the reluctance drive, because it can take so many forms, each with its own advantages and disadvantages. Since the papers concentrate mainly on a form using bifilar windings, I will use this for comparison.

Considering the motor, there can be no doubt that the reluctance motor is much less complex than the DC commutator motor, but the induction and reluctance motors seem to have similar complexity. The stator of the induction motor has a distributed winding, whereas the reluctance stator has bifilar salient-pole windings with some adjacent turns having between them, twice and probably 2.5 times (because of leakage reactance), the DC link voltage. A 415 V 3-phase supply produces a DC link voltage of about 600 V, so the adjacent turn voltage could be 1500 V, presenting some insulation problems. The rotor of the reluctance motor has no squirrel cage necessary for the induction motor but it does need rotor position switches which the induction motor does not have.

Comparing controllers, the phase controller for the DC motor is clearly the simplest, and the induction and reluctance controllers have similar complexity. Both need rectifiers, smoothing and interference suppression when an AC supply is used, and although the reluctance controller has less main switching devices than the induction controller, they are subjected to higher (2.5 × DC link) voltage. This, at present, rules out the use of single transistors in such controllers on

415 V 3-phase supplies. However, transistors are known to be the most cost-effective solution for induction controllers up to about 10 kW, giving the induction controller some advantage in this range.

The cost of wiring between controller and motor is often significant for industrial drives where the controllers tend to be sited together and the motors at various parts of the production line. Since there are many more wires required between motor and controller for the reluctance drive, this extra cost must be taken into account.

Interesting to note is why the reluctance drive is able to deliver high output power over a wide speed range. Although the drive is efficient, is still generates much heat. Referring to Fig. 12 of Paper 795B, at 750 rev/min the reluctance drive delivers 15 kW with 84% efficiency. I am informed that the controller is 98% efficient, which leaves 2.5 kW to be dissipated by the motor. The 9.25 kW, 86% efficient induction motor used for comparison has to dissipate only 1.5 kW at full load. This it does, presumably, with a class B temperature rise, whereas I understand the reluctance motor was tested for a class F rise. Even this extra 25% temperature rise does not account for all the heat dissipated by the reluctance motor, so it must be concluded that at least some of the high performance of the reluctance drive stems from its superior ability to dissipate heat. This important conclusion can be justified only if the method of cooling the reluctance motor is identical to that of the reference induction motor, so I would be grateful if the authors could comment on the equivalence of cooling.

Prof. J.V. Byrne (*University College, Dublin*): I would like to outline the switched-reluctance motor work at University College, Dublin.

We started in 1968 with a machine designed never to saturate. The specific output was appallingly low. By 1969, we had learnt that saturation could be exploited to remove the specific output penalty [E].

With saturation, tangential forces are, ideally, doubled and have a linear rather than a square-law dependence on current [F, G].

In exploiting these ideas in a machine, 'good' saturation is that of the variable-area constriction zone; 'bad' saturation is bulk saturation. I agree with the remark of Professor Harris that no good purpose is served by restricting the flux, beyond shaping the static torque characteristic. In the 5 kW, 2-phase machine exhibited here in 1976 [H], the low packing factor in the leading half of the rotor pole had just this limited purpose, and was not put there to give 'abnormal' levels of saturation, as Paper 795B would suggest.

In 1976, we began a collaboration with Joseph Lucas Ltd. aimed at developing a 10 kW mains-powered drive. The target specification was for a reversible machine, with constant-power speed range of 8.75:1. By bending the magnetic design in the direction of more flux and less MMF than would be typical of classical steppers, a speed range of 6:1 (1150–7000 rev/min) at constant power, constant voltage, was found to be possible.

The configuration is 3-phase, 6/4 pole, all poles have taper. Dimensions are 30 cm dia., 19 cm stack. Each phase winding is energised from the 0–500 V variable DC link, via two series

thyristors. De-energisation is via two diodes. Bifilar windings were not favoured at these voltage levels. The control system acts first to raise the link voltage, and then to relatively advance the angles of energisation and de-energisation. A specially developed electromagnetic transducer provides analogue speed and shaft-angle signals.

Static torque measurements show a roughly linear dependence on current. The variation with angle is within + 20%, − 30% of the mean (for 120 Nm mean). Power measurements show a machine efficiency of 89% around base speed, falling to 85% to 5000 rev/min. We observed, and were caught out by, the increasing iron losses with speed, mentioned by Professor Lawrenson. They cut the constant-power speed range to about 5:1.

The machine is capable of 20 kW, over a reduced speed range, with its present somewhat lossy core. A preliminary measurement shows over 90% efficiency at 18.6 kW.

Anyone testing these machines at high power will be conscious of noise. Do the authors agree that they are inherently more noisy than the conventional machines?

G.B. Smith (*British Rail*): Any transport organisation operating electric traction units must always be interested in new ideas in motors. However, the reality of economic factors, particularly in the present financial climate, must be questioned before an idea can be taken forward. A prime requirement of a traction system is that the equipment cost be a minimum. The installed cost must be low and so must the operating cost expressed in terms of good energy utilisation and minimal maintenance need. At first sight, the reluctance motor systems described in the papers appear to offer good economic characteristics plus improved performance when compared with similarly rated existing motor systems.

This year, a feasibility study was undertaken by the Leeds and Nottingham teams for BRB. The study has provided basic design information for a traction-motor system based on an extrapolation from proven systems of below 20 kW rating. Work has started on the construction of a motor system of about 200 kW to operate from a supply having DC third-rail characteristics. The extrapolation is a large one but the high standard of work done by the Leeds and Nottingham teams so far, gives me confidence that the reluctance motor in the doubly-salient form will live up to expectations.

J. Merrett (*Mullard Ltd*): The authors report having made FHP motors of this type; they may be useful for domestic washing-machine drives. These require high torque at low speed, and a constant horsepower characteristic similar to that classically used for traction. Have the authors any experience in this field? If so, how do total system costs compare with present solutions?

If the doubly-salient reluctance motor is suitable for a washing-machine drive, this would seem to be a good application to pursue as numbers are large, and so would provide valuable experience in large-scale production to help to reduce motor costs. A new gate turnoff switch, recently released on the market, would seem to provide an excellent means of power control. It would enable the commutation circuit represented, in Fig. 5 of the paper by Ray and Davis, by

$D_{4,5,6}$, $T_{4,5}$ and L and C, to be omitted, with obvious cost reduction. Have the authors any observations?

When induction and synchronous motors are supplied from variable-frequency inverters, there can be stability problems at low speed and light load. Have the authors found equivalent problems with their motor systems? As these systems are dependent on rotor position sensing for their operation plus being essentially synchronous in operation, I would expect some stability difficulties. How serious is this likely to be?

F. Devitt (*University College, Dublin*): It would be interesting to know if the authors have experienced any difficulties in maintaining balance between the different phase currents in the motor. Current unbalance implies uneven distribution of load and losses between phases, and this is obviously undesirable.

Two possible causes for current unbalance are as follows:

(*a*) Difficulty in accurately positioning the optical-switch rotor position transducers. Possible high sensitivites of current with respect to angle of ignition would accentuate any positioning errors of the position transducers.

(*b*) Fig. 7 of Paper 795B shows core flux waveforms for a 6/4 pole motor, where the pole polarities are three adjacent north poles with three adjacent south poles opposite. This arrangement gives 'rippled DC' flux in two stator core segments (Fig. 7*d*) , and the other four stator core segments have 'alternating' flux, as in Fig. 7*c*.

Closer examination of these waveforms show that, for a phase-switching sequence 1, 2, 3, phases 2 and 3 will each cause flux direction reversal in two segments, whereas phase 1 causes flux direction reversal in four segments, i.e. the stator-core iron losses due to switching phase 1 are twice those due to switching phases 2 and 3. This unbalance would increase with stator pole number.

If the poles were arranged with alternate north and south poles, the core loss per phase switching (for *each* phase) would be equal to that for phase 1 above: i.e. the total stator core loss would be higher, but would be evenly divided between phases.

Have the authors made any loss comparisons for these two pole arrangements, and did these influence the choice of pole arrangement used?

In Paper 8300P, the authors indicate that quadratic interpolation of i in ψ is done twice, first on inverting the input table, and secondly as needed during the integration procedure. Why was this first interpolation necessary? Is it true that the input table inversion is performed only once for any particular machine, and hence any computer time used is not significant?

Would the authors please indicate what iteration step length was used for the computer time and accuracy quoted in the paper for constant-speed operation?

Prof. R.J.A. Paul (*University College of North Wales*): The authors are to be congratulated on focusing attention on the significance of the variable-speed switched-reluctance motor as an attractive rival to the DC motor or variable-speed induction motor. It is implied in the papers of Lawrenson *et al.*, and Ray and Davis that the overall systems viewpoint of the interrelationship between electronic-drive circuits and machine design is of primary importance. It would be interesting to know if the authors could give some general design guidelines on such aspects as magnetic saturation and resulting stator-drive currents.

The high efficiency of the new machine is surprising, in that only a section of the iron is used at any instant to produce torque. Could the authors give a simple explanation for this result?

With reference to Section 3.2 of the paper by Lawrenson *et al.*, it is not clear why the authors have restricted their attention to the special case of one tooth per pole. On p. 258 it is stated that, for practical designs, $\beta_r > \beta_s$. It would be helpful if further clarification were given on this point, since, if we could have more than one tooth per pole, it could be argued that the optimum design is given when the stator-tooth width is equal to the rotor-tooth width.

K.K. Schwarz (*Laurence, Scott & Electromotors Ltd*) (*communicated*): It is important to analyse the salient features of any variable-speed drive system to obtain a better understanding of its potentialities. The arrangement described in the papers has some unusual features:

(*a*) the machine uses magnetic linkage, instead of the more usual current linkage

(*b*) the control apparatus uses a switching mode rather than frequency conversion, but

(*c*) the ultimate problem of loss-free variable speed, i.e. commutation, remains substantially the same. Unlike the case of normal industrial exploitation, the starting point is a fixed voltage DC busbar, which in itself means a basic reduction in the relative equipment requirements.

The possible comparisons with other thyristor power conversion drives [I, J] are 2-fold:

(*a*) a variable-voltage DC motor drive with armature and field control

(*b*) a variable-frequency drive utilising a squirrel-cage, reluctance or permanent-magnet motor.

However, the main point is that the battery-fed automotive traction equipment requires a very delicate compromise on weight and losses in order to get the best overall efficacy, pay load, range, speed, acceleration etc. Whatever the method of control, there are several vital parameters on which further comment would be welcome. These relate to the losses, a term which is preferred to 'efficiency' because of its connotation with cooling problems, i.e. how these were measured and how the correlation with the thermal design worked out. The effect of the oscillatory torque on the drive and the oscillatory current on the battery are also of relevance in any evaluation.

The single-motor drive, which was the subject of the papers, is, of course, a special case of the multimotor drive, which is common in traction problems; comments on possible arrangements here would be welcomed, i.e. parallel operation on the electrical and/or mechanical side.

I think it would be unfair to ask questions at this stage about the extension to a mains-fed generalised variable-speed drive system. The voltage/current characteristics currently available from semiconductors must play a large part in the economical assessment, which any system has to absorb — no particular principles being involved. In the medium term, if one is thinking of widening the application of the system, this may have quite a crucial effect on the potential of this most interesting development.

M.F. Mangan (*Chloride Technical Ltd*): I must first declare my interest in this work, since it is being funded by Chloride Technical, with assistance now from the Department of Industry. In view of my closeness to the work, I am here principally to hear the views of the other contributors on the developments of this very simple and very versatile drive system. However, I would like to ask the authors what prospects they see for further development and to ask them to comment on the use of this drive for applications other than battery-vehicle drives.

A.F. Anderson (*NEI Parsons Ltd*): I would like to congratulate the authors on their presentation of some very interesting and, in my opinion, significant results; they have provided a stimulating blend of electromagnetics, power electronics, economics and engineering design.

There is some historical justice in finding that motors designed in the early 1840s — dismissed as hopelessly inefficient — are the antecedents of the present generation of SR motors. Fig. C shows a motor of the type used by Robert Davidson, manufacturing chemist of Aberdeen, for driving a battery-powered vehicle weighing six tons at four miles an hour on the Edinburgh and Glasgow Railway in 1842 [K, L]. The motor had two U-shaped electromagnets, which were switched on alternately, and which attracted in turn one of the three axial iron bars placed round the periphery of a wooden drum. It is, I think, relevant to this discussion to outline the limitations of this machine, as follows:

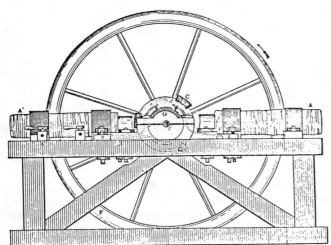

Fig. C *Davidson's electric motor*

(*a*) The motor had a wooden frame that was not stiff enough to withstand the large out-of-balance pull imposed by the electromagnets. Because he did not have the money to rebuild the machine, Davidson was reduced to the expedient of increasing the airgap to the thickness of a man's hand. In the circumstances, it is surprising that the machine worked at all.

(*b*) The rotor-iron circuit was solid and therefore subject to excessive eddy-current loss.

(*c*) The current pulses were of a fixed mark/space ratio of 1:1, and it seems likely that there was no provision for advancing or retarding the commutation.

(*d*) There was no means of energy recovery at switchoff. 'One curious phenomenon connected with the motion of this new and ingenious instrument', wrote the *Edinburgh Witness*, 'was the extent and brilliancy of the repeated electric flashes which accompanied the action of the machinery'.

Most of these limitations can now be seen to have been the result of an inadequate understanding and are not inherent in the switched-field machine.

The largest switched-field machine built in the last century was built for traction purposes by Charles Grafton Page, a US Patent examiner, in 1851 [M]. It was an axial machine, looking somewhat like a steam engine, and had an output of at least 8 HP. Page's experiment cost the US Government $30 000 and Page $6 000; it was a technical fiasco and financially ruinous to Page. His funding had been obtained on the basis that zinc was a waste product of lead mining and that the electromagnetic locomotive would provide an outlet for the zinc. Page forgot that Grove cells had to have two electrodes, one of zinc and the other of platinum. Whereas zinc was cheap, platinum was expensive. He spent 39 times as much on platinum as he did on zinc. This illustrates the importance of looking at the cost of the whole system and not just part of it; a point that the authors have stressed tonight.

My own interest in switched-field machines stems from the years 1961 to 1965, when I was working with Dr. Arthur Cruickshank at what was then Queens College, Dundee. I mention it in part answer to an earlier speaker who asked: 'why has the switched-field machine not been looked at before now?' At that time thyristors were just becoming available. We confined our attention to the idea of producing the simplest possible 2-path, switched-field machine that would run at variable speed from a 3-phase supply. We ended up with a naturally commutated envelope cycloconvertor [N]. The firing of the thyristors was done digitally and both current pulse length and voltage were made speed dependent.

The machine could be run open or closed loop, but when run open loop was subject to interference between supply and switching frequencies which was troublesome at high speeds. I suspect that this could still be a problem with any SR motor run from a rectified AC source unless special care was taken with smoothing. The authors might like to comment on this. We did not carry our work further, mainly because we felt that forced commutation was necessary and that, at the time, the economics of going to the additional complexity would rule the drive out of court. We were never in any doubt as to the potential of the SR motor, given cheaper and less cumbersome electronics.

R. Bourne (*South East London College*): At the top of p. 263, the authors state that 'the same airgap diameter was retained (to facilitate comparisons), even though these dimensional choices are unfavourable to the SR motor'.

Would the authors say what change they would like to make to this dimension and whether any such change would not also improve the performance of the induction motor?

Prof. P.J. Lawrenson, J.M. Stephenson and N.N. Fulton (*in reply*): We are grateful for the full and interesting remarks and thank discussers for several generous comments. In the long history of switched-reluctance motors, which has been so usefully illuminated by different contributions, the explanation for a relatively rapid advance over recent years must be attributed in great measure to the availability of suitable switching elements, but the importance of conception and design of the system, motor plus electronics, as an entity, cannot be overstated.

In this, we have been fortunate in being able to draw to an interactive focus our own previous experience of reluctance machines, stepping motors and their electronic drives, field analysis and dynamic modelling, together with our companion authors' exceptional experience with power and control circuitry and their appreciation of machine behaviour. We have also been extremely fortunate in benefiting from major financial support from industry since 1976, and we wish to acknowledge our debt, particularly to Dr. M.F. Mangan, Dr. B.D. Edwards and Mr. G. Cooper, whose foresight, courage and stimulus have greatly speeded both technical progress and commercial acceptance of switched-reluctance motor systems.

In reply to contributors who raised questions about the nature of the operating characteristics and the possible applications and ratings of switched-reluctance motor (SRM) systems, the following comments may be helpful:

(*a*) The system is inherently suited to 4-quadrant operation, and experience to date indicates that regeneration/braking performance can be arranged to match motoring performance. In any particular application, however, the design would be chosen to provide an appropriate balance in quality of performance between the different regimes.

(*b*) Although some early important applications have been to electric vehicle propulsion schemes, we see the SRM system as a variable-speed drive of general applicability. Numerous industrial applications look attractive, with robustness of motor and electronics, controllability, suitability for hazardous environments, high speeds and competitive costs being favourable points. Use in certain domestic appliances, including washing machines, as mentioned by Mr. Merrett, also looks attractive, and we have some work in hand in this area. For this, the natural series characteristic of the motors, giving high torque at low speed, is an advantage.

(*c*) With regard to size, we have experience of sizes between fractional and tens of horsepower in the laboratory and up to hundreds of kilowatts in design studies, including the rail application mentioned by Mr. Smith. Throughout this range, the performances appear very attractive and we broadly agree with Prof. Harris in his views on the upper limit of size to which SRM systems might be applied.

(*d*) In trying to put the SRM system into a frame of reference, comparison with the inverter-fed induction motor is most easily made and, both on performance aspects and cost, the SRM drive appears to have a clear advantage. In response

to Mr. Schwarz, the SRM system operating from the 3-phase mains will include a rectifier, but the above comparison takes cognisance of this. The results presented in the paper, although taken from a battery-fed system, are compared with induction-motor performance data for operation at fixed speed on sinusoidal mains and, therefore, are flattering to the induction motor. Under variable-frequency conditions, induction-motor performance would be substantially degraded.

An important operational feature is that, in contrast to the inverter-fed induction motor, the SRM system always has a motor winding in series with a main device, thus limiting possible fault currents.

(*e*) Variable-speed drives, in the main, remain DC-motor based, but even here the SRM system appears to be competitive in a number of cases (it is as flexible in control as the separately excited DC motor) and this competitiveness will improve further as electronic costs continue to fall and motor costs continue to rise.

Turning to more particular questions, we would first like to say that Prof. Harris' analysis of SR motor torque capability (and copper loss) as compared with the induction motor is most interesting. We accept his analysis and suggest that it goes a long way to answering Prof. Paul's question about high efficiency, notwithstanding the restricted iron usage in the machine. The possibility of further increasing torque output by using multiple teeth per pole is certainly one justifying exploration but, as noted, it leads to higher operating frequencies and losses. The authors have no experience to report on this nor, of any comprehensive kind, on the parameters influencing iron loss more generally; but use of 'thin' and low-loss lamination material is desirable for 'higher' frequency applications.

In connection with the possible use of unequal stator and rotor arcs and of pole combinations such as 10:4, we were concerned to indicate possible useful design options available. In this context it is important to recognise the need to achieve the best possible combination of efficiency and specific output for a particular speed range and that this may lead to parameter values which do not follow simply from a consideration of maximising static torque production.

Our statement on mutual inductance was based on the work of Acarnley [O] to the effect that mutual inductance is undesirable between phases in VR motors using unidirectional excitation.

On the subject of power factor and 'energy ratio', we are again much in sympathy with Prof. Harris. We were concerned in the paper to address a question which our experience tells us is always asked and to bring out the fact that the conventional concept of power factor is not applicable. The crucial question is, what rating of semiconductor switches must be used? In determining this, several other factors have to be taken into account, but our experience with the SR motor compares very favourably with that on induction-motor drives.

The question concerning the claimed beneficial effects of 'abnormal' saturation levels first propounded by Jarret *et al.*, and which has been with us for many years, was discussed by both Prof. Byrne and Harris. We are glad to see that they are in agreement with our own views, reflected in our designs, particularly that no good purpose is served by restricting flux in the pole faces at the air gap surface, beyond possibly shaping the static torque characteristic. With this latter exception, the design of the magnetic circuit of the SR motor is governed, in general terms, by essentially the same rules as apply to more conventional machines — directed to minimising (consistent with MMF and other considerations) overall or 'bulk' saturation, and maximising the working airgap levels of flux density. In permanent-magnet machines, the use of flux concentrators is well known.

Prof. Byrne's description of his SR motor and operating experience is most interesting. We are grateful for his contribution and the support it provides for the capabilties of SRM systems; we would like to acknowledge his sustained contribution to the subject.

The problem of noise is one which naturally comes to mind in view of the pulsating nature of the electromagnetic forces, and we have had some motors which have been noisy. Others, however, have been very quiet and, overall, we have every reason to believe that noise levels will be quite acceptable for most applications.

Dr. Ward's points about cost of wiring when motor and electronics are separated, and also about the motor's ability to dissipate heat are well made. Both he and Mr. Schwarz rightly emphasise 'losses'. Two important features of the SR motor are that there are no rotor copper losses to be dissipated and that the stator configuration is one which makes for very efficient cooling. However, the precise prediction of losses and their distribution demand (as do other aspects) a considerable sophistication of approach which belies the simplicity of the motor. It is necessary to think only of the effort expended over the years on losses in induction machines and then to note in the SR motor the complex waveforms of current and flux and their variation, not only with time but also with position in the machine, to appreciate something of the nature of the problem.

We have not given particular consideration to multimotor drives, but we do of course agree with Mr. Schwarz on the need for balance of design for single-motor battery-fed systems. The continuation of our traction project is evidence that the SR motor can offer an attractive combination of weight and losses.

On other points we would reply briefly as follows:

(a) The type of stability problem mentioned by Mr. Merrett in connection with synchronous and induction machines on a variable-frequency supply does not have a parallel in SR motors which, in this respect, behave like DC machines. Other forms of instability in the system can arise (as in any control system) but are not expected to be troublesome in practice.

(b) We have not encountered any difficulties in maintaining phase current balance and, though questions of accuracy of position transducer location have to be carefully addressed, they do not seem to pose any difficulties in practice.

(c) There appears to be no way of exploiting segmented magnetic circuits parallelling that in rotating-field machines.

(d) Possible interference and similar problems referred to by Dr. Anderson seem to us to be no different in the SRM system from those encountered in other electronically controlled drives, and no special difficulty has emerged or is anticipated.

(e) The point about the choice of airgap diameter (Mr. Bourne) is implied elsewhere in the discussion in that an overall better design, as between copper and iron, could in general be achieved in the SR motor with other than the induction-motor gap diameter.

(f) Pulsations in the drive torque do exist but, in practice, are surprisingly unobtrusive.

(g) In reply to the question concerning the quadratic interpolation described in Paper 8300P, it is true that the input table is inverted once only for any particular machine. This inversion produces a table of currents at (regular) values of flux linkage suitable for use in the numerical integration procedure. Quadratic interpolation is used at each step of the numerical integration to find the current corresponding to the flux linkage.

(h) In reply to Dr. Bolton, the parameter values relating to Fig. 6 of paper 8300P are $V_S/V_R = 1$, $L_{max}/R_1 = 0.057$, $V_S/L_{max} = 938$, $\omega = 25\,\pi$ rad/s, $\phi = 60°$.

R.M. Davis and W.F. Ray (*in reply*): We have some sympathy with Prof. Harris' comments on energy ratio and power factor, although we question the usefulness of these concepts in evaluation of the SR motor system performance. It would appear that peaky current waveforms ought to be penalised by a poor power factor, yet a motor operating with less peaky waveforms may prove suboptimal when considered as part of the complete system, including power electronics and control and the practical constraints which go with them.

Viewing the motor inverter as a system, one could justify considering the inverter DC terminals as those at which power factor is meaningful. A severe ripple current here would increase the kVA compared to the kW. This has important effects for a battery supply whose internal resistance is not negligible, and which will dissipate considerable power unnecessarily if the ripple current is excessive. The electrolytic capacitor bank used with this, and most other voltage-fed inverter drives has the important task of by-passing most of the ripple current. However this kW/kVA ratio mainly reflects the electrolytic capacitance rather than the inverter-motor performance.

In general, power factor is a useful concept for quantifying the performance of machines operating with sinusoidal voltages and/or currents. The SR motor operates with neither and it is doubtful whether the concept of an equivalent power factor is at all helpful in evaluating its performance. What really matters is the system efficiency.

The choice of pole ratios is not entirely the province of the motor designer; once again a system outlook must be applied. When forced commutation is considered, an even number of phases make resonant reversal of the commutation capacitor unnecessary if the motor phases are arranged alternately [P]. Also, an excessive number of phases must bring substantial, even overwhelming, advantages if the penalties in the more expensive inverter are to be outweighed. We feel that the fewer phases the better, provided that the drive can meet its specification.

The mutual inductance between phases can introduce EMFs. into dormant windings (see Fig. 18 of Reference P), perhaps as a result of commutations in adjacent phases, and these can cause problems. Once again, the interactions between motor design and power electronic design need careful consideration from the system viewpoint.

The contribution by Dr. Anderson is very welcome in reminding the authors that research interest in the SR motor goes back many years, and leads to the question posed by Dr. Bolton on why a successful SR drive was not developed earlier. Had the authors known the reason they also would perhaps have done it earlier! One author has already experienced the frustration of developing a new technology too soon, only to have it await the development of an associated technology to make it an economic proposition. We have been fortunate to have the silicon technology, both in computing and in power devices, which, in conjunction with the development of stepper motor technology and the difficulties faced by the PWM inverter fed induction motor to establish a commercial position, lead the authors to believe that the project has reached its success at the right time. The authors feel that the recognition of the following five important features have contributed significantly to this success:

(a) the necessity to keep the motor construction simple (some rather exotic forms of SR motor have been considered)

(b) the necessity to minimise the power electronic components required, so as to make the system viable

(c) the necessity for computer models to predict system behaviour — a considerable task

(d) the advantage of building up motor-phase current in advance of its torque productive period

(e) the necessity for close collaboration between motor and power electronic design.

Concerning regeneration, a worthy battery-vehicle drive must recover kinetic energy and minimise friction brake maintenance. However, high efficiency in motoring is more important than efficient regeneration, and the system design has been angled this way. Testing in regeneration has taken place at most speeds and loads in the range, yielding efficiencies (motor shaft to inverter DC terminals) only a few percent worse than for motoring at the same speed and power. The system would appear to offer interesting possibilities as a very high-speed generator, where the mechanical strength of the rotor is of prime importance; or as a wind-driven generator, where the speed is very variable or intermittent.

Concerning unipolar drives for other machines, it should be remembered that the SR system is possible by virtue of the motor's ability to operate efficiently from unidirectional current pulses, and the marriage between the thyristor and SR motor is consequently a happy one. To attempt a unipolar drive for a motor which requires fundamentally an alternating MMF seems to be a shotgun marriage by contrast! Since the SR motor and its inverter together have shown that they can outperform an induction motor on pure sinewaves, the effort seems hardly worthwhile.

The authors found themselves agreeing with many of Dr. Wards comments on induction motors compared to SR motors and on voltage levels in bifilar SR motors. Bifilar motors/ inverters have the advantage of only one main switching device per phase and are most relevant when operating from relatively low supply voltages especially when efficiency is important. The bifilar arrangement can have problems with high-voltage supplies, and is certainly inconvenient and perhaps impractical where the motor and inverter are widely separated. However, Prof. Byrne suggested a suitable inverter configuration for a nonbifilar motor, and the authors agree that, for high voltage supplies, this has important advantages. Further comments on this subject are now published [P]. Prof. Byrne's contribution of his experiences and success with this type of motor is very welcome and supports the claims of the authors.

Despite the possibility [P] of two main switching devices per phase for a single winding SR motor, the authors hold the view that the inverter for the SR motor is still simpler than a PWM inverter for an induction motor, on the grounds that, with thyristor circuits, one commutation circuit can be arranged to service all phases, which is extremely difficult for a thyristor PWM inverter. That transistors circumvent this problem is accepted, offering a simpler and more economic PWM inverter, but that does not mean that thyristors are not the best choice for SR motors. (It should be remembered that the peak/mean current ratio for SR motor waveforms is likely to be greater than for PWM.)

The use of gate turnoff devices as suggested by Mr. J. Merrett to simplify the circuit of Fig. 5, Paper T481P, is an interesting suggestion. Low-power drives tend to operate with current waveforms which peak at the commutation point, and it would thus be a requirement that the GTO devices have this turnoff capability. In other respects, the GTO device has the advantages of high forward voltage (certainly sufficient for operation from rectified single-phase supply), high dV/dt capability (easing snubbing problems) and control via short-duration gate pulses.

The question of instability was also raised. First, it is necessary to distinguish between damped oscillations arising from a flexible load shaft and genuine instability. The former type of oscillation can be excited by torque pulsations at the resonant frequency, but we do not regard this as true instability. We have found no evidence of genuine instability for drives taking their firing angles directly from the shaft position transducers. However, if these transducer signals are processed to modify dynamically the firing angles, then instability can occur, and we have experienced it.

Concerning the effect of magnetic saturation on the motor-phase currents as raised by Prof. Paul, the paper [P] recently published provides some guidance for chopping behaviour at low speeds. Since a high degree of saturation results in an incremental inductance which is lower than the minimum (out) inductance, the current excursions in chopping which become necessary to satisfy minimum on and minimum off times (inverter constraints) are considerably enlarged; to meet a prescribed torque, the peak currents can easily reach unacceptably high values if too high a specific output is sought.

Estimation of single-pulse waveforms with motor saturation

requires a nonlinear motor model; this was outside the scope of the paper [P]. However, the reaching of a satisfactory compromise on the degree of saturation, as reflected in specific motor output and in inverter component ratings, is an important part of the system design.

The measurement of power at the motor terminals when the voltages and currents are 'rich in harmonics' is, as Dr. Schwarz knows, very difficult. We have therefore, treated the motor and convertor as a unit measuring DC input power to the convertor and shaft output power. Convertor losses have been estimated as device conduction losses (including switching losses) and snubber losses; the remaining losses are assumed to occur in the motor. The separation of I^2R and iron losses is not easy as R is increased somewhat above its DC value by skin effect.

Concerning the possibility of a multimotor drive, the authors have so far only considered the SR drive as a single-motor drive. A multimotor traction application would certainly require individual inverters to permit wheel skimming and consequential differences in diameter. It is important to appreciate that in a multimotor drive fed from one inverter, only one motor can provide the rotor position transducer signals. All other parallel motors would operate as open loop stepper motors, taking up load angles (hopefully without instability) to suit their particular loads. The applied voltages and timings, dictated by the motor with the shaft transducer and its load, could well be quite wrong for the 'slave' motor, resulting in excessive currents. Although it would be unwise to rule out this possibility, we foresee many problems for a single inverter multimotor drive.

Finally, in answer to Dr. Anderson's question regarding beat effects between mains rectifier ripple and inverter ripple — the authors consider this to be in common with all rectifier fed inverter drives, and expect to find a similar solution.

References

A HARRIS, M.R. HUGHES, A., and LAWRENSON, P.J.: 'Static torque production in saturated doubly-salient machines', *Proc. IEE*, 1975, **122**, (10), pp. 1121–1127

B HARRIS, M.R., ANDJARGHOLI, V., LAWRENSON, P.J., HUGHES, A., and ERTAN, B.: 'Unifying approach to the static torque of stepping-motor structures', *Proc. IEE*, 1977, **124**, (12), pp. 1215–1224

C HARRIS, M.R., and FINCH, W.J.: 'Estimation of static characteristics in the hybrid stepping motor'. Proceedings of 8th conference on incremental motion control systems, IL, 1979, pp. 292–306

D SHEPHERD, W., and ZAND, E.: 'Energy flow and power factor in nonsinusoidal circuits' (Cambridge University Press, 1979)

E BYRNE, J.V. and LACY, J.G.: British Patent 1321110, 1973, based on Irish Patent Application 872, 25th June 1969

F BYRNE, J.V.: 'Tangential forces in overlapped pole geometries incorporating ideally saturable material', *IEEE Trans.*, 1972, MAG-8, pp. 2–9

G O'CONNOR, W.J.: 'Magnetic forces in idealised saturable-pole configurations', *IEE Proc. B, Electr. Power Appl.*, 1980, **127**, (1), pp. 29–53

H BYRNE, J.V., and LACY, J.G.: 'Characteristics of saturable stepper and reluctance motors'. *IEE Conf. Publ. 136*, 1976, pp. 93–96

I BUCKLEY, N.A.: 'High performance motors for high speed battery operated vehicles', *LSE Eng. Bull.* 1978, **14**, pp. 22–25

J BADER, Ch., and STEPHAN, W.: 'Comparison of electric drives for road vehicles', *ETZ-Arch.*, 1977, 98, pp. 22–26

K *Penny Mechanic and Chemist*, 23rd Sept., 1843

L ANDERSON, A.F.: 'Robert Davidson — father of the electric locomotive'. Proceedings IEE history of electrical engineering conference 1975

M POST, R.C. The Page locomotive — federal sponsorship of invention in mid-19th century America. *Technol. Culture*, 1973, **13**, pp. 140–169

N ANDERSON, A.F., and CRUICKSHANK, A.J.O. 'A.C. electric motor having reluctance-type rotor' British Patent 1114561, 1968

O ACARNLEY, P.P.: 'Analysis and improvement of the steady-state performance of variable-reluctance stepping motors'. Ph.D. thesis, Leeds University, 1977

P DAVIS, R.M., RAY, W.F., and BLAKE, R.J.: 'Inverter drive for switched reluctance motor: circuits and component ratings' B, *Electr. Power Appl.*, 1981, **128**, (2), pp. 126–136

DC 106 B

Erratum

LAWRENSON, P.J., STEPHENSON, J.M., BLENKINSOP, P.T., CORDA, J., and FULTON, N.N.: 'Variable-speed switched reluctance motors', IEE Proc. B, Electr. Power Appl., 1980, 127, (4), pp. 253–265:

The following changes are necessary:

p. 253, list of authors: for the penultimate author, read with an accent as J. Čorda

p. 254, column 2, line 5: after 'controlling' read 'these' to conclude line

p. 256, Fig. 4: read as subcaptions ———— current, ———— flux, — · — inductance

p. 257, column 2, line 24: for permenance read permeance

p. 258, column 1, line 33: for $\beta_s = \beta_r$ read $\beta_s = \hat{\beta}_s$

p. 258, Fig. 5: in lower half of figure read $\beta_s < \beta_r$

p. 260, Eqn. 22: for p read P

p. 260, Fig. 8: against vertical axis for K_L read k_L

p. 261, title of Section 5.1: insert comma after *linkage*

p. 263, Fig. 10: read as subcaptions ———— nonlinear method, ———— linear method, $L_{max}/L_{min} = 2$, — · — linear method, $L_{max}/L_{min} = 8$

p. 265, column 1, line 12: et seq.: read. . . 'between those responsible for designing and developing the electronic circuits and those responsible for the motor'

p. 265, Reference 12: read author as ČORDA, J.

ETC85 B

1.2 - Motor Design

A Review of Switched Reluctance Machine Design

N.N. Fulton and J.M. Stephenson

To be published in International Conference on Electric Machines, 1988

ABSTRACT

The last decade has seen an upsurge of interest in switched reluctance (SR) machines. Drives based on SR machines are now available commercially and are selling worldwide; many companies are known to be in the development stages of producing drives for applications covering a wide spectrum of the variable-speed market; and many researchers in different centres of learning are investigating different aspects of the behaviour of these unusual machines.

The published literature on SR machines includes a number of items which relate to the techniques used in approaching their design. This Paper reviews the bibliography of published methods of design, noting their points of commonality and of divergence, and discusses their importance.

Dr Fulton is the General Manager of Switched Reluctance Drives Ltd, Leeds, UK.

Dr Stephenson is a Senior Lecturer in the Dept of Electrical & Electronic Engineering, University of Leeds, UK.

N N FULTON, BSc, PhD, CEng, MIEE & J M STEPHENSON, BSc, PhD, CEng, MIEE
S R Drives Ltd, U K   Leeds University, U K

A REVIEW OF SWITCHED RELUCTANCE MACHINE DESIGN

1. INTRODUCTION

The last two decades have seen the initial growth and then the rapid spread of interest in doubly-salient, switched reluctance (SR) systems. Since many workers have pursued independent lines, this paper presents a timely review of the different design methods which have been adopted for the SR machine.

The object of the paper is to review published design information for doubly-salient, variable-reluctance machines which are electronically commutated in response to information supplied from a rotor position transducer (RPT). Although there is much published work, spanning many years, on the design of variable reluctance and hybrid stepping motors, a closer examination shows that there are significant differences in approach between these machines and the SR machines under review. Whereas stepping motors are predominantly current-fed, low-power, positioning devices which operate under open-loop regimes, SR machines are typically used in voltage-fed, high-power, variable-speed drives which normally operate under closed-loop control. Work by the authors, their colleagues and others has shown that these differences are so significant as to require a distinct approach to the problem of machine design.

Section 2 gives a brief review of the features of an SR machine. This leads, in Section 3, to considering the design problem and, in Section 4, to considering the different ways in which the published literature may be classified. Sections 5-7 discuss the design methods under the chosen headings and Section 8 draws together the conclusions. For the sake of brevity, papers are referenced by principal author only.

2. PRINCIPLES OF SR MACHINES

The basic features and principles of SR machines have been delineated by various authors, one of the first, and most comprehensive, papers being that by Lawrenson [1]. The machines are characterised by saliencies on both stator and rotor. This structure leads to stator phase windings which have an inductance which is dependent on rotor position. If the reluctance of the magnetic circuit is current independent, then the inductance has the trapezoidal profile as shown in many papers [eg 1]. When supplied by constant current, any one phase produces a torque variation with rotor angle, as shown in Fig 2 of Reference 1; when supplied from a constant-voltage bus, the drive can be controlled by altering the rotor angle at which the windings are switched on and off. This provides a wide variety of torque/speed profiles, as will be discussed below.

3. THE DESIGN PROBLEM

Some families of electrical machines have been intensively studied for many years, eg synchronous, induction and dc machines. The result of this work is that the design of these machines is, in broad terms, well understood and well documented. Formulas linking, say, geometrical dimensions to equivalent circuits are readily available, so that many aspects of the performance of the machine can easily be calculated. Simple equations ("output equations") are often derived to allow rapid performance estimation.

With SR systems, however, such methods are not readily applicable. There are the apparent complications of different pole combinations, different phase numbers, ranges of pole arcs from which to choose, etc. These parameters pose a completely new set of problems to the designer. On closer examination, there is a further notable complication in that the performance of an SR machine depends, often to a significant extent, on the form of power converter through which it is supplied. It follows that methods of analysis and design must be capable of taking these influences into account.

It is important, when considering the overall task of designing such machines, to differentiate between the relatively simple design problem of a fixed-speed motor operating from a 3-phase sinusoidal supply, and the much more complex problem of a variable-speed machine supplied from an electronic converter. In the second case, one has to consider: the operation of the machine over a wide range of torque and speed, usually with consequent changes in cooling; the interaction between parameters which affect the specific output of the machine yet which may not be optimised in isolation from, say, switch device ratings and costs; the control strategy of the converter; the dynamic response of the load and the dynamic response requirement of the complete drive system; and many, many other interactive effects.

The utopian dream is to develop a design method which, when supplied with the specification for a drive and any overriding requirements, will optimise every variable in the machine and produce the optimum design. If such a method has been developed, it has not yet been published! If experience (over many decades) with other families of electrical machines is a guide, it will be many years yet before that goal is approached. Nevertheless, some routes towards that goal are now being mapped out.

4. METHODS OF CLASSIFICATION

Since the literature is very diverse, there is a number of different approaches which could be adopted for the classification.

Some approaches centre on the physical construction of the machine:
        single or multiple teeth per pole
        single or multiple stack
        cylindrical (radial airgap) or disc-type
        (axial airgap)
        internal or external rotor
        linear or rotational motion
        etc.

Although perhaps useful for classifying the physical and performance-related aspects of the drive, these approaches do not yield much insight into design methods. A more helpful approach is to divide the work into methods based on:
(a) linear methods, ie based on the analysis of circuits with current-independent parameters.
(b) non-linear methods, ie based on circuits with current-dependent parameters.
(c) finite-element solutions of the magnetic fields, leading either to a direct solution of performance or to the calculation of parameters for circuits.
This latter grouping is used in Sections 5-7 below as the principal classification of the available information.

5. LINEAR METHODS

One of the earliest papers using linear analysis of the machine is by Koch [2]. Based on the now well-known 3-phase machine with 6 stator and 4 rotor poles, he considers the residual flux at switch-off to be very important. He uses a coefficient based on the residual flux to show how the lamination geometry should be chosen for different load characteristics. However, since the paper is based on the supposition that the iron is ideal ($\mu_r \to \infty$) and since the analysis omits the effect of currents flowing during the period of decreasing inductance, the validity of the conclusions (which arise from the substantial quantity of analysis in the paper) is doubtful.

The difficulties of Koch were avoided by Ray and Davis [3,4] in their derivations of simple expressions for the current waveforms in the machine so that the converter-component ratings can be estimated. This approach again depends on linearising the inductance, but (correctly) allows the voltage to be switched at any point in the cycle. This enables control strategies to be examined with sufficient accuracy that component ratings can be established. This work, however, was not aimed principally at solving design problems in the machine, but rather at design aspects of the converter which had not previously been considered in any detail.

A similar approach, based on "flat-topped" phase currents was used by Miller [5] and a further paper following his work is by Krishnan

122

[6]. Although valid for the specified conditions, these analyses are very restricted in their application, as one of the features of SR machines is that they have widely varying phase current waveshapes.

While all these approaches have some qualitative value, reliance on them for quantitative information is unwise.

6. NON-LINEAR METHODS

The phenomenon of saturation in SR machines is both inherent and important. It is also complex, with both local and bulk saturation effects being in evidence. Such a situation cannot be satisfactorily handled by the use of a simple 'saturation factor', as often applied to other machines.

Most authors who have considered the design of SR machines in any detail have used a non-linear design method of some sort. Most of their papers, however, have only a few comments on design methods or parameters, and no coherent design method is visible. In this group are contributions by: Unnewehr [7] on disc-type motors; by Vallese [8] on a bifilar-wound 3-phase motor; by Thornton [9] on a drive for a forklift truck; by Amin [10] on a large traction drive; and by Pollock [11] on a simple analysis which leads to a system with a high phase number.

A very general, qualitative approach to the design problem is contained in work by Kamerbeek [12], who has considered in broad terms the scaling question and has compared SR to other types of machines over a wide range of sizes. He concludes that reluctance machines will show superiority over many other types at relatively high powers.

The deliberate introduction of saturation into the magnetic circuit, originally proposed by Jarret [13], has consistently been advocated by Byrne and co-workers. They have made a significant contribution to the bibliography on SR systems, ranging from general papers on field patterns [14] and simulation [15] to exploitation of the saturated regions in design [16] and profiling of torque/angle relationships [17,18]. In all of these, it seems that the airgap must be kept as small as possible to allow the optimum exploitation of the deliberately saturated regions. In these publications, however, there is little material which bears directly on the design problem.

Also following the work of Jarret, Bakhuisen [19] discusses, mainly in isolation from the rest of the machine, the question of saturation in the iron and concludes that it is beneficial. Lawrenson [1], by contrast, concludes that the introduction of magnetic constrictions is not helpful. The difficulty of achieving a sound appreciation of the supposed merits of saturation in SR machines is discussed in a current paper [20].

Bearing in mind the superficial similarity between SR drives and stepping motor drives, it is not surprising that some workers have developed machines with more than one tooth per pole. The design techniques which they adopt are usually based on the ratios of tooth width to tooth pitch and of airgap length to tooth pitch. Relevant papers include those on basic theory by Harris [21,22] and Ertan [23], and those directed at practical designs, eg by Bausch [24], Welburn [25] and Finch [26]. The last paper describes a design technique which has been used by its authors as the basis of work reported in a variety of other publications [eg 27,28,29]. This method is undoubtedly a valuable approach for multiple-teeth-per-pole systems and is capable of producing good results in the limited number of applications for which it has been used. However, it is not well-suited to the single-tooth-per-pole structures which make up the majority of SR implementations reported up to the present time.

As would be expected, there is a group of papers by different authors which describe unusual configurations or methods of operation of SR systems. Among these are the work of: Bolton [30] and Chan [31] on low-power, low-cost, 1-phase machines with external, solid iron rotors; Finch [32] on linear actuator systems; Franceschini [33] on control by fixed angles; Lang [34] on servo applications with the use of bias windings; etc. None of these has anything significant to offer on the general problem of design.

An obvious route to choose in solving the design problem is to make use of the readily available power of modern computers by writing a set of programs to solve whatever design equations are chosen. This, of course, begs the question of which equations to use, but this is discussed below. Recent papers by Miller [35] and Fulton [36] show quite different approaches, though essentially based on the same analytical method of computing the machine performance. Miller implements an interactive, editor-based, pc system which uses many default parameters to produce an estimate of the machine performance; while Fulton uses a more traditional, rigorous approach to produce a more detailed and sophisticated result taking into account not only the machine geometry but also the power converter parameters. The latter approach has been in use for some years and its accuracy has been verified over a wide range of sizes, powers, speeds and supply voltages.

As noted above, the method of analysis has to be chosen before the design can proceed. While some have attempted to use output equations based on linearised theories [eg 28], there are only two methods which have received wide support. Pickup [37], Byrne [15] and Blenkinsop [38] all used formulations of the non-linear differential equations which required evaluation of differential coefficients, with consequent limitations on accuracy. Stephenson [39] subsequently published a method which does not use differential coefficients and, while requiring less input data, is more accurate. This method has been adopted by several others [eg 35,36] and a recent independent assessment of the two approaches [29] has confirmed the superiority of the Stephenson method.

7. FINITE-ELEMENT ANALYSIS

As computing power has become greater and more accessible, attention has been turned to the application of finite-element (FE) analyses to SR geometries. The principal contributions in this area are by Bausch [40], Arumugam [41,42] and Dawson [43]. There have also been some other, mainly University-based, projects where FE methods have been applied [eg 29] and it is believed that some industrial concerns are using their existing FE packages to investigate the qualities of SR machines.

The MAGNET package from Infolytica has been applied by Arumugam and Dawson. References 41 and 43 deal with one-tooth-per-pole designs for 3-phase and 4-phase machines respectively: Reference 42 deals with a two-teeth-per-pole,

3-phase machine. Though some information is given regarding design parameters, these papers principally report on the application of a general-purpose electromagnetic FE package to SR structures in order to calculate the torque.

Perhaps because it was not concerned solely with SR structures, an earlier paper by Simkin et al. [44] seems to have gone unnoticed by many, even though it included the application of both 2-D and 3-D FE methods to a structure typical of an SR machine. Other workers, eg Metwally [29] are known to have applied the Rutherford Laboratory PE2D software in the same manner and with similar results.

Though some success is evident with FE methods, there are clearly many problems yet to be overcome in the areas of meshing, solving for the field at rotor angles which introduce asymmetry to the geometry, etc. In addition, the need to cater for three dimensions in the field solution (as shown in Reference 44) is likely to bring a very great increase in complexity and in computing requirements. Even recognising the great advances being made in computing power, it is therefore likely that it will be some considerable time before FE methods can compete with circuit-based methods for general design work. There may well be, however, a place for FE techniques in the near future in the examination of particular parts of the doubly salient structure.

8. CONCLUSIONS

This paper has reviewed all the papers known to the authors which have a direct bearing on design methods for SR machines. Though it is difficult to draw detailed conclusions from such a diverse collection of publications, general features emerge.

While linear (ie current independent) methods are perhaps useful for qualitative assessments, they do not produce useful quantitative results. Several authors have shown that it is possible to apply sophisticated FE analyses to the SR structure. There is no evidence that such analyses are suitable for routine design work, though they must be considered to be of value for solving isolated problems in the hope of achieving 'generic' solutions which can then be applied more widely.

Non-linear, circuit-based techniques have been

shown to be the basis of various design methods, some of which have been successfully implemented in computer program suites. The authors and their colleagues pioneered this approach to SR motor design and have applied it over a very wide range of sizes and types. The review provides evidence that this approach has emerged as the preferred technique for SR machine design.

9. ACKNOWLEDGEMENTS

The authors acknowledge the contributions of their colleagues over many years. This paper is copyright of Switched Reluctance Drives Ltd and their permission to publish is gratefully acknowledged.

10. REFERENCES

1. "VARIABLE-SPEED SWITCHED RELUCTANCE MOTORS" Lawrenson, PJ, Stephenson, JM, Blenkinsop, PT, Corda, J and Fulton, NN. IEE Proc, Vol 127, Pt B, No 4, July 1980, pp 253 - 265 and "DISCUSSION ON VARIABLE-SPEED SWITCHED-RELUCTANCE MOTOR SYSTEMS" IEE Proc, Vol 128, Pt B, No 5, September 1981, pp 260 - 268

2. "THYRISTOR CONTROLLED PULSATING FIELD RELUCTANCE MOTOR SYSTEM", Koch, WH, Elec Machines & Electromechanics, Vol 1, 1977, pp 201 - 215

3. "INVERTER DRIVE FOR DOUBLY SALIENT RELUCTANCE MOTOR: ITS FUNDAMENTAL BEHAVIOUR, LINEAR ANALYSIS AND COST IMPLICATIONS", Ray, WF and Davis, RM, Elec Power Applications, December 1979, Vol 2, No 6, pp 185 - 193

4. "INVERTER DRIVE FOR SWITCHED RELUCTANCE MOTOR: CIRCUITS AND COMPONENT RATINGS", Davis, RM, Ray, WF and Blake, RJ, IEE Proc, Vol 128, Pt B, No 2, March 1981, pp 126 - 136

5. "CONVERTER VOLT-AMPERE REQUIREMENTS OF THE SWITCHED RELUCTANCE DRIVE", Miller, TJE, Trans IEEE on Ind Appl, Vol IA-21, 1985, pp1136-1144

6. "DESIGN PROCEDURE FOR SWITCHED RELUCTANCE MOTORS", Krishnan, R, Arumugam, R, and Lindsay, JF, Trans IAS Meeting, Denver, Sept 1986, pp 858 - 863

7. "AN AXIAL AIR-GAP MOTOR FOR VARIABLE SPEED APPLICATIONS", Unnewehr, LE and Koch, WH, IEEE Trans, 1974, PAS-93, pp 367 - 376

8. "VARIABLE-RELUCTANCE MOTORS FOR ELECTRIC VEHICLE PROPULSION", Vallese, FJ and Lang, JH, SAE International Congress & Exposition, Detroit, Michigan, February/March 1985, Paper No 850201

9. "DEVELOPMENT OF A SWITCHED RELUCTANCE DRIVE FOR A FORK LIFT TRUCK", Thornton, R, IEE Colloquium on Electric Vehicle Electronics and Control, November 1986, London, Paper No 9

10. "AN OPTIMUM DESIGNED EXAMPLE OF A 350KW, 1500V, 3000RPM VARIABLE RELUCTANCE MOTOR FOR ELECTRICAL TRACTION PURPOSE", Amin, E, Proc of Symposium on Electrical Drive, Cagliari (Italy), Sept 87, pp 347 - 351

11. "AN INTEGRATED APPROACH TO SWITCHED RELUCTANCE MOTOR DESIGN", Pollock, C, and Williams, B W, Proc of EPE Conference, Grenoble, September, 1987, pp 865 - 870

12. "SCALING LAWS FOR ELECTRIC MOTORS", Kamerbeek, EMH, Philips tech Rev 35, No 4, pp 116 - 123

13. "MACHINES ELECTRIQUES A RELUCTANCE VARIABLE ET A DENTS SATUREES", Jarret, J, Tech Mod 1967, 2, pp 78 - 80

14. "SATURABLE OVERLAPPING RECTANGULAR POLES: towards a functional relationship between force, overlap distance, mmf and saturation polarization", Byrne, JV and O'Connor, WJ, IEEE Trans on Magnetics, Vol Mag-11, No 5, September 1975, pp 1547 - 1549

15. "SATURABLE VARIABLE RELUCTANCE MACHINE SIMULATION USING EXPONENTIAL FUNCTIONS", Byrne, JV and O'Dwyer, JB, Proc Int Conf on Stepping Motors & Systems, Leeds, September 1979, pp 11 - 16

16. "CHARACTERISTICS OF SATURABLE STEPPER AND RELUCTANCE MOTORS", Byrne, JV and Lacy, JG, Proc EMDA Conf, London, July 1982, Conf Pub 213, pp 93 - 96

17. "DESIGN AND PERFORMANCE OF A SATURABLE VARIABLE RELUCTANCE SERVO MOTOR", Byrne, JV and Devitt, F, MOTOR-CON, October 1985, pp 139 - 146

18. "A HIGH PERFORMANCE VARIABLE RELUCTANCE DRIVE: A NEW BRUSHLESS SERVO", Byrne, JV, McMullie, MF and O'Dwyer, JB, MOTOR-CON, October 1985, pp 147 - 159

19. "IS SATURATION A BLESSING IN DISGUISE?", Bakhuizen, AJC, Niesten JG and Thoone, MLG, Proc IEE, Vol 125, No 5, May 1978, pp 407 - 410

20. "TORQUE PRODUCTION AND ENERGY CIRCULATION IN IDEALISED CURRENT FED S R MOTORS", Stephenson, JM, El-Khazendar, MA, Stroud, RJ, ICEM 88

21. "STATIC TORQUE PRODUCTION IN SATURATED DOUBLY-SALIENT MACHINES", Harris, MR, Hughes, A and Lawrenson, PJ, Proc IEE, Vol 122, No 10, October 1975, pp 1121 - 1127

22. "UNIFYING APPROACH TO THE STATIC TORQUE OF STEPPING-MOTOR STRUCTURES", Harris, MR, Andjargholi, V, Lawrenson, PJ, Hughes, A and Ertan, B, Proc IEE, Vol 124, No 12, December 1977, pp 1215 - 1224

23. "A NEW APPROACH TO THE PREDICTION OF STATIC TORQUE CURVE OF SATURATED VR STEPPING MOTORS", Ertan, HB, Hughes, A, Lawrenson, PJ and Harris MR, Proc 8th Annual Symposium on Incremental Motion Control Systems & Devices, Illinois, 1979, pp 169 - 179

24. "PERFORMANCE OF THYRISTOR-FED ELECTRIC CAR RELUCTANCE MACHINES", Bausch, H and Rieke, B, Int Conf on Electric Machines, 1978, Brussels, pp E4/2-1 - E4/2-10

25. "ULTRA HIGH TORQUE MOTOR SYSTEM FOR DIRECT DRIVE ROBOTICS", Welburn, R, pp 19-63 - 19-71

26. "VARIABLE SPEED DRIVES USING MULTI-TOOTH PER POLE SWITCHED RELUCTANCE MOTORS", Finch, JW, Harris, MR, Musoke, A and Metwally, HMB, Incr Motion Cont Sys Dev, Illinois 1984, pp 293 - 301

27. "A REVIEW OF THE INTEGRAL HORSEPOWER SWITCHED RELUCTANCE DRIVE", Harris, MR, Finch, JW, Mallick, JA and Miller, TJE, IEEE Transactions on Industry Applications, Vol IA-22, No 4, July/August 1986, pp 716 - 721

28. "SWITCHED RELUCTANCE MOTOR EXCITATION CURRENT: SCOPE FOR IMPROVEMENT", Finch, JW, Metwally, HMB and Harris, MR, Proc of PEVD Conference, Birmingham, November 1986, IEE Conf Pub 264, pp 196 - 199

29. "MULTI-TOOTH PER POLE VARIABLE RELUCTANCE MOTORS AND THEIR USE AS VARIABLE SPEED DRIVES", Metwally, HMB, PhD Thesis, University of Newcastle upon Tyne, UK, 1985

30. "LOW-COST, RELUCTANCE DRIVE SYSTEM FOR LOW POWER, LOW SPEED APPLICATION", Bolton, HR and Pedder, DAG, Proc EVSD, IEE Con Pub No 179, 1979, pp 88 - 92

31. "SINGLE-PHASE SWITCHED RELUCTANCE MOTORS", CC Chan, IEE Proceedings, Vol 134, Pt B, No 1, January 1987, pp 53 - 56

32. "LINEAR DOUBLY-SALIENT MAGNETIC COUPLERS", Finch, JW, Proc ICEM, Budapest, 1982, pp 1021 - 1024

33. "PARAMETER ESTIMATION OF A STEADY-STATE SWITCHED RELUCTANCE MACHINE MODEL", Franceschini, G, Pirani, S, Rinaldi, M, and Tassoni, C, Proc of Symposium on Electrical Drive, Cagliari (Italy), Sept 87, pp 329 - 338

34. "POWER OPTIMAL EXCITATION OF VARIABLE RELUCTANCE MOTORS", Lang, J H and Thornton, RD, Electric Machines & Electromechanics, Vol 2, 1978, pp 123 - 135

35. "PC CAD FOR SWITCHED RELUCTANCE DRIVES", Miller, T J E, and McGilp, M, Proc of EMD Conference, London, 16-18 Nov 1987, IEE Pub No 282, pp 360 - 366

36. "THE APPLICATION OF CAD TO SWITCHED RELUCTANCE DRIVES", Fulton, NN, Proc of IEE EMD Conf, London 1987, Pub. No 282, pp 275 - 279

37. "METHOD OF PREDICTING THE DYNAMIC RESPONSE OF A VARIABLE RELUCTANCE STEPPING MOTOR", Pickup, IED, and Tipping, D, Proc IEE, Vol 120, No 7, 1973, pp757-765

38. "A NOVEL, SELF-COMMUTATING, SINGLY-EXCITED MOTOR", BLENKINSOP, P T, Ph D Thesis, University of Leeds, UK, 1976

39. "COMPUTATION OF TORQUE AND CURRENT IN DOUBLY SALIENT RELUCTANCE MOTORS FROM NONLINEAR MAGNETISATION DATA", Stephenson, JM and Corda, J, Proc IEE, Vol 126, No 5, May 1979, pp 393 - 396

40. "MAGNETIC FIELDS IN TRACTION MACHINES FOR ELECTRIC CARS", Bausch, H and Kobler, HJ, Int Conf on Electric Machines, 1978, Brussels, pp G1/1-1 - 11

41. "MAGNETIC FIELD ANALYSIS OF A SWITCHED RELUCTANCE MOTOR USING A TWO DIMENSIONAL FINITE ELEMENT MODEL", Arumugam, R, Lowther, DA, Krishnan, R and Lindsay, JF, IEEE Transactions on Magnetics, Vol Mag-21, No 5, September 1985, pp 1883 - 1885

42. "FINITE-ELEMENT ANALYSIS CHARACTERISATION OF AN SR MOTOR WITH MULTITOOTH PER STATOR POLE", Lindsay, JF, Arumugam, R and Krishnan, R, IEE Proc, Vol 133, Pt B, No 6, November 1986, pp 347 - 353

43. "SWITCHED-RELUCTANCE MOTOR TORQUE CHARACTERISTICS: FINITE-ELEMENT ANALYSIS AND TEST RESULTS", Dawson, GE, Eastham, AR, and Mizia, J, IEEE Transactions on Industry Applications, Vol IA-23, No 3, May/June 1987, pp 532-537

44. "THREE-DIMENSIONAL NONLINEAR ELECTROMAGNETIC FIELD COMPUTATIONS, USING SCALAR POTENTIALS", Simkin, J and Trowbridge, CW, Proc IEE, Vol 127 Pt B, 1980, pp 368-374

The Application of CAD to Switched Reluctance Drives

N.M. Fulton

Electric Machines and Drives Conference, London, December 1987
IEE Conference Publication No. 282, pp. 275-279

1. INTRODUCTION

In the last few years, variable-speed drives based on switched reluctance (SR) machines have become commercially available. The growing awareness of the high levels of performance which these drives produce as a matter of routine have led many centres of research and many companies, both large and small, to begin to explore the benefits and potential of these systems.

The SR machine is, however, unusual in the eyes of designers who have been used to conventional machines which have benefited from a long history of continual development. It does not fall into the camp of a 'dc' type of machine, where a constant field flux can be 'seen' to be interacting with armature currents: nor does it have sinusoidal fluxes and currents which can be visualised by travelling fields and vector diagrams. Moreover, there is a marked contrast between the extreme simplicity of the structure of the SR machine and the sophisticated nature of the procedures required for its analysis and, particularly, its design.

The design of the machine requires a completely new approach and the modelling techniques required are not in common use. Methods fall into two basic categories: those treating the variation of the winding inductance with rotor angle as being current independent ('linear' methods); and those recognising the dependence of the inductance on current because of the saturating nature of the iron ('non-linear' methods). The linear methods, eg [1], are useful for qualitative examination of the effects of the design parameters and, in some cases, for initial design work on power converters. However, because of the high levels of saturation usually encountered in cost-effective motor designs, it is necessary to employ non-linear analysis of the system in order to predict machine performance (and hence converter requirements) with sufficient accuracy for commercial designs. However, the extreme non-linearity of the problem makes it essential, for both routine design work and exploration of the characteristics of SR systems, to implement the chosen method of solution in the form of computer programs. This paper describes the suite of programs which has been developed over a number of years by Switched Reluctance Drives Limited for the electromagnetic design of a wide range of SR machines.

2. DEVELOPMENT OF MODELLING PHILOSOPHY

Although the bibliography on SR systems displays an ever increasing number of items relating to experimental results, development projects and applications, there are relatively few papers concerning analysis and modelling. Methods of analysing stepping motors have been developed by many authors over a period of years and, since there are features common to stepping and SR motors, it is not surprising that some authors have developed SR analyses by extending stepping motor design techniques. This is the derivation of the approach adopted by Finch et al [2], but his method is based on the assumption of idealised, abruptly saturating iron. While useful for initial design work and for small perturbations about a known point where

coefficients are established, this method (and those similar) are unlikely to be sufficiently rigorous to give confidence over the wide range of machine size which will be discussed below.

Stephenson and Corda [3] have given a comprehensive review of possible approaches to the solution of the non-linear machine and demonstrate that, by careful formulation of the flux-linkage equations for one phase of the SR machine, accurate and economical numerical solution of the model is possible. Calculation of the performance of a polyphase machine is achieved by firstly solving a single-phase, non-linear model and then using superposition to incorporate the contributions of the other phases. Using their method, the SR machine can be described by:

- a statement of the number of independent phase windings
- the inductance variation (with rotor angle and current) of one phase winding, ie the 'inductance profile'
- the resistance of a phase winding.

Given these basic characteristics, the performance of the machine can then be computed (in principle at least) from a knowledge of the applied voltage and the switching angles.

While the inductance profile can be measured on an existing machine, some method of synthesising it is required for a design procedure. Unfortunately the inductance profile is a most complex relationship between current, flux and angle and there are no simple methods of establishing it. Corda and Stephenson [4] have published a method for estimating the inductance in the minimum and maximum inductance positions. Although the method takes account of the magnetisation characteristic of the iron (and hence is a 'non-linear' method), it has two short-comings. Firstly, the value of inductance in the minimum inductance position is current *independent*, ie the influence of iron saturation is ignored. This is inadequate for many design situations where, because of the high specific outputs being sought, the minimum inductance *is* current dependent. Secondly, and more seriously, the method does not allow for the third (axial) dimension of the machine. Practical experience (though strangely not in accord with some recent computational work [5]) has shown that the third dimension is significant and this significance is confirmed by other workers [7].

Finite-element (FE) methods have, in recent years, been applied to many design problems where it is important to know the shape and/or magnitude of the magnetic field in and around an iron structure. The doubly-salient structures used in SR geometries are obvious candidates for such an approach. Several FE software packages are now available for solving a range of magnetostatic problems and at least two of these have been applied to SR geometries to provide the phase inductance at specified rotor angles. Such methods are usually separated into 2-dimensional and 3-dimensional categories, each having a variety of mathematical techniques associated with them. In the former category, the work reported in References 5 and 6 report on the application of the Infolytica Magnet package to one- and two-teeth-per-pole structures. The first of these presents no experimental results and the second notes that the computational effort required precludes the satisfactory production of torque/angle curves. Trowbridge, Simkin and others at Rutherford Laboratory have made valuable contributions to the 3-dimensional category and Reference 7 illustrates the application of their TOSCA program to the calculation of the inductance of a 6-stator pole / 4-rotor pole structure. This gives good results over a range of currents and, since that publication, the methods have been further refined. However, in general, it seems that the commonly available FE packages are too cumbersome for routine design work in which many iterations will be required for a single design, each iteration requiring the inductance to be calculated at several currents (of the order of 10-15) and many rotor angles (of the order of 20-30). It is apparent that considerable skill is required for the correct meshing of the geometry, particularly at those rotor angles where the rotor position is not symmetrical with the stator geometry. Fortunately, developments are proceeding in this field also and accurate yet fast and economical FE methods may well become available within a relatively short time as a combined result both of developments in FE modelling and of the general increase of computing power.

The model developed by SR Drives and implemented in a CAD suite is a comprehensive one and, while based on the work described in References 3 and 4, it has been greatly extended both to overcome the shortcomings of Reference 4 and to provide comprehensive design information for a range of geometries for SR machines having one tooth per pole. The structure and scope of the suite will now be described and examples given of the options which are available to the designer.

3. STRUCTURE OF SUITE

The basis of the model is the solution of the flux-linkage/current/angle ($\psi/\theta/i$) equations and the suite is structured around a file which holds this data, as shown in Fig 1. The data can either be measured (on an existing machine for which further computation is required) or be synthesised from the geometry of the laminations and frame. Having obtained this data (and having viewed it in a variety of ways, if desired), two main routes are available: one allows calculation of the static torque/angle curves for a specified variety of currents and phase connections; the other allows calculation of the running performance of the drive at a given speed, taking account of the converter circuits, switching device characteristics, supply voltage(s), switching angles, current limits, etc. In both cases, both hardcopy and graphical output (in some cases, in colour) are available from a range of purpose-written post-processor programs. Fig 1 shows only a summary of the suite, which now numbers around fifteen programs and twenty associated utility routines. The suite has been written in FORTRAN77 and uses two graphical packages, GHOST80 and GINOSURF. It is generally structured to run on a mainframe and has run both on an ICL 1906 and an Amdahl 580 using the IBM operating system. One complete pass through the suite to calculate one load point takes typically around 5 seconds of CPU time.

4. DESIGN PARAMETERS

The range of parameters which has to be considered by an SR designer is large and it is appropriate to list the most important of these here - the effect of some of these has been discussed in a previous publication [8]. From the viewpoint of the modelling, the most significant of all the parameters is probably the phase number. Not only does its value determine the routines used for superposing the time-displaced solutions for the individual phases, but it also touches on the type of circuit being used. Further, for 2-phase systems, the use of stepped-gap rotors [9] involves a considerable complication of those

sections of the suite which handle the inductance calculation and $\psi/\theta/i$ synthesis. However, by careful construction of the suite, it is possible to reduce the number of places where phase dependency is dominant.

It is taken for granted that the front end of the suite will be able to handle a wide variety of lamination profiles. This can be achieved by a carefully chosen combination of parameters numbering between 15 and 25, depending on the particular style of lamination. This has been shown to be successful over a range of machine sizes.

The model includes the effect of the voltage-current characteristic of the converter devices and makes due allowance for these when solving the performance equations and calculating estimated device losses. The number of different circuit options which must be taken into account is quite large and includes circuits with:

- transistor-type devices where the on-state voltage is current dependent
- thyristor-type devices where the on-state voltage is current independent
- singly-wound or bifilar machine windings
- 'H-circuit' configuration of the windings [10]
- forced commutation of the main switches

The model incorporates a group of parameters which are used to control the model after the inductance profile has been established. These include:

- supply voltages (since the energy need not be returned to a sink of the same voltage as the source)
- ratio of turns in main and secondary windings (if a bifilar option is being considered)
- friction and windage coefficients to describe fan losses
- the switch-on, conduction and freewheeling angles
- the chopping levels set to control the currents at low speeds
- minimum-on and minimum-off times for the switching devices (which are used to override the chopping level controls where necessary)
- commutation capacitor parameters

Manipulation of the hardcopy and graphical output is, as is normal, achieved by a separate set of parameters which control the level of detail supplied to the designer.

5. EXAMPLES OF THE MODELLING

While it is not possible, in a short paper of this nature, to give a comprehensive review of the capabilities of this type of CAD package, the examples shown below will serve to illustrate the range of results which can be obtained.

The inductance profiles which are generated from the geometry of the machine can be viewed in a variety of ways, as illustrated in Figs 2-4. The variation of inductance with rotor angle over a rotor pole pitch is shown with current as a parameter in Fig 2. These curves are those of a 25kW machine and the effect of saturation of the iron is seen in the marked fall in inductance as the current is increased. (This is an indication of the difficulty which is encountered by any linear method which assumes current independence.) These inductance values can be displayed in the form of flux-linkage against current, with rotor angle as a parameter, as shown in Fig 3. A third method of displaying the data, Fig 4, is helpful in visualising the effect of saturation. The data shown in Fig 4 relates to a 6kW, 2-phase design and the effect of a stepped airgap on the shape of the inductance curves is seen at low values of current.

There is, of course, little point in producing inductance profiles for projected machines unless the accuracy can be expected to be sufficiently high to enable meaningful results to be achieved from the modelling process. Fig 5 shows the computed and measured values of the maximum inductance of a 20kW motor having a lamination diameter of 330mm and demonstrates the reliability of the modelling over a wide range of currents. The accuracy of computation of the minimum inductance is much more sensitive to the particular lamination geometry employed but, over a range of designs up to 100kW, is of the order of 1-5% and this gives very adequate results.

Turning now to the calculation of static torque, Fig 6 shows torque/angle curves computed for one phase of a 4-phase machine. While it is not particularly difficult to produce static torque curves having a shape which 'looks' correct, the most severe test of accuracy comes when the values of torque at the phase-to-phase transitions are computed. This torque is known as the ϵ-torque and is critically dependent on the precise shape of the sides of the torque/angle curve. Fig 7 shows this ϵ-torque plotted against current for a 4-phase, 50kW, 750 rev/min machine and demonstrates the good accuracy which can be achieved. When modelling drives having converter circuits which allow two phases to be energised together during chopping, it is important to be able to model the combining of the phase torques over the correct rotor angles. Fig 8 shows this technique applied to a 3-phase, 1kW machine.

The accuracy of solution of phase currents has already been published [3] and need not be discussed further here but it should be noted that the high level of accuracy demonstrated is necessary for good estimates of developed torque, rms currents, device ratings, etc. Typical current waveforms, for a range of switch-on angles, are shown in Fig 9.

A final example of the graphics facilities of the CAD suite is shown in Fig 10. This shows an isometric plot of the surface generated by the current waveform in a forced-commutated system when the switch-on angle is varied. This particular system incorporated a resonant-reversal technique in the commutation circuits and their influence is seen in the tail-end of the current waveforms.

6. CONCLUSIONS

It has been shown that, despite the unusual and varied nature of their structure and modes of operation, SR machines can successfully be modelled in a way that is both economical in computing time and sufficiently accurate for commercial design work. The CAD suite which has been developed over a number of years has proved its value over a wide range of machine designs, some of which have been demonstrated in this paper.

7. ACKNOWLEDGEMENTS

The Author is indebted to colleagues at SR Drives Ltd, particularly Dr J M Stephenson, for their contributions and helpful discussions over a long period. Copyright of this paper belongs to Switched Reluctance Drives Ltd and their permission to publish is gratefully acknowledged.

8. REFERENCES

1. Ray, WF, and Davis, RM, 1979, *EPA, Vol 2, No 6*, 185-193

2. Finch, JW, Harris, MR, Metwally, HMB, and Musoke, A, 1985, *Proc EMDA Conf, IEE No 254*, 134-138

3. Stephenson, JM, and Corda, J, 1979, *Proc IEE, 126 No 5*, 393-396

4. Corda, J, and Stephenson, JM, 1979, *Proc Int Conf Stepping Motors and Systems*, 50-59

5. Arumugam, R, Lowther, DA, Krishnan, R, Lindsay, JF, 1985, *Trans IEEE, Mag-21 No 5*, 1883-1885

6. Lindsay, JF, Arumugam, R, and Krishnan, R, 1986, *Proc IEE, Vol 133 Pt B No 6*, 347-353

7. Simkin, J, and Trowbridge, CW, 1980, *Proc IEE, Vol 127 Pt B No 6*, 368-374

8. Fulton, NN, Lawrenson, PJ, Stephenson, JM, Blake, RJ, Davis, RM, and Ray, WF, 1985, *Proc EMDA Conf, IEE No 254*, 130-133

9. El-Khazendar, MA, and Stephenson, JM, 1986, *Proc ICEM*, 1031-1034

10. Bausch, H, and Rieke, B, 1976, *Proc ICEM*, 128-1 - 128-10

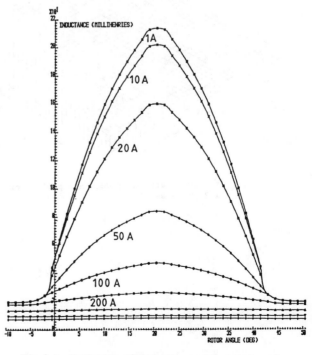

Fig 2 : Variation of inductance with rotor angle, current as a parameter

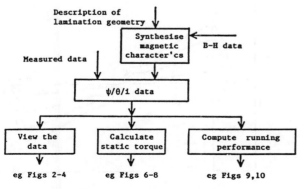

Fig 1 : Basic structure of CAD suite

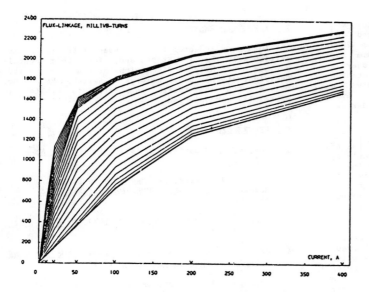

Fig 3 : Typical flux-linkage/current curves,
rotor angle as a parameter

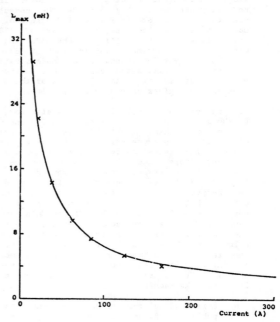

Fig 5 : Inductance, in the fully aligned position,
as a function of current
------ Calculated   x Measured

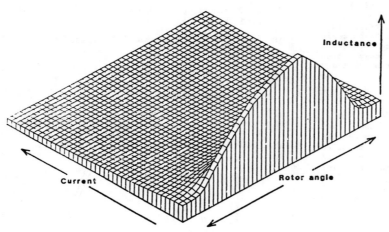

Fig 4 : Isometric view of inductance profile of
a 2-phase machine

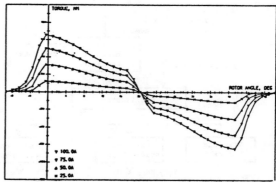

Fig 6 : Typical torque/angle curves for one phase
of a 4-phase machine

130

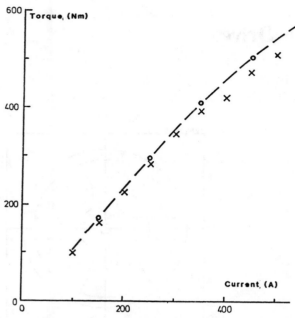

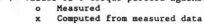

Fig 7 :   Values of ε-torque plotted against current
        o Measured
        x Computed from measured data

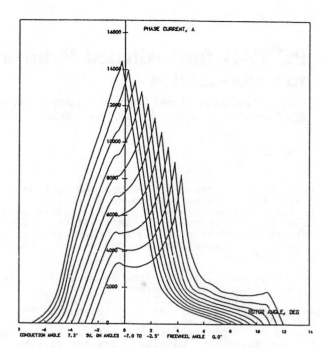

Fig 9 :   Typical current waveforms

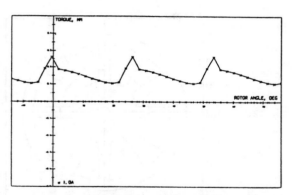

Fig 8 :   Superposition of torque/angle curves to
           model 2-phase-on operation of a 3-phase m/c

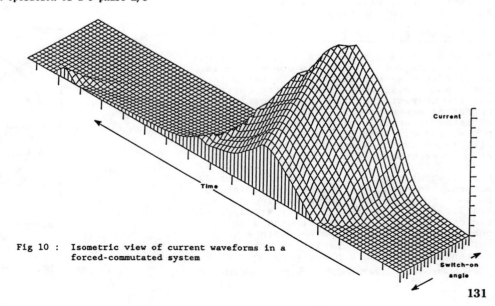

Fig 10 :   Isometric view of current waveforms in a
            forced-commutated system

PC CAD for Switched Reluctance Drives

T.J.E. Miller and M. McGilp

Electric Machines and Drives Conference, London, December 1987
IEE Conference Publication No. 282, pp. 360-366

INTRODUCTION

This paper describes a PC-based CAD package
PC-SRD for switched reluctance drives, written
in TURBO PASCAL. Graphics are used extensively
to bring out the characteristics of the drive
quickly and clearly. Using PC-SRD, a practical
design to any specification can be produced in
a few minutes.

PC-SRD is not limited to the motor design. It
also determines the waveforms necessary to
rate and select appropriate semiconductor
switches (transistors or GTO's being the
natural choice in most cases). It provides
loss and efficiency data, as well as the means
for designing control algorithms for current
chopping and switching angles in all parts of
the speed/torque range, including generating
or braking. The user can vary any design
parameter of the motor, power converter, or
control, and user-supplied steel magnetization
curves may be used.

PC-SRD uses algebraic approximations to
achieve rapid execution. These are constructed
so as to ensure a standard of accuracy
consistent with normal motor design
procedures. For very precise design, the
program structure permits the introduction of
more accurate data, e.g. magnetization curves
obtained by measurement or finite-element
analysis.

PC-SRD is written in Turbo Pascal and was
developed on an AMSTRAD PC1510 costing less
than £1000. Ideal for preliminary design and
evaluation purposes, it is also capable of
producing 'serious' designs. Some examples are
in operation at Glasgow University's Power
Electronics Laboratory.

BRIEF DESCRIPTION OF THE SR DRIVE

The SR drive is a brushless dc drive without
permanent magnets, [1-4]. In concept it is
similar to a single-stack VR stepper with
phase and amplitude control of the currents,
but its characteristics and application
potential are closer to those of brushless dc
PM motor drives. It also competes with the
p.w.m. induction motor drive, offering higher
efficiency over a wider speed range, possibly
at a lower cost.

Fig. 1 shows cross-sections of different SR
motors. Permissible combinations of stator
poles : rotor poles / phases include
6:4/3,12:8/3 and 6:2/3; 8:6/4; 4:2/2; 10:4/5;
and others, [4]. A phase winding usually
comprises two opposite pole windings in
series. Motoring torque is produced if
current i is fed to a phase winding when a

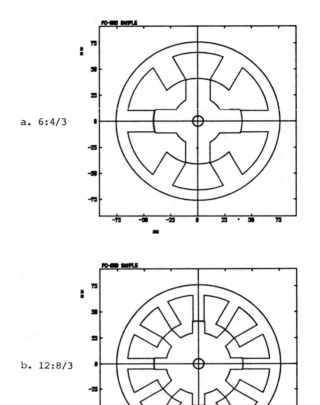

a. 6:4/3

b. 12:8/3

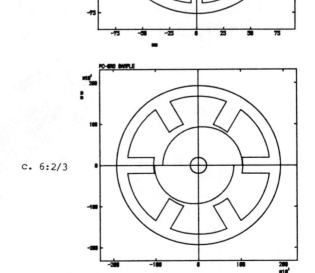

c. 6:2/3

Fig. 1 Typical SR motor geometries

pair of rotor poles is approaching alignment with the corresponding pair of stator poles. If there were no magnetic saturation the torque could be accurately expressed by the equation

$$T = \tfrac{1}{2}i^2 dL/d\theta \qquad (1)$$

where L is the phase self-inductance and θ is the rotor angle measured with respect to an arbitrary reference. The rate of change of inductance with θ must be positive to produce motoring torque. This condition is satisfied when the rotor poles are approaching the aligned position. When the rotor poles are separating from the stator poles $dL/d\theta$ is negative and if current is still flowing a negative or retarding torque will be produced. Therefore the pulses of phase current must be accurately phased relative to the rotor position. The obvious way to accomplish this is by means of a shaft angle transducer such as a slotted disc with optical interrupters; or Hall-effect sensors and a magnet ring. Signals from the sensor are used to 'commutate' the switches in the controller. In generating or braking mode, the current pulses are retarded to coincide with the separating of the rotor and stator poles.

Equation 1 also shows that the direction of current is immaterial. Unlike ac or PM brushless motors, the SR motor requires only unidirectional currents. This principle enables the use of controller circuits having only n or n+1 switches for an n-phase motor. Fig. 2 shows a 6-transistor circuit for a 3-phase machine. This circuit has 2 switches per phase, but it is simple to control; it has no restrictions on the number of phases conducting, or the conduction angles; and it requires no auxiliary components, [6].

The SR motor has been under development for several years especially in the UK and the USA and is nearing the point of sufficient maturity to appear in commercial applications. Compared with the brushless dc PM motor the SR drive may cost less, but it is likely to be noisier and less efficient. It has a high torque/inertia ratio and can withstand much higher temperatures than PM (ceramic) motors. It has no short-circuit current and no open-circuit voltage, which make it safer for use in critical applications where motor faults could be hazardous. Operation with a reduced number of phases is also possible in both generating and motoring.

The SRM is geometrically similar to the stepper motor and requires similar design techniques. However, stepper design usually ends with the static torque characteristics. For the SR motor it is necessary to go beyond this stage in designing for efficient operation with continuous rotation over a wide speed range. PC-SRD provides this. Most of the basic design principles are published in the References. Here the focus is on a synthesis of these principles in CAD software to produce practical designs quickly and effectively.

PRELIMINARY SIZING

The description of PC-SRD follows the sequence of a typical design. The starting point is usually the speed/torque curve of the load. Many other important factors will be included later, but the speed/torque curve provides a basis for initial sizing. In conventional ac and dc motor design the product of electric and magnetic loadings is often used to give an idea of the rotor D^2L required, [5]. These loading parameters are not very helpful when dealing with the SR motor, mainly because the flux is not constant. Instead, the output equation can be written as

$$T = 2\pi r^2 l\sigma \qquad (2)$$

where r is the rotor radius and l is the stack length. σ has the units of force per unit area and is proportional to the product of the conventional electric and magnetic loadings, but it can also be interpreted as the electromagnetic shear stress averaged over the cylindrical rotor surface. Equation 2 is valid for all electric machines. The rotor volume required to produce a given torque is inversely proportional to σ, which is thermally limited. Machines vary considerably in the efficiency with which they can produce shear stress. Some produce a high value of σ with low power losses, while others are not so good. The efficiency depends on the scale; on the quality and quantity of materials used; and on the torque-producing principle. SR motors are generally comparable with induction motors in this respect. A value of 1 kN/m² is typical of small nonventilated machines. Highly rated servomotors may reach 5–15 kN/m², and liquid-cooled or very large machines may attain at least ten times this value.

It is easy to determine σ for an existing motor whose torque and rotor dimensions are known. For similar cooling, this is often a convenient starting point for a new design, even when the new motor is of a new type (like the SRM) and the old motor is conventional.

The choice of a starting value for σ only determines the product r^2l. Another criterion is needed to get separate values for r and l. This is most easily done by assigning a starting value for the rotor l/r ratio. A large l/r ratio increases efficiency and torque/inertia ratio, but might be more expensive because of the need to maintain tolerances. The rotor critical speed also limits the l/r ratio. In the SR motor most of the lateral stiffness of the rotor is provided by the shaft, whose diameter is limited by the electromagnetic design.

The first task performed by PC-SRD is to read user-supplied values for T, σ, and l/r. It immediately returns separate values for l and r, together with rough estimates of the envelope (frame) dimensions of the stator; the weight and volume of electromagnetic materials; and the torque/weight, torque/volume, and torque/inertia ratios. These are crude estimates based on the assumption that the stator has twice the rotor diameter, and the overall length is 1.4 times the stack length. These preliminary estimates are rough-and-ready, but they save later effort by ensuring from the start that the design is within reasonable engineering limits.

Up to this point the design would be the same for any type of motor, with minor changes in the fixed ratios described above. In a dc commutator motor, for example, the overall length would be closer to 2 times the stack length, owing to the commutator length.

Motor Cross-section

Next PC-SRD generates a set of starting dimensions sufficient to define a lamination set. These are based on 'typical' proportions, scaled by the l and r calculated in the initial sizing. They do not embody any special properties, but follow well-established rules such as the 'feasible triangles' for pole arcs [4]. The user can change the geometry immediately, or proceed with the design.

Next, the cross-section is plotted, Fig. 1. This plot helps the designer to develop an eye for proportions that lead to particular performance characteristics (such as high efficiency or low iron loss). It also helps to judge the manufacturability of the lamination.

Winding design. The user next specifies the speed and the dc supply voltage, V. PC-SRD returns a recommended number of turns per pole. The user responds with a suggested current-density. PC-SRD calculates the cross-section of copper required and returns this value, expressed as a percentage of the available slot area. The user must decide whether this slot copper ratio is practical for the winding technique he plans to use. If it is too high he can reduce the current density or modify the geometry to make more slot area available. At each iteration of this interactive procedure, PC-SRD calculates the unsaturated inductance ratio between the aligned and unaligned positions of the rotor, [7]; this is important in determining the torque-producing capability, and must be kept as high as possible.

Dynamic Calculation for One Stroke

A 'stroke' is similar to one step of operation of a stepper motor, but the term 'stroke' or 'working stroke' is preferred here because of

the importance of the phasing of the turn-on and turn-off angles of the phase current. Adoption of the term from internal-combustion engine parlance is justified by the analogous operation.

Principle of calculation: 'dynamic design' or 'design by simulation'. [8] PC-SRD determines the torque and other important parameters dynamically by simulating the operation of the motor through one complete stroke, using Stephenson and Corda's technique, [9]. This is the only technique capable of determining the torque under dynamic conditions, because it is necessary to know the current waveform, which is calculated as part of the simulation. Techniques that are based solely on the static magnetization curves can only determine the torque if a current waveform is assumed, [10]. These methods provide an upper bound for the available torque, but only a fraction of this can be developed at speed.

Stephenson and Corda's method requires the magnetization curves for the aligned and unaligned positions of the rotor and for several intermediate positions. They used interpolating polynomials to approximate these curves on a piecewise basis, [9]. The simulation proceeds by stepwise integrating the phase flux-linkage u with respect to time:

$$u = \int (v - Ri)dt \qquad (3)$$

The value of v throughout one timestep dt is assigned the value +V, -V or 0 depending on the conduction states of the two phaseleg

switches (Fig. 2). At each timestep the most recent value of i is determined from the last value of u by solving (by interpolation) the following equation

$$u = u(i,\theta) \qquad (4)$$

which represents the magnetization curve for the current value of θ. In PC-SRD the rotor speed is assumed constant, so that θ is always known from the current value of t (time). The programming and interpretation are simplified if t is eliminated by integrating with respect to θ instead of t; equation (3) then becomes

$$u = \int (v - Ri)d\theta/\omega \qquad (5)$$

where w is the rotor speed in rad/sec.

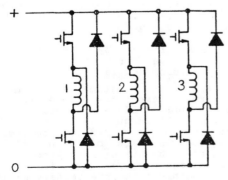

Fig. 2 Converter circuit for 3-phase motor

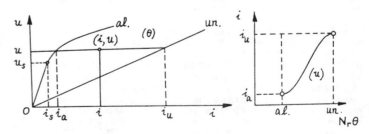

Fig. 3 Representation of magnetization curves

a. Flux-linkage u vs. current i, with θ as parameter
b. Interpolation for phase current i, with u as parameter

PC-SRD uses the following principle to achieve very fast execution of this procedure.

Representation of magnetization curves. The magnetization curve for the aligned rotor position is represented not by a piecewise series of low-order polynomials, but by two segments only. The lower segment is a straight line; the upper, a quadratic curve. The gradient is continuous across the junction. This is illustrated in Fig. 3a. The unaligned curve is a straight line whose gradient is the unsaturated, unaligned phase inductance L_u, [7,11]. In this position there is no saturation unless the current level is extremely high. (For small motors, this is well above the thermal limit).

The aligned and unaligned curves are determined as accurately as possible by one of three alternative methods:

(1) Calculation by the program;
(2) Measurement;
(3) Calculation by numerical technique such as the finite-element method.

Method 1 is used when starting a design from scratch. The necessary procedures [11] are included in PC-SRD. Method 2 is useful after the motor has been built; for example, where the program is being used to design the controller. Method 3 can be used to check or improve the results of method 1, especially when a very accurate design is needed.

Once determined, the aligned magnetization curve is curve-fitted to the straight and parabolic segment-pair. The magnetization curves for the intermediate rotor positions are not calculated explicitly. The following algorithm is used instead. Consider a typical point during the simulation. u has just been updated from equation 5, and θ is known. Corresponding to the present value of u is a current ia on the aligned magnetization curve, and a current iu on the unaligned curve. These currents are shown in Fig. 3a. ia is determined from u as follows:

$$i_a = u/L_{au} \qquad\qquad u < u_s \quad (6a)$$

$$i_a = i_s - a/m^2 + \{u - (u_s - 2a/m)\}^2/4a \qquad u > u_s \quad (6b)$$

where $u_s = L_{au} \cdot i_s$ and a is a parameter of the fitted aligned curve. i_u is determined from u as follows:

$$i_u = u/L_u \qquad\qquad (7)$$

The current corresponding to the actual rotor position is determined by a sinusoidal interpolation between i_a and i_u:

$$i = i_a + \{1 - \cos N_r\theta\}(i_u - i_a)/2 \qquad (8)$$

The nature of this interpolation is shown in Fig. 3b. In effect, the magnetization curve, expressed with i as the dependent and u the independent variable, is taken to be modulated by a sinusoidal function of rotor position between the two extremes of the aligned and unaligned positions. If the aligned and unaligned curves are accurate,

then for correctly-phased current pulses the energy converted during one stroke is bounded by these two curves and will be fairly accurate, as will the average torque. But the current waveform will be approximate. The structure of the program is such that accurate magnetization curves can be provided ab initio, if required.

The aligned and unaligned static magnetization curves can be plotted at this stage, permitting the user to verify the inductance ratio; the general level of inductance; and the effect of saturation, which limits the area between the curves. Since this area determines the available torque per ampere, the choice of steel can be partially evaluated here. Later, of course, iron losses must be examined also.

Practical Procedure; Phase Current Waveform

The user specifies the turn-on and turn-off (commutation) angles for the controller switches. A full treatment of the choice of these angles is beyond the scope of this paper, but the values used are typical. (See [4]). In broad terms, turn-on should be a few degrees before the approaching rotor poles begin to overlap the stator poles, and commutation should occur a few degrees before alignment. In fine-tuning these angles, adjustments are needed as a function of both speed and torque to get the best performance, [12]. Alternatively, a simple fixed-angle control can be used, as in brushless dc PM motors, [2].

The user must also specify the mode of current control during the conduction interval. Of several options, only one is described here. The current is assumed to be chopped between upper and lower reference levels specified by the user. The practical implementation of this algorithm would require a wide-bandwidth current transducer and a current-regulating comparator with hysteresis. Simpler, less expensive schemes are of course possible.

With this data PC-SRD proceeds to calculate its way through one complete stroke, terminating when the current reaches zero after commutation. Fig. 4a shows a typical phase current waveform calculated at a low speed with chopping control. Fig. 4b shows a higher speed, where the current is unable to reach the chopping level because, with the same switching angles, the time for which the voltage is applied is now much shorter and the self-e.m.f. is increased by the increased speed. If there were no magnetic nonlinearity, this effect could be expressed as follows:

$$V = du/dt = d/dt\ (Li)$$

$$= L\ di/dt + i\ dL/dt$$

i.e.,

$$V = \omega L\ di/d\theta + i\omega\ dL/d\theta \qquad (9)$$

The self-e.m.f. (the second term) is proportional to ω. So is the 'reactance' ωL. Consequently the rate of rise of current with respect to rotor position, di/dθ (i.e., the

135

initial gradient of Fig. 4) decreases as the speed increases. Ultimately this limits the current and limits the torque available at high speed. Operation in the chopping mode is noisier and less efficient than in the single-pulse mode, unless the chopping frequency is maintained well above the audible range.

Fig. 4c shows the current waveform in the generating mode. In this case the current-pulse is retarded past the aligned position, permitting the freewheeling current through the diodes to grow as a result of the negative self-e.m.f. (When the poles are separating $dL/d\theta$ is negative; see equation 9).

'Indicator Diagrams'

Fig. 5a shows the static magnetization curves for the aligned and unaligned positions. Superimposed is the trajectory of the point (i,u) during the stroke corresponding to Fig. 4a. The energy converted into mechanical work is equal to the area enclosed within this trajectory. and Fig. 5 can be regarded as an indicator diagram by analogy with steam engine operation. There is no electromagnetic equivalent of the Carnot cycle, but the diagram does show that for a given current the available conversion energy is limited by the magnetization curves for the two extreme positions. This underlines the need for a high aligned inductance, a low unaligned inductance, and the highest possible level of saturation.

In Fig. 5b (corresponding to Fig. 4b) the unused magnetic energy returned to the supply through the diodes can be identified as the area between the trajectory and the u-axis. This is a significant fraction of the conversion energy, and underlines the need for a storage capacitor in the dc supply. In Fig. 5c (generating) the net electrical energy generated per stroke is equal to the area enclosed. The area to the left of the trajectory, lying between the trajectory and the u-axis, is the excitation energy that is supplied through the switches when they are on. In this case the peak, mean, and r.m.s. diode currents are much larger than the corresponding switch currents.

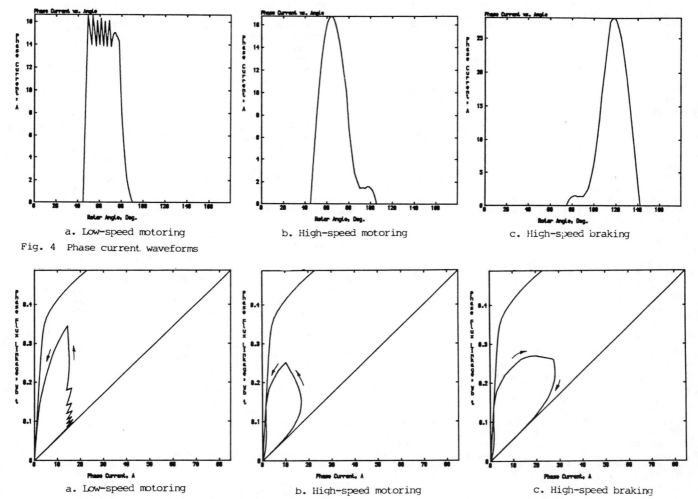

a. Low-speed motoring   b. High-speed motoring   c. High-speed braking

Fig. 4 Phase current waveforms

a. Low-speed motoring   b. High-speed motoring   c. High-speed braking

Fig. 5 Trajectory of the point (i,u) during one stroke

136

Torque

The average electromagnetic torque is given by equation 10 for a q-phase motor:

$$T = qN_r W/2\pi \qquad (10)$$

where W is the conversion energy enclosed by the trajectory [7]. From this the friction and windage loss and the iron loss are subtracted. Strictly speaking, the iron loss is an electrical loss that should be accounted for by the equivalent circuit and the associated electrical equations. However, there is no established equivalent-circuit representation of iron losses for machines of this type. Instead, the iron losses are estimated from the waveform of flux-density in the separate parts of the magnetic circuit. This waveform is generated from the flux-linkage,

which is calculated by equation 5. The flux-density in each section of the magnetic circuit is then determined in inverse proportion to its cross-section area, and the losses determined by scaling from the loss data for the steel. More precisely, iron losses would modify the indicator diagram, reducing the enclosed area. The simulation of this effect is left for future research.

The average torque is written to the output file along with several other useful parameters.

It would be useful to be able to calculate the instantaneous torque as a function of rotor position, including the contributions from all phases. This would require the calculation of incremental changes in stored field energy (or coenergy). The necessary functions can be derived from equation 8 together with equations 6 and 7, but this has been left for a future enhancement of the program.

DC supply current

The instantaneous DC supply current can be derived from a single phase current waveform by adding phase-shifted replicas thereof into a separate array. The result is shown in Fig.

6 for the three cases a, b, and c of Figs. 4 and 5. To a first approximation, the dc component is supplied by (or to) the rectifier or battery (or to a dynamic braking resistor). The remaining harmonics are shunted through the storage capacitor in parallel with the dc supply. The waveforms of Fig. 6 underline the need for sufficient ripple-current capability in addition to the energy storage already discussed.

Printed output

In addition to the graphical output PC-SRD also provides printed output as shown in Fig. 7.

PROGRAMMING CONSIDERATIONS

PC-SRD comprises a number of Pascal modules and was written using Turbo Pascal, [13]. The graphics were implemented with the Turbo Pascal Graphix Toolbox, [14]. Both Turbo Pascal and the Turbo Pascal Graphix Toolbox are commercial software products of Borland International, Inc. Development was done on an AMSTRAD PC1512HD10 personal computer with a CGA color graphics monitor running under MS-DOS 3.2.

CONCLUSION

PC-SRD provides a convenient design and evaluation tool that can be run more or less interactively on an inexpensive PC with very fast execution. It cannot compete in accuracy with the type of design software that could be implemented on a mainframe or workstation, but it is adequate for sizing purposes and is capable of providing sufficient data on which the controller can be designed, including the current and voltage stresses on the main semiconductor switches and the control strategy for the switching angles. For many applications the level of detail in this design will suffice for the electromagnetic part of the overall system design, the greater part of which is usually in the mechanical and electronic details and in planning how to manufacture it.

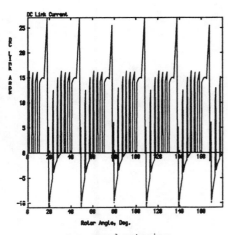

a. Low-speed motoring

Fig. 6 DC Supply current

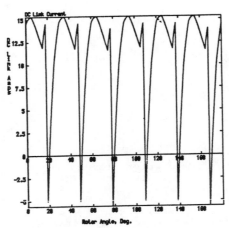

b. High-speed motoring

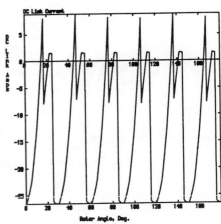

c. High-speed braking

```
30 JUN 1987 12:24        Glasgow University SR Design Package        Ver  32
------------------------------------------------------------------------------
 PC-SRD SAMPLE

--------------------------------------------------------- DIMENSIONS ( mm. )
Rotor  R0    23.000
       R1    41.000      4  Poles          Arc 32.000 Deg   Gap           0.250
Stator R2    66.000      6  Poles          Arc 29.000 Deg   Stack lgth 90.000
       R3    76.000

Stckng.fact. 0.960                         Steel = Losil 400
----------------------------------------------------------------WINDING DATA
Wire Dia.      1.500 mm. 3  Phases         75 Turns/Pole
C.S. Areas :
   1 Wire    = 1.77E+00 Sq.mm              Resistance/Phase =    0.542 Ohm
   Slot Cu   = 1.33E+02 Sq.mm              Temp.              = 105.000 C
Slot fill    = 3.04E-01
M.L.T.       = 2.79E+02 mm.                Inductance/Phase = 1.02E-01 H Al
Lgth o/ends  = 1.38E+02 mm.                                   5.74E-03 H Un
Copper wt.   = 4.34E+00 lb                              Ratio = 17.8
----------------------------------------------------------------CONTROL DATA
Voltage             160.0                  Current regulator settings
R.P.M.             3000.0                  14.0/16.0 A

Turn-on  angle      45.0 deg. BTC          Supply Res.      0.001 Ohm
Turn-off angle      12.0 deg. BTC          Transistor RQ    0.600 Ohm
Dwell angle   =     33.0 deg.              Transistor VQ    0.200 V
Stroke angle  =     30.0 deg.              Diode      VD    0.600 V
O/lap starts  =     30.5 deg. BTC          Phase freq. =    200.0 Hz
-----------------------------------------------------------------PERFORMANCE
Torque             = 3.63E+00  N-m         Efficiency   =   89.3 %
Shaft Power        = 1.14E+03  W

Losses: Copper     = 9.17E+01  W
        Iron       = 4.54E+01  W
                                           Deg. C / W        0.80
        Total      = 1.37E+02  W           Temp. rise    = 109.8 C

CURRENTS                         PEAK           MEAN           R.M.S.
Winding                         16.007          4.870           7.511
Transistor                      16.007          4.193           7.315
Diode                            8.970          0.677           1.164
DC Link (Supply)                                9.179
DC Link (Capacitor)                                             5.499

RMS Current Density  =    2742.3 A/Sq.in. =    4.3 A/Sq.mm.
------------------------------------------------------------------------------

--------------------------------------------------------SUPPLEMENTARY OUTPUT
WEIGHTS: Copper =     4.3 lb               Inertia       =  1.03E-03  kg-m2
         Iron   =    15.5 lb
         Total  =    19.8 lb
Resistivity     =     2.28 E-08 Ohm-m      Temp. fact. =  1.334
Sigma           =     0.55 psi

Pk Switch volts = 160.6     V

End of Design
```

Fig. 7 Typical printout of PC-SRD

REFERENCES

1. Miller, T.J.E., "Small motor drives expand their technology horizons", IEE Power Engineering Journal, September 1987.

2. Miller, T.J.E. and Jahns, T.M., "A current-controlled switched reluctance drive for f.h.p. applications", Conference on Applied Motion Control, Minneapolis, USA, June 10-12, 1986.

3. Harris, M.R., Finch, J.W., Mallick, J.A., and Miller, T.J.E., "A review of the integral-horsepower switched reluctance drive", Trans. IEEE, IA-22, No. 4, July/August 1986, pp 716-721.

4. Lawrenson, P.J. et al, "Variable-speed switched reluctance motors", Proc. IEE. 127, Pt. B, No. 4, July 1980, pp 253-265.

5. Harris, M.R., Discussion on Ref. 4, ibid., 128, Pt. B, No. 5, September 1981, pp 260-262.

6. Bass, J.T., Miller, T.J.E., and Ehsani, M., "Development of a unipolar converter for variable-reluctance motor drives", Proc. IEEE Industry Applications Soc. Annual Meeting, Toronto, 1985, pp 1062-1068

7. Miller, T.J.E., "Converter volt-ampere requirements of the switched reluctance motor drive", Trans. IEEE, IA-21, 1985, pp 1136-1144.

8. Miller, T.J.E., "Design of brushless drives using time-domain techniques", Proc. 22nd Universities Power Engineering Conf., Sunderland, April 1986.

9. Stephenson, J.M. and Corda, J., "Computation of torque and current in doubly salient reluctance motors from nonlinear magnetization data", Proc. IEE, 126, No. 5, May 1979, pp 393-396.

10. Harris, M.R. et al, "Static torque production in saturated doubly-salient machines", Proc. IEE, 122, No. 10, October 1975, pp 1121-1127.

11. Corda, J. and Stephenson, J.M., "Analytical estimation of the minimum and maximum inductances of a doubly-salient motor", Proc. Int. Conf. Stepping Motors and Systems, September 1979.

12. Bose, B.K., Miller, T.J.E., Szczesny, P.M. and Bicknell, W.H., "Microcomputer control of switched reluctance motor", Trans. IEEE, IA-22, No. 4, July/August 1986, pp 708-715.

13. Borland International, "Turbo Pascal", Scotts Valley, CA, 1985.

14. Borland International, "Turbo Pascal Graphix Toolbox", Scotts Valley, CA, 1985.

Computation of Torque and Current in Doubly-Salient Reluctance Motors From Nonlinear Magnetization Data

J.M. Stephenson and J. Corda

Proceedings IEE, Vol. 126, No. 5, May 1979, pp. 393-396

Indexing terms: Reluctance motor, Stepping motors

Abstract

The paper presents a new method for accurately calculating the torque and current of a variable-reluctance stepping motor from measured or computed flux-linkage/current/rotor position data. The new method is more accurate and uses less input data than previously known procedures, and is computationally efficient. It can be applied to both steady-state and transient problems.

List of principal symbols

i = current
J = moment of inertia
l = incremental inductance
L = inductance
R = resistance
t = time
T = torque
V = voltage
W' = coenergy
ϕ = angle of complete cycle
ψ = flux linkage
ω = angular velocity
θ = rotor displacement

1 Introduction

The need for accurate predictions of the performance of variable-reluctance motors and other related forms of doubly-salient switched reluctance motors has given rise to a number of methods of calculation in recent years.[1-6] The importance of these computational methods lies in their use in

(a) using measured inductance or flux-linkage data to enable detailed examination of behaviour over a wide variety of simulated operating conditions without the limitations associated with drive circuits, transducers or instrumentation
(b) using basic measured data modified as required to test the sensitivity of performance to hypothetical design changes
(c) using data computed by numerical field analysis, enabling, for example, purely computational studies of design to be made to meet specified performance requirements.

To predict performance (average and peak torques and currents, speed, efficiency etc.) it is necessary to solve the differential circuit equations for the appropriate switched conditions and, for transient performance, the mechanical equations also. The electromechanical nature of the motor is reflected in the variations of phase-winding inductances with rotor displacement. (Mutual inductances with other phase windings are often very small and are usually neglected.) To calculate the performance, it is simply necessary to integrate the set of 1st-order circuit and mechanical differential equations.

The problem centres on the measured or computed relationships between the flux linkage of the phase winding and its current and the angle of the rotor displacement. The degree of magnetic saturation to be handled is frequently high, which results in severe nonlinearity

in flux-linkage/current curves. The flux-linkage/angle relationships cannot, in general, be taken either as linear or as simple sinusoidal functions.

The various approaches to the problem are summarised below, but it is first necessary to state the equations of the basic machine circuit which is illustrated in Fig. 1. In this Figure, L represents the inducatance of one phase winding of the motor. Flux linkage is found by measurement or magnetic-field analysis as a function of exciting current i and rotor angular displacement θ over the complete range of current and cycle of inductance. The voltage equations describing the circuit in Fig. 1 are as follows:

$$\pm V = Ri + \frac{d\psi(\theta, i)}{dt} \tag{1}$$

In these equations, the $+$ve sign corresponds to the state when the controlling switch S is closed and $V = V_S$, $R = R_1$, and the $-$ve sign corresponds to the state when S is open and $V = V_R$, $R = R_2 + R_1$.

Eqns. 1 can be expressed in the following alternative forms, the choice depending on the method used for handling the nonlinear data:

$$\frac{di}{dt} = \frac{1}{\frac{\partial \psi}{\partial i}(\theta, i)} \left\{ \pm V - Ri - \omega \frac{\partial \psi}{\partial \theta}(\theta, i) \right\} \tag{2}$$

$$\frac{di}{dt} = \frac{1}{l(\theta, i)} \left\{ \pm V - \left[R + \omega \frac{\partial L}{\partial \theta}(\theta, i) \right] i \right\} \tag{3}$$

$$\frac{di}{dt} = \frac{1}{L(\theta, i)} \left\{ \pm V - \left[R + \omega \frac{dL(\theta, i)}{d\theta} \right] i \right\} \tag{4}$$

The mechanical equations are as follows:

$$\frac{d\theta}{dt} = \omega \tag{5}$$

$$\frac{d\omega}{dt} = \frac{1}{J} \{ T(\theta, i) - T_{mech} \} \tag{6}$$

Singh[1] used equations in the form of eqn. 3. The functions $l(\theta, i)$, $L(\theta, i)$ and $T(\theta, i)$ were approximated by fundamental cosinusoidal components in θ and by linear functions in i, the derivatives $\frac{\partial L}{\partial \theta}(\theta, i)$ being obtained analytically. Having approximated the function $l(\theta, i)$ by an analytical function, the function $L(\theta, i)$ and $T(\theta, i)$ could have been found analytically as $L(\theta, i) = \{ \int_0^i l(\theta, i) di \}/i$ and

Paper 8300P, first received 2nd May and in revised form 23rd November 1978

Dr. Stephenson and Mr. Čorda are with the Department of Electrical & Electronic Engineering, University of Leeds, Leeds, Leeds LS2 9JT, England

140

$T(\theta, i) = \partial\{\int_0^i L(\theta, i)idi\}/\partial\theta$, thus avoiding more measurements. The probable reason why this was not done is the poor approximation for $l(\theta, i)$.

Pickup and Tipping[2,3] used equations in the form of eqn. 2. Flux linkage was measured at equally spaced angles and currents. In their first method, the function $\psi(\theta, i)$ was approximated by polynomials in both θ and i. In the second method, this function was approximated by Fourier cosine series in θ and polynomials in i. The derivatives $\frac{\partial\psi}{\partial\theta}(\theta, i)$ and $\frac{\partial\psi}{\partial i}(\theta, i)$ and torque $T(\theta, i) = \partial\{\int_0^i \psi(\theta, i)di\}/\partial\theta$ were then found analytically.

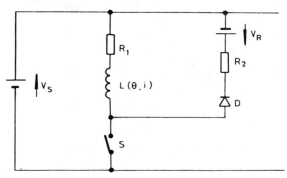

Fig. 1
Basic machine circuit

Byrne and Dwyer[4] also used equations in the form of eqn. 2. Flux linkage $\psi(\theta, i)$ was measured at equally spaced angles and currents. The curves $\psi(\theta, i)|_{\theta = const}$ were each fitted by a sum of three exponentials. The values of $\frac{\partial\psi}{\partial i}(\theta, i)|_{\theta = const}$ and $\int_0^i \psi(\theta, i)di|_{\theta = const}$ were obtained analytically and values of $\frac{\partial\psi}{\partial\theta}(\theta, i)|_{i=const}$ and $T = \partial\{\int_0^i \psi(\theta, i)di\}/\partial\theta$ were obtained numerically. The intermediate values of $\frac{\partial\psi}{\partial i}(\theta, i)$, $\frac{\partial\psi}{\partial\theta}(\theta, i)$ and $T(\theta, i)$ were found by linear interpolation in θ and i.

Blenkinsop[5] has used equations in the form of eqn. 4. Inductance $L(\theta, i)$ was measured at equally spaced angles and unequally spaced currents. Then, using polynominal fitting for the curves $L(\theta, i)|_{\theta = const}$, more values of $L(\theta, i)$ at equally spaced currents were obtained. The values of coenergy $W'(\theta, i) = \int_0^i L(\theta, i)idi|_{\theta = const}$ were found analytically, and the values of $\frac{dL(\theta, i)}{d\theta}|_{i=const}$ and $T(\theta, i) = \frac{\partial W'(\theta, i)}{\partial\theta}|_{i=const}$ were obtained numerically.

Acarnley[6] used equations in the form of eqn. 3. Incremental inductance $l(\theta, i)$ was measured at equally spaced angles and currents. The function $l(\theta, i)$ was approximated by polynomials in i and by Fourier series in θ. Then, the inductance $L(\theta, i)$ was found analytically as $L(\theta, i) = \psi(\theta, i)/i = \{\int_0^i l(\theta, i)di\}/i$. The derivative $\frac{\partial L}{\partial\theta}(\theta, i)$ and torque $T(\theta, i) = \partial\{\int_0^i L(\theta, i)idi\}/\partial\theta$ were then obtained analytically.

The voltage equations, eqns. 2, 3 and 4, contain coefficients $\frac{\partial\psi}{\partial i}(\theta, i)$, $\frac{\partial\psi}{\partial\theta}(\theta, i)$, $\frac{\partial L}{\partial\theta}(\theta, i)$ or $\frac{dL(\theta, i)}{d\theta}$, and the errors introduced by fitting the curves $\psi(\theta, i)$ or $L(\theta, i)$ are increased by the process of differentiation. Also the amount of data required in these methods is large for highly saturated machines. (This directly increases the computing time required for producing the data by numerical field analysis or the time required to make the number of measurements required.)

The fundamental difficulty of obtaining accurate values for the differential of a function specified only by tabulated points is well known. For example, an apparently good polynomial fit of flux linkage ψ with current i can give rise to significant errors in $\frac{\partial\psi}{\partial i}(\theta, i)|_{\theta = const}$. The situation is even worse in the present case, because when such polynomials are used to define the values $\psi(\theta, i)|_{\theta = const}$ for any value of i at different values of θ, even larger errors can arise in the differentiation with respect to θ (i.e. in evaluating $\frac{\partial\psi}{\partial\theta}(\theta, i)|_{i=const}$).

The essential feature of the new method described below is that, unlike the methods described above, it is based on using nonlinear differential equations which do not require the evaluation of the differential coefficients; consequently, the results are more accurate and require less input data.

2 New formulation and data handling

2.1 Voltage equations

Using voltage equations in the form

$$\frac{d\psi(\theta, i)}{dt} = \pm V - Ri \tag{7}$$

eliminates $\frac{\partial\psi}{\partial\theta}(\theta, i)$ and $\frac{\partial\psi}{\partial i}(\theta, i)$ as coefficients with the associated problem of error amplification. An additional advantage is that, in general, the maximum rate of change of flux linkage is not so rapid as the maximum rate of change of, for example, current, so that the step length in the integration of the model equations can be relatively long with consequential saving in computing time.

2.2 Method of solution

The given data defining the magnetic nature of the machine are stored as a table of values of $\psi(\theta, i)$. It is very desirable that the number of values to be specified is as small as possible, and it has been found that the curves of $\psi(\theta, i)|_{\theta = const}$ (being monotonic and having as a family their maximum rates of change in the same region of i) can be represented with sufficient accuracy and considerable economy by using unequally spaced values of current. This is illustrated in Fig. 2. A quadratic interpolation method has been found to be very effective in deriving intermediate values of flux linkage (see below). However, the curves $\psi(\theta, i)|_{i=const}$ are not monotonic (being cyclic), so that the quadratic interpolation cannot be simply applied, and as a family do not lend themselves so readily to economy of specifying data. Equally spaced angles have therefore been used in the table of input data $\psi(\theta, i)$.

The solution of eqns. 7, to obtain current waveforms and the motor performance, requires the definition of the magnetic behaviour of the motor in the form of a table $i(\theta, \psi)$ to enable the value of i to be updated after each step of the numerical integration of the model equations (i.e. after each integration step, to find a new value of ψ, it is necessary to find the corresponding current which is then inserted in the right-hand side of eqns. 7). It may be that the supposed difficulty in generating this table is the reason why eqns. 7 have not been used before. This table $i(\theta, \psi)$ (with a sufficient number of equally spaced angles and flux linkages to ensure sufficiently accurate interpolation) is obtained by inverting the input table. The values of $i(\psi)|_{\theta = const}$ at equally spaced flux linkages are found by using quadratic interpolation through three successive points (Fig. 3) because it has been found that zone II of the parabola fits very well with the curves $i(\psi)|_{\theta = const}$ for relatively large spaces between the currents i_{j-1}, i_j, i_{j+1}.

Thus, using this fast and very accurate method, the table $i(\theta, \psi)$, which is the inverse of the input table $\psi(\theta, i)$, is obtained. (The corresponding curves are shown in Fig. 4.) The intermediate values of $i(\theta, \psi)$ required in the course of numerical integration of the model equations are found by linear interpolation in θ and quadratic interpolation in ψ.

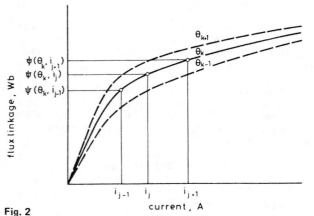

Fig. 2
Flux linkage as a function of current for 3 different rotor angles, illustrating the use of unequally spaced current values for tabulating the data

$j = 1, 2, \ldots, N_i$
$k = 1, 2, \ldots, N_\theta$

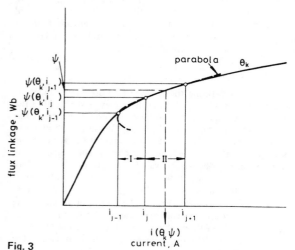

Fig. 3
Flux linkage as a function of current for a fixed rotor angle, illustrating the use of quadratic interpolation to find a current i, corresponding to a flux ψ using 3 tabulated points

$i(\theta_k, \psi) = A\psi^2 + B\psi + C$

The differential equations are solved by the Runge-Kutta 4th-order integration method, the values of the flux linkage and the current being found after each step of integration.

2.3 Torque computation

To compute the torque $T(\theta, i)$, the table of coenergies $W'(\theta, i)$ is required. $W'(\theta, i) = \int_0^i \psi(\theta, i) di|_{\theta = const}$ is found by numerical integration. (An increased number of values of $\psi(\theta, i)$ at equally spaced currents are computed using the method of quadratic interpolation.) $T(\theta, i) = \dfrac{\partial W'(\theta, i)}{\partial \theta}\bigg|_{i = const}$ is obtained by numerical differentiation using Stirling's formula.

142

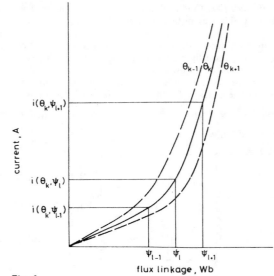

Fig. 4
Current as a function of flux linkage for 3 different rotor angles

These are the derived inverse functions of those shown in Fig. 2, but the flux linkages are tabulated at equal intervals
$l = 1, 2, \ldots, N_\psi$

The table of $T(\theta, i)$ is used for transient problems with the intermediate values of $T(\theta, i)$ being found by linear interpolation in θ and quadratic interpolation in i. The steady-state average torque per complete cycle of inductance is computed from the closed area in the ψ-i plane described during a complete cycle:

$$T_{avr} = \frac{\oint \psi di}{\phi}$$

3 Accuracy and efficiency of method

The choice of the number of elements in the input table $\psi(\theta, i)$ is made according to two tests of accuracy, consisting of the curves $T(\theta, i)|_{i = const}$ with the rotor stationary and $i(\theta)$ at constant speed, under the most onerous conditions, i.e. those involving the highest degree of saturation. The authors have found that for a machine with a sharply profiled airgap and high saturation, values of flux linkage specified at 11 equally spaced angles and 8 unequally spaced currents (distributed as shown in Fig. 5) are adequate.

Figs. 6 and 7 show the excellent degree of agreement achieved between measured and computed results using the new method. The greatest error in the current is 4% of the maximum current, and in the torque is 3% of the maximum torque. These results are particularly good in view of the very high degree of saturation occurring, which can be seen by relating the maximum current in Fig. 6 ($0.45\,\hat{i}$, where $\hat{i}$ is the maximum stored value of current) to the appropriate magnetisation curve in Fig. 5. Table 1 gives additional evidence of the accuracy of the method in the form of a comparison of predicted and measured average torque and efficiencies at three different loads at base speed. (Core loss was small for the operating conditons shown. Pickup and Tipping included an allowance for core loss by equivalent resistances in parallel with the windings. Such resistances have not been used in the new method described above, but could easily be added to the model. Transistor and diode voltage drops were, however, included.) The computing time for constant speed operation is 5 s (2 s for inverting the input $\psi(\theta, i)$ table and 3 s for integrating the model equations and for computing the performance). An ICL 1906A computer was used.

Table 1
PREDICTED AND, IN BRACKETS, MEASURED TORQUE AND EFFICIENCY

Torque, Nm	9·61 (9·50)	6·95 (7·00)	4·53 (4·77)
Efficiency, %	57·2 (55·2)	65·4 (64·7)	71·3 (70·2)

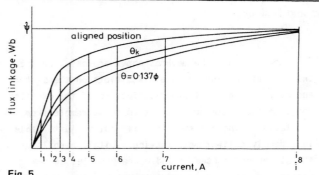

Fig. 5

Flux linkage against current for 3 different rotor angles (including the fully aligned position) showing the distribution of the 8 unequally spaced current values

$k = 1, 2, \ldots, 11$

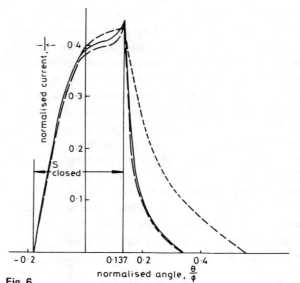

Fig. 6

Measured and computed current/angle responses at constant speed

—————— computed response using new method
– – – – computed response using Blenkinsop's method
– — — measured response

The problems associated with methods using the equations of the form of eqns. 2–4 above are also illustrated in Figs. 6 and 7. The method used in the computation of these curves was that due to Blenkinsop.[5] It can be seen that the results for torque at constant current (Fig. 6) are not seriously in error, nor indeed is the current curve until high degrees of saturation are reached. However, the polynomial fitting (5th-order was used here) of $L(\theta, i)|_{\theta = const}$ introduces serious errors at high flux densities. Indeed the flux continues to increase in value for a tenth of a cycle following switch off, and the calculated average torque is 29% high. The error in the rate of decay of current results in a large error in the input power (48% low) and the combined errors result in an apparent efficiency of 137%.

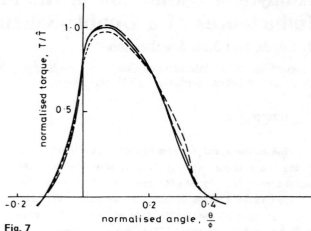

Fig. 7

Measured and computed torque/angle results at constant current

$$\hat{T} = \frac{\hat{\psi}\hat{i}}{\phi}$$

—————— computed results using new method
– – – – computed results using Blenkinsop's method
– — — measured results

4 Conclusions

The methods for predicting the performance of doubly-salient reluctance motors from computed or measured $\psi(\theta, i)$ data have been reviewed and basic weaknesses under highly saturated conditons have been identified.

The model equations have been reformulated in a more suitable form, which avoids the use of differentials of the input data $\psi(\theta, i)$, but which necessitates the generation of an inverse table of data $i(\theta, \psi)$. This is efficiently achieved using quadratic interpolation; at the same time it allows the specification of unequally spaced currents and hence the minimum quantity of input data.

Very good agreement between predicted and measured responses of phase current, average torque, efficiency and static torque under highly saturated conditions has been demonstrated using much reduced input data and requiring only very short computing times.

5 Acknowledgments

The authors are indebted to their colleagues N.N. Fulton and A. Hughes for stimulating discussions, and wish to acknowledge the value of the work of P.T. Blenkinsop which provided the background for the development of this method.

J. Čorda is indebted to the University of Leeds for financial support.

6 References

1 SINGH, G.: 'Mathematical modelling of step-motor'. Proceedings of the incremental motion control system and devices', University of Illinois, USA, pp. 59–148
2 PICKUP, I.E.D., and TIPPING, D.: 'Method for predicting the dynamic response of a variable reluctance stepping motor', *Proc. IEE*, 1973, 120, (7), pp. 757–765
3 PICKUP, I.E.D., and TIPPING, D.: 'Prediction of pull-in rate and settling-time characteristics of a variable-reluctance stepping motor and effect of stator-damping coils on these characteristics', *ibid.*, 1976, 123, (3), pp. 213–219
4 BYRNE, J.V., and O'DWYER, J.B.: 'Saturable variable reluctance machine simulation using exponential function', Proceedings of the international conference on stepping motors and systems, University of Leeds, July 1976, pp. 11–16
5 BLENKINSOP, P.T.: 'A novel, self-commutating, singly-excited motor'. Ph.D. thesis, University of Leeds, Oct. 1976
6 ACARNLEY, P.P.: 'Analysis and improvement of the steady-state performance of variable-reluctance stepping motors'. Ph.D. thesis, University of Leeds, Oct. 1977

Analytical Estimation of the Minimum and Maximum Inductances of a Double-Salient Motor

J. Corda and J.M. Stephenson

Proceedings of the International Conference on Stepping Motors and Systems, University of Leeds, September 1979, pp. 50-59

1. INTRODUCTION

The minimum and maximum inductances corresponding to the two extreme rotor positions, when the axis of excited pole is aligned with the rotor interpolar axis and when it is aligned with the rotor pole axis, are very important parameters in determining the behaviour of doubly-salient motors. This paper presents a simple method for calculating these parameters in terms of the geometric proportions of the machine including the effects of the distribution of the exciting coils. The magnetic configuration of a doubly-salient motor of typical dimensional proportions is considered and the results obtained by this method are compared with those found by numerical field solution and experimentally. An allowance for fringe flux at the ends of the core is made.

2. GENERAL APPROACH

A great deal of work has been done on analytically determining the magnetic permeance and force between toothed structures of stepping motors. Much of this springs from the work of F W Carter[1] based on the Schwarz-Christoffel transformation. Some of Carter's unpublished work has been modified and used as a basis for a very comprehensive numerical analysis of identically double-slotted structures by Mukherji and Neville[2]. This approach has been recently adapted for convenient application to stepping motors by Ward and Lawrenson[3]. Jones'[4] Schwarz-Christoffel transformation, to evaluate permeance and forces, was applicable to identical double-toothed structures where the gap is small compared to the tooth dimensions. Chai[5] has developed permeance formulae based on the assumption of the simple pattern of the field between toothed structures which consists of straight lines segments and concentric circular arcs. It is important to note that Jones' and Chai's results for the permeance are in very good agreement when the air-gap length is small compared to the other dimensions.

However, all the methods above are based on the analysis of rectangular teeth and slots on the stator and rotor and neglect the distribution of the exciting coils on the teeth or assume them to be remote from the teeth. These assumptions may not be justifiable in a doubly-salient motor having excitation coils on the stator poles (teeth) and where profiles of the stator and rotor poles and interpolar spaces are not rectangular.

The agreement of Jones' and Chai's results suggests that the assumption that field lines consist of straight line segments and circular arcs might be used even in the presence of distributed coils where the air-gap length is small compared with the other dimensions. Thus in the method for estimating the minimum inductance described below it is assumed that the field lines conform to this assumption and that the magnetic permeability of the iron is infinite with the flux lines entering the iron surface perpendicularly. The assumption of infinite permeability of the iron is realistic in this position, because the air paths of the field lines are very long and the mmf drop in the iron is small compared to that in the air. However, in the maximum inductance position the air-gap is small and the iron becomes highly saturated. A classical magnetic circuit analysis is therefore used in this position.

magnetic configuration corresponding to the minimum inductance position for a motor of typical dimensional proportions. It is assumed that the field inside the machine is plane-parallel with respect to the plane of the figure and can therefore be considered in two dimensions. The field effects at the ends of the machine (in the third dimension) are considered separately. Since it is assumed that the magnetic permeability of iron is infinite, the magnetic potential of the iron except for the poles of the excited phase are zero. The field lines may be devided into two groups: those which pass to the rotor and those

3. ESTIMATION OF THE MINIMUM INDUCTANCE

Fig 1 shows a sketch of the field pattern of the

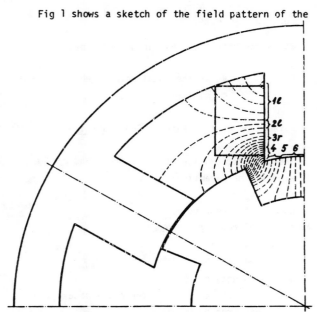

Fig. 1   Sketch of the field pattern in minimum
          inductance position

which do not.   This latter group consists of the
lines which lead from the excited stator pole side to
the stator back of core (path 1ℓ) and lines which lead
from the stator pole side to the adjacent pole (path
2ℓ).   The first group consists of the lines which
lead from the stator pole side to the rotor pole surface (path 3r), lines which lead from the stator pole
side to the rotor pole side (path 4), lines which lead
from the stator pole face to the rotor pole side (path
5) and lines which lead from the stator pole face to
the rotor interpolar surface (path 6).

It should be noted that each of the field lines
in the paths 1ℓ, 2ℓ and 3r is linked with a different
number of turns.   (Such partial linkages have not
been treated in the previously published methods for
determination of inductances of doubly-salient
machines.)   The lines of path 4, 5 and 6 are approximately linked with all the turns and are assumed to
link perfectly.

In order to allow an analytical solution it is
assumed that the conductors are uniformly distributed
over the coil cross-section and the field lines are
approximated as follows: It is clear from the path
length and linked amp-turns that the flux-linkage of
the path 2ℓ is very small in comparison with the flux-

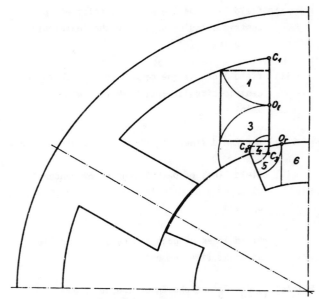

Fig. 2   Approximated flux paths

linkage of the paths 4, 5 and 6.   This also applies
to the flux-linkages of the path 1 and the outer
region of the path 3r, which are small compared with
the flux-linkages of the paths 4, 5, 6 and the inner
region of the path 3r, either because of the amount of
amp-turns or because of the length of the field lines.
Therefore, in calculating the total inductance, a
large error will not be incurred even if a gross
approximation is made to the mapping of the lines of
path 2ℓ.   Thus the path 2ℓ may be associated with
paths 1 and 3r forming together paths 1 and 3
(Fig 2).   The analytical determination of the position of the point O_ℓ is based on the condition that
the ratio of the lengths of the lines from the point
O_ℓ is equal to the ratio of amp-turns linked by them.
The results which are obtained by such rather extensive calculations show that point O_ℓ lies very near
to the mid-point of the pole side.   On the basis of
this argument and the explanations given above about
the values of the flux-linkages in the outer regions
of the paths 1 and 3, the simple assumption that the
point O_ℓ is at the mid-point of the coil side is
justified.   Obviously, these approximations are
valid for calculating the total inductance whereas
for calculating separately the leakage inductance and
the inductance between the stator and the rotor, it
would not be justified.   Fig 2 shows the simplified
field map inside the machine in which it is assumed
that:

i) field lines of the path 1 consist of concentric circular arcs with the centre at the point C_1,

ii) field lines of the path 3 consist of the concentric circular arcs with the centre at the point C_3

iii) the field lines of path 4 are as given below,

iv) field lines of path 5 consist of concentric circular arcs with the centre at the point C_5, and

v) field lines of path 6 consist of parallel straight line segments.

Points C_1, C_3, C_5 are chosen on the basis that

approximated field lines are perpendicular or near to perpendicular at the iron surface and that this field form is preserved when the dimensions of the magnetic configuration are changed. The position of the transition point 0_r is determined similarly to that of the transition point 0_ℓ. (This is shown below.)

Since the magnetic configuration in the minimum inductance position is symmetrical with respect to the pole axis and with respect to the axis perpendicular to it, then flux-linkage of one pole coil is

$$\psi = 2(\psi_1 + \psi_3 + \psi_4 + \psi_5 + \psi_6)$$

where ψ_1, ψ_3, ψ_4, ψ_5, ψ_6 are flux-linkages of the paths 1, 3, 4, 5, 6 respectively. The total inductance of one phase is

$$L_0 = 2 \frac{\psi}{i}$$

because there are two poles per phase each of which has a coil with N/2 turns, where N is number of turns per phase.

$$L_0 = \frac{4}{i} (\psi_1 + \psi_3 + \psi_4 + \psi_5 + \psi_6)$$

$$= 4 (L_1 + L_3 + L_4 + L_5 + L_6)$$

$$= N^2 \mu_0 \ell_F (P_1 + P_3 + P_4 + P_5 + P_6) \qquad (1)$$

$$= N^2 \mu_0 \ell P_0 \qquad (2)$$

where,

$$P_j = L_j / \left[\mu_0 \ell_F \left(\frac{N}{2} \right)^2 \right], \quad (j = 1, 3, 4, 5, 6)$$

ℓ_F is the effective core length (see below),

$$P_0 = \frac{\ell_F}{\ell} \sum_j P_j \text{ is the total normalised equivalent minimum permeance.}$$

3.1 Parameters of the Magnetic Configuration

The magnetic configuration is described by the following set of parameters

number of stator poles N_s,
number of rotor poles N_r,
stator pole arc s,
rotor pole arc r,
air-gap length g,
air-gap length of rotor interpolar space g_i,
rotor diameter d,
stator outside diameter d_0
back iron width c,
core length ℓ,
number of turns per phase N.

Let upper case letters represent normalised dimensional parameters, ie

$$G = \frac{g}{d_0}, \quad G_i = \frac{g_i}{d_0}, \quad D = \frac{d}{d_0}, \quad C = \frac{c}{d_0}, \quad L = \frac{\ell}{d_0}$$

The other angles and dimensions represented in Fig 3 can be represented in terms of the above parameters as follows

$$\phi = \frac{2\pi}{N_r}$$

$$\delta = \frac{2\pi}{N_s}$$

$$\gamma = \text{ang } H_2 C_1 E \simeq \frac{\pi}{2} - \frac{\delta}{2}$$

$$P = \frac{p}{d_0} = \left(\frac{D}{2} + G \right) \sin \frac{s}{2}$$

$$W = \frac{w}{d_0} = \left(\frac{D}{2} + G \right) \tan \frac{\delta}{2} - P$$

$$V = \frac{v}{d_0} = \left(\frac{1}{2} - C - \frac{\frac{D}{2} + G}{\cos \frac{\delta}{2}} \right) \frac{1}{\cos \frac{\delta}{2}}$$

$$Y = \frac{y}{d_0} = \frac{D}{2} + G - \frac{D}{2} \cos \frac{\phi - r}{2}$$

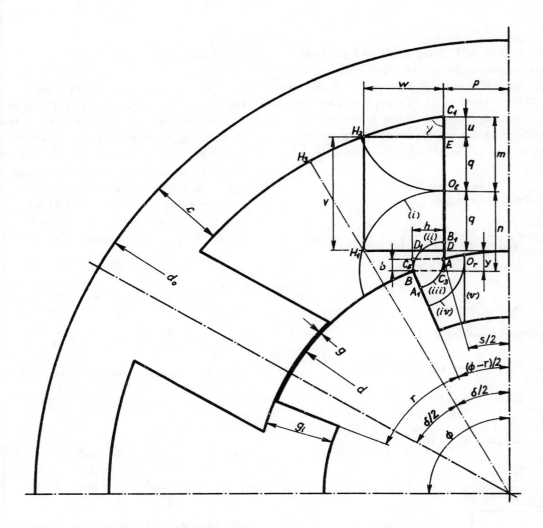

Fig. 3   Dimensional description of magnetic configuration

$$H = \frac{h}{d_0} = \frac{D}{2} \sin \frac{\phi - r}{2} - P$$

$$B = \frac{b}{d_0} = \left(\frac{D}{2} + G\right) \cos \frac{s}{2} - \frac{D}{2} \cos \frac{\phi - r}{2}$$

$$Q = \frac{q}{d_0} = \frac{V}{2}$$

$$U = \frac{u}{d_0} = \frac{W}{\tan \gamma}$$

$$M = \frac{m}{d_0} = U + Q$$

$$N = \frac{n}{d_0} = Y + Q$$

3.2   Calculation of Minimum Inductance Components

a)   Component L_1

Fig 4 shows detail of the magnetic configuration
for flux-path 1 where the broken line represents the
approximated back of core surface.   The flux-linkage
of path 1 is small in comparison with the flux-

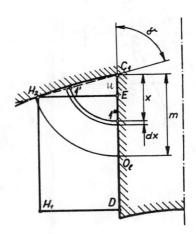

Fig. 4   Detail with flux path 1

147

linkages of paths 3, 4, 5 and 6, because the linked amp-turns become smaller as the field lines are nearer to the corner between the back of core and the pole. This is taken to justify the approximation that in calculating this particular inductance the small empty space between the pole winding and the back of core may be considered filled with pole winding turns.

The elementary flux-path 1'1" is linked with

$$\frac{\text{area } C_1 1'1"C_1}{\text{area } C_1 H_2 H_1 D} \left(\frac{Ni}{2}\right) \simeq \frac{\frac{1}{2} \gamma x \, x}{w v + \frac{1}{2} w u} \left(\frac{Ni}{2}\right) \text{ amp-turns}$$

The field on the elementary flux-path 1'1" is

$$H_x = \frac{1}{\gamma x} \left(\frac{\frac{1}{2} \gamma x \, x}{w v + \frac{1}{2} w u} \left(\frac{Ni}{2}\right) \right)$$

The elementary flux-linkage is

$$d\psi_x = \left(\frac{\frac{1}{2} \gamma x \, x}{w v + \frac{1}{2} w u} \left(\frac{N}{2}\right) \right) \mu_0 H_x \ell_F dx$$

Hence, L_1 is

$$L_1 = \frac{\int_{(o)}^{(m)} d\psi_x}{i} = \mu_0 \ell_F \left(\frac{N}{2}\right)^2 \frac{\gamma \, m^4}{4 \, w^2 (2v + u)^2}$$

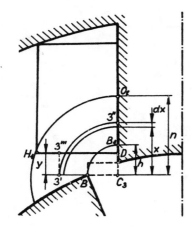

Fig. 5   Detail with flux path 3

or in the form of normalised equivalent permeance

$$P_1 = \frac{L_{1\ell}}{\mu_0 \ell_F \left(\frac{N}{2}\right)^2} = \frac{\gamma \, M^4}{4W^2 (2V + U)^2} \qquad (4)$$

For the machine with proportions shown in Fig 3,

$$P_1 = 0.04 = 0.0185 \sum P_j.$$

b)   Component L_3

Fig 5 shows details of the magnetic configuration for flux-path 3.   The amp-turns linked by field lines decrease and the length of the lines increases the further they are from line BB_1.   Thus the greater the radius of the line, the smaller the contribution of flux-line to the value of inductance component L_3. This reasoning supports the approximation that the rotor pole surface may be represented by a straight line at right angle to the stator pole side through the point C_3.   The elementary flux-path is linked with

$$\frac{wv\text{-area } D3"3"'D}{wv} \left(\frac{Ni}{2}\right)$$

$$\simeq \frac{wv + \text{area } D3"'3'C_3D - \text{area } C_3 3"3'C_3}{wv} \left(\frac{Ni}{2}\right)$$

$$= \frac{wv + xy - \frac{x^2 \pi}{4}}{wv} \left(\frac{Ni}{2}\right) \text{ amp-turns}$$

Using the procedure shown in Section 3.2(a)

$$P_3 = \frac{L_3}{\mu_0 \ell_F \left(\frac{N}{2}\right)^2} = \frac{2}{\pi} \left\{ \ln \frac{N}{H} + \frac{2(N-H)Y}{WV} \right.$$

$$- \frac{(N^2 - H^2)}{4(WV)^2} (\pi WV - 2Y^2) - \frac{(N^3 - H^3)Y\pi}{6(WV)^2}$$

$$\left. + \frac{(N^4 - H^4)\pi^2}{64(WV)^2} \right\} \qquad (5)$$

For the machine with proportions shown in Fig 3,

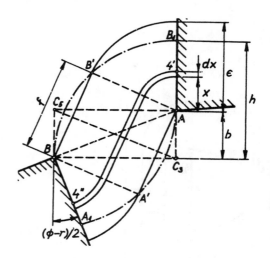

Fig. 6   Detail with flux path 4

$$P_3 = 0.51 = 0.246 \sum P_j.$$

148

c) Component L_4

Fig 6 shows details of the magnetic configuration for flux-path 4 between the semi-broken lines B_1B and AA_1. However, the field lines B_1B and AA_1 are not concentric circular arcs with centres C_3 and C_5 and therefore lines between them cannot be simply represented by concentric circular arcs with centres C_3 and C_5. Therefore path 4 is approximated by the lines which consist of concentric circular arcs with centres A and B and parallel straight line segments which are perpendicular to the diagonal C_3C_5 and of length

$$f \simeq 2b \cos(C_3 \, AA') \simeq 2b \, \frac{\phi-r}{2}$$

The radius

$$e = AB' = A'B = f/\tan(B'AB) \simeq f/\tan(\phi-r)$$

Since area DB_1D_1D (Fig 3) is small in comparison with area DEH_2H_1D, then it may be assumed that all field lines are linked with all amp-turns.

$$P_4 = \frac{L_4}{\mu_0 \ell_F (\frac{N}{2})^2} = \frac{2}{\phi-r} \ln \frac{2\tan(\phi-r) + \pi - (\phi-r)}{2\tan(\phi-r) + \pi - 2(\phi-r)} \qquad (6)$$

For the machine with proportions shown in Fig 3, $P_4 = 0.51 = 0.236 \sum P_j$.

d) Components L_5 and L_6

Fig 7 shows the magnetic configuration of flux-paths 5 and 6. If the field lines of path 5 consist of concentric circular arcs with centre C_5, and the field lines of path 6 consist of parallel straight line segments, then the transition point O_r on the stator pole surface, is determined from the condition mentioned above, ie the length of the circular arc (field line) O_rO_1 is equal to length of straight line segment O_2O_r. The angle which corresponds to the circular arc may be approximately taken to be $\pi/2 - (\phi-r)/2$. Therefore, the condition that $O_rO_1 = O_2O_r$ is

$$t(\frac{\pi}{2} - \frac{\phi-r}{2}) = g_i \quad \therefore \; t = \frac{2 \, g_i}{\pi - (\phi-r)}$$

The field lines of paths 5 and 6 are linked with all amp-turns. Hence

$$P_5 = \frac{L_5}{\mu_0 \ell_F (\frac{N}{2})^2} = \frac{2}{\pi - (\phi-r)} \ln \frac{2 \, G_i}{H[\pi - (\phi-r)]} \qquad (7)$$

$$P_6 = \frac{L_6}{\mu_0 \ell_F (\frac{N}{2})^2} = \frac{P + H}{G_i} - \frac{2}{\pi - (\phi-r)} \qquad (8)$$

For the machine with proportions shown in Fig 3,

$$P_5 = 0.55 = 0.255 \sum P_j \text{ and}$$

$$P_6 = 0.55 = 0.255 \sum P_j.$$

3.3 Fringe Flux at the Ends of Core Effective Core Length

The inductance component due to field at the ends of core of the machine has not been considered in any previous publication on doubly-salient motors. The mathematical treatment of the fringe flux at the ends of core is extremely complicated, particularly for doubly-salient machines, and would require a 3-dimensional numerical field solution for a rigorous

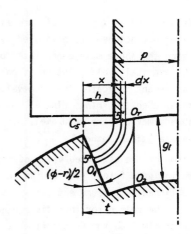

Fig. 7 Detail with flux paths 5 and 6

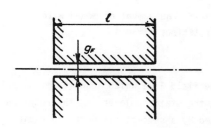

Fig. 8 Axial model of magnetic configuration with fictitious air-gap

treatment. This problem is the subject of continuing investigation but a simple method for making a very rough estimation of the contribution to the minimum inductance by the fringing flux is given below.

In Fig 8 the stator and rotor are represented by a pair of opposing faces with flanks extending to infinity and with a fictitious uniform air-gap, g_F, the length of which is chosen to be equal to the mean length of the lines (i), (ii), (iii), (iv) and (v) (Fig 3) (where the fringing is the most pronounced).

$$g_F = \frac{1}{5}\left[\frac{\pi}{2}n + \frac{\pi}{2}h + (\frac{\pi}{2} - \frac{\phi - r}{2})h + g_i + g_i\right] \quad (9)$$

It is assumed that the flux in the gap is uniform (ie fits the 2-dimensional field solution) to the end of the core and that the fringe flux at the ends of core is linked with all the turns.

The model in Fig 8 is symmetrical about the axis ss and therefore, the permeance of that model is equal to one half of the permeance of the model shown in Fig 9.

In the model in Fig 9 the fringing field is unlimited, ie the problem involves an infinite fringe. Carter[1] uses the Schwarz-Christoffel transformation to obtain an expression for fringeflux (effective core length) emphasizing that it is a matter of judgement how much of the computed fringe flux should be taken as effective. However, Carter's expressions for this case are very complex and are not suitable for simple analytical solution.

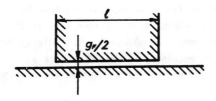

Fig. 9   'Half'-equivament of model with fictitious air-gap

If the field line which leads from the point O_ℓ (Fig 3) on the stator pole end to the rotor end is approximated by a semi-circle with the radius equal to n (see above), then the model in Fig 9 may be approximated by the model shown in Fig 10.

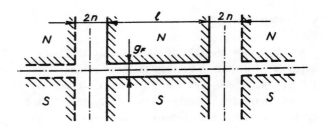

Fig. 10   Approximation of model shown in Fig. 8

Then the effective core length which takes into account fringe flux for this model is

$$\ell_F = (\ell + 2n)\sigma - 2n = \ell + 2n(1-\sigma) \quad (10)$$

where σ is Carter's coefficient

$$\sigma = \frac{2}{\pi}\left\{\arctan(\frac{2n}{g_F}) - \frac{g_F}{4n}\ln\left[1 + (\frac{2n}{g_F})^2\right]\right\} \quad (11)$$

For the machine with proportions shown in Fig 3 and ratio ℓ/d_0 equal to unity $\ell_F/\ell = 1.17$ which means that the flux-linkage at the ends of core at the minimum inductance position is 14.5% of the total flux-linkage.

3.4 Comparison Between Computed and Measured Results

The value of the minimum inductance on actual machine with typical proportions was computed using the method presented above and by the 2-dimensional boundary integral numerical method developed by Trowbridge and Simkin at the SRC Rutherford Laboratory.

When the fringe flux at the ends of core is excluded, the difference between the estimated value of inductance using the above method and that computed by 2-dimensional numerical field solution is only + 3%. Such excellent agreement indicates that the value of minimum inductance, when the fringe flux at the ends of core is excluded, may be very accurately estimated using the simple analytical method.

The comparison of the estimated value of minimum inductance which includes fringe flux at the ends of core with that obtained by measurement shows a difference of -14%. The reason for this discrepancy is the very rough approximation for fringe flux at the ends of core. (If, for example, the value of n in

Fig 10 is doubled then the difference is reduced to
-9%.) If this difference is expressed as a percentage of the inductance component due to the fringe
flux* it is 50%, ie the estimated value of inductance
component due to fringe flux is half the measured
value. Using this empirically obtained factor the
corrected value of minimum inductance is

$$L_o = L_{2D} + 2L_F = L_{2D} + 2(\frac{\ell_F}{\ell} L_{2D} - L_{2D})$$

$$= L_{2D} (2 \frac{\ell_F}{\ell} - 1) \qquad (12)$$

where L_{2D} is the estimated value of minimum inductance when the fringe flux is excluded and
L_F is the estimated value of inductance component due to fringe flux.

Work is continuing on this problem of estimating
the contribution of the fringing field and it is
hoped to have available soon results of a
3-dimensional numerical field solution.

4. ESTIMATION OF THE MAXIMUM INDUCTANCE USING B-H CURVE

In the maximum inductance position the mmf drop
in the iron is considerable, even in the case when
the iron is not saturated, in comparison with mmf
drop in the air-gap (the air-gap length is small
compared with the rotor diameter). When the iron
becomes saturated the mmf drop in the iron increases
nonlinearly and may exceed the mmf drop in the air-
gap. Therefore the mmf drop in the iron and the nonlinearity of B-H curve of iron must be taken into
account for the estimation of the maximum inductance.

To estimate the maximum inductance L_i by a
simple analytical method the following assumptions
are made:

i) When a phase winding is excited the magnetic
circuit is treated as a simple '2-pole'
pattern (Fig 11);

ii) There is no flux leakage, ie all flux
passes from the stator to the rotor and back;

---

* The inductance component due to fringe flux can be
obtained as the difference between the measured
value of inductance and the one obtained by
2-dimensional numerical field solution

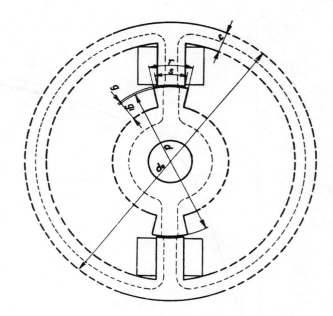

Fig. 11   Simple '2-pole' pattern

iii) The flux is linked with all the turns;

iv) The flux is uniformly distributed in the
cross-section normal to the field lines.

In the maximum inductance position the field
pattern is symmetrical about the axis of the excited
phase and therefore only one half of the magnetic
circuit, which carries one half of the total flux,
need be considered. Further simplification is to
split this half of the magnetic circuit into the following parts connected in the series:

two stator poles (subscript 's') with
cross-section: $a_s = \frac{d}{2} (\sin \frac{s}{2})\ell$
length   : $\ell_s = 2 (\frac{d_o}{2} - c - \frac{d}{2} - g)$

two air-gaps (subscript 'g') with
cross-section*: $a_g = [\frac{1}{2} (\frac{D}{2} + g)s + (1 - \sigma)i]\ell$
where   : $i = \frac{1}{2} \frac{D}{2} (r-s)$
                 $\sigma = \frac{2}{\pi} [\text{arc } \tan(\frac{i}{g}) - \frac{g}{2i} \ln (1 + (\frac{i}{g})^2)]$
length   : $\ell_g = 2g$

---

* Due to fringing effects the effective cross-section
in the air-gap is bigger than the stator pole
cross-section. To make some allowance for this
the air-gap cross-section is increased by introducing Carter's coefficient

151

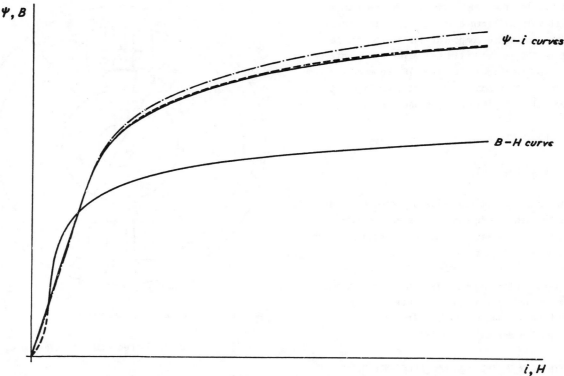

Fig. 12   B-H curve and ψ-i curve in maximum inductance position

——————— measured
— — — — computed by numerical field solution
——·—·— estimated by simple analytical method

two rotor poles (subscript 'r') with

cross-section**: $a_r = a_g$
length   : $\ell_r = 2g_i$

rotor body (subscript 'b') with

cross-section : $a_b = (\frac{d}{2} - g_i)\ell$
length   : $\ell_b = \frac{1}{2}(\frac{d}{2} - g_i)\pi$

stator yoke (subscript 'y') with
cross-section : $a_y = c\ell$
length   : $\ell_y = \frac{1}{2}(d_o - c)\pi$

(The numerical results are related to the machine with the parameters given above.)

The mmf equation is

$$Ni = \frac{B_g}{\mu_o}\ell_g + H_s\ell_s + H_r\ell_r + H_b\ell_b + H_y\ell_y$$

** It is assumed that the rotor pole cross-section is constant along the rotor pole and equal to effective cross-section in the air-gap (a_g)

Having given the value of flux-linkage, the values of flux densities B_g, B_s, B_r, B_b and B_y can be calculated ($B = \frac{\psi}{Na}$) and using the B-H curve of the appropriate steel the values for H_s, H_r, H_b and H_y can be found. Using the mmf equation the value of current corresponding to a given value of flux-linkage can be found and hence the maximum inductance if $L_i = \frac{\psi}{i}$. The normalised equivalent maximum permeance is given by

$$P_i = \frac{L_i}{\mu_o \ell N^2}$$

Bearing in mind that the above simple analytical method treats a very simplified field pattern, the error which is within 5% in the highly saturated region of ψ-i curve (in terms of inductance) is very good (Fig 12).   In the nonsaturated region the estimated and measured results are very close. Extremely good agreement between the measured ψ-i curve of the maximum inductance position and the one obtained by using 2-dimensional numerical field solution shows that the fringe flux at the ends of core is negligible compared with the flux within machine. It is very interesting to note that the fringe flux at the ends of core in the maximum inductance position is smaller (in absolute value) than the one in the minimum inductance position.

5. CONCLUSIONS

A method for the analytical estimation of the minimum and maximum inductances has been given which is valid for realistic saturation in the iron. The method is sufficiently simple to allow it to be programmed on a programmable calculator.

The estimation of the minimum inductance is based on the assumption that the field lines consist of circular arcs and straight line segments. The distribution of the winding is allowed for. The estimated value of minimum inductance, when the fringe flux at the ends of core is excluded, agrees very well (within 3%) with the result obtained by 2-dimensional numerical field solution. An allowance based on a rough approximation for fringe flux at the ends of core has been made. The estimated value of total minimum inductance is 14% lower than measured value. An empirical formula for the corrected value of minimum inductance has been suggested.

In the maximum inductance position the machine is treated as a simple 2-pole pattern and it is assumed that the flux is linked with all turns and there is no flux leakage. The B-H curve of the iron is used in calculating the maximum inductance. The difference between measured and estimated values of the maximum inductance is within 5% in the saturated region of the ψ-i curve and is insignificant in the nonsaturated region.

6. ACKNOWLEDGEMENTS

The authors wish to thank Dr N N Fulton and other colleagues in the Department of Electrical and Electronic Engineering of the University of Leeds for helpful discussions and in the SRC Rutherford Laboratory for their assistance in producing numerical field solutions. J Čorda is indebted to the University for scholarship support.

7. REFERENCES

1. F W Carter; 'The magnetic field of the dynamo-electric machine', Jour IEE, 1926, pp 1115-1138

2. K C Mukherji, S Neville; 'Magnetic permeance of identical double slotting', Proc IEE, Vol 118, No 9, Sept 1971, pp 1257-1268

3. P A Ward and P J Lawrenson; 'Magnetic permeance of doubly-salient air-gaps', Proc IEE, Vol 126, No 6, June 1977, pp 542-543

4. A L Jones; 'Permeance model and reluctance force between toothed structures', Proc of 5nd symposium on incremental motion control system and devices', University of Illinois, May 1976, paper H

5. H D Chai; 'Permeance model and reluctance force between toothed structures', Proc of the 2nd symposium on incremental motion control system and devices, University of Illinois, May 1973, paper K

Saturable Variable Reluctance Machine Simulation Using Exponential Functions

J.V. Byrne and J.B. O'Dwyer

Proceedings of the International Conference on Stepping Motors and Systems, University of Leeds, July 1976, pp. 11-16

1. INTRODUCTION

This paper relates to predicting the dynamic behaviour of a V.R. motor described by magnetization data ψ (i,θ) and circuit resistance.

The method of solution of the equations is that described by Pickup and Tipping[1], but the function ψ (i,θ) is stored, not as polynomials, but as sums of exponentials fitted to the test data by a variation of standard least-squares analysis.

This simulation method has been applied, after difficulties were found with polynomials, to a 6 kW power step motor[2] characterized by a highly non-linear set of magnetization curves which do not have symmetry with respect to angular displacements about the maximum permeance position.

2. REPRESENTING THE INDIVIDUAL MAGNETIZATION CURVES BY EXPONENTIAL FUNCTIONS

2.0 Advantages of exponential functions

The advantages have been recognized[3,4] in the context of the B-H characteristic. They include mathematical simplicity, with ease of differentiation and integration, the fact that exponentials are a "natural" fit to typical magnetization curves and reliably model the derivatives as well as the curves.

2.1 Form of Function used, and fitting methods.

Three exponential terms and a linear term were used, the latter taking care of, inter alia, the linear machine characteristic in the maximum reluctance position.

$$\psi = a_o i + a_1 \left\{ 1 - \exp(-\alpha_1 i) \right\}$$
$$+ a_2 \left\{ 1 - \exp(-\alpha_2 i) \right\}$$
$$+ a_3 \left\{ 1 - \exp(-\alpha_3 i) \right\} \qquad (1)$$

where ψ is flux-linkage and i is current.

Fig. 1 shows the character of the data to be fitted.

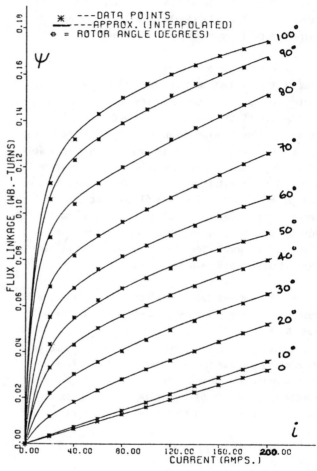

Fig. 1. Magnetization Characteristic of Motor

Various curve fitting techniques were considered. The simplest of these, discussed by MacFadyen et al.,[3] has the disadvantage that only one point is read from the magnetization curve for each coefficient to be evaluated. Therefore all measured data would not be utilized unless a very large number of terms were taken. For this reason the method was not considered appropriate.

The first attempt at fitting used a non-linear least squares analysis to obtain each of the coefficients a_0, a_1, a_2, a_3, α_1, α_2, and α_3. The aim was to explore the family of curves to find the range of exponential "frequencies" α which would be needed. Considerable time was spent in setting up these equations but finally it was discovered that it would be practically impossible to get a satisfactory solution. The reason for this is the unorthogonal nature of the exponential functions. This means that there are many minima of the sum of the squares and hence, in order to get a stable iteration scheme it is necessary to choose initial estimates very close to the final results. In fact a solution was found for only one angle.

With this in mind, and because many different sets of co-efficients were found to give a curve of the required accuracy, it was decided to fix the "α" coefficients and determine the "a" coefficients of Eqn. 1. In this case a linear least squares analysis can be used[5].

2.2 Least squares fitting of exponential series

In standard least-squares analysis the coefficients 'a' of Eqn. 1 are given by

$$\bar{a} = \left\{ Z^T Z \right\}^{-1} Z^T \bar{\psi} \qquad (2)$$

where $\bar{\psi}$ is the vector of flux-linkage data and the Z matrix is

$$\begin{bmatrix} i_1 & 1 - e^{-\alpha_1 i_1} & \cdots & 1 - e^{-\alpha_3 i_1} \\ i_2 & 1 - e^{-\alpha_1 i_2} & \cdots & 1 - e^{-\alpha_3 i_2} \\ \vdots & \vdots & & \vdots \\ i_n & 1 - e^{-\alpha_1 i_n} & \cdots & 1 - e^{-\alpha_3 i_n} \end{bmatrix} \quad (3)$$

i_1, i_2, i_3 . . . constitute the vector of current data.

This standard approach gave accuracy of 3% or better in the range 10-100% of maximum current but gave errors of up to 80% in initial slope. To impose more strongly on the solution the influence of data points near the origin, two modifications, in addition to the addition of dummy data points, were made.

The initial slope of the curve at zero current was inserted as additional data. This required that one extra row be added to the Z matrix, Eqn. 3. It is given by the elements

$$\frac{\partial}{\partial a_i} \left(\frac{\partial \psi}{\partial i} \right)_{i=0} ,$$

which from Eqn. 1 are

$$1 \qquad \alpha_1 \qquad \alpha_2 \qquad \alpha_3$$

The curve fitting of Fig. 1 was done in this way, with the same set of α throughout.

2.3 "Least squares of fractional error" fit

This second modification was directed at minimizing the sum of the fractional errors squared as distinct from the actual errors. The function to be minimized is

$$\sum_{u=1}^{n} \left[\frac{\psi_u - f(i, a)}{\psi_u} \right]^2 ,$$

which gives as solution for the 'a' coefficients, in place of Eqn. 2.

$$\bar{a} = \left\{ \left[\frac{Z}{\psi} \right]^T \cdot \left[\frac{Z}{\psi} \right] \right\}^{-1} \left[\frac{Z}{\psi} \right]^T \left[I \right] . \quad (4)$$

Z as before is given by Eqn. 3, and $\left[\frac{Z}{\psi} \right]$ is obtained by dividing each element of the Z matrix by the appropriate ψ value such that

$$\left[\frac{Z}{\psi} \right]_{ij} = \frac{Z_{ij}}{\psi_i} .$$

The final I in Eqn. 4 is a vector of ones.

The machine actually simulated had more evenly-spaced ψ-i curves than those of Fig. 1, and the data was extended to 400A. The following table shows how the modified fit (Eqn. 4) gave better percentage error than the standard fit (Eqn. 2) and how the results varied with the choice of alphas.

ALPHAS			Average % Error (whole field)	
			Eqn. 2	Eqn. 4
.1	.05	.01	2.22	1.68
.2	.08	.01	3.45	2.08
.3	.1	.03	3.39	1.83
.4	.1	.04	3.86	1.84

The maximum error in initial slope was below 2.5% for the last set of alphas.

The results show that acceptable values could be obtained for a fairly wide range of "α" coefficients. The most critical coefficients seemed to be the most rapidly rising ones. Values of .2 or .3 gave the best results for this. The other two coefficients were much less critical. But if these two were selected too close together interactions between the two take place and both components fluctuate simultaneously for changes in angle. Also, if the slowest rising component was less than about 0.005, interactions between it and the linear term take place in a similar manner. The actual coefficients used were 0.3 0.1 0.03.

3. REPRESENTATION OF THE FAMILY OF MAGNETIZATION CURVES

The curve fitting technique outlined in section 2 gave a set of four 'a' coefficients for each of 37 angles spanning 180 degrees at which the ψ/i curve was obtained. A means is required of representing the magnetization characteristic for intermediate θ (angle) values. Polynomials up to the 5th order were tried without success to fit the 'a' coefficients to θ.

Finally, instead of fitting the family of curves, two simple linear interpolation schemes were applied to give intermediate values of $\frac{\partial \psi}{\partial i}$, $\frac{\partial \psi}{\partial \theta}$ and $\frac{\partial W}{\partial \theta}$, the actual quantities appearing in equations 5 and 6.

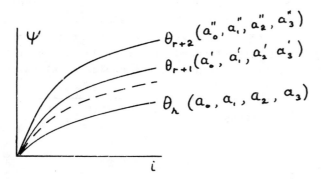

Fig. 2. Interpolation for ψ and $\frac{\partial \psi}{\partial i}$

In Fig. 2 the intermediate curve is taken to have 'a' coefficients.

$$a_i(\theta) = \frac{\theta - \theta_r}{\theta_{r+1} - \theta_r} a_i(\theta_{r+1}) + \frac{\theta_{r+1} - \theta}{\theta_{r+1} - \theta_r} a_i(\theta_r),$$

which is used to give, from Eqn. 1,

$$\frac{\partial \psi}{\partial i} = a_o + a_1 \alpha_1 e^{-\alpha_1 i} + a_2 \alpha_2 e^{-\alpha_2 i} + a_3 \alpha_3 e^{-\alpha_3 i}$$

Such a linear interpolation of ψ between measured curves means that constant values of $\frac{\partial \psi}{\partial \theta}$ and $\frac{\partial W'}{\partial \theta}$ are generated, leading to steps in both of those functions. The effect was reduced to a ramp transition by another application of linear interpolation: the value of $\frac{\partial \psi}{\partial \theta}$ and $\frac{\partial W'}{\partial \theta}$ calculated for the interval between measurements was taken to hold at the centre of that interval, and the values for intermediate points were calculated using interpolations between the centre points of adjacent intervals.

4. MACHINE EQUATIONS

These are, in the form required for numerical solution[1],

$$\frac{di}{d\theta} = \left\{ v - ri - \omega \frac{\partial \psi}{\partial \theta} \right\} \frac{1}{\partial \psi / \partial i} \qquad (5)$$

$$\frac{d\theta}{dt} = \omega$$

$$\frac{d\omega}{dt} = (1/J) \left\{ T(i,\theta) - T_{mech} \right\}$$

$$T(i,\theta) = \left. \frac{\partial W'}{\partial \theta} \right|_{i\ constant} \qquad (6)$$

$$W' = \int_0^i \psi \, di'.$$

W' is the co-energy and i' is a dummy variable. Eddy currents are neglected.

It was of interest to investigate steady state operation as a power motor with windings sequentially connected to a d.c. voltage source and sink. Then, in Eqn. 5, ω is constant and v is a function of time or angle.

Eqn. 5 was solved by Runge-Kutta integration procedure. Flux-linkage and torque (From Eqn. 6) were obtained by the interpolation scheme of section 3. The mechanical work in a complete cycle of the variables was computed from the trajectory area in the ψ-i plane[2]

$$W_{mech} = \oint i \, d\psi \qquad (7)$$

as well as from the torque-angle integral.

Copper losses, and average power at source and sink terminals were also calculated.

5. DESCRIPTION OF POWER V-R MOTOR

With appropriate design, saturation in V-R motors gives augmented torque and specific output.[6,7,8]. In particular, torque approaches the value

$$T = i \frac{\partial \psi(i,\theta)}{\partial \theta} \,,$$

varying directly as i, and of magnitude twice that of a linear device.

The 2-phase motor of Fig. 3 was laid out to exploit this effect in an economical brushless variable-speed drive system.

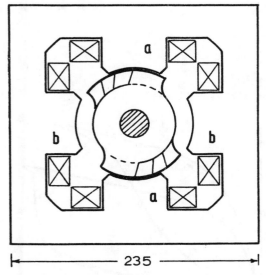

Fig. 3 Power V-R Motor

The rotor has elongated poles with graded iron density in the leading half. Excitation of one stator phase gives a step length of 90 degrees. The phase windings require to be energised sequentially with unidirectional currents, with provision for reversing only the voltage.

The iron is dimensioned to saturate only in the vicinity of the pole overlap zone. There is therefore a displacement-dependent constriction in the flux path. The air-gap is made as small as mechanical tolerances will allow. The magnetization characteristic

in consequence approaches that of a variable saturable reactor rather than that of a variable inductor.

6. SIMULATED AND MEASURED RESULTS

The machine windings were switched by thyristors to a variable d.c. voltage source and by diodes to a variable d.c. voltage sink. The switching instant, in relation to shaft angle, was adjustable. One of the objects of the simulation was to investigate how "ignition advance" affects the shape and enclosed area (representing work) of the operating trajectory in the ψ-1 plane.

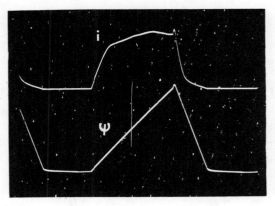

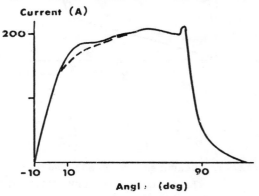

Fig. 4. Measured Current and Flux-linkage.
Simulated current waveform.
2088 RPM, 20/-40V, Ignition - 10°

Figs. 4, 5 compare computed and measured quantities for the test condition 2088 RPM, source/sink voltages 20/40V, ignition advanced 10° on maximum reluctance position, source "on-time" 90 degrees. The blips seen are associated with a commutation capacitor, whose voltage, as well as that of the semiconductor switching elements, was taken into account in Eqn. 5.

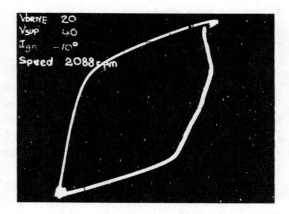

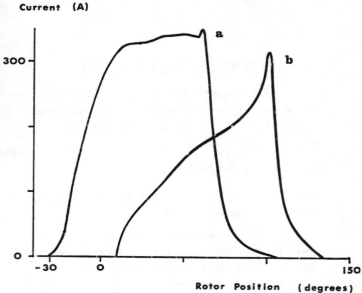

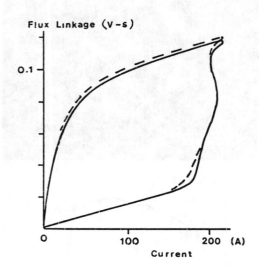

Fig. 5. Measured and Predicted ψ-i
Trajectory. 2088 RPM, 20/-40V,
Ignition -10°

Many such comparisions were made, up
to 8 kW mechanical output and 4000 RPM. The
only systematic deviation, found with
ignition advanced about 30 degrees, was in the
initial rate of rise of current. In these
cases the poles were overlapping at switch-on,
and the computed initial di/dt was too low.
The effect is visible in the trajectory 'a'
of Fig. 6, where the measured current is
shown dashed. It is thought that eddy
currents, neglected in the analysis, were
responsible for the deviation.

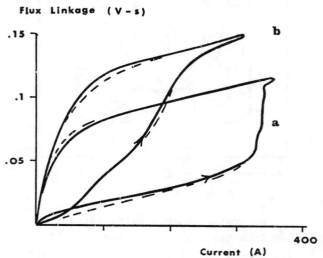

Fig. 6. Predicted Current Waveforms and
ψ-i trajectories.
(a) 2900 RPM 27/-45V, Ignition Advanced 30°
(b) 2181 RPM 25/-55V, Ignition Retarded 10°

158

The waveforms 'a' of Fig. 6 show the effect of adequate "ignition advance" in squaring up the trajectory and the current-time waveform. With over-retarded ignition ('b') the current rises throughout the working stroke and starts accelerating away, from generator action, just before commutation to the sink is effected.

The following extract from computer print-out shows the effectiveness of the interpolation scheme for torque.

Speed	2960 RPM
Null voltage angle	20°
Ignition advance	10°
Drive voltage	25.8
Suppression voltage	46.0

Angle	Current	Torque
59	195.4	11.84
62	193.8	11.68
65	191.9	10.90
68	191.3	10.33
71	182.7	10.26
74	160.0	8.92
77	139.6	7.57
80	120.9	6.32

Power from source	3662.2 watts
Power to sink	394.7
Power output (from Torque)	2601.5
" " (from $\int id\psi$)	2599.9
Copper loss	131.1
Semi-conductor device loss	538
Overall efficiency	79.61%

7. REFERENCES

1. Pickup, I.E.D.: "Method of predicting the dynamic response of a variable-reluctance stepping motor", Proc.IEE, 1973, 120, pp. 757-765.

2. Byrne, J.V. and Lacy, J.G.: "Characteristics of saturable stepper and reluctance motors", IEE Conference on Small Electrical Machines, London, 30-31 March 1976.

3. MacFadyen, W.K. et al.: "Representation of magnetization curves by exponential series", Proc. IEE, 1973, 120, pp.902-904.

4. El-Sherbiny, J.K.: "Representation of the magnetization characteristic by a sum of exponentials", IEEE Trans. on Magnetics, 1973, MAG-9, pp. 60-61.

5. Draper and Smith: "Applied Regression Analysis", Wiley, 1966.

6. Byrne, J.V.: "Forces in saturable electromechanical devices", Ph.D. Dissertation, University College Dublin1973.

7. Byrne, J.V. : "Tangential forces in overlapped pole geometries incorporating ideally saturable material", IEEE Trans. on Mag. 1972, MAG 8, pp. 2-9.

8. British Patent No. 1 321 110.

Limitations of Reluctance Torque in Doubly-Salient Structures

M.R. Harris, V. Andjargholi, A. Hughes, P.J. Lawrenson, and B. Ertan

Proceedings of the International Conference on Stepping Motors and Systems,
University of Leeds, July 1974, pp. 158-168

1. INTRODUCTION

1.1 General

This paper is concerned with the problem of determining torque-current and limiting-torque characteristics for a motor which essentially utilizes reluctance-type forces produced between teeth on both stator and rotor. A wide range of variable-reluctance and hybrid stepping motors produce torque by this mechanism, and their performance can be assessed with the aid of this theory - as, for example, in the companion paper, ref., 6. However, one type of stepping motor (common in the smaller sizes) employs a permanent-magnet rotor which in principle can be (and often is) cylindrical, and such motors require a different, though related, theory which will not be given here.

In an earlier examination of the general problem[7], a number of features have been pointed out which will briefly re-stated and expended before proceeding to a new theoretical formulation. The first, and perhaps most important, is that the machines in question are typically designed to work with heavily saturated regions of iron ; these regions occur in any teeth of the stator and rotor, the heads of which overlap across the airgap at a particular angular rotor position.

To put the matter another way, the airgap in typical designs is sized so that rated current produces an m.m.f. between stator and rotor which greatly exceeds that needed to drive a saturation level of flux density

across the airgap only. (This statement is less true for motors of the smallest size, in which - as usual - heating considerations severely limit the m.m.f. that can be designed into the machine.) The gap size is usually set at a minimum consistent with mechanical convenience.

This trend, which appears to have evolved empirically, is easily explained : more torque is produced by a motor of given tooth number, tooth/slot ratio and rated m.m.f., when the airgap is reduced so that heavy saturation results, than when it is large enough for a simple linear theory to be applicable. By "linear theory" will be meant hereafter one in which saturation of the iron regions is ignored and the profile of the stator and rotor teeth is treated as a magnetic equi-potential in analysing the airgap field pattern. Such an approach is exemplified in the work of Mukherjee and Neville[2], who present accurately calculated permeance functions for aligned and unaligned tooth positions respectively, for a range of relative proportions of tooth, slot and gap (hereafter referred to as "gap geometry"). A main purpose of this paper will be to show that, perhaps surprisingly, use can still be made of such linear permeance data in heavily saturated conditions to provide a good prediction of characteristics in respect of average torque. Before this can be done, however, some extension of existing theoretical treatment is needed, and this will be covered in section 2.

It follows from what has been said so far

that analyses aiming to predict torque/rotor-angle characteristics by linear approaches based on harmonic components of permeance variation (e.g. refs. 4 and 5), whilst useful, must be treated with caution. An indication of how far the saturated field-pattern associated with two partially overlapped teeth departs from that which a linear theory would predict is provided by Byrne[3]. Ref. 3 also provides an explanation of why the saturated machine produces more torque than the comparable unsaturated one. It is not the purpose of this paper to discuss that matter in detail, but in energetic terms it may be viewed as springing from the fact that in a saturated system the magnetic co-energy can greatly exceed the stored energy, whilst in the linear condition energy and co-energy are equal.

In the present work attention will be focussed, not on torque-angle characteristics, but only on average torque - i.e. the average value exerted over the full angular movement of the rotor for which the torque is of one sign ; the development of the theory demands this restriction of scope. However, for many calculations a knowledge of the average torque is of great assistance. Moreover a wide range of practical machines have a quite similar torque-angle characteristic, so that it is not difficult for example to form a fair estimate of the peak (or "holding") torque from a knowledge of the average torque.

In the next section of this introduction the "fundamental machine" - i.e. the basic machine geometry with which this paper is concerned - will be defined. After that, the results that are already known to follow from a linear theory will be critically reviewed, to provide a background for the present work. A non-linear form of analysis is then developed in section 2.

1.2 Fundamental machine

The following geometry is described as "fundamental", since it characterises a basic and simple machine, with which the torque

characteristics of any stepping motor may be usefully compared. The fundamental machine is a simple torque motor with no stepping capability, and so practical stepping motors must always differ from it with some consequent loss of torque. Their torque-current characteristics can however be related to those of a fundamental machine, using the concepts of "utilization factor" and "excitation factor" (see ref., 6).

The fundamental machine has the gap geometry already analysed by Mukherjee and Neville[2], and Chai[5]. It is as follows :

Both stator and rotor have N teeth of equal angular pitch. An m.m.f. F is applied between them, of constant sign, i.e. homopolar. The airgap is considered in equivalent "unrolled" form, so that stator teeth are parallel-sided and infinitely deep in respect of the calculation of (linear) airgap permeance. In both stator and rotor, tooth width = t, slot width = s, and slot pitch = λ, where

$$\lambda = t + s \qquad (1)$$

The airgap between any two overlapping stator and rotor teeth = g.

The relative position of stator and rotor in which teeth are perfectly aligned (stable, zero torque) will be termed the "in" (i) position, and the perfectly unaligned situation (tooth opposite slot, unstable, zero torque) the "out" (o) position. In the following it is to be understood that the term "permeance" relates to the results of linear theory. The total machine permeances in the two positions, P_i and P_o, are related to the corresponding permeance functions per tooth pitch of ref. 2, P_1 and P_2, as follows :

$$P_i = \mu_o N P_1 L \qquad (2)$$
$$P_o = \mu_o N P_2 L \qquad (3)$$

where L is the active core length.

The radius at the airgap (taken to be negligibly different for stator and rotor) is R, so that

$$\lambda N = 2\pi R \qquad (4)$$

and we may define the active rotor volume

$$V_r = \pi R^2 L \qquad (5)$$

1.3 Linear theory and its main results

The aim here is to review briefly, and generally without proof, the main results that follow from a simple linear analysis for average torque $\overline{T}$, (the average value exerted between the i and o positions). The main discrepancies between these results and those found in practice for saturated machines are then identified.

It is relatively straightforward[2,7] to express $\overline{T}$ in terms of the gap geometry and permeance functions, and so obtain the result

$$\overline{T} = \frac{2 V_r B^2}{\mu_o} \frac{(P_1 - P_2)}{(\lambda/g)^2} \qquad (6)$$

The flux density B is the value occurring between any two well-overlapped teeth, i.e.

$$B = \frac{\mu_o F}{g} \qquad (7)$$

Saturation can now be accounted for after a fashion by putting $B = B_s$, a limiting value which could arguably be taken somewhere between 1.7 and 2.1 T. We thus obtain an expression for the maximum value of average torque that the fundamental machine can produce, $\overline{T}'$, where

$$\overline{T}' = \frac{2V_r B_s^2}{\mu_o} \left\{ \frac{(P_1 - P_2)}{(\lambda/g)^2} \right\}_{max} \qquad (8)$$

According to this theory, the maximum torque is produced when the gap geometry is so proportioned as to maximise the bracketed function which we will denote F_1 ; i.e.

$$F_1 = \frac{P_1 - P_2}{(\lambda/g)^2} \qquad (9)$$

The behaviour of F_1 as t, s and g, are varied is analysed in ref., 2.

The main results that can be obtained from this general approach are now summarised and compared with those found in practice.

(1) According to eqn. 8 once F is sufficient to drive B close to B_s, then the torque developed should become constant and independent

162

of the slot number N. Experimental results in section 2 however will show that torque continues to increase rapidly with current well beyond the point where the teeth are strongly saturated in the i position. Also, as can be seen from data in ref., 6, there can be a correlation between torque and tooth number at certain levels of excitation.

(2) As the maximum possible value[2] of F_1 is $F_1' = 0.0195$, a value for $\overline{T}'$ can be found by substituting $B_s = 2.1$ T (say) and F_1' into eqn., 8. The result is

$$\frac{\overline{T}'}{V_r} = 1.37 \times 10^5 \quad \text{n-m/m}^3 \qquad (10)*$$

Experiment tends to show that this limiting value of specific torque (i.e. per unit rotor volume) is of the right order, though perhaps slightly low. On the other hand the optimum gap geometry that is implied is misleading (see point 3) and the theory leading to eqn. 10 is open to some objection (point 4).

(3) The highest value of F_1 (F_1') occurs[2] when $\lambda/g = 8.05$ and $t/\lambda = 0.38$. Practical stepping motors are commonly designed with small airgaps, so that λ/g is of the order of 50-100 and t/λ is in the range 0.35 - 0.40. According to linear theory it should be possible to increase $\overline{T}'$ for such machines simply by increasing the airgap, so bringing λ/g towards the value 8.05. In fact (as shown in section 3.3) maximum torque reduces as the gap increases.

Whilst the predicted value of t/λ appears to be of the right order, it must be appreciated that it is only obtained analytically in combination with the λ/g ratio of 8.05. If λ/g is to be as high as 50, the theory predicts that the best t/λ ratio would be very nearly 0.5 ; this conflicts with common design practice in modern machines.

(4) Some criticism of eqn., 8 has been made[7] on the grounds that whilst it holds the highest

---

* 72 gf-cm   = 1 oz-in.

  98.1 micro n-m = 1 gf-cm.

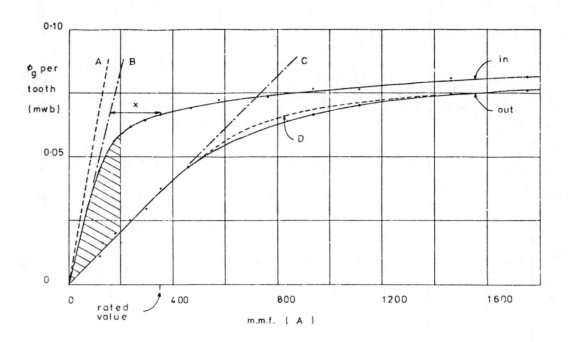

Fig 1   Flux per tooth vs m.m.f.

flux-density in the gap to a value B_s, the flux-density in the body of the tooth can rise to unrealistically high values due to accumulation of flux within the tooth from the tooth-sides. A modified linear theory (although still not totally satisfactory) avoids this difficulty and gives a different function to be maximised, F_2, where

$$F_2 = \frac{P_1 - P_2}{P_1^2} \left(\frac{t}{\lambda}\right)^2 \qquad (11)$$

However, the value of $\overline{T}'$ predicted by this theory is only about half that of eqn., 10, and is substantially different from that found experimentally.

In summary, whilst a linear theory is capable of indicating very broadly the correct order of magnitude of limiting specific torque, it has slight success in predicting the optimum gap geometry. Moreover it proves to be of little use in predicting torque-current characteristics - although that matter has not been demonstrated here.

2.   NON-LINEAR THEORY
2.1   General

To introduce an improved non-linear theory we may look at fig. 1, which shows experimental results obtained with a machine specified in section 3.1 which is similar in nature to the fundamental machine. The results are obtained from search coils wound on rotor teeth, and show the relationship between gap flux per tooth and excitation m.m.f. for the i and o positions respectively.

At sufficiently low m.m.f. the flux-m.m.f. relationship is linear ; the extrapolated line C is indistinguishable from that calculated by linear permeance theory[2] for the o position, whilst for the i position the extrapolated line B differs slightly from the calculated permeance line, A. This discrepancy is almost certainly due to difficulty in measuring the true effective value of g, however, and we may assume that the slope of the tangents B and C corresponds with the linear permeance values in the i and o positions respectively.

163

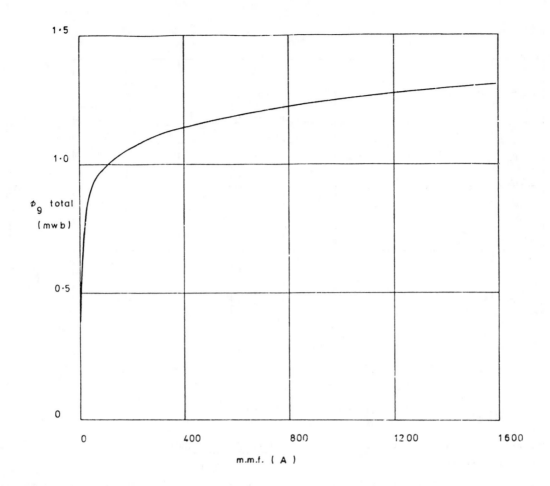

Fig 2   Total flux vs 'magnetising m.m.f.' ('x' in Fig 1)

It will be seen that within the range of rated m.m.f. the i-curve becomes markedly non-linear, but the o-curve remains linear. This variation of saturation-level between the two positions at a given current is characteristic of the general problem.

A well-established[1] and basic result for non-linear systems (involving no approximation) relates the average torque exerted between two positions to the change in system co-energy at constant current. Thus, if A is the shaded area (in wb - A) on fig. 1, mutliplied by N to represent the whole machine, it can be shown that the average torque at an m.m.f. of 200 A is given by

$$\overline{T} \;=\; \frac{N\,A}{\pi} \tag{12}$$

and similarly for any other level of m.m.f.

164

2.2 Theory

For almost all relative positions of stator and rotor, it is evident that the field patterns in the gap and slot areas alter markedly as the level of exciting m.m.f. increases into the heavily saturating range. Careful consideration shows however that this statement is less true in respect of the particular (symmetrical) i and o positions. A justification for this statement will not be detailed, but it is shown below to be supported by experimental evidence.

It is therefore postulated that in the i and o positions (and only in these) the problem can be modelled to a quite good approximation as follows. The total m.m.f. F is broken into two components : the first component (F_g) acts on an airgap which retains its linear permeance values

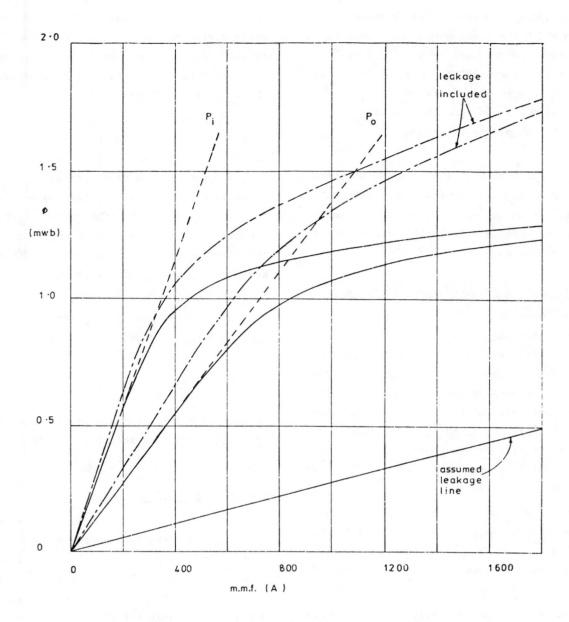

Fig 3   Theoretical diagram of flux vs m.m.f. for increased airgap

P_i and P_o ; and the second component (F_m) is absorbed in driving flux through the iron paths which complete the magnetic circuit.   We shall term F_g and F_m the "gap" and "magnetising" components respectively.

Thus, in fig. 1, the intercept x gives the values F_m at any particular level of gap flux, and this is re-plotted in fig. 2 to give the curve of total gap flux ϕ_g against F_m.

According to this model, having derived the ϕ_g - F_m curve from the experimental ϕ_g - F

curve for the i position, it should now be possible to predict the ϕ_g - F curve for the o position from the relationship

$$F = F_g + F_m$$
$$= \frac{\phi_g}{P_o} + F_m \qquad (13)$$

where F_m is to be evaluated at the level ϕ_g on fig., 2.   This provides an acid test of the validity of the theoretical assumptions.

Referring back to fig. 1, comparision between theory and experiment is seen to be remarkably good. The broken line D shows the extent to which the theoretical curve departs from the experimental one. Predictions of $\bar{T}$ based on this theory will be in error to the extent of the area enclosed between D and the experimental o - curve, and this is small compared to the whole area between the i and o curves.

In the present work the relationship between F_m and ϕ_g is obtained from the experimental i - curve. In subsequent work we shall examine methods by which this relationship can be theoretically predicted.

2.3 Prediction of torque-current curves and limiting torque value

The method for predicting torque-current curves is as follows. It is described with reference to fig. 3, which is a theoretical diagram of the ϕ_g - F curves in the i and o positions for the experimental machine, assuming the airgap to be enlarged by grinding. Thus the calculated P_i and P_o values are modified compared to fig. 1, but it is assumed that the ϕ_g - F_m curve is still as shown in fig. 2.

First the P_i and P_o lines are drawn, and then intercepts of magnetising mmf are added to each line at appropriate flux levels. This procedure yields the predicted ϕ_g - F curves for and o positions (solid lines).

To determine the theoretical average torque at a particular current it is necessary to find the co-energy area bounded by the two curves and vertical line at that current. (This may be done graphically or by computing.) Torque is then evalutated from equation 12.

In section 3.3 predicted torque-current curves for two different airgap sizes are compared with experimental results, and the comparison found to be encouraging.

Note that equation 12 is equally valid in relation to the co-energy area between curves of total flux (including leakage) versus m.m.f. An arbitrary leakage line is drawn on fig. 3, and

this used to modify the ϕ_g - F curves to the chain-dotted curves. These are intended simply to show the nature of the ϕ - F characteristics that could have been obtained by measurement at the machine terminals.

The theory outlined so far can also be used to predict the limiting (highest) value of average torque, $\bar{T}'$. Suppose that it can be said to a good approximation that saturation ultimately prevents ϕ_g exceeding a certain value ϕ_s. It is then easy to show that the total area between the ϕ_g - F curves for i and o positions (which is the co-energy area appropriate to infinite current) is equal to the area enclosed in the triangle bounded by the P_i, P_o and ϕ_s lines.

If the gap geometry is such that this triangular area is maximised, the maximum value being $\hat{A}$, then

$$\bar{T}' = \frac{N\hat{A}}{\pi} \qquad (14)$$

In section 4 this approach is developed into a new and interesting theory for optimum gap geometry.

2.4 Secondary effects

The fundamental machine is a theoretical concept, and there are certain differences between it and the practical machine used for the experimental work. These differences are listed here, but no attempt will be made to assess the magnitude of their effect on the torque characteristics.

Field end-effects

The test machine is heteropolar, having four reversals of m.m.f. around the gap circumference. whereas the fundamental machine is homopolar. This disturbs the field pattern surrounding some of the teeth in the machine. (The usual field fringing at the two axial ends of the core - though less important - must be included here.)

Finite slot depth

It is difficult to be sure whether the slots are effectively infinitely deep at the higher levels of test current.

166

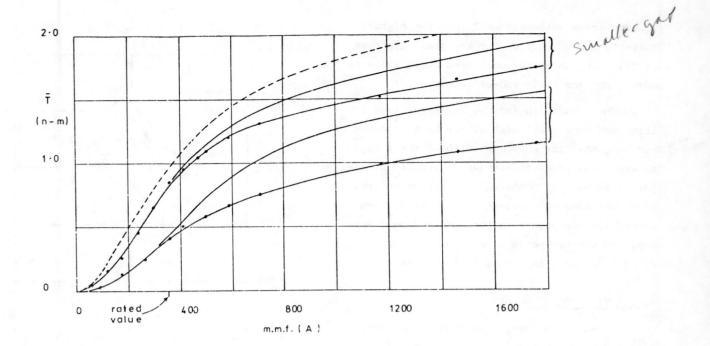

smaller gap

Fig 4 Average torque vs m.m.f.

Tooth taper

The teeth of the experimental machine taper slightly in the manner which is common in stepping-motor design, and is thought to provide some increase of torque. The fundamental machine, however, has parallel-sided teeth. The taper certainly enables the flux accumulated at the root of the tooth to exceed by 10% that which would be possible if its section were the same as the tooth head, when forced at very high current levels.

3. TORQUE-CURRENT CURVES

3.1 Specification of experimental machine

Tests were performed on one stack of a multi-stack motor with the following specification:

$N_s = N_r = 16$; $L = 1.43$ cm; $R = 1.5$ cm; $V_r = 10.1$ cm^3; $g = 0.0114$ cm (small-gap test), 0.0313 cm (large-gap tests); $t/\lambda = 0.42$; m.m.f. at rated current = 350A.

3.2 Flux-current characteristics

These characteristics, shown in fig. 1,

have already been discussed. They were obtained by search coils wound at the root of the rotor teeth, embracing several teeth but avoiding regions in which field end-effects are likely to modify the results. The small gap dimension quoted in section 3.1 is the apparent value obtained from the tangent-line B.

3.3 Torque-current characteristics

Tests were performed, measuring static torque over a range of displacement angles of the rotor and excitation current. From these, values of $\bar{T}$ were calculated at different currents, and these are plotted as the experimental curves of fig. 4. The higher curve is for the small gap, and the lower for the large gap.

Bracketed with each curve is the corresponding prediction from the (independent) experimental evidence of section 3.2, using the theory and method outlined in section 2.3. Note that these tests have been carried to very high current levels. The broken line is the prediction based on the measured (not apparent) airgap, and is not thought to be of great interest.

Up to rated current, predicted and measured

167

torques agree extremely well. At the highest
current, the prediction for the small-gap case
is 11% high, which is considered good, allowing
some error for field end-effects.

The prediction for the large-gap case is
less good (about 35% high at the highest current)
but note that the curves still have the right
nature. The predicted torque is reduced as the
gap is increased in contrast to linear theory.
It is not clear at present why the predictive
method is less good with the much enlarged air-
gap, and whether perhaps certain secondary
effects are more important in this case.

4. LIMITING TORQUE VALUE

We will now develop an analytical expression for $\bar{T}'$ from equation 14, in terms of permeance factors and gap geometry.

Let ϕ_s be an ultimate value of ϕ_g, assumed
to represent total saturation. It is then easy
to show that

$$\hat{A} = \frac{\phi_s^2}{2} \left(\frac{1}{P_o} - \frac{1}{P_i} \right) \tag{15}$$

We can relate ϕ_s to B_s, a saturated value of B,
where

$$\phi_s = 2\pi RL \left(\frac{t}{\lambda} \right) B_s \tag{16}$$

Bringing together the results of equations 2,3,
5,15 and 16, and substituting into 14 we finally
obtain

$$\bar{T}' = \frac{2V_r B_s^2}{\mu} \left(\frac{t}{\lambda} \right)^2 \left(\frac{1}{P_2} - \frac{1}{P_1} \right) \tag{17}$$

It is instructive to compare this equation with
equation 8 for the linear theory.

There is now a third function for which
an optimum can be sought, ie.

$$F_3 = \left(\frac{t}{\lambda} \right)^2 \left(\frac{1}{P_2} - \frac{1}{P_1} \right) \tag{18}$$

The behaviour of this function with variation of
gap geometry is plotted in fig. 5, and the result
is striking. At all values of λ/g, F_3 reaches
its maximum at $t/\lambda = 0.42$ (by coincidence

168

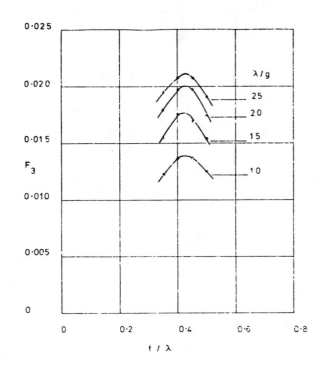

Fig 5 Variation of function F_3 with t/λ and
λ/g

precisely the designed value for the experimental machine). Moreover, as λ/g increases
(ie. g is made smaller at fixed λ) there is a
steady increase in F_3 and therefore $\bar{T}$. These
results compare well with established empirical
knowledge.

Finally, substituting into equation 17
$B_s = 2.1$ and $F_3 = 0.024$ (an approximate value
corresponding to $\lambda/g = 50$) an expression is
obtained for limiting specific torque

$$\bar{T}'/V_r = 1.69 \times 10^5 \quad n\text{-}m/m^3 \tag{19}$$

Equation 19 provides an asymptote for the small-
gap curve of fig. 4 of 1.7 n-m, and it is evident
that the measured $\bar{T}$ slightly exceeds this value.
Tooth taper is the probable explanation for this;
one would expect some increase of $\bar{T}$ to be associated with the slightly greater flux that the
tapered tooth is known to permit.

5. COMPARISON OF DIFFERENT MOTORS

The purpose of this section is to compare

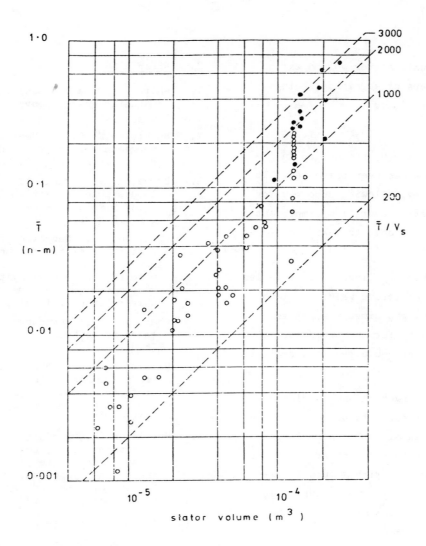

Fig 6   Average torque vs stator volume for doubly salient motors

o plain v.r.,   ● hybrid

the theoretical estimate of $\bar{T}'/V_r$ with data obtained from a wide range of manufactured doubly-salient motors. This requires that calculations for theoretical torque per unit rotor volume be converted into equivalent estimates of torque per unit stator volume. This matter will not be examined in detail, but a fairly careful study suggests that

$$(\bar{T}/V_s)_{\substack{\text{practical} \\ \text{motor}}} = K(\bar{T}'/V_r)_{\substack{\text{fundamental} \\ \text{machine}}} \quad (20)$$

where $1/40 \geqslant K \geqslant 1/80$

The range of K covers doubly-salient motors of various type. Equation 20 provides an expression for the torque per unit stator volume that can be achieved in a high-performance motor; it does not imply that all motors will achieve it.

From equation 19, it follows that no doubly salient motor should be able to produce a specific torque (referred to stator volume) greater than about 4000 n-m/m^3. The display of data in fig. 6 confirms this conclusion. (Figures for average torque are generally not given by manufacturers. The values used in fig. 6 are estimated, usually from holding-torque data by assuming that the peak/average ratio of the torque-angle curve is the same as that for a sine-wave.)

It is interesting to see that the values of

average torque per unit stator volume extend over nearly two orders of magnitude, being lower at the smaller sizes.

6. CONCLUSIONS

A new theory of torque production in doubly-salient motors has been presented. It is a non-linear theory, allowing properly for the effects of magnetic saturation (important in stepping motors) but able to make use of linear permeance data.

It is better than any linear theory in its prediction of limiting torque value, optimum t/λ ratio and optimum gap dimension. Its qualitative predictions for the dependence of $\bar{T}$ on excitation and gap geometry are correct in all respects.

Torque-current curves can be predicted with good accuracy at currents up to about the rated value. With an airgap of normal size, the agreement remains fairly good at six times rated current. However, with a much increased airgap and at six times rated current, the predicted torque is 35% higher than the measured value. The reasons for this discrepancy are not yet clear, but it may be that the slots are no longer effectively infinitely deep with the enlarged gap.

The general approach offers a much improved analytical model, and better physical understanding of the basic mechanism reluctance torque in strongly saturated doubly-salient devices.

7. REFERENCES

1. White, D. C. and Woodson, H. H., "Electro-magnetic energy conversion", Wiley 1959, Chapter 1.

2. Mukherjee, K. C. and Neville, S., "Magnetic permeance of identical double slotting", Proc. I.E.E., 118, 9, Sept. 1971, p 1257.

3. Byrne, J. V., "Tangential forces in over-lapped pole geometries incorporating ideally saturable material", Trans. I.E.E.E., Mag-8, No. 1, March 1972, p 2.

4. Chai, H. D., "Magnetic circuit and formulation of static torque for single-stack permanent-magnet and variable-reluctance motors", 2nd Annual Symposium, Incremental Motion Control Systems and Devices, University of Illinois, 1973 paper E.

5. Chai, H. D., "Permeance model and reluctance force between toothed structures", Ibid, paper K.

6. Harris, M. R. and Andjargholi, V., "Torque production in the hybrid motor", Conference Proceedings, Stepping Motors and Systems, University of Leeds, July, 1974.

7. Harris, M. R., Andjargholi, V., Hughes, A., Lawrenson, P. J., Ertan, B., "Limitations on specific torque in stepping motors", 3rd Annual Symposium, Incremental Motion Control Systems and Devices, University of Illinois, May 1974.

Note   Inclusion of this paper in these proceedings does not preclude submission elsewhere in extended and modified form.

Switched Reluctance Motor Excitation Current: Scope for Improvement

J.W. Finch, H.M.B. Metwally and M.R. Harris

Power Electronics and Variable-Speed Drives Conference, London, 1986
IEE Conference Publication No. 264, pp. 196-199

Summary

The optimum current waveform of a switched reluctance motor drive can be computed using an efficient numerical technique. A full nonlinear model of the motor magnetic circuit is used, with torque per rms ampere of excitation current being taken as the optimisation criterion. An idealised 50/50 square wave of current yields a convenient reference standard of torque per ampere. Results are presented for a prototype 2-teeth per pole motor. Simple voltage switched excitation is shown to give a high standard of performance. Possible improvements are discussed, relative to the necessary driving voltage requirements.

INTRODUCTION

The 'switched reluctance motor' (SRM) can take various forms but typically has construction geometry similar to a variable-reluctance (VR) stepping motor, with single or multiple teeth per stator pole and with each pole excited by a single coil as Lawrenson et al (1) and the present authors (2) have described. Pairs of stator coils are connected to form a phase winding, with the whole 3- or 4-phase machine winding being excited from an electronically switched inverter.

The SRM requires only unipolar current and is suited to operation on switched voltage supply. Very satisfactory performance can thus result from use of an inverter circuit which is simpler than that required for an induction motor drive as discussed by Davis et al (3).

Although simple square-wave voltage excitation gives good SRM operation, it is natural to ask what the optimum current waveform would be to yield the highest return in terms of torque. Such an optimum exists and can be expressed in analytical form, but is in fact most simply found by the use of widely available numerical optimisation routines. The approach is applicable to any SRM, working at any stipulated rms current, and takes full account of magnetic non-linearity (saturation). The answer provides a valuable measure, against which one may assess the effectiveness of any particular excitation scheme, including the usual simple switched voltage supply. It represents an absolute upper limit, for a given design of motor, to the improvement possible with any advanced waveform shaping techniques.

OPTIMISATION PROCEDURE

Performance criterion - torque per ampere

Several criteria for judgement or objective functions exist for a SRM, depending on the drive characteristics that are important for a particular application. Torque per rms ampere of excitation current (T/I) is chosen here, as it indicates the effectiveness of torque production within the motor, for given ohmic loss. Optimisation of this ratio, at fixed rms current, is in practice close to finding a true minimum of total motor losses for given output, and hence to maximising efficiency. Highest efficiency is the most generally applicable goal, in system design.

A convenient reference standard of torque per ampere is provided by considering an idealised 50/50 square wave of current, with the same rms value as that under consideration. This square wave is assumed to switch on and off at successive zeros of the per-phase static torque/angle diagram. It has been previously used in the development of a design method for SRMs (2), which is an extension of an established method for stepping motors developed by some of the present authors (4,5) that has now been used in industrial design for some years. The value of T/I for the square wave can be easily found, as a function of rms current. Both the theoretical optimum current waveform, and any practical excitation waveform, can then be related to this 50/50 wave through a dimensionless coefficient K_r which also forms part of the SRM design method (2) and a simple measure of waveform improvement thus established.

K_r may be slightly greater than unity, by up to perhaps 20%, and this reflects the improved running torque that can be obtained with a current of practical waveform, compared to the idealised 50/50 wave of equal rms value. This favourable excitation condition is always due to the practical current waveform being 'peakier' than the square wave, and the peak coinciding with the most torque-productive part (in torque per amp terms) of the torque/angle curve. For any excitation wave, K_r is an important indicator of its effectiveness in terms of torque per ampere.

Determination of optimum waveform

Saturation effects in a practical machine will cause the optimum current waveform to vary with its rms value. The chosen numerical method seeks the optimum waveform subject to the constraint of fixed rms current, and the rms value is then varied in steps to determine the corresponding waveform variation. To perform the optimisation the range of rotor displacement angle yielding positive motoring torque (0° to 180°E) is subdivided into a series of N equal segments. The current is assumed to have a constant value within each segment. Large values of N closely define the waveshape, but increase the computational effort. The value N=10 has been found satisfactory for most purposes.

EXPERIMENTAL SRM

The motor used for all the results quoted later is a prototype 6-pole 12/10 slotted SRM and has been described elsewhere (2). It was constructed in a standard TEFV D100 metric frame, with the same stack length and gap as an 8-pole induction motor. Measured torque/angle curves at three currents are shown in Fig. 1. Other SRMs may of course have significantly different characteristics (and correspondingly differing optimum current waveforms) but these are reasonably representative of typical gap geometry and of saturation influence. The general trends in behaviour discussed later are consequently widely applicable, both to single-tooth (1) and multi-tooth-per-pole designs.

OPTIMUM WAVEFORM

Current variations

Once the reliability of the computation and necessary level of discretisation were established, as described by the present authors (6) all the results were then obtained for fixed N = 10 and a range of values of rms current. Fig. 2 shows the optimum waveform at 3 A, a current close to a possible continuously rated value for this motor, without piped ventilation, at 750 rpm. These results are summarised in Fig. 3, which shows both optimum torque per ampere and T/I calculated for the 50/50 wave of equal rms value. The ratio $(T/I)_{opt}/(T/I)_{50/50}$ is plotted in Fig. 4, and this gives in effect the largest value of K_r that is theoretically possible at given current, = $\hat{K}_r$.

Optimum T/I reaches a peak of 7.3 Nm/A at an rms current of about 2.9 A, corresponding to a computed load torque of 21 Nm. The curve of T/I for the 50/50 wave has similar shape, believed characteristic of all SRMs. K_r ranges from almost 1.5 at low rms currents to a value approaching unity at high currents. This fall occurs because as the rms current increases, the optimum current waveform broadens, approaching more closely the square wave.

For this SRM, T/I peaks in the neighbourhood of rated operating condition. This is in fact true assuming any constant, reasonably torque-effective current waveform. Moreoever the T/I curve typically has a rather broad peak. It is believed that the same can be said of most SRMs and that it is a valuable natural design feature.

Also shown in Fig. 3 are the experimental values of T/I obtained with simple switched voltage excitation with on-times of 33% and 40%. These yield values for K_r of 1.05 and 0.95 respectively. This demonstrates how closely similar practical waveforms usually are, in terms of torque per ampere effectiveness, to the 50/50 square wave, for which of course K_r = 1, and brings out the usefulness of the 50/50 wave as a comparative standard in design work. The computed and experimental results show good agreement in T/I. Experimental values of running torque tend to be a little lower than the computations would predict, typically by about 5%, and this is ascribed to unbalance between phases, measurement error, and other second order effects.

Required driving voltage

From the outset of this work on optimum current, its main purpose was seen as demonstrating that an optimum current waveform existed, and deriving this optimum so that it might serve as a standard of assessment for practical waveforms. It was not expected that the optimum would be practically desirable, because its demands on peak voltage would considerably increase the cost of the excitation system. The required driving voltage waveform has been estimated via a smoothing and differentiating process. Its most important features are two pulses of large magnitude and short duration: a first positive pulse giving rapid current build-up, and a second (lower) negative pulse to commutate the current. The peak voltage required at 3 A and 750 rpm is 4 kV, with about -2.5 kV for commutation over a slightly longer time. These voltages can be compared with the 600 V (approx) rail voltage of a square wave inverter, suitable to drive the experimental motor in rated condition at 750 rpm.

SWITCHED-VOLTAGE EXCITATION

Base operating point

In this section, the effectiveness of simple switched-voltage supply for the experimental 12/10 SRM is briefly discussed by considering one operating condition specified (by simulation) as follows. 33% on-time, 600 V dc 480 V rms, 2.35 A rms, 750 rpm, switch-on angle = 186°E before alignment, 15 Nm output torque. This equals the 8-pole IM rating for this frame. (The SRM can in fact be thermally rated at higher torques, including the 4-pole IM rating of 20 Nm, for which it retains good performance characteristics.)

The experimental value of T/I is 6.25, comparing closely with a value of 6.39 obtained by simulation in which due allowance is made for both friction torque and an additional current component due to core loss. It is simpler to make comparisons with the optimum waveform without this friction and core-loss allowance, as core loss with the optimum waveform is unknown. At an electromagnetic torque of 15.2 Nm the switched-voltage value of T/I then becomes 6.66 and the optimum 7.1. Thus, switched-voltage excitation is only 6.2% less effective than the optimum in current utilisation for the same torque. However, at longer percentage on-times, a decrease in T/I typically occurs and there may be somewhat greater potential for improvement as discussed below. At the operating point specified above the current ratio (switched-voltage : optimum) is 1.066, implying an ohmic loss ratio of 1.136. The higher current for the same torque might produce an extra temperature rise equivalent to 3% increase of resistance (at the most) so that ohmic loss would be 17% greater for switched-voltage waveform, compared with the optimum.

Control variables - percentage on-time

Variation of on-time is one possible means for controlling an SR drive. Limiting the discussion to simple switched voltage excitation of an existing motor at a given speed, three variables remain for use by the control system. These are : dc link voltage within the converter, the rotor position angle at which excitation is applied to the motor windings (switch-on angle, here specified relative to the aligned position), and the rotor angle at which commutation is initiated. Sometimes it is convenient to specify the latter by the difference between the two angles expressed as a percentage of the whole switching cycle (i.e. the percentage on-time). Over an often surprising range of operating conditions a fixed percentage on-time, typically in the region of 33%, usually offers a high standard of performance. With variable dc link voltage this simple control can offer constant torque characteristics over a range of speeds around the chosen base. If the voltage is fixed for reasons of converter economy, characteristics rather akin to those of the series connected dc motor are given, suiting many traction applications.

If the dc voltage is fixed and constant torque characteristics are needed over a moderate speed range around the base value, percentage on-time can be varied.

It is a notable feature of SR drives that this simple control usually provides good performance over the constant-torque speed range. However, it is of interest to consider what variation in performance quality does occur. Fig. 5 effectively shows this, illustrating how T/I changes with percentage on-time relative to the 33% operating condition discussed earlier. These results were actually obtained, for convenience, at fixed speed by varying the dc voltage. This is roughly equivalent to operating at variable speed with constant torque and voltage, though the second order effects of friction and core loss variation are surpressed.

It can clearly be seen in Fig. 5 that variations in T/I are not rapid, reaching just over 8% at 45% on-time. The variations from 25-37% on-time are less than 2%. However, the value of T/I at 45% on-time gives a current ratio of 1.087 to the nominal (1.159 cf optimal), which would imply 34% more ohmic loss at this on-time than the optimal wave, even without a further correction for increased resistance with the higher loss. This larger loss appears to offer much greater potential for improvement than the nominal case discussed earlier, but investigations suggest that higher voltage from the converter is required for any substantial improvement. The higher voltage, if available, can be used to good effect simply to shorten the current pulse - thereby effectively returning the operating condition to a shorter on-time.

CONCLUSIONS

There exists a theoretical optimum current waveform for any SRM, which can be numerically determined. This optimum waveform sets a theoretical upper limit to the potential for improvement of torque at stipulated ohmic loss, in any given SRM, which is a valuable reference in system design.

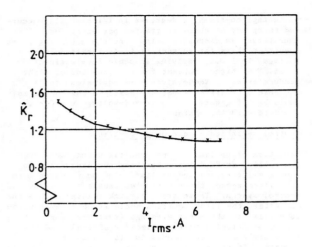

Figure 4   Variation of T/I ratio, optimum : 50/50 square wave, with current.

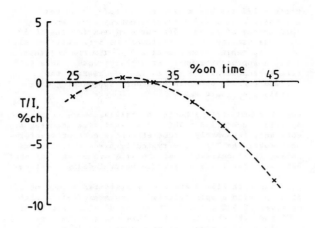

Figure 5   Percentage change in T/I with on-time, relative to 33% value.

A particular prototype small SRM has been studied, but the general nature of the results obtained is believed to be relevant to a wide range of machines. Simple switched-voltage excitation with 33% on-time was found to give about 17% more ohmic loss in one condition (corresponding to rated 8-pole IM torque) than the optimum. Longer on-times give more apparent scope for improvement, but exploitation of this difference would require higher driving voltage. In fact the voltage waveform required to drive the optimum current involves short-duration peak voltages of about six times the switched-voltage supply rail. Unless substantial over-voltages are available, the scope for improvement of current waveform is restricted.

ACKNOWLEDGEMENTS

This work forms part of a larger project supported by the U.K. Science and Engineering Research Council. Thanks are due to GEC Small Machines for technical advice and the gift of experimental machine parts. NEI Electronics provided valuable assistance with inverters. Grant support from the Egyptian Government to one of the authors (HMBM) is also acknowledged.

REFERENCES

1. Lawrenson, P.J., Stephenson, J.M., Blenkinsop, P.T., Corda, J., and Fulton, N.N., 1980, IEE Proc. B EPA, 127, (4), pp 253-265.

2. Finch, J.W., Harris, M.R., Metwally, H.M.B., and Musoke, A., 1985, Proc. IEE Conf. Electrical Machines - Design and Applications, London, pp 134-138.

3. Davis, R.M., Ray, W.F., and Blake, R.J., 1981, IEE Proc. B EPA, 128, (2), pp 126-136.

4. Harris, M.R., Andjarargholi, V., Lawrenson, P.J., Hughes, A., and Ertan, B., 1977, Proc. IEE, 124, (12), pp 1215-1224.

5. Harris, M.R. and Finch, J.W., 1979, Proc. 8th Symp. Incr. Motion Control Sys. & Devices, pp 293-306.

6. Finch, J.W., Metwally, H.M.B., and Harris, M.R., 1986, paper submitted IEE Proc. B EPA.

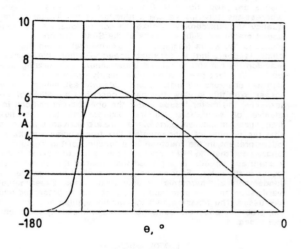

Figure 2   Optimum current waveform at rms current of 3 A, smoothed by curve fitting from discretised version, N = 10.

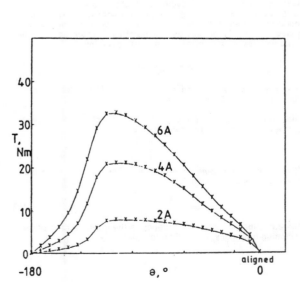

Figure 1   Measured torque/angle curves at three currents.

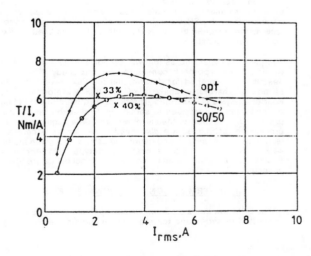

Figure 3   Torque per ampere for optimum and 50/50 square wave, variation with current.

Estimation of Switched Reluctance Motor Losses

P. Materu and R. Krishnan

Dept. of Electrical Engineering, Virginia Polytechnic Institute and State University, Blacksburg, VA 24061
IEEE-IAS Proceedings, Pittsburgh, PA, October 1988

Abstract

The losses in a switched reluctance motor (SRM) consist mainly of core losses and stator copper losses. The copper losses are proportional to the square of the r.m.s. current whereas the core losses are a function of the excitation frequency and flux density. Unlike the conventional a.c. variable speed machines, different parts of the SRM core are not subjected to the same frequency of flux reversals when the machine is operating at constant speed. Further, the current waveform is not sinusoidal and is dependent on operating conditions. The fact that this machine mostly operates in varying degrees of saturation further complicates the estimation of core losses. With this in view, this paper describes a method of determining the motor losses under the operating conditions inclusive of saturation. General expressions for the flux frequencies in different parts of the core are derived. It is shown that the number of (fundamental) flux frequencies simultaneously present in the machine is a function of the number of phases and that the SRM core losses are a function of excitation current and speed. It is proved that most of the core losses occur in the rotor. Fourier analysis is applied to separate the fundamental and harmonic components of losses. Experimental verification on a 6/4 pole laboratory prototype is presented and discussed. The measurement techniques applied and some of the problems encountered in the course of this study are briefly described.

I. INTRODUCTION

The losses in a switched reluctance motor (SRM) mainly comprise of iron and copper losses. Estimation of core losses is complicated by the fact that the frequency of flux reversals is different for different parts of the core. The complexity increases as the number of phases increases. An approximate but simple procedure for estimating these losses has been developed and is described in this paper. First, a general procedure for determining the flux waveforms for any pole combination is presented. Fourier analysis is then used to extract the a.c. components of flux. The core losses corresponding to each component are then determined. It is shown that a number of harmonics are significantly present especially in the rotor. It is also shown that the rotor losses are significantly higher than stator core losses at all frequencies. Data acquisition and processing techniques are used in determining the r.m.s. and average values of the phase currents. These methods are described and compared to d.c. meter measurements. A computer program for estimating the motor losses and other steady-state parameters of the SRM has been developed. A 6/4 pole prototype motor designed by one of the authors [2] was used for experimental verification.

The paper is organized as follows: section I gives a brief introduction to the approach considered for this study. Section II describes the method for determining the flux waveforms. In section III, harmonic analysis of the flux waveforms using Fourier analysis is described. The procedure used in predicting motor losses is described in section IV. Simulation results are presented in section V. Measurement and data acquisition and processing techniques are discussed in section VI. A comparison between measured and estimated data is also given in this section. A 6/4 pole 3 h.p. laboratory prototype SRM (Appendix 1) was used for experimental verification. Conclusions are given in section VII.

II. DETERMINATION OF FLUX WAVEFORMS

In the SRM, windings on diametrically opposite stator poles are connected in series and therefore simultaneously excited. Fig.

1 shows the basic circuit of the SRM drive. The voltage equation for one phase is given by

$$V = R\,i + \frac{d\Phi}{dt} \tag{1}$$

where V is the d.c. supply voltage, R is the stator winding resistance per phase and Φ is the flux linkages.

Since the resistive drop is much smaller than the supply voltage above 0.1 p.u. speed, the rate of change of flux linkages with respect to time can be considered to be approximately linear. The flux can therefore be approximated by triangular waveforms, positive slope corresponding to the time when the phase winding is excited and negative slope corresponding to

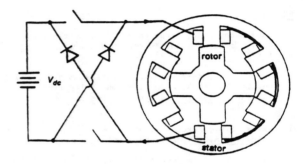

Fig. 1. Basic circuit of the SRM drive (6/4 pole)

the regeneration period during which a negative voltage equal to the supply voltage is applied across the phase winding. Since only the stator winding is excited, the flux waveforms in the parts of the core are determined by the switching sequence in the stator. The normal switching sequence for an m-phase motor is to excite the phases successively, that is, 1,2,3,...,m-1,m; 1,2,3..etc. In this case, each commutation step constitutes one working stroke and each cycle constitutes a switching period, T_s. Since a phase winding is excited whenever a rotor pole tends to alignment, the number of cycles (periods) for a stator phase per revolution equals the number of rotor poles. The switching period per phase is therefore given by

$$T_s = \frac{2\pi}{\omega_r\,N_R} = \frac{\lambda_r}{\omega_r} \tag{2}$$

where ω_r is angular speed of the rotor, N_R is number of rotor poles and λ_R is rotor pole pitch.

The stroke period T_w depends on the number of of phases. Each phase is excited once per cycle so that

$$T_w = \frac{T_s}{m} \tag{3}$$

Thus for any pole combination, the flux frequency in the stator poles is determined from T_s as

$$f_{ph} = \frac{1}{T_s} = \frac{\omega_r\,N_R}{2\pi} \tag{4}$$

The d.c. supply switching frequency is then m-times the phase switching frequency and is therefore given by

$$f_s = \frac{m \, \omega_r \, N_R}{2\pi} \qquad (5)$$

These two frequencies provide the basis for constructing the flux waveforms in all parts of the core.

It has been observed earlier that the period and form of the flux variations in the core differs from part to part [1]. Since the exciting currents are usually unipolar, the flux variations in the stator poles are unidirectional. This is however not the case in the other parts of the core. The flux waveforms in these parts are examined below.

A. Flux waveforms in the stator core

During commutation (switching off one phase and switching on another), only some sections of the stator back iron experience flux reversal. Thus, except for the 4/2 pole combination, the stator core flux in the SRM consists of multiple frequencies, the number of frequencies depending on the number of phases. Neglecting leakage effects, the flux waveforms in the stator core for an m-phase SRM can be constructed from the stator pole flux waveform. Table 1 shows the flux polarities in the 2m stator core segments of a general m-phase SRM for one switching period. This sequence is repeated in every period. A segment is defined as the part of the stator core between two adjacent stator poles. A transition from plus to minus represents reversal of flux whereas a plus (or a minus) represents a unipolar triangular flux pulse. An illustration for a 6/4 pole SRM is given in Fig. 2.

It is observed that the flux, in the segments bounded by the poles corresponding to phases 1 and 2, is always unipolar irrespective of the number of phases. The flux waveform in these segments is obtained by a direct sum of the stator pole flux waveforms. Note that the stator pole flux waveforms of adjacent phases (poles) are phase-shifted by one stroke period T_w. An illustration for a 6/4 pole SRM is given in Fig. 3 (a) where Φ_{s1}, Φ_{s2} and Φ_{s3} represent the flux linkages of stator phases 1, 2 and 3 respectively, and Φ_{b1} represents the flux linkages in the stator core segments with non-reversing flux (type 1 segments). The resulting waveform has a large d.c. component with a low amplitude a.c. ripple of frequency f_s. The amplitude of the ripple depends on the conduction angle and the level of excitation.

The other segments in the stator core experience one reversal in each switching period. However, these reversals do not occur at the same time. Thus each reversal constitutes a different frequency. There will therefore be an additional m-2 flux frequencies in the back iron. Again, the corresponding flux waveform is constructed by direct summation of the stator phase waveforms with the appropriate polarities and phase. Fig. 3(b) illustrates this for a 6/4 pole SRM where Φ_{b2} represents the flux linkages in the stator core segments with bipolar flux (type 2 segments). Since these flux waveforms are bidirectional, they are responsible for most of the iron losses in the stator. The number of different flux frequencies occuring in the stator core of a m-phase SRM can be obtained using [2]

$$c = \text{INTEGER}\left[\frac{m+2}{2}\right] \qquad (6)$$

where c is the number of frequencies. So, a 4/2 pole SRM will have 1 frequency; a 6/4 pole 2 frequencies, an 8/6 pole 3 frequencies etc. The frequencies of these flux waveforms will be given by

$$f_c = k \times f_{ph} \qquad 1 \leq k \leq m \qquad (7)$$

TABLE I

GENERALIZED CHART OF FLUX POLARITIES IN THE SRM STATOR CORE

Segment / Time	Stroke number						
	1	2	3	....	m-1	m	m+1
	T_w	$2T_w$	$3T_w$	....	$(m-1)\,T_w$	T_s	$(m+1)\,T_w$
Flux polarities							
1	+	+	+	....	+	+	+
2	+	+	+	....	+	-	+
3	+	+	+	....	-	-	+
m-1	+	+	-	....	-	-	+
m	+	-	-	....	-	-	+
m+1	-	-	-	....	-	-	-

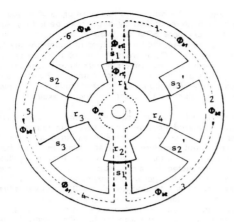

Stroke #1 (phase 1 ON)

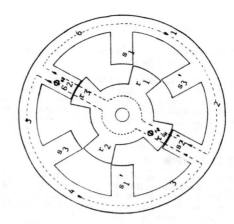

Stroke #2 (phase 2 ON)

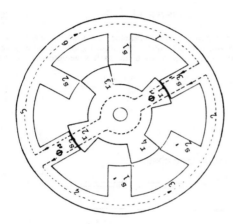

Stroke #3 (phase 3 ON)

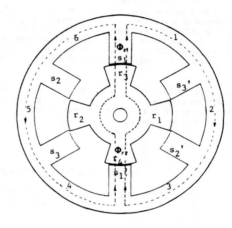

Stroke #4 (phase 1 ON again)

Fig. 2. Flux in a 6/4 pole SRM during one switching period

B. Flux waveforms in the rotor

1) The rotor poles

For an m-phase SRM with N_R rotor poles, each pair of rotor poles sets into alignment with a stator pole at a periodic interval given by

$$T_R = T_w \left(\frac{N_R}{2} \right) \qquad (8)$$

Further, the flux polarity in the pole reverses every half a revolution of the rotor. Thus, over a complete revolution, the flux waveform describes a bipolar envelope with a frequency

$$f_{R1} = \frac{f_{ph}}{N_R} \qquad (9)$$

In addition, from equation (8), there will be a carrier frequency

$$f_{R2} = \frac{1}{T_R} = \frac{2}{T_w N_R} \qquad (10)$$

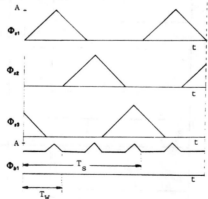

Fig. 3 (a) Construction of flux linkages waveforms in type 1 (unipolar flux) stator core segments

178

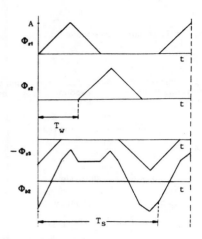

Fig. 3 (b) Construction of flux linkages waveforms in type 2 (bipolar flux) stator core segments

The rotor pole flux waveform can therefore be synthesized from the stator pole waveforms by observing the phase relationship stated in equation (8). Fig. 4(a) shows an example for a 6/4 pole SRM where Φ_{r1} represents the flux linkages in one pair of rotor poles.

2) The rotor core

Flux is established in the rotor core every time a stator phase is excited. Further, all the flux through the rotor poles also flows through the rotor core. The flux waveform for this part is therefore constructed by direct addition of the rotor pole flux waveforms. The flux in the rotor core, similar to the flux in the rotor poles, is bipolar and has the same fundamental frequency. Fig. 4(b) shows an illustration for a 6/4 pole SRM. In summary, the fundamental flux frequencies in an m-phase SRM are

Core part	Frequency
Stator poles:	$f_{sp} = f_{ph}$
Rotor poles:	$f_{rp} = f_{ph}/N_R$
Rotor core:	$f_{rc} = f_{ph}/N_R$
Stator core:	$f_{sc} = k \times f_{ph}; \quad 1 \le k \le m$

III. HARMONIC ANALYSIS

In the previous section, it was shown that the stator pole flux is unipolar and that in general, the flux waveforms in the core parts are not sinusoidal and therefore contain multiple harmonic frequencies. It is therefore necessary to determine the loss-causing sinusoidal components of various frequencies in each of the waveforms presented in section 2. This has been accomplished using Fourier Series analysis as described in the following subsection.

A. Stator waveforms

Since the stator core flux waveforms are constituted by direct addition of the stator pole waveforms, it is sufficient to perform Fourier Series expansion on the stator pole flux waveforms only. The harmonic spectrum of the core fluxes can be obtained by point-by-point addition of the pole waveforms for each harmonic. The analysis is first performed for the general case where the rates of rise and fall of fluxes are not equal such as would be in the case of dissipative regeneration. The origin of the waveform is set to be at the peak point of the flux waveform as shown in Fig. 5 such that an even function is achieved when $t_r = t_f$. The general Fourier expansion for period T_s is given by

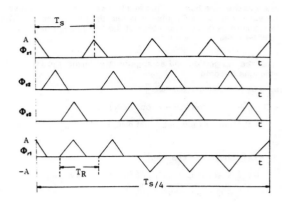

Fig. 4. (a) Construction of flux linkages waveforms in the rotor poles

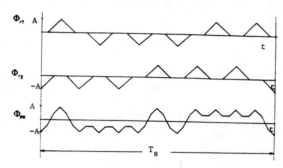

Fig. 4. (b) Construction of flux linkages waveforms in the rotor core

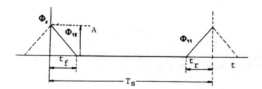

Fig. 5. Flux linkages waveform in the stator pole

$$\Phi_{s1}(t) = \frac{a_0}{2} + \sum_{n=1}^{\infty} \left[a_n \cos n\omega t + b_n \sin n\omega t \right] \quad (11)$$

where

$$a_0 = \frac{2}{T} \left[\int_{-t_r}^{0} \Phi_{11}(t) dt + \int_{0}^{t_f} \Phi_{12}(t) dt \right] \quad (12)$$

$$a_n = \frac{2}{T} \left[\int_{-t_r}^{0} \Phi_{11}(t) \cos n\omega t \, dt + \int_{0}^{t_f} \Phi_{12}(t) \cos n\omega t \, dt \right] \quad (13)$$

$$b_n = \frac{2}{T} \left[\int_{-t_r}^{0} \Phi_{11}(t) \sin n\omega t \, dt + \int_{0}^{t_f} \Phi_{12}(t) \sin n\omega t \, dt \right] \quad (14)$$

where

$$\omega = \frac{2\pi}{T} \quad (15)$$

$$\Phi_{11}(t) = A \left[1 + \frac{t}{t_r} \right] \quad (16)$$

$$\Phi_{12}(t) = A \left[1 - \frac{t}{t_f} \right] \quad (17)$$

and t_r is the time taken for the flux to rise to the maximum value hereafter called the ON time, t_f is the time taken for the flux to fall to zero, A is the peak value of the flux linkage function and n is the order of a harmonic.

After the integration and arrangement of terms, the Fourier coefficients become

$$a_n = C_1 \left[\sin n\omega t_r + \sin n\omega t_f \right]$$
$$+ C_2 \left[t_f + t_r - t_f \cos n\omega t - t_r \cos n\omega t_f \right]$$
$$- C_2 \left[n\omega t_r t_f (\sin n\omega t_r - \sin n\omega t_f) \right] \qquad (18)$$

and

$$b_n = C_1 \left[\cos n\omega t_r - \cos n\omega t_f \right]$$
$$+ C_2 \left[t_f \sin n\omega t_r - n\omega t_f t_r (\cos n\omega t_r + \cos n\omega t_f) \right]$$
$$- C_2 t_r \sin n\omega t_f \qquad (19)$$

and

$$a_0 = \frac{A}{T} (t_r + t_f) \qquad (20)$$

where

$$C_1 = \frac{2A}{n\omega T} \qquad (21)$$

and

$$C_2 = \frac{2A}{(n\omega)^2 T t_r t_f} \qquad (22)$$

For the case when the recovered energy is fed back to the dc supply, $t_r = t_f$ and the stator pole flux linkage function becomes an even function. In this case, $b_n = 0$ and a_n simplifies to

$$a_n = 2 C_1 \sin n\omega t_c$$
$$+ 2 C_2 \frac{t_r t_f}{t_c} \left[1 - \cos n\omega t_c - n\omega t_c \sin n\omega t_c \right] \qquad (23)$$

and

$$a_0 = \frac{2 A t_c}{T} \qquad (24)$$

The Fourier expansion for other phases can be derived from $\Phi_{s1}(t)$ by phase shifting as given below

$$\Phi_{s2}(t) = \frac{a_0}{2} + \sum_{n=1}^{\infty} a_n \cos n\omega(t + T/m)$$
$$+ \sum_{n=1}^{\infty} b_n \sin n\omega(t + T/m) \qquad (25)$$

$$\Phi_{s3}(t) = \frac{a_0}{2} + \sum_{n=1}^{\infty} a_n \cos n\omega(t + 2T/m)$$
$$+ \sum_{n=1}^{\infty} b_n \sin n\omega(t + 2T/m) \qquad (26)$$

$$\Phi_{sm}(t) = \frac{a_0}{2} + \sum_{n=1}^{\infty} a_n \cos n\omega(t + \frac{m-1}{m} T)$$
$$+ \sum_{n=1}^{\infty} b_n \sin n\omega(t + \frac{m-1}{m} T) \qquad (27)$$

As stated earlier on, the stator core flux spectrum can be constructed from a direct point-by-point summation of the stator pole waveforms for each harmonic. The functions can also be expressed in terms of the rotor position by substituting

$$t = \frac{\theta}{\omega_r} \qquad (28)$$

and setting $T = T_s$ as defined in equation (2).

B. The rotor

The rotor pole flux waveform is an odd function and therefore contains only sine terms. The corresponding Fourier expansion is of the form

$$\Phi_r(t) = \sum_{n=1}^{\infty} b_n \sin \alpha t \qquad (29)$$

where

$$b_n = \frac{4}{T} \int_0^{T/2} \Phi_r(t) \sin \alpha t \qquad (30)$$

$\Phi_r(t)$ is described by 2m functions given by

$$\Phi_{2k-1}(t) = -\frac{A}{t_c} (t - t_{k-1}) \qquad (31)$$

and

$$\Phi_{2k}(t) = -A + \frac{A}{t_c} (t - t_k); \qquad k = 1,2,3...2m \quad (32)$$

t_k are the points of discontinuity shown on Fig. 6 and T is the rotor flux period given by $1/f_{R1}$. The equations for determining the Fourier coefficients in the rotor are given in Appendix 2. Similar to the stator core flux spectrum, the rotor core flux spectrum is obtained by a point-by-point addition of the rotor pole waveforms for each harmonic. In all subsequent derivations, it is assumed that $t_r = t_f = t_c$.

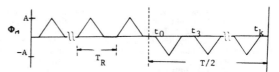

Fig. 6. Flux linkages waveform in the rotor pole

IV. CALCULATION OF LOSSES

A. Iron losses

Manufacturers of lamination core steel provide data showing the variation of core loss (in watts per pound or kilogram) as a function of flux density and frequency. The base frequency for the SRM is determined from the desired operating speed. Using Fourier analysis the harmonic frequencies and their relative amplitudes are determined. To determine the contribution of each harmonic to the core loss, the peak amplitude is used as a weighting factor for the flux density in the respective part of the motor. Thus, for a harmonic of order n, the weighted peak flux density is given by

$$\hat{B}_n = \frac{\hat{A}_n}{A} \hat{B} \qquad (33)$$

where A is the peak value of the flux linkages function used in the Fourier analysis (equation (17)), $\hat{A}_n$ is the peak value of the n-th order harmonic from Fourier analysis, $\hat{B}$ is the flux density in the core determined by the methods described in [3] and $\hat{B}_n$ is the peak flux density for harmonic of order n.

By choosing A = 1, the expression is further simplified. Once the flux density and frequency are known, the corresponding loss factor (watts/pound) is looked up from the loss characteristic and multiplied by the appropriate weight of iron.

A computer program for the analytical estimation of the flux density from motor dimensions and excitation data has been developed [3]. Briefly, the flux density is estimated in an iterative process using

$$N I_p = \sum^J H_j I_j \qquad (34)$$

and

$$\Phi = B_x N A_x \qquad (35)$$

where N is the number of turns per phase, i_p is the peak value of phase current, H_j is the field strength in flux path 'j' of length I_j and B_x is the flux density in part 'x' of the core with cross-sectional area A_x.

For a given peak current, the applied ampere-turns are fixed. N and A_x are known from design data. Starting from an initial value of flux linkage, the values of flux density B for different parts of the magnetic circuit are calculated using equation (35). The corresponding H values are obtained from the fitted B-H curve of the core material. These values are substituted into the right-hand side of equation (34) and compared against the left-hand side. If the difference between the two sides is greater than a pre-defined tolerance, the process is repeated until this tolerance limit is satisfied. The value of B corresponding to this condition is then used in equations (32) and (33). The method has been verified using Finite Element Analysis [4].

In most cases the loss characteristics given by the manufacturer are logarithmic. The variation of core loss as a function of frequency can be fitted using linear logarithmic polynomials of the form

$$P = W \left[\log^{-1} \{ a \log(f) - b \} \right] \qquad (36)$$

where a and b are constants representing the slope and y intercept, respectively, W is the core weight and P is the loss factor (watts/lb). The iron weight for each part is calculated from motor dimensions. It is to be noted that whereas the whole of the stator and rotor iron is continuously active, only a pair each of the stator and rotor poles is active at any instant. In determining the value of H in equation (34), the B-H characteristic of the core material has to be fitted. Several B-H curve fitting methods have been proposed in the literature [5,6,7]. The spline method is used in this study and for a discussion on the merits and demerits of the method refer to [3]

Once the loss factors and the iron weights are known, the iron losses in each section are calculated according to

$$P_i = \sum P_k \times W_k \qquad (37)$$

where P_i is the total core losses (watts), P_k is the loss factor in part 'k' of the core (watts/lb) and W_k is the weight of part 'k' contributing to the loss.

B. Copper losses

Copper losses constitute a considerable part of SRM losses particularly at high excitation levels. Accurate estimation of SRM copper losses at all speeds is complicated by the fact that the current waveform in the SRM is not sinusoidal. The waveform is dependent on operating conditions, particularly the excitation current, speed and the switching strategy. For the case of flat-topped currents, the current waveform can be approximated to a rectangular block. With this assumption it has been shown that the r.m.s. current per phase, assuming unity duty cycle is given by [1]

$$I = \frac{i_p}{\sqrt{(m)}} \qquad (38)$$

where i_p is the peak value of the stator current per phase.

The copper losses are generally given by

$$P_{cu} = m I^2 R_{ph} \qquad (39)$$

where R_{ph} is the resistance per stator phase and I is the r.m.s. value of the phase current.

For a precise prediction of the copper losses, the stator current waveform has to be accurately determined. A method for the accurate prediction of the SRM current waveform has been described by the authors elsewhere [8]. The method requires the phase inductance as an input. This can be predicted using Finite Element Analysis [4] or analytical methods [9]. A computer program has been developed for the purpose. The r.m.s. current is calculated from the output data using numerical integration.

5. SIMULATION RESULTS

A. Harmonic spectra of flux waveforms

Fig. 7(a) to (e) show the harmonic spectra in a 6/4 pole SRM for a switching period of 10 ms, and a conduction period of 2.7 ms. This corresponds to a speed of 1500 r.p.m. and a conduction angle of 24.3 °. The peak value of the flux linkage function is normalized to A = 1. The first five harmonics are shown in each case. It is observed that:

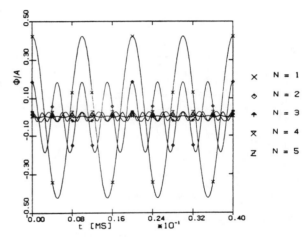

Fig. 7 (a) Harmonic spectrum of the stator pole flux

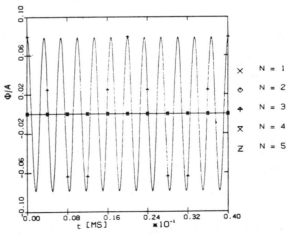

Fig. 7 (b) Harmonic spectrum of the flux in type 1 stator core segments

181

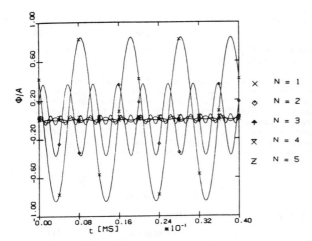

Fig. 7 (c) Harmonic spectrum of the flux in type 2 stator core

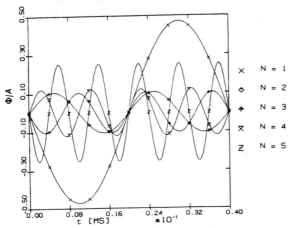

Fig. 7 (d) Harmonic spectrum of the rotor pole flux

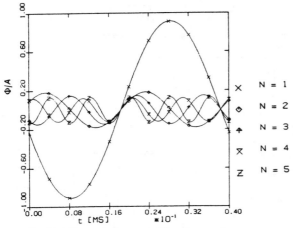

Fig. 7 (e) Harmonic spectrum of the rotor core flux

•Harmonics of all orders are significantly present in the rotor and stator.

•The fundamental component in the stator poles has a peak amplitude less than half of the overall peak. This was observed to be the case at all frequencies.

•The fundamental component in the core segments with non-reversing flux is less than 10 % of the overall peak. The value increases when the conduction angle is decreased.

•In the rotor core and poles, all harmonics are very significant. In this particular case, the peak amplitude of the 12-th harmonic for example, is 14.2 % of the peak amplitude of the fundamental. The corresponding figures for other parts of the core are 13.75 % for the rotor poles, 0.58 % for the stator poles and .29 % for the stator core. This implies that the core losses in the rotor are higher than in the stator. Taking the weighted harmonic content, the resulting error when all harmonics higher than order 12 are neglected is less than 1 % in all parts of the core. The weighted harmonic content is calculated from

$$g = \sqrt{\sum_{n=j}^{k} \left[\frac{F_n}{F_1 \, n} \right]^2} \qquad (40)$$

where F_n is the effective value of the n-th harmonic given by

$$F_n = \sqrt{\left[\sum_{t=0}^{T} f_n(t)^2 \; \frac{n \, \Delta t}{T} \right]} \qquad n = 1, 2, \dots k \quad (41)$$

T is the period of the fundamental frequency in the respective core part. The overall effective value (including all harmonics) in a given core part is calculated as

$$F = \sqrt{\left[\sum_{n=1}^{k} [F_n]^2 \right]} \qquad (42)$$

B. Effect of excitation level

The flux linkages in the SRM are proportional to the level of excitation. Increase in stator currents results in increased flux linkages leading to core saturation. The losses in the core will therefore be a function of excitation level. Variation of excitation level occurs every time the load is changed or during a commanded change in current such as would occur during chopping. This variation also occurs as the rotor moves from the unaligned to the fully aligned position. Variation in the excitation level is simulated by varying the peak amplitude A of the flux linkage function. Fig. 8 shows the change in the r.m.s. value of the a.c. component of flux linkage when A is varied from 0.1 to 1.0. The conduction angle and switching period are kept constant at 2.7 ms and 10 ms respectively. It is observed that the effective flux linkages in all parts of the core fall as A is decreased. The rotor core and the stator core segments with reversing flux show higher sensitivity to these changes.

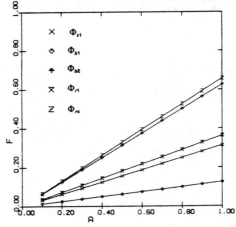

Fig. 8. Variation of effective value of flux linkages with excitation level

182

C. Effect of conduction angle and switching period

Variation of conduction angle is one of the control strategies adapted for the SRM operation and is generally used in multi-quadrant operation. Fig. 9(a) shows the variation of the r.m.s. value of the flux linkages in the different parts of the core when the conduction angle is varied from 9° to 45° with the peak amplitude A = 1 and a switching period of 10 ms. The effective flux linkages fall as the conduction angle is decreased. Again, the rotor core is observed to be more sensitive to these variations. By inference, the rotor losses increase as the conduction angle increases. The stator core segments with non-reversing flux show an interesting characteristic. For values of conduction angle smaller than T_w, the effective value of flux falls in inverse to the conduction angle, becoming zero at the threshold of overlap. Beyond this point it increases again while the effective values in the other core parts fall. This implies that the loss component due to this frequency is high for small values of conduction angle and also when the conduction overlap between phases is large. In the 8/6 pole SRM, conduction overlap is inevitable and therefore this loss component is significant.

Fig. 9(b) shows the effect of varying the switching period while maintaining a constant conduction angle. The simulation is done for a speed range of 300-2500 r.p.m. It can be observed that, except for the stator segments with non-reversing flux, the effective flux decreases as the speed decreases. In the non-reversing flux stator segments, the flux linkage is observed to

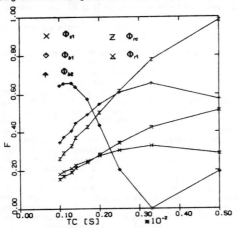

Fig. 9 (a) Effect of conduction period on flux linkages

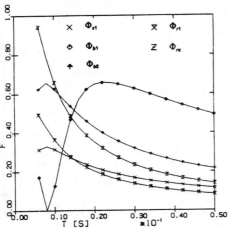

Fig. 9 (b) Effect of switching period on flux linkages

be more dependent on the duty cycle; at low values of duty cycle (ratio between conduction angle and switching period), the effective value is high. Due to its high frequency, this flux component has significant contribution to the core losses at low speeds.

D. Effect of Chopping

When operating in the constant torque region, it is necessary to maintain a constant current. To achieve this, the converter is operated in the chopping mode whereby the current is maintained at a desired average value. This is normally implemented by chopping the current between two levels or by using a PWM scheme. Chopping introduces a high frequency flux component which results in increased core loss. The chopping frequency depends on the magnitude of the specified current excursions, the magnitude of the average current and the rotor position. Taking the 6/4 pole SRM whose details are given in Appendix 1 and assuming a commanded average current of 9 A, a 20 % chopping window would result in current excursions between 10 A and 8 A. At low speed, this would result in 2.6 % excursions in flux linkages and a frequency of about 160 Hz. The core losses in the motor will therefore vary accordingly. Copper losses will also vary due to changes in the r.m.s. value of current as well as increased resistance due to skin effect, particularly if the chopping window is too small or, in the case of a PWM scheme, if the the modulation frequency is too high.

E. Separation of Stator and Rotor Core Losses

The losses corresponding to each part of the core were estimated using the procedure described in the previous section. All calculations of flux density are based on the fully aligned position. All harmonics up to the 8-th order were included. Losses were estimated for various values of phase current covering the entire load range of the prototype 3-h.p. SRM, inclusive of saturation. Fig. 10 (a) shows a variation of the stator and rotor (cores and poles) losses as a function of the excitation current. As predicted, the rotor losses are observed to be over 17 % higher than the stator losses over the whole load range. In deep saturation (10 A), the rotor losses (poles and core) constitute 58.5 % of the overall iron losses of the motor. The losses were predicted for a speed of 1500 r.p.m. and a conduction angle of 24.3°, from the same data used in determining the harmonic spectrum.

The core, copper and overall losses are shown in Fig. 10(b) for the same operating conditions as in Figure 10(a). At very low excitation, the core losses are dominant, representing about 68 % of overall losses at 1 A excitation. However, as the current increases, the copper losses increase quadratically and exceed the core losses.

F. Effect of speed on losses

At high speeds, the harmonic frequencies will cause considerably higher losses in the motor. Fig. 11 shows the variation of core losses with excitation current for the two speeds of 1500 and 5000 r.p.m. Harmonics of order 9 or higher have been neglected. As stated earlier, the error obtained when harmonics of order 12 or higher are neglected is less than 1 % in all core parts. It is observed that the core losses in deep saturation increase by a factor of six as the speed changes from 1500 r.p.m. to 5000 r.p.m. At this operating condition, the core losses constitute 19 % of the overall losses. The copper losses also increase due to skin effect.

VI. MEASUREMENTS AND EXPERIMENTAL RESULTS

A 6/4 pole 3 h.p. laboratory prototype SRM was used for experimental verification. The motor was coupled to a d.c. dynamometer for measuring the output power. As mentioned earlier, accurate measurements of motor currents can not be obtained with ordinary a.c. or d.c. meters because the current waveforms are neither sinusoidal nor pure d.c. and they vary with operating conditions namely, speed, saturation level and conduction angle. To alleviate the problem, a Data Acquisition System was used. In order to ensure uniformity in the accuracy

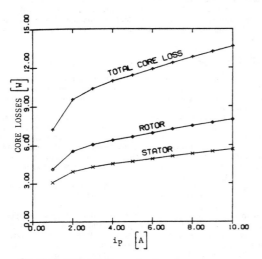

Fig. 10 (a) Core losses as a function of peak stator current

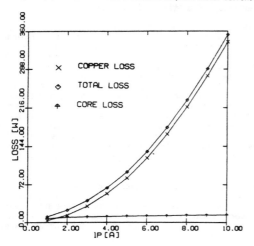

Fig. 10 (b) Core, copper and total losses as a function

of peak stator current

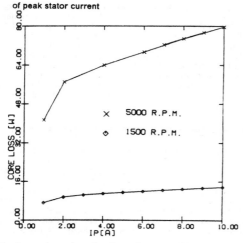

11. Comparison of predicted core losses at 1500 and 5000 r.p.m.

of the data, 1024 data points per period were acquired. All measurements were made at a fixed voltage of 180 V d.c. The r.m.s. current was calculated from the acquired data using

$$I = \sqrt{\left[\frac{A_I}{T_s}\right]} \qquad (43)$$

The average current is given by

$$\bar{I} = \frac{A_A}{T_s} \qquad (44)$$

where A_A is the area enclosed between the current waveform and the time axis and A_I is the area enclosed between the square of the current waveform and the time axis and T_s is the switching period. Areas A_I and A_A were calculated using Simpson's rule. Simpson's rule was chosen because the error by this method is proportional to the fourth power of the time step used and the third derivative of the function at the end points. Since the current function is smooth at both ends and the time step is 1/1024 of a period, the error is bound to be low. The input power to the motor is calculated from the average current and the d.c. supply voltage. The difference between this value and the output (shaft) power of the motor gives the total losses in the motor. By subtracting the copper losses, the friction and windage losses and the stray-load losses from this value, the core losses are obtained. The copper losses are calculated using equation (39) and the r.m.s. current calculated above. A friction and windage losses versus speed characteristic was obtained by driving the SRM at no load using a permanent magnet d.c. motor of a known torque constant and stator resistance. The stray load losses were estimated to be 6 % of the total motor losses [11].

Two types of current waveforms were investigated; a peaky and a flat-topped current waveform [10]. The peaky waveform is obtained by operating with a large advance angle. In this case, the advance angles were 11.95° and 0° and the conduction angles were 25.27° and 18.05° for the peaky and flat-topped currents, respectively. The conduction angle for the flat-topped current has to be small otherwise the curent becomes peaky just before turn-off due to the low inductance slope. Fig. 12(a) and (b) show oscillograms of the current and gating signal in one phase for the peaky and flat-topped waveforms respectively. Table II presents a comparison between average current readings obtained using a d.c. meter and the Data Acquistion system described earlier. For the flat-topped current, the difference between the two readings is less than 10 % for all the readings whereas for the peaky current, up to 15 % error is observed. Fig. 13(a) and (b) shows the variation of core and copper losses with excitation for the two types of current waveforms. Data samples for the same are shown in Tables III and IV. The phase resistance is 3.4 Ω. It is shown that operating with a peaky current waveform results in relatively higher copper losses when compared to the flat-topped current. The error obtained when the r.m.s. current is estimated using equation (38) and the appropriate duty cycle is observed to be larger for the peaky current. This accounts for the observed low copper losses compared to the predictions. In order to obtain accurate predictions, the exact current waveform has to be used. A method for predicting the current waveform in a SRM has been developed by the authors and is described elsewhere [8]. The method has been used to construct the flux linkages waveform from the peaky current waveform. The resulting waveform is shown in Fig. 14 for an operating point of 1095 r.p.m, a peak current of 5.1 A and a conduction angle of 25.3 °. Close agreement between the assumed and the actual waveforms is observed. The inductance in this case has been determined using an analytical method developed by the authors [3]. The predicted core losses for this current waveform are 13.3 W whereas the measured value is 16.2 W. The difference between the measured and the predicted value can be attributed to the change in the magnetic properties of the core material due to the stresses induced during cutting of laminations. Another possible source of error is imperfections in stacking the laminations. The combined effect of these errors has been observed to result in increasing the core losses by as much as 25 % [11].

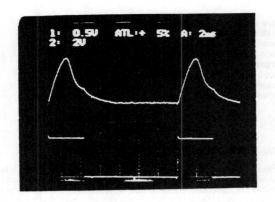

Fig. 12 (a) Phase current (top trace) and gate signal (lower trace) waveforms for a peaky stator current. Current scale: 2 A/div.

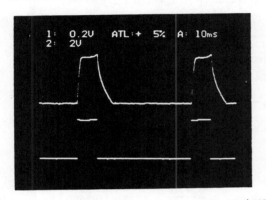

Fig. 12 (b) Phase current (top trace) and gate signal (lower trace) waveforms for a flat-topped current. Current scale: 0.8 A/div.

TABLE II

COMPARISON BETWEEN D.C. METER AND DATA ACQUISITION

SYSTEM MEASUREMENTS

Flat-topped current			Peaky current		
Peak current [A]	Average current [A]		Peak current [A]	Avarage current [A]	
	D.C Meter	Data Acqsn.		D.C Meter	Data Acqsn.
1.07	0.25	0.230	2.67	0.42	0.441
1.54	0.35	0.328	3.28	0.55	0.546
1.83	0.42	0.397	4.04	0.60	0.694
2.13	0.50	0.464	4.94	0.90	0.888
2.54	0.57	0.542	5.66	1.05	1.051

TABLE III

SAMPLE DATA FOR PEAKY STATOR CURRENTS

Peak current [A]	R.M.S. current by DAS [A]	R.M.S. current by [38] [A]	Iron losses [W]	Copper losses [W]	Speed [rpm]	Pcu Pin [%]
2.67	0.870	1.30	20.6	7.7	2324	4.5
3.28	1.067	1.59	11.3	11.6	1806	5.2
4.04	1.356	1.96	7.3	18.8	1458	7.1
4.94	1.712	2.40	15.5	29.9	1150	9.0
5.66	2.000	2.66	22.9	40.8	1004	10.6

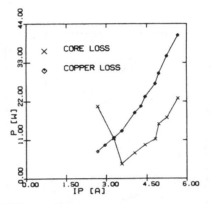

Fig. 13 (a) Measured losses for a peaky stator current waveform

VII. CONCLUSION

A method for determining the losses in a switched reluctance motor has been described. Fourier analysis has been used to determine the harmonic spectra of the fluxes in different parts of the core. It is is shown that harmonics of all orders exist and have a very significant effect on the core losses particularly in the rotor core. The maximum core losses occur in the rotor. It is also shown that, unlike conventional machines, core losses

TABLE IV

SAMPLE DATA FOR FLAT-TOPPED STATOR CURRENTS

Peak current [A]	R.M.S. current by DAS [A]	R.M.S. current by [38] [A]	Iron losses [W]	Copper losses [W]	Speed [rpm]	Pcu Pin [%]
1.07	0.446	0.618	5.6	2.0	955	3.4
1.54	0.629	0.889	9.6	4.0	729	4.8
1.83	0.768	1.057	10.4	6.0	615	5.9
2.13	0.888	1.230	10.5	8.0	550	6.9
2.54	1.051	1.466	10.70	11.3	486	8.0

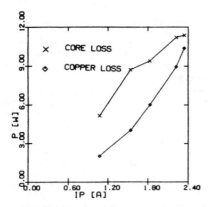

Fig. 13 (b) Measured losses for a flat-topped stator current waveform

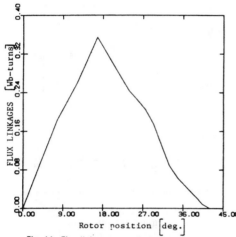

Fig. 14. Flux linkages waveform constructed from measured stator current

are not constant but vary widely with stator current, conduction angle and speed. Core losses increase considerably as the level of excitation and speed increase. Conduction angle modulation produces very significant changes in the flux spectrum in the motor. The high frequency component in the stator core is very sensitive to this change particularly during overlap and at low duty cycle. At low excitation, core losses are higher than copper losses; however, as the current increases, copper losses dominate. The problems associated with measurement of motor currents have been discussed and a solution is proposed. It is observed that the values of average current measured using a d.c. current analog meter differ by as much as 15 % from those obtained using a digital Data Acquisition System (DAS). Since the DAS is based on the actual current waveform, it is proposed as the most reliable method for measuring the SRM currents.

APPENDIX 1

Main dimensions and parameters of the SRM

Number of stator poles	6	
Number of rotor poles	4	
Stator pole arc	24	degrees
Rotor pole arc	36	degrees
Stator outer diameter	194	mm
Bore diameter	122.0	mm

Core length	60.37	mm
Shaft diameter	28.56	mm
Rotor pole height	27.25	mm
Stator back-iron thickness	10.15	mm
Airgap length	0.25	mm
Number of turns per phase	536	
Supply voltage	300	V
Rated peak current	10	A

APPENDIX 2

Equations for the Fourier coefficients in the rotor

These equations are developed for the general case where t_f is not equal to t_r. For the 6/4 SRM, the rotor pole flux linkage function is described by six linear functions of the form shown in equations (31) and (32). t_0 to t_8 are the points of discontinuity shown in Fig. 6. The corresponding Fourier coefficients are:

$$b_{n1} = \frac{-4A}{n^2 \omega^2 T t_r} \left[\sin n\omega t_1 - \sin n\omega t_0 \right]$$

$$\frac{+4A}{n^2 \omega^2 T t_r} \left[(n\omega t_1 \cos n\omega t_1 - n\omega t_0 \cos n\omega t_0) \right]$$

$$-\frac{4A t_0}{n\omega T t_r} \left[\cos n\omega t_1 - \cos n\omega t_0 \right] \qquad (1)$$

$$b_{n2} = \frac{4A}{n\omega T} \left[\cos n\omega t_2 - \cos n\omega t_1 \right]$$

$$+\frac{4A}{n^2 \omega^2 T t_f} \left[\sin n\omega t_2 - \sin n\omega t_1 \right]$$

$$-\frac{4A}{n^2 \omega^2 T t_f} \left[(n\omega t_2 \cos n\omega t_2 - n\omega t_1 \cos n\omega t_1) \right]$$

$$+\frac{4A t_1}{n\omega T t_f} \left[\cos n\omega t_2 - \cos n\omega t_1 \right] \qquad (2)$$

$$b_{n3} = \frac{-4A}{n^2 \omega^2 T t_r} \left[\sin n\omega t_4 - \sin n\omega t_3 \right]$$

$$\frac{+4A}{n^2 \omega^2 T t_r} \left[(n\omega t_4 \cos n\omega t_4 - n\omega t_3 \cos n\omega t_3) \right]$$

$$-\frac{4A t_3}{n\omega T t_r} \left[\cos n\omega t_4 - \cos n\omega t_3 \right] \qquad (3)$$

$$b_{n4} = \frac{4A}{n\omega T} \left[\cos n\omega t_5 - \cos n\omega t_4 \right]$$

$$+\frac{4A}{n^2 \omega^2 T t_f} \left[\sin n\omega t_5 - \sin n\omega t_4 \right]$$

$$-\frac{4A}{n^2 \omega^2 T t_f} \left[(n\omega t_5 \cos n\omega t_5 - n\omega t_4 \cos n\omega t_4) \right]$$

$$+\frac{4A t_4}{n\omega T t_f} \left[\cos n\omega t_5 - \cos n\omega t_4 \right] \qquad (4)$$

$$b_{n5} = \frac{-4A}{n^2 \omega^2 T t_r} \left[\sin n\omega t_7 - \sin n\omega t_6 \right]$$

$$\frac{+4A}{n^2 \omega^2 T t_r} \left[(n\omega t_7 \cos n\omega t_7 - n\omega t_6 \cos n\omega t_6) \right]$$

$$-\frac{4A t_6}{n\omega T t_r} \left[\cos n\omega t_7 - \cos n\omega t_6 \right] \qquad (5)$$

$$b_{n6} = \frac{4A}{n\omega T} \left[\cos n\omega t_8 - \cos n\omega t_7 \right]$$

$$+ \frac{4A}{n^2\omega^2 T t_f} \left[\sin n\omega t_8 - \sin n\omega t_7 \right]$$

$$- \frac{4A}{n^2\omega^2 T t_f} \left[(n\omega t_8 \cos n\omega t_8 - n\omega t_7 \cos n\omega t_7) \right]$$

$$+ \frac{4At_7}{n\omega T t_f} \left[\cos n\omega t_8 - \cos n\omega t_7 \right] \qquad (6)$$

$$b_n = b_{n1} + b_{n2} + b_{n3} + b_{n4} + b_{n5} + b_{n6} \qquad (7)$$

The flux linkage functions for the two pairs of rotor poles are are obtained as

$$\Phi_{r1} = \sum_{n=1}^{\infty} b_n \sin n\omega t \qquad (8)$$

and

$$\Phi_{r2} = \sum_{n=1}^{\infty} b_n \sin n\omega(t + T_w) \qquad (9)$$

The rotor core flux linkages function Φ_{rc} is obtained by a point-by-point summation of the two functions given in equations (8) and (9) above.

REFERENCES

[1] R. Krishnan, R. Arumugam and J. F. Lindsay, 'Design procedure for switched reluctance motors', Conf. Record, IEEE-IAS Annual Meeting, Colorado, pp.858-863, Oct. 1986

[2] P. J. Lawrenson et al., 'Variable speed switched reluctance motors', Proc. IEE, Vol.127, Pt. B, No. 4, pp. 253-265, October 1980

[3] R. Krishnan, S. Aravind and P. Materu, 'Computer Aided Design of electrical machines for variable speed applications', Proc. IEEE-IECON'87, Vol. 851, pp. 756-763, Cambridge, Nov. 1987

[4] R. Arumugam, D.A. Lowther, R. Krishnan and J.F. Lindsay, 'Magnetic field analysis of a switched reluctance motor using a two-dimensional finite-element model', Trans., IEEE Magnetics, Vol. mag-21, No. 5, pp 1883-1885, Sept. 1985

[5] J. R. Brauer, 'Simple equations for the magnetization and reluctivity curves of steel', Trans., IEEE Magnetics, Vol. MAG-11, No. 1, pp 81, 1975

[6] S. K. El-Sherbiny, 'Representation of the magnetization characteristics by a sum of exponentials', Trans, IEEE Magnetics, Vol. MAG-9, No. 1, pp. 60-61, March 1973

[7] F.C. Trutt, E. A. Erdelyi and R. E. Hopkins, 'Representation of Magnetization Characteristic of DC Machines for Computer Use', Trans, IEEE-PAS, Vol. 87, pp. 665-669, March 1968

[8] P. Materu, R. Krishnan and H. Farznehfard, 'Steady state analysis of the variable speed switched reluctance motor', Proc. IEEE-IECON'87, Vol. 854, pp. 294-302.,Cambridge, November 1987

[9] J. Corda and J. M. Stephenson, 'Analytical estimation of the minimum and maximum inductances of a double-salient motor', Proc. Int. Conf. on Stepping Motors and Systems, Leeds, pp. 50-59, Sept. 1979

[10] T. J. Miller, 'Converter volt-ampere requirements of the switched reluctance motor drive' Conf. Record, IEEE-IAS Annual Meeting, Chicago, pp. 813-819, October 1984

[11] E. Levi, 'Polyphase motors- A Direct Approach to their Design', John Wiley & Sons, 1984

Sensitivity of Pole Arc/Pole Pitch Ratio On Switched Reluctance Motor Performance

R. Arumugam and J.F. Lindsay

Dept. of Electrical & Computer Engineering, Concordia University, Montreal

R. Krishnan

Dept. of Electrical Engineering, VPI & SU, Blacksburg, VA
IEEE-IAS Proceedings, Pittsburgh, PA, October 1988

Abstract

In this paper, the sensitivity of the pole arc/pole pitch ratio of the stator and rotor on the performance of a switched reluctance (SR) motor is investigated. An analytical method based on "magnetic flux path" and a two dimensional finite element analysis are used for the study. The method of sensitivity study is performed by comparing the average torque developed for different stator as well as rotor pole arc/pole pitch ratios and choosing the ratio combination that produces the greatest value of average torque.

Introduction

In recent years a number of papers on SR motors has been published in the literature [1]-[14]. Only a few of them address the design aspects [1, 6, 9, 10]. The design philosophy published so far, is based on the knowledge of variable reluctance stepper motor designs. In these papers, the pole arc/pole pitch ratio has been either derived from the permeance values assuming parallel sided teeth and slots or taken to be slightly less than that used for variable reluctance stepper motors. Unlike these variable reluctance stepper motors, SR motors have smaller, unequal but even numbers of poles on the stator and rotor. Moreover, SR motors can be used for higher power applications requiring larger sizes. Hence, the change in motor performance due to variations of the stator and rotor pole arc/pole pitch ratios will be of interest to the SR motor designer. With the above considerations the present study has been undertaken.

An analytical method based on the lumped magnetic circuit model at different judiciously selected sections of the magnetic circuit has been proposed by Corda and Stephenson [15]. This method is used, as a first approximation, to determine the suitable pole arc/pole pitch ratio that enables the SR motor to develop the greatest value of the average torque.

The advantage of using finite element analysis for electromagnetic field analysis and particularly for the SR motor, in which high levels of saturation are encountered, has already been reported.[16,17] A two dimensional finite element analysis is used in this study to calculate the energy stored in the motor when its rotor is in the aligned and unaligned positions. The coenergy change between the above two positions of the rotor is computed from which the average torque developed is determined. As in the analytical method, the best choice for the pole arc/pole pitch ratio is determined by considering the average torques for different pole arc/pole pitch ratio combinations.

Basis of Sensitivity Study

In SR motors the air-gap geometry plays a vital role on their performance. A suitable choice of the air-gap length, pole width and pole height is necessary for a better design of the motor. Therefore, attention has been directed to optimize the air-gap geometry so that an SR motor with a greater value of torque is designed. The radial length of the air-gap is made as small as mechanically possible so that the torque developed is maximum when all other air gap parameters are held constant. In this study it is assumed that the height of the pole is fixed. The pole arcs on the stator and rotor are changed in steps of about 0.05 of a pole pitch, starting from a pole arc/pole pitch ratio of 0.25 up to 0.55. Values below 0.25 and above 0.55 are not required in this study as will become evident from the trend of the results which follow.

Due to the nonlinear nature of the magnetic fields in the SR motor under normal operating conditions, the virtual displacement principle provides the most convenient method of calculating the average torque. The flux linkages are calculated for different excitation currents when the rotor is in the aligned and unaligned positions. The coenergy change is obtained from which the average torque is calculated. The range of pole arc/pole pitch ratios which produces the greatest average torque is chosen as the preferred values for a better design.

Methods of torque calculation

In this study, two methods have been used for torque prediction. One is an analytical method which is used to calculate the aligned and unaligned flux linkages for various pole arc/pole pitch ratios. The other is a finite element analysis. The following sections give a brief description of these methods.

Analytical method

The analytical method describes the determination of the minimum and maximum inductances when the configuration of the motor is known. The minimum inductance is calculated using the assumption that the magnetic fields in the interpolar and air-gap regions consist of straight line segments and circular arcs. Unlike earlier methods, this one takes into account the actual distribution of the winding on the stator poles. The flux linkages when the rotor is in the unaligned position are obtained assuming that the minimum inductance remains constant for the range of excitation used in the analysis. When the rotor is in the aligned position there is considerable mmf drop in the stator and rotor cores compared with that in the air-gap. Hence, the conventional magnetic circuit analysis is used to determine the flux linkages.

The application of the technique proposed by Corda et al. may lead to large errors in the minimum inductance calculation unless care is exercised. The formulae derived for permeance components are based on the assumption that the ratio of pole arc/pole pitch is 0.5 for both the stator and rotor. Also, the windings are taken to be extended up to the middle of the stator interpolar regions. Hence a suitable modification of the formulae is required when this technique is used for a particular motor configuration and winding arrangement.

In the maximum inductance position the cores are saturated even at lower excitations due to the shorter air-gap length. Hence the mmf drops in various sections of the core and the nonlinear magnetization characteristic of the core are taken into account when the flux linkages vs current characteristic of the motor is computed.

The data for the magnetization characteristic of the iron (M-19 steel) is taken from the manufacturer's data sheets. The characteristic is divided into 20 line segments and each segment is represented using cubic spline polynomials. For flux densities beyond the range of available values, a linear extrapolation is used.

In order to determine the flux linkages vs current characteristic in the aligned position, the flux linkages are assumed and the excitations required to establish the assumed flux linkages are calculated. From the known number of turns, the flux and flux densities in various sections of the magnetic circuit are computed. Using the magnetization characteristic and the flux densities, the mmf drops in the different sections of the core are determined. The excitation current is obtained as the ratio of the total mmf to the number of turns.

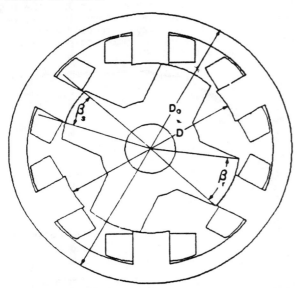

Figure 1: Typical SR motor configuration

The flux linkages vs current characteristic for the aligned and unaligned positions are stored in a data file. The area between these curves is obtained using numerical integration. Having obtained the coenergy change for the chosen ratios of pole arc/pole

pitch in the stator and rotor, the average torque developed is calculated.

The procedure used to determine the optimum pole arc/pole pitch ratio is as follows:

1 Set the pole arc/pole pitch ratio of the stator to 0.25.

2 Set the pole arc/pole pitch ratio of the rotor to 0.25.

3 Determine the flux linkages vs current characteristics for the aligned and unaligned positions.

4 Calculate the change in coenergy and hence the average torque developed.

5 Increment the pole arc/pole pitch ratio of the rotor by 0.05.

6 If the rotor pole arc/pole pitch ratio is greater than 0.55, then go to step 7. Otherwise, go to step 3.

7 Increment the stator pole arc/pole pitch ratio by 0.05.

8 If the stator pole arc/pole pitch ratio is greater than 0.55, then go to step 9. Otherwise, go to step 2.

9 Tabulate the average torque developed as a function of the stator and rotor pole enclosures for different excitations.

10 Choose the pole arc/pole pitch ratios corresponding to the greatest average torque for a given current.

It can be realized that there are 49 combinations of pole arc/pole pitch ratio when both the stator and rotor pole arcs are varied in this manner. For each combination, one flux linkages vs current characteristic is obtained for both aligned and unaligned positions. When five excitation currents are considered for each combination, there are 245 average torque values computed. It is presumed that this will give sufficient information on the nature of torque development for the changes in stator and rotor pole widths. Any number of average torque values can be calculated if a suitable number of excitations is considered. On the other hand, when finite element analysis is used, separate problems are to be set up and solved for each excitation and pole arc/pole pitch ratio combination in the aligned and unaligned positions. Thus, for the chosen five excitation currents 490 problems are solved and post-processed to obtain the coenergy for both aligned and unaligned positions. The coenergy is directly computed in the finite element analysis and hence the flux linkages vs current characteristic need not be plotted to determine the average torque.

Finite Element Analysis

The finite element formulation used for field analysis has already been described in [16]. The configuration of the SR motor considered for investigation is shown in Fig. 1.

189

For each stator pole arc/pole pitch ratio, the rotor pole arc/pole pitch ratio is changed from 0.25 to 0.55 in steps of 0.05. Thus, seven finite element analysis models are developed for the aligned position, each model having provision to change the rotor pole arc/pole pitch ratio. Since, the field solutions are obtained for five chosen excitations, 35 problems are set up for each stator pole arc/pole pitch ratio. Similarly, for the unaligned position, seven different models have been used, each time solving and postprocessing 35 problems.

The coenergy for each pole arc/pole pitch ratio combination and current, when the rotor is in a given position, is directly computed as described earlier. The change in coenergy and the average torque are determined.

Results

The average torque values calculated by the analytical method are given in Table I. To illustrate the nature of the results that lead to these values, a typical flux linkages vs current characteristic for a pole arc/pole pitch ratio of 0.4 on the stator and 0.35 on the rotor is shown in Fig. 2. The variation of the average torque with changes in stator pole arc/pole pitch ratio is plotted for different rotor pole arc/pole pitch ratios and is shown in Fig. 3. Keeping the pole arc/pole pitch ratios for the stator and rotor the same, average torque values are calculated for various excitation currents. The resulting average torque variation for different pole arc/pole pitch ratios is shown in Fig. 4.

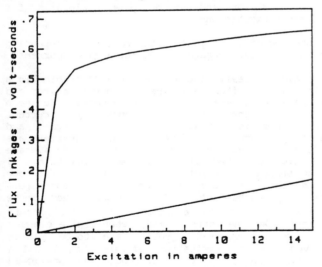

Figure 2: Flux linkages vs excitation current
(Analytical Method)

The average torque is proportional to the change in coenergy values when the rotor moves from the unaligned to the aligned position. Hence, the results obtained by the finite element method are presented in terms of coenergy itself. Excitation currents of 1, 2, 4, 8 and 12 amperes are used for the investigation. The results for 1 A excitation are not presented as they are very similar to the 2 A results. Fig. 5 shows the variation of coenergy change with varying stator and rotor pole arc/pole pitch ratios when the excitation is

2 A. The results for excitations of 4, 8 and 12 amperes are shown in Figs. 6, 7 and 8, respectively.

Conclusions

The changes in stator pole arc greatly influence the average torque compared with the changes in rotor pole arc. At lower excitation currents, when the saturation in the core is not appreciable, the developed torque increases invariably with increase in stator pole arc for a given rotor pole enclosure. Instead, at higher excitations, the increase in average torque with stator pole enclosure is less and at higher stator pole arcs the average torque decreases. It is particularly evident at a steady excitation of 12 A where the average torque is less for a stator pole arc/pole pitch ratio of 0.55 compared with that for 0.5. When the stator pole arc/pole pitch ratio is less than 0.35, there is distinctly less developed torque at all excitations. Hence, it may be recommended that the stator pole arc/pole pitch ratio be chosen in the range of 0.35 to 0.5. Higher pole arc/pole pitch ratios for the stator may be limited by considerations of winding space and the necessary clearance between the windings.

At all excitation currents, the average torque value increases with increase in rotor pole arc, reaches a peak value and then decreases when the stator pole arc is held constant. It is observed that there is no increase in torque developed when the rotor pole arc/pole pitch is increased beyond 0.45. Higher excitation currents produce maximum values of average torques when the ratio is slightly less than 0.4, whereas, at lower currents the average torque values reach peaks when the ratio is slightly above 0.4. Considering the nature of currents, normally encountered in SR motors, choosing the rotor pole arc/pole pitch ratio around 0.4 may produce the highest torques. The average torque is less for a rotor pole arc/pole pitch ratio of 0.25 compared with that for higher ratios when the stator pole arc/pole pitch ratio is in the range of 0.35 to 0.5.

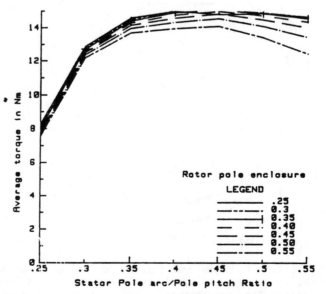

Figure 3: Average torque vs stator pole arc/pole pitch ratio (Excitation at 12 A)

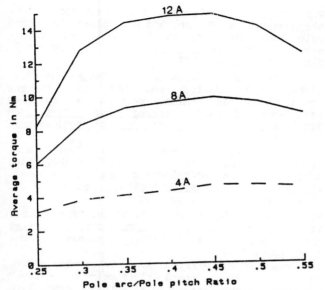

Figure 4: Average torque vs pole arc/pole pitch ratio
(Stator and Rotor ratios are equal)

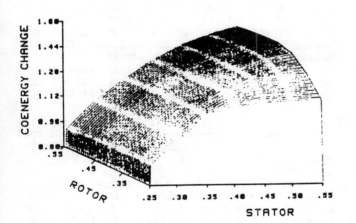

Figure 5: Average torque vs pole arc/pole pitch ratios
of stator and rotor (Excitation at 2A)

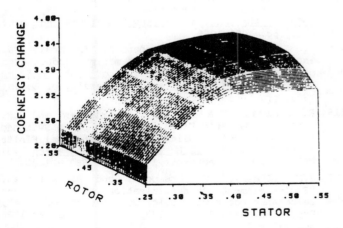

Figure 6: Average torque vs pole arc/pole pitch ratios
of stator and rotor (Excitation at 4A)

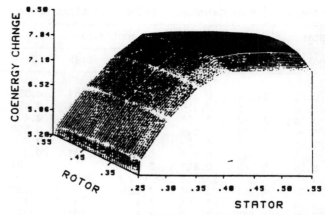

Figure 7: Average torque vs pole arc/pole pitch ratios
of stator and rotor (Excitation at 8A)

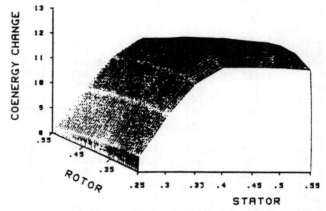

Figure 8: Average torque vs pole arc/pole pitch ratios
of stator and rotor (Excitation at 12A)

It can be concluded that the pole arc/pole pitch
ratio on the stator and rotor of an SR motor need not
be the same. The range of values that may be used for
the pole arc/pole pitch ratio of the rotor can be 0.3
to 0.45 and that of the stator 0.35 to 0.5.

Acknowledgements

This research forms part of a project receiving
financial support from the Natural Science and
Engineering Research Council of Canada (NSERC). The
authors are also grateful for the computational support
made available by Infolytica Corporation of Montreal.

191

TABLE I
Average Torque Calculated Using Analytical Method

Stator pole enclosure	Rotor pole enclosure				
	0.30	0.35	0.40	0.45	0.50
Excitation = 1 A					
0.25	0.354	0.355	0.357	0.357	0.357
0.30	0.415	0.416	0.418	0.418	0.418
0.35	0.472	0.475	0.476	0.476	0.476
0.40	0.524	0.528	0.530	0.530	0.530
0.45	0.562	0.629	0.686	0.732	0.769
0.50	0.561	0.629	0.685	0.731	0.768
0.55	0.560	0.628	0.683	0.728	0.764
Excitation = 2 A					
0.25	1.254	1.256	1.257	1.257	1.257
0.30	1.434	1.436	1.437	1.437	1.435
0.35	1.607	1.610	1.611	1.609	1.606
0.40	1.719	1.723	1.723	1.720	1.715
0.45	1.782	1.876	1.945	1.997	2.034
0.50	1.779	1.872	1.940	1.990	2.024
0.55	1.774	1.865	1.931	1.978	2.007
Excitation = 4 A					
0.25	3.158	3.149	3.141	3.133	3.123
0.30	3.872	3.868	3.862	3.853	3.840
0.35	4.123	4.118	4.109	4.096	4.077
0.40	4.342	4.338	4.326	4.308	4.282
0.45	4.424	4.522	4.588	4.629	4.645
0.50	4.406	4.501	4.561	4.593	4.596
0.55	4.380	4.469	4.521	4.542	4.526
Excitation = 8 A					
0.25	6.010	5.964	5.919	5.875	5.824
0.30	7.255	8.312	8.273	8.226	8.166
0.35	9.353	9.313	9.266	9.206	9.124
0.40	9.722	9.675	9.614	9.533	9.422
0.45	9.810	9.873	9.888	9.855	9.764
0.50	9.732	9.779	9.771	9.704	9.559
0.55	9.619	9.645	9.606	9.493	9.273
Excitation = 12 A					
0.25	8.136	8.019	7.915	7.810	7.690
0.30	12.816	12.727	12.635	12.528	12.391
0.35	14.514	14.414	14.301	14.162	13.975
0.40	15.030	14.913	14.770	14.585	14.331
0.45	15.102	15.090	15.009	14.846	14.568
0.50	14.922	14.874	14.740	14.500	14.102
0.55	14.666	14.570	14.367	14.023	13.454

References

[1] P.J. Lawrenson, J.M. Stephenson, P.T. Blenkinsop, J. Corda, and N.N. Fulton, "Variable speed switched reluctance motors", Proc.IEE, Vol 127, No.4, April 1980, pp. 253-265.

[2] P.J. Lawrenson, J.M. Stephenson, and J. Corda, "Switched reluctance motors for traction drives", Proceedings of the International Conference on Electrical Machines, Athens, Greece, 1980, pp. 410-417.

[3] W.F. Ray, R.M. Davis, J.M. Stephenson, P.J. Lawrenson, R.J. Blake, and N.N. Fulton, "Industrial switched reluctance drives - Concepts and performance", Proc. of IEE Conf. on Power Electronics and Variable Speed Drives, London, 1984, pp. 357-360.

[4] J.R. French, "Switched reluctance motor drives for rail traction: relative assessment", Proc.IEE, Vol 131, Pt.B, No.5, Sept. 1984, pp. 209-219.

[5] W.F. Ray, R.M. Davis, P.J. Lawrenson, J.M. Stephenson, N.N. Fulton, and R.J. Blake, "Switched reluctance motor drives for rail traction: a second view", Proc. IEE, Vol 131, Pt.B, No.5, Sept. 1984, pp. 220-225.

[6] N.N. Fulton, P.J. Lawrenson, J.M. Stephenson, R.J. Blake, R.M. Davis, and W.F. Ray, "Recent developments in high performance switched reluctance drives", Second Int. Conf. on Electrical Machines - Design and Applications, Sept. 1985, pp. 130-133.

[7] W.F. Ray, P.J. Lawrenson, R.M. Davis, J.M. Stephenson, N.N. Fulton, and R.J. Blake, "High performance switched reluctance brushless drives", IEEE Trans. on Industry Applications, Vol IA-22, No.4, July/Aug. 1986, pp. 722-730.

[8] W.F. Ray, R.M. Davis, and R.J. Blake, "The control of SR motors", Proc. of the Conference on Applied Motion Control 1986, Minneapolis, Minnesota, June 1986, pp. 137-145.

[9] J.W. Finch, M.R. Harris, A. Musoke, and H.M.B. Metwally, "Variable speed drives using multi-tooth per pole switched reluctance motors", Proc. 13th Annual Symposium on Incremental Motion Control Systems and Devices, May 1984, pp. 293-301.

[10] J.W. Finch, M.R. Harris, H.M.B. Metwally, and A. Musoke, "Switched Reluctance motors with multiple teeth per pole: Philosophy of design", Second Int. Conf. on Electrical Machines - Design and Applications, Sept. 1985, pp. 134-138.

[11] T.J.E. Miller, "Converter volt-ampere requirements of the switched reluctance motor drive", IEEE Trans. on Industry Applications, Vol IA-21, No.5, Sept/Oct. 1985, pp. 1136-1144.

[12] B.K. Bose, T.J.E. Miller, and P.M. Szczesny, and W.H. Bicknell, "Microprocessor control of switched reluctance motor", IEEE Trans. on Industry Applications, Vol IA-22, No.4, July/Aug. 1986, pp. 708-715.

[13] M.R. Harris, J.W. Finch, J.A. Mallick, and T.J.E. Miller, "A review of the integral horsepower switched reluctance drive", IEEE Trans. on Industry Applications, Vol IA-22, No.4, July/Aug. 1986, pp. 716-721.

[14] J.T. Bass, M. Ehsani, T.J.E. Miller, and R.L. Steigerwald, "Development of a unipolar converter for variable reluctance motor drives", IEEE Industry Applications Society Annual Meeting, Oct. 1985, pp. 1062-1068.

[15] J. Corda, and J.M. Stephenson, "An analytical estimation of the minimum and maximum inductances of a double salient motor", Proc. Int. Conf. on Stepping motors and systems, University of Leeds, 1979, pp. 50-59.

[16] R. Arumugam, D.A. Lowther, R. Krishnan, and J.F. Lindsay, "Magnetic Field Analysis of a Switched Reluctance Motor Using a Two-Dimensional Finite Element Model", IEEE Trans. on Magnetics, Vol MAG-21, No. 5, September 1985, pp. 1883 - 1885.

[17] J.F. Lindsay, R. Arumugam, and R. Krishnan, "Finite-element characterization of a switched reluctance motor with multitooth per stator pole", Proc. IEE, Part B, Vol 133, No 6, November 1986 pp. 347 -353.

1.3 - Finite-Element Analysis

Analysis of Variable Reluctance Motor Parameters Through Magnetic Field Simulations

Karl Konecny

MOTOR-CON Proceedings, 1981, 2A2.1-11

ABSTRACT

The static characteristics of a variable reluctance motor can be described by its phase resistance, inductance, and torque output. The latter two are nonlinear functions of current levels and rotor position. The prediction of these characteristics is vital for optimal motor design and is best accomplished through magnetic field simulation. For low current excitation, an analogous magnetic circuit technique yields accurate and general results. For high current excitation which saturates the magnetic materials, a finite element technique is used to solve the field equations. Motor symmetries are accounted for to reduce the magnitude of the problem. Excellent agreement between measured and predicted torque at all current levels is achieved.

High performance motion control systems place rigorous demands on variable reluctance motors. Designing a V.R. motor for these applications requires accurate knowledge of the magnetic fields that relate motor geometry and motor performance. These relationships can be investigated through extensive prototyping or accurate magnetic field simulation.

Extensive prototyping is impractical as it is very costly and time consuming. Also, many of the magnetic field characteristics of interest are difficult to measure. Magnetic field simulation directly yields predictions of flux linkages, field energy, and torque. This information allows the designer to shape torque curves, control inductance, minimize local saturation, investigate various lamination materials, and determine parameter sensitivities.

In this paper, two techniques of field simulation are presented. For low current levels, the motor is modeled by an analogous magnetic network circuit. Each segment of the rotor, stator, and various airgaps are replaced by a "resistor" of equivalent reluctance and the coils are modeled as voltage sources. The result is an equivalent network circuit from which the various flux linkages within the motor can be determined. For higher current levels which induce magnetic saturation in the rotor or stator, the analogous magnetic circuit technique breaks down. To extend the analysis into these excitation levels, a finite element technique is employed to solve for the magnetic potential vector. All material non-linearities are accounted for. These modeling techniques can be used in conjunction to optimize motor designs.

ANALOGOUS MAGNETIC NETWORK CIRCUIT

The accuracy of the analogous network circuit technique depends upon the inclusion of all significant flux paths in the network and the accurate determination of the reluctance of each path.

The modeling of the soft iron portions of the motor is straight forward. Each tooth and intertooth segment of the rotor and stator are replaced by individual resistors. The reluctance of each segment is determined from Equation (1)

$$R = \frac{L}{\mu_s A} \tag{1}$$

where L is the segment length, A is the segment cross sectional area, and μ_s is the permeability of the constituent material.

The reluctance of the various airgaps are functions of rotor position and are determined from apriori field estimations. Flux paths are assumed to exist between each stator tooth and the neighboring rotor teeth as well as between adjacent stator teeth (Figure 1). The geometry of each flux path is determined from the following assumptions:

1) Each flux path consists of parallel bundles of flux tubes of constant differential area dS.

2) The flux tubes follow arcs of circles and straight lines, maintaining continuity of slope.

3) Flux tubes enter and leave iron segments at right angles to the material surface. This assumption very nearly satisfies the boundary condition that the normal component of the magnetic induction be continuous across material interfaces.

4) Flux tubes do not cross each other and span the shortest distance, consistent with assumptions 2 and 3, between iron segments of opposite or neutral polarity.

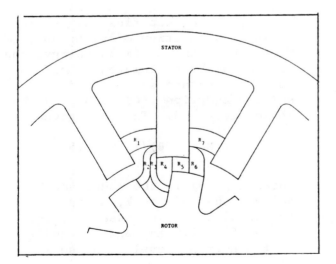

Figure 1 - Flux Path Estimations

The reluctances are now calculated by integrating across the component flux tubes of each path

$$\frac{1}{R} = \int_s \frac{\mu_o \, dS}{L(S)} \qquad (2)$$

where $L(S)$ is the flux tube length, dS is differential area, and μ_o is the permeability of free space. Often an air gap will contain several parallel flux paths, the net reluctance is determined from the parallel addition law

$$\frac{1}{R} = \sum \frac{1}{R_i} \qquad (3)$$

The goal of this work is to transform the geometry of the motor into an equivalent network circuit as shown in Figure 2. Equations of the form

$$\sum N_i I_i = \sum R_j(\theta) \, \phi_j(\theta) \qquad (4)$$

are constructed for each loop in the network where $\sum N_i I_i$ is the sum of all current linking each loop, $R(\theta)$ is the reluctance of the jth flux path, and $\phi_j(\theta)$ is the jth flux. Equation (4) represents a set of simultaneous equations that are linear in terms of the loop flux values. These equations are solved at incremental rotor positions.

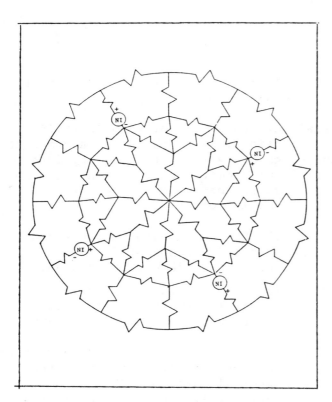

Figure 2 – Analogous Magnetic Network Circuit

Once the loop flux values are known as a function of rotor position; inductances, field energy, and torque can be calculated. Self and mutual inductances are found by solving Equation (4) with only one phase energized. They are given by

$$L_{jk}(\theta) = N_j \, \phi_j(\theta)/I_k \qquad (5)$$

where $L_{jk}(\theta)$ is the inductance between phase j and k, N_j is the number of turns in phase j, $\phi_j(\theta)$ is the flux linking phase j and I_k is the current level in phase k.

Field energy is found by summing the contribution of each energized phase

$$W(\theta) = \sum \tfrac{1}{2} N_j \, I_j \, \phi_j(\theta) \qquad (6)$$

output torque is found through numerical differentiation

$$T(\theta) = \frac{-\partial W}{\partial \theta}\bigg|_{I} = 1/2 \sum N_j I_j \frac{[\phi_j(\theta) - \phi_j(\theta + \Delta\theta)]}{\Delta\theta} \qquad (7)$$

The analogous magnetic network scheme allows the designer a method to predict static torque curves, inductance values, and leakage factors for a specific motor design. The technique also yields some general results that apply to most conventional V.R. motors.

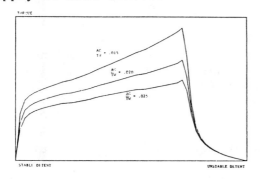

Figure 3 - Torque vs. Position for Various Air Gap to Tooth Width Ratios

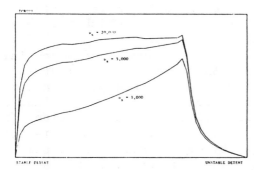

Figure 4 - Torque vs. Position for Various Rotor-Stator Permeabilities

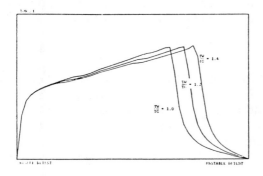

Figure 5 - Torque vs. Position for Various Stator Tooth Width to Stator Tooth Gap Ratios

Figure 3 shows predicted torque curves for various rotor-stator air gap lengths. As expected, the torque increases as the air gap decreases but this effect is greatest at positions far from the stable detent point. Figure 4 shows predicted torque curves for various permeabilities of rotor and stator constituents. The torque increases as the permeability increases but this effect is greatest near stable detent.

Figure 5 illustrates the effect of stator tooth width on the torque curves. As the teeth are widened, the peak torque drops and the torque curve broadens. Intra-stator leakage also increases as stator teeth are widened.

FINITE ELEMENT METHOD

V.R. motors are generally operated with some degree of induced magnetic saturation in the rotor or stator. The analogous magnetic network circuit modeling technique is not easily extended to encompass these material non-linearities. A finite element technique is used to analyze the magnetic fields within the motor for energization levels which induce saturation.

The finite element method directly yields the magnetic induction vector ($\vec{B}$) as opposed to flux (ϕ), an induction resultant, which the analogous network circuit technique produces. Thus the finite element method easily admits techniques to account for saturation, which depends upon induction levels.

The finite element method of magnetic field simulation relies on three equations from magnetostatics. The first states the fact that the magnetic induction vector ($\vec{B}$) can be represented by the curl of a magnetic potential vector ($\vec{A}$)

$$\vec{B} = \nabla \times \vec{A} \qquad (8)$$

For a V.R. motor, which can be modeled by its cross section, the potential vector is normal to the cross section. Second, the magnetic field vector ($\vec{H}$) is an increasing function of the magnetic induction vector

$$\vec{H} = \nu(B^2)\ \vec{B} \qquad (9)$$

where $\nu(B^2)$ is the inverse of the local permeability which, for non-linear materials, is a function of induction level. Finally, the three quantities are related through the minimization of Equation (10)

$$\int_V \int_0^{|B|} \vec{H} \cdot d\vec{B} - \vec{J} \cdot \vec{A}\ dV \qquad (10)$$

where $\vec{J}$ is the current density vector and the integration is carried out over the volume of the motor.

The geometry of the motor is discretized into a set of similar elements (Figure 6)[1]. Only one quarter of the motor need be considered due to symmetry considerations discussed later. Each element only admits a finite number of nodes at which the normal component of the vector potential is allowed to vary. The potential values throughout the element is determined from interpolation functions. Equation (10) is evaluated numerically, imposing Equations (8) and (9). The minimum value is determined by differentiating Equation (10) with respect to each nodal value and setting the result to zero. This process yields a set (one for each nodal degree of freedom) of non-linear algebraic equations as represented in Equation (11).

$$[S(B^2)]\ \{A\} = \{J\} \qquad (11)$$

where $[S(B^2)]$ is a reluctivity matrix which is a function of the induction level through variations in the permeability, $\{A\}$ is the nodal potential vector, and $\{J\}$ is the current distribution vector.

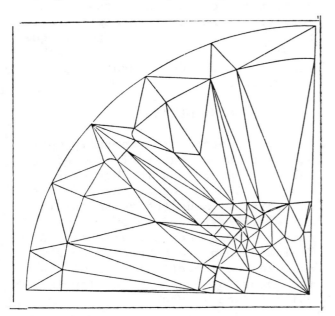

Figure 6 - Finite Element Mesh, 132 Elements, 265 Degrees of Freedom

---

[1] For a complete description of the finite element method see Ref. 1.

It is desired to know the potential values as a function of current, therefore Equation (11) is best solved at incremental levels of current excitation using a Newton-Raphson convergence scheme. Once the magnetic potential is known, the magnetic induction is found from Equation (8). The flux passing between two points is given by Equation (12)

$$\phi = \int_S \vec{B} \cdot d\vec{s} = \oint_L \vec{A} \cdot d\vec{L} = L(A_2 - A_1) \tag{12}$$

where L is the stator length and A_i is the potential value at point i.

The field energy is found through Equation (13).

$$W = \int_V \int_0^J \vec{A} \cdot d\vec{J} \; dV \tag{13}$$

Torque is the derivative of field energy with respect to rotor position holding current levels constant

$$T = \int_V \int_0^J \frac{\partial \vec{A}}{\partial \theta} \cdot dJ \; dV \tag{14}$$

The quantity $\frac{\partial \vec{A}}{\partial \theta}$ can be found as follows. Differentiating Equation (11) with respect to rotor position yields.

$$[S] \; \{\frac{\partial A}{\partial \theta}\} + [\frac{\partial S}{\partial \theta}] \; \{A\} = 0 \tag{15}$$

[S] and {A} are known and $[\frac{\partial S}{\partial \theta}]$ can be calculated from geometrical considerations (Appendix A), thus Equation (15) can be solved for $\frac{\partial A}{\partial \theta}$.

This finite element procedure is carried out at various rotor positions which yield torque and flux curves as functions of both current and position. The technique also yields information about leakage and local saturation.

IMPLEMENTATION

In the work presented, six noded, isoparametric, three sided elements are used (Figure 7). These provide two advantages over common three noded elements. Three nodes per side allow parabolic element boundaries which match the curved portions of the motor geometry well. This is particularly important in the rotor stator gap region. Failure to accurately match the true curvature of the tooth tips result in erroneous torque calculations. Six noded elements also allow quadratic interpolation of the potential vector. This produces a linear variation of the magnetic induction which admits localized saturation within elements. Thus, a lower elemental density is required to accurately model the magnetic fields.

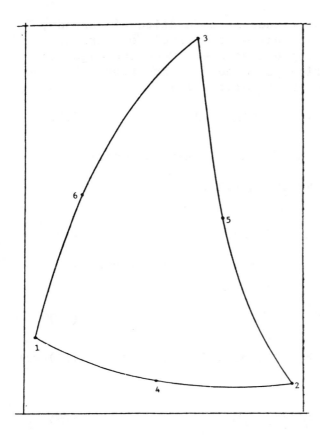

Figure 7 - 6 Noded, 3 Sided,
Isoparametic Element

Motor modeling through finite elements yields large scale systems of algebraic equations. The magnitude of the problem can be reduced through symmetry considerations. Only the portion of the motor cross/ section which, through symmetry operations, defines the total geometry and current distribution of the motor need be discretized. The specific symmetries are accounted for through constraints of the potential vector along the symmetry boundaries.

When a motor is at a stable or unstable detent position, several mirror planes exist. If the current distribution does not change signs in the mirror image, the nodal potentials along the mirror plane are unconstrained. This is equivalent to restricting the induction vector to be normal to the mirror plane. If the current distribution changes signs in the mirror image, the mirror plane potential values are set to zero. This is equivalent to restricting the induction vector to be parallel to the mirror plane.

For any other rotor stator orientation, mirror planes do not exist. An n fold rotation axis does though, where n is the number of stator teeth per phase. Thus only 1/nth of the motor cross section needs be considered. The remainder of the geometry is defined through n - 1 successive 360°/n clockwise rotations of the discretized portion about the rotor axis. This symmetry operation results in the left boundary of the model rotating into the right boundary.

If the current distribution maintains the same sign during the rotation, the magnetic potential also maintains the same sign. Thus nodal potential values on the left boundary must be constrained to equal the potential at corresponding points on the right boundary. This is accomplished by first defining nodes on the rotation boundaries at corresponding positions. A mapping function is then generated which maps corresponding nodes into the same degree of freedom. The resulting number of equations in Equation (11) will be N - M where N is the total number of defined nodes and M is the number of nodes lying on the left symmetry boundary.

If the current distribution reverses through each successive rotation, so will the potential vector. Thus potential values at corresponding points of the rotation boundaries are equal in magnitude but opposite in sign. This is accounted for by defining the same mapping function as in the previous case and introducing an elemental symmetry matrix that transforms calculated potential values $\{A\}_C$ into actual $\{A\}_A$ potential values

$$\{A\}_A = [D] \ \{A\}_C \tag{16}$$

For most elements $[D]$ is the identity matrix, that is, the actual potential values are directly calculated. For elements that lie along the left symmetry boundary $[D]$ is again a diagonal matrix with elements equal to 1 for nodes lying off the left boundary and -1 for nodes on the left boundary.

RESULTS

Computer programs were generated at the Hewlett-Packard Corporate Laboratories which implement both the analogous magnetic network circuit technique and finite element method of field simulation of V.R. motors. The Warner Electric SM-024-0035-AB V.R. motor was modeled. The finite element mesh required 100 to 150 elements and 200 to 300 nodes, depending upon rotor position. Only one eighth of the motor cross section needed to be considered for the stable and unstable detent orientations, one quarter of the motor was discretized for general orientations. An example is shown in Figure 6. The resulting magnetic induction is sketched in Figure 8 for a one phase on, two AMP excitation. Note that over 30 percent of the flux leaks back through secondary paths. The simulation was performed at various rotor positions. The predicted static torque curves are plotted along with measured torque curves in Figure 9.

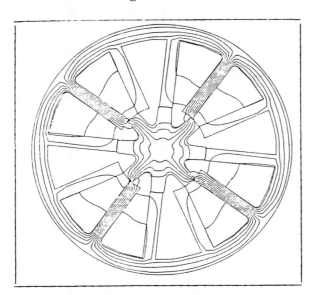

Figure 8 - Simulated Magnetic Induction, Warner Electric SM-024-0035-AB, θ = 14°, Phase Current = 2 AMPS

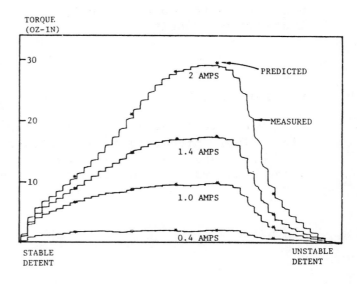

Figure 9 - Predicted and Measured Static Torque Curves, Warner Electric, SM-024-0035-AB

CONCLUSION

The analogous magnetic network circuit and finite element modeling techniques can be used in conjunction as an effective tool in motor design. Accurate magnetic field simulation allows the designer to predict and shape static torque curves, determine inductance levels, and control leakage and saturation.

The analogous network circuit method is particularly useful for a first cut analysis. A computer program for a particular geometry is easily written and implemented. As geometric dimensions of the motor are explicit parameters in the reluctance equations, they are easily varied and optimized. The refined design is then discretized and analyzed with the finite element techniques. This gives the designer a handle on saturation and leakage levels. The motor design can be fine tuned to minimize these problems.

The field simulation techniques presented can be extended to model linear V.R. motors and permanent magnet motors. Linear V.R. motors are axially symmetric and best analyzed in the (r, ϕ, z) cylindrical coordinates. For these motors, the magnetic potential vector $\vec{A}$ points in the ϕ direction. To extend the finite element method into this realm, the integrations in Equation (10) must be performed in the cylindrical domain and the cylindricity of the magnetic potential vector must be accounted for in Equation (8).

To include permanent magnets in a finite element model, the constitutive equation (Equation (9)) must be changed to

$$\vec{H} - \vec{H}_C = \nu(B^2)\vec{B} \tag{17}$$

where $\vec{H}_C$ is the coercive force vector. The added term contributes an equivalent current density

$$\vec{J}_e = -\nabla \times \vec{H}_C \tag{18}$$

to equation (10).

Variable reluctance motors offer many advantages to motion control systems. Their high torque to inertia ratio, high reliability, low cost and lack of brushes make them extremely attractive. Optimizing motor designs through accurate field simulations can only extend their usefulness.

202

APPENDIX: A DERIVATION OF $\frac{\partial [S]}{\partial \theta}$

The elemental contribution to the global reluctivity matrix is expressed as

$$[S]_e = \int_A \nu [C]^T [C] \, dA \qquad (A1)$$

where [C] is the matrix function that relates magnetic induction $\vec{B}$ to nodal potential values.

$$\vec{B} = [C] \{A\} \qquad (A2)$$

The integration is carried out over the element's cross sectional area.

The matrix [C] is expressed as

$$[C] = [J] [CL] \qquad (A3)$$

where [CL] depends only upon local coordinates and [J] is the transformation between local and global coordinates, thus [CL] is independent of element geometry. Equation (A1) is now expressed as

$$[S]_e = \int_{L_1} \int_{L_2} \nu [CL]^T [J]^T [J] [CL] \, DJ \, dL_1 \, dL_2 \qquad (A4)$$

where DJ is the determinant of $[J]$ and the integration is expressed in terms of local coordinates, a convenient form for numerical calculations.

Differentiating Equation (A4) with respect to rotor position yields

$$\frac{\partial [S]_e}{\partial \theta} = \int_{L_1} \int_{L_2} \nu \left[[CL]^T \left(\frac{\partial [J]^T}{\partial \theta} [J] + [J]^T \frac{\partial [J]}{\partial \theta} \right) [CL] \, DJ \right.$$

$$\left. + [CL]^T [J]^T [J] [CL] \frac{\partial DJ}{\partial \theta} \right] dL_1 \, dL_2$$

$$+ \int_{L_1} \int_{L_2} \frac{\partial \nu}{\partial \{A\}} \frac{\partial \{A\}}{\partial \theta} [C]^T [C] \, DJ \, dL_1 \, dL_2 \qquad (A5)$$

The second integral reduces to

$$\left[2 \int_{L_1} \int_{L_2} [C]^T \vec{B} \frac{\partial \nu}{\partial B^2} \vec{B}^T [C] \, dL_1 \, dL_2 \right] \frac{\partial \{A\}}{\partial \theta} \qquad (A6)$$

The bracketed matrix is identical to the incremental reluctivity matrix used in the Newton-Raphson solution scheme.

All that remains to be calculated are derivatives of the components of [J] with respect to rotor position. The component J_{ij} is an explicit function of x_k and y_k, the nodal coordinates within the element. Therefore

$$\frac{\partial J_{ij}}{\partial \theta} = \frac{\partial J_{ij}}{\partial x_k} y_k - \frac{\partial J_{ij}}{\partial y_k} x_k \tag{A7}$$

where $\dfrac{\partial J_{ij}}{\partial x_k}$ and $\dfrac{\partial J_{ij}}{\partial y_k}$ are non-zero only for nodes that lie on the rotor.

REFERENCES

1. Zienkiewicz, O.C., The Finite Element Method, 3rd edition, McGraw Hill, 1977.

2. Silvester, P., and Chari, M., "Finite Element Solution of Saturable Magnetic Field Problems", IEEE Transactions on Power and Systems, Vol. PAS-89, No. 7, Sept./Oct., 1970.

3. Lorrain, P., and Corson, D., Electromagnetic Fields and Waves, 2nd edition, W.H. Freeman and Co., San Francisco, 1970.

Switched Reluctance Motor Torque Characteristics: Finite Element Analysis and Test Results

G.E. Dawson, A.R. Eastham and J. Mizia

IEEE Transactions, Vol. IA-23, No. 3, May/June 1987, pp. 532-537

Abstract—The switched-reluctance motor is a simple and robust machine which is finding application over a wide power and speed range. To properly evaluate the motor design and performance and the effectiveness of different control schemes, an accurate model is required. The finite-element method can be used to predict the performance of the switched-reluctance motor as it can account for the salient pole geometry of the stator and rotor and the nonlinear properties of the magnetic materials. The computed results are shown to compare favorably with test results from a 7.5-kW Oulton switched-reluctance motor.

I. INTRODUCTION

IN RECENT YEARS there has been a growing interest in the design, performance, control, and application of switched-reluctance motor (SRM) drives [1]-[5]. Because of its simplicity and controllability, the motor is being evaluated for applications ranging from low-power servomotors to higher power traction drives [6], [7]. Switched-reluctance motors with power capabilities from 4 to 22 kW are presently commercially available for variable-speed applications. Experience indicates that system efficiencies of these drives are greater than comparable controlled induction motor drives over a wide speed range.

The switched-reluctance motor is a simple motor concept consisting of a salient pole structure on both the stator and the rotor (Fig. 1). The motor excitation is provided to a concentrated winding on each stator pole, with the rotor being a passive laminated steel structure. The excitation of diametrically opposite stator poles, when the rotor poles are nearby, creates a torque tending to align the stator and rotor poles. When the number of stator and rotor poles differ, the sequential switching of excitation from one set of stator poles to the next, in synchronism with the rotor position, results in a nearly constant torque. The synchronization of the turn-on of the excitation with rotor position can be accomplished with simple rotor position feedback.

To evaluate properly the switched-reluctance motor design and performance and the effectiveness of different control schemes, a reliable model is required. The finite-element technique [8], [9] can be conveniently used to obtain the magnetic vector potential values throughout the motor in the presence of complex magnetic circuit geometry and nonlinear properties of the magnetic materials. These vector potential values can be processed to obtain the field distribution, torque,

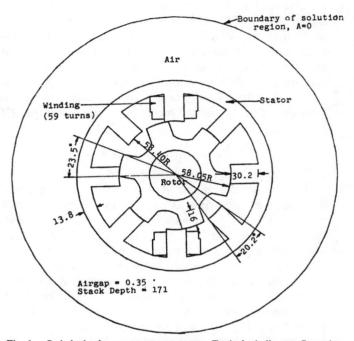

Fig. 1. Switched-reluctance motor geometry. Typical winding configuration shown for only one of four phases. Dimensions in mm for 7.5-kW Oulton machine.

and flux linkage. This paper will report on the use of the finite-element method to obtain the field distribution and torque at different operating values of rotor position and current. The computed results will be shown to compare favorably with experimental results. The flux linkage results will also be presented as this information is useful for field-network models [10].

II. FINITE-ELEMENT METHOD

Assuming a z-directed magnetic vector potential A_z, the two-dimensional magnetostatic problem is described by the nonlinear Poisson's equation

$$\frac{\partial}{\partial x}\left(\gamma \frac{\partial A}{\partial x}\right) + \frac{\partial}{\partial y}\left(\gamma \frac{\partial A}{\partial y}\right) = -J_z \qquad (1)$$

where γ is the magnetic reluctivity and J_z is the source current density. To solve (1), the following additional assumptions were made in the CAD-MAGNET package: 1) problem quantities are invariant with respect to the z coordinate, 2) the iron regions are isotropic, with a single-valued $B - H$ curve, and 3) the magnetic field is confined within the motor and a surrounding annulus of free space with the outer periphery being treated as a line of zero vector potential.

The finite-element analysis was performed on the switched-reluctance motor geometry shown in Fig. 1, which closely represents the 7.5-kW Oulton motor cross section. Discretized models of the motor were established for rotor and stator pole angular displacements θ, ranging from 0° to 26° in 2° increments. Fig. 2(a) shows the discretization of the complete region for $\theta = 18°$. The air gap between the vertical pair of stator and rotor poles (0.35 mm) was finely discretized in two layers with the remaining portions of the air space being more coarsely discretized in one layer (see Fig. 2(b) for the discretization of the air gap region only). This provided good resolution of the coenergy density of the motor. The approximate size of the discretized model (for one θ angle) was 2300 elements.

III. EXPERIMENTAL TORQUE RESULTS

The static torque performance of a 7.5-kW Oulton switched-reluctance motor was obtained over an angular displacement range from 0° to 30° for winding excitation currents of 4-, 6-, 8-, and 10-A dc. The rotor was blocked at different angles, and for a given excitation the torque value was obtained from a shaft torque transducer.

The "zero" angle was determined by energizing one of the phases with dc current and observing the position at which the rotor came to rest. At the zero point, a dead band occurred in the torque of about 2°. The experimental results of these tests are shown in Fig. 3. Fig. 4 shows the maximum torque as a function of winding excitation current. The maximum values of torque tend to lie at pole angular displacements of 18–20° over the range of excitation and can be seen to be proportional to the current squared as would be expected for a nonsaturating magnetic circuit.

IV. COMPUTED TORQUE RESULTS

The electromagnetic torque of the switched-reluctance motor was computed from the rate of change of coenergy with respect to angular displacement $W'(\theta, i)/\partial\theta$. The coenergy, in general, can be written as a function of the displacement angle θ and the current excitation i as follows:

$$W'(\theta, i) = \int_V \bar{A} \cdot \bar{J} \, dV - \int_V \int_0^B \frac{B'}{\mu(B')} \, dB' \, dV \quad (2)$$

where J is the source current density, B is the flux density obtained from $\nabla \times A$, μ is the permeability of a given material, and V is the volume of the region obtained from integration over the two-dimensional region of interest multiplied by the stack depth of the machine.

This equation was programmed as a user-defined verb in the postprocessing package of the MAGNET finite-element program. The torque was calculated from the derivative of the coenergy curves—(2)—with respect to the angle θ.

206

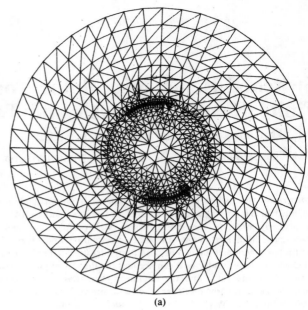

(a)

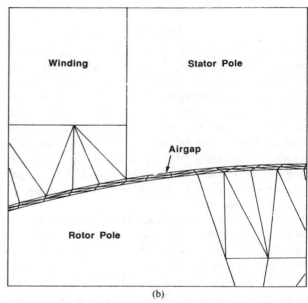

(b)

Fig. 2. Discretization of switched-reluctance motor for stator–rotor pole displacement of 18°. (a) Complete region. (b) Zoom (400×) of airgap region.

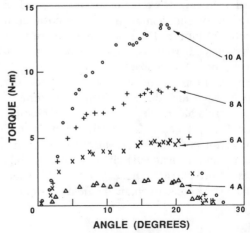

Fig. 3. Test results of torque-versus-displacement angle for winding currents of 4, 6, 8, and 10 A.

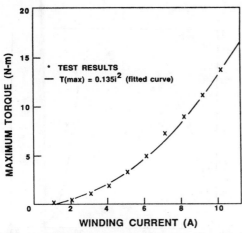

Fig. 4. Test results of maximum torque versus winding current. Results are compared to fitted curve; $T(\text{max}) = 0.135i^2$.

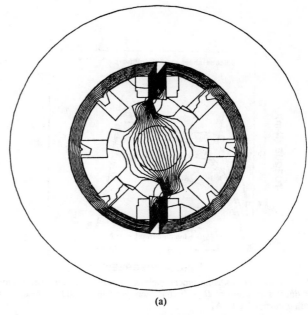

(a)

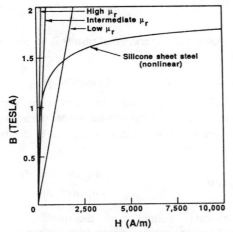

Fig. 5. $B - H$ characteristics for linear and nonlinear magnetic materials.

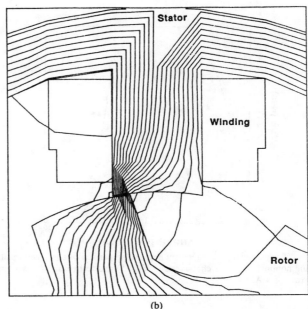

(b)

Fig. 6. Field distribution for displacement of 18° and winding current of 10 A. (a) Complete region. (b) Zoom of stator-pole-airgap-rotor-pole region.

To determine the sensitivity of the static torque characteristic to the permeability of the ferromagnetic material of the stator and rotor laminations, the coenergy was evaluated for the machine model with a range of angular displacements from 0° to 26°, in 2° increments, using low, intermediate, and high relative permeabilities. The actual values used ($\mu_r = 795$, 2789, and 5700) correspond to the approximate nonsaturating permeabilities of cast iron, cast steel, and silicon steel, respectively (Fig. 5). The coenergy was computed for four winding currents (4, 6, 8, and 10 A), and the torque was calculated from the derivative of the coenergy curves with respect to the angle θ.

Typical results of the finite-element method computations for an angle displacement of 18° and winding excitation current of 10 A are presented in Figs. 6 and 7. The field distribution is shown in Fig. 6 with the overall field distribution presented in Fig. 6(a) and the detailed stator-pole-airgap-rotor-pole field distribution presented in Fig. 6(b).

The computed torque values were obtained from the derivative of the coenergy curves generated from (2). Fig. 7 shows the effect of increasing the linear relative permeability and the effect of a nonlinear $B - H$ characteristic as compared

with a linear $B - H$ characteristic having the same low field μ_r.

The higher the material permeability, the larger the value of torque at all angular displacements for a given winding current. This observation illustrates the importance of the magnetic properties of the laminations on the torque performance of the motor.

The assumed magnetic properties of the nonlinear silicon sheet steel are shown in Fig. 5. A comparison between computed and measured values of torque versus angle is shown in Figs. 8–10 and Fig. 7 for winding excitations of 4, 6, 8, and 10 A, respectively. In view of an uncertainty in the precise $B - H$ characteristic of the SRM laminations and the sensitivity of the torque to permeability, the agreement

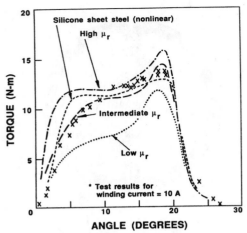

Fig. 7. Torque-versus-displacement angle. Comparison of computed results with linear and nonlinear $B - H$ characteristics (Fig. 5) and test results for winding current of 10 A.

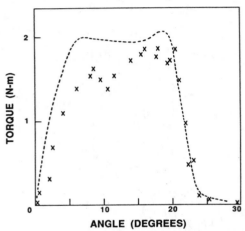

Fig. 8. Torque-versus-displacement angle. Comparison of computed results using nonlinear $B - H$ characteristic (Fig. 5) and test results for winding current of 4 A.

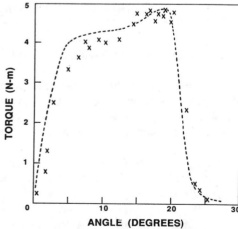

Fig. 9. Torque-versus-displacement angle. Comparison of computed results using nonlinear $B - H$ characteristic (Fig. 5) and test results for winding current of 6 A.

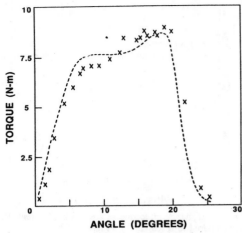

Fig. 10. Torque-versus-displacement angle. Comparison of computed results using nonlinear $B - H$ characteristic (Fig. 5) and test results for winding current of 8 A.

between computed and experimental results is as good as could be expected.

V. FLUX LINKAGE

The flux linkage $\Psi(\theta, i)$ for a phase winding varies with phase current i and cyclically with rotor position θ. The voltage v applied to the winding is related to Ψ at any instant by

$$v = Ri + \frac{d\Psi(\theta, i)}{dt} = Ri + \frac{\partial \Psi}{\partial i}\frac{di}{dt} + \frac{\partial \Psi}{\partial \theta}\frac{d\theta}{dt} \qquad (3)$$

where the resistance voltage drop, the transformer voltage, and the speed voltage terms appear in the right side of (3), respectively.

Unlike in many electrical machines, $\partial\Psi/\partial i$ and $\partial\Psi/\partial\theta$ are not constant, and, therefore, it is not possible to establish a fixed parameter equivalent circuit. The nonlinear function $\Psi(\theta, i)$ has been approximated by an analytical function [11] and used in conjunction with state variable equations to simulate the performance of the switched-reluctance motor and to investigate control schemes. It is also possible to take advantage of the finite-element solution of the magnetic vector potentials over the motor cross section to generate flux linkage curves as a function of angle θ and winding current i.

For a phase coil with axis in the y direction (using Cartesian coordinates), the flux linkage through one turn may be evaluated from the vector potential as

$$\Psi(\theta, i) = D \int_{x_1}^{x_2} B_y \, dx = D \int_{x_1}^{x_2} \frac{\partial A}{\partial x} \, dx$$

$$= D[A_2(x_2, y) - A_1(x_1, y)] \qquad (4)$$

where D is the stack width. Points x_1 and x_2 and the entire geometry for this derivation are shown in the schematic in Fig. 11. Equation (4) can be further developed to obtain

$$\psi(\theta, i) = D \frac{N}{S} (\alpha_1 - \alpha_2) \qquad (5)$$

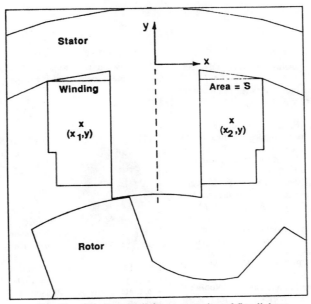

Fig. 11. Winding detail for computation of flux linkage.

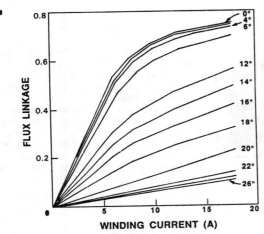

Fig. 12. Computed flux linkage as function of winding current and displacement angle using nonlinear $B - H$ characteristic (Fig. 5).

where

$$\alpha_{1,2} = \int_S A(x, y) \, ds_{1,2}$$

over windings 1 and 2; S is the winding cross section, and N is the number of turns of the winding.

Equation (5) was programmed into the post-processing package of the finite-element program to obtain a family of characteristics (Fig. 12) for flux linkage as a function of displacement angle θ and winding current excitation i. It can be seen that a magnetic saturation effect exists in the motor for small angles θ and higher excitation currents i. These results compare favorably with the flux linkage characteristics reported by others.

VI. Conclusion

The computed and experimental results presented in this paper have shown that the finite-element method is capable of predicting the static torque-angle characteristic of the switched reluctance motor. The method can account for the salient pole geometry and nonlinear magnetic material properties and their effects on the torque and flux linkage. It was shown that it is desirable to use high permeability ferromagnetic material to gain the maximum torque for a given geometry.

Having calculated the static torque-angle characteristic for a single pair of poles (or phase) of the machine, the pole switching sequence and the transient response of the excitation can be added to determine both the time average torque and the torque harmonics of the machine. Thus the finite-element method is an invaluable tool for magnetic design and performance calculations of switched-reluctance motors. In addition, the flux linkage results, in conjunction with state variable equations, can be used to obtain a field-network model for the evaluation of power conditioning units and control schemes.

Acknowledgment

The following individuals should be recognized: Mark Baker, Larry Pattison, and Chris Pallen for obtaining the test results, and Greg Ducharme for his assistance in the preparation of the figures.

References

[1] P. J. Lawrenson, J. M. Stephenson, P. T. Blenkinsop, J. Corda, and N. N. Fulton, "Variable-speed switched reluctance motors," *Proc. Inst. Elec. Eng.,* vol. 127, pt. B, no. 4, pp. 253–265, July 1980.

[2] W. F. Ray, R. M. Davis, and R. J. Blake, "The control of SR motors," in *Proc. Conf. Applied Motion Control 1986,* Minneapolis, MN, June 1986, pp. 137–145.

[3] P. J. Lawrenson *et al.,* "Controlled-speed switched-reluctance motors: Present status and future potential," in *Proc. Drives/Motors/Control 1982,* Leeds, 1982, pp. 23–31.

[4] J. V. Bryne, M. F. McMulin, and J. B. O'Dwyer, "A high performance variable reluctance drive: A new brushless servo," in *Motorcon 1985 Proc.,* 1985, pp. 147–160.

[5] T. J. E. Miller and T. M. Jahns, "A current-controlled switched-reluctance drive for FHP applications," in *Proc. Conf. Applied Motion Control 1986,* Minneapolis, MN, June 1986, pp. 109–117.

[6] M. R. Harris, J. W. Finch, J. A. Mallick, and T. J. E. Miller, "A review of the integral horsepower switched-reluctance drive," in *Proc.*

209

IEEE Industry Applications Soc. Ann. Meeting, Toronto, ON, Oct. 1985, pp. 783–789.

[7] W. F. Ray *et al.*, "Switched reluctance motor drives for rail traction: A second view," *Proc. Inst. Elec. Eng.,* vol. 131, pt. B, no. 5, pp. 220–225, Sept. 1984.

[8] A. R. Eastham, K. Ananthasivam, G. E. Dawson, and R. Ong, "Linear induction motor design and performance evaluation by complex frequency and time stepping finite element methods," in *Proc. Int. Conf. Evaluation and Modern Aspects of Induction Machines,* Torino, Italy, July 1986, pp. 42–48.

[9] R. Arumugam, D. A. Lowther, R. Krishnan, and J. F. Lindsay, "Magnetic field analysis of a switched reluctance motor using a two dimensional finite element model," *IEEE Trans. Magn.,* vol. MAG-21, pp.1883–1885, Sept. 1985.

[10] T. A. Nyamusa and N. A. Demerdash, "Transient analysis of partial armature short circuit in an electronically commutated permanent magnet motor system using an integrated nonlinear magnetic field-network model," presented at the IEEE/PES 1986 Winter Meeting, New York, Feb. 1986, paper 86 WM 228-1.

[11] M. Ilić-Spong, R. Marino, S. Peresada, and D. G. Taylor, "Nonlinear control of switched reluctance motors in robotics applications," in *Proc. Conf. Applied Motion Control 1986,* Minneapolis, MN, June 1986, pp. 129–136.

Graham E. Dawson (S'66–M'69) was born in North Vancouver, BC, on November 11, 1939. He received the B.A.Sc., M.A.Sc., and Ph.D. degrees from the University of British Columbia in 1963, 1966, and 1970, respectively.

In 1969 he joined the Department of Electrical Engineering, Queen's University at Kingston, as an Assistant Professor. He was promoted to Associate Professor in 1975 and to Professor in 1981. His electrical engineering research activities have been associated with the transportation industry, and he has current interest in the computer-aided design and performance of rotary and linear traction motors and energy management of transportation systems.

Dr. Dawson is a Registered Professional Engineer in the Province of Ontario and a member of the Canadian Society for Electrical Engineering.

Anthony R. Eastham (M'75–SM'83) received the B.Sc. degree in physics from the University of London, London, England, in 1965, and the Ph.D. degree from the University of Surrey, England, in 1969.

After research work at Plessey Telecommunications Limited and at the University of Warwick, he joined the Canadian Institute of Guided Ground Transport, where he coordinated a group that technically defined, component-tested, and assessed high-speed Maglev in Canada. He is now a Professor of Electrical Engineering at Queen's University in Kingston, ON, Canada, having joined the faculty in 1978. His research activities include innovative urban and high-speed transportation, linear electric drives, and electromagnetic analysis.

Dr. Eastham is a Registered Professional Engineer in the Province of Ontario.

Jerzy T. Mizia was born in Poland in 1944. He received the M.Sc. degree from the Jagiellonian University in 1965 and the Ph.D. degree from the Academy of Mining and Metallurgy (Kracow) in 1970, both in physics.

His research activities include theoretical solid-state physics and electromagnetic engineering, including finite-element analysis of electromagnetic devices. He is currently employed as a Research Engineer in the Department of Electrical Engineering, Queen's University, Kingston, ON, Canada.

Finite-Element Analysis Characterization of a Switched Reluctance Motor with Multi-tooth per Stator Pole

J.F. Lindsay, R. Arumugam and R. Krishnan

Proceedings IEE, Vol. 133, Pt. B, November 1986, pp. 347-353

Indexing trems: *Motors, reluctance, Reluctance motors, Mathematical techniques*

Abstract: The paper describes the use of a 2-dimensional finite-element field analysis (FEA) to obtain $N\phi/i/$ rotor position characteristics of a doubly salient switched reluctance motor. The stator and rotor have even but unequal numbers of poles. Unlike the single tooth per pole construction of the most common doubly salient pole motors, the motor considered for analysis has two teeth per stator pole. The increase in torque developed due to the multitooth stator pole configuration has been reported by Finch *et al.* in May 1984. An accurate knowledge of flux pattern inside the motor is essential to a machine designer for performance prediction. The complex configuration and material nonlinearities of the motor makes analytical methods of magnetic field analysis difficult. The numerical method (FEA) used gives a better understanding of the flux distribution inside the motor. The terminal inductance per phase, the flux linkage of each stator pole winding, and the components of the leakage inductances are determined for different rotor positions and excitation currents. The average torque developed by the motor is also determined under current forced conditions.

1 Introduction

Switched reluctance motor drives have drawn the attention of investigators in recent years [1–4]. Their construction is simple and robust. The power density of such drives is high and they operate with high power factor and efficiency. The control circuits are simple with a minimum number of switching devices compared with induction motor controllers [5]. There is no rotor I^2R loss because there are no windings on the rotor.

The doubly salient switched reluctance motor resembles the variable reluctance stepping motor in construction. The stator and rotor of the motor have even but unequal numbers of salient poles. Only the stator poles carry windings. Each concentrated winding on the stator poles, when connected in series with the diametrically opposite pole winding constitutes one phase of the motor. Each phase winding is excited at appropriate times such that the motor develops torque in the required direction. Rotor position sensors are used to generate control signals and thus switch the phase currents.

The airgap length of the motor is made very small so that the motor can be operated under saturated conditions. This results in a higher coenergy change as the rotor moves from the maximum to minimum reluctance position. Hence more torque is developed than would be the case with unsaturated conditions [6].

The doubly salient switched reluctance motor operates on the principle that the rotor poles tend to align themselves along the axis of minimum reluctance when the appropriate stator windings are excited. As the torque is developed due to the change of reluctance in the motor,

Paper 4885B(P1), received 24th March 1986

Prof. Lindsay and Mr. Arumugam are with the Department of Electrical Engineering, Concordia University, 1455 de Maissoneuve Blvd. West, Montreal, Quebec, Canada H3G 1M8. Dr. Krishnan was formerly with Gould Research Labs., Rolling Meadows, IL, USA. He is now with Virginia Polytechnic Institute and State University, Blacksburg, VA 24061, USA

the direction of current flow in the excited winding is immaterial. This results in unipolar operation of the motor with only one switching device in each phase. The number of switching operations for each revolution of the motor depends on the number of phases and the number of rotor poles. The direction of rotation of the motor depends on the sequence of switching the stator phase windings.

The speed of the motor can be varied by altering the switching frequency when the torque is held constant. The developed torque at constant power can be controlled by the instant and duration of switching. Thus the switching angles are controlled to achieve the desired speed and torque and to stabilise the operation of the motor. At low speeds current chopping is often used to limit the peak current.

The design of the switched reluctance motor becomes complicated due to the complex geometry of the motor. The reluctance variation of the motor has an important role on the performance of the motor. Hence an accurate knowledge of the flux distribution inside the motor for different excitation currents and rotor positions is essential for the prediction of performance of the motor. The motor is highly saturated under normal operating conditions and hence the nonlinear magnetisation characteristic of the core must be taken into account when the flux distribution is computed. All these aspects are included in a finite-element analysis.

For convenience, the position when a stator tooth is opposite a rotor tooth, such that the reluctance is minimum, is defined as the aligned position. The unaligned position is defined as that when the stator tooth is opposite the rotor slot such that the reluctance is maximum.

In earlier investigations, the performance of the motor was predicted based on the concept of mean static torque produced during the rotor movement from the maximum to minimum reluctance position when one stator phase winding is excited [2–4]. They make use of the permeances calculated for aligned and unaligned positions. In such permeance calculations the tooth and slot are assumed to be parallel sided, infinitely deep and the excitation source is remote from the tooth tip. An analytical estimation of the aligned and unaligned position permeance values from the motor geometry has been reported by Corda and Stephenson [7] for the doubly salient motor. All the above methods have considerable error in predicting permeance in the unaligned position which plays a critical role in determining the motor performance.

Stephenson and Corda [8] have reported an efficient method to predict the performance of a doubly salient reluctance motor from the measured or calculated magnetisation data. The finite-element analysis of the machine provides an alternative approach to the prediction of performance by directly calculating the coenergy at different rotor positions for a given excitation.

For the present study, it has been assumed that each phase of the motor is excited with rectangular blocks of current. In the case of a voltage fed motor, there would be the additional problem of modelling and predicting the dynamics of the build up and decay of the current.

A two-dimensional finite-element analysis of a switched reluctance motor with one tooth per stator pole has already been presented [9] by the authors. This accounts for the motor configuration, nonlinearities in the magnetisation curve of the motor steel laminations, and the location of the excitation source. In this paper, the finite-element analysis is used to study a switched reluctance motor having two teeth per stator pole. The flux linkages of an excited stator pole winding, the terminal inductance of the motor, the leakage inductances and the torque developed by the motor for various excitation currents and rotor positions are determined.

2 Motor specifications

The motor specifications considered for study are the same as those used by Finch et al. [1]. The motor has six stator poles with two teeth per pole. The rotor has ten projecting poles. Fig. 1 shows the geometry of the stator and rotor laminations in the aligned position. The winding on each stator pole is concentrated and has 0.9 mm diameter wire. The diametrically opposite pole windings are connected in series.

The main dimensions of the motor are:

stator core outer diameter = 16.51 cm
stator bore diameter = 9.3 cm
length of iron core = 10.8 cm
stator pole neck = 2.45 cm
back of core = 1.25 cm
length of airgap = 0.255 mm
rotor pole pitch = 2.91 cm
tooth width = 1.02 cm
diameter of the shaft = 2.858 cm
number of turns per pole = 222

The stator and rotor core are made of non-oriented silicon steel laminations. The magnetisation curve is taken from the manufacturer's data sheets for M-19 steel.

212

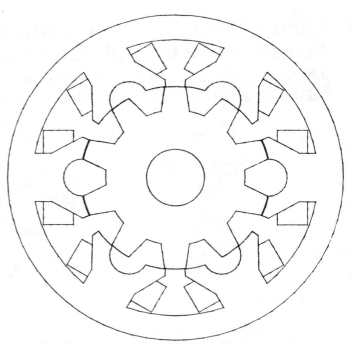

Fig. 1 *Motor configuration in aligned position*

3 Finite-element analysis

To determine the magnetic field distribution inside the motor the following assumptions are made:

(*a*) The magnetic field outside the motor periphery is negligible and hence the outer periphery of the motor surface can be treated as a zero magnetic vector potential line.

(*b*) The magnetic materjals of the stator and rotor are isotropic and the magnetisation curve is single valued; i.e. hysteresis effects are neglected.

(*c*) The magnetic vector potential A and the current density vector J have only z-directed components

(*d*) The magnetic field distribution inside the motor is constant along the axial direction of the motor

(*e*) The end effects are neglected.

The magnetic flux density B in a magnetic material can be given as

$$B = H/\gamma \qquad (1)$$

where H is the magnetic field intensity and γ is the reluctivity of the magnetic material. From Ampere's law

$$\text{curl } H = J \qquad (2)$$

where J is the current density vector.
From eqns. 1 and 2

$$\text{curl } (\gamma B) = J \qquad (3)$$

By defining the magnetic vector potential A as

$$B = \text{curl } A \qquad (4)$$

and substituting eqn. 4 in eqn. 3

$$\text{curl } (\gamma \text{ curl } A) = J \qquad (5)$$

which may be regarded as more general because the problem region may include nonzero current densities. The reluctivity is field dependent and thus eqn. 5 is nonlinear.

With the assumptions made above, eqn. 5 can be written, in rectangular co-ordinates, as

$$\frac{\delta}{\delta x}\left(\gamma \frac{\delta A}{\delta x}\right) + \frac{\delta}{\delta y}\left(\gamma \frac{\delta A}{\delta y}\right) = -J \tag{6}$$

The solution of eqn. 6 yields the magnetic vector potential A inside the motor by using appropriate boundary conditions. The magnetic flux density components can be given as

$$B_x = \frac{\delta A}{\delta y}; \qquad B_y = -\frac{\delta A}{\delta x} \tag{7}$$

The motor outer surface and the bore contour of the shaft are flux lines of zero magnetic vector potential. The homogeneous Dirichlet boundary condition ($A = 0$) defines the nodes along these flux lines.

When the rotor is in the aligned and unaligned positions, there exists symmetry along the maximum and minimum permeance axes, respectively. Hence the axes of maximum permeance in the aligned condition and minimum permeance in the unaligned condition are the flux lines of zero vector potential.

When only one phase winding of the motor is excited, the vector potential is a periodic function of three stator pole pitches in the case of a 6-stator pole motor. This implies that the vector potential along the nodes on one side of the shaft is equal in magnitude but opposite in sign to that on the corresponding nodes lying on the opposite side of the shaft.

The tangential component of the magnetic field intensity vector H and the normal component of the magnetic flux density vector B are continuous along the iron-air boundary of the motor.

The nonlinearity of the magnetisation curve of the iron and the complex geometry of the stator and rotor do not permit analytical solutions for the magnetic vector potential. However, numerical methods provide a convenient approach. In the variational method, the solution to the problem is obtained using an interpolation technique by minimising the nonlinear energy functional

$$F = \int_R \left[\int_0^B H \cdot dB - \int_0^A J \cdot dA\right]dR \tag{8}$$

where R is the problem region of integration.

The entire problem region R is subdivided into triangular finite elements. The elements are defined such that the sides of the triangles coincide with the boundary of each material. The finite element subdivision for one quarter of the motor, when the rotor is in the aligned position, is shown in Fig. 2.

Owing to the symmetry when the rotor is in the aligned and unaligned positions, one quarter of the motor is modelled for calculating the solution. The nodes on the periphery of the stator, the shaft periphery, the maximum permeance axis for the aligned position, and the minimum

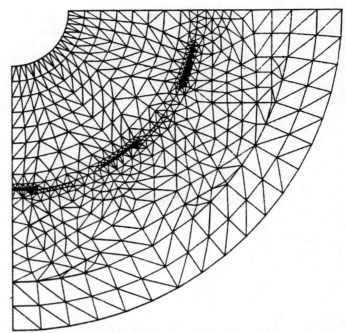

Fig. 2 *Finite element subdivision of the motor model: aligned position*

permeance axis for the unaligned position are defined to have zero vector potential. The Neumann boundary condition is applied at the nodes on the horizontal axis of the model.

As the periodicity condition repeats after three stator pole pitches, a model of one half of the motor is required for other positions of rotor. The homogeneous Dirichlet boundary condition is used along the shaft and motor peripheries. Binary constraints are used at the nodes on the horizontal axis of the motor model.

To give some measure of the position dependence of the torque developed by the motor, it is modelled for different rotor positions in addition to the aligned and unaligned positions. The rotor is considered to be displaced by 4°, 8° and 12° from the aligned position. Calculations for more intermediate positions of the rotor would result in better prediction of the torque during its movement from the unaligned to aligned position. The study is limited to only five rotor positions due to the computational effort for each rotor position.

The field analysis has been performed using a Magnet CAD package [10] which is based on the variational energy minimisation technique to solve for the magnetic vector potential. The vector potential has been postprocessed to obtain desired results such as the energy stored in the field, flux density, flux distribution, inductance, torque etc. The equipotential plot obtained represents the flux density distribution in the motor.

4 Results

The flux density distribution inside the motor for the aligned and unaligned positions of the rotor and for the intermediate positions of 4°, 8° and 12° displaced from the

213

aligned position are shown in Figs. 3–7, respectively. These plots are obtained for a steady current excitation of 8 A in one phase winding. The flux plots obtained for other excitations are not shown. It can be observed that some of the flux lines cross the airgap to the rotor and complete their path through the adjacent pole winding back into the stator. It has been observed that the flux densities in the excited stator poles and the torque producing rotor poles are higher compared with those in the back of rotor and

stator cores for a given excitation. The flux densities in other parts of the motor are very much less. For a steady excitation of 8 A, the back of the cores of rotor and stator do not saturate whereas the excited stator poles and the torque producing rotor poles are heavily saturated.

A plot of the flux linkages of one phase winding for varying excitation current is shown in Fig. 8 in normalised form. The steady excitation current (18.1 A) that causes saturation in the motor iron when the rotor is in the unaligned position, and the corresponding flux linkage (1.18 Vs) are taken as the base values. The normalised flux

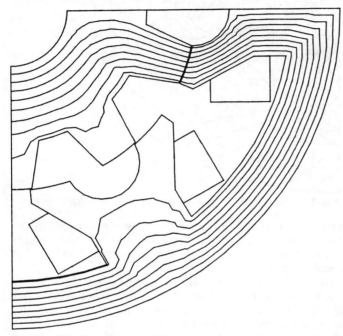

Fig. 3 *Flux plot: aligned position*

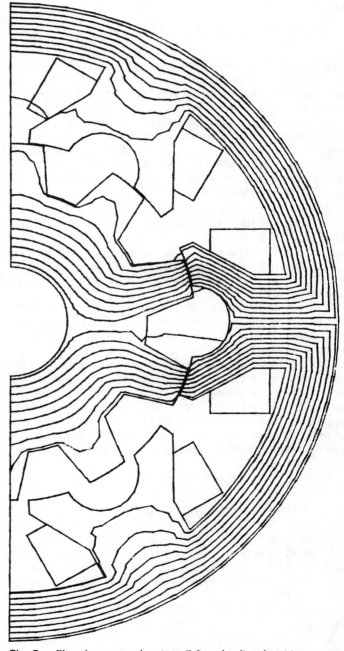

Fig. 5 *Flux plot: rotor pole axis at 4° from the aligned position*

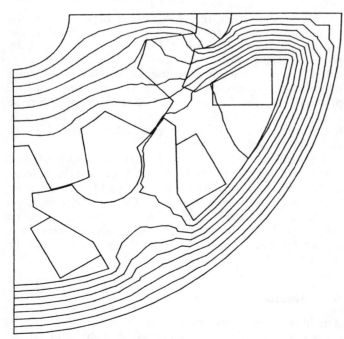

Fig. 4 *Flux plot: unaligned position*

214

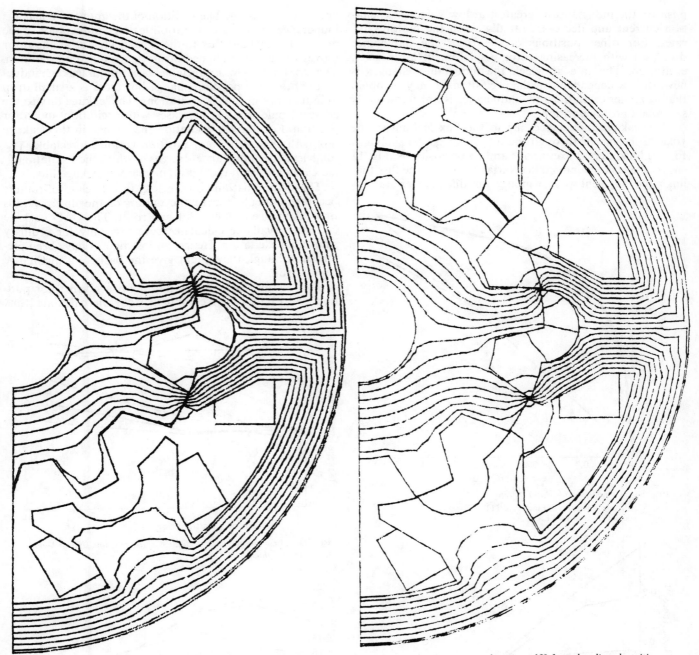

Fig. 6   *Flux plot: rotor pole axis at 8° from the aligned position*

Fig. 7   *Flux plot: rotor pole axis at 12° from the aligned position*

linkages obtained through the finite-element analysis show higher values than those of the experimental values reported by Finch *et al.* [1] when the rotor is in the aligned position. The flux linkages in the unaligned position are the same as those reported by Finch *et al.* until the beginning of saturation, beyond which the finite-element analysis shows a smaller value; however, the difference is within 2%. In spite of the same physical dimensions of the motor, the magnetic material used in the finite-element analysis may not be the same as that used by Finch *et al.*

Thus the difference in the material B/H characteristics shows itself in the flux linkage calculations.

The inductance has been defined as the ratio of flux linkages to the exciting current. Values based on this definition are presented in Figs. 9–11, although the nonlinear characteristic of the magnetic circuit limits their use. It should be noted that it is the flux linkages that are determined and used in all calculations.

Fig. 9 shows the terminal inductance variation for different rotor positions and excitation current. In the aligned

215

position the inductance is greatest at low values of excitation current and decreases rapidly when the motor saturates. For other positions of the rotor the inductance decreases with increasing rotor displacement for a given excitation. The rate at which the inductance changes is nevertheless dependent on the excitation. It may be noted that at higher excitations the inductance is not particularly sensitive to rotor position.

The leakage inductance due to the flux leaking to the yoke of the motor through the interpolar gap is shown in Fig. 10 for different excitations and rotor positions. Fig. 11 shows the leakage inductance variation due to the flux that links the adjacent pole winding for different excitations

and rotor positions. The addition of the above two leakage inductances constitutes the total leakage inductance of the motor. The leakage flux to the yoke decreases as the rotor moves towards the aligned position, for a given excitation. This does not show appreciable variation in leakage inductance with rotor position change but there is a small drop in inductance at higher excitations. The leakage flux to the adjacent pole winding increases with excitation and rotor position. For a given excitation the change in this leakage flux with change in rotor position is not appreciable. The total leakage inductance decreases when the excitation is increased although there is an increase in leakage flux.

The determination of core loss in this class of motors becomes difficult due to the nature of the flux reversals that take place in the stator and rotor. The CAD package currently available calculates core loss only for ordinary sinusoidal excitation. Hence, no loss figures have been presented although they might give the basis of a qualitative comparison.

The energy stored in the motor for a given rotor position and excitation is computed from the calculated flux

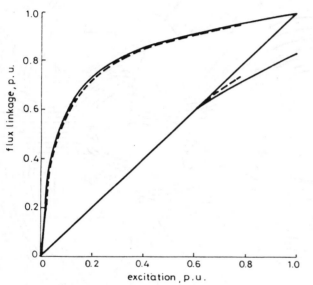

Fig. 8   *Comparison of Flux Linkages*

- – – – experimental results of Finch *et al.* [1]
———— calculated values by FEA

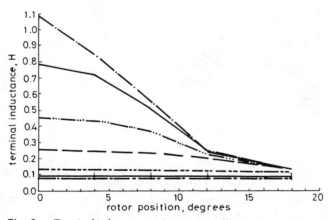

Fig. 9   *Terminal inductance against rotor positions*

—··— 0.5 A
——— 2 A
—···— 4 A
— — — 8 A
—·—· 18 A
—+—+ 27 A
—··—· 36 A

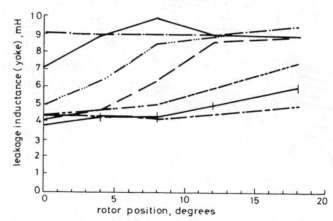

Fig. 10   *Leakage inductance against rotor position, due to leakage flux to yoke of one phase*

See Fig. 9 for key

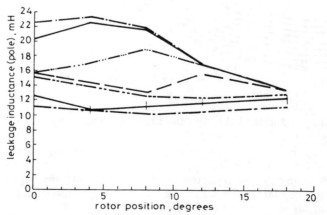

Fig. 11   *Leakage inductance against rotor position, due to leakage flux to adjacent pole winding of one phase*

See Fig. 9 for key

216

densities in each element of the motor model and the magnetisation characteristic of the material used. The torque developed by the motor is determined as the ratio of the change in coenergy when the rotor moves from one position to the other and the angle between the two positions. From the flux linkage against excitation curves shown in Fig. 12, for different rotor positions, it is observed that the

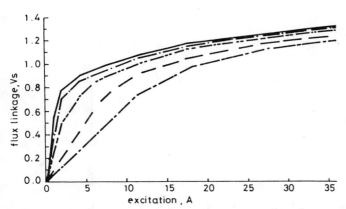

Fig. 12 *Total flux linkages against excitation of one pole winding*

——— aligned
—·—· 4°
——··— 8°
– – – – 12°
—·—·— unaligned

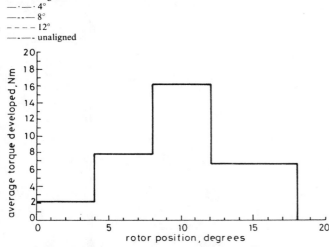

Fig. 13 *Average torque against rotor position for 4 A excitation*

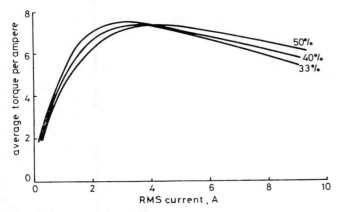

Fig. 14 *Average torque per ampere*

change in flux linkages for the last 4° of implied movement into the aligned position is considerably smaller than that for the other 4° increments in rotor position. The rate of change of coenergy is therefore reduced significantly when the rotor pole is near the aligned position resulting in minimal additional developed torque. In contrast, the rate of change is greater when there is partial overlap of the pole faces. Hence the current could perhaps be switched off just before the aligned position provided, of course, that the next pole is in a position to develop torque.

The average torque developed between the rotor positions, i.e. 0°, 4°, 8°, 12° and 18° from the aligned position, at which the finite-element analysis was performed has been calculated. The values obtained show clearly that the developed torque is a function of rotor position. However, when plotted in Fig. 13 for a steady excitation of 4 A, the torque has been shown as constant over these intervals, for convenience. Depending on the duration of excitation the average torque developed per RMS ampere varies. Fig. 14 shows the calculated values of the average torque per ampere for rectangular blocks of excitation current having durations of 33, 40 and 50% starting at angles of 2°, 1.6° and 0°, respectively, from unaligned position. Although the average torque increases as the duration is increased, the maximum value of the torque per ampere decreases. Thus, for a particular application, if the torque per ampere is a more important criterion than the torque per unit volume, the duration of excitation can be chosen accordingly taking due account of the likely increase in pulsating torque. The average torque also varies with the angle of switching. The general pattern of the torque per ampere characteristics shows substantial agreement with the experimental values reported by Finch *et al.* [1]. Differences are inevitable due to the assumption of rectangular blocks of current and a different magnetic material in the computation.

5 Conclusion

A 2-dimensional finite-element analysis of a switched reluctance motor has been given. This paper presents an alternative method of predicting the doubly salient switched reluctance motor performance by directly computing the coenergy in the motor at different rotor positions. The field pattern obtained gives a better understanding of the flux distribution in the motor.

The dependence of terminal inductances and leakage fluxes on rotor position and excitation has been demonstrated. In general, there does not seem to be any simple relationship. The one exception is the total leakage flux which is almost independent of rotor position for a given excitation. The terminal inductance does not have a linear dependence on rotor position with the result that $dL/d\theta$ is not constant.

Even with constant excitation, the developed torque is inevitably a function of position and thus the motor produces a pulsating torque. This is particularly evident in Fig. 13 where the average values of the developed torque over the intervals in rotor position are shown. A pulsating torque is also present with standard excitation arrangements. As in all variable reluctance motors the developed torque maximises before full alignment is attained.

The value of inductance for this configuration seems somewhat high compared with that for a 3-phase motor with six stator and four rotor poles [9] and could inhibit the growth of current. When the increased losses due to the higher switching frequency of the two teeth per stator pole arrangement are also taken into account, this motor may be considered unsuitable for high-speed operation.

6   Acknowledgment

This project has been carried out with the aid of a grant from the Natural Sciences and Engineering Research Council of Canada. The authors are also indebted to Infolytica Corporation Ltd., Montreal, Canada who provided the computational facilities.

7   References

1 FINCH, J.W., HARRIS, M.R., MUSOKE, A., and METWALLY, H.M.B.: 'Variable speed drives using multitooth per pole switched reluctance motors', Proceedings of 13th Annual Symposium on Incremental Motion Control Systems and Devices, May 1984, pp. 293–301

2 LAWRENSON, P.J., STEPHENSON, J.M., BLENKINSOP, P.T., CORDA, J., and FULTON, N.N.: 'Variable-speed switched reluctance motors', IEE Proc. B, Electr. Power Appl., 1980, 127, (4), pp. 253–265

3 HARRIS, M.R., HUGHES, A., and LAWRENSON, P.J.: 'Static torque production in saturated doubly-salient machines', ibid., 1975, 122, (10), pp. 1121–1127

4 HARRIS, M.R., ANDJARGHOLI, V., LAWRENSON, P.J., HUGHES, A., and ERTAN, B.: 'Unifying approach to the static torque of stepping-motor structures', ibid., 1977, 124, (12), pp. 1215–1224

5 DAVIS, R.M., RAY, W.F., and BLAKE, R.J.: 'Inverter drive for switched reluctance motor: circuits and component ratings', IEE Proc. B, Electr. Power Appl., 1981, 128, (2), pp. 126–136

6 BYRNE, J., and LACY, J.C.: 'Electrodynamic system comprising a variable reluctance machine'. US Patent 3956678, 11th May 1976

7 CORDA, J., and STEPHENSON, J.M.: 'An analytical estimation of the minimum and maximum inductances of a double salient motor', Proceedings of International Conference on Stepping motors and systems, University of Leeds, UK, 1979, pp. 50–59

8 STEPHENSON, J.M., and ČORDA, J.: 'Computation of torque and current in doubly salient reluctance motors from nonlinear magnetisation data', Proc. IEE, 1979, 126, (5), pp. 393–396

9 ARUMUGAM, R., LOWTHER, D.A., KRISHNAN, R., and LINDSAY, J.F.: 'Magnetic field analysis of switched reluctance motor using a 2-dimensional finite element model', IEEE Trans., 1985, MAG-21, (5), pp. 1883–1885.

10 Magnet CAD package: user manual. Infolytica Corporation Ltd., Montreal, Canada

1.4 - Variations on the Classical SR Motor

Variable-Speed Drives Using Multi-tooth per Pole Switched Reluctance Motors

J.W. Finch, M.R. Harris, A. Musoke and H.M.B. Metwally

13th Incremental Motion Control Systems Symposium,
University of Illinois, Urbana-Champaign, IL, 1984, pp. 293-302

I. INTRODUCTION

There has been considerable recent interest in the possibility of using a doubly-salient reluctance motor with electronically switched stator windings as a variable speed drive. This 'switched reluctance motor' (SRM) has already been shown[1,2] to be capable of challenging the well established induction motor/inverter drive. A design method for stepping motors has been developed by the present authors[3-5] and shown to yield good accuracy of torque prediction, whilst being convenient and simple in use. The present paper employs this method, with some simple extensions, to make performance predictions for various SRMs, providing at the same time good insight into the fundamentals of motor behaviour. The method is verified by experimental results obtained from a prototype SRM. Overall motor performance is shown to be extremely competitive in comparison to equivalent induction motors, and this leads to a general discussion of SRM capabilities.

In a practical drive, phase switching of the SRM will be controlled via a simple shaft-mounted position transducer, so that stator windings are excited at appropriate times to ensure continuous rotation; the motor and inverter are thus locked together in operation. This combination tends to exhibit characteristics similar to a series d.c. motor, suitable for traction applications. As load falls motor speed increases to a high value, whilst at high load (such as climbing a steep gradient) the speed falls until the developed torque is sufficiently large. The shaft transducer makes the drive 'robust' in application, since abrupt increases in load torque cannot cause the loss of synchronism that is possible with open-loop operation. Motor operation without the transducer is still possible, however, and the basic test results presented here were so obtained.

The characteristics of the drive can be radi-

cally altered by simple changes in control strategy. For example, speed holding accuracy of high order can be achieved with a high stability oscillator; in this mode, torque control at constant speed is obtained by varying the period of excitation (or on-time), with either chopped or square voltage waveform. In conjunction with this, the rotor angle at switch-on (switching angle) may be controlled, principally in order to minimise rms current for given torque. The switching angle for any phase may be measured to any convenient datum, a good choice being the aligned (tooth-opposite-tooth) position, which is the stable equilibrium position for continuous excitation of the one phase. If, however, the on-time is kept fixed, then varying switching angle may be visualised as selecting that portion of the whole (positive and negative) torque/angle curve which is developed by the motor during excitation of each phase. By this means, it is possible in principle to make the net output torque zero for any given rms current, simply by adjusting switching angle so that equal positive and negative torques are produced on average during steady rotation.

The SRM operates well on switched voltage supply. Unlike the induction motor, a square wave of voltage is a better excitation waveform (though still not strictly optimum) than a sinewave. Moreover, like all VR stepping motors, only unipolar current supply is required, so that only one switch may be needed in series with each phase winding. These features mean in summary that the SRM operates well from an inverter circuit of simpler design than is needed for the induction motor, and this matter has been extensively discussed elsewhere[1,2,6].

Apart from a general discussion of SRM operation, and the presentation of a design method for predicting operating torque at given current, it is a major purpose in this paper to explore the differences in performance characteristics that arise, as a consequence of varying the number of teeth that are carried on each stator pole. This matter has not been examined in any detail elsewhere, and SRM designs up to the present time[1,2,6] are generally one-tooth per pole. We may note immediately, however, that the choice of tooth number per pole is a design feature of fundamental importance, and that the one-tooth design is certainly not optimum for all possible ranges of operating speed.

II. BASIC DESIGN METHOD

An estimate for SRM running torque, T_r, is conveniently obtained by simple extension of a torque prediction method for stepping motors, which has proved to be a helpful design approach. The following expression for T_r is obtained:

$$T_r = K_r\, K_u\, K_e\, \overline{T}_{\ell s}\, V_r \qquad (1)$$

Since the basic design method is documented elsewhere[4,5] only a brief review is attempted here, concentrating on the SRM application. V_r is active rotor volume, equal to $\pi D^2 L/4$, with L being the effective active rotor length. $\overline{T}_{\ell s}$ is a theoretical limiting value of mean torque between successive torque zeros, expressed per unit rotor volume, obtained by assuming idealised 'abruptly saturating' magnetisation characteristics (see later, Fig. 1). K_e represents the proportion of $\overline{T}_{\ell s}$ which would be achieved by an idealised square wave of phase current which has the same rms value as the applied current and which is assumed to switch on and off at successive zeros of the per-phase static torque/angle diagram; K_e inherently allows for the differences between practical and idealised magnetisation characteristics. K_u represents the fraction of the rotor that is effectively utilised for torque production at any one time, with the given excitation pattern. K_r, which is usually slightly greater than unity, allows for the improved running torque obtained with a current of practical waveform, compared to the idealised 50/50 square wave of equivalent rms value.

The fact that K_r may exceed unity (and usually does with optimum switching strategy), is due to the actual current waveform being 'peakier' than the square wave, and the peak coinciding with the most torque-productive part (in torque per amp terms) of the torque/angle curve.

Per-tooth magnetisation characteristics are depicted in Fig. 1, which defines the main quantities involved in the analysis. One important quantity is the critical mmf, F_c, required for the idealised unaligned tooth flux to reach the saturation value of gap flux per tooth pitch, ψ_{ts}. It is very helpful to express excitation mmf and flux per tooth in normalised forms, $(F) = F/F_c$ and $(\psi_t) = \psi_t/\psi_{ts}$ respectively, which gives the usual benefits associated with any per-unit representation. The range of numerical values of the variables is greatly reduced; moreover, a single set of typical normalised magnetisation characteristics will often be adequate for at least preliminary work, with the resultant additional errors usually proving to be surprisingly small.

An expression for F_c can be shown[4] to be

$$F_c = 1.671 \times 10^6\ t/P_2 \qquad (2)$$

where t is the tooth-width and P_2 is the usual dimensionless function relating to gap permeance per tooth pitch in the unaligned position. P_2, and P_1 (the aligned equivalent), are available in convenient form for design use if the geometry is rectilinear with infinite-depth slots. In that case the tooth width, t, toothpitch, λ, and air-gap, g, completely define the geometry. Once F_c is known, we can immediately estimate the order of magnitude of mmf needed to obtain a torque which is a sizeable fraction of $\overline{T}_{\ell s}$. SRMs, being intended as drives, are often large machines by comparison with most stepping motors. It is advantageous to keep g as small as possible and λ/g may be large in these motors, values over 100 being common. In such cases, if t/λ is fixed and significantly less than 0.5 (which is commonly the case), then P_2 becomes nearly independent of g, since from Eqn. 2

$$F_c/\lambda = 1.671 \times 10^6\ (t/\lambda)/P_2 \qquad (3)$$

It transpires that F_c/λ is an important ratio; for typical t/λ values and with large values of λ/g, we find that F_c/λ lies in the range 100 to 120 A/mm.

The co-energy areas S_1, S_2, and S_3 as shown in Fig. 1 govern the average torque developed between the aligned and unaligned positions. When these areas are normalised, by dividing by the whole triangular area OXY, a simple expression for

K_e results:

$$K_e = (S_1) - (S_2) + (S_3) \qquad (4)$$

III. MOTOR SPECIFICATION AND DESIGN CALCULATIONS

Prototype Motor

The prototype 6-pole 12/10 slotted SRM has the lamination profile indicated in Fig. 2, which shows the motor in the unaligned position for the pole-pairs on the vertical axis. It has 2 teeth on each stator pole with the windings on diametrically opposite poles connected to form one phase of the 3-phase stator. Each pole carries 222 turns of 0.9 mm wire in 6 sections of 37 turns, which may be series or parallel connected; all currents quoted later assume series connection of all 12 sections per pole-pair to form a phase winding, although tests were actually performed with various series/parallel combinations.
Basic motor dimensions are:
(All lengths quoted are measured in mm)
Stator OD = 165.1, bore = 93, g = 0.255;
L_s = 108, L_r = 114.3; t = 10.2, λ_r = 29.1.
Hence, t/λ = 0.35, λ/g = 115.
Stator pole neck = 24.5, back of core = 12.5.

This motor forms part of a continuing development and test programme. Its static characteristics have been carefully measured and will be used here to verify the design method. It has been constructed in a standard TEFV D100 metric frame, with the same stack length and air-gap as an 8-pole induction motor, to facilitate comparisons.

Design Coefficients

The measured terminal magnetisation characteristics are shown in normalised form in Fig. 3 as full lines. These characteristics naturally include any inter-pole leakage flux which occurs. This contribution, often small in practice, is included within the permeance estimates discussed later. The areas enclosed within these curves for particular values of exciting winding mmf are directly proportional to co-energy change, and thence to the estimated average torque. At typical values of mmf, this estimate differs by less than 2% from the average torque measured directly on the torque/angle curve between torque zeros,

providing clear confirmation of the theoretical method. Values of K_e are also obtained from the characteristics of Fig. 3, and are given in Fig. 4 as functions of the mmf per tooth pitch, F/λ. If F is the mmf produced in the pole winding by I, the crest value of the rms-equivalent 50/50 square-wave current, then the corresponding value of K_e can be used in Eqn. 1 for torque. Note that the rms value of this square-wave current is $I_r = I/\sqrt{2}$.

Since it is the total output torque that is normally required, it is correct to define K_u to include contributions from all phases. For the 3-phase 12/10 motor, for each phase that is excited 4 out of 10 rotor teeth are torque-productive for half the time, but 3 phases contribute to torque. Hence in this case, K_u = (4/10) x (1/2) x 3 = 0.6. Correspondingly, a 3-phase 6/4 motor would have K_u = (2/4) x (1/2) x 3 = 0.75. The factor K_r can be derived from a detailed knowledge of waveforms within the machine, but at best is likely in practice to exceed unity by a small margin. A nominal value of 1.05 is acceptable for the present work.

Permeance Values

The accuracy of the design method is obviously dependent on the reliability of the permeance coefficients, P_1 and P_2. In much stepping motor work, permeance data for idealised rectilinear geometry is sufficiently accurate to yield torque estimates within 5-10%. Better permeance estimates are commonly required for SRMs, particularly because the low tooth number increases the significance of field end-effects and (notably in the 12/10 motor) stator slots are appreciably less than infinitely deep. This matter will be briefly discussed here, but whilst it is important for good accuracy it should not be allowed to obscure the essential simplicity of the basic design method. For the aligned position, the air-gap permeance is dominated by t/g and end-effects are small. Rectilinear values for P_1 are therefore usually satisfactory, but it is often worth applying a correction factor for the long iron paths (even at unsaturated flux levels) in series with the gap, to yield an effective value P_{1e}. A simple procedure is to determine an iron coefficient $P_i = \mu_r(W_i/\ell_i)$ where W_i and ℓ_i are the effective width and length of the iron path per pole, and μ_r is the effective unsaturated relative permeability of the iron (typically 4000-6000).

Then, $1/P_{1e} = 1/P_1 + 1/P_i$. Great accuracy is unnecessary since $\overline{T}_{\ell s}$ is not strongly affected by P_1, if the ratio P_1/P_2 is large.

Drop of mmf in the iron has negligible effect on the linear value of unaligned permeance, but other modifications to the simple 2-D rectilinear data are often needed. Methods of correcting for field end-effects, and for shallow stator slots, are next briefly discussed. So far as finite slot depth is concerned, two-dimensional field computations may be used to estimate the effective values, or a simple 'assumed flux path' approach[7] adopted. Used with care the latter method can be sufficiently accurate, but errors are possible, particularly at low values of λ/g; in the SRM, its use has proved acceptable.

A similar method may be used to estimate the magnitude of three-dimensional field end effects; this is reported elsewhere[7] and may be briefly described as follows. The end fields are assumed to behave as if the length of the motor formed an effective tooth in a supposed repeating two-dimensional structure. Permeance contributions are then estimated to find the end contribution to P_2.

The measured permeances in the 12/10 motor are $P_{1e} = 37.0$ and $P_{2e} = 5.56$. Since $t/\lambda = 0.35$ and $\lambda/g = 115$, the rectilinear data would give $P_1 = 43.6$ and $P_2 = 4.4$. A 2-D value of P_2 which allows for the shape of the teeth and the finite depth of the slots, using the assumed flux path method, is 4.75. The estimate of 3-D end effects (allowing also for the fact that the rotor is longer than the stator) shows that the ends contribute an extra 14.5% to permeance, so that the estimated P_2 becomes $4.75 \times 1.145 = 5.44$, which is just over 2% lower than the measured value. It is not claimed that such accuracy by the assumed flux path method is always possible; indeed, the end-effects deduced from measurement on a one-tooth per pole motor were found to be double the estimate.

Estimated Running Torque

Assuming now the measured permeances for the 12/10 motor, we obtain $\overline{T}_{\ell s} = 131000\,\text{Nm/m}^3$, $P_{2e}/P_{1e} = 0.1503$, and $F_c = 3066\text{A}$. If $F/\lambda = 25\text{A/mm}$, which is in fact a modest value for this motor, K_e is found to be 0.23. Assuming $K_r \doteq 1.05$, an estimate may be formed through Eqn. 1 of running torque, and it

is interesting to express this in specific form, per unit rotor volume. The value obtained is in fact $19000\,\text{Nm/m}^3$, which happens to be closely comparable with that for the 8-pole IM in the same frame (though note that the rotor volumes are different in the two machines). At the very high level of excitation of $F/\lambda = 105\text{A/mm}$ (i.e. $F = F_c$, assuming this were permissible) about $60000\,\text{Nm/m}^3$ might be achieved.

Consider for interest's sake, a 3-phase 6/4 one-tooth per pole design in the same frame. For convenience, the proportions of a previously published design[7] are used as an illustration. In this design $t_r > t_s$ by a considerable margin, almost 1.5:1. Assuming the same gap as the 12/10 motor gives $P_1 = 94.3$ reducing to $P_{1e} = 79.1$ when corrected for iron mmf drop, whilst $P_{2e} = 5.60$ (coincidentally almost exactly the measured value for the 12/10 motor). To strike a closer comparison with the 12/10 motor, we will assume instead that $t_s = t_r$, and the best estimates of permeances then become $P_{1e} = 78.8$, $P_{2e} = 4.9$. For this case, $F_c = 7809\text{A}$, $F_c/\lambda = 112\text{A/mm}$, $t/\lambda = 0.33$, so that $\overline{T}_{\ell s} = 146000\,\text{Nm/m}^3$, with a possible specific running torque of about $85000\,\text{Nm/m}^3$. After detailed calculations have been made of the mmf drop at various flux levels in the iron circuit of the 6/4 motor, a curve of K_e versus F/λ is eventually obtained, shown plotted in Fig. 4. It is clearly seen that despite the radically different nature of these two designs, very similar curves result for K_e, when expressed in this normalised form. The measured values of (S_2) are shown in Fig. 5 as a function of (F) for the 12/10 motor, and the estimated curve for the 6/4 design is also indicated.

Comparing the 6/4 and 12/10 designs, assuming both motors operating at the same F/λ-value, we note that the running torque obtained will be rather higher in the 6/4 case, since both $\overline{T}_{\ell s}$ and K_u are larger (146 cf 133 and 0.75 cf 0.6). However, it is important to note that the actual value of F/λ available will not be the same, being greatly reduced in the 6/4 motor, since λ is much larger. It is a very distinctive feature of doubly-salient motors[3-5,8] that the normalised excitation mmf can be increased simply by the expedient of increasing the number of teeth per pole, without paying any penalty in increased ohmic loss. The cost to be paid in a particular drive will be in the increased switching frequency required for a given shaft speed (in inverse ratio to the above excitation benefit), and therefore increased core loss.

A brief comparison with the IM may be noted. Rated running torque per unit rotor volume in an IM might range from a typical value of 16000 Nm/m^3 for an 8-pole 1 kW motor to a value approaching 36000 Nm/m^3 for a modern 4-pole 10 kW design, TEFV, with class B rise. The figure would only rise to perhaps 85000 Nm/m^3 for a 5 MW motor. Values of ultimate running torque deduced earlier thus show that SRMs are at least in principle comparable to IMs in torque capability up to very large sizes. However, it remains for detailed future work to assess which type of motor, IM or SRM, is most favourable for a given drive application at a particular power level.

IV. EXPERIMENTAL PERFORMANCE CHARACTERISTICS

The value of torque per unit current is a convenient measure of the effectiveness of a given machine. The measured value of average torque has been converted to this basis ($T_r/I_r = 1.5\ \overline{T}/I_r$) in Fig. 6. The calculated values using the basic design method are slightly lower than these measured values, with an average error of 2% for values of I_r between 2 and 7 A. Also shown in Fig. 6 are two sets of experimental results for the 12/10 prototype at 750 rpm, with fixed on-times of 33% and 40% respectively. These are open-loop tests in which the motor adopts its own switching angle as the load torque is gradually increased. Consequently K_r varies markedly with load, reaching maxima of just above and just below unity in the two cases.

Efficiency was carefully measured using wide-bandwidth electronic wattmeters and is shown in Fig. 7 for the same tests. It should be noted that both tests show an improvement in efficiency of 10% or more over the 72% measured on the 8-pole IM, a most encouraging result. In normal closed-loop control, switching angles corresponding to the peak of the efficiency curve would be maintained.

A full programme of tests is in progress, but it is already clear that the improvement in efficiency can be used to yield a higher running torque than the 15 Nm rated value of the 8-pole IM. Simulation studies indicate that over 20 Nm is possible, which exceeds the rated torque of the 4 pole IM.

V. CONCLUSIONS

A method of predicting running torque of a SRM has been discussed. It is capable of acceptable accuracy and, being based on co-energy change, correctly represents the important effects of magnetic saturation. It employs dimensionless curves that apply to a wide range of machines, and is simple and quick in use.

SRMs have been shown to be, at least in principle, competitive in torque production with IMs up to large sizes. It remains for future work to establish detailed comparisons. Multi-tooth per pole SRMs have a large excitation-mmf advantage over single-tooth designs. This feature means that multi-tooth designs may well be preferred, particularly for lower speed application where their increased switching frequency and core loss are no disadvantage. Preliminary tests on a prototype motor have supported this view, and show the substantial efficiency advantage over the IM.

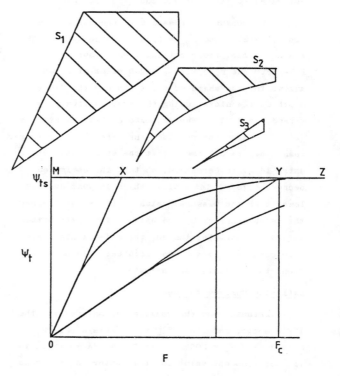

Fig. 1. Per-tooth magnetisation characteristics

VI. REFERENCES

1. Lawrenson, P.J., Stephenson, J.M., Blenkinsop, P.T,, Corda, J., and Fulton, N.N., 'Variable-speed switched reluctance motors', IEE Proc. B, Electr. Power Appl., 1980, 127, (4), pp 253-265.

2. Ray, W.F. and Davis, R.M., 'Inverter drive for doubly salient reluctance motor: its fundamental behaviour, linear analysis and cost implications', IEE J. Electr. Power Appl., 1979, 2, (6), pp 185-193.

3. Harris, M.R., Hughes, A., and Lawrenson, P.J., 'Static torque production in saturated doubly-salient machines', Proc. IEE, 1975, 122, (10), pp 1121-1127.

4. Harris, M.R., Andjargholi, V., Lawrenson, P.J., Hughes, A., and Ertan, B., 'Unifying approach to the static torque of stepping-motor structures', Proc IEE, 1977, 124, (12), pp 1215-1224.

5. Harris, M.R., and Finch, J.W., 'Estimation of static characteristics in the hybrid stepping motor', Proc. Eighth Annual Symposium on Incremental Motion Control Systems and Devices, 1979, pp 293-306.

6. Davis, R.M., Ray, W.F., and Blake, R.J., 'Inverter drive for switched reluctance motor: circuits and component ratings', IEE Proc. B, Electr. Power Appl. 1981, 128, (2), pp 126-136.

7. Corda, J., and Stephenson, J.M., 'Analytical estimation of the minimum and maximum inductances of a double-salient motor', Proc. International Conference on Stepping Motors and Systems, Leeds, 1979, pp 50-59.

8. Harris, M.R., 'Discussion on Variable-Speed Switched-Reluctance Motor Systems', IEE Proc. B, Electr. Power Appl., 1981, 128, (5), pp 260-269.

Fig. 2. Lamination profile for the prototype motor

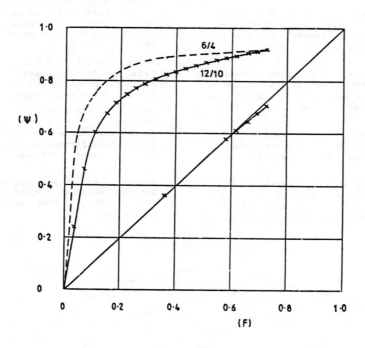

Fig. 3. Normalised magnetisation characteristics, measured for the 12/10 prototype, assumed for the 6/4 design

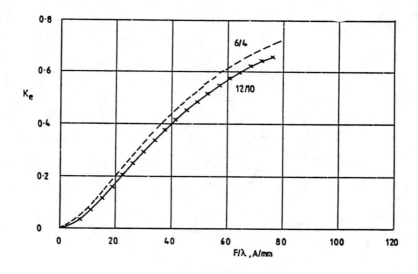

Fig. 4. Variation of K_e with F/λ, derived from flux measurement on the 12/10 prototype, calculated for the 6/4 design

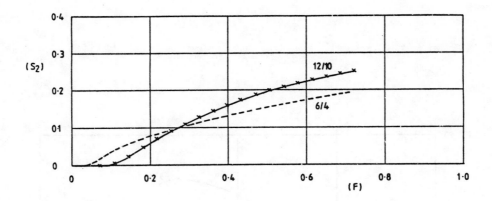

Fig. 5. Variation of (S_2) with (F), derived from flux measurement on the 12/10 prototype, calculated for the 6/4 design

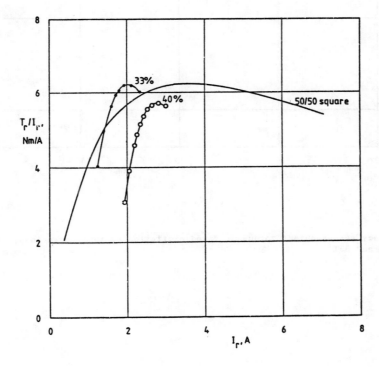

Fig. 6. Running torque per amp against rms current, deduced for idealised 50/50 square-wave current, measured open-loop at 750 rpm and fixed 40% excitation pattern on 12/10 motor, ditto with fixed 33% pattern

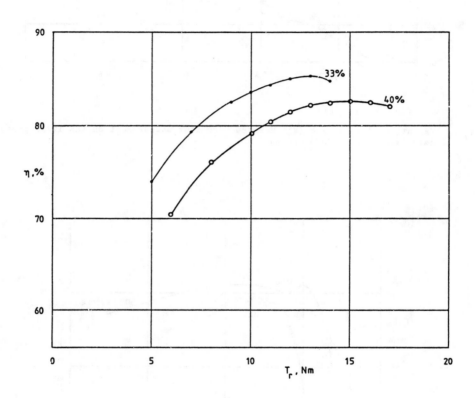

Fig. 7. Efficiency versus output torque at fixed excitation pattern, open-loop, 750 rpm and 40%, 750 rpm and 33%

228

Analysis and Optimization of the 2-Phase Self-Starting Switched Reluctance Motor

M.A. El-Khazendar and J.M. Stephenson

International Conference on Electrical Machines, Munich, September 8-10, 1986, pp. 1031-1034

1. Introduction

The Switched Reluctance Motor (SRM) system is the most recent development in the field of variable speed drives /1/2/ and such a system is illustrated in Figure 1. It consists of a simple form of a reluctance motor, position transducer, switching circuit, and controller. The torque is developed by successively switching the current to the appropriate phase winding according to the position of the rotor with respect to the stator. At least 3 phase windings, distributed on an appropriate number of stator poles, with a related number of rotor poles (without any windings), will give a reversible self-starting drive. Both 1-phase /3/ and 2-phase SRMs /4/ are attractive due to the advantage of minimising the dominant cost of the electronic circuit, but they suffer from being not inherently self-starting or being unidirectional when modified for self-starting. The model presented in this paper deals with a 2-phase SRM with a stepped-air-gap as a means of achieving the starting capability. The stator has 4 salient poles and the rotor 2.

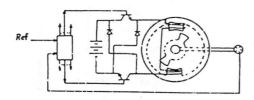

Fig 1 A scheme of SRM system

Figure 2 shows a 2-pole rotor with stepped air-gap. The effect of stepping the air-gap is to extend the region of positive inductance variation such that at any rotor position there is a $+\frac{dL}{d\theta}$ for either phase winding. This effect is demonstrated in Figure 3.

Switched reluctance motors are among those types of machine which normally operate with a high level of saturation, so it can hardly be expected to represent the machine with a linear model in which saturation is neglected. To predict machine performance accurately, iron saturation must be taken into account, resulting in a more complicated model. The mathematical equations are non-linear and numerical methods must be used to solve such a non-linear model. The linear model is more suitable, compared with the non-linear model, for a broad understanding of the combined effects of the different parameters on the machine characteristics. This will lead to a simple physical view of the behaviour of the quite complicated situation of stepping the air-gap.

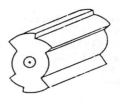

Fig 2 Stepped air-gap rotor

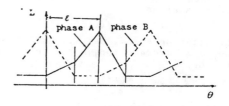

Fig 3 Inductance variation with step

2. Analysis for flux and current

In deriving the model the following assumptions are made:

(i) Saturation, fringing, mutual coupling and winding resistance are neglected.

(ii) Rotor angular velocity is constant.

(iii) The conduction angle must not be greater than half the angle of one complete cycle of inductance variation.

(iv) One step in the air-gap per rotor pole is considered.

(v) Quantities will be normalised as follows:

$$\bar{\theta} = \frac{\theta}{\theta_S} \quad , \quad \bar{t} = \frac{t}{\tau} = \bar{\theta} \quad , \quad \bar{L} = \frac{L}{L_m}$$

$$\bar{\psi} = \frac{\psi}{V\tau} \quad , \quad \text{and} \quad \bar{i} = \frac{i}{V\tau/L_m}$$

The voltage equation when the winding is switched on is

$$\frac{d\psi}{dt} = V = \text{constant}$$

whence in normalised form:

$$\bar{\psi}(\bar{t}) = \bar{t} \qquad (1)$$

$$\bar{i}(\bar{t}) = \bar{t}/\bar{L}(\bar{t}) \qquad (2)$$

Similarly in the switch off period before current zero

$$\bar{\psi}(\bar{t}) = -\bar{t} + 2\bar{t}_{con} \qquad (3$$

$$\bar{I}(\bar{t}) = (-\bar{t} + 2\bar{t}_{con})/\bar{L}(\bar{t}) \qquad (4)$$

where t_{con} is the time through which the switching devices are on.

$\bar{L}(\bar{t})$ is the normalised inductance as a function of normalised time and machine parameters. These parameters are: α_1, α_2, K_d, the switch-on angle in its normalised form (K_c), the switch-off angle in its normalised form (K_e), and the position at which both flux and current decay to zero (x).

Fig 4 shows the inductance variation in its normalised form. This figure has been divided into seven distinct regions. For each region an expression of $\bar{L}(\bar{t})$ as a function of normalised time and the previously defined parameters is obtained.

3. Analysis for torque

The average torque is calculated from the change of system coenergy when the rotor moves through an angle equal to the rotor pole pitch. In mathematical form the average torque per phase is given by:

$$T = \frac{1}{\phi} \oint \psi(i)di \qquad (5)$$

Substituting normalised values of both ψ and i, the total average torque for an m phase machine will be:

$$T = \frac{m(V\tau)^2}{\phi L_o} F \qquad (6)$$

$$\text{where; } F = \frac{1}{\alpha_1} \oint \bar{\psi}(\bar{I})d\bar{I} \qquad (7)$$

This integration can be evaluated by defining the locus of the flux linkage as a function of current for each region of normalised inductance variation with the limits of the integration determined by the three parameters: K_c, K_e, X.

Expressions of the function F for all possible combinations of K_c, K_e, and X have been derived for a stepped air-gap machine with one step per rotor pole (19 possibilities for the motoring mode).

As an illustration we will consider only one of these combinations:
Switch-on in region 3 $-1 < K_c < 0$
switch-off in region 4 $0 < K_e < 1-K_c$
flux & current decay to zero in region 6 $0 < X < 1$

The function F is given by:

$$F = F_1 + F_2 + F_3 + F_4 + F_5$$

where:

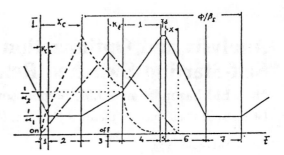

Fig 4 (—) Inductance variation (-··-) Flux-linkage (-·-) Current waveform. $(K_{con} = K_c + K_e + 1)$

It is seen that F is a function of five dimensionless parameters, the mechanical parameters (α_1, α_2, K_d) and the controlling parameters (K_c, K_e)

$$\text{i.e } F = f_1(\alpha_1, \alpha_2, K_d, K_c, K_e) \qquad (8)$$

4. Analysis for Optimisation

The effect of these parameters on the torque production of a given machine will be studied in such a way that both magnetic and electric characteristics are well under control by imposing two constraints upon the optimisation process, these are :
the peak flux density (β), and r.m.s. current density (J).

R.m.s. phase current I is given by :

$$I = \frac{V\tau}{L_o} \sqrt{\frac{\beta_s}{\phi}} R \qquad (9)$$

$$\text{where } R = \sqrt{\frac{1}{\alpha_1} \oint \bar{I}^2 d\bar{t}}$$

$$\text{or } R = f_2(\alpha_1, \alpha_2, K_d, K_c, K_e) \qquad (10)$$

$$J = NI/2A_w \qquad (11)$$

where A_w is the estimated available copper area per pole side which can be shown to be given by

$$A_w = W_1 D^2 \qquad (12)$$

where W_1 is a function of δ, β_s, S_r, S_c and β_s

$$\therefore J = NI/(2D^2 W_1) \qquad (13)$$

$$F_1 = \frac{\alpha_2 - (\alpha_1 - \alpha_2)K_c}{\alpha_1 - \alpha_2} \left\{ \frac{\alpha_2}{\alpha_1}(1 + K_c) + \frac{\alpha_2}{\alpha_1 - \alpha_2} \ell n \left[\frac{\alpha_2 - (\alpha_1 - \alpha_2)K_c}{\alpha_1} \right] \right\}$$

$$F_2 = \frac{1 - (\alpha_2 - 1)(1 + K_c)}{\alpha_1(\alpha_2 - 1)} \left\{ \frac{\alpha_2 K_e [1 - (\alpha_2 - 1)(1 + K_c)]}{1 + (\alpha_2 - 1)K_e} - \frac{\alpha_2}{\alpha_2 - 1} \ell n [1 + (\alpha_2 - 1)K_e] \right\}$$

$$F_3 = \frac{\alpha_2 + (\alpha_2 - 1)(K_c + 2K_e)}{\alpha_1(\alpha_2 - 1)} \left\{ \frac{(1 - K_e)[\alpha_2 + (\alpha_2 - 1)(K_c + 2K_e)]}{1 + (\alpha_2 - 1)K_e} + \frac{\alpha_2}{\alpha_2 - 1} \ell n \left[\frac{1 + (\alpha_2 - 1)K_e}{\alpha_2} \right] \right\}$$

$$F_4 = \frac{K_d}{\alpha_1} \left(K_c + 2K_e - \frac{K_d}{2} \right)$$

$$F_5 = \frac{[(\alpha_1 - 1)(K_c - K_d + 2K_e) - \alpha_1]}{\alpha_1 - 1} \left\{ \frac{K_c - K_d + 2K_e}{\alpha_1} + \frac{1}{\alpha_1 - 1} \ell n \left[\frac{\alpha_1 - (\alpha_1 - 1)(K_c - K_d + 2K_e)}{\alpha_1} \right] \right\}$$

230

Substituting the value of I from (9) into (13) gives :

$$V_T = \frac{2}{N} \sqrt{\phi/\beta_s} \; L_o D^2 J \; W_1 / R \qquad (14)$$

As the maximum flux linkage occurs at the instant of switch-off, then

$$\hat{\phi} = V_T(K_c + K_e + 1), \quad N\hat{B}A_p = V_T(K_c + K_e + 1)$$

But $A_p = S_r D \ell \beta_s / 2$

$$\therefore V_T = N\hat{B}S_r D \ell \beta_s / 2 (K_c + K_e + 1) \qquad (15)$$

Now, it is required to get an expression for the average torque in which both β and J are controlled.

Multiplying equations (14) and (15) gives:

$$(V_T)^2 = (D^3 \ell)(J\hat{B}) \left(\frac{S_r L_o \sqrt{\beta_s \phi} \; W_1}{R(K_c + K_e + 1)} \right) \qquad (16)$$

Dividing equation (14) by equation (15) gives:

$$\beta = J \left(\frac{4\mu_o}{\beta_s \sqrt{\beta_s/\phi}} \; \frac{P_o D}{S_r} \; \frac{W_1(K_c + K_e + 1)}{R} \right) \qquad (17)$$

where P_o is the normalised minimum permeance and is given by:

$$P_o = L_o / (N^2 \mu_o \ell) \qquad (18)$$

Substituting $(V_T)^2$ from equation (15) into equation (6), the average torque equation becomes:

$$T = 2D^3 \ell W_2 \qquad (19)$$

Where W_2 is the torque function and is given by:

$$W_2 = (\sqrt{\beta_s/\phi} \; S_r \; W_1) \left(\frac{F}{R(K_c + K_e + 1)} \right) J\hat{B} \qquad (20)$$

i.e. $W_2 = f_3(S_r, \delta, \beta_s, S_c, \alpha_1, \alpha_2, K_d, K_c, K_e, J, \hat{B}) \qquad (21)$

So, the average torque produced from a machine with stator outer diameter (D) and core length (ℓ) is directly proportional to W_2 which is a function of 11 parameters.

It is very time consuming to maximise W_2 with all these parameters, so it is helpful to pick up the less dependent variables and treat them as constants throughout the following analysis by assigning a realistic value for each of them.

5.
Effect of machine parameters on torque function

Numerical optimisation techniques to search for the optimum parameters are beyond the scope of this paper but a graphical routine will be developed to study the effect of some important parameters on the torque function W_2.

The procedure is summarised in the following steps:

(i) Specify both the limiting current density according to thermal considerations (this value depends on motor size, speed and method of cooling) and the peak value of the flux density according to realistic magnetic considerations.

(ii) Compute the torque function W_2 from equation (20) for the given speed and J for all the possible combinations of K_c, K_e, X, recording only those values for which β given by equation (17) is less than a given value.

(iii) Plot a curve representing the value of the function W_2 against the conduction angle. Repeating this procedure for different values of the studied parameter will give a clear picture of the effect of this parameter on the characteristics of this type of machine.

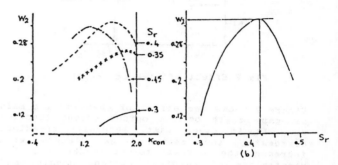

Fig 5 Effect of split ratio on torque function W_2

In Figure 5a a torque function W_2 is drawn against the conduction angle coefficient for four different split ratios (0.3, 0.35, 0.4, 0.45). Each point on those curves is specified by the two constraints $J = 5 A/mm^2$ and $\beta < 2$ T. Figure 5b shows that with $S_r = 0.42$ the torque function W_2 is maximised with the rest of parameters held constant.

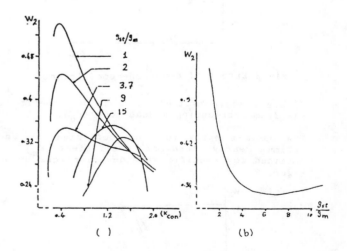

Fig 6 Effect of starting air-gap length

The effect of the air gap step size is depicted in Figure 6. In (a), a set of curves have been drawn for different g_{st}/g_m (1,2,4,9,15). In (b), the relation between the maximum obtainable torque function W_2 and the ratio g_{st}/g_m is shown.

231

It is quite clear from this figure that introducing a step as a means of self starting from any position results in derating the machine output torque. Unfortunately, the optimum starting air-gap length from the point of view of starting performance is approximately the worst for running performance.

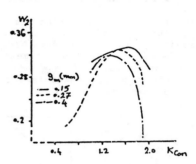

Fig 7 Effect of air-gap length

Figure 7 shows the effect of changing the main air-gap length on the output torque function W_2. It is clear that, generally speaking, decreasing the main air-gap length will increase the output torque under the two constraints of constant current density and peak flux density, but the percentage increase in the output torque reaches an insignificant value when the main air-gap length is decreased to a small value which would present a mechanical difficulty.

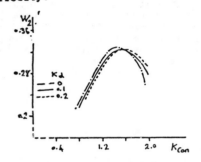

Fig 8 Effect of dead-zone coefficient

Figure 8 suggests that there is no benefit at all from introducing a dead zone (K_d).

The optimum early turn-on coefficient K_c and optimum early turn-off coefficient K_e are obtained for specific machine parameters using Figure 9.

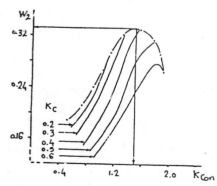

Fig 9 Effect of switching on earlier

232

6. Conclusions

A linearised model for a switched reluctance motor has been developed and the presence of the step in the air-gap is taken into consideration. A graphical routine based upon analytical solutions with both peak flux density and current density under control is used to study the effect of parameters on the characteristics of the SR machine.

The conclusions drawn from this study can be summarised as follows:

1- There is no advantage to be gained by introducing a dead zone.

2- There is no need to decrease the main air-gap length to the value which represents mechanical difficulties as the percentage increase in the output torque then is insignificant.

3- The 2-phase SRM is inflexible in its design due to the conflicting requirements between its starting and running performance.

7. List of main symbols:

B	Peak flux density
g_m	Main air gap length
g_{st}	Starting air gap length
i	Instantaneous current
I	R.m.s. winding current
J	R.m.s. current density
K_c	Early turn-on coefficient
K_d	Dead zone coefficient $(\beta_m - \beta_s)/\beta_s$
K_e	Early turn-off coefficient
L	Inductance of phase winding
N	Number of turns per phase
S_r	Split ratio
t	Time
T	Torque
V	Voltage
α_1	Ratio of maximum to minimum inductance
α_2	Ratio maximum to intermediate inductance
β_m	Main rotor pole arc
β_r	Rotor pole arc
β_s	Stator pole arc
β_{st}	Starting rotor pole arc
ζ	Efficiency
θ	Rotor angle
ψ	Flux linkage
ω	Angular velocity
τ	β_s/ω

8. References

1) RAY, W. F., LAWRENSON, P. J., DAVIS, R. M., STEPHENSON, J. M., FULTON, N. N., BLAKE, R. J. : 'High performance switched reluctance brushless drives'. Proc, IEEE, IAS Conference, Toronto, 1985

2) CORDA, J., : 'Speed control of switched reluctance motors', ICEM, Budapest, 1982

3) CHATRATANA, S., BOLTON , H.R.. PEDDER, D.A.G., : 'Investigation into small single-phase switched reluctance motors', Proc. IEE, Conf. on Small and Special Electrical Machines, Sept. 1981

4) EL-KHAZENDAR, M.A., : 'The Design of 2-phase Switched Reluctance Motors', PhD thesis, University of Leeds, 1982

Microprocessor-Controlled Single-Phase Reluctance Motor

J.C. Compter

Drives/Motors/Controls, Brighton, 1984, pp. 64-68
Published by Peter Peregrinus Ltd., London

1. INTRODUCTION

This paper deals with a brushless electronically controlled single-phase reluctance
motor. The motor has a high speed
capability, due to its very robust rotor,
and requires only one electronic power
switch in its control circuitry. The latter
feature considerably reduces the cost of
the drive system.
Research on polyphase reluctance motors has
been done by Prof. Lawrenson et al. (see
refs. 1, 2 and 3). The polyphase operation
of these motors, however, entails the use
of many electronic power switches in the
control circuitry and the cost of these
switches is evidently an obstacle to the
successful commercial use of polyphase
motors in consumer products.

As our study is aimed particularly at the
applicability of a reluctance motor in
consumer products it concentrates on the
single-phase reluctance motor. However the
choice of a single-phase motor involves a
starting problem and a strongly pulsating
torque, which means that the motor is not
suitable for applications that require
constant torque or speed, as for example in
video or audio equipment.
Typical applications of this motor are to
be found in domestic appliances, and in
this area we consider this motor a serious
competitor of the widely used A.C. series
motor, especially because the speed ranges
of both motors are compatible.
To solve the starting problem we use a
microprocessor, which also facilitates
control of the torque-speed curve over a
wide range.

2. THE MOTOR AND THE ADDITIONAL COMPONENTS

It is well-known characteristic of the
reluctance motor that the optimum switch-on
and switch-off angle depend closely on the
motor speed and required torque
production. As a consequence a rather
sophisticated control has to be used in
combination with a position sensor as shown
in figure 1. This figure gives the motor
and its control circuitry.

From the rotor position signal the speed is
obtained by measuring the time the rotor
needs to rotate over a certain angle.
Assuming a constant rotor speed this angle
might be e.g. 180 degrees. We choose this
angle because then a simple and relatively
cheap sensor can be used to get a pulse
signal every 180 degrees. This gives the
control the additional task of measuring
the time between two successive sensor
signals to determine the speed. For the

further discussion it is important to note,
that the sensor pulse is produced when the
rotor attains the in-line position.

The electrical circuit belonging to figure
1 is given in simplified form in figure 2.

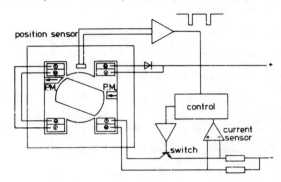

Fig. 1. Motor control.

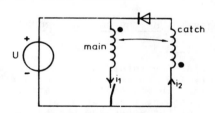

Fig. 2. Electrical circuit.

Bifilar wound main and catch coils are
used to recuperate the energy stored in the
magnetic field when the electronic switch
is opened. Both coils have the same number
of turns.
The current in the coils of the motor is
determined by measuring the voltage drop
across two resistors connected in series
with the main and catch coils of the
motor. The current in both coils has to be
sensed, because due to a negative value of
$\partial \phi / \partial \theta$ the current in the catch coil can
rise above values allowed by the switch.
The switch should not be closed when a
current overload occurs.

The start from standstill is a well-known
problem of the single-phase reluctance
motor (see e.g. ref. 4). Let us suppose
that the rotor is in line with the stator
poles at standstill. Excitation of the
coils is of no use, because the rotor will
remain in line and will not rotate. To ensure that this situation does not occur we
propose placing two permanent magnets in
the stator bore, which give the rotor a defined position at standstill, as shown in
figure 1. Excitation of the coils at standstill will always produce an initial
rotation clockwise. It will be the task of

233

the control unit to accomplish a good start, a good start meaning that the rotor starts rotating in the desired direction from the very first moment.

In the appendix is shown that some attention has to be given to the magnets, because a flux linkage exists between the magnets and the coils.

The solution with "parking" magnets is found to be suited to loads with a negligible static friction, e.g. a vacuum-cleaner. For applications with a significant static friction other solutions are investigated.

We propose the use of a microprocessor as control unit because it offers the following advantages compared with other solutions such as discrete electronics, or analog and digital integrated circuits :
- small size;
- low energy consumption (hence a small power supply);
- flexibility during the experimental phase (no new wiring when changes are necessary);
- complicated algorithms and features are easily introduced.

The single-chip microprocessor type 8048 has been chosen for its low cost of production.

3. DESCRIPTION OF THE CONTROL PROGRAM

3.1. Input signals

The sensor for the rotor position gives a signal every 180 degrees. The processor has to acknowledge this signal immediately, because it contains the essential information of the speed and the rotor position. From this information the correct moment to open or close the electronic switch has to be calculated. For this purpose we feed the sensor signal to the interrupt input of the processor. To record the time between two interrupts, which is related to the actual speed, we use the 8-bit timer of the processor.

A signal will be given by the current sensor when the current approaches the maximum allowable value for the power electronics.

This signal has a high priority, because to ignore it would lead to damage. Owing to the lack of a second interrupt input this signal has to be fed to another input port. Because of the inductive character of the motor coils the current cannot rise to unacceptable values within e.g. 20 μs, and this allows us to connect the current detector with an input port of the processor. This means that the program should verify whether or not the current exceeds the maximum value by a repetitive check of this port. In our program this is done every 25 μs.

3.2. Steady state

The basic idea of the control is given in figure 3, starting with the opening of the switch. The measured time between the two preceding signals of the sensor ($T_{i,n}$) is used within the time $T_{c,n+1}$ to determine with a "table look-up" procedure the new values of the time intervals $T_{a,n+2}$ and $T_{b,n+2}$; this is done every 32 sensor pulses for reasons of control loop stability. Next the reception of a sensor pulse is awaited (the time interval $T_{w,n+1}$). When a new pulse is received from the sensor the control program waits for the time $T_{a,n+2}$, closes the switch and opens it again after the time interval $T_{b,n+2}$.

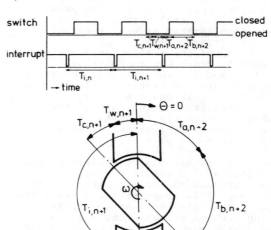

Fig. 3. Principle of control (steady state).

A software timer is required to realize the times T_a and T_b, because the type 8048 processor contains only one on-board timer and it is the task of this timer to measure the time between the interrupt signals. We have developed for this purpose a procedure whose execution time is adjustable by means of the value of a variable.

In this procedure a regular check is made to determine whether the current exceeds the maximum allowable value and, if it does, the switch is opened for a short time.

The new values of T_a and T_b are calculated within the time between the opening of the switch and the reception of a new interrupt, because here the processor has not other tasks, as shown in figure 4.

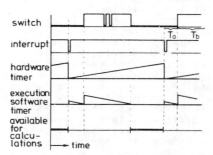

Fig. 4. Time available for calculations.

3.3. Start procedure

After the command to start the motor the program has to check whether the rotor is rotating. If it is, then the speed has to be determined by measuring the time between two successive interrupt pulses and a jump has to be made to the part of the program pertaining to this speed (see figure 5). This check is necessary because we can not use the procedure for a start from standstill.

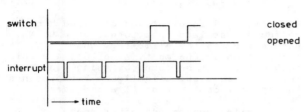

Fig. 5.   Rotating rotor at the start.

When no rotation is found, the required initial values are given to a number of variables, e.g. $T_{b,1}$, $T_{a,2}$ and $T_{b,2}$, and the switch is closed for the time $T_{b,1}$ (see figure 6). The time intervals mentioned above result in a maximum acceleration for the motor used and are found by experiment.

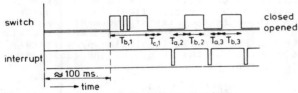

Fig. 6.   Start from stand-still.

The switch will be opened a short time if the current exceeds the maximum allowable value. The value of $T_{b,1}$ is such that in practice the rotor has reached the position where the sensor produces an interrupt when the interval $T_{b,1}$ is passed. Once the interrupt is received, then the switch remains open for the time $T_{a,2}$ and is closed again for the time $T_{b,2}$.
The same time values are used again after the reception of the second interrupt ($T_{a,3} = T_{a,2}$ and $T_{b,3} = T_{b,2}$).
After the third pulse the "table look-up" procedure is used to obtain new values for T_a and T_b by means of the measured time between the preceding interrupts. This will be repeated every five interrupt pulses until the time interval is less than 2.56 ms (corresponding speed : 11,720 r.p.m.).

3.4. Error conditions

Until now we have assumed that the sensor pulse was received during the time the processor was waiting for it to occur. Supposing a very fast acceleration of the rotor in the case of a small load torque, the interrupt will occur at an earlier moment than expected by the program, e.g. in the time interval when the switch is closed or within the time taken to calculate new values of T_a and T_b. We distinguish between two situations, namely when this phenomenon occurs before the second pulse is finished (case A) and when it occurs in the time afterwards (case B).

Case A, interrupt during :
$T_{b,1}$: open switch, replacement of the predefined $T_{a,2}$ by
$T'_{a,2} = T_{a,2}-1.28$ ms and
$T'_{b,2} = T_i/2$ (see figure 7).
$T_{c,1}$: open switch, replacement of the predefined $T_{a,2}$ by
$T'_{a,2} = T_{a,2}-1.28$ ms and
$T'_{b,2} = T_i/2$.

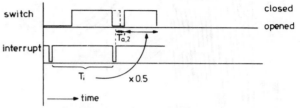

Fig. 7.   Reception of an interrupt with in the time interval $T_{b,1}$.

$T_{b,2}$: open switch, replacement of the predefined $T_{a,3}$ by
$T'_{a,3} = T_{a,3}-1.28$ ms and
$T'_{b,3} = T_i/2$.

During the start we cannot afford the loss of a pulse for the electronic switch, because this increases the risk of a stillstanding rotor. The control program has to obtain new values of T_a and T_b by using the available information and this has to be realized in a minimum of time, because the processor has also the task of guarding the T_a by means of a software timer. So a calculation requiring a long execution time is not possible. As the speed is higher than expected, the new time interval T_a has to be smaller than the predefined value; the same argument holds for the pulse width. The actions mentioned above are crude, but they require a mininum of time.

Case B, when an interrupt occurs within the following intervals the program proceeds to :

$T_{b,n}$: open switch and
for $T_{a,n}$ 2.56 ms - $T_{a,n+1}=T_a$,$T_{b,n+1}=T_i/2$
for $T_{a,n}$ 2.56 ms - new calculation for
$T_{a,n+1}$ and $T_{b,n+1}$

$T_{c,n}$: new calculation of $T_{a,n+1}$ and $T_{b,n+1}$
(see figure 8).

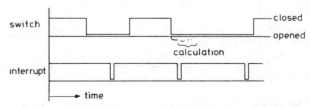

Fig. 8.   Reception of an interrupt within the time taken up by calculations.

The aim of these procedures is to ensure that the the interrupt is not received at a wrong moment a second time. A more detailed description of the program can be found in ref. 5.

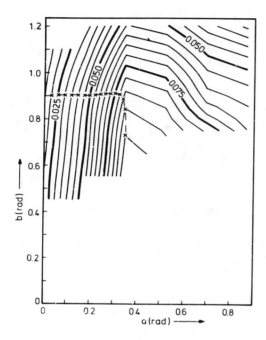

Fig. 9. The measured torque at 20,000 r.p.m. and 310 V supply (Nm).

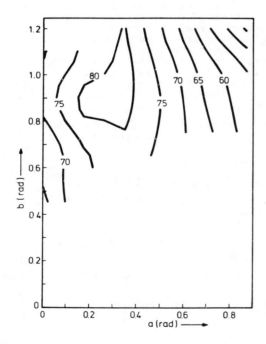

Fig. 10. The measured efficiency at 20,000 r.p.m. and 310 V supply (%).

4. TORQUE AND EFFICIENCY

An example of the influence of the switch-on and switch-off angle (a and b respectively) is given in figures 9 and 10, which show the torque production and the efficiency of the motor in figure 11 at 20,000 r.p.m. The relations between the time intervals T_a and T_b and the angles a and b are :

$$a = \pi/2 - \omega \cdot T_a \qquad (1)$$
$$b = \pi - \omega \cdot (T_a + T_b) \qquad (2)$$

From figure 9 combinations of a and b can be found that lead to a torque of 0.05 Nm. With a view to the efficiency one combination is the most favourable and this combination can be found by using figure 10. The dashed curve in figure 9 represents the optimum combination of a and b at 20,000 r.p.m. with a 310 V supply. Together with similar curves for other speeds a number of different torque versus speed curve can be realized in such a manner that a maximum efficiency is obtained for every occurring combination of speed and torque. The related values of T_a and T_b have to be stored in the memory of the micro-processor to be used by the "table look-up" routine. If more than one torque versus speed relation is required, one has to store the same number of tables and select one of these tables by means of an external switch. The practical limit for the number of tables is the size of the memory in the processor used. The 8048 has a memory of 1 kByte; the program length is approximately 600 bytes and the length of one table is 160 bytes.

5. EXPERIMENTAL RESULTS

Results of measurement on the experimental motor are given in the tables 1 and 2. The supply voltage is 310 V, the current is limited to 4.2 A and the reaction torque acting on the stator is used to determine the torque production. The bearing losses are not taken into account in determining the efficiency, nor the losses in the low power electronics. The experimental motor is not provided with forced cooling.

Speed (r.p.m.)	10,000	15,000	20,000
Torque (Nm)	0.184	0.164	0.100
Efficiency (%)	71	78	77
Output power (W)	193	260	210
a (rad)	0.17	0.35	0.35
b (rad)	0.52	0.52	0.52

Table 1. Measured maximum torque values, ambient temperature 35°C.

Speed (r.p.m.)	10,000	15,000	20,000
Torque (Nm)	0.119	0.133	0.073
Efficiency (%°)	77.5	85	81
Output power (W)	125	209	153
a (rad)	0.03	0.17	0.35
b (rad)	0.90	0.70	1.05

Table 2. Measured maximum efficiency values, ambient temperature 35°C.

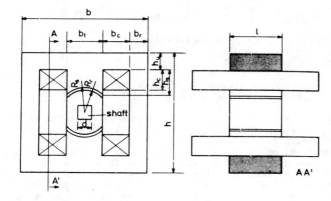

b=70, D_t=18.1, b_c=15, b_r=11, h_u=10, h_c=10, h_s=13, h=70, l=29.5, d=8
R_s=15.00, R_r=14.90
Number of turns of the main and catch coil : 314
Wire diameter : 0.5
Laminateel iron : 0.3 mm

Fig. 11. Dimensions of the motor (mm).

6. CONCLUSIONS

The electronically controlled single-phase reluctance motor is as a potential replacement of the A.C. series motor in domestic appliances. Correct operation of this type of motor requires adaptation of the switch-on and switch-off angle to the speed and it is shown that a microprocessor can do this job well. The program allows the use of a very simple sensor, which produces just two pulses per revolution. Two "parking" magnets and a special start procedure in the control unit solve the starting problem for loads with a low static friction (e.g. a vacuum-cleaner). For loads with a high static friction other solutions are under investigation.

7. APPENDIX

Assuming linear behaviour of the magnetic circuit and the absence of eddy-currents we can write the following formula for the torque T :

$$ T = i_i \frac{d\phi_{pm}}{d\theta} + T_{pm} + \frac{1}{2} i_i^2 \frac{dL}{d\theta} \qquad A.1 $$

The symbols refer to :
i_i : current carried by the main or catch coil (i_1 or i_2).
ϕ_{pm} : the flux linked with the main and catch coil when i_i equals zero. This flux is produced by the permanent magnets.
T_{pm} : the torque produced by the field of the permanent magnets when i_i equals zero.
L : the selfinductance of the main and catch coil.
θ : the rotor position, which equals zero at the in-line position.

The contribution to the mechanical output of component 2 equals zero when one revolution is considered. Component 3 represents the reluctance torque and component 1 is the hybrid part. It holds in our circuit that

$$ i_i \geqslant 0 \qquad A.2 $$

A positive contribution of the hybrid part to the torque T is consequently developed when $d\phi_{pm}/d\theta > 0$ as long as $i_i > 0$. We conclude that a correct design of the parking magnets leads to an increased torque production compared with the same motor without magnets and assuming the same current values as a function of the rotor position. In practice, however, increased eddy current and hysteresis losses will be found and the current will be influenced by the presence of the magnets. The increased iron losses are hardly accessible to numerical or analytical calculations and for this reason we give in table A.1 the maximum measured efficiency of the experimental motor at 15,000 r.p.m. and 120 V supply voltage. The difference between case A and B is the reversal of the voltage induced by the magnets; this is realized by changing direction of the current in the main and catch coils. Case C gives the efficiency without magnets.

case torque Nm	A eff. %	B eff. %	C eff. %
0.010	72	73	76
0.015	76	78	78
0.020	79	79	80
0.025	79	80	80
0.030	79	80	80
0.035	78	80	80
0.040	78	80	80

Table A.1. Efficiencies found for a motor provided with magnets (the cases A and B) and without magnets (case C).

8. REFERENCES

1. Lawrenson, P.J. and J.M. Stephenson, P.T. Blenkinsop, J. Corda, N.N. Fulton, Variable-speed switched reluctance motors. IEE Proc. B, Vol. 127 (1980), p. 253-265.
2. Davis, R.M. and W.F. Ray, R.J. Blake, Inverter drive for switched reluctance motor : Circuit and component rathings. IEE Proc. B, Vol. 128 (1981), p. 126-136.
3. Chappell, P.H. and W.F. Ray, R.J. Blake, Microprocessor control of a variable reluctance motor. IEE Proc. B, Vol. 131 (1984), p. 51-60.
4. Chatratana, S. and H.R. Bolton, D.A.G. Pedder, Investigations into a small single-phase switched reluctance motor. IEE, Conf. Publ. 202 (1981), Small and Special Electrical Machines, p. 99-102.
5. Compter, J.C. Microprocessor-controlled reluctance motor. Thesis, Eindhoven University of Technology, 1984, The Netherlands.

Performances of a Multi-Disk Variable Reluctance Machine

J.P. Bastos, R. Goyet, J. Lucidarme, C. Quichaud and F. Rioux-Damidau

International Conference on Electrical Machines, Budapest, 1982, pp. 254-257

INTRODUCTION

Multidisc variable reluctance machine have the potential of producing very much high powers and torques per unit volume than conventionally designed machines, in particular because such a structure gives a very important gap surface (1)(2). The performance greatly depends upon the geometry chosen in the useful region, where the torque is generated. In order to optimize the structure, one has to calculate the magnetic field distribution in the machine as a function of the various dimensions. This calculus cannot be made analytically. We have therefore developed a two dimensional modelisation based on a finite elements computer programme and which not only takes into account the saturation in the iron but also allows us to describe an anisotropy in the ferromagnetic material (this is particularly important when the machine teeth are composed of sheets or grain orientated steels). Two different forms of the teeth have been considered : rectangular and trapezoidal. A first three-dimensional description has been based on our two dimensional results.

The use of the finite elements variational method for our problem has soon been described (3)(4) as well as some results (5). These results are here synthesised with the new one recently obtained, that permits to choose the better structures for the machine.

2. THE MACHINE

Its structure is given on figure 1.

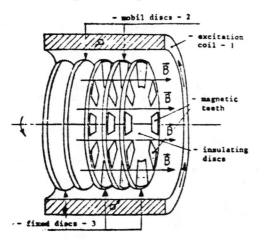

FIGURE 1 : General structure of the multidisc variable reluctance machine

A general magnetic field $\vec{B}$ is generated by the excitation coil (1). In the central region, are placed two interleaved series of identical discs, made of an insulating material. These discs feature a large number of magnetic teeth. The first series (2) is rigidly connected to the axis of the rotor ; the second one (3) is fixed to the stator.

In this device, the flux ϕ depends on :

- the current in the coil

- the angle giving the relative position of the stationary and the rotating discs.

The reluctance and flux variations are obtained by the rotation of the first series. It is minimal when the teeth are in "conjunction" (the mobil teeth are in front of the fixed ones), and it is maximal when they are in "opposition". Current I is injected in the coil as the teeth are in opposition and they are attracted. When they reach the "conjunction" position the current is annuled ; by inertia, they came again in opposition and a new cycle can occur.

One can draw the diagram $\phi(I)$ (cf. figure 2) and the hachured area represents the electrical energy for one cycle :

$$W = \int I d\phi$$

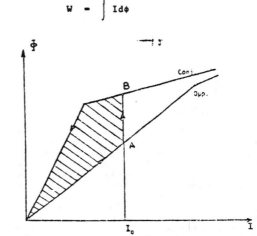

FIGURE 2 : Flux ϕ as a function of the current I

For a given frequency, the performance of the machine therefore depends upon the area W, and is as much great as W is. We thus have to optimise W as a function of the dimensions of the discs and of the teeth.

3. THE MODEL

The calculus of the magnetic field distribution in the discs region will permit to determinate the characteristics of the machine. In a first approach, we have used a bidimensional model. As the machine has a periodic structure, the study can be limited to the "elementary domain" represented in figure 3. It consists of a part of a stationary disc and a part of a rotating disc ; the radial dependence of the domain width is neglected. The geometric shape of the domain is defined by the normalised parameters given by figure 3.

The normalized parameters are :

$$e = \frac{E}{L}, \quad \text{the relative air-gap}$$

$$\lambda = \frac{\ell_1}{L}, \quad \text{the external shape}$$

$$s = \frac{\ell_2}{\ell_1}, \quad \text{the air-iron ratio}$$

$t = \dfrac{\ell_3}{\ell_2}$, the tooth slope

(t = 1 for a rectangular tooth).

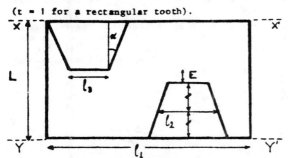

FIGURE 3 : The elementary domain

The current I circulating in the coil creates a magnetic field $\vec{H}$ in the elementary domain. As no current is present, one can write :

$$\vec{H} = -\vec{\nabla}V$$

where V is a scalar potential. The potential difference V_0 between the lines XX' and YY' (see figure 3) is directly proportional to I ; V_0 is the Ampere-turn number of the coil over an elementary domain.

We can calculate, for a given V_0, the value of $\vec{H}$ everywhere by using the finite elements method (5), and so determine the flux through the domain. In the calculus, we use "normalized" units (3)(5) that facilitate the results application. The magnetic field used in the model is $\bar{h} = \bar{H}/H_c$, where $\bar{H}$ is the real field and H_c the value at which the material begins to saturate. The material is said not saturated when $|\bar{h}| \leq 1$, and saturated in the other way.

4. THE IRON PERMEABILITY FUNCTION

The "normalized" permeability used is $\mu_r = \mu/\mu_0$, where μ is the real one and $\mu_0 = 4\pi.10^{-7}$ H/m. The chosen permeability function is :

$$\mu_r(h) = \mu_r(h=0)e^{-Gh^3/3} \quad \text{for} \quad |\vec{h}| \leq 1$$

and

$$\mu_r(h) = 1 + \frac{\mu_r(h=1) - 1}{h} \quad \text{for} \quad |\vec{h}| \geq 1$$

$\mu_r(h=0)$ and G depend upon the magnetic material of the teeth.

In the first experimented prototype (6), the teeth were made of a Fe-Co alloy. For this material we choose $\mu_r(h=0) = 1000$ and G = 0.9986. Figure 4 shows the real permeability and the theoretical one ; we see the agreement is good.

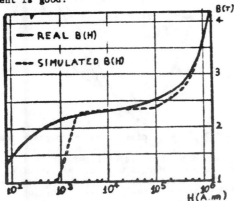

FIGURE 4 : The permeability function

5. ENERGY POWER PER CYCLE

The "normalized" energy per cycle obtained from the calculus (corresponding to $W = \int Id\phi$, fig.2) is called $A.\lambda^2$ and its relationship with W is (5) :

$$W = \mu_0 H_c^2 L^2 A\lambda^2$$

and the volumic power P is

$$P = \mu_0 H_c^2 \frac{v}{L} A$$

where v is the mobile tooth speed.

The expression of P shows that, for a given v :

- the discs must be narrow (small L values)

- the saturation induction of the teeth material (proportional to H_c) must be high

- the function $A(e,s,\lambda,t,V_0)$, proportional to the hachured area in figure 2 must also be high. Therefore, for a given V_0, we have to optimize A as a function of the elementary domain geometric shape.

6. THE CALCULUS

Two aspects of the machine design have been closely considered :

- the influence of the nature of the magnetic material, that can be isotropic or anisotropic. An anisotropic modelisation has been created for stacked iron sheets (3)(4).

- the influence of a teeth shape (rectangular and trapezoïdal teeth have been considered). Concerning the rectangular teeth shape, we take the following geometric parameters :

$$\lambda = 1, 2, 3, 5$$
$$e = 0.04, 0.07, 0.10$$
$$s = 0.17, 0.25, 0.33, 0.42, 0.5$$

and for the trapezoïdal ones, the same λ and e and :

$$s = 0.5, 0.6, 0.7$$
$$t = 0.5, 0.75, 1.$$

The performance of over 200 different combinations of these parameters have been evaluated to being varied from non-saturated case to saturated one. This calculus cover a large range of different possible structures.

Figures 5 illustrate some magnetic field configuration that we obtained.

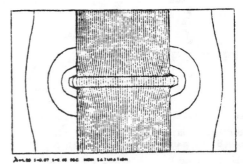

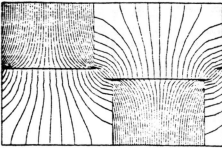

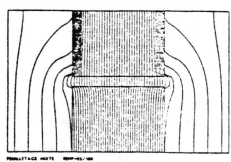

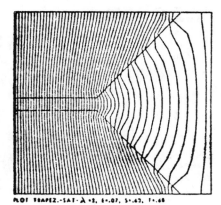

FIGURE 5 : Different magnetic field configurations.
(a) and (b) rectangular teeth. (c) teeth
made of vertically (upper one) and horizontally (lower one) stacked iron sheets.
(d) trapezoïdal teeth

7. THE FINAL RESULTS

We have plotted A as a function of the different parameters. Figures 6 show typical diagrams. By comparison
of all these curves, we conclude that :

- the gap must be as small as possible

- the best operation of those machines occurs when
  teeth are saturated (see fig. 7)

- the teeth made of stacked iron sheets present no

240

real interest.

- the air-iron ratio s must be about .42 for rectangular teeth.

- the trapezoïdal teeth are better than the rectangular ones. If we choose an air-iron ratio s ∿ .6 and
  t ∿ .6 and if the iron is greatly saturated, the
  performance can be about 1.5 the better obtained with
  the rectangular teeth.

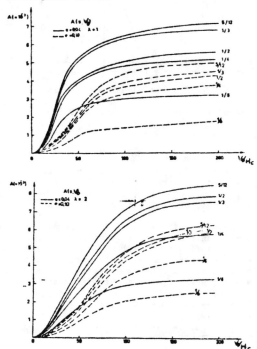

FIGURE 6 : The normalised power A as a
function of the potential V_0

8. TRIDIMENSIONAL EXTENSION OF THE MODEL

Figure 7 shows the tridimensional elementary domain
corresponding to a real machine. We can see that the
bidimensional parameter $\lambda = \ell_1/L$ is proportional to
the radius R, but that $s = \ell_2/L$ and $e = E/L$ are constants. For a given e and V_0, the study shows that
there is

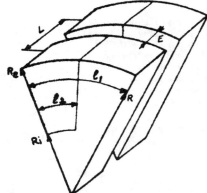

FIGURE 7 : The tridimensional elementary domain

only one optimal geometric configuration characterized
by s_M and λ_M. As λ varies with R, the global performance of the machine will be less than calculated with
λ_M. We have estimated the loss as a function or the
radius R_M for which λ is chosen equal to λ_M and therefore obtained the better value for R_M (5). A typical

machine has R_i/R_e = .5 and R_M/R_e must be = .8 and the performance is about .75 the optimal performance.

9. CONCLUSION

The calculus presented in this paper, which has been confirmed by some experiments (7), has permit to determine the optimum form of the variable reluctance machines in a bidimensional approximation.

REFERENCES

(1) C.RIOUX, Théorie générale comparative des machines électriques établie à partir des équations du champ électromagnétique. R.G.E. 79, 415, 1970.

(2) C.RIOUX, J.LUCIDARME, Les machines à réluctance variable polydiscoïdes. Journées d'étude sur l'Electrotechnique avancée. DRET Paris, mai 1981.

(3) J.P.BASTOS, Calcul des performances intrinsèques des machines à réluctance variable polydiscoïdes par une méthode d'éléments finis. Thèse de Docteur-Ingénieur, Université Pierre et Marie Curie, Paris, 1980.

(4) J.P.BASTOS, C.JABLON, G.QUICHAUD, F.RIOUX-DAMIDAU. Modélisation magnétique et numérique par éléments finis de feuilletages ferromagnétiques, Rev.Phys. Appl. 15, 1625, 1980.

(5) J.P.BASTOS, R.GOYET, J.LUCIDARME, Performances intrinsèques des machines à réluctance variable à disques imbriqués. Rev.Phys.App. 15, 45, 1980.

(6) J.NUTA, C.A.BLEIJS, C.GLAIZE, R.GOYET, Tests of an axial air-gap variable reluctance motor supplied by a solid state converter. Electric machines and Electromech. 6, 361, 1981.

(7) R.GOYET, Contribution à l'étude des machines à réluctance variables à disques imbriqués. Thèse de Doctorat d'Etat. Université Pierre et Marie Curie, Paris, 1981.

Experiment, As a Generator, of a Multi-Disk Reluctance Machine Having a 200kW Nominal Power at 500RPM

R. Goyet, C. Rioux, J. Lucidarme, R. Guillet, D. Griffault, and C. Bleijs
International Conference on Electrical Machines

ABSTRACT

Experimental results shown in the present paper concern a three phases variable reluctance machine of which structure is particular : the three phases are set in a line, as magneticly independent modules. Each phase has two rotor disks and three stator disks. There are 24 magnetic poles on a disk, this fixs nominal frequency and speed (200 Hz and 500 r.p.m.). The machine is associated with a multichanel data acquisition system located on the rotor. One chanel gives instantaneous torque from strain gages sticked on the output shaft.

Experiment relates to some static tests and main results obtained when the machine works as a generator with a resistance-capacitance load; these results concern power factor, efficiency and torque. A special presentation in the flux-ampere turn plan allows to bring out the extreme possibilities of the machine. The results, connected with more general studies, confirm the possibility to get, with this type of machine, very good performances expecially concerning torque to weight ratio. [1] [2]

INTRODUCTION

Each phase of the above introduced machine has the elementary structure of figure 1 The three phases are one beside the other. Latterly a such machine has been build and tested as a motor, with several electronic devices [3] [4]. In the present paper torque and power levels are somewhat bigger (4.000 mN. 200 kW - 1600 kg) and experiment is restricted to generator working. Thus current and voltages stay close to sinewaves and it is easy to analysis electromagnetic characteristics of the machine. Interferences are also smaller than with thyristor devices. This is a good point for a data acquisition system functionning. The system used there is described in the last section of the paper.

Electromagnetic characteristics of a reluctance machine are well introduced by the elementary cycle described in the flux ampere turn plan. This cycle stands between two static curves corresponding to the two extreme positions of the rotor, conjunction and opposition. These are given on figure 2.

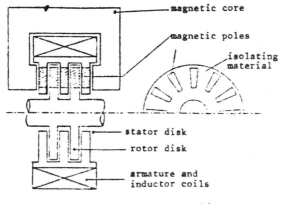

Figure 1

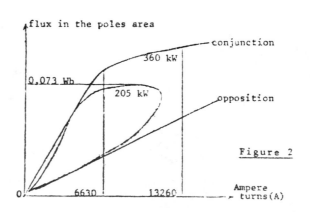

Figure 2

Reprinted by kind permission of the authors.

Flux is this observed between outer and inner radius of the poles. It gives permeances corresponding to the poles area (13,6 and 4 μH). Inductances from the coil are superior because of the leakage fields around the poles, extreme values are 15,5 and 5,3 μH per turn.

The area between the two static curves and a constant ampere turn straigt line gives the maximum available energy per cycle. At 200 Hz frequency and 13260 A there is a theorical power of 360 kW, somewhat greater than nominal power of the machine (200 kW).

ROTATING TESTS AS GENERATOR

Tests are carried out with separated armature and inductor coils. The 3 armature coils are set in a star circuit with connected neutral conductor. The 3 inductor coils are set in series with a current feed. Figure 3

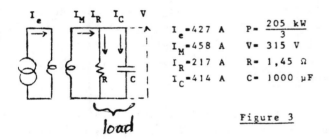

$I_e = 427$ A	$P = \dfrac{205 \text{ kW}}{3}$
$I_M = 458$ A	$V = 315$ V
$I_R = 217$ A	$R = 1,45$ Ω
$I_C = 414$ A	$C = 1000$ μF

Figure 3

gives wiring, values of the load; and main results in a test where the output power is 205 kW at 500 r.p.M. On each phase the load consists of a parallel connecting of resistance and capacitance. With a sine wave the corresponding power factor is 0.48 at the working frequency (200 Hz). The oscillogram of figure 4 shows how armature current stands away from the sine wave.

Another way of seeing electromagnetic functioning of the generator is to consider the flux ampere turns cycle. This is replaced on figure 2 between curves corresponding to conjunction and opposition. Then it appears the three essential following remarks :

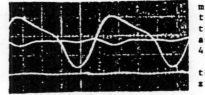

Armature current — measured total torque. average: 4100 mN — torque zero

Figure 4

- Working of the generator is very little saturated, this confirms that intrensic possibilities of the machine are greater than the adopted nominal power.
- Cycle stands away from the classical ellipse get with sinewave conditions.
- Cycle also keeps close to the two extreme curves, specially at the proximity to the origin. This results from the load adaptation to the generator and has a great effect upon the instantaneous torque. This last point is developed in the following section.

INSTANTANEOUS TORQUE

Numerical computation of single phase electromagnetic torque has been carried out. Figure 5. shows the three elementary curves and their algebraic sum.

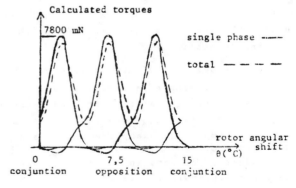

Figure 5

This sum is compared to experimental torque measured on the output shaft of the machine (figure 4). The measure is provided by the data transmission system farther described.

Calculated torque depends on the two independant variables, rotor angular shift θ and flux ϕ , and on the reluctance function $R(\theta,\phi)$. In saturated conditions single phase torque T derives from the expression :

$$T(\theta,\phi) = \int_0^{\phi} \frac{\partial R}{\partial \theta}(\theta,\phi)\ \phi'd\phi'$$

where function $R(\theta,\phi)$ determination results from no load tests. Then T (θ) is obtained from the knowledge of $\phi(\theta)$ relation.

Direct components of calculated and measured torques are of course identical (4100 mN) On the other hand, in experiment, alternating component is about five times smaller than in calculation. This results of the filter effect of inertias and elasticity of the shaft-line. This is an interesting point with regard to mechanical working of the machine : the output torque is almost constant.

DATA ACQUISITION SYSTEM

The Torque of figure 4, as also many other mechanical thermic and electric variables have been achieved by means of a special data acquisition system. This ensures the transmission of 16 independant measures in the digital form from rotor to stator.. This transmission utilizes two signals : one is the succession of measured values, these are writen on 9 bits and available at the output of serial registers embarked in the rotor; the second signal is a synchronizing pulse which allows to reconstitute clock signal and measurements sequences. The device is characterized by the following points :

- At the begining of acquisition system, each signal to measure is shaped in a universal wheastone bridge. After that data transfer is performed in the usual way : preamplifier, multiplexer , amplifier, sample-and-hold and A/D converters.
- Inductive effects in the strain gages (and their supplywires), as also thermal drifts of the amplifiers are eliminated by the use of a unique chopping method that relys on two separate measurements. The first is made when the bridge is under supplied with a measuring current, the second is made when the bridge is unactivated. Each mesurement is achieved from the difference between the two corresponding digital numbers.

- Logical driving system allows different choices in the programmation of the multiplexer. Thus one measurement can be done several times among the 16 elementary cycles.
- Data and clock are transfered from rotor to stator via rotating transformers. These transformers are fed with an 6 MHz oscillator located on the rotor.

The second part of the acquisition system located on the stator, is more common. Measures are replaced in parallel way and send to oscilloscopes.

CONCLUSION

A multi disk reluctance machine has been associated to a data acquisition system. Present paper concerns the working as a generator and main experiment results. Firstly it shows the feasibility of the machine at a significant power and torque level. Then it gives some characteristics of the electro-magnetic and mechanical behavior of the machine. Indeed working as a generator is useful step in the knowledge of a prototype : currents and voltages are near from sinewaves and the measurements interpretation is easier. Paper insists on the output torque aspect. It shows stepped down ondulations. This is a good point for indusrial applications as motor drived by electronics.

REFERENCES

I Aspects préliminaires de la théorie des machines électriques comportant des matériaux ferromagnétiques.
C. RIOUX. Rev. Phys. Appl. 15 (1980) 1505-1515

2 Performances intrinsèques des machines à réluctance variable à disques imbriqués.
J.P. BASTOS, R. GOYET, J. LUCIDARME.
Rev. Phys. Appl. 15 (janvier 1980) 45-54

3 An axial air gap reluctance motor for variable speed applications .Unnewehr LE. Koch W.M.
IEEE Trans. Power. Appl. Syst. Pas 93 n° (1974) 367-373.
4 Series commutated SCR controllers for variable speed reluctance motor drives .Unnewehr
IEEE Power electronics PESC (June 1973) 180-191.

Electrical Control of a Linear Reluctance Motor Prototype

R. Goyet, R. Gheysens, J. Lucidarme, D. Matt, and C. Rioux

European Power Electronics Conference, Grenoble, France, September 1987, pp. 1055-1059

ABSTRACT

This paper describes the experimentation of an original linear reluctance motor driven by means of a transistorized inverter.

After briefly explaining the structure of the motor, stress is laid on the inverter drive. This is obtained from moving part position and P.W.M. current control ; it has a great effect on the performance of the whole device.

KEYWORDS

Reluctance, Multi-airgap, Longitudinal-field, Actuator, Linear, Electric-jack.

1) Introduction

Orsay Laboratory of Electrotechnology has been studying reluctance machines for about ten years. It has developed axial multi air-gap structures which are capable of high power per unit weight.

The principle of air-gap area multiplication [1] has formed the basis for the actual design of an original linear mortor [2,3] having exceptional performances. The idea was in fact to build a motor of the jack type, able to compete, in terms of specific forces, with fluid pressure devices, and offering with respect to these a number of advantages, one of which being in particular the possibility of being able to operate in step by step mode or in position servo-control.

2) Description of the prototype. Power supply

The actuator is schematically represented by Figure 1. It is a structure having alternately fixed and mobile (1) magnetic teeth, excited by a longitudinal field created by windings (III). The useful volume is divided into fractions so as to offer a very large active area (26 air-gaps).

Fig. 3: Linear reluctance actuator

I: Moving part

1: Magnetic tooth
2: No magnetic support

II: Stator

3: No magnetic support

III: Coil

IV: Magnetic yoke

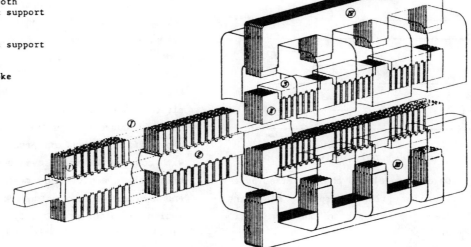

Figure 1. Linear reluctance actuator

The mass of the prototype is less than 2 kg.

Since it is a machine intended for intermittent operation, we may get the fixed and moving parts to rub and thus ensure guiding, with the value of the air-gap being consequently limited to the functional clearance. The course chosen arbitrarily is of 4 cm.

The whole system exhibits two characteristic positions with respect to the excitation field : when the teeth are staggered (opposition), the flux is maximum, when they are face to face (conjunction), the flux is minimum.

The actuator consists of three elemental modules associated in line and shifted by electric $2\pi/3$ so as to obtain a three-phase machine.

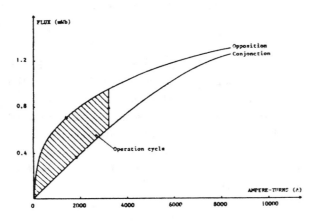

Figure 2. Flux-ampereturns curves

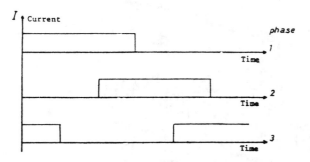

Figure 3. Current supply in the three phases

The operation of a module is characterized by the cycle described in a flux ampereturns plane inscribed between the opposition and conjunction curves (Figure 2).

The ideal mode of power supply, maximizing the thrust force, consists in injecting a constant current into the phase when the inductance of the

246

latter is increasing, and in zeroing this current when the phase inductance is decreasing, which leads to current pulses with a filling factor of 1/2 (Figure 3).

3) Experimental process

The simplified experimental process aims at revealing the performances in terms of useful force which are, as we shall see, beyond all common measure with what is usually obtained from an electric linear motor.

We choose to lift a load formed by a bucket filled up with lead shot and to set it back again with softness (Fig. 4).

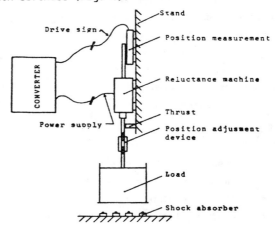

Figure 4. Testing stand

4) Description of the power supply and measuring means

The driving inverter consists of three unidirectional half-bridges based on bipolar transistors (Figure 5), operating in self-guided mode. In this configuration, there exists no coupling between phases. The basic control of the transistors is performed by a special-purpose circuit (UAA 4002 Thomson Efcis).

For the current regulation, we choose to test two different chopping processes : one version with constant current level and constant extinction time, achieved by means of a monostable, and another version using the pulse width modulation technique provided by a dedicated circuit (TL494 Texas Instruments). In both cases, the upper transistor of a half-bridge serves only to switch the current, the lower transistor performing the chopping around the prescribed value.

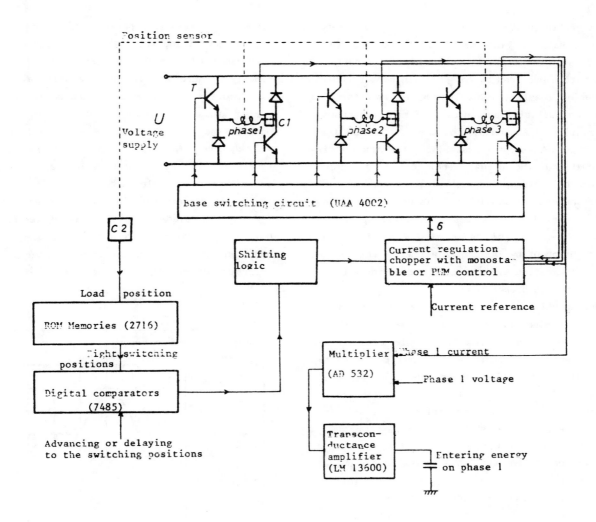

Figure 5. Power electronic control and measurement

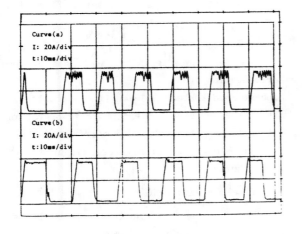

Figure 6. Current with monostable chopper (a)
or PWM regulation (b)

Table 1. Main experiment results

Load	(kg)	110	60	60
Current reference	(A)	33	33	23
Electric energy entering in the three windings	(J)	89	64	54
Potential energy at the top of the motion	(J)	33	20	18
Joule effect losses	(J)	47	26	16
Eddy currents and friction losses	(J)	9	18	20
Lifting up time	(s)	.18	.1	.13
Mean speed of the up motion	(m/s)	.17	.3	.23

The particularity of the device lies in the manner the positional information is being processed. A first incremental device using the finest information provided by the position sensor (0.01mm) gave no result chiefly because of imperfections in the variation periodicity of magnetic characteristics of the phases, as the fabrication sizes of the teeth-support assemblies, the primary difficulty in the actual mechanical construction, could not be complied with. The accuracy obtained for the commutation instants is essential. Indeed, we observed that a desynchronization, even a small one, of the order of 0.2 mm ($\frac{1}{15}$ step), brought about by failing to allow for not keeping the fabrication sizes, caused the performances to drop in vast proportions.

Consequently, we opted for the utilization of memories directly addressed by the position with a 0.1 mm ($\frac{1}{30}$ step) resolution. The memories are programmed from magnitudes measured on the machine after fabrication. These magnitudes are relative to the position and to the variation of the machine magnetic characteristics. A set of comparators enable transistor firing advances or delays to be effected with respect to the theoretical reference instant available in the memory output, the purpose of this being to correct the influence of the current rise and fall times.

The experimental reported in this paper is a lifting up followed by a setting back again with softness. It consists of a succession of transitory states which requires special measures. For instance, the electric energy input to the windings is got from instantaneous power by means of a controlled current transconductance amplifier and a capacitor.

The time-dependent position of the load is merely obtained through digital-analogue conversion of the sensor output.

5) Experimental results

Figure 6 shows the current during lifting of a 110 kg load. Curve (a) corresponds to the monostable chopper and curve (b) to PWM regulation. Both have similar performances during lifting up but the monostable solution cannot ensure constant current during descending motion. This results from the fact that the current does not naturally lessen under the reference during the free wheel operations in down motion. Table 1

gives main measured parameters during three significant rising motions with two current references and two different loads.

Finally, Figure 7 indicates the vertical position and the input energy in one phase winding during up and down motions. These curves correspond to the first test of Table 1, that with a 110 kg load. Magnetic energy recovery during free wheel operations and kinetic energy recuperation are perceptible in the figure.

Tests with current shiftings with respect to the referene revealed the necessity to perform a 0.1 mm ($\frac{1}{30}$ step) turn-off and turn-on advance of the transistors. The gain estimated under particular conditions of the experimentation is of 10 %.

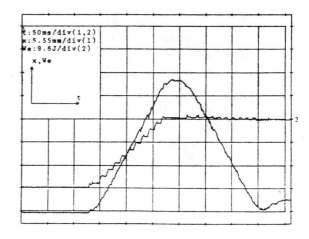

Figure 7. Energy and position during the motion

6) Conclusion

The two characteristic magnitudes of the experimentation are :
- highest load tested : 110 kg (the absolute limit of the prototype is estimated to be 180 kg)
- maximum speed attained : 0.3 m/s.

The force P per unit cross-sectional area S of the machine is :

$$P = \frac{Mg}{S} = 12.5 \times 10^5 \ Pa$$

The obtained value is close to the upper operating limit of pneumatic equipment.

Knowing that an order of magnitude of per unit area forces in an electric machine is situated around 0.4×10^5 Pa, we gained a factor 30 with respect to this value (of course, the calculation is based on referring the forces onto an area S which is representative only of the machine dimensions). It should be recalled that this gain could be achieved only through the position sensor-processing Electronics association which enables the almost unavoidable mechanical imperfections of the machine to be made up for.

The possible applications of an actuator, such as the one just described, are naturally very numerous. Let us simply say that this actuator might advantageously replace all similar pneumatic devices that do not allow an adequate control of positioning outside mechanical abutments, as well as the rival electromechanical devices using a rotating motor associated with a screw-nut system, these having lowe specific performances, especially from the point of view of displacement speeds.

BIBLIOGRAPHY

[1] Performances intrinsèques des machines à réluctance variable à disques imbriqués.
J.P. BASTOS, R. GOYET, J. LUCIDARME
RPA Janvier 1980.

[2] Moteur électrique polyphasé à réluctance variable.
Brevet n° 8514363 déposé le 27 septembre 1985.
Titulaire : CNRS
Inventeurs : C. RIOUX, R. GUILLET, J. ROCHE, J. LUCIDARME.

[3] Etude de deux structures originales de machine à réluctance variable polyentrefer,
- une machine tournante vernier,
- un actionneur linéaire à haute performance.
Thèse de Doctorat, Paris VI, Avril 1987.

1.5 - Early Milestone Papers

Tangential Forces in Overlapped Pole Geometries Incorporating Ideally Saturable Material

J.V. Byrne

IEEE Transactions, Vol. MAG-8, No. 1, March 1972, pp. 2-9

Abstract—The tangential force in overlapped-pole and attracted-slab geometries typical of electromechanical devices is examined for the case of magnetic or dielectric material having an ideally square saturation characteristic. Simple force expressions for large overlap are presented, these indicating in general a doubling of force as compared with the linear case, and a linear rather than square-law variation of force with driving potential difference (magnetic or electric). The approach for the overlapped-pole geometry, based on the physically nonrealizable idealization of ideally square $B–H$ material characteristic, yields results correlating remarkably well with experimental measurements.

INTRODUCTION

A WIDE RANGE of electromechanical devices exploits the tangential forces acting between overlapping poles or teeth, or between pairs of poles or plates and an inserted slab or plunger. A device with a single overlap zone is the most elementary and is repeated many times in conventional rotating machines.

In general, the tangential force can be derived only from a knowledge of the field spatial distribution, which is a function of overlap. For the linear case, exemplified by infinitely permeable poles separated by an air gap, the Schwartz–Christoffel transform is widely applied [1]–[3]. Where nonlinearity of the materials must be considered, iterative methods using digital computers [4], [5] now enable the field distribution to be computed.

Very simple overlap-independent force expressions apply to certain magnetically or electrically linear geometries, provided sufficient overlap exists. The force on a slab of linear magnetic or dielectric material attracted into the space between a pair of ideally permeable poles or plates is [6], [7]

$$F_x = \tfrac{1}{2}(\mu - 1)\mu_0 H_x{}^2 yz, \quad \text{magnetic case} \quad (1)$$

$$F_x = \tfrac{1}{2}(\epsilon - 1)\epsilon_0 E_v{}^2 yz, \quad \text{electric case} \quad (1')$$

where μ is slab relative permeability, y is thickness, and z is transverse length. For a pair of partially overlapped infinitely permeable poles or plates separated by a vacuum medium, the tangential force, derivable by energy [8], Maxwell stress tensor [6], [9], and Schwartz–Christoffel transform [1] methods, is

$$F_x = \tfrac{1}{2}\mu_0 H_v{}^2 yz, \quad \text{magnetic case} \quad (2)$$

$$F_x = \tfrac{1}{2}\epsilon_0 E_v{}^2 yz, \quad \text{electric case.} \quad (2')$$

Manuscript received August 3, 1970; revised July 8, 1971.
The author is with the Department of Electrical Engineering, University College, Dublin, Ireland.

Dvoracek [1] derives (2) as the limiting case of the force-per-tooth side when the ratio of air-gap length to slot width approaches zero. He notes that for this condition only those teeth of a dc machine which are partially overlapped with poles can give a torque contribution.

A characteristic of linear devices is that "with constant currents (voltages in the electric case) the energy supplied by the electric sources during a displacement is divided into two equal parts. One half is converted into mechanical work; the other half increases the stored energy of the system" [10], [11]. This accounts for the factor $\tfrac{1}{2}$ in (1) and (2). The concept is attributed [10] to Kelvin who quotes [12] the relevant extract from unpublished notes dated 1851. It will be clear that if the stored energy increments can be reduced, a more complete conversion of electrical to mechanical energy will be achieved.

An ideally square "step-function" shaped material characteristic ($B–H$, or $D–E$) is not physically realizable. Nevertheless, as in magnetic amplifier analysis [13], it is a useful idealization. There, it leads to the concept of the ideal saturable wound core as a switching element having no energy storage or other conventional inductive properties. A square magnetization characteristic ($M–H$) is in principle realizable and describes with varying approximation a range of materials from crystalline iron to silicon steels. Real ferroelectric materials have $P–E$ characteristics less ideally square than their ferromagnetic counterparts.

The purpose of this paper is to show that very simple force laws paralleling (1) and (2), but indicating a doubling of force as compared with the linear case, apply to certain overlap geometries incorporating saturable material appropriately idealized. This does not appear to have been noted in previous literature.

ATTRACTED SLAB GEOMETRY

General Approach

A slab of homogenous material, having B a single-valued function of H, is attracted (Fig. 1) into the space between a pair of ideally permeable poles maintained at a magnetic potential difference $\mathfrak{F}$ (amperes). The case corresponds with that of a slab of dielectric moving into the space between capacitor plates. Using the lumped-parameter energy approach, the treatment is essentially the same whether or not the slab material is linear, the force being given by the rate of increase of coenergy with respect to displacement, at constant MMF [6], [11].

Provided the slab is inserted sufficiently far so that the fringing flux distribution is stabilized and a region of sensi-

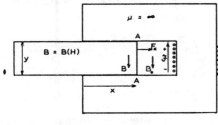

Fig. 1. Attracted slab.

bly uniform flux density exists somewhere in the overlap zone, the energy and coenergy changes are simply calculable, being associated with an extension of the uniform field zone [14]. While not crucial to the argument, it may be noted that this zone will extend right up to the leading face of the slab at AA: the fact that the field quantities will tend to be tangential to the leading face can be seen by considering the insertion of an ideal flow boundary there; the equipotentials will have the same uniform distribution on the vacuum side as on the iron side, and thus the flow boundary could be removed.

The force on the slab is therefore

$$F_x = \frac{dW'}{dx} = \frac{d}{dx} \int_V w' \, dv$$

the integral being taken over all space. Since the fringing flux was supposed independent of displacement,

$$F_x = (w_1' - w_0')yz, \qquad \text{N} \qquad (3)$$

where W' is coenergy of the device (joules), w' is coenergy density, and w_1' and w_0' are coenergy densities in the uniformly magnetized regions inside and ahead of the slab. In terms of the field quantities and material properties in these two regions, the force is

$$F_x = yz \left[\int_0^H B \, dH - \int_0^H B_0 \, dH \right]$$
$$= \mu_0 yz \int_0^H M \, dH, \qquad \text{N} \qquad (4)$$

where H is the same in the two regions. The difference between the flux densities in the two regions $B - B_0$ is μ_0 times the magnetization density $M(\text{A} \cdot \text{m}^{-1})$. The force for the corresponding electric case is, with P the polarization density $(\text{C} \cdot \text{m}^{-2})$,

$$F_x = yz \int_0^E P \, dE, \qquad \text{N}. \qquad (4')$$

Linear Case

For linear materials with $M = (\mu - 1)H$ or $P = (\epsilon - 1)\epsilon_0 E$, (4) and (4') give (1) and (1'), which, for comparison with the saturated case, will be written

$$\frac{F_x}{z} = \frac{1}{2} \mu_0 M \mathfrak{F}, \qquad \text{N} \cdot \text{m}^{-1} \qquad (5)$$

$$\frac{F_x}{z} = \frac{1}{2} PV, \qquad \text{N} \cdot \text{m}^{-1} \qquad (5')$$

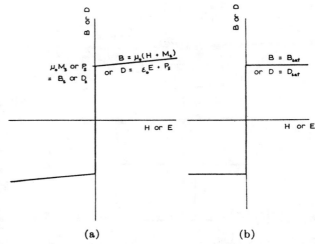

Fig. 2. Ideal B–H or D–E material characteristics. (a) Ideally square M–H or P–E relation. (b) Ideally square B–H or D–E relation.

where $\mathfrak{F}$ and $\dot{V}$ are the driving potential differences in amperes and volts, respectively, and M and P are magnetization and polarization densities in the uniform field zone, respectively.

Expressed as normal pressures referred to the leading face of the slab, (1) and (1') give, in conformity with alternative approaches via volume force density [6], [7],

$$p = \frac{1}{2}(\mu - 1)\mu_0 H^2 = \frac{1}{2}\mu_0 MH, \qquad \text{N} \cdot \text{m}^{-2} \qquad (6)$$

$$p = \frac{1}{2}(\epsilon - 1)\epsilon_0 E^2 = \frac{1}{2}PE, \qquad \text{N} \cdot \text{m}^{-2}. \qquad (6')$$

A method of measuring the dielectric constant of liquids depends on the pressure of (6') causing the liquid to be drawn up between parallel plates [7].

Saturated Case

For slab material having the ideally square M–H (or P–E) characteristic of Fig. 2(a), (4) and (4') give

$$\frac{F_x}{z} = \mu_0 M_s \mathfrak{F} = B_s \mathfrak{F}, \qquad \text{N} \cdot \text{m}^{-1} \qquad (7)$$

$$\frac{F_x}{z} = P_s V = D_s V, \qquad \text{N} \cdot \text{m}^{-1} \qquad (7')$$

where B_s and D_s are defined (Fig. 2) as the flux densities at the threshold of saturation. The corresponding pressures, referred to the leading face of the slab, are

$$p = \mu_0 M_s H = B_s H, \qquad \text{N} \cdot \text{m}^{-2} \qquad (8)$$

$$p = P_s E = D_s E, \qquad \text{N} \cdot \text{m}^{-2}. \qquad (8')$$

The forces and pressures of (7) and (8) for a slab of material having the ideally square M–H or P–E characteristic of Fig. 2 are exactly double those of (5) and (6) for the linear case, given the same driving potentials and magnetization or polarization densities. This is in conformity with the fact that the field energy is not a function of displacement, the energy density in the uniformly saturated part of the slab being the same as that in the vacuum ahead of the slab.

253

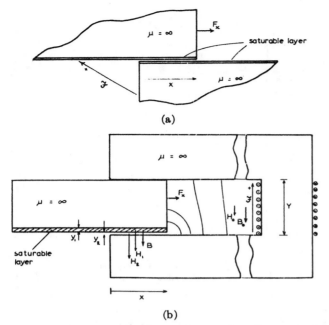

(a)

(b)

Fig. 3. Saturable material in surface layer. (a) Overlapping-pole geometry. (b) Attracted-slab geometry.

OVERLAPPING POLE GEOMETRY—CASE 1: SATURABLE MATERIAL LOCATED IN SURFACE LAYER

Here, the saturable material is supposed to be in slablike layers attached to one or both pole faces, the poles being elsewhere infinitely permeable, or conducting (Fig. 3(a)). This geometry is perhaps most relevant to the electric field case, where ferroelectric material might conveniently form a thin layer on a conducting backing plate. For greater generality, a series air gap will be included in the treatment. The overlap zone is then a sandwich of linear and saturable materials between equipotential boundaries, and the fields remote from fringing effects will be normal to these boundaries. The case also represents the attracted-slab geometry of Fig. 3(b), with the pole spacing Y taken very large. The treatment by the energy method includes, as before, the linear case.

For the geometry of Fig. 3(b) the force is, in the same way as (3),

$$F_x = \frac{d}{dx} W' = (w_1' y_1 + w_2' y_2 - w_0' Y) z \quad (9)$$

where w_1' and w_2' are coenergy densities in the two materials in the uniform field regions of the overlap zone and y_1 and y_2 are the thicknesses. The term w_0' is the coenergy density in the uniform field region well ahead of the leading face where of course the field is complicated by fringing. The force per transverse meter z is therefore, from (9),

$$\frac{F_x}{z} = y_1 \int_0^{H_1} B\, dH_1 + y_2 \int_0^{H_2} B\, dH_2 - Y \int_0^{H_0} B_0\, dH_0. \quad (10)$$

If the first integral refers to material having the ideally square M–H characteristic of Fig. 2(a), the remaining

254

integrals referring to media where $B = \mu_0 H$, then (10) gives

$$\frac{F_x}{z} = \mu_0 \left(M_s + \frac{1}{2} H_1 \right) H_1 y + \frac{1}{2} B H_2 y_2 - \frac{1}{2} B_0 H_0 Y$$

$$= \mu_0 \left(M_s + \frac{1}{2} H_1 \right) \mathfrak{F}_1 + \frac{1}{2} B \mathfrak{F}_2 - \frac{\mu_0}{2} \frac{\mathfrak{F}^2}{Y} \quad (11)$$

where $\mathfrak{F}$ is the total driving potential difference and $\mathfrak{F}_1$ and $\mathfrak{F}_2$ are its components across the saturated and linear layers of the sandwich. For large Y, assumed henceforward, the last term of (11) becomes negligible, and the forces for the geometries of both Fig. 3(a) and (b) are

$$\frac{F_x}{z} = \mu_0 \left(M_s + \frac{1}{2} H_1 \right) \mathfrak{F}_1 + \frac{1}{2} B \mathfrak{F}_2 \quad (12)$$

$$\frac{F_x}{z} = \left(P_s + \frac{1}{2} \epsilon_0 E_1 \right) V_1 + \frac{1}{2} D V_2. \quad (12')$$

For the ideal case of zero air gap ($y_2 = 0$), $\mathfrak{F}_2 = 0$, $\mathfrak{F}_1 = \mathfrak{F}$, so that (12) and (12′) give

$$\frac{F_x}{z} = \mu_0 \left(M_s + \frac{1}{2} H_1 \right) \mathfrak{F} \quad (13)$$

$$\frac{F_x}{z} = \left(P_s + \frac{1}{2} \epsilon_0 E_1 \right) V. \quad (13')$$

If, instead of an ideally square M–H or P–E characteristic, the ideally square B–H or D–E characteristic of Fig. 2(b) is assumed for the saturable layer, (12) takes the simpler form

$$\frac{F_x}{z} = B_{sat} \mathfrak{F}_1 + \frac{1}{2} B \mathfrak{F}_2 \quad (14)$$

$$\frac{F_x}{z} = D_{sat} V_1 + \frac{1}{2} D V_2 \quad (14')$$

where B is B_{sat} and D is D_{sat} if saturation level is reached. The potential "dropped" across the saturated layer is twice as effective in producing force as that "dropped" across the linear layer. For zero air gap, the second term disappears. If the saturable material is below saturation and hence infinitely permeable, it merges with the poles proper, and the first term disappears yielding (2) and (2′) already known for the linear case.

OVERLAPPING POLE GEOMETRY—CASE 2: SATURABLE MATERIAL IN DEPTH

Fringing and Leakage Fluxes Neglected

For the geometries already considered, the energy approach gives solutions, e.g., (4) and (10), valid for any "soft" material characteristic, the linear and ideally square characteristics being special cases. For the geometry of Fig. 4, only the saturated case with ideally square B–H or D–E materials will be considered, this imposing a simplifying constraint on the field distribution.

An ideally square B–H characteristic leads to the following properties (two-dimensional symmetry is assumed).

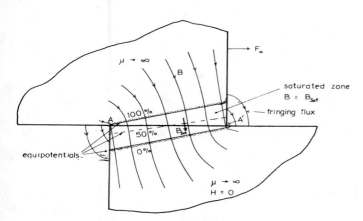

Fig. 4. Overlapping poles of material having ideally square B–H characteristic.

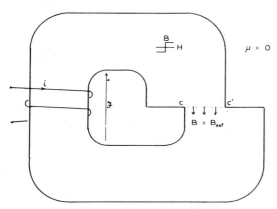

Fig. 5. Iron circuit with constriction. External medium impermeable to flux.

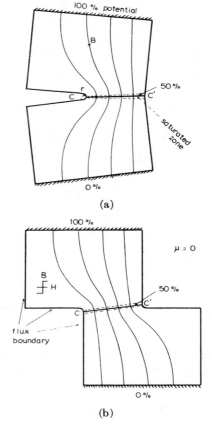

(a)

(b)

Fig. 6. Symmetrical pole pair. External medium impermeable to flux.

1) The energy density is everywhere zero in the material.
2) A convergent–divergent flux tube must include a saturated region at its waist if there is finite magnetic potential difference between its wider parts.
3) In a saturated region $B = B_{\text{sat}}$, and since div $\bar{B} = 0$, the flux lines cannot converge or diverge. Adjacent stream lines will have a common center of curvature.
4) For isotropic material in a saturated region there will be no H component normal to the B direction. Because of 3), the equipotential surfaces are planes. The boundary surfaces between saturated/unsaturated material, being equipotentials, are also plane.
5) In a convergent–divergent flux path driven by a finite MMF the throat or smallest section will be saturated in the normal direction, provided the surrounding medium is nonconductive to flux.

The truth of property 5) may be seen in several ways. Most generally, the flux will dispose itself for minimum energy and/or maximum flux-linkage. Referring to Fig. 5, assume that remanent flux density B_{sat} has been set up normal to the constriction section cc', this giving the maximum possible flux, and zero stored energy in the system. (This flux could be thought of as set up by inserting a premagnetized thin slab at cc'.) The subsequent application of an MMF $\mathfrak{F}$, cannot deflect the flux density

at cc' from its original normal direction or magnitude because such deflection would reduce the surface integral $\int_s B_n \, dA$ over cc', corresponding with energy transfer from the device to the electrical source. There being no stored energy in the device, this is impossible.

An alternative approach to 5), valid for a symmetrical pole pair, is illustrated in Fig. 6. The iron outside the saturated zone will be supposed to have very high, but not infinite, permeability so that the flux distribution is determinate. The problem may be visualized in terms of a two-dimensional resistive analog with ideally saturable Teledeltos paper, unfortunately not available. In Fig. 6(a), because of the symmetry of the convergent–divergent flux pattern, the surface cc' bisects the saturated zone, constitutes the 50-percent equipotential, and is normal to the flux-density vector over its entire area. Furthermore, and in contradistinction from the same case in a linear medium, the flux density at this surface has everywhere its saturated value and is therefore uniform. In these conditions of both potential and flux-density uniformity over the surface cc', the lower block may be turned around to give the configuration of Fig. 6(b), without effect on the flux distribution in either block. A small fillet radius r has been assumed so as to avoid difficulties in visualizing the bisecting plane cc' of the saturated zone in Fig. 6(a). For decreasing r, the plane cc' of Fig. 6(b) has decreasing tilt and in the limit corresponds with the sliding interface AA' of Fig. 5.

255

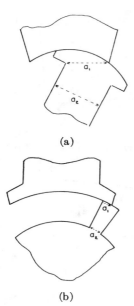

(a)

(b)

Fig. 7. Geometries exhibiting jump in constriction location. a_1: variable constriction. a_2: fixed constriction.

Device stored energy in the absence of fringing flux is nil, and the coenergy $\mathfrak{F}\Phi_{\text{sat}}$ is proportional to the constriction section a_c

$$W' = \mathfrak{F}B_{\text{sat}}a_c.$$

For devices giving linear and rotary motion, respectively, the force and torque are

$$F_x = \mathfrak{F}B_{\text{sat}}\frac{da_c}{dx} \qquad (15)$$

$$T = \mathfrak{F}B_{\text{sat}}\frac{da_c}{d\theta}.$$

In the examples of Fig. 7 the variable constriction a_1 becomes the fixed constriction a_2 as rotation proceeds, the torque thereafter being nil. For the geometry of Fig. 4 the force for all overlap from zero to full is, from (15),

$$\frac{F_x}{z} = \mathfrak{F}B_{\text{sat}}, \qquad \text{N} \cdot \text{m}^{-1}. \qquad (16)$$

Effect of Fringing Flux

The presence of fringing flux in the geometry of Fig. 4 will cause the flux density in the interface AA' to deviate from the normal direction it would otherwise have, the waist of the convergent–divergent flux pattern being no longer defined by the interface.

A rough indication of factors affecting the tilting of the saturated constriction zone is obtained by mentally removing the sharp corners at A and A' in Fig. 4 and supposing some of the fringing flux to be accommodated in fillets as already shown in Fig. 6(b). The fillet size is supposed to increase with $\mathfrak{F}$ to represent the corresponding growth in fringing flux. For poles with such tip extensions having a circular fillet profile of radius $\frac{1}{2}d$ and total

256

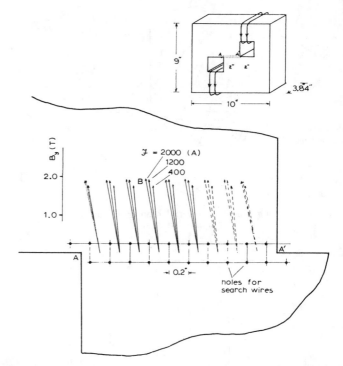

Fig. 8. Lamination stack simulating overlapped poles. Flux density measured in the plane AA'.

overlap x, (15) easily gives

$$\frac{F_x}{z} = \mathfrak{F}B_{\text{sat}}\left(1 + \frac{d^2}{x^2}\right)^{-1/2}$$

which approaches (16) for large x. The constriction surface is tilted $\tan^{-1}(d/x)$ from the sliding interface AA', the angle increasing with MMF and decreasing with overlap.

The deviation of the B field from a normal direction at the interface surface has been measured using the lamination stack shown in Fig. 8 to represent a pair of overlapped poles. The steel is a conventional power transformer silicon iron,[1] the grain direction being set normal to the interface. Search holes 0.76 mm in diameter, spaced 0.2 in apart in square formation straddle what would be, in a machine, the sliding interface. The field distribution here, measured by the reversal method with a Grassot fluxmeter, is shown for a range of driving MMF. The behavior approximates to that suggested for an ideally square B–H material in that the B field remains remarkably uniform and parallel to itself over most of the overlap zone and shows a tilt increasing with driving MMF.

In explanation of the relative constancy of the B field magnitude, it is suggested that the H field in a saturated overlap zone will not increase as fast as $\mathfrak{F}$ because of the thickening of the zone itself with $\mathfrak{F}$.

The fringing flux distribution and the corresponding stored energy may be expected to depend on the tilt angle of the saturated zone. If the overlap length in relation to

[1] Unisil 46, British Steel Corporation.

the driving MMF is small so as to give large tilt, it will not be possible to ignore the variation of the fringing flux accompanying an incremental change in overlap. For increasing overlap, as the tilt angle and its rate of change decrease, the change in fringing flux for an incremental displacement eventually becomes a negligible fraction of the flux increment due to the increase in interface area.

As overlap increases, therefore, the flux and coenergy increments for a displacement at constant MMF include diminishing contributions from fringing flux changes and progressively approach the values

$$d\Phi = B_{sat}z\,dx$$

$$dW' = \mathfrak{F}B_{sat}z\,dx$$

appropriate to the iron-borne flux and coenergy increments as the cosine of the tilting angle approaches unity. The force for large overlap is therefore

$$\frac{F_x}{z} = \mathfrak{F}B_{sat} \qquad (17)$$

which is the same as (16) for the case with fringing neglected. It is of interest that the cosine of the tilt angle measured on the test rig of Fig. 8 was in the range 0.98–1.0.

For the case where one pole only is ideally saturable, the other being infinitely permeable, the discussion is simpler, but again leads to (17). At the interface, there will be no tilt of the B vector from the normal direction in the saturable material because of the boundary constraint $H_x = 0$ imposed by the infinitely permeable pole.

The preceding arguments are not sensitive to the location of magnetizing windings, provided these are rigidly attached to one or both poles.

FORCES AT SMALL OVERLAP

As indicated in the Introduction, solutions for the forces at small overlap require, in general, a detailed knowledge of the field distribution. For the nonrealizable idealization of a square B–H (or D–E) material characteristic, sufficient information on the field distribution at zero overlap is already available. Unfortunately, the idealization is here least valid because of the intense fields to be expected at the overlap threshold. However, we obtain a guide to the character of the force-overlap variation at small overlap.

In Fig. 9(a), for the pair of poles at zero overlap, the plane AA' is by symmetry an equipotential surface having magnetic scalar potential, say,

$$\psi AA' = 0, \qquad A.$$

The surfaces CC' and DD' of the unsaturated parts of the poles have the potentials $\pm 0.5\mathfrak{F}$. In the saturated zones, shown hatched, the B field is everywhere parallel to itself and, from symmetry, has the tilt angle $\pi/4$. The surface $0C$ of the saturated zone is, for the fringing field, a flux source having normal component

$$B_n = \frac{1}{\sqrt{2}}B_{sat}.$$

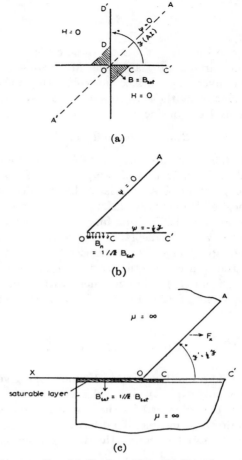

Fig. 9. Zero overlap. (a) Configuration. (b) Boundary conditions. (c) Model giving same boundary conditions.

The field in the space bounded by $A0CC'$ (Fig. 9(b)) is uniquely determined by the values set for $\mathfrak{F}$ and B_{sat}. It could be modeled by resistive analog plotter, constant currents being fed into a series of segments along $0C'$, C marking the last segment which can be so fed without its voltage exceeding that representing the potential $0.5\mathfrak{F}$. The remaining boundaries would be equipotentials.

The same field is generated by the geometry of Fig. 9(c), for which the force component F_x is known from energy considerations. The infinitely permeable upper pole has surface $0A$ inclined at $\pi/4$ to the sliding interface. The lower pole, also infinitely permeable, has a thin surface layer of ideally square B–H material saturating at $B_{sat}' = B_{sat}/\sqrt{2}$. The force F_x on the upper pole (from (17), which is applicable because the fringing flux is not displacement dependent) is

$$F_x = \frac{\mathfrak{F}}{2}\frac{B_{sat}}{\sqrt{2}}z. \qquad (18)$$

This must correspond [9] with the x component of the normal surface stress $B^2/2\mu_0$ integrated over the surface $A0$ since the surface $0X$ contributes nothing to the translational force on the upper pole. The force normal to $A0$,

257

being $\sqrt{2}$ times greater than (18) is

$$F_{A_0} = \frac{\mathfrak{F}}{2} B_{sat} z. \qquad (19)$$

Reverting to Fig. 9(a), the total force on either pole is given by the surface integral of the Maxwell stresses in vacuo taken over any surface enclosing the body (9). From (19), since only the surface AA', assumed to extend to infinity, need be considered,

$$F = \mathfrak{F}B_{sat}z, \qquad \text{normal to } AA'. \qquad (20)$$

Equation (20) may also be derived by assuming an incremental interpenetration of the pole tips of Fig. 9(a) and supposing the surplus material to be transferred to a region of negligible coenergy density. If AA' remains a plane of symmetry, the fringing flux, for reasons already given, will not vary for displacements made at constant MMF. Coenergy changes will be associated with the increment in the iron constriction section, located in AA', and the force from (15) is in agreement with (20).

The component of (20) in the direction of proposed overlap is

$$\frac{F_x}{z} = \frac{1}{\sqrt{2}} \mathfrak{F}B_{sat}, \qquad \text{at zero overlap.} \qquad (21)$$

The force at zero overlap is $1/\sqrt{2}$ times that given by (17) for large overlap. By an analogous argument, the same result (21) is obtained for the case where one of the poles is infinitely permeable. The derivation of (20) and (21) assumes remote location of the driving winding.

Measurements with Overlapped C Cores

Of interest is the extent to which (17) and (21) apply to real situations. The device geometry (Fig. 10(a)) departs from the ideal chiefly in that the poles have limited length both transversely and in the direction of motion.

The measured forces (Fig. 10(b)) are average values computed from a family of device magnetization characteristics by dividing coenergy increment by displacement. In this way it was possible to eliminate air gaps and have the ground (machined) faces of the cores in physical contact. The width of the hysteresis loop was so small in relation to the driving MMF that the magnetization characteristic could be considered single valued. This energy method is least satisfactory where there is rapid force variation with overlap, such as near zero overlap. At zero overlap, the force reduction shown, to about 90 percent of the plateau level, is the mean of results ranging from 85 to 95 percent of plateau level obtained by direct measurement of torque in a rotary geometry.

With a real material (here a conventional transformer silicon iron), there is an inherent difficulty in setting a value B_{sat} for prediction of force by (17). For ideally square M–H material, the value $\mu_0 M_s$, corresponding to the threshold of saturation, might be selected. Here, quite arbitrarily, the magnetization characteristic has been linearized to yield a corresponding value supposed to occur at $H = 0$; at full overlap the core flux density

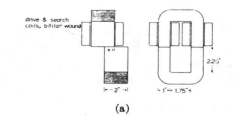

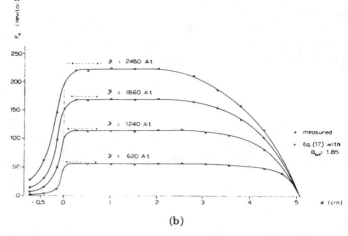

(b)

Fig. 10. Overlapped C cores. Force variation with MMF and overlap.

TABLE I

$\mathfrak{F}$ per Overlap (A)	2480	1860	1240	620
Force (N)				
Measured	224	170	115	56.6
Equation (17) (with B_{sat} = 1.85)	233	175	116.5	58.2
Measured force/$\mathfrak{F}z$ (T)	1.78	1.80	1.82	1.80

varied from 1.87 to 1.93 T over the four-to-one range of MMF for which the force is shown, giving

$$B_{sat} = 1.85, \qquad \text{T.}$$

The forces, from (17) using this value of B_{sat}, are shown in Fig. 10(b) beside the measured values. For the plateau regions the forces are given in Table I. The last line shows that (17) would very closely describe the device in the chosen MMF range if the value 1.8 (rather than 1.85) had been selected for B_{sat}. The force variation with driving ampere-turns is remarkably linear.

Force has also been measured directly, using C cores of smaller core section 0.625-by-0.75 in. There was difficulty in maintaining consistent small air gaps due to the large normal force between the pole surfaces. Good correlation with (14) was indicated as the air gap was varied.

The extent of the force plateau diminishes with increasing $\mathfrak{F}$, as might be expected in a geometry of limited pole length; with increasing overlap the fringing flux boundary surfaces foreshorten, and also, the core sectional area becomes inadequate to carry the combination of iron-borne and fringing flux.

258

The force at zero overlap is closer to the plateau level than a comparison of (21) and (17) would suggest. This is attributable, at least in part, to the poor validity here of the ideally square B–H assumption, as already noted. The geometry is, in any case, unsuitable for confirming 21), in the derivation of which the magnetizing winding was presumed to be remote and the magnetic material isotropic.

DISCUSSION AND CONCLUSIONS

Simple equations have been presented describing the tangential force at large overlap for typical electromechanical device geometries incorporating ideally saturable material. Because the diversion of half the electrical input energy into field storage, a characteristic of linear devices driven at constant magnetic or electric potential difference, does not take place, the forces are correspondingly augmented. If the whole of the driving potential difference is developed across ideally square M–H (or P–E) material in the case of the slab geometry, or across square B–H (or D–E) material instead of an air gap in the case of the overlapping pole geometry, the forces, for equal flux densities, are exactly doubled. A direct rather than a square-law variation of force with driving potential difference is observed.

For geometries incorporating nonlinear material in slab or layer form, treatment by the energy method is straightforward. Where the saturable material extends in depth throughout a pole or poles, the derivation given depends on the assumption of a physically unrealizable ideally square B–H or D–E material characteristic. The resulting simplifying constraints on flux distribution permit simple solutions for force at large and zero overlaps. The experimental measurements on steel cores indicate that the solution for large overlap has remarkably good validity in a real situation, describing the force almost down to zero overlap, whereas the solution for zero overlap gives less than the observed force. In explanation, it may be sup-posed that at small overlap the real flux densities depart most from the ideal saturation levels assumed, this compensating in part for force reductions associated with tilting of the flux direction away from a normal direction to the interface. As a result, the plateau level of force is reached after a surprisingly small initial overlap.

It would be interesting to model by digital computer methods the overlapped pole case with square M–H material characteristic so that the error involved in the square B–H idealization could be assessed. The force expressions, although derived for idealized situations, have application to practical electromechanical devices, including rotating electric machinery.

REFERENCES

[1] A. I. Dvoracek, "Forces on the teeth in dc machines," *IEEE Trans. Power App. Syst.*, vol. PAS-64, pp. 1054–1061, Feb. 1963.
[2] F. C. Williams and R. S. Mamak, "Electromagnetic forces in slotted structures," Inst. Elec. Eng., London, Monogr. 456 U, 1961.
[3] K. J. Binns, "Pole-entry flux pulsations," Inst. Elec. Eng., London, Monogr. 469U, 1961.
[4] E. A. Erdélyi and E. F. Fuchs, "Nonlinear magnetic field analysis of dc machines," *IEEE Trans. Power App. Syst.*, vol. PAS-89, pp. 1546–1554, Sept.–Oct. 1970.
[5] P. Silvester and M. K. V. Chari, "Finite element solution of saturable magnetic field problems," *IEEE Trans. Power App. Syst.*, vol. PAS-89, pp. 1642–1651, Sept.–Oct. 1970.
[6] H. H. Woodson and J. R. Melcher, *Electromechanical Dynamics*. New York: Wiley, 1968, pp. 444–454.
[7] R. Becker and F. Sauter, *Electromagnetic Fields and Interactions*. Glasgow, Scotland: Blackie, 1964.
[8] E. B. Moullin, *Principles of Electromagnetism*. London: Oxford, 1932, p. 180.
[9] C. J. Carpenter, "Surface-integral methods of calculating forces on magnetized iron parts," Inst. Elec. Eng., London, Monogr. 342, Aug. 1959.
[10] V. Karapetoff, "Mechanical forces between electric currents and saturated magnetic fields," *AIEE Trans.*, vol. 46, pp. 563–569, May 1927.
[11] A. E. Fitzgerald and C. Kingsley, *Electric Machinery*. New York: McGraw-Hill, 1961, pp. 43–45, 57–60.
[12] W. Thomson, *Reprint of Papers on Electrostatics and Magnetism*. London: Macmillan, 1872, pp. 440–443.
[13] G. M. Attura, *Magnetic Amplifier Engineering*. New York: McGraw-Hill, 1959.
[14] G. R. Slemon, *Magnetoelectric Devices: Transducers, Transformers, and Machines*. New York: Wiley, 1966, pp. 16–18.

Compatible Brushless Reluctance Motors and Controlled Switch Circuits

B.D. Bedford

U.S. Patent No. 3,679,953, July 1972

[54] **COMPATIBLE BRUSHLESS RELUCTANCE MOTORS AND CONTROLLED SWITCH CIRCUITS**

[72] Inventor: **Burnice D. Bedford**, Scotia, N.Y.

[73] Assignee: **General Electric Company**

[22] Filed: **Nov. 6, 1970**

[21] Appl. No.: **87,565**

[52] **U.S. Cl.**..............................**318/138**, 318/138, 318/166, 318/254
[51] **Int. Cl.**..**H02k 29/00**
[58] **Field of Search**...................318/138, 166, 254, 685, 696

[56] **References Cited**

UNITED STATES PATENTS

3,374,410   3/1968   Cronquist 318/138
3,127,548   3/1964   Van Emden...........................318/138
3,486,096   12/1969   Van Cleave 318/138
3,530,347   9/1970   Newell..................................318/138

Primary Examiner—Gene Z. Rubinson
Attorney—Paul A. Frank, John F. Ahern, Julius J. Zaskalicky, Donald R. Campbell, Frank L. Neuhauser, Oscar B. Waddell and Joseph B. Forman

[57]   **ABSTRACT**

Brushless reluctance motors are compatible with energization by the simple rectilinear voltages easily produced by motor control circuits having a small number of mechanical or solid state switches. These motors with two or three opposing pairs of main and feedback stator windings, and suitable variable or constant gap nonpolarized rotors, require a like number of controlled switches and uncontrolled rectifiers actuated by a rotor position sensor to generate square wave voltages on which the reluctance motors operate efficiently.

11 Claims, 8 Drawing Figures

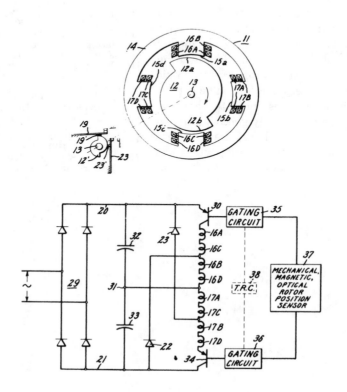

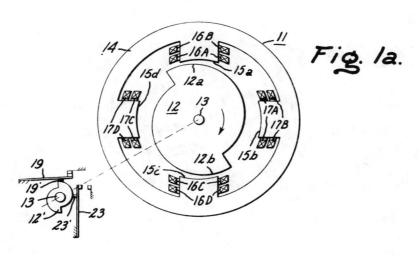

Fig. 1a.

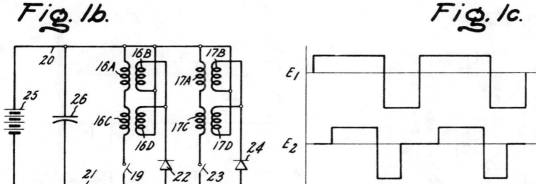

Fig. 1b.

Fig. 1c.

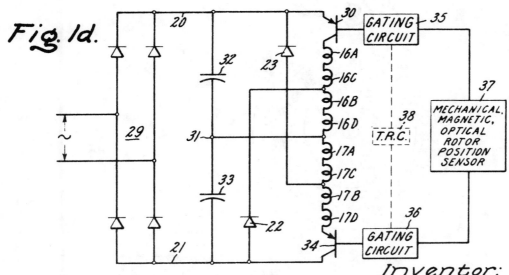

Fig. 1d.

Inventor:
Burnice D. Bedford,

by Donald R. Campbell
His Attorney.

261

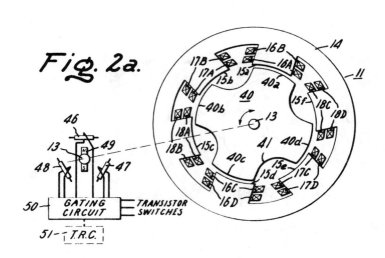

Fig. 2a.

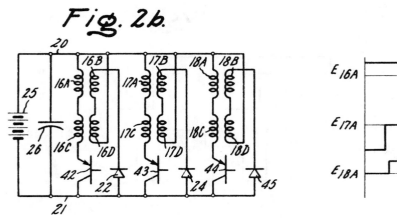

Fig. 2b.

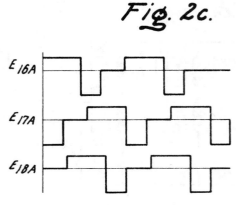

Fig. 2c.

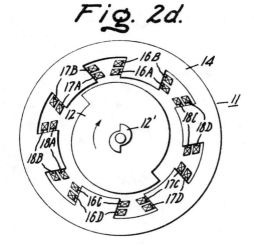

Fig. 2d.

Inventor:
Burnice D. Bedford,
by Donald R. Campbell
His Attorney.

262

COMPATIBLE BRUSHLESS RELUCTANCE MOTORS AND CONTROLLED SWITCH CIRCUITS

This invention relates to brushless reluctance electric motors energized from a direct current supply, and more particularly to brushless motors designed for efficient operation by motor control circuits using a few controlled switches that produce simple rectilinear voltage wave shapes. These brushless motors are suitable for adjustable speed operation.

In the brushless or commutatorless direct current motor, the drive coils carrying the magnetizing current which results in the production of torque are located on the stator housing, rather than on the movable rotor as in the conventional direct current motor. The rotor in a brushless motor is commonly a permanent magnet rotor, and the circuit energizing the drive coils uses controlled switches to control the flow of current in the drive coils in a sequence to produce continuous rotation of the rotor. Consequently, the commutators and brushes utilized on the conventional motor to supply current to the armature windings are eliminated, together with such undesirable features as for example the need to replace brushes, arcing between commutator segments, and repair of worn commutator segments. In order to reduce the cost of the control circuit, and thus minimize the cost of the combination of the brushless motor and control circuit, it is advantageous to choose a circuit that requires only a few controlled on-off switches, preferably solid-state switches such as transistors or silicon controlled rectifiers. These simple control circuits, however, produce only square or stepped voltage wave shapes, and the particular rectilinear wave shape that is generated depends on the circuit configuration, the number of switches, and the sequence and timing of the closing of the switches. A particular brushless motor that operates satisfactorily with one of these voltage wave shapes will not operate as well with another simple wave shape. Although brushless motors operated by motor control circuits comprising only a few controlled switches have appeared in the prior art, there is inadequate recognition in the prior art of the motor design needed to operate efficiently with a particular square or stepped voltage wave shape that can be produced by an inexpensive control circuit. The alternative approach of using a conventional motor design and modifying the motor control circuit to produce a more complex voltage wave shape to meet the more severe motor requirements frequently results in a higher cost for the combined motor and control circuit.

As will be further described in this application, details of the impressed voltage and back electromotive force voltage are not critical in a reluctance motor, and such a motor is suitable to be energized by an extremely simple control circuit which in the preferred embodiments of the invention comprises only two or three controlled switches. As a continuation of the general subject matter, another concurrently filed application by the same inventor assigned to the same assignee, Ser. No. 87,484, filed Nov. 6, 1970 discloses and claims other motor control circuits with up to six controlled switches that easily produce other square and stepped voltage wave shapes, together with an appropriate design of a permanent magnet or variable reluctance brushless motor having a similar back emf wave shape, whereby efficient motor operation is assured.

Therefore, an object of the invention is the improved combination of a brushless or commutatorless electric motor and a simple motor control circuit employing only a few controlled switches to produce rectilinear wave shapes, wherein the brushless motor is designed to operate efficiently on the voltage and current wave shapes produced by the particular control circuit.

Another object is the provision of a family of compatible brushless reluctance motor and motor control circuits employing a small number of mechanical or solid-state controlled switches that are useful in a variety of small or medium size motor applications.

Yet another object is to provide a brushless reluctance motor and control circuit therefor having an interrelated design such that the combination is inexpensive and suitable for adjustable speed operation.

In accordance with the invention, the combination of a compatible brushless reluctance motor and motor control circuit includes a stator member supporting a plurality of spaced sets of opposing stator windings for generating overlapping magnetic fields of opposite polarity in torque producing relation to a rotatable magnetically nonpolarized ferromagnetic rotor. Preferably each set of stator windings includes bifilar closely coupled main and feedback windings. An energizing circuit applies simple rectilinear, or substantially rectilinear, voltages of both polarities to each set of stator windings under the control of only one controlled switch device for each set. When main windings with associated feedback windings are used, the controlled switch is in series with the main windings and an uncontrolled rectifier is effectively in series with each feedback winding to discharge the stored magnetic energy in the main winding magnetic circuit. This circuit generates square wave energizing voltages. Control means responsive to the instantaneous rotor position renders conductive each controlled switch for desired intervals of conduction in a sequence to produce continuous torque to rotate the rotor in a given direction. A double spiral rotor creating a variable air gap is used in a motor with one direction of rotation, while a symmetrical four pole rotor with constant air gap is used in a reversible motor.

The foregoing and other objects, features, and advantages of the invention will be apparent from the following more particular description of several preferred embodiments of the invention as illustrated in the accompanying drawings wherein:

FIG. 1a is a schematic end view of a variable reluctance brushless motor having a double off-center spiral rotor, the stator winding coils being shown in cross section here and in the similar views of other motors to follow, and further including a diagrammatic representation of an extension of the motor shaft to which is attached a cam for actuating mechanical control circuit switches;

FIG. 1b is a schematic circuit diagram of a motor control circuit using two mechanical switches for energizing the motor of FIG. 1a;

FIG. 1c shows two different voltage wave shapes easily produced by the FIG. 1b circuit;

FIG. 1d is a schematic diagram of a modification of the motor control circuit for the FIG. 1a motor using solid state transistor switches that can be operated by time ratio control principles to control the speed of the motor;

FIG. 2a is a schematic end view of a variable reluctance brushless motor having a symmetrical rotor for reversible operation, in which the motor shaft position is sensed by magnetic Hall generator sensors to switch the solid-state control circuit switches;

FIG. 2b is a compatible control circuit for the FIG. 2a brushless motor that employs three transistor switches;

FIG. 2c is a series of three waveform diagrams illustrating the voltage wave shapes generated in the three pairs of stator windings by the FIG. 2b control circuit; and

FIG. 2d is a view similar to FIG. 2a of a modified form of the variable reluctance brushless motor having a double spiral rotor for one direction of rotation that can also be used with the FIG. 2b control circuit.

The variable reluctance brushless motor illustrated in FIG. 1a is a relatively small motor that can be used in a variety of applications requiring only one direction of rotation, such as in a small electric appliance. This motor operates efficiently on the voltage and current wave shapes produced by a motor control circuit comprising only two controlled switches and two feedback diodes. As a special feature, the motor has feedback windings which aid commutation and improve the efficiency by returning magnetic energy to the power circuit. The motor comprises in general an annular stator member 11 within which a nonsymmetrical variable reluctance rotor 12 rotates on a shaft 13. The variable reluctance rotor 12 is constructed of an appropriate ferromagnetic material such as soft iron in a shape characterized by two off-center spiral surfaces 12a and 12b, each occupying 180° and having a high point with respect to the center of the rotor that terminates at an inwardly directed shoulder connecting to the low point of the other spiral.

The stator 11 comprises an annular frame 14 provided with four inwardly directed, equally spaced salient pole structures 15a–15. A bifilar stator winding is disposed about each of the stator poles. Main windings 16A and 16C are wound respectively about the diametrically opposite stator poles 15a and 15c, and (see also FIG. 1b) are connected in series circuit relationship and wound in opposite directions so as to produce, when energized, magnetic poles of opposite polarity. Feedback windings 16B and 16D wound about the same two poles are closely coupled with the respective main windings 16A and 16C but are connected in parallel circuit relationship to effectively obtain a 2:1 turns ratio. In the same manner, bifilar windings 17A and 17B are associated with stator poles 15b, and windings 17C and 17D with stator pole 15d. In the motor control circuit, FIG. 1b, a mechanical switch 19 is connected in series with the first pair of series connected main windings 16A and 16C between direct current power supply terminals 20 and 21. A feedback diode 22 is connected in series with each of the feedback windings 16B and 16D between the pair of supply terminals. Similarly a second mechanical switch 23 is connected in series with the other pair of series connected main windings 17A and 17C, and a second feedback diode 24 is connected in series with the parallel combination of feedback windings 17B and 17D. The power source of the motor control circuit is conveniently a battery 25 across the terminals of which is connected a filter capacitor 26.

The mechanical switches 19 and 23 are closed and opened in dependence upon the position of the rotor 12 and in a sequence to obtain continuous torque in one direction. As is shown schematically in FIG. 1a, the rotor position is sensed mechanically by a small mechanical cam 12' having the same double spiral shape as the rotor 12 which is fastened to the motor shaft 13 or an extension thereof with a leading angle of orientation. Cam riders 19' and 23' are mounted at right angles to one another and actuate the respective mechanical switches 19 and 23, which can be simple spring contacts as shown or can be modern sealed switches such as the reed switch. Contacts 19 and 23 are opened and closed twice during each revolution of rotor 12, mechanically displaced by exactly 90° or some other selected angle. As will be explained in greater detail later, it is preferred to use solid-state switches in place of the mechanical switches 19 and 23, and to use a magnetic or optical sensor for sensing the rotor position in place of the mechanical cam 12', but the use of the mechanical equivalent in this first embodiment of the invention serves to clarify its operation.

As is well known, the principle of operation of a reluctance machine is that the torque tends to rotate the rotor so as to minimize the reluctance offered by the air gap to the magnetomotive force developed by the stator windings. Consequently, rotor 12 tends to rotate in a clockwise direction until the air gap between a pole face and the double spiral rotor 12 is at a minimum, and in order to maintain continuous torque, of course, the other set of stator windings are energized to develop the desired flux density level in advance of the rotor actually reaching the minimum air gap position. Upon applying voltage to a particular set of stator windings, small currents are induced at the adjacent surfaces of the moving rotor 12 to create a magnetic pole of opposite polarity. Since the direction of the current in the main windings cannot be changed, the polarity of the magnetic poles induced on rotor 12 by each set of opposing windings during each complete revolution changes from north to south, but this presents no problem because the rotor is made of soft laminated iron. In the operation of the motor shown in FIG. 1a, closing of contacts 19 by mechanical cam 12' energizes windings 16A and 16C for a predetermined angle of rotation of rotor 12. When switch 19 is opened by cam 12', the potential across closely coupled feedback windings 16B and 16D is such that feedback diode 22 is forward biased and is conductive to limit the transient voltage due to the opening of

mechanical switch 19 and to return a considerable amount of magnetic energy to the power circuit. The return of energy aids commutation and greatly improves the efficiency of this type of motor. After rotor 12 has rotated approximately one-quarter of a revolution, switch 23 is closed to energize the opposite pair of main windings 17A and 17C, following which feedback diode 24 is rendered conductive to dissipate the energy in feedback windings 17B and 17D.

The FIG. 1b motor control circuit is capable of producing the two idealized voltage wave shapes shown in FIG. 1c. The continuous voltage wave shape E_1 and noncontinuous voltage wave shape E_2 are obtained depending upon the adjustment of the timing of the opening and closing of switches 19 and 23 with respect to the cam position. To maintain continuous torque, the voltages applied to the opposing sets of stator windings can be overlapping. The application of continuous voltage wave shape E_1 to an opposing pair of main stator windings may, depending on the particular operating conditions and physical parameters, result in the production of negative torque. Although it is possible to operate an electric motor with negative torque, the more desirable voltage wave shape E_2 is ordinarily used because the generation of positive torque is more easily obtained. The delay between two complete waves is at least equal in angular measurements to the stator pole face length. The turns ratio of the feedback windings is selected to achieve fast flux decay with only a reasonable increase in induced voltage. The parallel feedback winding arrangement shown in FIG. 1a effectively achieves a 2:1 turns ratio, but with good coupling in a manner easily wound when manufacturing. The same or another turns ratio is achieved in a conventional manner by reducing the number of turns and connecting the two feedback windings in series with its respective feedback diode. Although fast flux decay is desirable, in practice the turns ratio is limited by the induced voltage applied to switches 19 and 23, assuming that the mechanical switches are replaced by low cost solid-state switches having low voltage ratings. It is also pointed out that rapid flux decay is permitted without excessive loss by the use of laminated rotor and stator members.

While not ideal, voltage wave shape E_2 is a good voltage wave shape for operating a variable reluctance motor. The back electromotive force voltage (back emf) of the motor can be a similar wave shape. Details of the voltage wave shapes in the variable reluctance motor are not critical since both pole area and flux density can change with little loss to match an impressed voltage wave shape. In the conventional electric motor, the applied voltage to operate the motor should have about the same wave shape as the back emf in order to avoid high circulating currents and losses. A reluctance motor, however, has no characteristic back emf and can run on a variety of wave shapes without producing excessive circulating currents and undesirable characteristics. Voltage wave shapes E_1 and E_2 are easy to produce with a few controlled switches, and operate the motor with good, if not ideal, efficiency. Other applied voltage wave shapes which give more torque are difficult to obtain with a control circuit having only a few control switches.

The motor control circuit illustrated in FIG. 1d is a modification of the FIG. 1b control circuit and is also compatible for use with the reluctance motor of FIG. 1a. This motor control circuit uses solid-state transistor switches, is powered from a single phase alternating current source, and has provision for time ratio control operation of the switches to change the voltage supply to the motor windings and thereby control the speed. The d-c voltage supply is provided by a full wave diode rectifier 29 connected to an a-c source. As in the previous control circuit, main windings 16A and 16C are connected in series circuit relationship with one another and a first transistor switch 30. In this circuit, however, feedback windings 16B and 16D are connected in series with one another, and are further in series with the two main windings 16A and 16C and transistor 30, the entire series circuit being connected between positive supply terminal 20 and the junc-

3,679,953

7

removing the gating signal to each transistor, rendering them nonconductive, the feedback diodes 22, 45, and 24 are respectively forward biased to return stored magnetic energy in the main winding magnetic circuits. In the counterclockwise direction of rotation, for the position of the rotor illustrated, stator windings 16A, 18A, and 17A are energized in overlapping sequence to rotate the rotor incrementally through one-sixth of a revolution to move rotor pole 40a adjacent to stator pole 15b. Thereafter the sequence is repeated for windings 16C, 18C, and 17C, and so on. In the manner already explained, the timing of the switching of transistors 42–44 is such that the magnetic flux in a stator pole is built up and exerts useful torque on the rotor, then is returned to the zero or reference level during the time the rotor pole face, because of the difference in length, is rotating adjacent the stator pole face without exerting torque. As before, the delay before again energizing that same pair of windings, to avoid negative torque, is sufficient to allow that rotor pole to move completely off of the stator pole. The sequence of operation to produce clockwise rotation of the rotor is obvious from this description.

FIG. 2d shows a modification of the motor of FIG. 2a wherein the symmetrical four pole rotor 40 is replaced by the double spiral rotor 12. The modified FIG. 2d motor, therefore, has only one direction of rotation obtained by energizing the stator windings in clockwise sequence, i.e., in the order of 18A, 17A, 16A, etc. It is desirable to overlap the application of voltage so that two of the three opposing pairs of windings are energized at the same time. Thus, at any given time four of the six stator windings are energized and supplying torque to turn the rotor. Voltage wave shape E₂, FIG. 1c, is preferred.

The remarks made before with regard to FIGS. 1a–1d as to the advantages of the variable reluctance motor described and the compatibility of the motor control circuit with the motor design apply equally as well to the six stator pole, three control switch motor control circuit forms of the invention shown in FIGS. 2a–2bd. Moreover, although the transistor and diode are uniquely suited for these simple square wave shape circuits, they can be replaced by other solid-state switching devices such as the silicon controlled rectifier, the triac or diac, if controlled for one direction of conduction, the silicon controlled switch, etc. These different controlled semiconductors, as well as some specific time ratio control circuits, are further discussed and illustrated in the Silicon Controlled Rectifier Manual, 4the Edition, available from the Semiconductor Products Department, General Electric Company, Electronics Park, Syracuse, New York, copyright 1967. Transistor gating circuits that can be employed are also described in the General Electric Transistor Manual, copyright 1964, available from the same address.

In summary, relatively simple motor control circuits using a small number of controlled switches to produce simple rectilinear wave shapes have been disclosed in conjunction with suitable reluctance brushless motors that can be operated with good efficiency by these wave shapes. The control circuit has as few as two or three mechanical or semiconductor controlled switches for generating a nonsymmetrical square wave, while the compatible reluctance motor has two or more pairs of bifilar main and feedback windings. The brushless reluctance motor and control circuit combinations are further suitable for variable speed operation by controlling the magnitude of the voltage supplied to the motor.

While the invention has been particularly shown and described with reference to several preferred embodiments thereof, it will be understood by those skilled in the art that the foregoing and other changes in form and details may be made therein without departing from the spirit and scope of the invention.

What I claim as new and desire to secure by Letters Patent of the United States is:

1. A compatible brushless reluctance motor and motor control circuit comprising

8

a stator member supporting a plurality of spaced sets of opposing stator windings for generating overlapping magnetic fields of opposite polarity in torque producing relation to a rotatable magnetically nonpolarized ferromagnetic rotor,

circuit means for applying simple substantially rectilinear energizing voltages to each of said sets of stator windings under the control of only one controlled switch device for each set of stator windings, and

control means responsive to the instantaneous position of said rotor for rendering conductive each controlled switch device for desired intervals of conduction in a sequence to produce continuous torque to rotate said rotor in a given direction,

wherein each set of stator windings comprises pairs of opposing closely coupled main and feedback windings, and said controlled switch device is effectively connected in series circuit relationship with the pair of main windings, and

said circuit means further includes a second switch device for each set of stator windings effectively connected in series circuit relationship with each feedback winding to discharge the stored magnetic energy associated with said main windings upon rendering nonconductive the respective controlled switch device.

2. The combination recited in claim 1 wherein all of said switch devices are solid-state devices.

3. The combination recited in claim 1 wherein there are no more than three sets of opposing stator windings equally spaced from one another, and

said rotor is shaped as a double spiral, whereby there is a variable air gap between the rotor and stator member.

4. The combination recited in claim 1 wherein there are three sets of equally spaced opposing stator windings, and

said rotor is shaped to have four circumferentially oriented arcuate pole faces, whereby there is a constant air gap between the rotor and stator member.

5. A compatible brushless reluctance motor and motor control circuit comprising

an annular stator member supporting a plurality of spaced pairs of diametrically opposing closely coupled main and feedback stator windings which generate overlapping magnetic fields of opposite polarity in torque producing relation to a rotatable magnetically nonpolarized ferromagnetic rotor,

circuit means for applying substantially square wave energizing voltages to each of said pairs of stator windings including a controlled solid-state switching device connected in series circuit relationship with each pair of main stator windings, and an uncontrolled rectifier effectively connected in series circuit relationship with each feedback winding of each pair of windings with a polarity to be forward biased to discharge stored magnetic energy in the magnetic circuits associated with the respective closely coupled main windings,

means for sensing the instantaneous position of said rotor, and

control means responsive to said rotor position sensing means for rendering conductive each controlled switch device for desired intervals of conduction in a sequence to produce continuous torque to rotate said rotor in a given direction.

6. The combination recited in claim 8 wherein there are only two pairs of opposing main and feedback windings approximately at right angles to one another, and wherein

said circuit means includes a pair of unidirectional voltage power supply terminals between which are connected each pair of main windings and series connected controlled switch device, and also each pair of feedback windings in circuit relationship with the respective uncontrolled rectifier, and

said rotor is shaped in a double spiral configuration.

265

tion point 31 between two series connected filter and voltage divider capacitors 32 and 33. The other set of main and feedback windings 17A--17D are similarly connected in series with one another and with a second transistor switch 34, the entire series circuit being connected between junction point 31 and negative supply terminal 21. Feedback diodes 22 and 23 are now respectively connected to the junctions between the pair of main windings and feedback windings in the appropriate set of windings and to the opposite supply terminal. To control the turn-on and turn-off of the two transistor switches 30 and 34, gating circuits 35 and 36 are provided and are under the control of a mechanical, magnetic, or optical rotor position sensor 37. Optionally, to achieve speed control, transistor switches 30 and 34 may be turned on and off rapidly during one-quarter of a revolution of the motor to reduce the voltage supply to the motor windings by time ratio control principles. To this end, gating circuits 35 and 36 can be under the control of a time ratio control circuit 38.

In the operation of the motor control circuit of FIG. 1d, it is assumed that rotor position sensor 37 initially actuates transistor gating circuit 35, which supplies a timed turn-on signal to the base electrode of transistor 30, rendering it conductive. The winding utilization is better in this circuit as compared to FIG. 1b, since current flows through all four stator windings 16A–16D into the junction point 31 between capacitors 32 and 33, the voltage at junction point 31 having an average value of half the supply voltage. If voltage control is required, time ratio control circuit 38 modifies the action of gating circuit 35, and during the short intervals of nonconduction of transistor 30, feedback diode 22 becomes forward biased to supply current to these stator windings. When rotor position sensor 37 signals that windings 16A–16D are to be de-energized and the other set of windings 17A–17D are to be energized, feedback diode 22 is also conductive to return stored magnetic energy to the supply terminals to achieve complete flux decay in windings 16A–16D. The other set of stator windings 17A–17D are then energized when gating circuit 36 turns on transistor 34, as previously explained. Feedback diode 23 similarly conducts current to these windings during the nonconducting intervals of time ratio control operation, and returns magnetic energy when transistor 34 is turned off for longer periods of time.

The variable reluctance motor tends to have the voltage-speed characteristics of a d-c series motor, hence reduction of the voltage supply to the motor windings serves effectively to decrease the speed of the motor. Other ways of achieving voltage control are to change the supply voltage in advance of the motor circuit, or to use a voltage control circuit such as a phase controlled rectifier in place of the diode rectifier 29. These voltage control techniques are more suitable for the FIG. 1b motor control circuit. Both motor control circuits use only two controlled switches and two feedback rectifiers, and are compatible in terms of low cost, the voltage wave shape produced, and the variable reluctance motor of FIG. 1a. These motor control circuits provide no way of reversing the current through the opposite pairs of main windings 16A, 16C and 17A, 17C, hence are not suitable for use with a permanent magnet rotor. As was mentioned, the details of voltage wave shapes are not critical in a variable reluctance motor as both pole area and flux density can change with little loss to match an impressed voltage wave shape. Moreover, the direct control of the time of switching by the mechanical cam 12' or rotor position sensor 37 makes the motor function as a d-c motor with no synchronizing problem. The motor energized by a square voltage waveform may not have a desirable torque angle characteristic for operation as an ordinary synchronous motor. The effective circumferential length of stator poles 15a–15, or the winding pitch of the stator windings, is not critical so long as the magnetic fields that are produced and the patterns of decreasing reluctance are overlapping.

The variable reluctance motor illustrated in FIG. 2a has a symmetrical rotor for reversible operation, and has a compatible motor control circuit employing only three controlled switches. This form of the variable reluctance motor has four equally spaced arcuate rotor poles, and six stator poles which are energized in the appropriate direction to produce overlapping clockwise or counterclockwise torque. The dimensions and spacing of the stator and rotor poles are such as to produce overlapping periods of decreasing reluctance in the stator winding magnetic circuits. The three pairs of opposing stator poles 15a–15f are wound with closely coupled bifilar main and feedback windings in the same manner as the motor of FIG. 1a. For this purpose, an additional set of series connected main windings 18A and 18C are provided together with the respective parallel connected feedback windings 18B and 18D. As will be further explained, the compatible motor control circuit of FIG. 2b produces the same square wave shapes as are illustrated in FIG. 1c. Either of voltage wave shapes E_1 or E_2 can be used, but voltage wave shape E_2 is preferred for the same reasons. The symmetrical rotor 40 is essentially circular with four arcuate cutouts or lobes 41 defining the four pole faces 40a–40d. The circumferential length of each rotor pole is greater than the stator pole circumferential length. Assuming that the rotor is rotating, the magnetic flux developed in a particular stator pole exerts pull upon the rotor during the time the rotor is advancing toward but not yet overlapping the stator pole. During the time that the rotor pole overlaps the stator pole, as determined by the difference in their circumferential lengths, the flux in the stator pole is reversed and driven back to zero. To prevent the development of negative torque, no voltages are applied to the stator winding until the rotor pole has advanced in its entirety away from that stator pole. This time period corresponds to the delay between two complete waves of voltage wave shape E_2, FIG. 1c. As a condition to generating no negative torque, the stator pole circumferential length should be about equal to the circumferential space between adjacent rotor poles.

The motor control circuit of FIG. 2b is similar to the FIG. 1b control circuit with the exception of the addition of the third set of windings and the replacement of the mechanical switches by transistor switches 42–44. An additional feedback diode 45 is in series with each of the feedback windings 18B and 18D. To provide gating signals for transistors 42–44 that are timed in dependence upon the rotor position, a plurality of magnetic sensors in the form of Hall elements or generators 46–48 are assembled about an extension of the motor shaft 13 and actuated by a permanent magnet 49 secured for rotation with shaft 13. Sensors of this type which operate on the Hall effect principle generate an output voltage between the two output terminals when a magnetic field is applied perpendicular to the face of the Hall element and an energizing control current is applied between the two input terminals, normally aligned with the longitudinal axis of the element. Further information on the Hall generator itself and its utility as a rotor position sensor in a brushless motor can be obtained from the prior art patents, as for example, U.S. Pat. No. 3,159,777 to E.W. Manteuffel, granted Dec. 1, 1964 and assigned to the General Electric Company. Although in the interest of simplicity it may be desirable to use six Hall generators to sense the rotor position at 60° intervals, it is also possible to use the three Hall generators 46–48 spaced at 120° intervals as illustrated, provided that the control current is reversed twice during each revolution of the permanent magnet 49. The output terminals of Hall generators 46–48 are connected to a gating circuit 50 for transistor switches 42–45. If desired, the gating circuit 50 may also be under the control of a time ration control circuit 51 to change the speed of the motor as previously explained.

Assuming that the variable reluctance motor is to rotate in a counterclockwise direction, the control circuit of FIG. 2b is operated to produce the square voltage wave shapes E_{18A}, E_{17A}, and E_{18A} illustrated in FIG. 2c. The wave shapes for the other main stator windings are similar. Square wave gating pulses are supplied from gating circuit 50 to transistor switches 42, 44, and 43 in that sequence, and it will be noted that the gating signals are overlapping. Immediately after

9

7. The combination recited in claim 5 wherein there are only two pairs of opposing main and feedback windings approximately at right angles to one another, and wherein

said circuit means includes a pair of unidirectional voltage power supply terminals between which are connected a pair of voltage divider capacitors, each pair of opposing main and feedback stator windings being connected in series with one another and with the respective controlled switch device between one power supply terminal and the junction of said voltage divider capacitors, the respective uncontrolled rectifier being connected between the other power supply terminal and the junction of said pair of main windings and pair of feedback windings, and

said rotor is shaped in a double spiral configuration.

8. The combination recited in claim 7 further including time ratio control means for rapidly turning on and turning off each controlled switch device to vary the magnitude of the voltage applied to said pairs of main stator windings.

9. The combination recited in claim 5 wherein there are only three pairs of opposing main and feedback windings equally spaced from one another, and wherein

said circuit means includes a pair of unidirectional voltage

10

power supply terminals between which are connected each pair of main windings and the respective series connected controlled switching device, and also each pair of feedback windings in circuit relationship with the respective uncontrolled rectifier, and

said rotor is shaped to have four circumferentially oriented arcuate pole faces.

10. The combination recited in claim 9 further including time ratio control means for rapidly turning on and turning off each controlled switch device to vary the magnitude of the voltage applied to said pairs of main stator windings.

11. The combination recited in claim 5 wherein there are only three pairs of opposing main and feedback windings equally spaced from one another, and wherein

said circuit means includes a pair of unidirectional voltage power supply terminals between which are connected each pair of main windings and the respective series connected controlled switching device, and also each pair of feedback windings in circuit relationship with the respective uncontrolled rectifier, and

said rotor is shaped in a double spiral configuration.

* * * * *

Compatible Permanent Magnet or Reluctance Brushless Motors and Controlled Switch Circuits

Burnice D. Bedford

U.S. Patent No. 3,678,352, July 1972

[54] **COMPATIBLE PERMANENT MAGNET OR RELUCTANCE BRUSHLESS MOTORS AND CONTROLLED SWITCH CIRCUITS**

[72] Inventor: **Burnice D. Bedford,** Scotia, N.Y.

[73] Assignee: **General Electric Company**

[22] Filed: **Nov. 6, 1970**

[21] Appl. No.: **87,484**

[52] U.S. Cl. **318/138,** 318/254
[51] Int. Cl. **H02k 29/00**
[58] Field of Search 318/138, 254, 696, 685

[56] **References Cited**

UNITED STATES PATENTS

3,482,156	12/1969	Porath	318/138
3,023,348	2/1962	Cox	318/138
3,127,548	3/1964	VanEmden	318/696
3,159,777	12/1964	Manteuffel	318/138

Primary Examiner—Gene Z. Rubinson
Attorney—Paul A. Frank, John F. Ahern, Julius J. Zaskalicky, Donald R. Campbell, Frank L. Neuhauser, Oscar B. Waddell and Joseph B. Forman

[57] **ABSTRACT**

Brushless motors have physical and electrical characteristics to be compatible with energization by the simple square or stepped wave voltages easily produced with inverter type motor control circuits having a small number of controlled switches responsive to a rotor position sensor. These motors employ a few pairs of opposing non-distributed stator windings arranged overlapping or non-overlapping with a predetermined winding pitch, and a constant gap magnetically polarized or non-polarized rotor with pole faces having related angular dimensions, to thereby produce rectilinear back emf voltages with approximately the same wave shape as the energizing voltages. The circuits can control the applied voltage to adjust motor speed.

9 Claims, 12 Drawing Figures

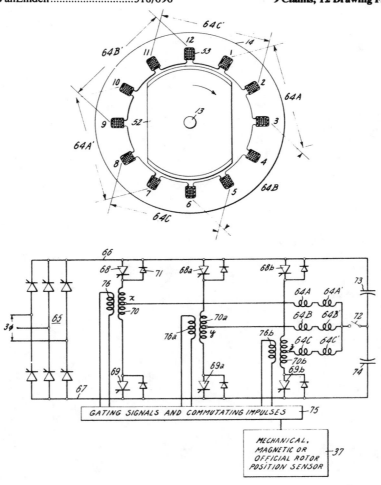

Fig.1a.

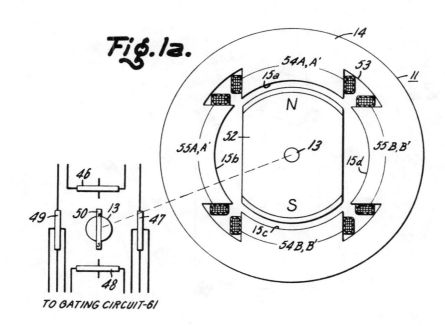

TO GATING CIRCUIT-61

Fig.1b.

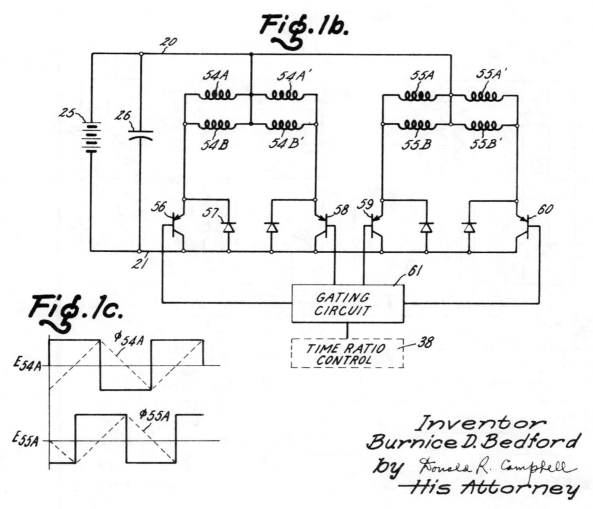

GATING CIRCUIT

TIME RATIO CONTROL — 38

Fig.1c.

Inventor
Burnice D. Bedford
by Donald R. Campbell
His Attorney

269

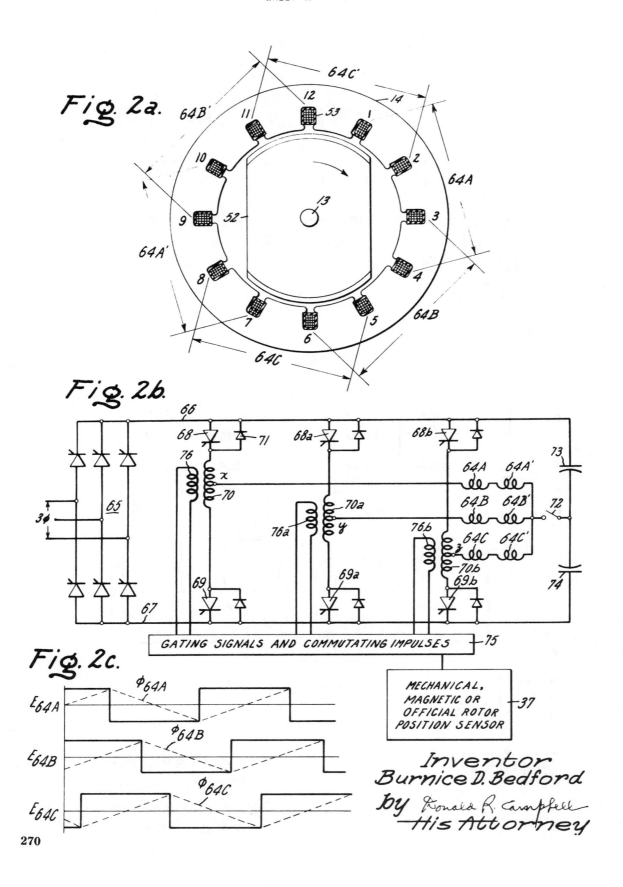

Fig. 2a.

Fig. 2b.

GATING SIGNALS AND COMMUTATING IMPULSES — 75

MECHANICAL, MAGNETIC OR OFFICIAL ROTOR POSITION SENSOR — 37

Fig. 2c.

Inventor
Burnice D. Bedford
by Donald R. Campbell
His Attorney

270

Fig.3a.

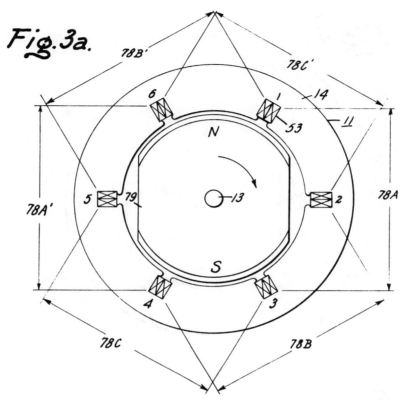

Fig.3b.

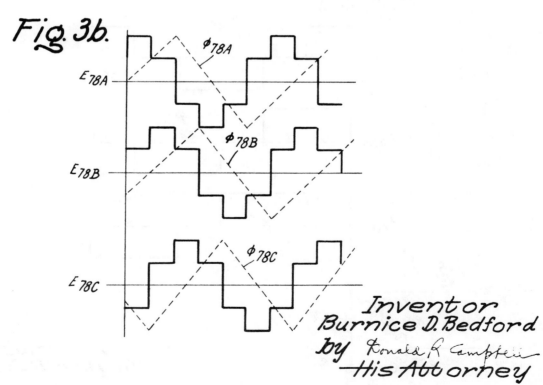

Inventor
Burnice D. Bedford
by Ronald R. Campbell
His Attorney

Fig. 4a.

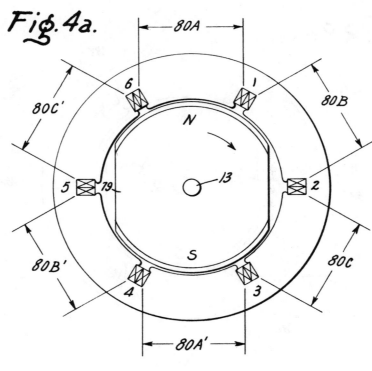

Fig. 4b.

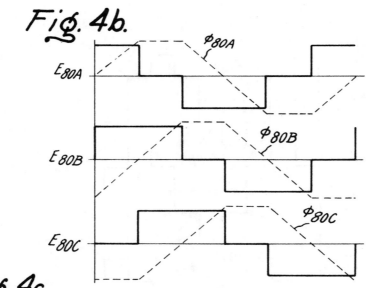

Fig. 4c.

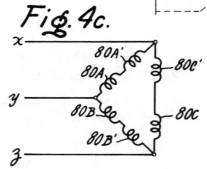

Inventor
Burnice D. Bedford
by Donald R. Campbell
His Attorney

272

Fig. 5.

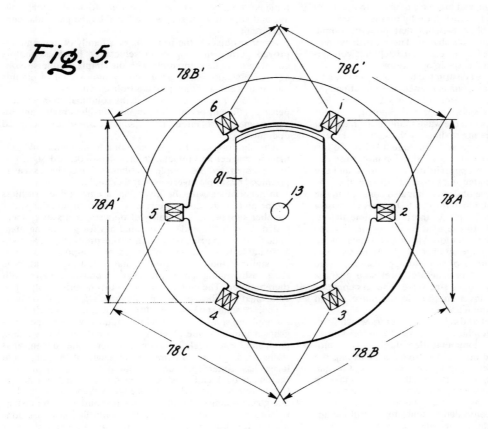

Inventor
Burnice D. Bedford
by   Donald R. Campbell
His Attorney

1

COMPATIBLE PERMANENT MAGNET OR RELUCTANCE BRUSHLESS MOTORS AND CONTROLLED SWITCH CIRCUITS

This invention relates to brushless permanent magnet and reluctance electric motors, and more particularly to brushless motors designed for efficient operation by motor control circuits using a few controlled switches that produce simple square or stepped voltage wave shapes. These brushless motors are specially designed to have similar back emf voltages and are suitable for adjustable speed operation.

Conventional motors are constructed to be operated by sine wave voltages. Motors traditionally classed as alternating current motors are designed to produce sine wave back electromotive force voltages since, as is well known, a motor operates most efficiently when the wave shape of the back emf and energizing voltage are approximately the same to thereby avoid high circulating currents and consequent losses. To this end, ordinary motors such as the a-c induction motor use distributed stator windings to approximate a sine wave and use stator poles that are rounded outwardly at either end of the pole face to attain the same objective. It is also common to use stator slots that are skewed with respect to the axis to cause a rounding of the generated emf. At the present time the attempt is frequently made to use a stepped wave inverter to energize such a conventional motor. Although it is possible to select an inverter with a large number of controlled switches that produce a great number of steps to approximate a sine wave, less expensive inverters generate fewer step changes that deviate substantially from a sine wave. The less complex square or stepped wave inverters thus do not make a good combination with an a-c induction motor.

The permanent magnet and reluctance motors which form the subject of the invention have similarities to motors known in the art as brushless or commutatorless direct current motors. In the brushless d-c motor, the drive coils carrying the magnetizing current which result in the production of torque are located on the stator housing rather than on the movable rotor as in the conventional direct current motor. The rotor is commonly a permanent magnet rotor, and the circuit energizing the drive coils uses controlled switches to control the application of voltage to the stator windings in a sequence to produce continuous rotation of the rotor. As a result, the commutators and brushes utilized on the conventional motor to supply current to the armature windings is eliminated together with such undesirable features as the need to replace brushes, arcing between commutator segments, and repair of worn commutator segments. In order to reduce the cost of the control circuit, and thus minimize the cost of the combination of brushless motor and control circuit, it is desirable to employ a simple control circuit using a small number of controlled switches to produce simple square or stepped voltage wave shapes. Although brushless motors operated by motor control circuits comprising only a few controlled switches have appeared in the prior art, there is inadequate recognition in the prior art of the need to match the physical and electrical parameters of the motor to the particular rectilinear energizing voltage that can be easily produced by a simple control circuit.

The compatible brushless motors and motor control circuits described in this application employ, as illustratory embodiments of the invention, motor control circuits with four to six controlled switches for producing simple square and stepped voltage wave shapes, together with appropriate designs of permanent magnet or reluctance brushless motors that generate a similar back emf wave shape, whereby efficient motor operation is obtained. As a continuation of the general subject matter, another concurrently filed application by the same inventor assigned to the same assignee, Docket Ser. No. 87,565, filed Nov. 6 1970, discloses and claims other brushless reluctance motors suitable to be energized by even simpler control circuits having as few as two or three controlled switches, wherein efficient operation is possible because reluctance motors have no characteristic back emf and can operate on a variety of wave shapes.

2

Accordingly, an object of the invention is the improved combination of a brushless or commutatorless electric motor and a motor control circuit employing only a few controlled switches to produce simple square or stepped voltage wave shapes, wherein the brushless motor is designed to have the same back emf wave shape and operate efficiently on the voltage and current wave shapes produced by the particular control circuit.

Another object is the provision of a family of new and improved permanent magnet or reluctance brushless motors constructed to be compatible with the simple rectilinear voltage wave shapes easily produced by motor control circuits comprising a small number of controlled switches.

Yet another object is to provide the combination of an inexpensive, efficient permanent magnet brushless motor and inverter type control circuit that is suitable for adjustable speed operation.

A further object is the provision of new and improved permanent magnet and reluctance brushless motors designed to be operated by the simple rectilinear wave shapes easily produced by simple inverter control circuits.

In accordance with the invention, a compatible brushless motor and motor control circuit includes an annular stator member supporting a plurality of opposing pairs of concentrated (non-distributed) stator windings for generating magnetic fields of opposite polarity in torque producing relation to a rotatable ferromagnetic polarized or non-polarized rotor. The stator member and rotor each have opposing arcuate pole faces establishing an approximately constant gap width therebetween. The motor control circuit comprises only a few controlled switches, preferably only one alternately conducting pair of solid state switches for each pair of stator windings, for applying simple substantially rectilinear wave shape energizing voltages to the pairs of stator windings. The winding pitch and location relative to one another of the concentrated stator windings, and circumferential length of the rotor pole faces, have interdependent angular dimensions to produce rectilinear back emf voltages in each pair of stator windings that have approximately the same wave shape as the applied energizing voltages. Control means responsive to the instantaneous rotor position renders the controlled switches conductive for desired intervals to energize the pairs of stator windings in a sequence to produce continuous torque to rotate the rotor in a given direction.

The invention is also directed to the brushless motor per se. In various forms of the invention, there are two or three pairs of overlapping and non-overlapping concentrated stator windings with a winding pitch of about 90°, 60°, and 120°, and the rotor pole faces selectively have one of these angular dimensions.

The foregoing and other objects, features, and advantages of the invention will be apparent from the following more particular description of several preferred embodiments of the invention, as illustrated in the accompanying drawings wherein:

FIG. 1a shows a diagrammatic end view of a permanent magnet brushless motor constructed in accordance with the invention with four 90° stator poles and concentrated stator windings, and two 90° rotor poles;

FIG. 1b is a schematic circuit diagram of a motor control circuit employing four controlled solid state switches that is compatible with the FIG. 1a motor;

FIG. 1c is a waveform diagram of the two phase square voltage waveforms produced by the motor control circuit of FIG. 1b, further showing in dotted lines the average magnetic flux characteristics produced by these applied voltages;

FIG. 2a illustrates a second embodiment of a permanent magnet brushless motor having six overlapping 90° concentrated stator windings, and 90° rotor poles;

FIGS. 2b and 2c show respectively a motor control circuit comprising six SCRs arranged in the form of a conventional full wave, three phase inverter circuit with the addition of a neutral switch which when closed and operated to generate the square wave voltage wave shapes and resultant magnetic flux curves shown in FIG. 2c is suitable to energize the FIG. 2a motor;

FIGS. 3a and 3b illustrates respectively another embodiment of the permanent magnet brushless motor with six overlapping 120° concentrated windings and a 120° rotor that is energized by the control circuit of FIG. 2b, the neutral switch assumed to be open, in a manner to produce the three phase stepped voltage wave shapes and resulting average magnetic flux characteristics shown in FIG. 3b;

FIG. 4a is still another embodiment of a permanent magnet brushless motor characterized by six 60° non-overlapping concentrated windings and a 120° rotor;

FIGS. 4b and 4c show respectively the quasi-square voltage wave shapes supplied to three adjacent stator windings of the FIG. 4a motor by the motor control circuit of FIG. 2b when modified to have delta-connected pairs of stator windings as illustrated in FIG. 4c; and

FIG. 5 is a modified form of the motor shown in FIG. 4a having six overlapping 120° windings and a 120° reluctance rotor, that is suitable for energization by the same voltage wave shapes as illustrated in FIG. 4b.

The compatible brushless motor designs and motor control circuits to be described are suitable for manufacture in the small to medium horsepower range. These motors are preferably made with permanent magnet rotors, shaped to create a constant air gap at the interface with the stator pole faces, but can also be made with reluctance type soft iron rotors having a similar shape. In some applications, the higher manufacturing cost of the permanent magnet motor is justified to save weight and power.

FIG. 1a shows a permanent magnet motor in a size suitable for operation from a battery source by a motor control circuit that uses four controlled switches for producing a two phase square voltage wave shape. The motor comprises in general an annular stator member 11 including a frame member 14 within which a permanent magnet rotor 52 rotates on a shaft 13. The permanent magnet rotor 52 has diametrically opposite 90° arcuate pole faces and two parallel sides, and preferably has laminated pole tips to prevent hysteresis and eddy current loss due to variations of the flux at the pole surfaces. The stator frame member 14 is provided with four equally spaced winding slots 53 shaped to have small openings to provide a good flux path for the rotor flux and to define stator segments or pole faces 15a–15d that have an almost 90° circumferential length. It will be noted that the entire length of each stator segment is arcuate, i.e., the ends are not rounded outwardly toward the outside of the motor. Two opposing pairs of stator windings are supported on the stator frame 14 within the slots 53, each with a 90° winding pitch.

Referring to the motor control circuit shown in FIG. 1b, stator windings 54A and 54A′ are essentially a single winding wound in two adjacent stator slots, while windings 54B and 54B′ are wound effectively as a single winding in the opposing stator slots. Electrically, the center tap junctions between windings 54A and 54A′ and between windings 54B and 54B′, are connected together and to positive supply terminal 20. The other free ends of the respective pairs of windings so formed are coupled to negative supply terminal 21 respectively through transistor switch 56 and inverse parallel connected diode 57, and through transistor switch 58 and its associated feedback diode. Supply terminals 20 and 21 are connected across a battery 25 and parallel filter capacitor 26. Since windings 54A and 54B conduct current in only one direction, they are wound oppositely to produce opposite magnetic poles, and the same is true of windings 54A′ and 54B′. The other pairs of the opposing windings 55A and 55A′, and 55B and 55B′, are displaced mechanically by 90°, but are otherwise identically arranged and connected with the use of the third and fourth transistors 59 and 60 and their respective inverse-parallel feedback diodes.

Gating circuit 61 for the four transistors is under the control of a mechanical, magnetic, or optical rotor position sensor, to initiate switching of the transistor switches as the motor rotates in dependence upon the instantaneous position of rotor 52. Specific gating circuits that can be used are given, for example, in the Transistor Manual, 7th Edition, copyright 1964, published by the General Electric Company and available from the Semiconductor Products Department, Electronics Park, Syracuse, New York. The rotor position sensor illustrated in FIG. 1a comprises a plurality of magnetic sensors in the form of Hall elements or generators 46–49 assembled at 90° intervals about an extension of the motor shaft 13 and actuated by a permanent magnet 50 secured for rotation with shaft 13. Sensors of this type which operate on the Hall effect principle generate an output voltage between the two output terminals when a magnetic field is applied perpendicular to the face of the Hall element and an energizing control current is applied between the two input terminals, which are usually aligned with the longitudinal axis of the element. Further information on the Hall generator itself and its utility as a rotor position sensor in a brushless motor can be obtained from the prior art patents, as for example, U.S. Pat. No. 3,159,777 to E.W. Mantcuffel, granted Dec. 1, 1964, and assigned to the General Electric Company. As has been indicated, a mechanical cam or an optical sensor, as is known in the art, can also be used to sense the instantaneous position of permanent magnet rotor 52 whereby gating circuit 61 provides gating signals for transistors 56, 58, 59, and 60 that are timed in dependence upon the rotor position. This direct control of the time of switching of the control switches makes the motor function much as a d-c motor with no synchronizing problems.

The permanent magnet motor of FIG. 1a is symmetrical and is operable in both directions. The compatible motor control circuit of FIG. 1b is controlled to apply to the opposing pairs of stator windings the two phase square wave voltage wave shapes illustrated in FIG. 1c for windings 54A and 55A, the wave shapes for the other pair of opposed windings being similar. The voltage wave shapes and average magnetic flux characteristics, shown in dotted lines, are idealized, and the discussion of the motor to follow assumes idealized parameters and characteristics. The motor air gap and control circuit switching times may depart slightly from the idealized situation to compensate for leakage flux and flux shift due to motor torque. The same remarks apply to FIGS. 2a–5. Each stator winding and its associated stator pole acts like a simple solenoid in that the magnetic flux characteristic produced by the application of a constant unidirectional voltage is linear. The magnetic flux characteristics ϕ_{54A} and ϕ_{55A} generated respectively by the application of square wave voltages E_{54A} and E_{55A} increase and decrease linearly in the manner shown in FIG. 1c. Switching the transistors in the control circuit to change the polarity of the applied voltage is timed such that the rotor 52 is centered on a particular stator pole when the flux in that stator pole is at a maximum. Accordingly, transistors 56 and 58 change state almost simultaneously, i.e., the conducting one is turned off and the non-conducting one is turned on, when rotor 52 is centered and stator poles 15a and 15c. In like manner, transistors 59 and 60 are operated as a complementary pair and change state when rotor 52 is centered on the other two stator poles 15b and 15d.

In a typical sequence of operations for counter-clockwise rotation, transistor 56 is turned on to establish current flow through stator windings 54A and 54B when a reference end of the rotor (the S pole) is centered on stator pole 15c. Looking only at the state of the flux in stator pole 15a, it is seen that the average flux is driven from its maximum negative value and crosses the zero ordinate just at the time that the leading edge of permanent magnet rotor 52 reaches the near end of stator pole 15a, adjacent to stator pole 15d. During the next one-quarter of revolution when the arcuate pole face of rotor 52 is moving into alignment with stator pole 15a, the average magnetic flux in this stator pole increases from about zero to its maximum. At this time the state of transistors 56 and 58 is changed, and transistor 58 is now conductive and energizes windings 54A′ and 54B′. During the succeeding one-quarter of revolution of rotor 52, the average flux in stator pole 15a is decreasing linearly from its maximum to zero, and the polarity of the flux changes just as the trailing end of rotor 52 leaves the end of stator pole 15a adjacent to stator pole 15b. The

5

magnetic flux in the orthogonally oriented stator pole 15d is tracing a similar characteristic but delayed by 90° as related to the rotation of the rotor, and so on for the other stator poles. Thus, continuous torque is exerted upon rotor 52.

When the FIG. 1a motor is used as a permanent magnet motor, the rotor flux is relatively constant and the motor tends to have the voltage-speed characteristics of a d-c shunt motor. If the permanent magnet rotor is replaced by a reluctance rotor, the motor runs as a variable reluctance motor and has the voltage-speed characteristics that tend to be more like those of series d-c motors. Magnetic saturation may be used to limit the flux and obtain a characteristic which is a compromise between that of a series and a shunt motor. Either the series or shunt motor characteristic is suitable for speed control by controlling the d-c voltage. To vary the d-c supply voltage, and thus change the speed of the motor, it is possible to use a time ratio voltage control circuit ahead of the motor to change supply voltage for the motor windings, or a single phase alternating current source can be used in conjunction with a phase controlled rectifier. The motor switching circuit of FIG. 1b, however, is well suited to include time ratio voltage control. To do this, a transistor switch that is conducting is turned on and off rapidly at a fast rate compared to the motor speed to control the average motor voltage and motor current. The associated feedback diode 57 is conductive during the short period when the transistor is turned off to achieve time ratio control. Thus, gating circuit 61 is optionally under the control of time ratio control circuit 38. The reduction of the voltage applied to the pairs of stator windings, as is well known, depends upon the ratio of the time it is conducting to the time it is non-conducting as it is being turned on and off rapidly.

The permanent magnet or reluctance motor and motor control circuit are compatible because the brushless motor constructed as shown in FIG. 1a generates approximately the same back emf voltage wave shape as is applied to the pairs of stator windings by the motor control circuit of FIG. 1d. This assumes that there is constant air gap between the rotor pole faces and the stator pole faces. To illustrate this, the back emf voltage wave shape generated in stator winding 54A will be traced (assuming an ideal motor) during one-half revolution of rotor 52. Looking only at one end of the rotor, as for instance, the north pole, the average magnetic flux generated in stator pole 15a increases linearly from zero to a maximum and then back to zero as rotor 52 rotates 90° clockwise from out of alignment into complete alignment with stator pole 15a, and then another 90° from complete alignment to out of alignment in the other direction. At this time the south pole of rotor 52 is beginning to move into alignment with stator pole 15a, and during the next one-half revolution the magnetic flux decreases linearly from zero to a minimum in the other direction and back to zero. Consequently, the back emf voltage generated in stator winding 54A has the same square voltage wave shape as that for the energizing voltage shown in FIG. 1c, and the phasing is the same. Of course, a suitable technique is used to assure that the motor has running current, such as by making the magnitude of the applied voltage greater than the back emf voltage and the IR drop involved in circulating current through the stator windings. Also, the switching times of the switches with respect to rotor position may be advanced to compensate for leakage reactance, or some departure from the idealized air gap may be made to compensate for leakage flux and flux shift due to load current or flux shift due to motor torque. To be compatible with two phase square wave voltage energization of the motor, and to avoid the excessive circulating currents and losses that are incurred when the back emf voltage wave shape is not the same as the applied voltage wave shape, it is seen that the FIG. 1a motor is constructed with non-overlapping approximately 90° concentrated stator windings, and a rotor with arcuate pole faces that have an effective circumferential length of about 90° also. The brushless motor constructed in this manner operates efficiently on applied voltages with a two phase square wave shape.

6

The second embodiment of the invention shown in FIG. 2a also uses a 90° permanent magnet rotor, however there are now six overlapping 90° concentrated stator windings, and different combinations of 30° stator segments forming 90° stator poles. The combinations of stator segments, it is understood, become a pole under the influence of winding magnetomotive force and in dependence on the position of the rotor. This brushless motor is energized by what can be referred to as three phase square voltage wave shapes, and this motor design produces a similar back emf wave shape. The three pairs of opposing 90° concentrated stator windings are identified as windings 64A and 64A', 64B and 64B', 64C and 64C'. A total of 12 stator slots 53 are located about the inner periphery of stator member 14, defining 12 equal stator segments, and each winding disposed in the stator slots overlaps the two adjacent windings by 30°. The three phase square voltage wave shapes required for efficient operation of the motor of FIG. 2a are shown in FIG. 2c, and can be produced by what can loosely be called a three-phase version of the FIG. 1b control circuit. Another compatible motor circuit for producing this type of voltage wave shape is the motor control circuit of FIG. 2b, which is energized by a three phase alternating current source, uses a phase controlled rectifier for input voltage control, and employs six silicon controlled rectifiers as the power switching devices arranged as a conventional three phase inverter circuit.

Referring to FIG. 2b, the input terminals of the motor control circuit are connected to a conventional full wave, phase controlled rectifier 65 comprising six SCR's. As is well known, varying the phase of which the SCR's are rendered conductive adjusts the magnitude of the d-c voltage produced at output terminals 66 and 67. The motor control circuit proper is a full wave, three phase inverter circuit. The first phase comprises two SCR's 68 and 69 connected in series with a current limiting and commutating winding 70, and further includes a feedback diode 71 for reactive current connected in inverse-parallel relationship across the load terminals of each thyristor. The center tap point x of commutating winding 70 is coupled directly to series connected stator windings 64A and 64A', which are optionally connected through a neutral switch 72 to the junction between two series connected voltage divider capacitors 73 and 74. The other two phases of the inverter circuit have an identical arrangement of components designated by the same numeral with an "a" or "b" suffix. The three pairs of opposing stator windings are wye-connected to one terminal of neutral switch 72, and to respective commutating winding center tap points x, y, and z in the three phases of the inverter circuit.

The silicon controlled rectifier is a triode reverse blocking thyristor that is rendered conductive when the anode is positive with respect to the cathode and when a gating signal is applied to the gate electrode. Thereafter the gate electrode loses control over conduction through the device and to commutate it off or render it non-conductive it is necessary to apply a reverse bias voltage or to reduce the flow of current through the device below the holding value for a determined time before reapplying forward voltage. Gating signals and commutating impulses for the thyristors are generated in a circuit 75 under the control of a mechanical, magnetic, or optical rotor position sensor 37 of the type previously discussed with regard to FIG. 1a. Suitable gating circuits that can be used are given in the Silicon Controlled Rectifier Manual, 4th Edition, copyright 1967, published by the General Electric Company, and available from the address previously given. The commutating impulses are coupled respectively to commutating windings 70, 70a, and 70b by closely coupled windings 76, 76a, and 76b. The commutating pulses generate reverse currents that oppose the flow of load current through a conducting SCR to commutate it off. In order to produce the square voltage wave shapes shown in FIG. 2c, it is necessary to close neutral switch 72. The operation of this type of inverter circuit, which operates on simple square wave gating signals, is well known as described for example in the book, "Principles of Inverter

Circuits" by Bedford and Hoft, John Wiley and Sons, Inc., New York, Library of Congress catalog card No. 64–20078, copyright 1964. Briefly, supplying a gating signal from circuit 75 to the gate electrode of SCR 68 to render it conductive applies a positive polarity square wave voltage to windings 64A and 64A', which are wound in opposite directions to produce opposite magnetic poles. Upon commutating off SCR 68 and turning on SCR 69, the polarity of the voltage applied to windings 64A and 64A' is reversed, thereby reversing the polarity of the stator magnetic poles produced when the flow of current through the windings changes direction. The other thyristors in the other phases of the inverter are operated similarly at the proper intervals to produce the 60° phase displaced square voltage wave shapes shown in FIG. 2c. Only voltage waveforms E_{64A}, E_{64B}, and E_{64C} representing the voltages applied to those windings are illustrated, the others being complementary.

Each opposing pair of stator windings is alternately energized with positive polarity voltage and negative polarity voltage at 180° intervals of rotation of permanent magnet rotor 52. The timing of application of voltage is the same as described with regard to FIGS. 1a–1c, that is, the change from one polarity to the other is made when an adjacent rotor pole is approximately centered on a particular concentrated winding and the stator pole defined by that winding. As is also the case with FIG. 1c, the flux at the instant at that particular stator pole is approximately at a maximum. As will be observed in FIG. 2c, where the average magnetic flux characteristics ϕ_{64A}, ϕ_{64B}, and ϕ_{64C} are illustrated, the magnetic flux in any particular pole increases and decreases linearly. The motor shown in FIG. 2a is, loosely speaking, a three-phase version of the two phase motor of FIG. 1a, and operates in a similar manner. The only difference is that the stator windings overlap by 30° on each side, and consequently, the total average magnetic flux in these overlapping stator pole regions is the sum of the flux produced by each stator winding operating individually. The flux due to the overlapping windings is of the same polarity during most of the time that any portion of the rotor is adjacent that stator pole portion, and thus is beneficial since it tends to smooth the torque. For example, the flux in the stator segment between stator slots 1 and 2 is the sum of that due to stator winding 64A and stator winding 64C'. At the time the flux in this stator segment produced by winding 64A is going through zero toward its maximum (the leading end of rotor 52 is adjacent stator slot No. 1), the flux generated by stator winding 64C' is increasing toward the same polarity maximum and does not begin to decrease until the leading end of rotor 52 is adjacent stator slot No. 2. The overlapping stator winding, therefore, is beneficial to the production of continuous torque.

The motor constructed as illustrated in FIG. 2a is compatible with the motor control circuit of FIG. 2b when connected and controlled to produce the voltage wave shapes shown in FIG. 2c because the motor operates efficiently with these applied voltage wave shapes, as has already been explained, while at the same time the back emf generated in the stator windings have the same wave shape. In the same manner as has been discussed with regard to FIGS. 1a–1c, the rotor flux acting on each opposing pair of concentrated stator windings increases linearly to a maximum and then decreases linearly to the same level as the rotor poles rotate 180° from out of alignment into complete alignment, and then from complete alignment to out of alignment in the other direction. This induces a square voltage wave shape or square back emf in each stator winding pair. Since the pairs of stator coils are physically displaced by 60°, the resulting back emf's are also phase displaced by 60° as related to the rotation of the motor, and this corresponds to the three phase square wave voltage energization of the stator windings illustrated in FIG. 2c. Adjustable speed operation of this motor and of the other motors to be described hereafter that are operated by the FIG. 2b control circuit is obtained, as previously mentioned, by control of the d-c supply voltage by the phase controlled rectifier. The volt-

age impressed on the stator windings is also changed by time ratio control operation of the thyristor switches. Moreover, the SCR's in this control circuit can be replaced by other suitable solid-state switches such as the transistor, diac, triac, etc.

The conventional three phase stepped wave voltage wave shapes shown in FIG. 3b are obtained when the neutral switch in the motor control circuit of FIG. 2b is opened and the inverter is operated according to the widely used mode of operation as described for instance in the aforementioned Bedford and Hoft book. In FIG. 2b, it is assumed that windings 78A and 78A' replace windings 64A and 64A', and so on. The compatible motor design for use with these voltage wave shapes is illustrated in FIG. 3a. The three pairs of opposing concentrated 120° stator windings 78A and 78A', 78B and 78B', and 78C and 78C', are located so as to overlap each adjacent winding by 60°. To this end there are six of the stator slots 53, equally spaced from one another, defining six equal stator segments. The permanent magnet rotor 79 has 120° arcuate rotor pole faces.

The three phase stepped voltage wave shapes E_{78A}, E_{78B}, and E_{78C}, illustrated in FIG. 3b, are phase displaced by 60° as related to the revolution of the motor. The corresponding magnetic flux characteristics ϕ_{78A}, ϕ_{78B}, and ϕ_{78C} produced by these energizing voltages are also shown in dashed lines in FIG. 3b. The application of voltage to a particular stator winding is timed so that the change from positive to negative polarity of applied voltage occurs when the flux produced by that stator winding is at a maximum and the rotor is centered on that particular winding or completely aligned with it. Subsequent step changes in the applied voltage are made at 60° intervals as related to the revolution of the motor. With the use of a 120° rotor and a 120° concentrated stator winding, there may be some negative torque exercised on the rotor during one-quarter of the time that any portion of a rotor pole face is opposite any portion of the corresponding stator pole face. This is because the rotor takes 120° to rotate into alignment with the stator pole, and 120° to rotate out of alignment, making a total of 240°, whereas the magnetic flux in any particular stator winding is changing at 180° intervals. During a 60° interval, then, negative torque may be exerted on the rotor, depending on such factors as the leakage reactance, etc., as is known in the art. However, some negative torque is not detrimental to good operation of the motor. In the manner already explained with regard to FIG. 2a, the use of overlapping stator windings in general is beneficial because of the smoother torque produced.

The back emf wave shape produced by a permanent magnet or reluctance motor constructed as shown in FIG. 3a is the same as the applied step voltage wave shapes, and hence the motor and control circuit are compatible. During any single 180° interval of travel of the rotor past a particular concentrated stator winding, such as winding 78A, and in view of the 120° circumferential length of the rotor pole faces, there is a period when one rotor pole is acting on the stator winding, a period when both rotor poles are acting on the stator winding, and a period when the other rotor pole is acting on the stator winding solely. Assuming that the leading edge of the north pole of rotor 52 is commencing to rotate clockwise into alignment with concentrated winding 78A, the aligned rotor poles at 60° intervas are: S 60°, N 60°, N 120°, N 60°, and S 60°. This plots out a magnetic flux characteristic acting on concentrated stator winding 78A that is identical to the curve ϕ_{78A} in FIG. 3b. The resulting induced back emf has the same wave shape as applied voltage E_{78A}.

The permanent magnet motor designs illustrated in FIG. 4a and FIG. 5a are both compatible for energization by the voltage wave shapes shown in FIG. 4b. These quasisquare voltage wave shapes are produced when the motor control circuit of FIG. 2b is operated with the neutral switch open with with the three pairs of stator windings delta-connected as shown in FIG. 4c. To obtain these quasi-square voltage wave shapes, in which the zero voltage intervals are half as long as the positive and negative voltage intervals, the thyristors are supplied with 180° square wave gating signals and are sequentially gated at 60° intervals in the manner further explained on page 267 of the aforementioned Bedford and Hoft book.

The compatible motor design of FIG. 4a uses the 120° permanent magnet rotor 79, but requires three pairs of opposing 60° concentrated stator windings 80A and 80A', 80B and 80B', and 80C and 80C' that are non-overlapping. The three applied voltage wave forms E_{80A}, E_{80B}, and E_{80C} illustrated in FIG. 4b are phase displaced by 60° as related to the revolution of motor, corresponding to the physical displacement of the opposing pairs of stator windings. As is observed from the magnetic flux characteristics ϕ_{80A}, ϕ_{80B}, and ϕ_{80C}, the flux is driven to a maximum of one polarity by the applied voltage, and remains at this maximum level during the zero voltage interval, and is driven down only by applying a voltage of the opposite polarity. The switching of the thyristors in the motor control circuit occurs at intervals about 30° before and after the center of the rotor pole face is at the center of a particular concentrated winding, since both the flux and the applied voltage are constant during the intervening 60° interval. Efficient operation of the motor is obtained as the magnetic flux in succeeding stator poles in a clockwise or counterclockwise sequence are driven to their constant maximum value at succeeding 60° intervals. No negative torque is produced since the entire 120° permanent magnet rotor 79 requires exactly 180° to rotate past every portion of a 60° stator pole.

The back emf voltages produced by the motor of FIG. 4a have the same quasi-square voltage wave shapes as the applied voltages illustrated in FIG. 4b. Going through the analysis for concentrated stator winding 80B, the rotor flux acting on stator winding 80B increases linearly, assuming clockwise rotation, as the leading edge of rotor 79 rotates from stator slot No. 1 to stator slot No. 2. For the next 60° rotation, the rotor flux is constant, while for the remaining 60° of rotation, the flux is decreasing linearly back towards zero. This is the same characteristic as the flux characteristics shown in FIG. 4b, and the induced back emf generated in stator winding 80B has the same wave shape as the applied voltage wave shape.

The motor design of FIG. 5 is also compatible with the FIG. 2b motor control circuit when operated to produce the quasi-square voltage wave shapes given in FIG. 4b. In place of the 120° permanent magnet rotor 79, the rotor 81 has a 60° circumferential length and is illustrated as being a soft iron reluctance rotor in place of the permanent magnet rotor. The three pairs of opposing concentrated 120° stator windings are identical to the stator winding arrangement shown in FIG. 3a, and for convenience are given the same identifying numerals. These concentrated 120° stator windings overlap each adjacent winding by 60°. The timing of the applied voltage to a particular stator winding is such that the magnetic flux in that stator pole, taken individually, is midway through its constant maximum magnitude at the time the rotor 81 is centered upon that particular stator winding. No negative torque is exerted upon rotor 81, and the resultant flux pattern produced by the overlapping pairs of stator windings is beneficial to the production of continuous for the reasons given in the discussion of FIG. 3a. The back emf generated by the FIG. 5 motor in any one stator winding is the same as the applied voltage wave shapes. As rotor 81 rotates past stator winding 78A, for example, the flux acting on stator winding 78A to produce an induced back emf increases to a maximum as the stator rotates 60°, is constant for the next 60°, and then decreases linearly for the final 60° of rotation. Consequently, the motor of FIG. 5 and the motor control circuit of FIG. 2b, operated to produce the applied voltage wave shapes shown in FIG. 4b, are compatible.

In summary, a family of permanent magnet or reluctance brushless motors are designed to have a simple approximately square or stepped wave shape back emf characteristic that matches the applied voltage wave shapes easily produced by a motor control switching circuit comprising a number of controlled switches. The switching circuits disclosed are known

inverter-type circuits using solid state switches that can if desired be controlled by time ratio control principles to vary the voltage impressed on the motor windings to adjust the speed of the motor. The brushless motors comprise an annular stator with opposing pairs of concentrated (non-distributed) stator windings wound with a predetermined winding pitch in non-overlapping or overlapping relationship. The permanent magnet or reluctance rotors have opposing arcuate poles, to define a constant or an approximately constant width gap between the rotor and stator poles faces, and a specified circumferential length. The motors are simple, but operate with good efficiency on rectilinear or substantially rectilinear wave shapes, as compared to conventional alternating current motors that have design features adapting them for sine wave energization.

While the invention has been particularly shown and described with reference to several preferred embodiments thereof, it will be understood by those skilled in the art that the foregoing and other changes may be made therein without departing from the spirit and scope of the invention.

What I claim as new and desire to secure by Letters Patent of the United States is:

1. A compatible brushless motor and motor control circuit comprising

an annular stator member supporting a plurality of opposing pairs of concentrated stator windings for generating magnetic fields of opposite polarity in torque producing relation to a rotatable ferromagnetic rotor, said stator member and rotor each having opposing arcuate pole faces to establish an approximately constant gap width therebetween,

a motor control circuit comprising only a few controlled switches for applying simple substantially rectilinear wave shape energizing voltages of both polarities to said pairs of stator windings,

the winding pitch and location relative to one another of said concentrated stator windings, and the circumferential length of said rotor pole faces, having interdependent angular dimensions to produce substantially rectilinear back emf voltages in each pair of stator windings that have approximately the same wave shape as the energizing voltage applied by said motor control circuit, and

control means responsive to the instantaneous position of said rotor for rendering conductive said controlled switches for desired intervals of conduction to energize said pairs of stator windings in a sequence to produce continuous torque to rotate said rotor in a given direction,

wherein there are only three opposing pairs of series connected concentrated stator windings,

said motor control circuit comprises a pair of alternately conducting controlled switches for each pair of stator windings, said controlled switches being solid state controlled devices,

said three opposing pairs of concentrated stator windings each having a winding pitch of about 90° and each stator winding overlaps the adjacent stator windings by about 30°, and

said rotor pole faces have a circumferential length of about 90°, whereby the brushless motor is suitable for square voltage wave shape energization.

2. A compatible brushless motor and motor control circuit comprising

an annular stator member supporting a plurality of opposing pairs of concentrated stator windings for generating magnetic fields of opposite polarity in torque producing relation to a rotatable ferromagnetic rotor, said stator member and rotor each having opposing arcuate pole faces to establish an approximately constant gap width therebetween,

a motor control circuit comprising only a few controlled switches for applying simple substantially rectilinear wave

11

shape energizing voltages of both polarities to said pairs of stator windings,

the winding pitch and location relative to one another of said concentrated stator windings, and the circumferential length of said rotor pole faces, having interdependent angular dimensions to produce substantially rectilinear back emf voltages in each pair of stator windings that have approximately the same wave shape as the energizing voltage applied by said motor control circuit, and

control means responsive to the instantaneous position of said rotor for rendering conductive said controlled switches for desired intervals of conduction to energize said pairs of stator windings in a sequence to produce continuous torque to rotate said rotor in a given direction,

wherein there are only three opposing pairs of series connected concentrated stator windings,

said motor control circuit comprises a pair of alternately conducting controlled switches for each pair of stator windings, said controlled switches being solid state controlled devices,

said three opposing pairs of concentrated stator windings have a winding pitch of about 120° and each stator winding overlaps the adjacent stator windings by about 60°, and

said rotor pole faces have a circumferential length of about 120°, whereby the brushless motor is suitable for stepped voltage wave shape energization.

3. A compatible brushless motor and motor control circuit comprising

an annular stator member supporting a plurality of opposing pairs of concentrated stator windings for generating magnetic fields of opposite polarity in torque producing relation to a rotatable ferromagnetic rotor, said stator member and rotor each having opposing arcuate pole faces to establish an approximately constant gap width therebetween,

a motor control circuit comprising only a few controlled switches for applying simple substantially rectilinear wave shape energizing voltages of both polarities to said pairs of stator windings,

the winding pitch and location relative to one another of said concentrated stator windings, and the circumferential length of said rotor pole faces, having interdependent angular dimensions to produce substantially rectilinear back emf voltages in each pair of stator windings that have approximately the same wave shape as the energizing voltage applied by said motor control circuit, and

control means responsive to the instantaneous position of said rotor for rendering conductive said controlled switches for desired intervals of conduction to energize said pairs of stator windings in a sequence to produce continuous torque to rotate said rotor in a given direction,

wherein there are only three opposing pairs of series connected concentrated stator windings,

said motor control circuit comprises a pair of alternately conducting controlled switches for each pair of stator windings, said controlled switches being solid state controlled devices,

said three opposing pairs of concentrated stator windings are non-overlapping windings each having a winding pitch of about 60°, and

said rotor pole faces have a circumferential length of about 120°, whereby the brushless motor is suitable for quasi-square voltage wave shape energization.

4. A compatible brushless motor and motor control circuit comprising

an annular stator member supporting a plurality of opposing pairs of concentrated stator windings for generating magnetic fields of opposite polarity in torque producing rela-

12

tion to a rotatable ferromagnetic rotor, said stator member and rotor each having opposing arcuate pole faces to establish an approximately constant gap width therebetween,

a motor control circuit comprising only a few controlled switches for applying simple substantially rectilinear wave shape energizing voltages of both polarities to said pairs of stator windings,

the winding pitch and location relative to one another of said concentrated stator windings, and the circumferential length of said rotor pole faces, having interdependent angular dimensions to produce substantially rectilinear back emf voltages in each pair of stator windings that have approximately the same wave shape as the energizing voltage applied by said motor control circuit, and

control means responsive to the instantaneous position of said rotor for rendering conductive said controlled switches for desired intervals of conduction to energize said pairs of stator windings in a sequence to produce continuous torque to rotate said rotor in a given direction,

wherein there are only three opposing pairs of series connected concentrated stator windings,

said motor control circuit comprises a pair of alternately conducting controlled switches for each pair of stator windings, said controlled switches being solid state controlled devices,

said three opposing pairs of concentrated stator windings have a winding pitch of about 120° and each stator winding overlaps the adjacent stator windings by about 60°, and

said rotor pole faces have a circumferential length of about 60°, whereby the brushless motor is suitable for quasi-square voltage wave shape energization.

5. A brushless motor suitable for energization by simple substantially rectilinear voltage wave shapes comprising

an annular stator member supporting a plurality of opposing pairs of series connected concentrated stator windings for generating magnetic fields of opposite polarity in torque producing relation to a ferromagnetic rotor mounted for rotation within said stator member, said stator member and rotor each having opposing arcuate pole faces to establish an approximately constant gap width therebetween,

the winding pitch and location relative to one another of said concentrated stator windings, and the circumferential length of said rotor pole faces, having interdependent angular dimensions to produce simple substantially rectilinear back emf voltages in each pair of stator windings that have approximately the same wave shape as the energizing rectilinear voltage wave shapes,

wherein there are three opposing pairs of concentrated stator windings each having a winding pitch of about 90° and each stator winding overlaps the adjacent stator windings by about 30°, and

said rotor pole faces have a circumferential length of about 90°, whereby the brushless motor is suitable for square voltage wave shape energization.

6. A brushless motor suitable for energization by simple substantially rectilinear voltage wave shapes comprising

an annular stator member supporting a plurality of opposing pairs of series connected concentrated stator windings for generating magnetic fields of opposite polarity in torque producing relation to a ferromagnetic rotor mounted for rotation within said stator member, said stator member and rotor each having opposing arcuate pole faces to establish an approximately constant gap width therebetween,

the winding pitch and location relative to one another of said concentrated stator windings, and the circumferential length of said rotor pole faces, having interdependent angular dimensions to produce simple substantially rectilinear back emf voltages in each pair of stator windings that have approximately the same wave shape as the energizing rectilinear voltage wave shapes,

wherein there are three opposing pairs of concentrated stator windings each having a winding pitch of about 120° and each stator winding overlaps the adjacent stator windings by about 60°, and

said rotor pole faces have a circumferential length of about 120°, whereby the brushless motor is suitable for stepped wave energization.

7. A brushless motor suitable for energization by simple substantially rectilinear voltage wave shapes comprising

an annular stator member supporting a plurality of opposing pairs of series connected concentrated stator windings for generating magnetic fields of opposite polarity in torque producing relation to a ferromagnetic rotor mounted for rotation within said stator member, said stator member and rotor each having opposing arcuate pole faces to establish an approximately constant gap width therebetween,

the winding pitch and location relative to one another of said concentrated stator windings, and the circumferential length of said rotor pole faces, having interdependent angular dimensions to produce simple substantially rectilinear back emf voltages in each pair of stator windings that have approximately the same wave shape as the energizing rectilinear voltage wave shapes,

wherein there are three non-overlapping opposing pairs of concentrated stator windings each having a winding pitch of about 60°, and

said rotor pole faces have a circumferential length of about 120°, whereby the brushless motor is suitable for quasi-square wave shape energization.

8. A brushless motor suitable for energization by simple substantially rectilinear voltage wave shapes comprising

an annular stator member supporting a plurality of opposing pairs of series connected concentrated stator windings for generating magnetic fields of opposite polarity in torque producing relation to a ferromagnetic rotor mounted for rotation within said stator member, said stator member and rotor each having opposing arcuate pole faces to establish an approximately constant gap width

therebetween,

the winding pitch and location relative to one another of said concentrated stator windings, and the circumferential length of said rotor pole faces, having interdependent angular dimensions to produce simple substantially rectilinear back emf voltages in each pair of stator windings that have approximately the same wave shape as the energizing rectilinear voltage wave shapes,

wherein there are three opposing pairs of concentrated stator windings each having a winding pitch of about 120° and each stator winding overlaps the adjacent stator windings by about 60°, and

said rotor pole faces have a circumferential length of about 60°, whereby the brushless motor is suitable for quasi-square wave shape energization.

9. A compatible brushless motor and motor control circuit comprising

an annular stator member supporting a plurality of symmetrically arranged opposing pairs of concentrated stator windings each simultaneously generating magnetic fields of opposite polarity in torque producing relation to a rotatable ferromagnetic rotor, said stator member and rotor having opposing arcuate pole faces to establish an approximately constant gap width therebetween,

a motor control circuit comprising only a few controlled switches for applying simple substantially rectilinear stepped wave shape energizing voltages of both polarities to said pairs of stator windings, wherein

the sum of the angular dimensions of the winding pitch of said concentrated stator windings and the circumferential length of one of said rotor pole faces totals approximately 240° and produces substantially rectilinear back emf voltages in each pair of stator windings that have approximately the same wave shape as the energizing voltage applied by said motor control circuit, and

control means responsive to the instantaneous position of said rotor for rendering conductive said controlled switches for desired intervals of conduction to energize said pairs of stator windings in a sequence to produce continuous torque to rotate said rotor in a given direction.

* * * * *

An Axial Air-Gap Reluctance Motor for Variable Speed Applications

L.E. Unnewehr and W.H. Koch

IEEE Transactions, Vol. PAS-93, 1974, pp. 367-376

ABSTRACT

The disc motor is a brushless, reluctance-principle rotating machine that has been developed to the prototype stage for traction and variable speed drive application. It is attractive for these and other applications because of a simple magnetic structure which has potential for very low assembly costs, and because it can be controlled over a wide range of speed and torque by thyristor circuits of comparable simplicity to the chopper circuit used with dc traction motors. The motor theory is briefly discussed and laboratory measurements of machine performance is presented.

INTRODUCTION

Electric drive systems for propulsion of railways and commercial road and off-highway vehicles have been in use for several decades. More recently a combination of very important advances in electronic components and magnetic materials and of new needs in transportation has led to the construction and application of electric motor drive systems suitable for a multitude of different propulsion purposes in the power range from 2 to over 3000 KW. Some sectors of this field of technology, notably the one involving dc commutator motors and their controllers, have already matured into lines of commercial products. Other sectors, in which commutator motors are impractical and brushless, high performance motors are desirable, are characterized by continuing innovation research and development. Such sectors involve very high speed rail passenger cars and locomotives, wheel drives for off-highway military vehicles and those electric automobiles and trucks potentially satisfying more ambitious expectations relative to low weight, high efficiency, zero maintenance and high reliability. The objective of this paper is to contribute to the technology of brushless, high performance drive systems. Specifically a brushless motor for use in a drive system supplied from a dc power source whose terminal voltage cannot be controlled by the operator will be described. This drive system is intended to meet principal performance and design requirements for vehicle traction applications.

Such requirements include: (a) Constant, rated power output capability over the range of 25% to 100% of maximum rated speed; (b) high torque capability over the range from zero to one quarter of maximum rated speed; (c) capability for regenerative motor braking; (d) rapid and smooth control of motoring and braking torques at all operating speeds without reliance on separate control of dc power source voltage; (e) high propulsion energy efficiency when operating the system in a representative vehicle driving cycle; (f) a high ratio of rated output power to total system weight; (g) ruggedness; high reliability.

Paper T 73 331-6 recommended and approved by the IEEE Rotating Machinery Committee of the IEEE Power Engineering Society for presentation at the IEEE PES Summer Meeting & EHV/UHV Conference, Vancouver, B.C., Canada, July 15-20, 1973. Manuscript submitted February 8, 1973; made available for printing April 24, 1973.

Systems Concept

The principal components of this drive system are a disc type variable reluctance motor, a solid state phase current controller and a rotor position sensor associated with the motor. These components and their interconnections for power and signal flows are shown in schematic form in Fig. 1 for a single phase motor embodiment. The essential principles of motor operation can be described as follows. The annular phase winding is linked with a closed path for magnetic flux passing through soft magnetic stator and rotor poles, the air gaps in between and magnetic stator yokes connecting the poles. In the rotor discs, soft magnetic, highly permeable pole sections alternate with nonmagnetic sections along the circumference. As the rotor turns, the permeance of the flux path enclosing the annular phase winding, rises and falls periodically and so does the phase winding self-inductance. Positive, that is "motoring" torque, is developed by this variable reluctance device whenever, during a period of rising permeance the phase current is nonzero. These inductance relationships are qualitatively illustrated in Fig. 2. It may be said that the magnetic field between stator and rotor pole surfaces and edges, which is associated with a given phase current and rotor position, is pulling the moving rotor pole sections into alignment with the stator poles. A single phase section of the variable reluctance motor is thus capable of developing a pulse train of positive torque, if and when the solid state controller can force current through the phase winding during the recurring intervals of rising winding self-inductance. The controller thus has to perform the difficult task of (a) rapidly forcing energy into the reluctance machine when the rotor position sensor indicates a rising inductance interval to be approaching, (b) maintaining this current during the same interval and (c) recovering all magnetic energy from the machine when the rotor approaches the interval of declining inductance and phase current is to be reduced to zero. These tasks have to be performed over an extensive range of variable speeds, including very low speed and stall conditions. For any controlled circuit to be accepted for vehicle drive application, it should be capable of accomplishing phase current turn-on and turn-off with very high energy efficiency and with the utmost economy in the use of electronic power handling components.

Review of Related Technology

During the past decade, extensive work has been carried out in the development of solid state device controlled, brushless drive systems. Substantial accomplishments have been made worldwide toward meeting different technological design and performance requirements in industrial, aerospace and vehicle propulsion applications of such drive systems.[1-4] Reluctance motor drive systems have received some attention[5], but not nearly as much as synchronous and induction motor systems. In order to relate why and how the authors' choice of a particular reluctance drive concept came about, the nature of some well known dc powered drive systems will be briefly reviewed.

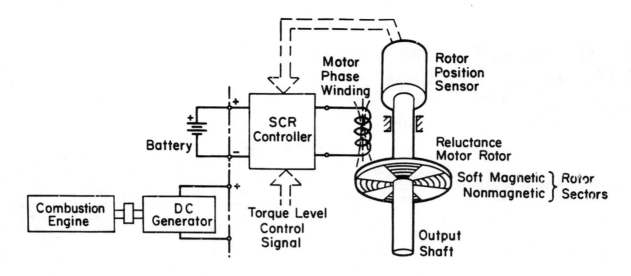

Figure 1. Schematic Diagram of Disc Motor System

Polyphase induction motors and synchronous motors fed from battery power sources through variable frequency solid state inverters have been reported and demonstrated in experimental road vehicles[6-9]. Both of these motor types develop revolving air gap fields internally and accordingly require for power input a set of AC armature currents whose fundamental wave components form a balanced system of polyphase, usually three-phase currents. Ac frequency and phase sequence have to be made variable in accordance with the requirements of vehicle motor speed; furthermore, the current amplitudes have to be controlled in accordance with the desired motor output torque and given motor speed. Various ac-dc three-phase inverters or combinations of independent current level controllers and frequency variable, current invariant inverters have been demonstrated which perform these functions[3,5,6,9,10,11,12,13]. Such inverters are generally designed in a three-phase bridge configuration. As may be seen from the circuit examples in Fig. 3, such bridge configurations contain six individually controlled SCR or transistor branches connecting each of the three motor winding terminals with each pole of the dc power source. The complexity of these branches and the required number of SCR's, diodes and capacitors vary. They can be quite high and can make such drive systems costly. High inverter branch complexity can also become a significant burden in the attainment of acceptable MTBF (mean time between failures) for the total drive system.

The variable reluctance motor drive system which is the subject of this paper differs in concept from the inverter driven three-phase revolving field systems. Consideration of induction and synchronous motors above and, in the following, of two different brushless servomotor systems should shed some light on the nature of and the reasons for the conceptual differences of the motors and electronic controllers to be described here.

A dc powered synchronous motor servo system used in aerospace applications has a permanent magnet rotor for field excitation and two to four stationary phase windings[14,15]. The phase winding currents are

supplied by a transistor dc-ac inverter which periodically reverses phase current polarity and is responding to a rotor position sensor and rotor speed feedback, as shown in Fig. 4. A position sensor, for example a Hall plate or an electro-optical device, is required for controlling the optimum phase angle of the drive signals for the power transistor switches. A design variation of such motors is conceivable according to which the permanent magnet rotor is replaced by a salient pole, soft magnetic, laminated rotor and the rotor position sensor is readjusted (Figure 5). The resulting three-phase reluctance motor would continue to require a solid state supply of three-phase currents, capable of periodically reversing the polarity of all phase currents. Polarity reversal is required for steady output torque and for sure motor startup because all inductances of the phase windings undergo cyclical changes, including sign reversals, as the soft magnetic rotor turns. For this servo motor, the most simple, still adequate inverter configuration would thus continue to be the three-phase transistor bridge working in conjunction with a three phase armature winding. Another well known type of servo motor, frequently used for position control in the stepping mode, also makes use of a soft magnetic rotor, of the principles of angularly displaced polyphase windings for achieving positive torque in any rotor position, and of phase winding current control as a function of rotor position sensor signals. However, unlike the previously discussed polyphase, revolving field type reluctance motor, derived from the permanent magnet rotor synchronous machine, this servo motor does not permit magnetic coupling among different phase windings to occur. It, therefore, does not have to be supplied with a balanced set of sign-reversing polyphase current waveforms such as unidirectionally pulsing phase currents, which are turned on and off as a function of rotor position and torque level. Such currents can be supplied from various other configurations of solid state controllers which are designed for pulsing currents and can, therefore, be assembled from circuit branches containing no more than one solid state power switching set of elements per phase, such as a transistor-diode combination or an SCR-diode and commutation branch combination. This amounts to only half the number of power switching devices per phase required in those corresponding bridge type inverters, which are associated with all brushless motors discussed above. The circuit of a representative servo drive system of this type[17,18] is shown in Fig. 6. Another remarkable example of a magnetically phase decoupled variable reluctance motor with SCR controllers has been built for a torpedo drive and reported in Ref. 21. In view of the high cost, the switching losses and the sensitivity to false triggering of SCR devices used in three-phase bridge inverters suitable for the power requirements of brushless vehicle traction motors, the simplified controller circuit characteristics applicable to stepper motors were recognized and put to use in the development of the drive system to be reported. The magnetically decoupled polyphase reluctance motor in its general concept permits the excitation with arbitrary current waveforms - alternating, unidirectionally pulsing, or combinations thereof - which can be economically and efficiently generated by several kinds of solid state controllers, without incurring detrimental interaction of electric or magnetic variables associated with the different motor phases. These controllers have been described elsewhere.[19]

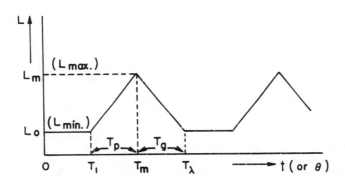

Figure 2.   Idealized Representation of
Disc Motor Inductance Variation

Motor Description

The term "disc motor" is derived from the geometry of the rotor and stator sections of this machine, both of which are of disc-shape and can be "stacked" together in proper sequence to form various combinations of phases. The resulting outline of the motor may be of small diameter and relatively long - almost cigar-shaped at one extreme, or of large diameter and of short length, somewhat pancake-shaped at the other extreme, with many variations in between. It thus has a great flexibility of size and shape and can be designed to match the unusual space requirements of automotive, truck and tractor applications.

This configuration was proposed independently by three different investigators,[20,21] including one of the authors. Previous applications include those as torpedo motor and as stepper motor for control applications. The traction motor application is more demanding than either of these earlier applications in terms of voltage, current, and torque levels over a wide speed range, and in its requirements for ruggedness and operating life time.

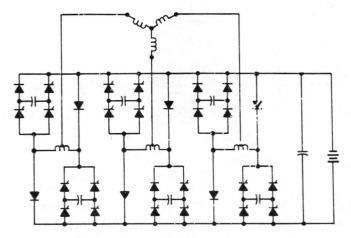

Figure 3.   Bridge Inverter Circuit

The disc motor is an axial air gap machine which accounts for its disc-shape structure. Figure 7 illustrates the cross section of a typical polyphase configuration made up of three magnetically uncoupled sections or "phases". Figure 8 illustrates a single phase prototype of this configuration upon which many hours of testing and evaluation have been achieved. Several other disc motor configurations have been developed and in some cases constructed;[22] Figure 9 illustrates a 2-phase prototype machine in which the rotor and stator magnetic members are essentially reversed from the version shown in Figure 8. Both of the configurations have the following general physical characteristics:

1. Solenoidal-type exciting (armature) windings. This is the simplest and potentially least expensive form of electrical winding, and also permits the use of Litz wire, flat conductors, and other unconventional conductors that result in lower "skin effect" and resistance losses.

2. The solenoidal coils are more amenable to forced cooling - either air or liquid - than conventional windings.

3. Motor torque is approximately proportional to effective or RMS current, as shown below.

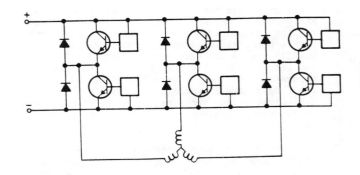

Figure 4.   Transistor Controller

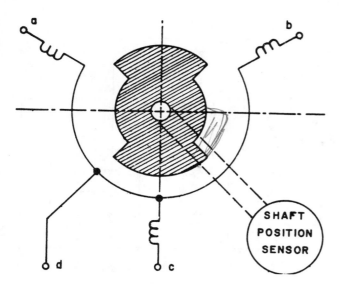

Figure 5. Polyphase Reluctance Motor

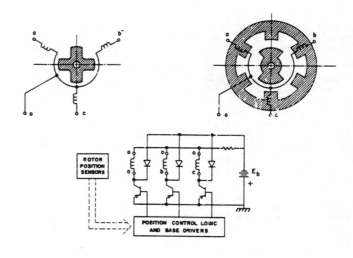

Figure 6. Reluctance Servo Drive

4. The magnetic poles or saliencies consist of rectangular or radial sectors spaced uniformly around both rotor and stator discs and separated by nonmagnetic material. The magnetic sectors may be laminated or formed from powdered iron to reduce eddy current losses. The machines illustrated in Figures 8 and 9 have magnetic sectors constructed of .004" silicon steel laminations.

5. The number of poles used is arbitrary up to a certain point determined by the frequency limitations of the controller and the leakage flux in the motor. Ford disc motors have been built with up to 29 poles in a 10.5 in. diameter rotor disc (Figure 9); at least 40 appear to be feasible.

6. When the machine is built for more than one phase, the phase windings are magnetically uncoupled. This permits the use of many phases or sections in the motor by stacking the sections in an axial direction. There is a practical limit usually determined by the resulting controller complexity. With more than one phase, the motor is self-starting.

7. The air gap of the disc motor can be increased by using more than one rotor disc per phase section. This requires the use of a dummy stationary disc between adjacent rotor discs. The motors illustrated in Figs. 7 and 8 have two rotor discs and one dummy disc per phase.

8. The nonmagnetic structural portions of both rotor and stator discs may be either electrically conducting or nonconducting materials. The former give a somewhat improved inductance variation but usually result in excessive eddy current losses and reduced motor efficiency. Discs for the configuration of Fig. 7 and Fig. 8 use a glass-reinforced phenolic material. These discs have a 5.6" diameter and speed rating of 5000 rpm. Burst speed during destruction tests has been measured at 17,500 RPM.

9. The principal cooling technique has been to force air through one end of a hollow shaft and radially outward through holes in the shaft past the exciting coils. Since most of the motor loss is in the coils, this is an effective technique. Water-filled conductors have also been evaluated.

284

Motor Theory

The principle of this particular type of reluctance machine is related to one of the earliest observations of mechanical forces made by men, the attraction between two magnetic materials. It is the basis for the operation of many types of magnetic relays, magnetic transducers, and the lift magnet. The disc motor differs from the conventional reluctance motor in that both its stator and rotor members have saliency. Its magnetic circuit is akin to that of the Flux-Switch Inductor machine, but the disc motor is singly-excited.

The basic theory of the disc motor has been developed in terms of both its field and circuit parameters, and several design techniques and associated computer programs have been prepared. Only a short summary of this development will be presented here in order to introduce some of the key design parameters. Nomenclature is defined in Appendix A.

The energy conversion process resulting from the mechanical translation within a magnetically-coupled system is well described in several texts.[23] For a singly-excited system and a system in which there are only two magnetic members - one stationary and one free to rotate about its axis - this process can be described in terms of a developed torque,

$$T = -\frac{\partial W_m}{\partial \theta} + i\frac{\partial \psi}{\partial \theta} \qquad (1)$$

where W_m is the magnetic energy, ψ the flux linkages, and θ the angular displacement between rotary and stationary members. Let $W_m = \frac{1}{2} i^2 L$ and $\psi = iL$, where L is some time varying inductance function such as that shown in Figure 2. Then eq. (1) can be expressed as

$$T = -\frac{1}{2} i^2\frac{\partial \dot{L}}{\partial \theta} - Li\frac{\partial i}{\partial \theta} + i^2\frac{\partial L}{\partial \theta} + Li\frac{\partial i}{\partial \theta}$$

$$= \frac{1}{(d\theta/dt)} \left[\frac{1}{2}i^2\frac{dL}{dt} \right] \qquad (2)$$

Assuming the magnetic material in the system to have infinite permeability, the energy exists entirely in the air regions between and surrounding the magnetic

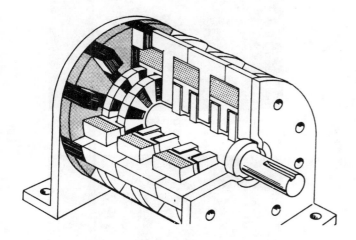

Figure 7. Disc Motor Crossection

Figure 8. 8 H.P., 5000 rpm Disc Motor

members. In the increasing inductance interval, T_p of Figure 2, the inductance function can be expressed as

$$L = L_{min} + Xt \qquad (T_1 < t < T_m) \qquad (3)$$

where X is the slope of the function.

Equation (2) is a nonlinear differential equation with time-varying coefficients. The term, $d\theta/dt$ is recognized as the instantaneous rotor speed. However since the disc motor is excited by a series of discontinuous current pulses, the developed torque must be examined first during the intervals of one pulse. For motor action, this interval is the period of increasing inductance, $T_p = T_m - T_1$; shown in Figure 2, or a portion of this period. Due to rotor, shaft, and load inertias, and since the time interval of a typical current pulse is generally very small, rotor speed may be considered constant during the interval. Mathematically, this can be thought of as a piecewise linearization of the speedtime characteristics, the increment of linearization being the duration of a current pulse. Therefore, let

$$\frac{d\theta}{dt} = 2\pi n$$

where n = rotor speed in revolutions/sec.

Substituting into eq. (2) gives

$$T = \frac{1}{2\pi n} \left[\tfrac{1}{2} i^2 X \right] \qquad (4)$$

which is the instantaneous torque developed by the disc motor.

To gain further insight into the nature of the disc motor torque, it is necessary to note the fact as shown in Ref. 23 that eq. (1) is valid <u>regardless of the manner in which i and θ vary</u> as long as the variation is compatible with the internal system constraints. That is, it is not necessary to know a precise analytical expression for the current, i, in eq. (4) to further evaluate the torque. Therefore, assume the current is zero at $t = T_1$ in Figure 2; reaches a maximum value, I_m, at $t = T_x$, where $T_1 < T_x < T_m$; and is forced back to zero at $t = T_m$. The average value of torque which can be measured at the motor

shaft is that value averaged over the time interval of a pole period, T_λ in Figure 2, since this value is repeated identically during each pole period of one rotor revolution. Therefore,

$$T_{ave} = \frac{1}{T_\lambda} \left[\int_{T_1}^{T_x} T \, dt + \int_{T_x}^{T_m} T \, dt \right]$$

$$= \frac{X}{4\pi n} \left[\frac{1}{T_\lambda} \left(\int_{T_1}^{T_x} i^2 dt + \int_{T_x}^{T_m} i^2 dt \right) \right] \qquad (5)$$

The expression within the brackets of eq. (5) can be recognized as the square of the RMS value of the current flowing in the motor exciting winding. Therefore,

$$T_{ave} = \frac{X}{4\pi n} I_{rms}^2 \qquad (6)$$

The electromagnetic developed power of the motor is equal to $2\pi n \, T_{ave}$, or

$$P_d = 1/2 \, X \, I_{rms}^2 \qquad (7)$$

The role and significance of the term, X, in eqs. (3) through (7) requires further discussion. This term represents the slope of the inductance or permeance variation of the motor: for example, under the idealized assumptions shown in Figure 2,

$$X = \frac{L_{max} - L_{min}}{T_p} \qquad (8)$$

This expression is valuable in discussing the theoretical capabilities of the disc motor and in developing design equations from which a practical configuration can be sized to satisfy a set of performance specifications.

However, as in the case of most machine equations derived from electromagnetic theory, several practical considerations can - and usually do - result in measured values that differ from those predicted by eqs. (6) and (7): (a) Obviously, the inductance of a practical configuration will have rounded corners rather than the sharp discontinuities shown in Fig. 2; (b) saturation of any portion of the magnetic circuit will alter the value of X; disc motor current

285

Figure 9.   20 H.P., 3600 rpm Disc Motor

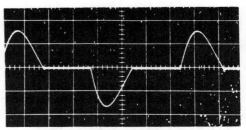

a.   Series Commutated Controller
100 amp./div., 200 μs/div.

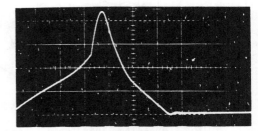

b.   Parallel Commutated Controller
200 amp./div., 1 ms/div.

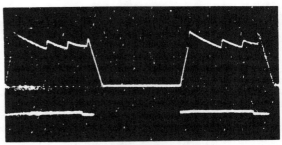

c.   Controlled Freewheeling
125 amp./div., 100 μs/div.

Figure 10.   Disc Motor Load Current

waves are often characterized by high peaks of short duration (as during capacitor discharge in an SCR parallel commutation condition), which can cause X to vary even during one current pulse; (c) disc motor current pulses are often purposefully caused to exist beyond the region of increasing inductance and extend over into regions of constant or even decreasing inductance, which results in an equivalent X factor based upon all three inductance slopes. (d) the motor mechanical loss torque is, of course, included in the eqs. (6) and (7), and should be considered in using these formulae. In general, it does not appear desirable to spend much effort controlling the wave forms of a disc motor, for this can only result in increased controller cost and complexity. And, eqs. (6) and (7) show that all current of any wave form existing during a condition of increasing inductance results in useful torque, with the exceptions of conditions of <u>stator</u> magnetic circuit saturation (it can be shown[24] that with <u>rotor</u> saturation, a torque proportional to the first power of the current is developed). Typical observed disc motor wave forms are shown in Figure 10 for various conditions of speed, torque, and controller configuration.

The RMS and average values of current functions are defined in terms of the electrical pulse width, T_o, and the mechanical pole period, T_λ, of Figure 2 as follows:

$$I_{rms} = I_{max} \sqrt{K_f \frac{T_o}{T_\lambda}} \qquad (9)$$

$$I_{ave} = I_{max} K_A \frac{T_o}{T_\lambda} \qquad (10)$$

In these equations, K_f and K_A are the usual form factors describing the shape of the pulse. For example, in a sine pulse, $K_f = 1/2$ and $K_A = 2/\Pi$; in a triangular pulse, $K_f = 1/3$ and $K_A = 1/2$; in a square pulse, $K_f = K_A = 1.0$.

These standard wave forms often represent reasonably accurate approximations of the actual motor coil pulses. However, in general, the motor current pulses consist of series of discrete sections of continuous waves. The discontinuities are caused by the near instantaneous switching action of the controller SCR's. The continuous current functions are determined by the source voltage and impedance and the circuit parameters in the usual manner. For example,

with the motor coil switched directly across a dc source, the circuit differential equation becomes,

$$E_b = i (R_c + X) + (L_{min} + Xt) \frac{di}{dt} \qquad (11)$$

where X is defined by eq. (8) and a constant rotor speed is assumed. For the initial condition, $i = 0$ and $L = L_{min}$ at $t = 0$, it can be shown[19] that the solution to eq. (11) is

$$i = \frac{E_b}{X + R_c} \left[1 - \frac{L_{min}}{L_{min} + Xt} \right]^{1 + \frac{R_c}{X}} \qquad (12)$$

For all operating conditions except at very low motor speeds, $R_c << X$, and eq. (12) becomes,

$$i \approx \frac{E_b}{X + R_c} \frac{Xt}{L_{min} + Xt} \qquad (13)$$

An example of this wave form is shown by the first portion of the current pulses in Figure 10b.

Structural Considerations

Assume that the electrical pulse period, T_o, is

equal to the mechanical period, T_p, and that $T_\lambda = \frac{1}{np}$. Then eq. (7) becomes

$$P_d = (1/2)np K_f (L_{max} - L_{min}) I_m^2 \qquad (14)$$

This expression shows the principal design parameters determining the power rating of a disc motor. The role of the current terms, I_m^2 and K_f, and the speed, n, are obvious and are characteristic of most machine power equations. Much of the motor design challenge rests with the remaining terms, p, L_{max}, and L_{min}, and these terms also greatly influence the requirements of the controller. The number of magnetic sectors per disc, or poles, p, along with the speed largely determine the switching frequency and pulse widths of the controller which in turn influence SCR switching losses, SCR dynamic ratings, and system efficiency and cost. The minimum inductance, L_{min}, determines the amount of residual magnetic energy that must be continually intercharged among load, capacitor, and source. The permeance coefficients that make up the inductance functions have been analyzed by magnetic circuit theory, by electrical analog techniques, and by iterative computer studies of magnetic field distributions, and are well defined.

A major laboratory effort has also been made to evaluate the magnetic characteristics of the unique magnetic circuits used in various disc motor configurations. At present, most magnetic members are constructed from slabs of laminated sheet steel. A special bonding technique had to be developed to give the laminated slab sufficient strength to withstand the subsequent cutting and grinding processes required to form the individual magnetic sectors used in the rotor and stator. There was much concern that both the bonding and machining processes might severely deteriorate many of the magnetic properties of the sectors. It has been shown[25] that surface grinding greatly deteriorates the "surface permeability" of mumetal down to a depth of about 100 μm, for example. Since this is a rather small distance compared to the dimensions of the motor sectors, it was felt that the bulk properties of the sectors should not be too badly affected. Also, saturation density and core loss are more important parameters than permeability in motor applications. Our measurements tend to confirm this supposition for both silicon steels and the 48% nickel alloy, Monimax, having lamination thickness of .004 in. However, the properties of a 27% cobalt alloy were deteriorated by over 50% due to the bonding-machining process. Techniques have since been evolved to use standard laminations and bonding techniques for the stator sectors, but the above cutting and grinding method is still used for the rotor sectors. It is hoped that these can eventually be replaced by powdered iron sectors.

The disc motor must be synchronized with the pulse frequency supplied by the controller by means of a position feedback circuit. This circuit consists of a rotor position-voltage transducer and appropriate logic for modifying the position voltage signal for use in the controlling of the power SCR's in the controller. An optical sensing system[26] has been used on the prototype motors and can be seen on the motor in Figure 9. Inductive, reluctance, and capacitive transducers have also been used but are generally more susceptible to electromagnetic noise resulting from SCR switching.

Iron loss and efficiency measurements have been performed during excitation by discontinuous, non-

TABLE I — Selected Test Data on Disc Motor (excluding controller losses)

Speed (r/min)	I_{rms} (A)	V_{rms} (V)	Load Power (W)	Load Torque (N.m)	Eff. %
500	205	42	416	7.9	46
1000	190	60	760	7.2	65
1200	164	68	750	5.9	79
1500	182	90	1070	6.8	75
1930	126	180	670	3.3	70
2500	162	110	1430	5.4	83
4050	136	195	1730	4.2	79
4300	142	200	1960	4.4	77
4700	80	98	780	1.6	75
4840	148	213	2450	4.8	80
5000	184	200	3750	7.1	86
5000	243	240	5950	11.3	87
5000	287	255	6200	11.8	88
5400	35	47	128	0.25	53

sinusoidal wave forms illustrated in Fig. 10 at power factors down to .02 by means of an electronic wattmeter.[27] This wattmeter was also valuable in the development of the disc motor excitation coils. These coils tend to be constructed of relatively few turns and of large cross section conductor. The disc motor current can contain many high frequency components due to the discontinuities and steeply rising wave fronts even though the actual pulse frequency is not unusually high. In order to minimize skin effect losses in the coils, several coil configurations were developed, including bundled conductors, aluminum foil coils, copper strip coils, and Litz wire. The Litz wire coils were found superior, and the ac/dc resistance ratio remained less than 1.05 for pulse widths down to 165 μs with a repetition rate of 1600 pps and a wave form similar to those in Figure 10-a.

In general, both of the prototype configurations shown in Figs. (8) and (9) have performed well under the usual laboratory evaluation techniques, which include conditions of temperature extremes, misalignment with the load dynamometer, sudden failure of control elements, mishandling, etc. The principal limitation in the present design has been due to the use of laminated phenolic material in the rotor discs. Discs of large diameter, such as those used in the motor of Figure 9, do not have sufficient flexural strength to withstand the high axial magnetic forces occurring during current pulses of high peaks.

Motor Performance

A number of complete variable-speed drive systems using the components described have been designed, assembled, and laboratory tested with dynamometer loads. Both battery and dc generator power supplies have been used in the tests. Three motor and five controller configurations have been tested under varying conditions of speed and load. The data given in Table I is for the single phase machine shown in Figure 8 which has 17-poles, 2 rotor discs, and runs at 5000 r/min and weighs 43 lbs. including base plate.

It is seen that the efficiency of the disc motor is relatively high over a wide range of speed and torque levels. This is in part due to the very low windage and bearing loss of this axial air gap design. The effects of magnetic saturation and the overlapping of the current pulses beyond the region of increasing inductance is apparent in the above measurements at high current levels. The inconsistancies in the efficiencies shown are due to: (a) the effects of different types of current wave shapes resulting from the use of several different controller configurations in the above tests; (b) although the motor configuration was the same during all tests reported above, different types of steel in both stator and rotor sectors and different bearing configurations were being evaluated throughout the tests; (c) the electronic wattmeter may have an error within ±3 % at these low power factors.[27]

Conclusions

A variable speed drive system has been developed which is particularly suitable for battery-powered traction applications. The core of this system is a brushless, axial air-gap reluctance machine controlled by an SCR capable of full motoring and regenerative operation over a wide speed range without the need of auxiliary voltage regulatory circuitry.

The motor is of simple construction, capable of long operation in an automotive environment. It has no brushes, slip rings, coil slots, or moving coils, and is excited by a simple solenoidal winding. The magnetic circuit consists of a series of alternate rotating and stationary discs amenable to mass production techniques. The number of magnetic sectors on these discs can be varied in order to make use of the maximum dynamic characteristics available in modern power SCR's. A Motor specific weight (lbs./h.p.) comparable to other brushless, variable speed motors can be achieved without resorting to unusually high rotor speed designs. The motor efficiency is higher over a wider speed range than in most brushless variable speed motors.

Many different component configurations and several complete systems have been designed, assembled, and laboratory tested. A few of the lab tests are included in this paper. Several of these configurations have given excellent performance characteristics and have survived many hours of normal and abnormal operating conditions. This evaluation indicates that this reluctance motor system does have the potential for low-cost, good efficiency, and highly reliable performance in many vehicular applications.

APPENDIX A - NOMENCLATURE

E_b — source voltage; volts

i, I — current; amperes

L — inductance, henries

n — rotor speed; revolutions/second

p — the number of magnetic sectors per disc

P_d — motor electromagnetic developed power; watts

R_c — motor coil resistance; ohms

t — the independent variable time; seconds

T — motor electromagnetic developed torque; newton-meters

T_o — electrical pulse width; seconds

T_p — mechanical period, or the time period during which motor inductance is increasing; seconds

T_λ — pole period, or the time period of a repeatable disc section; seconds

Wm — magnetic energy; joules

X — motor equivalent load resistance; ohms

θ — angle between rotor and stator references; radians

λ — pole pitch = $2\Pi/p$ radians

ψ — flux linkages; webers

REFERENCES

(1) D.A. Bradley et al., "Adjustable-Frequency Inverters and Their Application to Variable Speed Drives," Proc. IEE (London), vol. III, no. 11, pp. 1833-1846, November 1964.

(2) H. Stemmler, "The Use of Static Converters to Vary the Speed of Three-Phase Drives Without Losses," Brown Boveri Review, vol. 54, no. 5/6, pp. 217-232, May/June 1967.

(3) K. Hermann, "Development of Inverters with Forced Commutation for ac Motor Speed Control Up to the Megawatt Range," Conference Record, 1967 International Electronics Conference, Toronto.

(4) J. J. Pollack, "Advanced Pulsewidth Modulated Inverter Techniques", IEEE Transactions on Industry and General Applications, vol. 1A-8, no. 2, pp. 145-159, March/April 1972.

(5) P. J. Lawrenson, "Theory & Performance of Polyphase Reluctance Machines", Proceedings of the IEE, August, 1964.

(6) J. T. Salilii, P. I. Agarwel, G. J. Spix, "Induction Motor Control Scheme for Battery-Powered Electric Car (GM-Electrovair I)", IEEE Trans. on Industry and General Applications, vol. IGA-3, no. 5, pp. 463-469, Sept./Oct., 1967

(7) R. W. Johnston, "Modulating Inverter System for Variable Speed Induction Motor Drive (GM Electrovair II), IEEE Trans. on Power Apparatus and Systems, vol. PAS-88, no. 2, pp. 81-85, Feb., 1969.

(8) Y. Miyake, "Electric Vehicles In Japan," SAE Preprint 690072, Jan. 13-17, 1969.

(9) S. Grimmin, "Japanese Spur Electric Car Efforts, Automatic News, March 20, 1972.

(10) S. B. Dewan, D. L. Duff, "Optimum Design of an Input-Commutated Inverter for ac Motor Control," IEEE Trans. on Industry and General Applications, vol. IGA-5, no. 6, pp. 699-705, Nov./Dec., 1969.

(11) S. Miyairi, Y. Tuneliero, "The Characteristics of Commutatorless Motor With SCR Inverter," Proceedings, International Conference on Magnetics, Washington D.C., April, 1968, 11-4.

(12) M. Yamaguchi, H. Fujiwara, "Analysis of Commutatorless Motors," Electrical Engineering in Japan, vol. 89, no. 9, pp. 35-43, 1969.

(13) T. Tsuchiya et al., "Basic Characteristics of Series Commutatorless Motors," Electrical Engineering in Japan, vol. 89, no. 9, pp. 71-76 (1969).

(14) N. Sato, V. V. Semenor. "Adjustable Speed Drive with a Brushless dc Motor," IEEE Transactions on Industry and General Applications, vol. IGA-7, no. 4, pp. 539-543, July/August 1971.

(15) W. W. Yates, R. E. Shamfer, "Brushless dc Torque Motor," NASA Report No. WAED 64.5SE, Contract No. NAS5-3934, Sept. 1964.

(16) W. Radziwill, "A Highly Efficient Small Brushless dc Motor," Philips Technical Review, vol. 30, p. 7-12, (1969/70).

(17) Sidney A. Davis, "Stepper Motors," Electromechanical Design, July 1964, pp. 109-119.

(18) T. R. Fredriksen, "Direct Digital Processor Control of Stepping Motors," IBM Journal, March 1967, pp. 179-188.

(19) L. E. Unnewehr, "Theory and Practice of Thyristor Series Commutation", Ford Scientific Research Staff Technical Report, SR72-72, June, 1972

(20) H. Ranseen, "Variable Reluctance Stepper Motor", U.S. Patent #3,005,118,(1959).

(21) P. French, A. H. Williams, "A New Electric Propulsion Motor," Proceedings AIAA Third Propulsion Joint Specialist Conference, Wash., D.C., July 1967.

(22) W. K. Heintz and L. E. Unnewehr, "Disc-Type Variable Reluctance Rotating Machine", U.S. Patent #3,700,943; Nov., 1972.

(23) D. C. White and H. H. Woodson; "Electromechanical Energy Conversion" (book), John Wiley and Sons, 1959.

(24) Jacques Bajer, "Machines Electriques Jarret", La Technique Moderne, February, 1967.

(25) I. Preece and R. Thomas; "The Effects of Various Stresses on Mumetal", IEEE Transactions on Magnetics, September, 1971.

(26) D. R. Hamburg; U.S. Patent #3,673,476; June 27, 1972.

(27) D. R. Hamburg and L. E. Unnewehr, "An Electronic Wattmeter for Nonsinusoidal Low Power Factor Power Measurement" IEEE Transactions on Magnetics, September, 1971.

Discussion

S. Noodleman (Inland Motor Corporation, Radford, Virginia 24141): The paper "An Axial Air-Gap Reluctance Motor for Variable Speed Applications" outlines a rather unique motor construction operated

Manuscript received August 1, 1973.

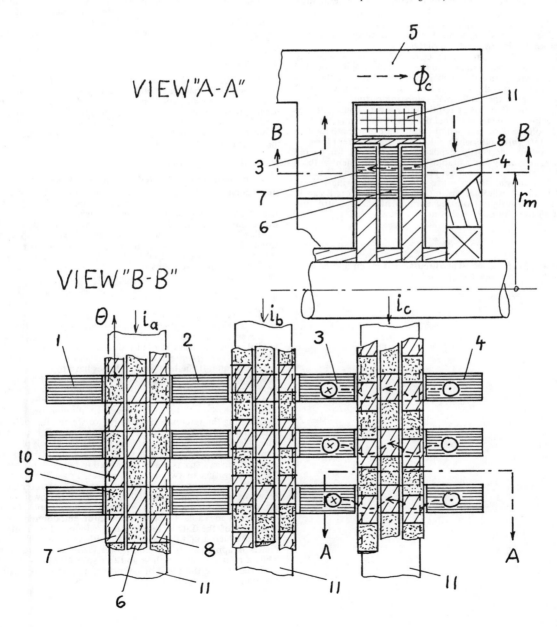

from a relatively simple drive system.

In reviewing this paper it would be helpful if there were an illustration showing how the magnetic flux is distributed between the rotating member and the stationary member and a clearer illustration of the configuration of the magnet coils used in generating this flux.

It is not clear exactly what is meant by a single phase machine since it would be expected that at least two windings properly displaced would be necessary in order to provide a field to generate the starting torque.

While the main feature of this paper appears to be the fact that this type reluctance motor permits the use of unidirectional SCR's for energizing each coil thus simplifying the electronics there is some question whether the motor performance for the motor illustrated is optimum in regard to the amount of magnetic material and copper required. If weight is the prime consideration the 43 lbs. indicated is not an optimum figure for iron, copper and appropriate permanent magnet material for a motor developing a torque of approximately 8 ft. lbs. The volume of the iron, copper and magnet material generally determine the torque capability of the motor while the ability to operate to the high speeds depend on the electronics to appropriately control the switching of the required currents and of the rotor to withstand the centrifugal forces generated at these speeds. It would be interesting to have information regarding the peak torque capability for the 17 pole, 2 rotor disc motor described and it would seem that the peak torques would be limited by saturation of the iron and internal losses.

The use of the relatively large number of poles appears necessary in order to obtain acceptable torques for the size and weight indicated but its use requires proportionally higher switching rates of the electronics for these higher speeds. This results in high frequency current that must be forced into the coils to develop the output torque and these generate the skin effect losses requiring Litz wire coils. This has to be balanced against more conventional designs using permanent magnets with 2 or 4 pole construction which would not require these high switching rates or Litz wire material to hold down eddy current losses.

Each type system necessarily needs to be analyzed for overall cost, performance, reliability and suitability to the requirement with all factors weighed to provide for the most practical drive system. As costs and reliability of electronic drivers are continually improving a good case can be made for considering the most efficient motor design in terms of material and weight (rotating permanent magnet structure with wound stator) and a suitable electronic driver.

L. E. Unnewehr and **W. H. Koch:** The discussion is much appreciated. Answers to four items questioned are given below.

Manuscript received October 9, 1973.

1. Illustration of Flux Path

A three-phase machine, abbreviated "3PM" (Fig. 7 of paper) is shown in two crossectional views. View A-A presents a plane section of phase c in the axial-radial plane. View B-B presents a cylindrical section of all three phases made at a radius r_m and shown unrolled for 3 of p identical pole groups. r_m = mean pole face radius. The electromagnetic components are: 1, 2, 3, 4 laminated steel stator pole "fingers"; 5 Phase c lam. steel stator yoke; 6 stationary disc containing p pairs of non-magnetic and lam. magn. sectors 9, 10; 7, 8 Rotor discs containing identical sectors as 6; 11 annular phase exciting winding. The rotor discs of phases a, b and c are shown in positions of, respectively, minimum, decreasing, and increasing phase winding inductance. The flux path of phase c, which is shown in a position for developing motoring torque, is indicated. In accordance with terms introduced in Fig. 2 of the paper, the inductance function $L_a(t)$ of phase a is characterized by $(T_m - T_1)/T_\lambda = (T_\lambda - T_m)/T_\lambda = 0.475$; $T_1/T_\lambda = 0.05$; $L_m = 125 \ \mu H$, and $L_o = 80 \ \mu H$. The period length T_λ corresponds to a rotor displacement $\theta_\lambda = 2\pi/p$ by one repeatable section, where p is the number of magn. rotor sectors. For the single phase section, abbreviated "1PS", (Fig. 8), p = 17. The minimum and maximum of $L_a(t)$ occurs at $t/T_\lambda = 0.025$ and 0.525, respectively. The rotor discs of phases b and c are displaced against those of phase a by $\pm 1/3 \ \theta_\lambda$ respectively. This causes $L_b(t)$ and $L_c(t)$ to have the same shape as $L_a(t)$ and to be shifted versus $L_a(t)$ by $\pm 1/3 \ T_\lambda$ respectively.

2. Motor Phase Number

In Fig. 7 a 3PM is illustrated and in Fig. 8 an experimental 1PS with p = 17 and having a total weight of 43 lbs. is shown. The 3PM is able to start in any rotor position and develops continuous running torque, if the phase currents i_a, i_b and i_c are each controlled to be nonzero during the rising inductance interval of magnetic phase sections a, b and c. The 1PS is able to start and develop motoring torque only during the interval $T_1 < t < T_m$.

3. Saturation and Torque

Torque is indeed limited by pole and yoke saturation, depending upon current and rotor position. A lower torque limit can be associated with a chosen limit on winding power losses. Taking into account 8 winding terms, and 4 air gaps, magnetically series connected, each air gap equal to 10^{-3}m, defining the saturation flux density for the steel used as $B_s = 1.4$ Tesla, the 1PS at small values of t/T_λ exhibits saturation at $i_s \simeq 550$ A. At $t = T_m$ the machine saturates for $i_s \simeq 460$ A. Now the maximum torque can be calculated form Eq. 2 for any assumed current waveform, after substituting the values of T_1, T_m, L_o and L_m given above.

4. Power, Utilization and Losses

A permanent magnet ("PM") synchr. motor may for comparable speed- and torque characteristics possibly be lighter than a disc motor. However, three observations on such a comparison should be made. (a) PM motors cause PM field induced stator iron losses at medium to maximum speed, which do not decrease along with any reduction of output torque. If the total energy losses have to be limited over a motor operating cycle containing significant medium to high speed running at only low output torque (this is the case with battery powered vehicles operating in representative driving cycles), high performance PM motors would probably not be competitive with disc motors in efficiency. (b_1) Considering the reference base for motor weight, it should be noted that the weight of 43 lbs of the 1PS contains only 22 lbs of active material (mag. lamin. and conductor copper). (b_2) A polyphase machine assembled by axially stacking several 1PS could utilize all stator rings of magn. "fingers" (parts 2 and 3) located in between phases twice, that is for carrying flux of two adjacent phases. For a 3PM this leads to a saving of $\simeq 23$ percent in active material weight with reference to the combined weight of 3 individual 1PSs. (c) Winding power losses are among the major factors limiting the power to weight ratio. The performance data given in Table 1 apply to a motor embodiment with a low fill factor k_f of the Litz wire excitation winding. For a different coil design, in which a k_f value associated with windings of conventional machines is approached, winding power losses are approximately halved. Thus a relatively high stall — and low speed torque — to weight ratio is achieved before the onset of saturation, at a power loss level comparing favorably with that attainable in PM motors.

Characteristics of Saturable Stepper and Reluctance Motors

J.V. Byrne and J.G. Lacy

IEE Conference Publication No. 136, Small Electrical Machines, London, March 1976, pp. 93-96

INTRODUCTION

Stepping type reluctance motors, now applied as positioning devices, were as "electromagnetic engines" the electric motors of the mid-nineteenth century. In that role they gave way to the conventional machines of today. A review of energy conversion processes in saturable versions of such machines indicates that, with the use of semi-conductor switches, they have interesting possibilities in the area of variable-speed drives.

The special characteristics of machines designed to exploit saturation are outlined. A 5kW d.c. traction drive is described.

ENERGY CONVERSION IN SATURABLE VARIABLE-RELUCTANCE DEVICES

Idealized Cases

It is helpful to start with non-realizable idealizations including "ideally-square" B-H material and the neglect of fringing flux.

Force and torque. For devices with negligible airgap Byrne (1) has shown that the torque is

$$T = \Im B_{sat} \frac{dA_c}{d\theta},\qquad(1)$$

where $\Im$ is the driving m.m.f., B_{sat} is the saturation flux-density and A_c is the constriction cross-section of the ferromagnetic path. By writing eqn.1 in the forms

$$T = \Im \frac{\partial \phi}{\partial \theta} = i \frac{\partial \psi}{\partial \theta},\qquad(2)$$

it can be seen that the torque is, for the same values of driving m.m.f. and flux-change with angle, twice that obtaining in a magnetically linear device, and is proportional to the first power of the current.

Kelvin's "law" and its evasion. Karapetoff (2) elevated Kelvin's finding to the status of a "law": at constant currents, only half the electrical input to a linear device is available as mechanical work; the other half is stored in the increasing inductance. This loss of output is avoided in the ideal saturable case, there being zero field energy density in soft square-loop material. All, instead of half, of the electrical input is converted to mechanical form.

Equivalent circuit. The equivalent circuit in the absence of motion is that of an ideal saturable reactor, viz., an open or closed switch. With motion, the varying constriction in the flux path acts as a flux-sink, accepting for all non-zero m.m.f. the flux $B_{sat} A_c$. The electric equivalent circuit is a voltage source/sink with

$$e = \frac{d\psi_{sat}}{d\theta} \omega_{mech} = B_{sat} \frac{dA_c}{d\theta} \omega_{mech},\qquad(3)$$

where ω is the angular velocity.

If the constriction section A_c increases linearly with motion, eqns.(3) and (1) indicate "back-e.m.f." proportional to speed only, and torque proportional to current, exactly as for a d.c. machine operated at fixed excitation.

Real Devices

The magnetization M, rather than the flux-density B, tends towards a saturated value. Fringing fluxes are unavoidable. Nevertheless the main features of the ideal case are preserved.

Force on overlapping rectangular poles. The tangential force for poles of infinite extent has been shown by Byrne and O'Connor (3) to lie in the range 96% to 100% of the value

$$F = \Im B_s z \qquad(4)$$

for all overlap. B_s is the flux-density at the knee of the square M-H characteristic, and z is the transverse depth.

Measurements with partially-overlapped C-cores of silicon iron (1) fitted eqn.(4) with B_s 1.8. With an airgap, the gap m.m.f. is only half as effective in force production as the part "dropped" across saturated iron.

Magnetization characteristics. Due primarily to unwanted fringing flux, the ψ-i curves are inclined rather than sharply saturating. Fig.1 is for an early version of the rotating machine shown in Fig.2. One is concerned with the total area of trajectory in the ψ-i plane representing mechanical work, as shown by Woodson and Melcher (4)

$$W_{mech}/ \text{cycle} = \text{trajectory area},\qquad(5)$$

and with torque uniformity over the working stroke, requiring, for equal motion increments, equal areas between successive magnetization curves. These areas are far from equal in Fig.1, resulting in the torque variation over the working stroke shown in Fig.5(a).

Unfortunately, the possible trajectory area does not increase proportionally with current, because of the way the magnetization curves converge at high current.

The characteristics of Fig.1 represent, approximately, a flux sink paralleled with a fixed reluctance. The corresponding electric equivalent circuit is a source of "speed voltage"

$$e = \frac{\partial \psi}{\partial \theta} \omega_{mech},\qquad(6)$$

in series with constant inductance

$$L = \frac{\partial \psi}{\partial i}\qquad(7)$$

Dynamic operation. The closed trajectory a,b,c of Fig.1 has three distinct stages, with character as shown in brackets:

rate dependent on the difference between the source and speed voltages, the incremental winding inductance and the circuit resistance. At high rotational speeds the slow growth of current due to the small difference in driving voltage may be compensated for by advancing the gating of T_1 and T_4 relative to the rotor position. When the winding current (magnetically sensed at the supply bus 1) reaches the desired value T_5 is gated thus commutating T_1. The winding current is now supplied through T_5 and C_1 re-charging C_1 to its original polarity. T_5 now self-commutates as D_1 comes into conduction the decaying winding current now being driven from the energy stored in the winding inductance. When the current now sensed at "freewheel" busbar 4 falls to 66% of its original peak value T_1, T_4 and T_7 are gated and the process repeats. When the rotor position signal indicates the end of the working stroke the logic first checks the state of T_1. If conducting it is first commutated by gating T_5 and when D_1 conducts T_2 is gated together with T_3 and T_7 as winding BB' commences its working stroke. The gating of T_2 also serves to commutate T_4 thus leaving winding AA' reverse connected across the supply through D_1 and D_2 when C_2 has re-charged. Unconverted energy is thus returned to the supply. This energy recuperation process can be seen as winding voltage reversals on the oscillograms the winding current of course being uni-directional. The system has an excellent part load efficiency characteristic mainly due to the absence of the parasitic loss paths normally associated with the inductive energy storage required for chopper operation. The full-load efficiency is 85%.

REFERENCES

1. Byrne, J.V., "Tangential forces in overlapped pole geometries incorporating ideally saturable material", IEEE Transactions on Magnetics, MAG 8, 1972, 1, 2-9.

2. Karapetoff, V., "Mechanical Forces between electric currents and saturated magnetic fields", AIEE Trans., 1927, 46, 563-569.

3. Byrne, J.V. and O'Connor, W.J. "Saturable overlapping rectangular poles: towards a functional relationship between force, overlap distance, m.m.f. and saturation polarization", INTERMAG 1975, Imperial College, London.

4. Byrne, J.V., and Lacy, J.G.,"Electrodynamic system comprising a variable reluctance machine", British Patent No. 1321110.

5. Byrne, J.V., and Lacy, J.G., "Compatible controller-motor system for battery-electric vehicle", Proc.IEE, 1970, 117, 369-376.

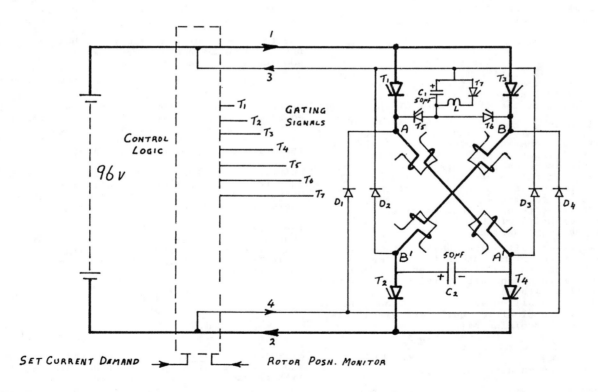

Figure 4 Chopper controlled traction system power circuit

292

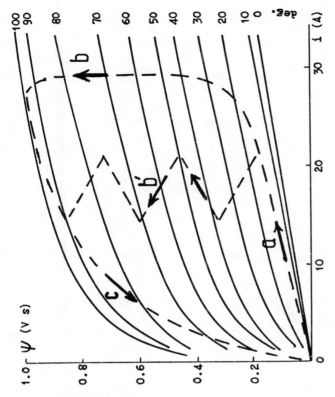

Figure 1 Machine magnetization characteristic before rotor refinement. Sample trajectories shown dotted.

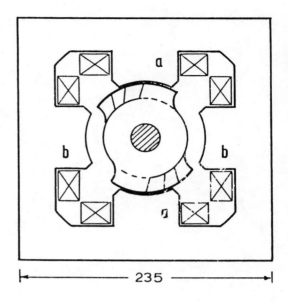

Figure 2 6kW saturable traction motor arrangement.

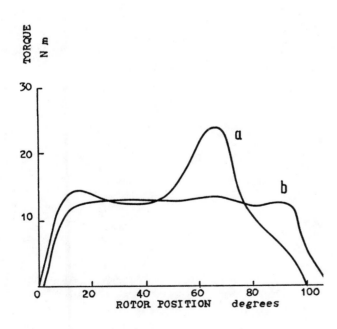

Figure 3 Torque variation in working stroke
(a) before rotor refinement
(b) after rotor refinement

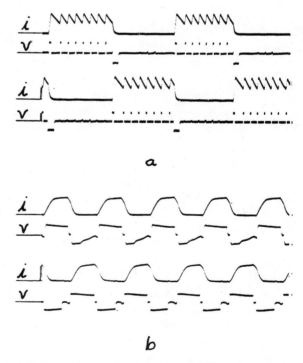

Figure 5 Current and voltage waveshapes
(a) low speed, showing chopping
(b) high speed, showing commutation only

Section 2
Converters and Controls

2.1 - Converters

Inverter Drive for Doubly-Salient Reluctance Motor: Its Fundamental Behaviour, Linear Analysis and Cost Implications

W.F. Ray and R.M. Davis

IEE Electric Power Applications, Vol. 2, No. 6, December 1979, pp. 185-193

Abstract: A variable-speed drive using a doubly salient variable reluctance motor, because of its ability to operate efficiently from unidirectional winding currents, enables the number of main switching devices in the inverter to be halved, yielding a very economical brushless drive. The drive is analysed using a linear model for the motor, yielding phase-current waveforms which enable an economical design for the thyristor inverter and provide guidance for a motor design which will optimise the performance/cost ratio of the drive as a whole. A control strategy for firing and turning off the inverter thyristors at appropriate rotor angles for a traction application also emerges from the analysis. Experimental waveforms exhibit shapes similar to those predicted, and the results of this analysis have been of great value in designing such a drive for a battery vehicle.

List of principal symbols

Normalised equivalents are printed in bold type, where appropriate.

i	=	motor phase current
i_p	=	current at commutation
i_{rms}	=	r.m.s. value of torque-producing current
I	=	unit current for normalisation
θ	=	rotor angle
$\Delta\theta$	=	angular duration of rising inductance = unit angle for normalisation
θ_0	=	thyristor firing angle
θ_p	=	thyristor commutation angle
θ_q	=	current extinction angle
θ_{cy}	=	angular repetition period
L	=	motor phase inductance
ΔL	=	$L_{max} - L_{min}$
b	=	$L_{min}/\Delta L$
τ	=	instantaneous torque
τ_m	=	mean torque per phase
P	=	mean power per phase
ω	=	rotor angular velocity
T	=	time duration of rising inductance
n	=	main/auxiliary winding-turns ratio
a	=	(angular duration of rising inductance)/(angular duration of falling inductance)

1 Introduction

The last two decades have seen the development of cheaper, faster and more powerful thyristors and a consequent growth of thyristor variable-speed drives, using both d.c. and a.c. motors. Inverter drives performing as well as Ward Leonard systems were claimed in 1964,[1] and since then publications on a.c. drives, using various circuits and modulation schemes for voltage control, have proliferated. Inverter-fed induction motors for rail traction have recently appeared.[2]

Paper T481P, first received 23rd August and in revised form 4th October 1979

Mr Ray and Mr Davis are with the Department of Electrical Engineering, the University of Nottingham, Nottingham NG7 2RD, England

An a.c. drive is economically viable compared to a rectifier d.c. drive only if the savings in capital and maintenance by replacing the d.c. motor with an a.c. motor exceed the cost difference between the variable-output-frequency power convertor and the controlled rectifier. Although a.c. drives using induction motors are becoming competitive in certain specialised applications, they have not yet been generally accepted.

The high convertor cost for the induction-motor drive arises from the double power conversion, a.c. → d.c. → a.c. (although with p.w.m. inverter schemes the first can be uncontrolled rectification), and from the large number of thyristors required: a p.w.m. inverter often requires 12 thyristors for a 3-phase bridge.[2,3] Although it is natural that the development of a.c. drives should have started with induction motors, their requirement of bidirectional phase currents does not readily match the unidirectional current capability of the thyristor. It is questionable whether the induction motor is the ideal partner for the thyristor in a cheap thyristor drive, and a system design approach, freed from the constraint of pseudosinusoidal alternating currents, could yield a motor-convertor system of greater simplicity and lower cost.

In 1974, Chloride Technical Ltd. asked Nottingham University to study the feasibility of a.c. drives for battery vehicles. From this study, it was found that an inverter feeding a doubly salient reluctance motor appeared more promising than an induction motor system. (The doubly salient reluctance motor should not be confused with the smooth-stator reluctance motor designed for sinewaves and requiring bidirectional currents.) The advantage of the doubly salient reluctance motor, in common with stepper motors generally, is its ability to operate with unidirectional current pulses, thus halving the number of main switching devices in the inverter.

Low-power stepper motors with a small step angle for positioning drives have received considerable attention and some high-power low-speed applications have been published.[4,5] However, few publications on high-power doubly-salient drives have appeared and these mainly concern the motor. Unnewehr and Koch[6,7] of Ford (USA) in 1973 gave details of a multistack disc-type motor and thyristor power converter which achieved an efficiency of 85% at 6 kW. However, both the motor and inverter were disagreeably complex; the latter, using 3 or more thyristors per phase, was presumably designed more for experimental flexibility than for low cost. In 1976 Byrne[8] of Dublin University gave details of the simulation and performance of a single-stack 2-phase motor with unidirectional rotation, following earlier patent specifications. However, the 2-phase inverter[9] used 7 thyristors and the main emphasis of Byrne's research was on motor design. Koch[10] also considered a single-stack 3-phase motor but without the merits of the current waveforms shown in this paper.

Work on switched reluctance motors at Leeds University has been in progress since 1972, and as a result of proposals arising from the above feasibility study, a combined programme of work at the two universities for battery-vehicle applications was sponsored by Chloride Technical Ltd. in 1975. The aims were to study the proposed motor-inverter system and to design and build a prototype which would prove the acceptabilities of the new system and whether it could compete with the well established chopper and separately excited d.c. motor.

2 Objectives of the paper

When designing variable-speed drives, the motor is generally of long-standing development and dictates the type of current or voltage waveforms which the inverter must supply. However, for a new system in which the development of both motor and inverter are taking place simultaneously, such as is described in this paper, it is important that the waveforms required by the motor should not demand excessive inverter-component ratings, especially as the inverter represents the majority of the cost. A motor designed for minimum cost will not necessarily result in the cheapest system for which some compromise between the motor requirements and inverter ratings must be made. It is essential that a system design approach is applied.

The object of this paper is, by using a linear model of the motor, to develop simplified expressions for the current waveforms so that the inverter-component ratings can be determined with sufficient accuracy, operating conditions and motor parameters may be selected which result in a good compromise of motor performance and inverter-component ratings, and a simple control system can be envisaged for a battery-vehicle application.

3 Basic system operation

Fig. 1 illustrates a 3-phase single-stack motor with three pairs of stator poles and four rotor poles, the simplest possible self-starting 3-phase arrangement giving zero mutual inductance between phase windings and minimum switching frequency for a given speed. Torque is developed according to the well-known stepper-motor principle that, when a phase is excited, the rotor is pulled in a direction of increasing flux linkage (or inductance) such that the adjacent rotor pole is pulled into alignment with the excited stator pole.

Fig. 2a shows a developed view of the motor for which the stator and rotor pole arcs have been arbitrarily taken as 30° and 45°, respectively. Fig. 2b shows the idealised (current-independent) variation of inductance $L(\theta)$ with rotor position θ for a phase 1, and Fig. 2c shows the corresponding variation of torque $\tau(\theta)$ for constant current excitation, torque being proportional to $dL(\theta)/d\theta$ and independent of current direction.

A mean motoring torque is produced by supplying each phase with pulses of current timed so as to coincide predominantly with the angular period for which $dL/d\theta$ is positive. If the current pulse overlaps the period for which $dL/d\theta$ is negative, then some counter productive torque is produced, but this may be unavoidable at high speeds

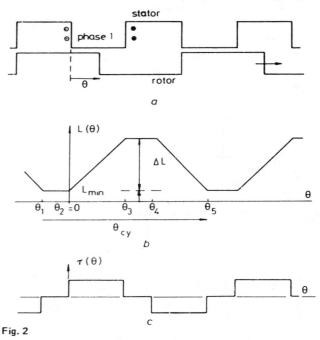

Fig. 2

a Developed view of the motor
b Variation of phase 1 inductance $L(\theta)$ with rotor position θ
c Variation of developed torque $\tau(\theta)$ with θ for a constant current in phase 1

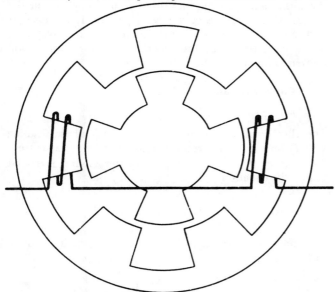

Fig. 1 *3-phase single-stack v.r. motor, showing the phase 1 winding*

owing to the increased angular duration of the pulse.

Fig. 3 shows the ideal (but impractical) current waveform for which only useful torque-producing current is allowed to flow. It implies that it is advantageous to build the current up and force it down as rapidly as possible at the appropriate rotor angles. Fig. 4 shows a circuit arrangement for a single phase, achieving this by utilising closely coupled windings on the motor, which, although incurring a slight motor penalty, gives considerable inverter advantage. Current is built up under the source voltage V_s by firing thyristor T_1 and, on turning off T_1 by a commutation such that the current pulses coincide predominantly with the angular period for which $dL/d\theta$ is negative (see Fig. 2), then negative torque is produced and the system regenerates, the mean current flowing back to the source through the diodes and auxiliary windings predominating over the current in the main windings and thyristors. Regeneration is achieved without any additional inverter components, which in many applications is a considerable system advantage.

Appropriate snubbing networks must be used, and leakage inductances and circuit losses will slightly modify the voltage waveforms but do not change the basic very simple behaviour of the inverter. During commutation of a given phase, short-duration current pulses will be introduced in the other phases which only slightly affect the mean torque produced. These could be eliminated by replacing the diodes $D_{4,5,6}$ by thyristors and eliminating T_4.[12]

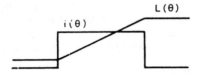

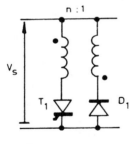

Fig. 3 *'Ideal' current pulse*

Fig. 4 *Switching circuit for one phase*

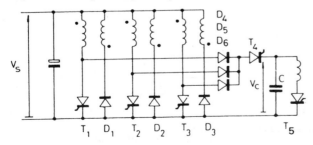

Fig. 5 *3-phase inverter circuit*

298

circuit (not shown), current transfers to the auxiliary winding in proportion to the turns ratio n and returns to the supply through D_1. The main winding sees a voltage $-n\,V_s$, which collapses the field, returning the energy to the supply, although torque is still produced in the same direction since a current in D_1 may be replaced by an equivalent current in T_1 from the viewpoint of flux linkage.

The winding arrangement has the advantage that no split supply or secondary voltage sink or storage for the return of energy is required. The inclusion of a second thyristor in parallel with the auxiliary winding to provide a free-wheeling path for the current, as used by Byrne,[9] although offering some advantage at low speeds where 'chopping' is required (see Section 5), is unnecessary for the inverter operation and merely adds to the cost.

With each motor phase driven by the arrangement of Fig. 4, there are a number of alternative commutation circuit arrangements which may be used to turn off the main thyristors, the simplest of which is shown in Fig. 5.[12] Assuming for simplicity that the windings have perfect mutual coupling, let thyristor T_1 be conducting and the commutation capacitor voltage be $V_c = -(1+n)\,V_s$. Firing T_4 will turn off T_1 (and T_2, T_3 if either of them are also conducting) and charge C to $V_c = +(1+n)\,V_s$ at which point D_1 becomes forward biased, clamps V_c at $+(1+n)\,V_s$ and transfers the phase 1 current to the auxiliary winding. After allowing sufficient time for T_4 to turn off, V_c may be resonantly reversed to $V_c = -(1+n)\,V_s$ by firing T_5.

The thyristors are fired by a controller at rotor angles according to a predetermined law, necessitating a rotor-position transducer. If the firing is delayed (or advanced)

There is a significant a.c. component as current is built up in the main windings and returned via the auxiliary windings, and since it is undesirable for the source to supply this component, electrolytic capacitance is required, as shown in Fig. 5. The r.m.s. current taken by this capacitance will depend on the degree of overlap and hence the positioning and duration of the current pulses in the phases. Once the source voltage V_s is fixed, the inverter costs are solely dependent on the motor current waveforms, and a simplified analytical technique for their estimation is given in the next section.

4 Current waveform analysis

4.1 Nonlinear equations

The flux linkage associated with a motor phase winding $\psi(\theta, i)$ depends on the rotor position θ and winding current i, and hence the voltage v applied to the winding is related to i by the equation

$$v = Ri + \frac{d\psi(\theta, i)}{dt} \qquad (1)$$

It is convenient to work in terms of the winding inductance $L(\theta, i)$ rather than flux linkage; putting $\psi(\theta, i) = iL(\theta, i)$ gives

$$v = Ri + \left\{ L(\theta, i) + i\frac{\partial L(\theta, i)}{\partial i} \right\} \frac{di}{dt} + i\frac{\partial L(\theta, i)}{\partial \theta}\omega \qquad (2)$$

where $\omega = d\theta/dt$.

The instantaneous gross torque $\tau(\theta, i)$ transmitted to the rotor is given by

$$\tau(\theta, i) = \frac{\partial W'(\theta, i)}{\partial\theta} \qquad (3)$$

where $W'(\theta, i)$ is the winding coenergy for a given rotor angle defined by

$$W'(\theta, i) = \int_0^i iL(\theta, i)\, di\,|_{\theta = constant}$$

In this application, the applied voltage $v(\theta)$ is dictated by the inverter under the control of the position sensor, and has a specified rectangular waveform given by

$$v(\theta) = V_s \qquad (\theta_0 < \theta < \theta_p)$$
$$v(\theta) = -nV_s \qquad (\theta_p < \theta < \theta_q)$$

where n is the winding-turns ratio, θ_0 and θ_p are controlled switching angles and θ_q is the angle at which the current becomes zero, as shown in Fig. 6.

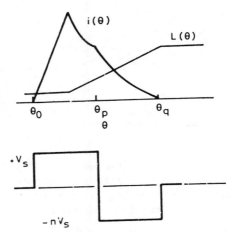

Fig. 6 *Typical current pulse with corresponding phase voltage*

4.2 Computational problems, objectives and assumptions

The inverter ratings required to produce specified mean torques τ_m over the operating-speed range depend on the current waveforms $i(\theta)$. It is desirable therefore to calculate $i(\theta)$ for various combinations of θ_0 and θ_p at different speeds ω to discover which control angle combinations minimise the inverter ratings. The effect of utilising different turns ratios n needs also to be examined.

Calculation of $i(\theta)$ depends on the solution of eqn. 2, which has been examined recently by Stephenson and Corda[11] who also mention several earlier methods. These are all based on a knowledge of $L(\theta, i)$ or $\psi(\theta, i)$ in the form of a table of values, which, if the motor has been previously built, may be obtained by static tests.

However, as previously stated, the motor design and the inverter ratings must be considered together to achieve an overall optimal system. In this case, the geometry of the motor is a further unknown variable to be included in the search, and $L(\theta, i)$ must be computed for a large number of alternative motor geometries. Since it is well known that the computation of $L(\theta, i)$ for a single specified geometry is very difficult and expensive, owing to the nonlinearity of the magnetic circuit, the cost of conducting an optimal search with variable motor geometry is at present prohibitive.

It is therefore expedient to simplify eqn. 2 by making the following assumptions:

(a) L follows the trapezoidal current-independent variation $L(\theta)$ shown in Fig. 2b

(b) The resistive voltage drop Ri is negligible. Ri must be small compared with the effective back e.m.f. $i\omega dL(\theta)/d\theta$ or compared with the reactive drop $L(\theta)\, di/dt$, if the motor efficiency is to be acceptable, and the subsequent error is well within that due to (a).

Further assumptions concerning inverter behaviour are:

(c) Commutation takes place over a negligible rotor angle, during which the motor phase current is constant.

(d) For the coupled windings previously described, the phase current $i(\theta)$ is the sum of the main winding current and $1/n$ times the auxiliary winding current, leakage inductance effects being negligible.

The simplified equations then yield $i(\theta)$ analytically, enabling r.m.s. current and mean torque to be calculated. Although $i(\theta)$ is only an approximation of the actual current waveform, it is sufficiently accurate to indicate the desirable design trends.

To the inverter designer the motor is now represented as an inductance profile $L(\theta)$, and he is able to calculate which profiles are likely to minimise his costs, thereby giving him the opportunity not only to determine the optimal control angles θ_0 and θ_p but also to influence the motor design.

4.3 Simplified equations

The simplified equations are

$$v(\theta) = kV_s = \omega\frac{d}{d\theta}\{iL(\theta)\} \qquad (4)$$

$$\tau(\theta, i) = \tfrac{1}{2} i^2 \frac{dL(\theta)}{d\theta} \qquad (5)$$

where V_s is the source or battery voltage, and $k = +1$ or $-n$ as defined in Section 4.1.

Consider the period of increasing inductance ($\theta_2 < \theta < \theta_3$ in Fig. 2b) during which positive torque is produced by the current i, and let $\theta_2 = 0$ for convenience. Then

$$L(\theta) = \Delta L\left(b + \frac{\theta}{\Delta\theta}\right)$$

and

$$\frac{dL(\theta)}{d\theta} = \frac{\Delta L}{\Delta \theta}$$

where ΔL is the change in inductance over the period $\Delta \theta$ of rising inductance as shown in Fig. 2, and

$$b = \frac{L_{min}}{\Delta L} \qquad (6)$$

4.4 Normalisation

It is convenient to normalise both rotor angle and current to give an analytic solution for normalised current i in terms of normalised angle θ which is independent of speed. This greatly simplifies the calculation of current waveforms, since a set of normalised waveforms may be computed which are (with appropriate scaling) applicable at any speed, as shown below.

Let unit angle be $\Delta \theta$ such that $\theta = \theta/\Delta\theta$ and $L(\theta) = \Delta L(b + \theta)$ for $0 < \theta < 1$. Then

$$kV_s = \omega \frac{\Delta L}{\Delta \theta} \frac{d}{d\theta} \{i(b + \theta)\}$$

and

$$\tau = \tfrac{1}{2} i^2 \frac{\Delta L}{\Delta \theta}$$

for $0 < \theta < 1$. Let unit current

$$I = \frac{V_s \Delta \theta}{\omega \Delta L} = \frac{V_s T}{\Delta L} \qquad (7)$$

where T is the time corresponding to the rising inductance period such that $i = i/I$, in which case

$$k = \frac{d}{d\theta} \{i(b + \theta)\} \qquad (8)$$

and

$$\tau = \frac{1}{2} \frac{V_s I}{\omega} i^2 \qquad (9)$$

Eqn. 8 has the solution

$$i = \frac{k(\theta - \theta_a) + i_a(b + \theta_a)}{(b + \theta)} \qquad (10)$$

for $0 < \theta < 1$, where (θ_a, i_a) is one known arbitrary point on the waveform. A suitable point to define a current pulse is (θ_p, i_p), since this point is common both to the waveform before commutation when $k = +1$ and after commutation when $k = -n$.

The above equations apply to the region of rising inductance for which the current waveforms are curves. For the regions of constant maximum or minimum inductance shown in Fig. 2b, the current waveforms are straight lines defined by

300

$$kV_s = \omega p \Delta L \frac{di}{d\theta}$$

which on normalisation becomes

$$k = p \frac{di}{d\theta} \qquad (11)$$

where $p = b$, $k = +1$ for $\theta_1 < \theta < \theta_2$ ($\theta < 0$), and $p = 1 + b$, $k = -n$ for $\theta_3 < \theta < \theta_4$ ($\theta > 1$). As for eqn. 8, eqn. 11 is independent of speed.

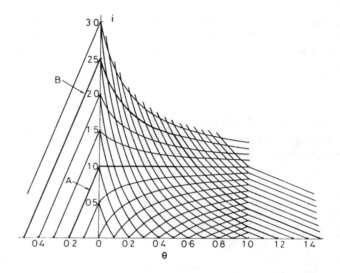

Fig. 7 *Current waveforms for winding ratio n = 1 and inductance ratio b = 0·2, showing two particular examples A and B*

4.5 Normalised current waveforms

Fig. 7 shows a set of normalised current waveforms, applicable at any speed for $n = 1$ and $b = 0.2$, from which, for a given choice of control angles θ_0 and θ_p or commutation point θ_p, i_p, the current pulse may be constructed. It has been assumed for convenience that the pulses do not overlap the falling inductance region ($\theta < \theta_1$ or $\theta > \theta_4$) although, were this to be so, the corresponding waveforms in this region could be calculated equally readily using the same normalisation as in Section 4.4., as shown in the Appendix.

Two examples of current pulses are shown in Fig. 7, of which pulse A is particularly instructive. If the thyristor is fired so that at the end of the minimum inductance period the normalised current i has reached unity, then eqn. 10 shows it will remain at unity throughout the rising inductance period until commutation. The motor back e.m.f. $i\Delta L/\Delta\theta$ exactly counterbalances the applied voltage $+V_s$, giving a zero value for $di/d\theta$.

It must be noted that the unit current I is inversely proportional to speed ω (eqn. 7). Hence, although waveform A may be applied at speeds of either 1000 rev/min or 2000 rev/min, the actual current values $i(\theta)$ for A at 2000 rev/min will be half those of the waveform at 1000 rev/min.

It must also be noted that, if resistive and iron losses are negligible, the thyristor conduction angle $\theta_p - \theta_0$ must equal the diode conduction angle $\theta_q - \theta_p$ for a turns ratio $n = 1$, where θ_q is the angle of current extinction. With reference to eqn. 1, the flux increase up to θ_p due to the winding voltage $+V_s$ must equal the flux decrease after θ_p due to $-V_s$.

5 Computation and control of torque and power

Torque and hence power are controlled by the switching angles θ_0 and θ_p shown in Fig. 6, and once θ_0 and θ_p are defined the normalised current waveform is defined irrespective of speed.

Provided no current exists in the falling inductance period ($\theta_4 < \theta < \theta_5$ in Fig. 2), increasing the 'on' angle ($\theta_p - \theta_0$) increases the size of the current pulse and hence the torque and power produced at a given speed.

Since the instantaneous torque τ at a given speed is proportional to i^2 (eqn. 9), the mean torque τ_m and power P developed per phase are given by

$$\tau_m = \frac{1}{2} \frac{V_s I}{\omega} i_{rms}^2 \qquad (12)$$

$$P = \tfrac{1}{2} V_s I \, i_{rms}^2 \qquad (13)$$

where

$$i_{rms} = \left\{ \frac{1}{\theta_{cy}} \int_0^1 i^2 \, d\theta \right\}^{1/2} \qquad (14)$$

i.e. the normalised r.m.s. value of the torque producing current over a complete angular cycle θ_{cy} of the rotor geometry. Eqn. 14 is extended if current also flows in the falling inductance region, as shown in Appendix 10.

Since I is inversely proportional to speed ω (eqn. 7), a particular normalised current waveform (such as the flat-topped waveform A of Fig. 7) with a particular value of i_{rms}, if applied throughout the speed range by keeping the control angles θ_0 and θ_p fixed, will result in the following variations of torque and power with supply voltage and speed:

$$\tau_m \propto \frac{V_s^2}{\omega^2}; \qquad P \propto \frac{V_s^2}{\omega}$$

Hence, if the control angles are kept constant, the torque/speed characteristic of the series motor naturally results, a characteristic that has been found in practice to suit traction applications.

The value of i_{rms} for a particular waveform defined by (θ_p, i_p) may be obtained by substituting for i in eqn. 14 using eqn. 10. The current i_{rms} is then a function of (θ_p, i_p), as shown in the Appendix.

If it is desired to maintain constant power over a range of speeds, then, with reference to eqn. 13, since I decreases inversely with speed, i_{rms}^2 must increase proportionately with speed. For example, to achieve the same power at 3000 rev/min as at 1000 rev/min the value of normalised

torque producing current i_{rms} at 3000 rev/min must be $\sqrt{3}$ greater than at 1000 rev/min, although the actual torque-producing current i_{rms} will be $\sqrt{3}$ times less. This is illustrated by the waveforms A and B in Fig. 7, for which $i_{rms} = 0.811/\sqrt{\theta_{cy}}$ for A and $i_{rms} = 1.4/\sqrt{\theta_{cy}}$ for B.

For a motor having the inductance profile of Fig. 2 with $\theta_{cy} = 3$, $L_{min} = 2$ mH, $\Delta L = 10$ mH and $V_s = 200$ V, the unit current $I = 100$ A at 1000 rev/min and 33.3 A at 3000 rev/min. From eqn. 13, waveform A produces 2.2 kW/phase at 1000 rev/min with a flat-topped current peak $\hat{i} = 100$ A, and B produces 2.2 kW/phase at 3000 rev/min with $\hat{i} = 83.3$ A.

The variation of i_{rms} with current pulse shape is illustrated in Fig. 8 in which the co-ordinates i_p, θ_p represent the current and rotor angle at commutation and define the entire pulse. For constant power, i_{rms} must be proportional to (speed)$^{1/2}$. The contours for i_{rms} have been calculated in Fig. 8 for $b = 0.2$ and $n = 1$; similar contours may be calculated for alternative values of inductance and winding-turns ratio.

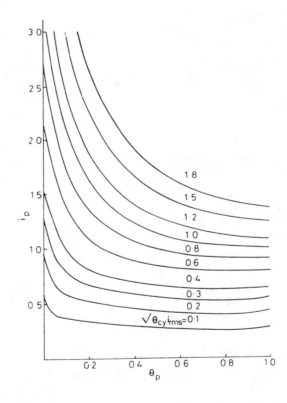

Fig. 8 *Variation of i_{rms} with commutation point, shown as contours of constant $\sqrt{\theta_{cy}} \, i_{rms}$ in the i_p-θ_p plane for winding ratio $n = 1$ and inductance ratio $b = 0.2$*

To achieve greater values of i_{rms} as the speed increases, the angular duration of the current pulse must be increased (by increasing $\theta_p - \theta_0$). A limit is reached when the current pulse occupies the whole of the angular cycle of rotor geometry θ_{cy}, in which case, since current is flowing during the falling inductance period, some counter-productive torque will be produced. For running at speeds greater than the speed corresponding to this limiting case, a decrease in power must be accepted unless continuous current is allowed to flow in the motor windings, in which case a different form of control is required.

As the speed is reduced to low values it is not feasible to maintain constant power, and at low speeds constant torque is generally attained, as shown by the typical traction characteristic of Fig. 9.

To achieve constant torque, it is necessary to maintain a constant torque-producing r.m.s. current, which is not feasible using a single commutation per pulse owing to the relatively large values of di/dt compared with the time T of the rising inductance period. A different mode of control is required, whereby the current is maintained at its desired value by chopping between two levels, as shown in Fig. 10.

Hence, in addition to the rotor position sensor for angular control at higher speeds, a current sensor is required for current control at lower speeds.

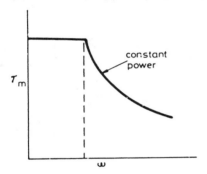

Fig. 9 *Typical torque/speed characteristic for traction*

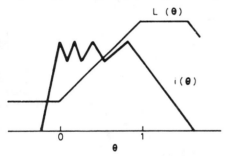

Fig. 10 *Typical low-speed current waveform*

6 Inverter design considerations

It has been shown that every current pulse waveform is uniquely defined by i_p and θ_p. A given power may be produced from a wide variety of waveforms, which will impose a correspondingly wide variation in the current ratings for thyristors, diodes and electrolytic capacitors. Voltage ratings are predominantly determined by the source voltage V_s and the turns ratio n, although leakage inductance causes an increase in the voltage seen by the commutating capacitors and by the thyristors above the zero-leakage value of $(1 + n) V_s$.

In seeking an optimum design for the system for a given application, the influence of motor inductance-profile changes on component ratings can be obtained, since an analytical formula for the normalised current waveform exists. By repeated calculation, desirable trends both for motor inductance profile and for current waveforms, which together optimise the system, can be identified. If a cost function for the system can be formulated, then the search can be directed to cost minimisation.

If transistors are used as the main switching devices, then the peak current $\hat{i}$ is likely to be the most sensitive parameter in determining cost at a given source voltage. Fig. 11 shows the variation of the ratio $\sqrt{(\theta_{cy})}\, i_{rms}/\hat{i}$ in the i_p-θ_p plane, and indicates that flat-topped current waveforms ($i_p = 1$) and late commutation angles θ_p are desirable.

If thyristors are used, whose cost is determined mainly by r.m.s. current rating, and which have the ability to handle high peak/r.m.s. ratios, then a different conclusion is reached, as follows. By putting $Ii_{rms} = i_{rms}$ in eqn. 13

$$P = \tfrac{1}{2}\, V_s i_{rms} i_{rms} \qquad (15)$$

which indicates that the actual current i_{rms} may be reduced by using a high value of normalised current i_{rms} (i.e. by reducing the unit current I for the motor by designing it for a higher inductance). Since 'peaky' waveforms, like B in Fig. 7, have larger values of i_{rms} compared to the flat-topped waveform A, the 'peaky' waveform is desirable,[13] with earlier firing and longer duration. The limit for this trend, as already stated, arises at top speed and full power, where the onset of continous current may cause problems.

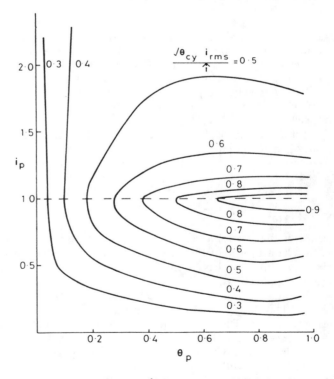

Fig. 11 *Variation of $i_{rms}/\hat{i}$ with commutation point, shown as contours of constant $\sqrt{\theta_{cy}}i_{rms}/\hat{i}$ in the i_p-θ_p plane for winding ratio $n = 1$ and inductance ratio $b = 0.2$*

7 Conclusions

Doubly salient reluctance motors offer an attractive alternative to induction motors in brushless variable-speed drives in that the inverter is only required to supply a unidirectional current to each motor phase, thereby halving the number of main switches required. The motor, even with bifilar windings, has the constructional simplicity of an induction motor, but with no rotor conductors, and is expected to be similar in cost, size and weight.* Owing to its symmetry, the motor is self starting in either direction, and regeneration arises naturally by appropriate positioning of the current pulses, thereby giving 4-quadrant operation without requiring any additional inverter components.

The power convertor was designed using thyristors because, at the power level required for a battery-vehicle drive, power transistors were found to be more costly. Although no direct cost comparison with an equivalent induction-motor drive has been made, the simplicity of the power convertor both in the number of power devices and in its ease of control suggests that the reluctance-motor drive will be substantially cheaper. Present experience indicates that the volumes of power convertor and motor are similar, and that the power convertor becomes smaller than the motor at power levels suitable for a vehicle drive.

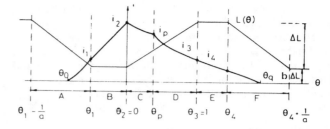

Fig. 13 *General current pulse showing regions for corresponding formula*

The linearised analysis presented in this paper shows that the motor phase currents follow simple functions of rotor angle, independent of speed, and that the current and angle at commutation define uniquely the normalised current waveform. The use of a linear motor model in the analysis has been justified on the grounds of providing:

(*a*) sufficiently accurate current waveforms for inverter-design purposes

(*b*) a control strategy for the inverter in terms of firing and commutation angles for various speeds and powers

(*c*) an indication of desirable inductance profiles for motor design whereby the performance and cost of the system as a whole may be optimised.

It must however be emphasised that, for a particular motor, inductance is a function of current as well as rotor angle, and current waveforms will show the effects of saturation at high flux values. Nevertheless, predictions from linear analysis greatly assisted the liaison between Nottingham and Leeds Universities in designing a motor at Leeds and its inverter at Nottingham for a cost-effective prototype system.

Fig. 12 shows actual current waveforms for an inverter-driven v.r. motor developing 2·8 kW/phase at 1000 rev/min and at 2250 rev/min. The main and auxiliary winding currents for one phase have been superimposed to show the effective phase current, and it will be seen that the pulses approximately follow the shapes predicted by linear theory. The main thyristor r.m.s. currents were 47 and 36 A, respectively.

The object of the paper has been to demonstrate the value of linear analysis of an inverter-fed v.r. motor for inverter-design purposes and for system-design optimisation. It has not been possible to present more experimental results at this stage but the authors anticipate further publications as the project proceeds, and a companion paper is currently being prepared (see footnote on previous page).

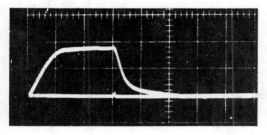

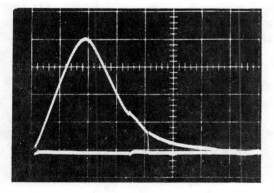

Fig. 12 *Actual motor phase current waveforms:*

a 1000 rev/min; vertical scale 50 A/div., horizontal scale 1 ms/div.
b 2250 rev/min; vertical scale 20 A/div.; horizontal scale 0·5 ms/div.

8 Acknowledgments

The authors are indebted to their colleagues at Leeds University, J.M. Stephenson, for his helpful comments during the conduction of the analysis in this paper, and Prof. P.J. Lawrenson and N.N. Fulton who, with Dr. Stephenson, designed and constructed the motor from which the waveforms of Fig. 12 were taken.

The authors wish to thank Chloride Technical Ltd. for its sponsorship of the project, for permission to release

* 'Variable-speed switched reluctance motors', a paper currently being prepared by P.J. Lawrenson, J.M. Stephenson, P.T. Blenkinsop, J. Corda and N.N. Fulton

the details given in this paper and for helpful discussion with M.F. Mangan.

The authors also wish to thank their colleague at Nottingham University, R.J. Blake, for designing and constructing the electronic controls for the drive, and Prof. R. L. Beurle for the use of the facilities of the Department of Electrical and Electronic Engineering.

9   References

1 BRADLEY, A., CLARKE, C.D., DAVIS, R.M., and JONES, A.: 'Adjustable-frequency inverters and their application to variable-speed drives', *Proc. IEE*, 1964, **111**, (11), pp. 1833–1846
2 BRENNEISEN, J., FUTTERLIEB, E., MULLER, E., and SCHULZ, M.: 'A new converter drive system for a diesel electric locomotive with asynchronous traction motors', *IEEE Trans.*, 1973, **IA-9**, pp. 482–491
3 GIBSON, J.P.: 'New inverter circuit suitable for high-current p.w.m. operation', *Proc. IEE*, 1976, **123**, (10), pp. 993–998
4 HENDER, B.S.: 'Watts on'. Proceedings of the 2nd international electric vehicle symposium, November 1971, pp. 51–55
5 BAUSCH, H., and RIEKE, B.: 'Performance of thyristor-fed electric car reluctance machines'. Proceedings of the international conference on electrical machines, Brussels 1978, p. E4/2-1
6 UNNEWEHR, L.E., and KOCH, W.H.: 'An axial air-gap reluctance motor for variable speed applications', *IEEE Trans.*, 1974, PAS-93 pp. 367–376
7 UNNEWEHR, L.E.: 'Series commutated SCR controllers for variable speed reluctance motor drives', IEEE PESC 73 record, pp. 180–190
8 BYRNE, J.V., and O'DWYER, J.B.: 'Saturable variable reluctance machine simulation using exponential functions'. Proceedings of the international conference on stepping motors and systems. University of Leeds, July 1976, pp. 11–16
9 BYRNE, J.V., and LACY, J.G.: 'Electrodynamic system comprising a variable reluctance machine'. British Patent 1321110, June 1970
10 KOCH, W.H.: 'Thyristor controlled pulsating field reluctance motor system', *Elect. Mach. Electromech.*, 1977, **1**, pp. 201–215
11 STEPHENSON, J.M., and CORDA, J.: 'Computation of torque and current in doubly salient reluctance motors from nonlinear magnetisation data', *Proc. IEE*, 1979, **126**, (5), pp. 393–396
12 DAVIS, R.M. and RAY, W.F.: 'Variable reluctance motors'. British Patent 13416/77, March 1977
13 RAY, W.F. and DAVIS, R.M.: 'Reluctance electric motor dirve systems'. British Patent 13415/77, March 1977

10   Appendix

General formula for normalised current waveforms for an asymetrical linearised inductance profile

For generality, the falling inductance region is taken to be of slope $-a\Delta L/\Delta\theta$, compared with $+\Delta L/\Delta\theta$ for the rising inductance region. The normalisation is based on the rising inductance region as given in Section 4.4 and all symbols are the same as previously used in the paper. The current pulse is taken to overlap the falling inductance regions as shown in Fig. 13, but to be discontinuous (i.e. $\theta_q - \theta_0 < \theta_{cy}$).

To calculate r.m.s. and mean currents for thyristors, diodes, and for torque production, it is advantageous to subdivide the current into the six regions A to F shown in Fig. 13, A to C representing thyristor conduction before commutation and D to F representing diode conduction

after commutation. For pulses of shorter duration than that shown in Fig. 13, the current in regions A, B, E or F and the values for i_1, i_2, i_3 or i_4 may be zero.

10.1   Equations for r.m.s. and mean currents

Using the symbols $SS = \int i^2 d\theta$ and $S = \int id\theta$ over the six regions A to F, we have the following expressions:
(a) Torque-producing current:

$$i_{rms} = \{(-a\,SS_A + SS_C + SS_D - a\,SS_F)/\theta_{cy}\}^{1/2}$$

(b) Thyristor and diode r.m.s. and mean currents:

$$i_{T_{rms}} = \{(SS_A + SS_B + SS_C)/\theta_{cy}\}^{1/2}$$
$$i_{T_{mean}} = (S_A + S_B + S_C)/\theta_{cy}$$
$$i_{D_{rms}} = n\{(SS_D + SS_E + SS_F)/\theta_{cy}\}^{1/2}$$
$$i_{D_{mean}} = n(S_D + S_E + S_F)/\theta_{cy}$$

(c) Interrelationship between $i_{rms}, i_{T_{mean}}$ and $i_{D_{mean}}$:
The mean power per phase P, previously defined by eqn. 13, is also given by

$$P = V_s I\,(i_{T_{mean}} - i_{D_{mean}})$$

Hence

$$i^2_{rms} = 2\,(i_{T_{mean}} - i_{D_{mean}})$$

10.2   Definition of the current pulse

The current pulse is defined by (i_p, θ_p) from which θ_0 and θ_q are obtained.

$$\theta_0 = \theta_p - i_p\,(b + \theta_p)$$
$$\theta_q = \theta_p + i_p\,(b + \theta_p)/n$$

Before the above integrals are evaluated it is necessary to determine the normalised current values at the region boundaries, as shown in Fig. 13.

$$i_2 = \{-\theta_p + i_p\,(b + \theta_p)\}/b \quad \text{for} \quad \theta_0 < 0$$
$$i_2 = 0 \quad \text{for} \quad \theta_0 \geqslant 0$$
$$i_1 = i_2 + \theta_1/b \quad \text{for} \quad \theta_0 < \theta_1$$
$$i_1 = 0 \quad \text{for} \quad \theta_0 \geqslant \theta_1$$
$$i_3 = \{-n\,(1 - \theta_p) + i_p\,(b + \theta_p)\}/(1 + b) \quad \text{for} \quad \theta_q > 1$$
$$i_3 = 0 \quad \text{for} \quad \theta_q \leqslant 1$$
$$i_4 = i_3 - n\,(\theta_4 - 1)/(1 + b) \quad \text{for} \quad \theta_q > \theta_4$$
$$i_4 = 0 \quad \text{for} \quad \theta_q \leqslant \theta_4$$

10.3   Equations for current and current integrals

(a) Regions C and D:

$$i = \frac{k(\theta - \theta_p) + i_p\,(b + \theta_p)}{(b + \theta)}$$

$$SS_{C,D} = k\,|k|\,\{(\theta_p - \lambda)(1 + GH^2) - 2(b + \theta_p)\,H\log_e G\}$$

$$S_{C,D} = |k|\,\{\theta_p - \lambda - (b + \theta_p)\,H\log_e G\}$$

where $G = (b + \theta_p)/(b + \lambda)$, and $H = 1 - i_p/k$.

For region C, $k = 1$ and $\lambda = 0$ for $\theta_0 \leqslant 0$, $\lambda = \theta_0$ for $\theta_0 > 0$.

For region D, $k = -n$ and $\lambda = 1$ for $\theta_q \geqslant 1$, $\lambda = \theta_q$ for $\theta_q < 1$.

(b) Region B

$$i = i_2 + \theta/b$$

$$SS_B = b\,(i_2^3 - i_1^3)/3$$

$$S_B = b\,(i_2^2 - i_1^2)/2$$

(c) Region E

$$i = i_3 - n\,(\theta - 1)/(1 + b)$$

$$SS_E = (1 + b)(i_3^3 - i_4^3)/3n$$

$$S_E = (1 + b)(i_3^2 - i_4^2)/2n$$

(d) Regions A and F

$$i = \frac{-\theta_1 + b\,i_1}{b - a\,(\theta - \theta_1)} \qquad \text{for region A}$$

$$i = \frac{-n\,(\theta - \theta_4) + (1 + b)\,i_4}{(1 + b) - a\,(\theta - \theta_4)} \qquad \text{for region F}$$

$$SS_{A,F} = k\,|k|\,p\,(G^2 - 1 - 2G\log_e G)/a^3$$

$$S_{A,F} = -\,|k|\,p\,(G - 1 - G\log_e G)/a^2$$

where for region A, $k = 1$, $p = b$ and $G = 1 + ai_1$, and for region F, $k = -n$, $p = 1 + b$ and $G = 1 - ai_4/n$

305

Inverter Drive for Switched Reluctance Motor: Circuits and Component Ratings

R.M. Davis, W.F. Ray and R.J. Blake

Proceedings IEE, Vol. 128, Pt. B., No. 2, March 1981, pp. 126-136

List of principal symbols

Normalised equivalents are printed in bold type, where appropriate

i	=	motor phase current
i_p	=	current at commutation
i_{RMS}	=	RMS value of torque-producing current
I	=	unit current for normalisation
I_{MC}	=	mean current in chopping mode
I_{RR}	=	resonant reversal current
θ	=	rotor angle
$\Delta\theta$	=	angular duration of rising inductance
	=	1 when normalised
θ_o	=	thyristor firing angle
θ_p	=	thyristor commutation angle
θ_q	=	current extinction angle
θ_{cy}	=	angular repetition period
$L(\theta, i)$	=	motor phase inductance
L_{max}, L_{min}	=	maximum, minimum value of L, respectively
ΔL	=	$L_{max} - L_{min}$
L_{inc}	=	incremental inductance
l	=	leakage inductance
λ	=	$\sqrt{L_{min}/l}$
b	=	$L_{min}/(L_{max} - L_{min})$
ω	=	motor speed, rad/s
ω_b, ω_t	=	base, top speed, respectively
P_r	=	motor rated power
P	=	power/phase
p	=	number of phases
n	=	main/auxiliary turns ratio
V_s	=	supply voltage (rated value = 150 V)
V_c	=	voltage across lower commutating capacitor
$\Delta V = \beta V_s$	=	voltage boost from l
ΔV_r	=	voltage boost after resonant reversal
V_p	=	initial reverse thyristor voltage
V_a	=	auxiliary (commutation) supply voltage
μ	=	$i_p V_s/P_r$ = commutation current/nominal input current
N_c	=	dimensionless commutation constant
t_{RVM}	=	reverse-voltage time, main thyristors
t_{RVC}	=	reverse-voltage time, commutation thyristors
t_{CM}	=	conduction time, commutation thyristors
T_c	=	chopping period
t_{RR}	=	resonant reversal time

1 Introduction

1.1 General

In a previous paper [1], two of the authors described in general terms the inverter drive for a switched reluctance motor. The motor's ability to operate efficiently from unidirectional current pulses synchronised with rotor movement

make it an ideal partner for an inverter using unidirectional current switches, e.g. thyristors. As each motor phase normally requires only one switch, compared to two, the inverter is simpler and cheaper than a PWM inverter for an induction motor. Linear analysis was used to predict approximate current waveforms, from which the inverter device ratings may be estimated in advance of motor manufacture: it also indicated those operating conditions likely to lead to an economic system design. The elements of a control philosophy emerged and it was concluded that the system offered the prospect of a competitive variable speed drive combining high efficiency, flexibility and reliability.

A companion paper [2] from Leeds University examines the motor behaviour in more detail, highlighting its nonlinear behaviour, its high specific output and efficiency and its constructional simplicity.

Both Nottingham and Leeds Universities have had papers accepted for the ICEM Conference in Athens, 1980 [3, 4]. Extensive references are listed elsewhere [1, 2].

The purpose of the present paper is to describe in more detail the configurations for power convertors for both 3-

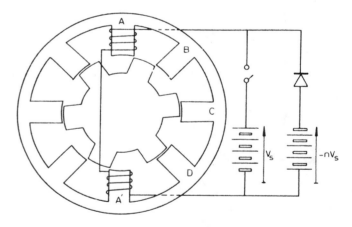

Fig. 1 *Power source/sink for one phase of SR motor*

and 4-phase motors having single and bifilar windings, and to detail the operation and ratings of a 4-phase bifilar system over a wide range of speed and power. The circuits to be described, because of their power ratings, use thyristors as the main device, but the commutation circuits can take a variety of forms [5]. Of particular interest is the possibility with the 4-phase motor of using dormant motor windings for the resonant reversal of the commutating capacitors in the low-speed chopping mode [6]; the power circuit configuration then avoids totally any need for resonant reversal in the single-pulse mode, which occupies the vast majority of the speed range [7]. This combination yields a very low-loss power convertor.

1.2 Summary of system behaviour

Before embarking on the above topics, it will help the reader if a summary of the previous paper [1] is presented. A 4-phase 8/6 motor is shown in Fig. 1 (in contrast to the 6/4 motor shown previously). Energising the A phase causes anticlockwise rotation, during which the phase inductance increases. The idealised variation of inductance with normalised rotor angle is shown in Fig. 2, which also shows a typical current

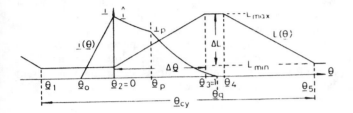

Fig. 2 *Linearised inductance profile and typical current waveform*

pulse and defines the various angles. Energy flows from the source V_s into the motor during the thyristor ON angle ($\theta_p - \theta_o$) and returns to the sink $-nV_s$ via the diode during the diode ON angle ($\theta_q - \theta_p$). The direction of the torque produced depends on whether the phase current coincides predominantly with rising or falling inductance.

The family of current waveform segments shown in Fig. 3 apply at all speeds, provided the current is expressed in a normalised form. The normalised current i is the ratio of the actual current to unit current I which is defined (eqn. 1) as that current which equalises the motor back EMF and the supply voltage V_s; eqn. 2 then follows.

$$ I = \frac{V_s \Delta \theta}{\omega \Delta L} \quad \text{and} \quad i = \frac{i \omega \Delta L}{V_s \Delta \theta} \quad (1), (2) $$

At high speed ω and for a given winding current i, i must be large, e.g. waveform X, requiring appreciable early turnon [8]. The top-speed full-power condition generally requires the current pulse ($\theta_q - \theta_o$) to occupy virtually the whole cycle θ_{cy}.

Waveform Y on Fig. 3 identifies a lower speed current waveform for which the supply voltage and motor back EMF are equal: it therefore has a flat top, corresponding to a normalised current of 1.

At very low speed, a given winding current will correspond to an extremely low value of normalised current i which can only be obtained at constant supply voltage by chopping the main thyristor on and off repeatedly, waveform Z. The value of the linear analysis is demonstrated in Section 3 where current ratings for inverter devices are predicted for the power electronics designer who will also find the linearised system behaviour easy to grasp.

Power at a given speed is controlled by adjusting θ_o and θ_p to give the best 'torque producing' current in the interval

$0 < \upsilon < 1$, bearing in mind that a high current when the poles are nearing full alignment only worsens saturation (see Fig. 9c of Reference 2).

At constant supply voltage and constant angles, the motor has a characteristic similar to a DC series motor. If, however, the angles are suitably adjusted for rated winding current at each speed, the characteristic lies somewhere between constant torque and constant power. It is thus possible, with a little sacrifice, to meet either a constant torque characteristic or a constant power characteristic; the latter is particularly suited to the chloride battery vehicle project [1], which requires 25% of the speed range at constant maximum torque (mostly short-time rated) and constant power up to 75% of top speed.

When evaluating inverter component ratings, both voltages and currents must be considered. Section 2 presents several circuits, with their implications for device voltages. Sections 3 and 4 examine current ratings and component values for both main and commutation components, and in particular consider the use of motor windings in place of separate components for resonant reversal during chopping. Section 5 describes a digital controller built to examine on an experimental test rig the variation of drive power, efficiency and behaviour with control angle settings θ_o and θ_p, and to establish optimal control laws for a battery vehicle application. Section 6 gives a selection of the waveforms obtained, illustrating the behaviour of the 4-phase inverter built for this purpose.

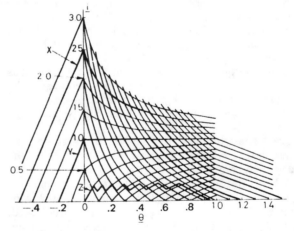

Fig. 3 *Normalised current waveforms against normalised rotor angle*

X typical high-speed waveform
Y typical midspeed waveform
Z typical low-speed (chopping) waveform

2 Inverter configurations and device voltage ratings

2.1 Motor windings

The motor windings must be arranged both to accept power from a source and to return power to it. A motor with a single winding per phase uses it for both functions, requiring either a 3-wire supply, Fig. 4a or a bridge configuration of thyristors and diodes, Fig. 4b. For a given winding voltage rating, Fig. 4a requires a thyristor with twice the voltage rating of Fig. 4b, but the latter requires two thyristors.

Bifilar windings avoid the need for a 3-wire supply, although retaining the advantage of only one thyristor and one diode per phase, Fig. 4c.

2.2 Inverter configurations

Where a 3-wire supply is available, the circuit of Fig. 4a can be useful provided the motor has an even number of phases to ensure equal loading on the halves of the 3-wire

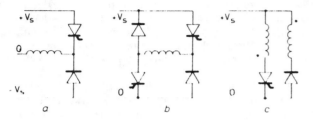

Fig. 4 *Circuits for energising one phase of SR motor*

supply. The circuit requires devices rated at $2V_s$ (thyristors) and $2V_s + V_a$ (diodes) where V_a is the reverse thyristor voltage for commutation. By arranging the phases with their thyristors connected alternately to $+V_s$ and $-V_s$, the need for the commutation circuit to be reset between main thyristor commutations is avoided, except in the chopping mode. It might be considered that this configuration would suit a battery-powered drive, since a mid-tapping can be obtained; but the winding is then rated for only half the total battery voltage, thus doubling the current and the forward-drop losses, which cannot be neglected when the total battery voltage is 180 V or less.

The bridge configuration, the circuit of Fig. 4b, has thyristors rated at only V_s, but requires two per phase: the two diodes are rated at $V_s + V_a$. It is a circuit whose merits are likely to be most important at high voltage and high power.

The bifilar configuration, the circuit of Fig. 4c, has thyristors rated somewhat in excess of $2V_s$, because of the small leakage inductance which exists even between bifilar windings. When, at the end of a commutation, the anode of the thyristor and the commutating capacitor (not shown) reach a potential $2V_s$, the diode becomes forward biased. It takes a few microseconds for the current to transfer from the main winding and commutation capacitor to the bifilar auxiliary winding, and during this time the capacitor charges by an additional voltage ΔV, which depends on the magnitude of the winding current at commutation. This extra voltage is helpful in the single-pulse mode where it increases the commutation capability as the winding current increases. Its effect is examined further in Section 4. It should be noted that the effective phase current waveform as shown in Fig. 3 for a bifilar motor is the sum of the currents in both windings.

Once again, by arranging the phases with their thyristors connected alternately to two supply terminals, $+V_s$ and 0, the need to reset the commutation circuit between main thyristor commutations is avoided except in the chopping mode, for motors with an even number of phases.

This configuration was chosen for the battery-vehicle drive because the winding is rated for the full battery voltage and only one device is in series with the winding, thus minimising the forward-drop losses. The high specific motor output reported elsewhere [2, 3, 4] has been achieved with bifilar windings.

Each of the above circuits requires a bank of electrolytic capacitors in parallel with the direct voltage supply. This is because the inverter, like any other voltage-fed inverter-motor drive, draws a considerable ripple current associated with the supply and return of energy to the windings. A larger number of phases tends to produce more overlap between one phase returning current and the next drawing current, so that the capacitor duty is eased. The circuit of Fig. 4a is the worst in this respect, as, for a 4-phase motor, one half of the supply has only two phases supplied by it, and the capacitor bank is repeated for the other half. The capacitor bank is rated to carry the AC component of current at the direct-voltage terminals of the inverter.

2.3 Commutation circuits

Where transistors are used for the main switching devices, commutation circuits are unnecessary. With thyristors, however, the commutation circuit(s) will vary considerably with the configuration of main devices.

The general approach has been, first, to apply a reverse voltage to the thyristor undergoing turnoff, rather than to drive a current through a reverse parallel diode. This increases certain voltage levels within the circuit, but reduces the need for resonant turn-off inductors and diodes. Secondly, a single commutation circuit has been designed to service all the main thyristors, using as few commutation thyristors as possible. Thirdly, the circuits have been designed where possible to avoid the need for extra components for resonant reversal of the commutating capacitor voltage.

For the circuits of Figs. 4a and b, the winding end potentials are clamped by the auxiliary diodes, and this makes it more difficult to obtain the necessary voltage on the commutating capacitor to reverse bias the thyristors. One solution, shown in Fig. 5, uses an auxiliary supply V_a to provide the

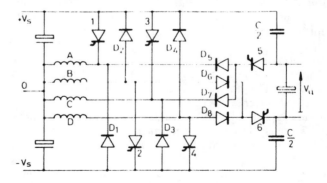

Fig. 5 *4-phase circuit based on Fig. 4a*

required reverse voltage for the 4-phase version of Fig. 4a. In the single pulse mode, the commutating capacitor is always

308

left with the correct polarity for the next commutation. In the chopping mode however, after turning off 1, say, by firing 5, the capacitors must be resonantly reversed before 1 can be turned-off a second time. This is achieved by firing 6 and using windings B and D for resonant reversal. As B is at or near minimum inductance, the resonant reversal generally occupies less than the 'on' time for 1 in chopping. In the case of a battery vehicle, the power for the auxiliary supply would be supplied via a DC/DC convertor, otherwise from a transformer/rectifier; the power rating is a small proportion of the drive power.

Because of the requirement of two thyristors per phase, the circuit configuration of Fig. 4b has not been developed, and the battery-vehicle project has concentrated on the bifilar winding configuration of Fig. 4c.

The 3-phase version of Fig. 4c appeared in the previous paper [1] (Fig. 5). That arrangement had all the windings connected to $+V_s$ and all the thyristors and diodes to 0 V. A resonant reversal circuit (L, T_5) was essential for the single-pulse and chopping modes. No auxiliary supply was needed as the transformer action of the bifilar windings provides ample commutating voltage.

A 4-phase version of Fig. 4c which uses separate resonant reversal in the chopping mode is shown in Fig. 6. This arrangement, although requiring two extra thyristors and a resonant

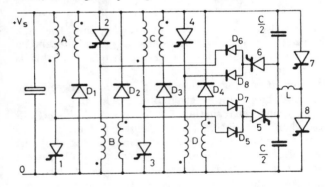

Fig. 6 *4-phase circuit based on Fig. 4c using separate resonant reversal*

inductor, has the advantage of providing a higher initial reverse voltage in chopping than in the single-pulse mode, enhancing the relative circuit capabilities in chopping, but imposing higher voltages on the commutation components and diodes.

When the motor windings are used for resonant reversal in chopping, the separate resonant reversal components in Fig. 6 are superfluous. Fig. 7 shows the 4-phase bifilar

circuit with winding resonant reversal. Design details and further description of the operation of this circuit are presented in Section 4.

In the Figures 5, 6 and 7, the commutation diodes D_5-D_8 can be replaced by thyristors, eliminating the need for thyristors 5 and 6.

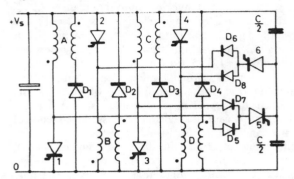

Fig. 7 *4-phase circuit based on Fig. 4c using winding resonant reversal*

2.4 Component voltage ratings

Table 1 summarises the demands of the several circuits in terms of device voltage ratings. In each case, the winding is rated for a supply voltage V_s, and hence the current ratings of corresponding devices are similar. Resonant reversal is assumed to be loss-free.

It is clear that the VA rating of the main thyristors is consistent for all circuits. The voltage ΔV for the bifilar circuits arises from the leakage inductance between the bifilar pair. The diodes are more highly rated for the bifilar circuit, as are the commutating capacitors. However, bifilar circuits normally exhibit a higher initial reverse voltage so that the commutating capacitor can be smaller.

3 Current ratings and inductance values

3.1 Perspective

The voltage ratings of the inverter devices depend mostly on the supply voltage, the winding turns ratio and its leakage inductance, and the circuit configuration. The device current ratings depend additionally on the drive power and the motor characteristics, and it is the latter which makes the overall system design difficult. Only by treating the system as a whole can the most economical design be achieved, and, as explained in the previous paper [1], the motor geometry is an essential but unknown factor when designing the system from scratch. A computer search for a global optimum system which incorporates all system interactions and includes motor nonlinearities is prohibitive at present.

Table 1: Voltage ratings for major inverter components

Circuit	Thyristors			Diodes		Electrolytics	Commutation circuit		Thyristor reverse voltage
	Main	Comm.	Res. rev.	Aux.	Comm.		Cap.	Aux. sup.	
Fig. 5	$2V_s$	$2V_s + V_a$	–	$2V_s + V_a$	$2V_s$	$V_s(2)$	V_s	V_a	$-V_a$ $-2V_s - \Delta V^*$
Fig. 6	$2V_s + \Delta V$	$4V_s + \Delta V$	$3V_s + \Delta V$	$4V_s + \Delta V$	$3V_s + \Delta V$	V_s	$3V_s + \Delta V$	–	$-V_s - \Delta V$
Fig. 7	$2V_s + \Delta V$	$3V_s + \Delta V$	–	$3V_s + \Delta V$	$2V_s + \Delta V$	V_s	$2V_s + \Delta V$	–	$-V_s - \Delta V$

*This reverse voltage applies only in chopping; in the single-pulse mode the reverse voltage is $-V_s - \Delta V$

The linear analysis presented previously [1] has been of considerable value in predicting current ratings because it permits a computer search embracing several trapezoidal inductance/angle profiles $L(\theta)$, each corresponding to a particular motor geometry but neglecting saturation. Each inductance profile is characterised by its normalised cycle duration θ_{cy}, its maximum and minimum inductances L_{max}, L_{min} and their normalised angular durations $(\theta_4 - 1)$ and θ_1 as shown in Fig. 2. The ratio L_{max}/L_{min} in practice is strongly saturation sensitive and varies dramatically between low and high currents [2]. For each profile, a constant value of L_{max}/L_{min} must be chosen by experience, representing its average value over the practical range of current. The computer search then predicts for selected speeds a range of device current waveforms with their associated control angles θ_o, θ_p, and the profile which minimises the device RMS currents while satisfying the drive requirements can be identified. This method has assisted in the system design and has provided a useful guide for the power electronic component ratings, which can be finalised later after practical tests.

3.2 Constraints at rated power
For a specified inductance profile, the normalised current waveform is defined by the control angles θ_o and θ_p (or the commutation values $i_p \theta_p$). For a particular combination θ_o, θ_p, the equations previously derived [1] enable values of normalised RMS current to be obtained for torque production (i_{RMS}) and for the main thyristors ($i_{T_{RMS}}$), diodes ($i_{D_{RMS}}$) and electrolytic capacitance ($i_{C_{RMS}}$).
The mean power per phase P is given by

$$P = \tfrac{1}{2} V_s I i_{RMS}^2 = \tfrac{1}{2} V_s i_{RMS} i_{RMS} \qquad (3)$$

so that for a specified P and V_s, the unit current I may be calculated and hence the actual currents $i_{T_{RMS}}$, $i_{D_{RMS}}$, $i_{C_{RMS}}$ obtained. This calculation may be repeated for alternative combinations of θ_o, θ_p.
Substituting in eqn. 1 gives

$$i_{RMS} = \frac{\omega^{1/2}}{V_s} \left(\frac{2P\Delta L}{\Delta \theta} \right)^{1/2} \qquad (4)$$

which indicates that to provide rated power over the speed range ω_b to ω_t it is necessary to select current pulses for these speeds for which

$$i_{RMS_t} = i_{RMS_b}(\omega_t/\omega_b)^{1/2} \qquad (5)$$

In general, the longer the relative duration $\{(\theta_q - \theta_o)/\theta_{cy}\}$ of the current pulse, the larger the value of i_{RMS} and the smaller the device ratings $i_{T_{RMS}}$ etc. required for the specified power. Linear analysis thus indicates that a current pulse which occupies the full cycle θ_{cy} at the top speed ω_t is beneficial.

The device ratings will then depend on the choice of control angles θ_o, θ_p for the base speed ω_b condition, giving a value i_{RMS_b} which is limited by the constraint that at the top speed ω_t a waveform with $i_{RMS_t} = i_{RMS_b} (\omega_t/\omega_b)^{1/2}$

must be available. For a specified inductance profile, control angles may therefore be chosen subject to the above limitation which minimise the device ratings $i_{T_{RMS}}$ etc. required for the specified power, and by repeating this for various alternative inductance profiles, the global minimum device ratings are located.

3.3 Constraints at constant torque
Linear analysis has also indicated the benefit of maximising θ_{cy} subject to the constraint of developing rated starting torque in any position. From a linear viewpoint, this constraint is $\theta_{cy} \leqslant p$, but in practice, saturation reduces the torque per unit current significantly as θ approaches 1, and a smaller value of θ_{cy} may be preferable.

It has been assumed above that device ratings are dependent on the base-speed condition. However, the commutation circuit ratings depend on the maximum current to be commutated and this may occur during the low-speed chopping range (Fig. 3) over which base-speed torque is available. If the chopping current waveform is taken to be a rectangular pulse of magnitude I_{MC} and duty cycle $1/p$, where p is the number of phases, occupying part of the rising inductance period, then, since its RMS value must equal that at base speed,

$$I_{MC} = i_{RMS_b}/p^{1/2} \qquad (6)$$

I_{MC} represents the mean current during chopping. The peak or commutation current will be somewhat greater, depending on the magnitude of the current excursions, and this is examined in more detail in Section 4.

The current during chopping is shared between the thyristor and diode for a given phase. Assuming constant current I_{MC}, and equal sharing, the thyristor and diode RMS currents when operating in the chopping mode are equal to $I_{MC}/(2p)^{1/2}$. Unless the current excursions during chopping increase this value significantly, the thyristor RMS current in chopping is less than at base speed and hence the latter has been used below for predicting the thyristor RMS rating, $I_{T_{RMS}}$.

However, to satisfy chopping (and regeneration) conditions, the diode RMS rating $I_{D_{RMS}}$ should be made equal to that for the thyristors $I_{T_{RMS}}$ and not be based on the base-speed motoring condition.

3.4 Example of computer predictions
An illustrative example of the power electronics design for a traction drive is now introduced and developed in Section 4. The drive has a rated output $P_r = 16$ kW between $\omega_b = 25\pi$ and $\omega_t = 75\pi$ rad/s, with constant torque below 25π rad/s and a chopping mode covering the lower part of the constant torque range. For the reasons stated in Section 2, a 4-phase bifilar motor was chosen with unity turns ratio. The supply voltage at rated power was taken as 150 V and the drive efficiency was chosen arbitrarily at 85% (shown later to be a good choice). The effective power per phase was thus $P = 4.7$ kW. The motor geometry was chosen as $\theta_1 = -0.85$, $\theta_4 = 1.15$ and $\theta_{cy} = 3$, and four values of $L_{max}/L_{min} = 21$, 11, 6, 4.33 were considered. For each value, the control

angles θ_o, θ_p which minimised the current ratings at 25π rad/s were identified and it was checked that a corresponding condition existed at 75π rad/s having a normalised torque-producing current $\sqrt{3}$ times larger. The results which satisfy the above are shown in Table 2, which compares theoretical predictions with the practical results for a 16 kW motor of corresponding geometry whose control angles were adjusted for rated power and maximum efficiency.

Examining Table 2 more closely, the agreement between the predicted and calculated values of θ_o and θ_p is noteworthy. The similarity of $\theta_p - \theta_o$ for all inductance ratios indicates a consistency of flux linkages and good agreement with the practical motor: the L_{min} consistency and agreement is impressive, and is of particular value in the commutation circuit design when the windings are used for resonant reversal. the RMS current ratings for the thyristors, and more noticeably for the diodes, vary with the L_{max}/L_{min} ratio as expected; a reasonable choice of this ratio must be made to reflect the degree of motor saturation, and 6 is an acceptable choice for this motor. The capacitor RMS current $i_{C_{RMS}}$ is seriously underestimated by the linear motor model, a direct result of the effects of saturation for the practical motor. It is also a reflection of the inadequacy of the linear model in predicting the actual winding current waveform, reported earlier at Fig. 10 [2]. The predicted and actual current waveforms each for rated power and base speed corresponding to Table 2 are shown in Fig. 8. The occurrence

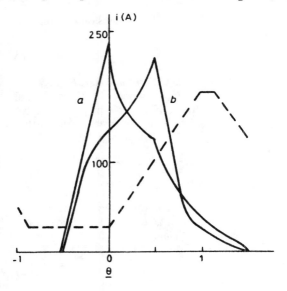

Fig. 8 *Current waveforms*

a Based on linear analysis and $L_{max}/L_{min} = 6$
b Obtained experimentally for the same power, showing similarity of peak currents in spite of differing waveshape

of the maximum current at commutation increases the electrolytic current $i_{C_{RMS}}$ since the current i_p always appears as a step change of electrolytic current.

The actual value of peak current, $\hat{i}$, for equal powers is, however, well predicted by linear theory, although the predicted timing at $\theta = 0$ is incorrect; it occurs at θ_p. Since the

predicted value of $\hat{i}$ at $\theta = 0$ gives a better indication of the actual current at commutation, the value of $\hat{i}$ will be used hereafter as i_p for commutation circuit design.

The approximate equality of the predicted and actual values of $\hat{i}$ for equal powers (Fig. 8) is not entirely fortuitous since, in order to produce the required torque with current pulses of the same duration, the magnitude of these pulses should be similar even if the peak occurs at different angles. To obtain an exact correlation between predicted and actual pulse shapes would, as has already been stated, necessitate analysis of the saturation behaviour which would be burdensome for the inverter designer.

3.5 Conclusions

With a judicious choice of the L_{max}/L_{min} ratio, the predictions based on a linear motor model have been shown to give useful guidance for power electronic circuit design with the exception of electrolytic ripple current. The following approximations for current ratings in terms of rated power P_r, voltage V_s and top speed ω_t, when $L_{max}/L_{min} = 6$ and $\omega_t/\omega_b = 3$, have proved useful.

$$i_{T_{RMS}} = i_{D_{RMS}} = i_{C_{RMS}} = 0.8\, P_r/V_s$$

$$I_{MC} = 1.5\, P_r/V_s; \quad \hat{i} = i_p = 2.1\, P_r/V_s$$

$$L_{min} = 0.27\, V_s^2/P_r\omega_t$$

L_{min} has been related to ω_t rather than to ω_b since the top speed limitation of a 'fully opened' current pulse tends to dictate the value of L_{min} and is generally independent of ω_b unless the ratio ω_t/ω_b is significantly greater than 3.

4 Commutation analysis for 4-phase bifilar inverters

This Section examines the commutation behaviour for the circuits of Figs. 6 and 7, the former using separate resonant reversal components and the latter using the motor windings. It develops relationships from which the capacitance value and commutation thyristor ratings may be estimated. However, for generality, the relationships are also expressed in terms of P_r, V_s and ω_t (using the formula given in Section 3.5) wherever appropriate. The particular application is that summarised in Table 2 with $P_r = 16$ kW, $V_s = 150$ V and $\omega_t = 75\pi$ rad/s.

The choice of commutation capacitance C is influenced by the chopping frequency (or current excursions) at low speed, the reverse voltage time t_{RVM} of the main thyristors and the voltage boost ΔV due to leakage inductance. Whether t_{RVM}, ΔV or chopping frequency dictates the value of C and whether under base speed or chopping conditions depends on the particular application.

For the battery-vehicle application, the supply voltage V_s can vary, rising to 210 V when regenerating, compared to the rated motoring condition of $V_s = 150$ V. Excess thyristor voltage in regeneration can be avoided by limiting the voltage boost either by increasing C or by accepting less regenerated power which is often preferred for a battery vehicle. To avoid confusion in the following analysis, V_s (= 150 V) is the supply voltage at rated power, although the implications of regener-

Table 2: Base speed motoring current ratings for a 150 V 16 kW 4-phase traction drive for various values of L_{max}/L_{min}. (To allow for low speed and regeneration conditions $I_{D_{RMS}}$ should be made equal to $I_{T_{RMS}}$)

		Theoretical values				
L_{max}/L_{min}		21	11	6	4.33	Practical values
$b = \dfrac{L_{min}}{\Delta L}$		0.05	0.10	0.20	0.30	
θ_o		0.6	0.58	0.55	0.54	0.5
θ_p		0.5	0.5	0.5	0.5	0.5
$\theta_p - \theta_o$		1.1	1.08	1.05	1.04	1.0
i_{RMS}		1.763	1.254	0.863	0.688	
i_p		2.0	1.8	1.5	1.3	
l	A	20.2	39.9	84.2	132	
I	A	242	231	232	239	220
i_{TRMS}	A	71	75	87	99	78
i_{DRMS}	A	10	19	34	47	35
i_{CRMS}	A	42	40	57	84	94
I_{MC}	A	71	100	145	182	160
L_{min}	mH	1.65	1.67	1.59	1.51	1.65

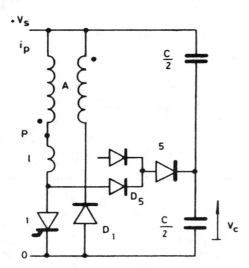

Fig. 9   *Circuit at commutation of thyristor 1.*

ation voltage are mentioned in Section 4.5 in assessing actual values. For low-speed operation where little power is drawn from the battery, the supply voltage should approach the nominal 180 V, but for starting with a discharged battery, chopping behaviour with $V_s = 150$ V is relevant.

4.1 Voltage boost ΔV caused by leakage inductance

Fig. 9 illustrates this effect for the commutation of thyristor 1. The current i_p to be commutated flows into C until, at $V_c = 2V_s$, D_1 turns on and clamps the point P at $2V_s$. The energy in the leakage inductance l transfers to C as i_p transfers to the auxiliary winding, giving the capacitor an additional voltage boost

$$\Delta V = i_p \sqrt{l/C} \tag{7}$$

where i_p is assumed constant during commutation. Let the commutation current i_p be defined in terms of the nominal input current P_r/V_s, i.e.

$$i_p = \mu P_r / V_s \tag{8}$$

the value of μ depending on whether the base speed or chopping condition applies. ΔV is now given more generally as βV_s and is related to μ by

$$\beta = \Delta V/V_s = 0.52\,\mu/\lambda N_c \tag{9}$$

where λ^2 is the ratio L_{min}/l and N_c is a dimensionless constant where

$$N_c = (\omega_t C V_s^2/P_r)^{1/2} \tag{10}$$

4.2 Main thyristor reverse voltage time: inverter, Fig. 6

As is well known, $t_{RVM} = C V_p/i_p$ where V_p is the initial reverse voltage at commutation. For single-pulse operation, no resonant reversal is required and the reverse voltage for the next commutation is $V_p = V_s + \Delta V$, giving

$$t_{RVM} = \frac{CV_s}{i_p} + \sqrt{lC} = \frac{N_c^2(1 + \beta)}{\omega_t \mu} \tag{11}$$

However, when chopping, the capacitor voltage is resonantly reversed using separate components to give $V_p = 2V_s + \Delta V$, assuming negligible reversal loss. Hence

$$t_{RVM} = \frac{2CV_s}{i_p} + \sqrt{lC} = \frac{N_c^2(2 + \beta)}{\omega_t \mu} \tag{12}$$

The total commutation time t_{CM} (i.e. the time for which the commutation thyristor conducts) during chopping is given by

$$t_{CM} = t_{RVM}\left[\frac{4V_s + \Delta V}{2V_s + \Delta V}\right] + \frac{\pi}{2}\sqrt{lC} = \frac{N_c^2(4 + 2.57\beta)}{\omega_t \mu} \tag{13}$$

312

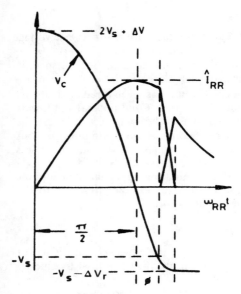

Fig. 10 *Capacitor voltage V_c (Fig. 9) and winding current during resonant reversal*

4.3 Main thyristor reverse voltage time: inverter, Fig. 7

For single-pulse operation, the inverter of Fig. 7 behaves identically to the inverter of Fig. 6 and t_{RVM} is given by eqn. 11.

In chopping, however, using motor windings for resonant reversal, the reverse voltage V_p is clipped at $V_p = V_s + \Delta V_r$ by the action of the auxiliary windings, where ΔV_r is the voltage boost associated with the resonant reversal current which is much less than ΔV due to the commutation current. Hence less than half the reverse voltage V_p for separate resonant reversal is achieved.

This behaviour is illustrated in Fig. 10. Since while chopping in phase A, phase B will be at minimum inductance L_{min}, and neglecting the very much smaller reversal current in phase D, the peak reversal current

$$I_{RR} = (2V_s + \Delta V)\sqrt{C/L_{min}}$$

and the current at transfer to the auxiliary windings is $\hat{I}_{RR} \cos \phi$. The voltage boost $\Delta V_r = I_{RR} \cos \phi \sqrt{1/C}$ and hence

$$\frac{\Delta V_r}{V_s} = (\cot \phi)/\lambda \quad \text{where} \quad \sin \phi = \frac{V_s}{2V_s + \Delta V} = \frac{1}{2+\beta}$$

The main thyristor reverse voltage time during chopping

$$t_{RVM} = \frac{C(V_s + \Delta V_r)}{i_p} = \frac{N_c^2}{\omega_t \mu}\left[1 + \frac{\Delta V_r}{V_s}\right] \qquad (14)$$

and the resonant reversal time

$$t_{RR} = \left(\frac{\pi}{2} + \phi\right)\sqrt{L_{min}C} + \frac{\pi}{2}\sqrt{IC}$$

$$= \frac{0.52 N_c}{\omega_t}\left(\phi + \frac{\pi}{2}\left(1 + \frac{1}{\lambda}\right)\right) \qquad (15)$$

Generally, $\beta(= \Delta V/V_s)$ lies in the range $0.75 < \beta < 1.5$ for which $0.37 > \phi > 0.29$, and λ lies in the range $10 < \lambda < 20$ for which $0.13 < \Delta V_r/V_s < 0.33$. Eqns. 14 and 15 may therefore be simplified with acceptable accuracy to

$$t_{RVM} \simeq \frac{1.2 N_c^2}{\omega_t \mu} \qquad t_{RR} \simeq \frac{1.04 N_c}{\omega_t} \qquad (16, 17)$$

The commutation time

$$t_{CM} = t_{RVM}\left(\frac{3.2}{1.2}\right) + \frac{\pi}{2}\sqrt{IC} = \frac{N_c^2(3.2 + 1.57\beta)}{\omega_t \mu} \qquad (18)$$

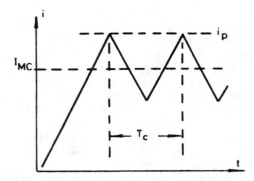

Fig. 11 *Phase current during chopping*

4.4 Chopping period T_c

Consider the zero speed current excursions of magnitude $2(i_p - I_{MC})$ about a mean value I_{MC} with an incremental inductance L_{inc} and an applied voltage $\pm V_s$ as shown in Fig. 11. The chopping period T_c is given by

$$T_c = \frac{4(i_p - I_{MC})L_{inc}}{V_s} = \frac{1.08(\mu - 1.5)}{\omega_t}\frac{L_{inc}}{L_{min}} \qquad (19)$$

During this period, the following times must be accommodated:
   (a) the main thyristor commutation time t_{CM}
   (b) the commutating thyristor reverse voltage time t_{RVC}
   (c) the resonant reversal time t_{RR}
   (d) the reverse voltage time for the resonant reversal thyristor, also taken to be t_{RVC}.
Hence it is a requirement that

$$T_c > t_{CM} + t_{RR} + 2t_{RVC} \qquad (20)$$

For the inverter of Fig. 7, it is logical to fire the resonant reversal thyristor and the main thyristor together, giving

$$0.5\,T_c > t_{RR} + t_{RVC} \qquad (21)$$

Assuming that chopping is controlled by current levels (i.e. the current excursion $2(i_p - I_{MC})$ is constant) then T_c is a minimum when the incremental inductance L_{inc} is a minimum. According to linear theory this occurs at the beginning of the phase when $L_{inc} = L_{min}$, and the chopping period should subsequently increase. However, a practical motor exhibits saturation, especially as θ approaches unity, and the chopping period consequently decreases.

313

To obtain the greatest mean torque for a given commutation capability it is necessary to depart from control of chopping by current levels alone and to use a combination of current levels, ontime and offtime control as described in Section 5.3. This results in chopping waveforms of variable current excursion (Fig. 16), but for the purpose of this Section, constant current excusions with $L_{inc} = L_{min}/2$ are assumed.

4.5 Predicted component values

The commutating capacitor C for the present application is chosen to limit the voltage boost to 150 V, which with a battery voltage of 210 V in regeneration, gives a maximum thyristor voltage of 570 V. At rated motoring power and base speed ($i_p = 224$ A), and for $\beta = \Delta V/V_s = 1$ and $\lambda = 15$ (a midrange value), eqns. 8–11 give $\mu = 2.1$, $N_c = 0.0728$, $C = 16$ μF and $t_{RVM} = 21.4$ μs; these values apply equally to Figs. 6 and 7. When regenerating, the power must then be restricted to ensure that i_p does not exceed 224 A.

With winding resonant reversal (Fig. 7) eqn. 17 gives $t_{RR} = 324$ μs and using 250 μs turnoff commutation thyristors, with $t_{RVC} = 300$ μs, eqn. 21 indicates a chopping period $T_c = 1250$ μs is appropriate. The corresponding commutation current for $L_{inc}/L_{min} = 0.5$ is $i_p = 218$ A ($\mu = 2.044$ from eqn. 19) giving a 73% peak-to-peak/mean ratio.

For $\mu = 2.044$, $\beta = 0.975$ (eqn. 9), $t_{RVM} = 13.2$ μs and $t_{CM} = 52$ μs (eqns. 16, 18). Main thyristors with 10 μs turnoff times are necessary.

The commutation thyristor ratings are based on a rectangular current waveform of 218 A peak, duty cycle 52/1250 = 0.042 while chopping, and, noting that a given commutation thyristor only services alternate phases, the equivalent RMS current is 32 A. The peak resonant reversal current $\hat{I}_{RR} = 45$ A.

With separate resonant reversal (Fig. 6), allowing $t_{RR} = 100$ μs, the chopping period could be reduced to $T_c = 800$ μs (eqn. 21), giving $i_p = 197$ A, $\mu = 1.85$ with a 47% peak-to-peak/mean ratio, thereby giving smoother torque. The values for t_{RVM} and t_{CM} are 35 μs and 76 μs, respectively (eqns. 12, 13). However, 15 μs main thyristors would be required since $t_{RVM} = 21.4$ μs at base speed. The commutation thyristor ratings would be based on 197 A peak and 0.095 duty cycle while chopping. For $t_{RR} = 100$ μs, a 64 μH inductor would be required with a reversal current of 216 A peak, 54 A RMS.

4.6 Other considerations

The relationships derived above illustrate how the device ratings can be estimated for a 4-phase bifilar wound motor for the specified application in advance of an actual motor design, and hence system costs may be estimated. A similar approach may be used for 3-phase systems. The operating conditions which dictate the ratings and capacitance value depend on the application, but the relationships should enable these conditions to be identified.

The above relationships ignore several important inverter characteristics for the sake of simplicity. In particular, di/dt-limiting inductors are required in series with the commutation

thyristors; these will reduce the main thyristor reverse voltage and influence the size of commutation capacitor.

At speeds below base speed, the constant torque specification requires increasingly high peak thyristor current in the single-pulse mode, beyond what has been predicted above, but usually with a short-time rating. Conditions at the lowest single-pulse speed and highest chopping speed must be included when considering commutation components, but the voltage boost makes the single pulse conditions less onerous than they would at first appear.

The final choice of commutating capacitance, chopping frequency, current thresholds etc. can at present only be made after tests on the actual motor.

5 Control electronics

An electronic control system for the experimental SR motor drive based on Fig. 7 is shown in Fig. 12. Digital techniques

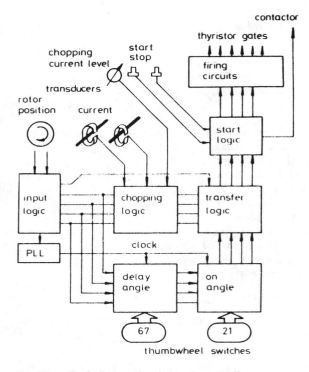

Fig. 12 *Block diagram for electronic controller*

were used to determine the switching angles in the single-pulse mode to ensure accurate and repeatable test conditions. Testing was necessary over a wide range of motoring and regenerating torques at all speeds in the single-pulse mode to identify those angle combinations which give efficient motor operation with acceptably small current ratings for the inverter components. At low speed, the control electronics operate the drive in the chopping mode, where thyristor switching is mainly under current control. The electronics also prime the commutating capacitor at switchon, and organise the mode change from chopping to single pulse and vice versa. Finally, it provides overall protection by tripping the battery contactors if the current becomes excessive. The electronics require transducer inputs from shaft position and winding currents, and manual inputs for start/stop, chopping current level and single-pulse switching angles. Its outputs are the thyristor gate signals and contactor control signals.

5.1 Rotor position transducer

The operative phase is determined by a rotor position transducer designed and built at Leeds University, using two slotted optical sensors which give two 6-pulse per revolution signals 90° phase displaced. Each change of state of these signals corresponds to the beginning of the rising inductance period for a phase, hence 4-phase logic signals can be obtained by combining the transducer signals with appropriate gates. To ensure reliable operation at very low rotor speeds, it is necessary to incorporate a large amount of hysteresis on the input logic.

In the single-pulse mode, the switching angles are determined using digital techniques developed from earlier work [9] at Leeds University. The major development involves a phase-locked loop to generate a high-frequency clock directly from the phase signals. The clock frequency of 720 pulses per revolution permits an angular precision of 0.5 degrees, which was sufficient for the system tests. The phase-locked loop chosen uses an edge-controlled flip-flop phase comparator and exhibits no tendency to lock on to VCO harmonics. Its minimum capture speed is 1.67π rad/s and it remains locked up to 133π rad/s. The capture time was adequately short for the machine inertia.

5.2 Phase-current transducers

At slow speeds the system operates in the chopping mode when thyristor switching is mainly under phase-current control; an isolated current feedback signal is thus required. This was provided by two magneto-resistor current transducers each serving a pair of alternate phases. A magneto resistor is a 2-terminal semiconductor device with a resistance proportional to the magnitude of perpendicular flux density passing through it for flux densities greater than 300 mT, and proportional to the square of flux density at lesser values. When the magneto-resistor is placed in the air gap of an iron core it measures the net current passing through the core. The transfer characteristic exhibits a central linear region bounded by two nonlinearities, one at low current caused by the magneto-resistor characteristics and one at high current caused by core saturation. To ensure a suitable bandwidth of not less than 5 kHz, the lamination thickness of the iron core used was 0.05 mm. The magneto resistor is biased with a constant current source such that the voltage across it is proportional to its resistance. The main winding current and the inverse of the auxiliary winding current for two motor phases are summed in the core to produce a composite analogue current signal which requires no further amplification. The nonlinearities and temperature dependence of the magneto-resistor current transducer were not found to cause problems in the experimental control system. Fig. 13 shows at (a) the current measuring system for one pair of phases, with a current transducer transfer characteristic at (b). Fig.

14 shows a waveform example in which the upper trace is the magneto-resistor output voltage V_{MR} and the lower trace is the main winding current measured with a coaxial shunt.

5.3 Slow speed operation: priming, chopping mode and mode charge

When the drive is started from rest, the start logic block in

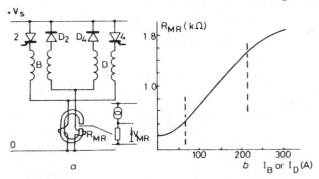

Fig. 13 *Magneto resistor*

a   Circuit for current measurement in two phases
b   Resistance against current characteristic

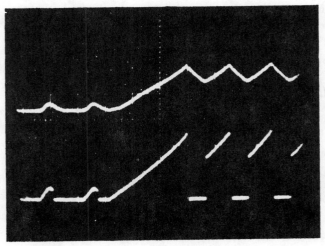

Fig. 14 *Magneto-resistor voltage waveform V_{MR} (upper trace) compared with thyristor current (lower trace) measured with coaxial shunt*

Calibrations: 2 V/cm, 100 A/cm, 1 ms/cm

Fig. 12 closes the battery contactors and 'primes' the commutating capacitor into the required state depending on the rotor position by sequentially firing the commutating thyristors. Control is then transferred to the chopping logic which fires the main and commutating thyristors in the operative phase by comparing the phase current with an upper and lower threshold respectively. The output torque is controlled by varying the current threshold levels.

With the exception of the first commutation of the incoming phase, there must be a resonant reversal of the commutation capacitor prior to each commutation: the dormant commutating thyristor is thus fired each time the main thyristor is refired. This control policy gives rise to a minimum

'ON' time for the main thyristor to accommodate the resonant reversal. When the machine is operating with a high degree of saturation the minimum 'ON' time will over-ride the control from the current transducer; at the end of the phase the minimum 'ON' time will also delay commutation until it is safe to do so. Thus it becomes necessary to incorporate a minimum 'OFF' time, which delays re-firing the main thyristor when current control is over-ridden, to ensure peak-current amplitude stability. The minimum OFF time can be increased manually to provide control of torque below that obtained with minimum threshold levels. The control system does not provide regeneration during the chopping mode.

At an appropriate speed within the constant torque range, control is transferred from the chopping mode to the single-pulse mode; this is handled by the transfer logic block. A frequency to voltage convertor driven from the clock signal and a comparator determine the mode change speed. The transfer of control must be sequenced to ensure that correct commutation is maintained. At mode change, firing pulses are blanked at the last commutation of the A phase and begin again with the D-phase firing pulse of the new control mode. The sequence is the same for the downwards transfer from the single pulse to the chopping mode. The output torque is zero for the duration of the transfer period, hence a hysteresis of approximately 12% is incorporated in the mode-change speed detector.

5.4 Operation in single-pulse mode
For single-pulse operation the switching angles are determined by digital angle counters which are fed by the high-frequency clock signal. The appropriate main thyristor is fired after a delay angle set into the counter by two decade digit switches

Fig. 15 *Integral motor-inverter construction*

and the commutation thyristor is fired after a second delay set by another two decade digit switches which define the on angle. The angular resolution is $\frac{1}{2}°$/digit with an angular range of 49.5° for both control angles. The control angles can be changed while the machine is running, by means of a synchronised data-loading system. The reference angle for the delay counter can be switched to any of the 4-phase

signals to give control over the complete machine inductance cycle. In single-pulse operation, the change from motoring to regeneration is achieved by simply altering the control angles, hence the magnitude and direction of torque for a large variety of switching angles can be established and an optimum control law identified.

5.5 Protection
The overall protection for the experimental SR drive system is provided by high-speed DC contactors in series with the battery supply. A combined signal from the two magneto-resistor current transducers is fed into a comparator which changes state if the current in any of the four phases exceeds a safe level, tripping the contactors. Some filtering is used so that the protection system does not respond to fast spikes of current, and once tripped the contactors cannot be reclosed until a manual reset button is operated.

5.6 Practical features
The electronic control system described was powered from a 12 V automotive battery supply and was built using CMOS logic on a 10 V logic level. CMOS was used because of its superior noise immunity and low power consumption. The thyristors were fired with single gate pulses isolated from the power electronics by means of pulse transformers.

6 Experimental results

6.1 Integral construction
The 4-phase bifilar motor has 16 winding ends. Some of these can be commoned, leaving 12 power connections from the motor. This, and the desirability of a single motor-inverter package for a battery vehicle led to the integral construction shown in Fig. 15. The system is force ventilated, not because it requires excessive cooling, but because a shaft fan designed for the required cooling at 25π rad/s would consume too much power at high speeds: a separate constant-speed fan is therefore preferable.

The entire inverter, including electrolytics, battery contactors, gate pulse transformers and current transducers is housed at the nondrive end of the motor. To convey an idea of size, the inverter is 380 mm long.

For the test results given below the commutation capacitance was 16 μF; the battery voltage varied from 150 V motoring to 210 V regenerating and was generally between 170 and 180 V for low-speed chopping.

6.2 Experimental results: chopping
In the chopping mode the current at commutation and refiring is determined by two comparator levels, provided that the thyristor ON and OFF times each exceed preset minima. The lower trace of Fig. 16 shows the phase current (main and auxiliary currents combined) preceded by six resonant reversal pulses required by the previous phase. Even at this low value of current (mean level $\simeq$ 50 A) corresponding to a measured output torque of 57 Nm, the chopping frequency does not decrease as θ approaches 0.75. The predicted torque for $I_{MC} = 50$ A is 63 Nm.

At higher current (upper trace, torque = 194 Nm) the chopping frequency increases and the minimum OFF time

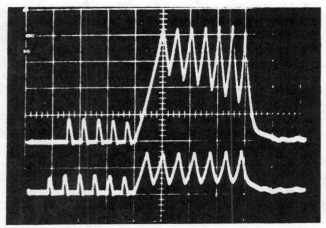

Fig. 16 *Phase current (chopping at 6.25 π rad/s) preceded by resonant reversal pulses*

Upper trace (194 Nm), current minima set by minimum offtime
Lower trace (57 Nm), current maxima and minima comparator-controlled
Calibrations: 53 A/cm, 3.3 ms/cm

forces the lower threshold to droop. The peak or commutation current is very similar to that predicted in Section 4.5 (220 A) but the chopping period is longer (1600, c.f. 1250 μs). This is because the minimum OFF time was set to a higher value (800 μs) during preliminary tests to give a greater margin for safe commutation. As a result, the current excursion (155 A) at the end of the chopping sequence is larger than predicted (116 A), the mean current is lower and the torque produced (194 Nm) is slightly below the 203 Nm equivalent to 16 kW at 750 rev/min. For 155 A excursions with 170 V supply, the incremental inductance is 0.9 mH but would be

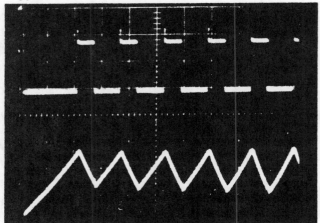

Fig. 17 *Main thyristor voltage and phase current (chopping at 3.33 π rad/s)*

Calibrations: 200 V/cm, 110 A/cm, 1 ms/cm, zero for current trace at bottom of graticule

slightly lower for 116 A excursions with the same peak. The resonant reversal pulses are 45 A peak as predicted.

Fig. 17 shows the phase current expanded for a mean chopping level of 100 A with the thyristor voltage above it. The measured torque was 122 Nm compared with 127 Nm for $I_{MC} = 100$ A. The reverse voltage (barely visible) is

approximately −170 V which is 40 V below that expected with 175 V supply. This loss of reverse voltage is mainly due to the effect of the *di/dt* inductor as current transfers from the main to the commutating thyristor.

6.3 Experimental results: 25 π rad/s
At base speed, the phase current and thyristor voltage are shown on Fig. 18, corresponding to an output power of 16 kW, and an efficiency of 84.5%. The upturn of current near the peak is the result of saturation, which also causes

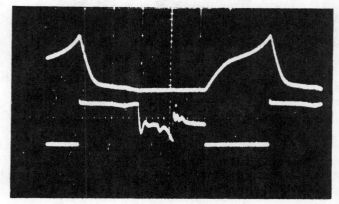

Fig. 18 *Phase current and main thyristor voltage, motoring at 25 π rad/s and 16 kW*
Calibrations: 110 A/cm, 200 V/cm, 2 ms/cm

a faster initial decay rate after commutation. The thyristor voltage waveform shows the thyristor ON time followed by diode conduction of slightly shorter duration. After diode recovery the winding is open circuit and responds to any magnetic unbalance. Just after the centre of the screen the auxiliary diode is seen to conduct very briefly following the commutation of the next-but-one phase, since the two phases are always commutated together.

6.4 Experimental results: 75 π rad/s
At this speed and rated power, the current pulse is expected to occupy virtually a whole cycle, and this is illustrated in Fig. 19. Although small, the presence of diode current is confirmed by the thyristor voltage trace which remains high (proving diode conduction) for almost 50% of the cycle. The commutation is visible at the end of each thyristor ON period and just before the ON period the effect of the commutation of the next-but-one phase can be seen. Some blurring of the former commutation is attributed to random 1 bit error on the digital counters which determine firing instants. The output power is 17.6 kW at 90.6% efficiency.

6.5 Further results
The above test results were taken from a wide range performed to identify the best control law. It is hoped to publish more details in the future.

7 Conclusions
Various inverter circuits and their requirements in terms of device voltage ratings have been considered for bifilar and

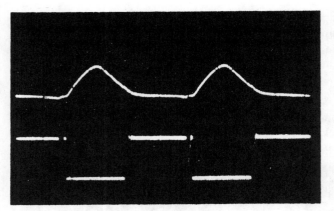

Fig. 19 *Phase current and main thyristor voltage, motoring at 75 π rad/s and 17.6 kw*
Calibrations: 110 A/cm, 200 V/cm, 1 ms/cm

non-bifilar switched reluctance motors. For the battery vehicle application, the 4-phase bifilar arrangement is preferred.

A design method for an inverter feeding such a motor and the prediction of ratings for its major components, based on currents calculated from a linear motor model, has been presented. Since the motor does not operate with familiar current waveshapes, e.g. sine waves, the design of the power electronics cannot proceed without the knowledge of current waveforms, even if only approximate. Good agreement with measurements obtained from a practical drive confirms that the linear model is helpful both in this respect and in providing a simple picture of system behaviour. In seeking an optimum system, due allowance for the effects of saturation in the motor must however be made.

The importance of designing the system as a whole is emphasised. The inductance profile which minimises device ratings also serves as a useful guide for the design of a real motor which will hopefully minimise overall costs.

The control electronics, designed specifically to facilitate accurate and repeatable test conditions using digitally controlled 'delay' and 'on' angles, has proved of great value as a development aid. It is not, however, representative of the control techniques required for a commercial application.

8 Acknowledgments

The authors wish to thank their colleagues [2] at Leeds University for their invaluable contribution to the project and, in particular, for designing and constructing the motor and its rotor position transducer, from which test data were taken.

Thanks are also due to Chloride Technical Ltd. for its sponsorship of the project, for permission to release the details given in this paper, and for helpful discussions with Dr. M.F. Mangan, and to Prof. R.L. Beurle for the use of the departmental facilities at Nottingham University.

9 References

1 RAY, W.F., and DAVIS, R.M.: 'Inverter drive for doubly salient reluctance motor: its fundamental behaviour, linear analysis and cost implications', *IEE J. Electr. Power Appl.*, 1979, **2**, (6), pp. 185–193

2 LAWRENSON, P.J., STEPHENSON, J.M., BLENKINSOP, P.T., CORDA, J., and FULTON, N.N.: 'Variable-speed switched reluctance motors', *IEE Proc. B, Electr. Power Appl.*, 1980, **127**, (3), pp. 253–265

3 DAVIS, R.M., RAY, W.F., and BLAKE, R.J.: 'An inverter drive for a switched reluctance motor'. International conference on electrical machines, Athens, Sept. 1980

4 LAWRENSON, P.J., STEPHENSON, J.M., FULTON, N.N., and CORDA, J.: 'Switched reluctance motors for traction drives'. *Ibid.*

5 DAVIS, R.M., and RAY, W.F.: British Patent 13416, 1977

6 *Idem*, British Patent 22892, 1978

7 *Idem*, British Patent 22893, 1978

8 *Idem*, British Patent 13415, 1977

9 STEPHENSON, J.M.: British Patent 22891, 1978

Development of a Unipolar Converter for Variable Reluctance Motor Drives

J.T. Bass, M. Ehsani, T.J.E. Miller and R.L. Steigerwald

IEEE Transactions, Vol. IA-23, 1987, pp. 545-553

Abstract—A new converter concept for driving the switched reluctance motor has been developed. This converter has only one switching device per phase, uses a unipolar dc supply, returns all the trapped energy to the source, and does not require bifilar windings; it is called a C-dump converter because the trapped energy is dumped in a capacitor and then returned to the dc source. The topology for several different C-dump converters is presented. In addition, the design and experimental results for a C-dump converter using a chopper to recover the energy dumped on the capacitor are presented.

I. INTRODUCTION

The widespread interest in the switched reluctance (SR) motor in recent years [1]–[3] is focused on the design and capabilities of the motor with relatively less attention on the electronic power converter [4]. The motor is a variable-reluctance (VR) stepper motor operated with controlled current and controlled conduction angles. It has no brushgear and no rotor windings, and the stator is simple and robust with short endwindings. The motor has been shown to be at least equal to the inverter-fed induction motor [2] in efficiency, torque per unit volume, copper utilization, and inverter kVA requirements, and substantial improvements are expected in the future.

The SR drive has additional advantages compared with the conventional adjustable-speed ac or brushless dc drives (including permanent-magnet motor drives). First, shoot-through faults are impossible [3]. This is true for all SR converter circuits because there is always a motor winding in series with each main power switching device. Second, there is a greater degree of independence between the phases than is possible in conventional ac or brushless dc drives. A fault in one phase (whether in the motor or in the converter) generally affects only that phase; the other phases can continue to operate independently. In ac drives a fault in one phase can disable two phases. Third, the SR motor has zero short-circuit current and zero open-circuit voltage, so that the secondary

Paper IPCSD 86-48, approved by the Static Power Converter Committee of the IEEE Industry Applications Society for presentation at the 1985 Industry Applications Society Annual Meeting, Toronto, ON, Canada, October 6–11. Manuscript released for publication November 7, 1986.

M. Ehsani is with the Department of Electrical Engineering, Texas A&M University, College Station, TX 77843.

J. T. Bass is with the Bass Engineering Company, Inc., P.O. Box 5279, Longview, TX 75608.

T. J. E. Miller is with the Department of Electronics and Electrical Engineering, University of Glasgow, Glasgow, Scotland, UK G12 8QQ.

R. L. Steigerwald is with Corporate R&D, General Electric Company, Building 37-463, 1 River Road, Schenectady, NY 12345.

IEEE Log Number 8613392.

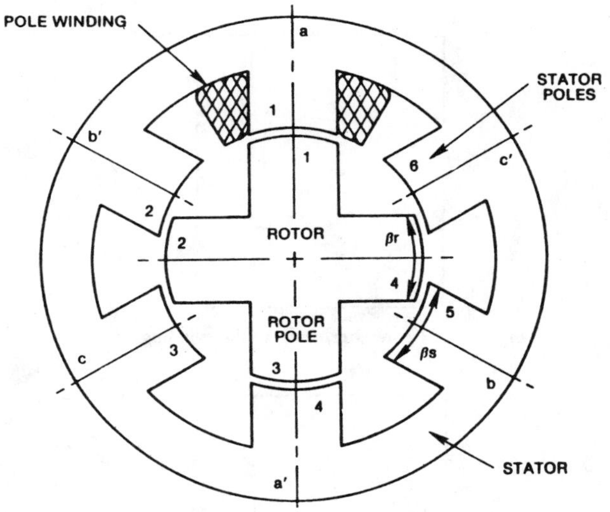

Fig. 1. Basic switched reluctance motor.

effects (such as overvoltages or overheating) of faults in either the motor or the converter tend to be benign.

Fig. 1 shows the cross-section of a small developmental SR motor. It has been realized [1], [3] that the reluctance motors requires only unipolar or unidirectional currents and this gives rise to the possibility of operating with only one switching device in series per phase, instead of two in series in each phase leg of an ac or brushless dc drive. Such circuits use fewer semiconductor devices than their ac counterparts, and they have only one forward voltage drop in series per phase, so that the power losses may, in principle, be lower than in conventional inverters. Other factors being equal, both of these factors should permit a reduction in the physical size of the converter and an increase in its reliability. The isolated current sensors employed in many ac drives are eliminated in many SR drives, which can use sensors referenced to the common dc rail.

The main classes of SR converter circuits are reviewed and a number of new ones, including a new unipolar circuit with only one main switching device per phase is described. This circuit employs a capacitor to recover the magnetically trapped energy in the off-going motor phase, and is therefore called the C-dump, or "capacitive suppression," circuit.

II. CONVENTIONAL SR CONVERTER TOPOLOGIES

Fig. 2 shows a SR converter circuit that is similar to the conventional six-step ac inverter, except that the motor windings are in series with the phase switches. The upper and

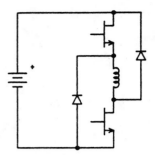

Fig. 2. SR converter with two switches per phase.

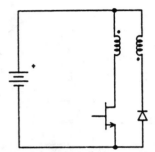

Fig. 3. SR converter using bifilar windings.

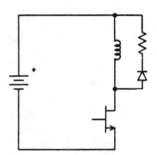

Fig. 4. SR converter with suppression resistor.

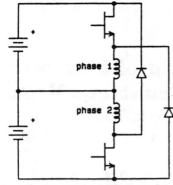

Fig. 5. SR converter using double-rail dc supply.

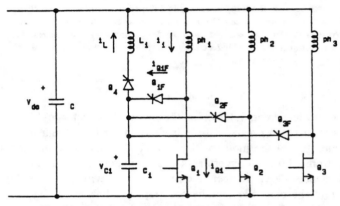

Fig. 6. C-dump converter with resonant energy recovery.

lower switches turn on and off together and have voltage and current ratings that are practically the same as those of equivalent ac inverter drives. Shoot-through faults are prevented by the motor winding and "lock-out" circuitry is not necessary.

Among the circuits that have been proposed to achieve unipolar operation with only one switch per phase, the simplest uses bifilar motor windings (Fig. 3). This circuit has been used extensively in stepper motor drives. The bifilar winding is undesirable because the coupling between the primary and secondary windings is always imperfect, and this leads to voltage spikes during current turn-off. In addition, the utilization of copper in the motor is less than in a comparable monofilar design, and the number of connections between the motor and the converter is doubled.

Fig. 4 shows a single-switch circuit that avoids the use of bifilar windings by commutating the phase currents into a common free-wheeling resistor ("suppression resistor" or "dump resistor"). The power losses in the resistor are inherently large in most cases, and this circuit is certainly not practical in integral-horsepower sizes. For maximum efficiency, commutation must take place as close as possible to the aligned position of stator and rotor poles, and a high resistance

is then needed to force the current quickly to zero. But this produces an unacceptably high peak voltage on the switching device.

Fig. 5 shows a derivative of Fig. 2 in which a split-level dc source is used. The circuit shown has four switches, four diodes, and four motor phases, giving one main switch per phase. Note that each of the upper phases freewheels into the lower dc source, and vice versa. However, a fault in any phase would unbalance the upper and lower levels. This circuit, therefore, does not have as much independence between phases as the previous ones.

III. C-Dump Converters

A. C-Dump SR Converter with Resonant Energy Recovery

The circuit shown in Fig. 6 is called the C-dump circuit because the trapped energy is dumped into a capacitor before being returned to the dc source by a resonant circuit. In addition, this circuit has only one switch per phase, and it uses a single rail dc supply. The waveforms describing the converter's operation are shown in Fig. 7. When (or just before) Q_1 switches off, the phase current i_1 that was flowing through Q_1 commutates to Q_{1F} and begins to charge capacitor C_1. The rising capacitor voltage forces an increasingly rapid decay of the phase current. At the end of this working stroke, all the trapped energy is stored on the capacitor. To recover this energy to the dc source, Q_4 is fired, and C_1 discharges its energy resonantly through $L1$ into the dc source capacitor C.

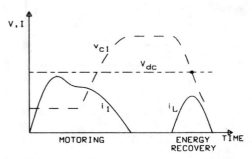

Fig. 7. Voltage and current waveforms from C-dump converter with resonant energy recovery.

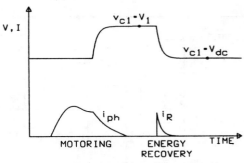

Fig. 9. Voltage and current waveforms from C-dump converter with damped energy recovery.

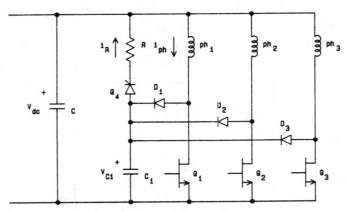

Fig. 8. C-dump converter with damped energy recovery.

Q_4 turns off when i_L reaches zero. However, during the resonant recovery, the freewheeling SCR's become forward biased when $v_{c1} = V_{dc}$. This explains why SCR's are used for freewheeling, instead of diodes.

The circuits in Figs. 2, 3, and 5 produce a -1 pu phase voltage to demagnetize the phase winding after a phase is switched off. As a result the air gap flux decreases linearly after the phase is turned off, causing the conduction times to be equal in the freewheeling diodes and phase switches. For the converter in Fig. 6, however, the flux continues to increase after the phase is switched off, until the dump capacitor voltage is equal to the supply voltage. As a result the phase current slope does not change drastically at the commutation point as it would with the conventional SR converters. After the capacitor voltage rises above the supply voltage, the flux starts decreasing at an increasing rate.

B. C-Dump SR Converter with Damped Energy Recovery

The difficulties of the circuit in Fig. 6 arise from the fact that v_{c1} falls, or "rings down," well below V_{dc}, making it necessary to use the free-wheeling thyristors. Instead, it is desirable to maintain the capacitor voltage v_{c1} above the supply voltage permanently. One possible means to do this is shown in Fig. 8. The resistor prevents v_{c1} from dipping below the supply voltage V_{dc}, but the resistor also dissipates part of the energy being recovered.

The proportion of energy lost in the resistor is a function of the supply voltage V_{dc}, and the capacitor voltage V_1 after the phase current decays to zero. The waveforms that describe the operation of the energy recovery circuit are shown in Fig. 9. It can be shown that the expressions for the recoverable energy given up by the capacitor and the energy lost in the resistor are given by (1) and (2), respectively:

$$\text{Recoverable Energy} = E_C = \frac{1}{2} C(V_1^2 - V_{dc}^2) \qquad (1)$$

$$\text{Energy lost in } R = E_R = \frac{1}{2} C(V_1 - V_{dc})^2. \qquad (2)$$

The proportion of the recoverable energy that is dissipated in the resistor is given by (3):

$$\rho = \frac{E_R}{E_C} = \frac{V_1 - V_{dc}}{V_1 + V_{dc}}. \qquad (3)$$

If V_1 approaches V_{dc}, most of the energy given up by the capacitor is recovered by the source. On the other hand, if V_1 is much larger than V_{dc}, all the energy is lost in the resistor. For a typical case where V_1 is twice V_{dc}, one-third of the energy dumped in the capacitor is lost in the resistor during the recovery cycle.

C. C-Dump SR Converter with Chopper Energy Recovery

The circuit in Fig. 8 provided a means of recovering part of the trapped energy and still permanently maintaining the dump capacitor voltage above the supply voltage. However, in many applications, the amount of energy lost during the recovery cycle would be unacceptable. The chopper shown in Fig. 10 provides a low-loss means of recovering the stored energy while permanently maintaining the capacitor above the supply voltage.

The amount of energy recovered during each cycle can be controlled with the chopper switch Q_4. When the chopper switch is closed, the dump capacitor C_1 begins to discharge. After the chopper switch is opened, the stored energy in L is returned to the source through the free-wheeling diode D. This is, therefore, the conventional buck, or step-down, converter. The capacitor voltage can be controlled with the converter and most of the energy is recovered.

IV. ANALYSIS AND DESIGN OF CHOPPER ENERGY RECOVERY

The C-dump circuit with chopper energy recovery was selected to be analyzed further and prototyped because it has

321

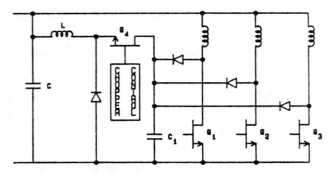

Fig. 10. C-dump converter with chopper energy recovery.

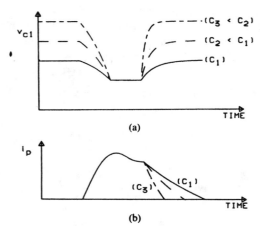

(a)

(b)

Fig. 11. Three dump capacitors. (a) Dump capacitor voltage. (b) Phase current.

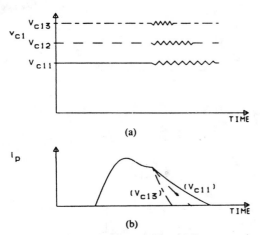

(a)

(b)

Fig. 12. Three mean dump capacitor voltages. (a) Dump capacitor voltage. (b) Phase current.

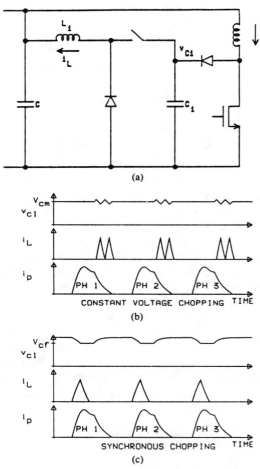

(a)

CONSTANT VOLTAGE CHOPPING

(b)

SYNCHRONOUS CHOPPING

(c)

Fig. 13. Constant voltage and synchronous chopping. (a) Dump capacitor voltage. (b) Reactor current. (c) Phase current.

an all-transistor implementation and is controllable from a simple strategy. This simple control strategy will be presented next.

A. Phase Current Control

One feature of this converter is phase current waveshape control after the phase is switched off. If the chopper switch is held off while the phase is being demagnetized, the dump capacitor voltage rises as the phase current decays. When a small dump capacitor is used, its voltage increases sharply, forcing an increasingly rapid decay of the phase current to zero. As the capacitor size decreases, the voltage rise increases forcing the current to zero faster. The plots in Fig. 11 show the waveshape for three different dump capacitors and the resulting capacitor voltage.

If the chopper is used to maintain the dump capacitor voltage at a constant value, the waveshape after turn-off can still be controlled. If the voltage is maintained at a high level, the phase current is forced to zero quickly. However, when the voltage is maintained at a low level, the phase current decays to zero much more slowly. The plots in Fig. 12 show the phase current for three different mean dump capacitor voltages. By controlling the waveshape in this manner, it may be possible to use the capacitor voltage as a secondary speed control.

B. Chopper Control

The aforementioned waveshape control indicates the possibility of two control modes. These modes are shown in Fig. 13

322

and will be referred to as constant voltage and synchronous control. The constant voltage mode uses a bang-bang control to maintain the dump capacitor voltage within a fixed band. The width or amplitude of this band is ΔV_c, and the center of the band is V_{cm}. The resulting chopping frequency is inversely related to ΔV_c.

The circuit in Fig. 14 is used to implement the constant voltage control. Since the reference voltage (αV_{dc}) is a

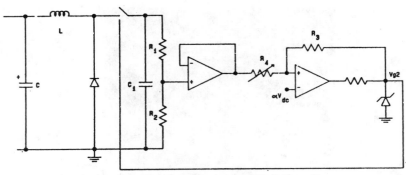

Fig. 14. Control circuit for constant voltage chopping.

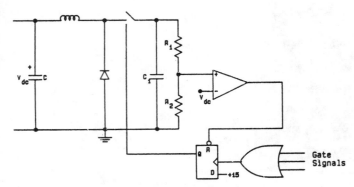

Fig. 15. Control circuit for synchronous chopping.

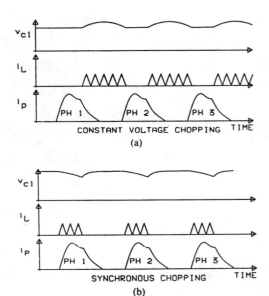

Fig. 16. Phase current using peak current limit. (a) Dump capacitor voltage. (b) Reactor current.

multiple of the dc link voltage, V_{cm} is a multiple of the supply voltage. Consequently, if the dc link voltage is varied to implement a torque or speed control, V_{cm} changes proportionally. The chopping band ΔV_c is adjusted by R_4.

If the synchronous control mode is used, the dump capacitor is discharged at the beginning of each working stroke to recover the energy dumped on the capacitor at the end of the preceeding working stroke. This recovery should be complete before the phase switch turns off. After phase is switched off, the capacitor voltage increases until the phase current is zero. The increase in the capacitor voltage is a function of the capacitor size, initial voltage V_{cf}, and the amount of trapped energy that is recovered.

The circuit in Fig. 15 is used to implement the synchronous control. When any phase switch is gated, the chopper switch is turned on by the D flip-flop. The comparator resets the flip-flop, which turns off the chopper switch when the capacitor voltage is equal to the desired V_{cf}. As in the previous control circuit, the desired V_{cf} is a multiple of the supply voltage.

In many cases, both the constant voltage and synchronous control produce a low duty cycle, low chopping frequency, and large peak reactor currents. This occurs because all the energy is being recovered by the source during only a small part of the phase period. For the constant voltage control, the energy is recovered only while the phase winding is being demagnetized. When synchronous control is used, all the energy is recovered in only one chopping cycle. In both cases, if the energy can be recovered in smaller quantities spread over a greater part of the period, the peak reactor current will be much lower. Consequently, the physical reactor size may

be smaller and the peak current rating of the switch can be decreased. The plots in Fig. 16 show the reactor current, phase current, and dump capacitor voltage when the peak reactor current is limited.

V. PERFORMANCE EVALUATION

A C-dump converter using a 10-μF dump capacitor and a 1.5-mH reactor was used to drive a 2.5-W SR motor from a 40-V dc bus. This prototype system was used to determine experimentally how the system efficiency using the C-dump converter compared to the 46.5-percent system efficiency obtained with the 2.5-W motor and the two-switch-per-phase converter shown in Fig. 2.

A. Constant Voltage Chopping

Several parameters affect the operation and efficiency of the chopper. The four that affect it significantly are the width of the chopping band, the mean dump capacitor voltage, the reactor size, and the dump capacitance. Therefore the constant voltage control was tested by holding three of these constant and by varying the remaining parameter. The results obtained

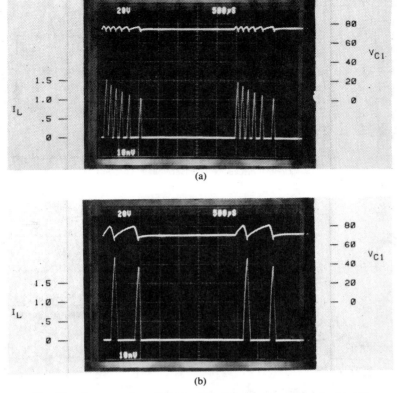

(a)

(b)

Fig. 17.   Constant voltage chopping with different chopping bands. (a) ΔV_c = 5 V. (b) ΔV_c = 15 V.

TABLE I
RESULTS FROM CONSTANT VOLTAGE CONTROL USING DIFFERENT CHOPPING BAND

ΔV_c (V)	P_{in} (W)	P_{out} (W)	Efficiency (%)	Speed (r/min)
1	6.0	2.47	41.2	1510
5	5.9	2.50	42.4	1520
15	6.1	2.45	40.2	1495

by varying experimentally one parameter at a time in this manner are presented in Fig. 16.

The oscillograms in Fig. 17 show the dump capacitor voltage and the reactor current for two different voltage bands. The experimental results are given in Table I. The variation in the amplitude of the chopping band does not have a significant effect on the system efficiency. However, the efficiency with a 5-V amplitude is slightly higher.

The oscillograms in Fig. 18 show the dump capacitor voltage, reactor current, and the phase current for two different mean dump capacitor voltages. The system efficiency and motor speed for four dump capacitor voltages are given in Table II. As discussed previously, the phase current after the phase switch turns off decays faster as the mean dump capacitor voltage increases, resulting in a decrease in the torque produced. This effect can be seen in Table II as the speed decreases when the capacitor voltage increases. In addition, the system efficiency decreases as the dump capaci-

324

tor voltage increases. However, at V_{cm} = 80 V which produces the −1 pu demagnetization voltage typical of conventional SR converters, the system efficiency compares favorably to the 46.5 percent obtained from the two-switch-per-phase converter.

The converter was tested with two different reactors. The inductances of the reactors were 1.5 mH and 2.0 mH and the resistances 2.4Ω and 0.9Ω, respectively. In addition, the converter was tested with a resistor in place of the reactor. The system efficiency for each reactor and the resistor are given in Table III. The large reactor was the most efficient for both V_{cm} levels. However, when V_{cm} was 120 V, there was only a slight difference between the two reactors. The primary reason for the system efficiency being higher with the large reactor is that resistance of the larger reactor is much lower, which reduces the I^2R losses. Therefore a tradeoff exists between the reactor costs and the system efficiency.

The system efficiency was measured using four different dump capacitor sizes. For each capacitor, measurements were taken at three different V_{cm} levels. The results are given in Table IV. For a given dump capacitor voltage, the efficiency is independent of the capacitor size over the range tested. Since small dump capacitors are desirable, this result is significant.

B. Constant Voltage Control with Peak Current Limit

The system was evaluated using the peak current limit with constant voltage control. The measurements were performed using a 2.0- and a 1.5-mH reactor. With each reactor, data

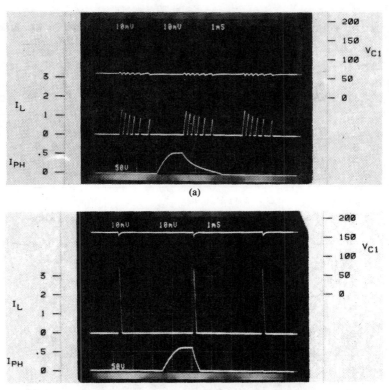

(a)

TABLE II
RESULTS FROM CONSTANT VOLTAGE CONTROL FOR DIFFERENT MEAN
DUMP CAPACITOR VOLTAGE

V_{cm} (V)	P_{in} (W)	P_{out} (W)	Efficiency (%)	Speed (r/min)
60	6.0	2.63	43.8	1580
80	5.8	2.53	43.6	1520
120	5.9	2.37	40.2	1440
160	6.0	2.30	38.3	1392

TABLE III
RESULTS FROM CONSTANT VOLTAGE CONTROL WITH DIFFERENT
REACTORS AND MEAN DUMP CAPACITOR VOLTAGE

Reactor	V_{cm} = 60 V			V_{cm} = 120 V		
	P_{in} (W)	P_{out} (W)	Efficiency (%)	P_{in} (W)	P_{out} (W)	Efficiency (%)
2.0 mH, 0.9 Ω	6.0	2.59	43.2	5.90	2.31	39.1
1.5 mH, 2.4 Ω	6.1	2.55	41.8	5.95	2.30	38.6
5 Ω	7.6	2.53	33.3	8.30	2.29	27.6

was obtained for 0.5- and 0.75-A peak current limits and with no current limits. The results of those measurements are given in Table V. The dump capacitor voltage, reactor current, and phase current for a peak current limit of 0.5 A are shown in Fig. 19. This oscillogram should be compared to the one in Fig. 18(a) since the energy recovery circuit is operating with the same parameters except for the peak current limit. Since the energy is recovered in small quantities, the dump capacitor voltage rises while the phase winding is being demagnetized. After the phase current decays to zero, the chopper circuit continues to discharge the dump capacitor until its voltage decays to the bottom of the constant voltage band.

The efficiency is constant for all three cases with each reactor. When the peak current limit is used, the I^2R losses decrease, but the switching losses increase. The data suggests that these two factors offset each other, and as a result, the efficiency remains constant. Therefore the primary advantage of the peak current limit is to reduce the reactor size. The peak energy stored by the reactor when no current limit is used is nine times larger than the peak energy stored when a 0.5-A peak current limit is used.

VI. CONCLUSION

The C-dump converter is an attempt to achieve efficient unipolar operation of the switched reluctance motor with only one switch per phase. The main phase switches are all referenced to the negative dc rail, and there are no "upper" devices in the phase legs. The C-dump converter has all of the main advantages of SR converters, including a high degree of independence between the phases and freedom from shoot-through faults. It avoids the use of bifilar windings or split-level dc supplies, and returns a high porportion of the trapped energy to the source instead of dissipating it in a suppression resistor as in many stepper motor circuits. The control of the

325

TABLE IV
RESULTS FROM CONSTANT VOLTAGE CONTROL WITH DIFFERENT DUMP CAPACITORS AND MEAN DUMP CAPACITOR VOLTAGE

Capacitor (μF)	V_{cm} = 60 V			V_{cm} = 80 V			V_{cm} = 120 V		
	P_{in} (W)	P_{out} (W)	Efficiency (%)	P_{in} (W)	P_{out} (W)	Efficiency (%)	P_{in} (W)	P_{out} (W)	Efficiency (%)
4	5.75	2.51	43.6	5.6	2.43	43.4	5.55	2.26	40.7
10	6.00	2.55	42.5	5.8	2.44	42.1	5.80	2.30	39.7
20	5.75	2.50	43.5	5.6	2.38	42.5	5.60	2.23	39.8
40	5.80	2.52	43.4	5.6	2.40	42.9	5.60	2.26	40.4

TABLE V
RESULTS FROM CONSTANT VOLTAGE CONTROL USING DIFFERENT PEAK CURRENT LIMITS

Current Limit (A)	L = 2.0 mH, 0.9Ω			L = 1.5 mH, 2.4 Ω		
	P_{in} (W)	P_{out} (W)	Efficiency (%)	P_{in} (W)	P_{out} (W)	Efficiency (%)
0.50	5.9	2.55	43.4	5.9	2.56	43.4
0.75	5.9	2.59	43.9	5.9	2.56	43.4
No Limit	6.0	2.60	43.3	6.1	2.56	42.0

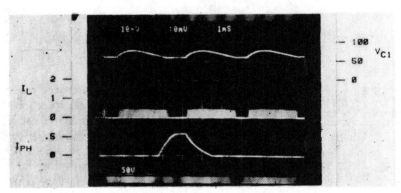

Fig. 19. Constant voltage chopping with peak current limit.

dump capacitor voltage is simple to implement, and the dump reactor current can be controlled by chopping the chopper transistor. This minimizes the required size of the dump reactor.

The C-dump converter is not strictly a one-switch-per-phase circuit because of the addition of the chopper transistor. In addition, there are the dump capacitor, reactor, and diode. There is an absolute minimum rating for all four of these components, and this is determined by the efficiency of the motor.

The volt-amperes that must be handled by the chopper are typically of the order of 20–30 percent of the power conversion in the motor, and there are losses associated with the energy recovery process, primarily in the chopper transistor and reactor. Nevertheless the tests have shown efficiencies very close to those attainable with all-silicon converters having two switches per phase. At lower voltages the C-dump converter can be expected to improve its efficiency relative to the two-switch-per-phase circuits.

The chopper transistor is not referenced to the negative dc rail, and this complicates its gate drive to some degree. If the chopper transistor fails open-circuit, the dump capacitor voltage will increase rapidly until the phase transistors are destroyed by overvoltage. This poses some interesting protection problems, although these are not necessarily worse than those of conventional ac drives.

REFERENCES

[1] T. J. E. Miller, "Converter volt-ampere requirements of the switched reluctance drive," presented at IEEE/IAS Annual Conference, Chicago, IL, October 1984.
[2] M. R. Harris, T. J. E. Miller, and J. Finch, "A review of the integral-horsepower switched reluctance drive," presented at IEEE/IAS Annual Conference, Toronto, ON., Canada, October 1985.
[3] P. J. Lawrenson, et al, "Variable-speed switched reluctance motors," in IEE Proc., vol. 127, July 1980, pp. 253–265.
[4] W. F. Ray and R. M. Davis, "Inverter drive for doubly salient reluctance motor: Its fundamental behaviour, linear analysis and cost implications," Electric Power Applications, vol. 2, pp. 185–193, December 1979.

Mehrdad Ehsani (S'80–M'81–SM'83) received the B.S. and M.S. degrees from the University of Texas at Austin in 1973 and 1974, respectively, and the Ph.D. degree from the University of Wisconsin—Madison in 1981, all in electrical engineering.

From 1974 to 1977 he was with the Fusion Research Center, University of Texas, as a Research Engineer. From 1977 to 1981 he was with Argonne National Laboratory, Argonne, IL, as a Resident Research Associate, while simultaneously pursuing his doctoral work at the University of Wisconsin—Madison in energy systems and control systems. Since 1981 he has been an Assistant Professor of electrical engineering at Texas A&M University, College Station, where he has founded a power electronics program and laboratories. His current research is in power electronics and their control systems.

Dr. Ehsani is a member of the IEEE Industrial Electronics Society, the IEEE Power Engineering Society, Sigma Xi; is Chairman of Papers Review Subcommittee of Static Power Converter Committee of IEEE Industry Applications Society; and is a Registered Professional Engineer in the State of Texas. He was the recipient of the Best Paper Award in Static Power Converters at the IEEE–Industry Applications Society 1985 Annual Meeting. He also is coauthor of a book on converter circuits for superconductive magnetic energy storage and is a contributor to the *IEEE Guide for Self-Commutated Converters* and other monographs.

James T. Bass received the B.S., M.S., and D.E. degrees from Texas A&M University, College Station, in 1980, 1983, and 1985, respectively, all in electrical engineering.

While working on his doctoral research project on variable-speed motor control, he spent a year at the General Electric Corporate Research and Development Center in Schenectady, NY. He is currently with the Bass Engineering Company in Longview, TX.

Dr. Bass is a member of the Industry Applications Society, the National Association of Corrosion Engineers, HKN, and $\Psi\beta\Pi$ honorary societies. He is the coauthor of six technical publications in signal and power electronics.

Timothy J. E. Miller (M'74–SM'82) is a native of Lancashire, UK. He was educated at Atlantic College and at Glasgow University, where he graduated in 1970 with honors in electrical engineering. He studied at Glasgow University under the Student Apprenticeship Scheme of Tube Investments Ltd. In 1973 he became a Research Fellow at the University of Leeds, where he received the Ph.D. degree in 1977 for his investigations of the electrical and mechanical properties of superconducting ac generators.

He spent two years as Assistant Research Engineer with International R&D Company in Newcastle-Upon-Tyne, UK, before joining the British Gas Engineering Research Station at Killingworth, Northumberland, as a Research Engineer working on the development of electronic instruments for the tracing of underground gas pipes. He then joined GEC Power Transmission Division at Stafford as Head of Transmission Systems Development. He moved to the General Electric Company in Schenectady, NY, in 1979, and was Manager of the Adjustable Speed Drives Program and the Power Electronics Control Program from 1983. While at G.E. he worked on the design of ac motors and generators, particularly permanent-magnet machines for both line-start and inverter-fed (adjustable-speed) applications; on the design and development of a wide range of switched-reluctance motor drives; and on drives and controls for a wide range of industrial, aerospace, and commercial products and applications. He also contributed to the development of controls for very large thyristor-switched capacitor compensators for power transmission systems. He was Manager of the Power Electronics Control Program at the Corporate Research and Development Center of G.E., Schenectady, NY, when he joined the University of Glasgow, Department of Electronics and Electrical Engineering to establish teaching and research in the fields of power electronics and motor drives. He is an Adjunct Professor at Union College in Schenectady, NY, and a short-course faculty member at the University of Wisconsin—Madison.

Dr. Miller is the author of a widely used text on reactive power control. He holds several US and UK patents, and is a Chartered Engineer in the UK and a Senior Member of the IEEE.

Robert L. Steigerwald (S'66–M'79–SM'85) was born in Auburn, NY. He received the B.S. degree in electrical engineering with distinction from Clarkson College of Technology, Potsdam, NY, in 1967 and the M.E.E. and Ph.D. degrees from Rensselaer Polytechnic Institute, Troy, NY, in 1968 and 1978, respectively.

Since joining the research staff of General Electric Corporate Research and Development in 1968, he has conducted research and advanced development of solid-state power conversion circuits employing thyristors, power transistors, and gate-turn-off thyristors (GTO's). He also has worked in the area of motor drives and computer simulation of power circuits and drive systems. His current interests include the development of high-frequency resonant power supplies and their controls, as well as the development of power-integrated circuits and their application to power electronic systems.

Dr. Steigerwald holds 39 patents and is a member of Eta Kappa Nu.

Converter Volt-Ampere Requirements of The Switched Reluctance Motor Drive

T.J.E. Miller

IEEE Transactions, Vol. IA-21, 1985, pp. 1136-1144

Abstract — This paper describes an algebraic, nonlinear analysis of the switched reluctance drive system. The analysis is intended to provide an understanding of the factors that determine the kVA requirements of the electronic power converter, and to determine the fundamental nature of the torque/speed characteristics.

The effect of saturation is given special attention. It is shown that saturation has the two main effects of increasing the motor size required for a given torque, and at the same time decreasing the kVA per horsepower (i.e., increasing the effective "power factor" by analogy with an ac machine). The kVA per horsepower is lower than predicted by simple linear analysis that neglects saturation.

The paper also develops the necessary conditions for a flat-topped current waveform by correctly determining the motor back-emf. It also explains why it is desirable to allow the phase current to continue (though with much reduced magnitude) even after the poles have passed the aligned position. The theory provides a formula for determining the required commutation angle for the phase current.

The paper provides the basis for an estimation of the kVA requirements of the SR drive. These requirements have been measured and also calculated by a computer simulation program developed at CRD.

INTRODUCTION

The switched reluctance motor is a variable reluctance (VR) stepping motor designed for efficient energy conversion. The motor is "doubly salient" and has different numbers of stator and rotor poles (8/6, 6/4, 12/8, and 6/2 being among the preferred choices) [1-4]. It is singly-excited, and there are no rotor windings or magnets or squirrel cage. Torque is produced exclusively by reluctance variation and is independent of the direction of the stator (phase) currents, giving rise to the possibility of truly "unipolar" operation in which only one main electronic switching device is required for each phase (compared with 2 required for all AC motor drives). A typical example of a motor cross-section is shown in Fig. 1.

The phase currents must be switched (from a dc power supply) in synchronism with the rotor shaft position in such a way that essentially the phase is excited when and only when a pair of rotor poles is approaching alignment with the stator poles of that phase. Shaft position sensors of the Hall-effect type and optical interrupters have been used to generate the switching reference signals. The conduction angles must be modulated as a function of both the speed and torque in order to stabilize operation and limit the peak current.

The SR motor drive has received a revival of interest and activity in the last 5-10 years, particularly in Europe where serious attempts have been made to commercialize it and numerous analytical papers have been published. With the notable exeption of the work of Ray and Davis [5,6] and, to some extent, that of Byrne [7], the published analyses concentrate on the motor and its capability, and there has been very little analysis of the power electronic requirements in terms of the kVA/hp of silicon devices required. This paper attempts to provide such an analysis.

Fig. 1. Cross-section of a typical switched reluctance motor with 6 stator poles and 4 rotor poles. The stator pole windings are connected in opposite pairs to form "phases" (3 in this case). There are no rotor windings or magnets.

The theoretical treatment is based on an idealized nonlinear energy conversion model that explains physically the factors controlling the voltampere requirement for a given power conversion. This is equivalent to the theory of power factor in pure ac machines, and indeed the same method can be applied to the ac machine but the mathematics would be unnecessarily cumbersome in their case. The kVA or voltampere requirements calculated using the idealized model are substantiated with computer simulation results and with measurements made on a laboratory machine of about 5 hp rating at 1800 rpm. The most important result is a comparison between the inverter voltampere requirements of induction-motor and switched reluctance motor drives.

The calculations and results in this paper are valid for essentially all inverter topologies that use either one or two switching devices per phase. If only one device is used in each phase, its voltage rating is typically twice that of each device in an inverter that uses two devices per phase. Its current rating, however, is the same. Inverters of both types are described in the literature [2-7]. An example of one that uses only one device per phase is shown in Fig 2. It has a bifilar winding wound such that freewheeling current returns to the dc source through the secondary after the main transistor is turned off. At turnoff the transistor collector voltage rises to twice the dc link voltage if the turn ratio is unity. (Actually the voltage is somewhat higher because of resistance and imperfect coupling between the bifilar windings. Snubbing may be required to limit the voltage to 2 p.u.)

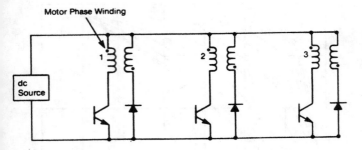

Fig. 2. Example of a switched reluctance inverter that uses only one main switching device per phase, givng a total of three main devices and only one forward voltage drop in series per phase.

SIMPLE NONLINEAR THEORY

Basic Relationships and Assumptions

In one revolution each phase conducts as many pulses of unidirectional current as there are rotor poles, so that there are qN_r pulses or "working strokes" per revolution. The shape of the current pulse varies with torque and speed, but an important special case is shown in Fig. 3, in which the current waveform is flat-topped. Since $di/dt = 0$ between A and C, the motor back-emf in this interval must be equal to the applied voltage V (resistance is neglected for the present).

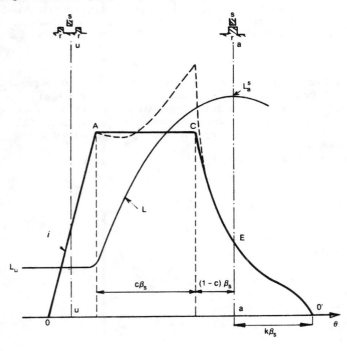

Fig. 3. Flat-topped current waveform of switched reluctance motor (OACEO'), superimposed on the inductance waveform L. The "unaligned" and "aligned" rotor positions are illustrated at the top of the diagram.

The electromagnetic energy W that is available to be converted into mechanical work in each working stroke is equal to the area enclosed by the trajectory of the operating point in the flux-linkage/current diagram, Fig. 4. The trajectory corresponding to Fig. 3 is shown with the heavy line, and the shaded area OACEO represents the converted energy, W. The points O, A, C ... correspond to those similarly labeled in Figure 1. Again, resistance is neglected. The average torque is given by

$$\text{Average torque} = \text{Work/stroke} \times \frac{\text{No. of strokes/rev}}{2\pi}$$

$$T_a = W \frac{qN_r}{2\pi} \tag{1}$$

The average power conversion P is given by

$$P = \omega T_a \tag{2}$$

where ω is the speed in rad/s. If iron losses and windage and friction are neglected, then T_a and P appear at the shaft.

The area W will be calculated from Figure 2 in terms of known inductance parameters, the peak current i, and the applied dc voltage V. The following assumptions are embodied in the structure of this diagram:

1. The minimum inductance L_u occurs in the "unaligned" position of the rotor and is constant.
2. The magnetization curve for the "aligned" rotor position is linear up to point D, with inductance L_a^s. Thereafter it continues with slope L_u so that DB is parallel to OA and separated

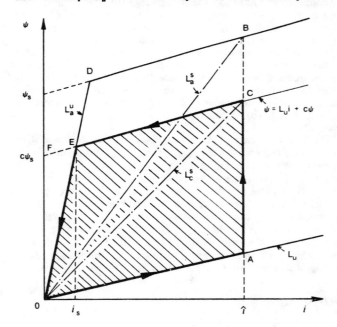

Fig. 4. Flux linkage/current characteristics of one phase of switched reluctance motor. ODB is for the "aligned" position, and OA for the "unaligned" position of the rotor. The shaded area represents the energy converted into mechanical work in each "stroke" or cycle.

from it by the fixed flux linkage ψ_s. The saturated inductance in the aligned position is L_a^s, and this is clearly a function of current i. The aligned saturation factor is defined as the ratio

$$\sigma = \frac{L_a^s}{L_a^u} \qquad (3)$$

The unsaturated inductance ratio is defined as

$$\lambda_u = \frac{L_a^u}{L_u} \qquad (4)$$

Note that $\lambda_u > 1$ and $\sigma < 1$.

3. The operating-point trajectory reaches the end of its flat-topped current interval at the commutation point C such that

$$AC = c \cdot AB = c\psi_s \qquad (5)$$

where $c < 1$. Note that AC represents the change of flux linkage between A and C, which is equal to the voltage-time integral between these points. After commutation, the operating point moves along CE, assumed parallel to AO, and then along EO as the flux linkage is forced down by the inverter. (It is assumed that the inverter is of such a design as to apply $-V$ to the phase winding between C and O).

The conditions in Fig. 4 correspond to early commutation, i.e., with $c < 1$. It is generally necessary to commutate early—that is, before the rotor reaches the aligned position at B. The phase must be essentially de-energized before this; otherwise a significant negative torque would be generated. The aligned position is first reached at E, and although the flux linkage still appears quite high, the stored magnetic energy (equal to the area to the left of the "current" magnetization curve) will be much smaller than at C, provided the unsaturated aligned inductance L_a^u is large. The negative torque generated between E and O is therefore small, permitting this section of the trajectory to be approximated by a straight line. It is important to note that early commutation precludes the conversion of the energy represented by area CBDE. This is typically of the order of 30% of the maximum theoretically available area OABDO, and must be accounted for in any design procedure.

The trapezoidal area OACEO in Fig. 4 is amenable to analysis, but it is artificial and strictly nonphysical. Its justification is that it is a reasonably good approximation to trajectories observed in the laboratory for flat-topped current waveforms, and it gives sufficiently accurate results to make it a reliable source of physical understanding and a good vehicle for trend analysis. An idea of how realistic it is can be obtained from Fig. 5, which shows a series of ψ/i curves for the model 5-hp motor. This drive was intended to operate with a flat-topped current of 51 A, and in the neighborhood of this current the slopes of the magnetization characteristics are all roughly parallel to L_u.

Analysis

At E,

$$L_a^u i_s = L_u i_s + c\psi_s \qquad (6)$$

and therefore

$$i_s = \frac{c\psi_s}{L_a^u - L_u} = \frac{c\psi_s}{L_u(\lambda_u - 1)} \qquad (7)$$

Along AC,

$$c\psi_s = V \frac{\theta_{AC}}{\omega} \qquad (8)$$

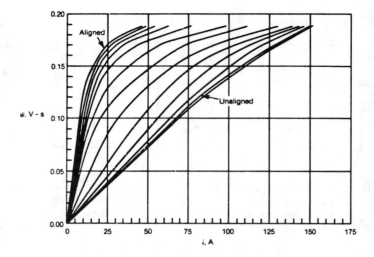

Fig. 5. Flux linkage/current magnetization curves computed by finite-element analysis for 5-hp switched reluctance drive. The curves are at 5° intervals of the rotor.

where θ_{AC} is the angular rotation of the rotor as the operating point moves from A to C. The analysis is much simpler if it is assumed that

$$\theta_{AC} = c\beta_s \qquad (9)$$

where β_s is the stator pole arc. This is equivalent to assuming that the phase inductance variation is restricted to an angular rotation of β_s, which is only approximately true. With this assumption,

$$c\psi_s = cV \frac{\beta_s}{\omega} \qquad (10)$$

which implies that the flux linkage varies linearly with rotor position if the phase current is constant. Although this appears to be a "wild" assumption, it is in fact a necessary condition for the existence of a flat-topped current waveform at constant speed. This means that the present analysis is meaningful only for flat-topped current waveforms.

The conversion energy W can now be evaluated as

$$W = OACEO = OACF - OEF$$
$$= c\psi_s i - \tfrac{1}{2} c\psi_s i_s = c\psi_s (i - \tfrac{1}{2} i_s)$$

At B,

$$L_a^s i = L_u i + \psi_s$$

and therefore

$$i = \frac{\psi_s}{L_a^s - L_u} = \frac{\psi_s}{L_u(\lambda_u \sigma - 1)} \qquad (11)$$

From equations 7 and 11,

$$i_s = i \cdot c \frac{\lambda_u \sigma - 1}{\lambda_u - 1} = i \frac{c}{s} \qquad (12)$$

where

$$s = \frac{\lambda_u - 1}{\lambda_u \sigma - 1} \qquad (13)$$

Substituting from equations 10 and 12 in the expression for W,

$$W = V \frac{\beta_s}{\omega} ci \left[1 - \frac{1}{2} \frac{c}{s} \right] \qquad (14)$$

The average electromagnetic power is therefore

$$P = Vi \frac{\beta_s q N_r}{4\pi} \cdot c \left[2 - \frac{c}{s} \right] = Vi \frac{\beta_s q N_r}{4\pi} \cdot Q \qquad (15)$$

and $T_a = P/\omega$.

This is the fundamental energy conversion equation for the SR drive, in which V is the dc source voltage and i is the peak flat-topped phase current at speed ω. It permits the fundamental inverter volt-ampere requirement to be estimated and compared with those of other drives.

Special Case of Nonsaturating Motor

A special case arises when there is no saturation; then $\sigma = 1$ and

$$W = V \frac{\beta_s}{\omega} i \cdot c \left[1 - \frac{c}{2} \right] \qquad (16)$$

If also $c = 1$, we get

$$W = \frac{1}{2} V \frac{\beta_s}{\omega} i \qquad (17)$$

This is illustrated in Fig. 6b. The quantity $V\beta_s i/\omega$ is the energy supplied by the inverter along AC as the rotor moves from a position of minimum inductance to the aligned position, traversing an angle β_s in time β_s/ω. ($V\beta_s/\omega$ is also the change in flux linkage during this interval). The quantity $V\beta_s i/\omega$ is equal to the area of the rectangle whose rightmost vertical side is AC. Since triangle OAC has the same base (AC) and height (i) as this rectangle, it has half the area; hence the coefficient of 1/2 in equation 17. The energy supplied by the inverter up to the point of commutation C is $W + R$, and the energy R is returned to the supply after commutation (or dumped into a resistor or capacitor, depending on the type of inverter). The "energy ratio" has been defined as $W/(W+R)$ and is a kind of generalized power factor which tells how much energy conversion (W) is obtained for a given input energy ($W+R$) in each working stroke. The energy ratio of the nonsaturating system in Fig. 6b is determined as follows:

$$W = \frac{1}{2} V \frac{\beta_s}{\omega} i - \frac{1}{2} (L_a^u - L_u) i^2 \qquad (18)$$

$$R = \frac{1}{2} L_a^u i^2 \qquad (19)$$

$$\frac{W}{W+R} = \frac{\lambda_u - 1}{2\lambda_u - 1} \qquad (20)$$

where $\lambda_u = L_a^u/L_u$ is the inductance ratio. With $\lambda_u = 6$ the energy ratio is 0.455. It improves with increasing λ_u, but can never exceed 0.5 in a nonsaturating motor.

Inverter Volt-Ampere Requirement and Energy Ratio of Saturating Motor

Figure 6a is drawn for a motor with the same unsaturated inductance ratio ($\lambda_u = 6$) and the same energy conversion W (and therefore the same torque) as in Fig. 6b. Saturation now limits the maximum flux linkage to a smaller value ($\psi_a < \psi_b$), and if the rotor displacement between A and C is unchanged (β_s), then the supply voltage V must be reduced compared with that in Fig. 6b. In other words, the saturating motor cannot accept as large a voltage impulse (voltage-time integral) as the nonsaturating motor. For the same energy conversion, W, the maximum current i must now be greater. Fig. 6a is drawn for a saturation factor $\sigma = 0.4$, which is typical. With $\lambda_u = 6$ this gives $s = 3.57$, so that from equation 14 with $c = 1$,

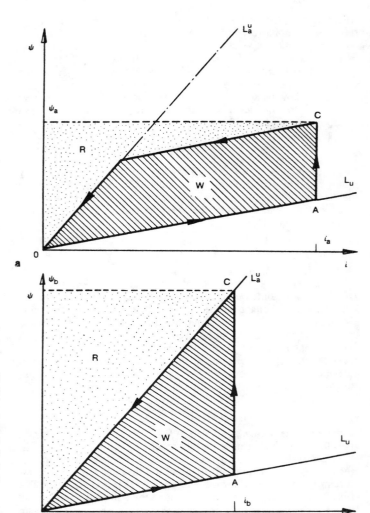

Fig. 6. Energy converted (W) and energy returned to supply (R) during one stroke. (a) Nonlinear saturating case. (b) Linear nonsaturating case.

$$W = 0.860 V \frac{\beta_s}{\omega} i \qquad (21)$$

This should be compared with equation 17 for the nonsaturating motor. Although i is greater, the volt-ampere product Vi is only $0.5/0.860 = 58\%$ of the value required in the nonsaturating motor, for the same torque and speed.

The ratio of the maximum current i for the two motors can be determined as follows. From equation 11

$$\psi_s = i_a L_u (\lambda_u \sigma - 1) \quad \text{so that}$$

$$W = \psi_s \left(i - \frac{1}{2} i_s \right)$$

$$= i_a^2 L_u (\lambda_u \sigma - 1) \left[1 - \frac{1}{2s} \right] \qquad (22)$$

Dividing by equation 18 (with $i = i_b$) gives

$$i_a/i_b = \frac{s}{\sqrt{2s-1}} \qquad (23)$$

331

where subscript a refers to the saturating motor and subscript b refers to the nonsaturating motor, as in Figs 6a and 6b, respectively. This equation is valid only if $c = 1$. In the example illustrated in Fig. 6, $s = 3.57$ and $i_a/i_b = 1.44$; i.e., the saturating motor requires a maximum current of 1.44 times that of the nonsaturating motor. However, its voltage requirement is lower by a greater factor, as will now be shown. Dividing equation 14 for the saturating motor (a) by equation 17 for the nonsaturating motor (b), and incorporating equation 23 gives

$$\frac{Va}{Vb} = \frac{1}{\sqrt{2s-1}} \qquad (24)$$

In the example with $s = 3.57$, $Va/Vb = 0.403$, i.e., the saturating motor requires a source voltage of only 40.3% of that required by the nonsaturating motor. The Vi products are in the ratio $1.44 \times 0.403 = 0.58$, which checks the previous result. In general from equations 23 and 24,

$$r = \frac{(Vi)a}{(Vi)b} = \frac{s}{2s-1} \qquad (25)$$

This result is valid only if $c = 1$.

Of course the saturating motor could be rewound with $1/0.403 = 2.48$ times as many turns, so that it would accept the same dc source voltage as the nonsaturating motor with the same switching angles. For the same torque, its peak current would then be only 58% of that required by the nonsaturating motor.

In the general case ($c < 1$) the ratio between the volt-ampere requirements of the saturating and the nonsaturating motors is determined from equation 14. In the nonsaturating case $\sigma = s = 1$ so that in general, for the same torque and speed,

$$r = \frac{(Vi)a}{(Vi)b} = \frac{s(2-c)}{2s-c} \qquad (26)$$

In the example with $\lambda_u = 0.6$, $\sigma = 0.4$, $s = 3.57$, and $c = 1$, this gives 0.58, which checks the earlier result.

The result expressed by equation 26 is fundamentally important. It says that saturation *decreases* the inverter volt-ampere requirement for a given torque and speed. If it were not for this effect, i.e., if there were no saturation, then the SR drive would have no chance of competing with the induction motor drive in terms of inverter kVA per horsepower. The result is intuitively surprising, but it is important to interpret it correctly. Notice that it says nothing about the dimensions of the motor; it merely states that if the motor saturates then the inverter volt-ampere requirement for a given torque and speed will be smaller than for a nonsaturating motor with the same values of L_a^u and L_u. While saturation is desirable from this point of view, at the same time it *reduces* the energy conversion capability of a motor of given dimensions, evaluated at a given peak current. This is obvious from Fig. 6; for the same value of W in both cases a and b, $i_a = 1.44 i_b$. Saturation therefore generally leads to a smaller inverter but a larger motor. The degree of saturation will influence the balance between the motor size and the inverter size.

The "degree of saturation" is difficult to define with precision (σ and s being together perhaps the most helpful approximation). However, it is helpful to think of it as the degree to which the trajectory of the operating point will approximate closer to Fig. 6a than to Fig. 6b. To minimize the sizes of *both* the motor and the inverter, W should be maximized and R should be minimized. This requires the largest possible inductance ratio λ_u, the highest possible saturation flux density, and the highest possible value of L_a^u. This requires a small airgap and iron with very high permeability.

These results are so important that they will be restated slightly differently. The effect of saturation is to reduce the energy conver-

sion capability of a motor of given dimensions, but at the same time it reduces the inverter volt-ampere requirement (for a given torque and speed) *by a greater factor,* leading overall to a reduced inverter kVA/hp.

In the foregoing discussion the "size" of the inverter has been represented on a per-phase basis by the volt-ampere product Vi. This measure is particularly meaningful for transistor inverters because transistors are sensitive to peak currents. For SCR inverters the rms phase current may be a more meaningful indication of inverter size. Unlike the peak flat-topped current, the rms current cannot be estimated by a simple theory.

Finally, the energy ratio for the saturating case is given in general by

$$\frac{W}{W+R} = \frac{c(2s-c)(\lambda_u\sigma-1)}{s+2c(\lambda_u-1)} \qquad (27)$$

In the example discussed above with $c = 1$, $\lambda_u = 6$, $\sigma = 0.4$, $s = 3.57$ and the energy ratio has the value 0.633, which can be verified graphically from Fig. 6a.

Motor Back-emf Required to Produce a Flat-Topped Current

Referring to Fig. 3, along AC we have $\Delta\psi = c\psi_s = V\beta_s/\omega = ciL_u(\lambda_u - 1)$ and therefore

$$V\frac{\beta_s}{\omega} = iL_u(\lambda_u\sigma-1) \qquad (28)$$

This equation defines the back-emf required to produce a flat-topped current i; it is equal to V. Equivalently, for a given dc source voltage V the equation defines the value of the flat-topped current, i. Note that if saturation is ignored (i.e., $\sigma = 1$) the estimate of V (or i) obtained from this formula will be incorrect by the factor s defined in equation 13.

Determination of the Commutation Angle

To produce positive motoring torque it is expected that the current pulse will coincide more or less exactly with the rising part of the phase self-inductance curve marked L in Fig. 3. There is no point in conducting current before this "torque zone." Also, the current must be brought essentially to zero by the time the poles are aligned (at "a" in Fig. 3); otherwise a significant negative torque component may be generated. The torque zone extends for a rotor displacement approximately equal to β_s, provided $\beta_s < \beta_r$ (which is normally the case).

The problem to be solved is to determine the angular position of the commutation point C. This point divides the torque zone into two sections: the first has an angle of $c\beta_s$, and during this time the current is assumed to be constant. The second has an angle of $(1-c)\beta_s$, and ends when the rotor and stator poles are aligned. Conduction continues thereafter for an angle equal to $k\beta_s$, where k is unknown. (It is assumed that $\beta_s = \beta_r$). Along CO',

$$V\frac{(1-c+k)\beta_s}{\omega} = L_u i + c\psi_s \qquad (29)$$

This merely expresses the fact that the negative voltage impulse supplied by the inverter after commutation must reduce the flux linkage to zero. The flux linkage in question is that which is achieved at point C in Fig. 4, and is given by the RHS of equation 29. Eliminating ψ_s by means of equation 11, and simplifying, we get

$$V\frac{\beta_s}{\omega}(1-c+k) = iL_u[1+c(\lambda_u\sigma-1)] \qquad (30)$$

Now divide equation 30 by equation 28, and rearrange to give an explicit formula for c:

$$c = \frac{(1+k)(\lambda_u \sigma - 1) - 1}{2(\lambda_u \sigma - 1)} \qquad (31)$$

Also of interest is the flux linkage ψ_E still in the winding at point E, where the poles are aligned. Ideally this should be zero, but in practice

$$\psi_E = \frac{k}{1 - c + k} \psi_C \qquad (32)$$

A numerical example shows the disadvantages of commutating early enough to make $k = 0$, i.e., to eliminate the current by the time the poles are aligned. With $\lambda_u = 6$ and $\sigma = 0.4$, setting $k = 0$ in equation 31 gives $c = 0.14$; equation 15 shows this to imply a low Q-factor and a large inverter volt-ampere product. If, on the other hand, conduction is permitted to continue throughout the negative torque zone, i.e., $k = 1$, then $c = 0.64$. Now equation 15 shows a much improved Q-factor, implying a smaller inverter volt-ampere product. Also in this case $\psi_E = 0.74 \psi_C$, which is consistent with observations in the laboratory. A negative torque is indeed produced, but it is negligible. The negative torque is completely swamped by positive torque produced by other phases and does not appear on the shaft. The polyphase induction motor exhibits precisely the same phenomenon: on a per-phase basis the torque makes a small negative excursion shortly before every current-zero; the total torque, however, is never negative in the steady state under balanced conditions (ideally it is constant).

Torque/Speed Envelope

In many drives it is required that the torque/speed envelope follow a constant-torque locus from standstill to "base speed" and a constant-power locus at speeds above base speed. At the higher speeds the torque is limited by the available voltage; the term "field weakening" is often used to refer to this speed range. At low speeds, and particularly for breaking away from standstill, the drive may be required to provide for a short time a torque greater than the normal maximum.

At low speeds it is characteristic of the SR drive that the available dc source voltage greatly exceeds the motor back-emf, so that chopping (with current feedback) is necessary to achieve a flat-topped current waveform. This being so, it is possible to work with $c = 1$ provided that thermal limits are not exceeded. Under these conditions, of course, equations 14 and 15 become invalid because V is applied for only a fraction of the time. The torque must therefore be expressed as a function of current only, and the result is

$$T_a = i^2 L_u (\lambda_u \sigma - 1) \frac{qN_r}{4\pi} \cdot c \left(2 - \frac{c}{s}\right) \qquad (33)$$

All the analysis in the previous sections has been for a flat-topped current waveform sustained by the dc source voltage V at the rated speed ω; by definition this is the base speed. If the commutation angle corresponds to a particular value c_b at base speed, then the torque at base speed is given by equation 33 with $c = c_b$.

At zero speed it is possible to work with an effective value $c = 1$. Therefore, for the same peak current the ratio between the torque at zero speed and at base speed is

$$\frac{T_0}{T_b} = \frac{2s - 1}{c_b(2s - c_b)} \qquad (34)$$

For example, if $\lambda_u = 6$ and $\sigma = 0.4$, as in earlier examples, $s = 3.57$, and with $c_b = 0.6$ this ratio is equal to 1.56. The SR drive inherently can produce more *average* torque at standstill than at its base speed, for the same peak current. Since the same peak current is attained in both cases the *peak* torque can be no greater at zero speed than at base speed; however, since the average torque is higher, the torque waveform must be smoother at zero speed. Evidently the standstill torque T_0 can be delivered throughout a substantial part of the speed range in which chopping is necessary to limit the peak current.

At speeds above base speed a simplified analysis was presented by Byrne and McMullin [7], but this is too approximate to be useful; it is better to analyze the high speed performance by computer methods.

COMPARISON OF SR AND INDUCTION MOTOR kVA REQUIREMENTS

Attention is confined to the volt-ampere requirement at "base" speed — that is, the maximum speed at which full torque and full voltage are achieved without chopping. This makes the analysis simpler and gives a meaningful comparison between the two drives. Only the "inverter" part of the converter is considered, and the evaluation is restricted to the controlled switches without considering freewheeling diodes.

For the IM drive operating from a six-step voltage source inverter the equivalent of equation 15 is

$$P = Vi \frac{3}{\pi k} \cdot PF \qquad (35)$$

where V and i have the same meaning as in the SR drive and PF is the fundamental power factor. The factor k is the ratio between the actual peak current and the peak of the fundamental component of current.

Converter Sized By Peak Current and Peak Voltage

S is defined as the volt-ampere product Vi multiplied by the number of switching devices. This measure of the converter volt-ampere requirement is particularly meaningful for transistor converters. For the SR drive the number of switching devices can be either one per phase or two per phase. If there is only one switching device per phase the peak voltage on each device (assuming perfect snubbing) is $2V$. If there are two switching devices per phase the peak voltage on each device is V. In both cases $S = 2qVi$, where q is the number of phases. The converter volt-ampere requirement per kilowatt is now obtainable from equation 15 as

$$\text{SR:} \qquad S/P = \frac{8\pi}{\beta_s N_r Q} \quad \text{kVA/kW} \qquad (36)$$

The induction motor is assumed to have three phases and two switching devices per phaseleg, giving $S = 6Vi$. From equation 35 the "specific peak voltamperes" or kVA/kW is given by

$$\text{IM:} \qquad S/P = \frac{2\pi k}{PF} \quad \text{kVA/kW} \qquad (37)$$

Typical Values

For the SR drive the preferred combinations of stator and rotor poles are 8/6 and 6/4. For the 8/6 motor $q = 4$ and $N_r = 6$. The optimum value of β_s is approximately 0.4 times the stator pole pitch. In this case $\beta_s = \pi/10$ rad. Equation 36 gives

$$\text{8/6 SR:} \qquad S/P = 13.3/Q \quad \text{kVA/kW} \qquad (38)$$

For the 6/4 motor $q = 3$ and $N_r = 4$. The optimum value of β_s is again approximately 0.4 times the stator pole pitch, i.e., 0.419 rad. Equation 36 gives

6/4 SR: $S/P = 15/Q$ kVA/kW   (39)

From equations 38 and 39 it appears that the four-phase SR motor has a volt-ampere requirement that is only 13.3/15 or 89% of that required by the three-phase motor, assuming that the same value of Q can be achieved in both cases. In other words, the four-phase motor has an approximately 10% better utilization of the converter volt-ampere capacity.

For the integral-horsepower industrial induction motor a typical value of k is 1.25 and a typical value of PF is 0.85. From equation 37,

IM: $S/P = 9.25$ kVA/kW $= 6.89$ kVA/hp   (40)

Comparing the 8/6 four-phase SR drive and the IM drive, from equation 7 with the typical parameter values substituted, the ratio of volt-ampere requirements is given by

$$\frac{SR \text{ kVA/kW}}{IM \text{ kVA/kW}} = \frac{1.44}{Q} \quad (41)$$

Design calculations (and measurements on the 5-hp model drive) suggest that the following values can be obtained at base speed and full load:

$\lambda_u = 8$

$\sigma = 0.3$

$c = 0.65$

With these values $s = 5.0$ and $Q = 1.22$. The SR drive therefore has a converter volt-ampere requirement that is $1.44/1.22 = 1.18$ times that of the induction motor drive for the same shaft power. The SR drive should be capable of operating with about $1.18 \times 9.24 = 11$ kVA/kW or about 8 kVA/hp of switching devices (based on peak current and peak voltage).

Measured kVA/kW of SR Drive

The 5 hp model SR drive built and tested at CRD was computed with the following parameters:

Shaft power $= 6.01$ hp $= 4.48$ kW
(windage and friction and iron losses neglected)
Peak phase current $= 43.6$ A
Dc link voltage $= 135.4$ V
kVA/kW $= 2 \times 4 \times 135.4 \times 43.6/4480 = 10.5$
kVA/hp $= 7.86$

This is 1.14 times the induction motor requirement and is actually better than predicted by the simple theory. (The computation was performed using a program that has been found to give accurate predictions of the phase current waveforms for the 5 hp drive.)

Converter Sized By RMS Current and Peak Voltage

For SCR converters the rms current is more significant than the peak current. For the six-step IM drive the rms phase current is somewhat larger than the rms of the fundamental component, and a ratio of 1.15 will be assumed to be typical. Since the rms of the fundamental component is $i/\sqrt{2}$, equation 40 becomes

IM: S/P (rms) $= 0.24 \times 1.15/\sqrt{2}/1.25$

$= 6.0$ kVA/kW $= 4.48$ kVA/hp

This is not yet a correct measure of the inverter size, however, because each switch conducts only on alternate half-cycles with a conduction angle of 180°, and this is equivalent to a reduction in the rms switch current by a factor of $\sqrt{2}$, for the same peak current. Therefore

IM: S/P (rms) $= 6.0/\sqrt{2} = 4.24$ kVA/kW

$= 3.17$ kVA/hp

For the SR drive the ratio between peak current and rms current is much harder to define, and it can vary between wide limits depending on the control strategy and the electromagnetic design of the motor. Computer studies show that the rms phase current typically lies between 0.3 and 0.5 times the peak current. Perhaps the most useful information that can be presented is the actual set of values for the model 5-hp drive.

For the 5 hp drive the rms phase current was 21.3 A with a peak of 43.6 A, giving a rms/peak current ratio of 0.49 and a volt-ampere requirement of $0.49 \times 10.5 = 5.1$ kVA/kW or 3.8 kVA/hp. This is 20% higher than for the induction motor drive.

CONCLUSIONS

(a) The converter volt-ampere requirement has been defined in two alternative ways. The first is the product of the peak current and peak voltage in each switch, multiplied by the number of switches. The second is similar, except that rms switch current is used instead of peak switch current. In both cases the requirement is expressed in kVA/kW or kVA/hp of shaft power (losses are neglected). The definition based on peak current will be an important factor in determining the cost of converters built with transistors and GTOs. The definition based on rms current will be more important when the converter is built with SCRs. In all cases the drive is assumed to be operating at "base" speed.

(b) The converter kVA requirement for the SR drive is higher than the theoretical value for the induction motor six-step converter. This is true for both the kVA/hp (peak) and the kVA/hp (rms). The relative requirements are summarized in Table 1. The figures in the table are normalized to the induction motor drive figures, which are calculated to be 6.9 kVA/hp (peak) and 3.2 kVA/hp (rms). The figures for the 5-hp model drive are about 10% optimistic because windage and friction and iron losses were neglected in deriving them.

Table 1

CONVERTER VOLT-AMPERE REQUIREMENTS OF SWITCHED RELUCTANCE AND INDUCTION MOTOR DRIVES

	Induction Motor	5-hp CRD SR Motor
kVA/hp (peak)	1.0	1.14
kVA/hp (rms)	1.0	1.20
Number of main switching devices	6	4
Peak/rms current ratio for each switching device	2.17	2.04

334

(c) The SR converter can be built with only one switching device per phase, giving four devices (SCRs or transistors) in a four-phase drive and only three in a three-phase drive. When discrete devices are used, this will permit a considerable reduction in inverter size and losses.

(d) The four-phase 8/6 SR motor is about 10% more efficient in utilizing the volt-ampere capacity of the converter than the three-phase 6/4 motor, assuming that the same value of the coefficient Q is obtained in both cases. Since Q can easily vary by amounts comparable to this 10%, it would appear that the two motors are roughly equal in their volt-ampere requirements.

(e) The nonsaturating SR motor requires more kVA/hp than the saturating motor. Analyses performed without proper treatment of the effects of saturation will overestimate the requirement for converter volt-amperes.

(f) The present analysis does not take account of second-order effects such as line transients and the effectiveness of snubbers, etc. These effects can greatly increase the amount of "kVA" (voltage and current capacity) that is finally built into a converter. The present analysis should be regarded as a baseline indication of the fundamental requirements.

(g) The kVA requirement of the SR drive is essentially the same whether it is designed with one switching device per phaseleg or with two. In the former case, the peak voltage experienced by the switches is twice that experienced by those in the two-per-phase circuits. However, the amount of space required will be less with only one device per phase. Also, the current passes through only one device instead of two, and this will lead to a higher efficiency. (In the induction drive the current also passes through two devices in series).

(h) Note that the peak/rms current ratio for the devices in the SR drive is lower than in the induction motor drive. The reason for this is the increased duty factor in the SR drive, which follows from the fact that each device conducts current in only one direction and conducts every time a rotor pole passes the poles of its associated phase winding. In the induction motor drive each device is idle for 180 electrical degrees in every cycle.

The theory provides a means for estimating the motor back-emf. This emf is only a fraction of the value calculated using the linear theory. If the linear theory is used to design an inverter, the power semiconductor devices will be seriously overdesigned, yet the motor will produce only a fraction (typically one-half) of its desired power.

The theory also provides a formula for estimating the motor power output for a given terminal volt-ampere product and indicates how far the machine must be driven into saturation to achieve this.

The energy ratio is derived for both saturating and nonsaturating motors. It is shown that the energy ratio of the nonsaturating motor cannot exceed 0.5, but that of the saturating motor can significantly exceed this.

The conditions necessary to achieve a flat-topped current waveform are discussed. This waveform provides a natural definition of base speed in the torque/speed diagram. However, it is shown that standstill torque capability is inherently greater than the torque at base speed, and a typical example shows a 56% difference, with no increase in peak current.

Finally, an explicit formula is given that enables the commutation angle to be determined. This formula shows why it is not only permissible but desirable to commutate late enough to cause the instantaneous torque of one phase to have a small negative excursion. This phenomenon has a parallel in the polyphase induction motor. In both motors the total torque of all phases together never becomes negative.

Acknowledgment

The author is indebted to colleagues J.T. Bass, R. Gerbetz, J.A. Mallick, E. Richter, J.A. Rulison, and F.G. Turnbull for their design and experimental work; and to J.C. Bunner for general guidance.

DEFINITION OF SYMBOLS

c Ratio between voltage-time integral applied to the phase winding at constant current through the main switching device, and the maximum change of flux linkage that can be obtained at the same current as the rotor moves from the unaligned to the aligned position

i Peak phase current

k Ratio of actual peak to fundamental peak phase current in six-step induction motor drive

N_r Number of rotor poles

Q Coefficient appearing in equation 1 that determines the effectiveness of the SR converter in utilizing volt-ampere capacity to deliver real power ($= c\,(2 - c/s)$)

q Number of phases

~~r Ratio of actual rms to fundamental rms phase current in six-step induction motor drive~~

s Ratio between the peak phase current and the current required to just saturate the magnetic circuit in the aligned position

V dc source voltage

β_s Stator pole arc, radians

λ_u Unsaturated inductance ratio ($= L_a^u / L_u$)

σ Saturation factor for aligned inductance ($= L_a^s / L_a^u$)

REFERENCES

[1] G. Charlish, "An Idea 100 Years Old Comes to Fruition," *Financial Times,* Technology Section, January 26, 1983.

[2] B.D. Bedford, U.S. Patent 3,678,352 (1972).

[3] B.D. Bedford, U.S. Patent 3,679,953 (1972).

[4] P.J. Lawrenson, J.M. Stephenson, P.T. Blenkinsop, J. Corda, and N.N. Fulton, "Variable-Speed Switched Reluctance Motors," *IEE Proc.* Vol. 127, Pt. B, No. 4, July 1980, pp. 253-265.

[5] W.F. Ray, R.M. Davis, "Inverter Drive for Doubly Salient Reluctance Motor: Its Fundamental Behavior, Linear Analysis and Cost Implications," *Electric Power Applications,* December 1979, Vol. 2, No. 6 pp. 185-193.

[6] R.M. Davis, W.F. Ray, and R.J. Blake, "Inverter Drive for Switched Reluctance Motor; Circuits and Component Ratings," *IEE Proc.,* Vol. 128, Pt. B, No. 2, March 1981, pp. 126-136.

[7] J. Byrne and M.F. McMullin, *Motorcon September 1982 Proceedings,* pp. 10-24.

2.2 - Controls

The Control of SR Motors

W.F. Ray, R.M. Davis and R.J. Blake

Conference on Applied Motion Control, Minneapolis, June 1986, pp. 137-145

ABSTRACT   The basic control strategy for a switched reluctance motor is explained. Performance is based on the ability to provide each phase with pulses of current during torque productive periods of rotation and to adjust their amplitude either by chopping or by selecting predetermined switching angles. The strategy is influenced by the type of power converter circuit and by the drive application; various alternatives are explored. The role of rotor position measurement and the relative merits of feedforward and feedback control are discussed. The exceptional controllability of the SR motor is justified and its potential as a servomotor examined. A typical control schematic is given.

1. Introduction

Although the switched reluctance (SR) variable speed drive had little publicity in the United States before 1984, it is now receiving a rapidly increasing interest from University and Industrial research groups. The authors, however, have had over ten years experience designing, building and operating several forms of SR drive, and have supplied the majority of the designs for the only present commercially available range marketed by TASC Drives Ltd. [1] Their previous publications have dealt in the main with the mechanism of torque production the associated current waveforms and various alternative power converter circuits; relatively less has been said of the associated control strategy and hardware, although this has also received considerable past study.

The ability of the SR drive to convert energy from a direct voltage source to a fully speed-controlled shaft with an efficiency which matches or exceeds other alternative drives is largely recognised. This merit is also true of its controllability, indeed efficiency and control ability are closely linked. The high efficiency is achieved through the integrated design of the complete drive and the optimisation of its control, in particular by arranging phase currents to be present during the most torque productive periods of rotation.

An SR motor has salient poles on both its stator and rotor. Each stator pole has a simple concentrated winding and there are no conductors on the rotor. Diametrically opposite windings are connected together either as a pair or in groups to form motor phases and for each phase a circuit with a single controlled switch is necessary and sufficient to supply a unidirectional current during appropriate intervals of rotor rotation. Fig. 1 shows the typical cross sectional arrangement for a 4-phase SR motor having 8 stator and 6 rotor poles.

In concept the SR motor and its switching power converter circuit is the simplest arrangement that can be conceived - no commutator, magnets, distributed windings, rotor conductors or alternating currents are required. In practice it matches or exceeds the specific output of a cage induction motor, benefitting from the virtual absence of rotor losses and from the cooling of the stator coils by air circulation due to the stirring effect of the rotor poles.

✳

Messrs. Ray and Davis are principally with the Department of Electrical and Electronic Engineering at the University of Nottingham, England.

In its internal appearance the SR motor is seen to be very similar to a stepping motor with a large step angle (15° in the example of Fig. 1). The phases are energised in sequence and, when each pair of stator poles is excited, the most adjacent pair of rotor poles is pulled towards alignment according to the well known principle of minimisation of reluctance. The rotor therefore 'steps' around in the opposite direction of rotation to that of the stator phase excitation. However, unlike the stepping motor, the SR motor is capable of smooth rotation at low speeds, of high efficiency throughout the speed range both at full and partial load torques, and does not exhibit instability. The reason for this substantially improved performance is, not least, the control strategy based on a knowledge of rotor position.

Fig. 2 shows for a constant current flowing in a particular phase the typical variation of torque with rotor position. While the rotor poles approach an excited pair of stator poles motoring torque is produced, but while the rotor poles recede braking torque and regeneration occur.

A stepping motor phase is ideally supplied with pulses of current of fixed magnitude and, for a particular speed of fixed time duration and duty. (For a 4-phase motor with dual-phase excitation the duty will be 50%, at least at low speed.) For the rotor to move in synchronism, the pulses of current must take a relative position with respect to the torque-angle profile such that the developed torque matches the load torque. At zero load the pulse would straddle the position of maximum pole alignment such that the developed motoring and braking torques cancel. With increasing load the relative position of the pulse advances to create a net motoring torque. The speed of the motor is dictated by the frequency of the pulsed excitation. Since the current is always at its maximum level the efficiency under part load will be particularly poor and in addition at low speed there will be a tendency to move discretely from one position of alignment to the next.

For an SR motor with rotor position sensing the current pulses are ideally timed to coincide with the positive (motoring) torque-productive arc of rotor movement for a particular pair (or set) of poles. The torque magnitude is controlled by adjusting the current amplitude. The method of adjustment may be different at high and low speeds and will be influenced by whether the direct voltage source can be varied. However, whatever method of adjustment of current amplitude is used, the ability to control the pulse angular position is of fundamental importance and is a major reason for maintaining high efficiency even at reduced torque.

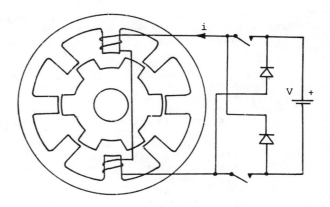

Fig. 1. 4-Phase SR Motor Showing Switching Circuit for one Phase.

In an earlier paper [2] the authors described a flexible control system which enabled the rotor angles at which the current pulse was initiated and collapsed to be independently set. The motor speed for this experimental rig was dictated by speed control of the loading system. By experimental variation the most efficient switching angles for different load torques and speeds could be established for a particular motor. This information has subsequently been stored either using digital or analogue memory in the SR controllers so that, for a given torque demand and speed the optimum switching angles are automatically selected. The control of SR motors is now receiving increasing attention and a variety of policies are emerging from companies other than SR Drives or TASC Drives such as GE [3] and Inland-Kolmorgan [4] as well as from Universities such as Nottingham, Leeds, Newcastle [5], Cork and MIT.

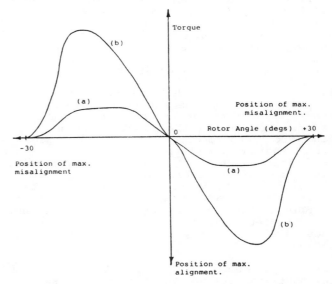

Fig. 2. Variation of Torque with Rotor Angle for a constant phase current for a typical 4-Phase SR Motor.
(a) Low current.
(b) High current.

The fundamental control features of pulse angular position and pulse amplitude are influenced by various factors such as:

Fixed or variable direct supply voltage

Single or 2/4-quadrant operation

The importance of high specific output or efficiency versus smooth torque

The value of the SR motor's stepping cability for positioning

The use of direct or indirect rotor position sensing

The method of signal processing: analogue/digital or microprocessor

The use of feed forward or feedback controls

It is the aim of this paper to examine how these factors influence the method of control of SR motors and to show how extraordinarily flexible the SR drive is from a control viewpoint.

2. Circuits

The essential features of the power switching circuit comprise, per phase, (i) a controlled switch to connect a direct voltage source to the winding to build up the current, (ii) an alternative path for the current to flow when the switch is turned off, generally through a diode, such that the winding experiences a reverse voltage so as to collapse the current. One or more direct voltage sources or sinks may be used.

Fig. 3 shows the 4-phase circuit used by TASC Drives Ltd. The dc link filter capacitor is split to produce a dual direct voltage source for the converter [6]. The current in Phase A is drawn from the top half supply when the GTO is on and returned to the bottom half supply when the GTO is off, but always flows into the capacitor mid-junction. The converse applies for Phase B.

It will be seen that an extra constraint is imposed on the control strategy: if the capacitor voltages are to remain stable at half the link voltage, then the combined mean current in Phases A and C must equal that of Phases B and D. This circuit is therefore only feasible for an even number of phases, but it retains the advantage of one controlled switch per phase. The imposed constraint implies that it is not possible to have independent control for each phase current but this is only of relevance at low speed.

It is possible to avoid the constraint without increasing the number of switches by using a motor with bifilar windings [2] but this is not attractive unless

the motor and power converter are in close proximity due to the substantially increased number of connections. The alternative circuit shown in Fig. 1 also avoids the constraint but at the cost of two switches per phase and more connections (although not as many as for the bifilar case).

A further alternative utilising the 4-phase motor winding circuit of Fig. 3, proposed by Bausch [7] for a low speed motor, is to leave the common or H-point of the windings disconnected from the supply. Only one switch per phase is required and the need of a split-supply is avoided, but at the penalty of the constraint that at any time instant the sum of the currents in Phases A and C must equal that of B and D. This contraint imposes a significant inflexibility on the control of the current pulses.

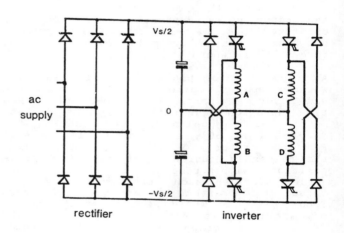

Fig. 3.   4-Phase SR Power Converter Circuit with Split Supply

Yet another possibility is the circuit proposed by Egan et. al. [4] and shown in Fig. 4. This utilises a unipolar supply and retains one switch per phase. However, when the switches are turned off the reactive energy is returned via diodes to a capacitive sink (C_2) and thence back to the dc link capacitor (C_1) using a flyback-converter. This circuit has the capability of independent control of each phase current but at the penalty of an extra storage capacitor and an extra switch. Furthermore, for motoring conditions the average current returned by the diodes can approach in value the average current drawn by the phase switches. For equal values of source and sink voltage (Vs_1 and Vs_2) the current rating for the flyback converter switch is typically twice that of the phase switches, and if braking duty is required the ratio can be significantly higher.

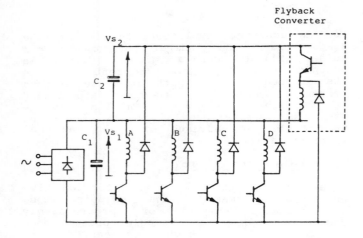

Fig. 4. 4-Phase SR Power Converter with auxiliary energy recovery circuit.

The circuit of Fig. 4 does however exhibit an extra degree of control in that the sink voltage Vs_2 may be varied by varying the duty cycle of the flyback converter. Other configurations of converter may be used for recovering the energy from C_2.

The direct voltage source is usually derived from a single-phase or 3-phase utility supply by diode rectification. In this aspect the "front end" of the SR power converter matches exactly that for a PWM induction motor drive, or any drive operating from a voltage-fed dc link. From the viewpoint of the utility supply, power factor is purely a function of the type of rectifier, the ac source impedance and the presence, if any, of a dc link choke, and is independent of the type of drive downstream of the link capacitor except in so far that the efficiency of the drive has a bearing on the magnitude of ac supply current drawn.

However, diode rectification does not permit regeneration without an additional fully-controlled line-commutated inverter bridge. Dynamic braking can of course be readily achieved using an energy dumping resistive chopper circuit. If the direct voltage source is battery supported then energy recovering is inherently available without the need for additional components; early SR drive developments by the authors used this feature for a battery vehicle which was fully 4-quadrant.

Very large SR drives may benefit from a variable direct voltage source at the cost of a front end controlled rectifier. This enables the amplitude of the phase current pulses to be controlled by the source voltage as well as by the duty of the phases switches, and therefore gives an extra degree of control. However, most of the authors' experience has sprung from fixed direct voltage sources, and the control by rotor-position related switching has been most adequate.

3. Modes of Control

The control objective is to provide each phase with current over all or part of the torque productive period of rotation, and to adjust the amplitude of the current such that the desired level of torque is obtained. This desired level may be dictated by an external feedback loop for controlling speed or position.

At low speeds the phase period is sufficiently long to enable shaping of the current pulse by repeated swiching within this period, which has been called "chopping" control. At higher speeds sufficient time may only exist for one on-off operation of the phase switch per phase period in which case the pulse shape will depend on the non-linear relationship of current flux-linkage and rotor angle. This has been called "single pulse" control.

3.1 Chopping Control

Fig. 5 shows a typical phase current waveform with chopping control. The pulse is timed to occupy the 30° torque productive period shown in Fig. 2 and controlled to have an approximately constant amplitude over this period by chopping the direct source voltage. Generally the phase switch is turned off when the current reaches some upper level and turned on again when the current has fallen by a set amount or after a set time. The sawtooth current excusion during chopping will reduce as the chopping frequency is increased, until, with very high chopping frequencies the current pulse will be relatively flat-topped. The torque magnitude is controlled by raising or lowering the upper level.

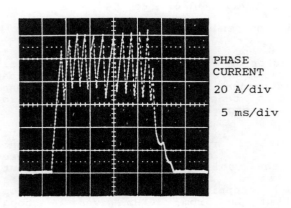

PHASE CURRENT

20 A/div

5 ms/div

Fig. 5. Typical phase current waveform in the chopping mode at 250 rev/min

If the current pulse is flat-topped the torque variation per phase will take the typical shape shown in Fig. 2. The torques from all the phases are then superimposed. There will obviously be some torque ripple due to the variation of torque with rotor angle. If very smooth low speed torque is required the current pulse shape can be adjusted, by modulating the chopping, to counteract the natural ripple torque. However, bearing in mind that the frequency of the torque pulsation is, for the 4-phase system illustrated, 24 times the shaft frequency, the rotor inertia, although considerably less than for a cage induction motor, is still sufficient to give smooth rotation. Without modulation the low speed torque variation is typically $\pm$ 15% of rated torque and the speed variation due to the ripple torque is very small. Experience with many months of reliable operation through a gearbox confirms that there are no practical problems for normal industrial drives.

The Inland SR servo drive system [4] uses modulation to eliminate ripple torque. The motor geometry and pole shape are designed to give a torque γ which is proportional to current i and varies sinusoidally with rotor angle θ ie

$$\gamma = k\,i\,\sin(6\theta)$$

(where k = constant) for one phase of an 8 stator/6 rotor pole motor.

The current is then modulated by chopping such that the pulse also takes a half-sinusoidal shape,

$$i = \hat{I}\,\sin(6\theta) \qquad (0 < \theta < \pi/6)$$

giving

$$\gamma = k\,\hat{I}\,\sin^2(6\theta)$$

Since the adjacent phases are displaced by 15°, and at any given time two phases are contributing torque, the total torque

$$\gamma_{tot} = k\,\hat{I}\,(\sin^2 6\theta + \cos^2 6\theta) = k\,\hat{I}$$

and is independent of rotor angle.

The absence of torque pulsations in the rotor is a help in reducing acoustic noise but it must be remembered that the individual stator poles still experience pulsating tangential and radial electro-magnetic forces.

The Inland system requires independent control of chopping for each phase and the ability to chop throughout the speed range which necessitates a high chopping frequency; with a top speed of 3000 rpm, 10KHz gives 16 chops per pulse which is sufficient for a half-sine modulation.

342

For a constant amplitude pulse, as the speed rises the number of chops per pulse reduces and a greater angle is occupied building the current up to its desired level. Both of these effects tends to reduce the torque per amp making it desirable to switch on in advance of the 'misaligned' poles position. The angle of advance can be made a function of speed.

3.2 Single Pulse Control

At a speed which may lie between 10% and 50% of maximum the chopping mode is either switched or progressively faded out, and the 'single pulse' mode prevails at all higher speeds. The current amplitude is now controlled predominantly by the switch conduction angle, but is also significantly influenced by the switch-on angle, both of which are functions of speed and torque for maximum efficiency.

At approaching maximum speed and maximum torque for that speed, the phase current occupies an arc of rotation which approaches 360° electrical (60° mechanical for an 8-6 motor), so that currents exist in at least two and frequently three windings simultaneously. To achieve maximum power from the motor for minimum phase current it is necessary to substantially advance the switch on angle in order to build the current up in advance of the back emf E which arises when the stator and rotor poles start to overlap.

The back emf is both current, speed and position dependent and is given by

$$E = i\,\frac{\partial L(\theta,i)}{\partial \theta} \cdot \frac{d\theta}{dt}$$

At high speeds the back emf may be greater than the source voltage! Fig. 6 shows a typical 'single-pulse' type current waveform at high speed for a general purpose industrial drive where, it will be seen, the current starts to fall due to the back emf effect before the switch off angle is reached [7].

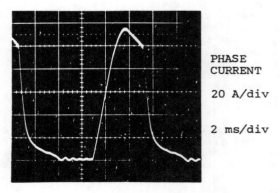

PHASE CURRENT

20 A/div

2 ms/div

Fig. 6. Typical phase current waveform in the single pulse mode at 1050 rev/min

4. Sensors

4.1 Rotor Position

Measurement of rotor position is essential if a high performance is to be obtained from the SR drive. The authors' experience has been to use a slotted vane attached to the shaft with simple optical or electromagnetic sensors. Using two sensors displaced by 15°, Gray-coded logic signals are obtained which subdivide the 60° repetition arc of a 4-phase 8-6 motor into 15° segments. The sensors are positioned such that for each phase two of these segments corresponds to the period of positive torque production. With the rotor stationary it is therefore known which pair of phases to energise in order to obtain either forward or reverse torque, and, with low speed rotation, at which instants to initiate and terminate chopping.

At higher speeds where the switch-on angle has to be progressively advanced, electronic interpolation is used to obtain a finer resolution of rotor position. For example, in industrial drives, a 4-phase ramp waveform generator is phase and amplitude locked to the Gray-coded logic signals. Digital (counting) interpolation has also been used in conjunction with microprocessor controls.

An alternative direct method for servo-motor applications [4] uses a purpose-designed resolver which provides precise angle information at all rotor positions. Ultimately the choice is based on ecomomic and performance factors.

Much current research [5] is directed at determining rotor position from winding currents and/or voltages, either from those normally occuring or from those specially injected for this purpose. One promising idea is to identify the positions at which the phase incremental inductance is a minimum (or a maximum) by measuring the current rise times with constant amplitude chopping. A high chopping frequency will be necessary to obtain sufficient resolution at higher speeds.

4.2 Current Sensing

Current sensing is necessary if current amplitude is controlled by raising or lowering the chopping level. For the 4-phase circuit of Fig. 3 the current in Phases A and C can be combined and measured in a single transducer, likewise for Phases B and D. Hall-effect sensors either direct or using flux-nulling have proved satisfactory.

5. Controllability

For the two modes of control previously described there will be predetermined control laws relating the switching angles to the demanded torque for a particular speed. In chopping the torque is mainly dependent on current amplitude, and it is possible to formulate a family of curves relating current level to torque for a range of speeds. In single pulse the torque is dependent on the angle of switch-on and switch-off and, once again, families of curves can be formulated relating these angles to torque and speed.

The controllers used by the authors have therefore two main features: firstly a memory either digital or analogue for the control laws, and secondly logic for translating this information into a control on-off signal for each phase switch. The method of torque control used is therefore essentially feed-forward rather than feedback, and has been very successful.

A typical control system is shown schematically in Fig. 7. The 4-phase reference signal from the rotor position sensors and decoding is used to initiate ramp waveforms which enable the single-pulse switch-on and switch-off instants to be determined from analogue signals from the control law memory. For chopping control, comparators are used to determine these instants by comparing the measured currents with the desired level from the memory. The digital signal from the position sensors is converted to an analogue speed signal using a frequency to analogue converter. An outer speed loop with at least integral and proportional control is used to set the torque demand signal.

For the battery vehicle controller developed at Nottingham University using an SR drive, an automatic speed contrl loop was inappropriate (the human driver providing the torque demand signal) but more sophisticated logic was necessary to interface the inputs from accelerator and brake pedals and to cater for conditions such as producing forward torque when rolling backwards. Satisfactory operation was obtained from a fully 4-quadrant controller, and a schematic diagram of the experimental control system is given in reference [9].

The control flexibility of the SR motor arises from the individual control of the torque contribution of each phase current pulse. It is possible that the current in Phase A, timed to produce maximum motoring torque, can be followed with a current in the next phase, B, to produce maximum braking torque, this change occuring in little more than 15° of shaft rotation.

343

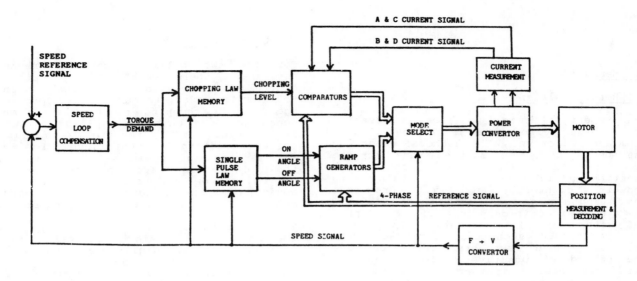

Fig. 7.   Schematic Diagram for a 4-Phase SR Control System

This response rate for a step change in torque coupled with the low rotor inertia offers excellent servo capabilities as has now been reported by Kollmorgan-Inland.

Table 1 gives relevant parameters for servo applications for two of the TASC 'Oulton' drives. The torque/inertia ratios for the respective motor size and power are impressive and compare with custom-built servomotors. Peak torque is less a function of the motor than of the ratings for the electronic switches supplying the current. The torque/inertia ratio can be improved by utilising a lower rotor/stator diameter (split ratio), at the penalty of reduced torque and power for a given stack size. Inland motors of approximately the same stack diameter and maximum speed require approximately 50% extra stack length to produce the same continuous torque rating but have lower rotor inertias.

TASC TYPE		D112	D132
Continuous Torque	(Nm)	25.5	47.8
Peak Torque	(Nm)	38.2	71.7
Maximum Speed	(rpm)	1500	1500
Continuous Power	(kW)	4.0	7.5
Stack Diameter D	(mm)	170	205
Stack Length L	(mm)	150	170
D^2L	(dm^3)	4.33	7.14
Rotor Inertia	(g.m^2)	5.0	12.8
Cont. T/J Ratio	(k.rad/S^2)	5.1	3.73
Peak T/J Ratio	(k.rad/S^2)	7.65	5.6

Table 1 : Ratings, size and accelerations for TASC general industrial SR motors

In contrast to the type of feedforward torque control described above, it has been suggested in a recent paper [3] that the provision of a torque loop within the speed control can make the response of the speed loop faster. Since it is not practical to measure torque directly, in this case dc supply current was measured and hence supply power, and knowing the speed the actual torque could be computed by division. However bearing in mind that the current must be measured downstream of the link capacitor and contains an appreciable ac component, the necessary filtering will result in significant time delays. It is therefore very unlikely that controlled torque changes can be made as rapidly as with feedforward control, apart from the additional stability problems that such a loop might introduce. In the case where an SR drive is to be used for a specific torque control application an external means of torque measurement would be available.

In the authors' experience, the response time of the speed loop is almost entirely dependent on the type of speed transducer and the amount of load inertia. For the industrial drives, a frequency to voltage conversion of the 24 pulse per rev position sensor signal is used with a 100:1 range of speed control. The transducer filter time constant is dictated by the low speed condition.

Fig. 8 shows a typical transient for speed and torque for a load of negligible inertia (compared with the rotor) and a friction coefficient corresponding to 50% rated torque at full speed. Faster responses could be obtained using a transducer with a higher number of pulses per rev.

6. Control Hardware

There are many ways to implement the control strategy described in this paper. Most of the early developments carried out by SRD, including the TASC Oulton drive, have used a combination of analogue and MSI digital circuitry to provide the necessary control functions. These control systems have provided excellent performance at a relatively low component cost. However the component count for these analogue/digital systems is fairly high and for this reason a higher level of integration is required.

The development of highly integrated control systems at SRD is currently moving in two directions. Firstly microprocessor-based systems are being developed to service a variety of application areas requiring a high level of control flexibility and programmability. These control systems provide full 4-quadrant control of the SR drive providing efficient operation throughout the torque/speed range and are capable of being 'programmed' to suit a range of motor sizes and torque/speed characteristic requirements. The flexible microprocessor-based control system is now in an advanced state of development. It is likely that some of the peripheral functions required by the microprocessor control system will eventually be implemented by a semicustom integrated circuit to provide a further reduction in component count.

The second direction is aimed at very high volume markets such as low power drives for the domestic appliance market. In this case a dedicated control system will be integrated on to a full custom chip requiring only a minimal number of external components to provide full closed loop control. The control strategy of such a dedicated chip would have some similarity with that described in this paper but would also include new techniques to provide the limited requirement of such drives in a form suitable for custom integration.

In the near future it will be possible to integrate both control functions and power switches on to one silicon chip. This technology of power integrated circuits will provide further cost and volume reductions for small drives.

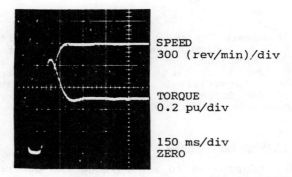

SPEED
300 (rev/min)/div

TORQUE
0.2 pu/div

150 ms/div
ZERO

Fig 8. Typical step response of speed and Torque for a 0 to 1500 rev/min speed reference demand, with low load inertion and 0.5 pu viscous friction.

7. Conclusions

The inherent flixbility of control of the switched reluctance drive has been illustrated in this paper. This flexibility is due to the nature of excitation whereby torque is produced by a sequence of individually controlled current pulses at intervals of typically 15° of rotor rotation. These pulses may be positioned relative to the rotation to obtain maximum efficiency (or to maximise whatever performance criteria is desirable) for any operating torque and speed. The use of direct rotor position measurement eliminates the instability or resonance phenomena associated with conventional stepper motor drives or other synchronous systems.

At low speeds modulation can be used to shape the current pulses to give smoother operation if this is necessary. This can be extended throughout the speed range at the penalty of a relatively high modulation frequency. A variety of power converter circuits exist, the relative merits of which depend on the supply voltage and the particular application. Braking is simply achieved by repositioning the current pulses.

A digital or analogue memory can be used to store the optimum control laws relating switching angles to demanded torque and operating speed. Torque is adjusted by a simple feedforward control, and an external feedback loop can be used to control speed. Torque adjustment is very rapid and together with the inherently low rotor inertia and robust construction makes the SR drive suitable for servo applications as well as the many other applications for which its high efficiency is more important.

The hardware implementation of the controller does not indicate undue difficulty either in a technical or commercial sense and with the increasing availability of large scale integrating competitive variable speed SR drives for applications as small as domestic appliances are quite feasible.

8. Acknowledgements

The authors wish to acknowledge the contribution made to the development of SR controllers by their colleagues D.M. Sugden, P.D. Webster, H.C. Lovatt and S.P. Randall and to the large contribution made by TASC Drives Ltd in producing the SR drives described in this paper.

9. References

1. Ray, W F et.al. "High Performance Switched Reluctance Brushless Drives", IEEE-IAS Conference Record, October 1985, p.p. 1769-1776.

2. Davis, R M, Ray, W F and Blake, R J "Inverter Drive for Switched Reluctance Motor : Circuits and Component Ratings", IEE Proc. Vol. 128, Pt.B, No. 2, March 1981, p.p. 126-136.

3. Bose, B K et.al. "Microcomputer Control of Switched Reluctance Motors", IEEE-IAS Conference Record, October 1985, p.p. 542-547.

4. Egan, M G et.al. "A High Performance Variable Reluctance Drive : Achieving Servomotor Control", Proc. Motorcon 1985.

5. Acarnley, P P, Hill, R J and Hooper, C W, "Detection of Rotor Position in Stepping and Switched Motors by Monitoring of Current Waveforms", IEEE Trans, Vol. IE-32, No. 3, August 1985, p.p. 215-222.

6. Ray, W F and Davis, R M, "Power Conversion Circuit", UK Patent GB 2105933.

7. Bausch, H and Rieke, B, "Speed and Torque Control of Thyristor-Fed Reluctance Motors", Proc. of Int. Conf. on Electrical Machines, 1976, p.p. I28-1 - I28-10.

8. Ray, W F and Davis, R M, "Reluctance Electric Motor Drive Systems", UK Patent GB 1591346.

9. Blake, R J et.al. "The Control of Switched Reluctance Motors for Battery Electric Road Vehicles",

Speed Control of Switched Reluctance Motors

J. Corda and J.M. Stephenson

International Conference on Electrical Machines, Budapest, 1982

SUMMARY

The paper outlines the analysis of an ideal electrically-switched doubly-salient reluctance motor. Expressions for torque and rms current are given in terms of a normalised set of electrical and mechanical parameters. Two methods of speed control are identified from the analysis and an elegant graphical method for choosing parameters to meet any required load characteristic is presented.

SPEED CONTROL OF SWITCHED RELUCTANCE MOTORS

Dr J Čorda
Faculty of Electrotechnics
University of Sarajevo
Yugoslavia

Dr J M Stephenson
Department of Electrical and Electronic Engineering
University of Leeds
England

LIST OF SYMBOLS

i = instantaneous current
L = inductance of phase winding
r,s = subscripts for rotor and stator
t = time
T = torque
V = voltage
α = L_i/L_o
β_s = stator pole arc
θ = rotor displacement
ϕ = rotor pole pitch
ψ = flux linkage
ω = rotor angular velocity

1 INTRODUCTION

A switched reluctance motor has salient poles
on both the stator and rotor with excitation
coils on the stator only. Torque is produced
by the pulling into alignment of stator and
rotor poles when current flows in the
appropriate coils. The currents in the coils
are switched electronically in sequence, and at
instants determined by the position of the
rotor, so as to produce continuous rotation.
One form of the motor with 4 'phases', i.e. 4
stator circuits, is illustrated in Fig. 1.

In previous papers[1-9] the authors and their
colleagues at Leeds and Nottingham Universities
have presented the principles of operation,
design considerations and practical results for
this class of motor and its associated power
electronics.

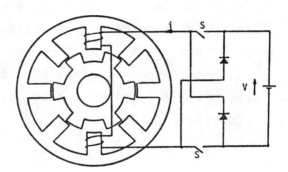

FIG 1 Motor and drive circuit

Analysis of a linearised, ideal motor is very
limited in its value for detailed motor design,
but is useful for gaining a broad understanding
of the influence of the many parameters. This
paper presents the application of linear analysis
to the design of a switched reluctance motor to
meet a given load torque/speed characteristic.

2 DERIVATION OF EQUATIONS

2.1 The Basic Model

The switched reluctance motor produces torque by
virtue of the inductance of the phase windings
varying with rotor position. In practice these
inductances vary in a complex manner with rotor
position and with the magnitude of the current
flowing in the winding. In order to make linear
analysis possible it is necessary to idealise
these and other motor characteristics as follows:
(i) Inductances are independent of current;
(ii) Flux fringing at pole tips is negligible,
     i.e. the phase inductance varies as shown
     in Fig. 2., This is drawn for a motor with
     rotor pole arc β_r > stator pole arc β_s.

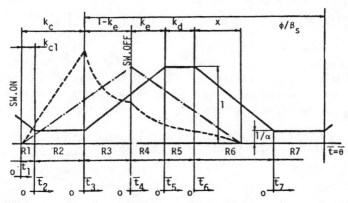

FIG 2 Normalised inductance profile (———) with flux-linkage waveform (—·—) and current waveform (—--—).

The inductance rises and falls linearly with increasing and decreasing overlap of stator and rotor poles. It has regions of constant maximum inductance because $\beta_r > \beta_s$ and minimum inductance when the stator pole is opposite the inter-polar slot on the rotor. A cycle of inductance variation is defined by a rotor pole pitch ϕ;

(iii) All power losses are ignored;

(iv) Switching is instantaneous;

(v) Angular velocity is constant.

With these assumptions the equation defining the behaviour of the circuit of Fig. 1 is

$$\pm V = \frac{d\psi}{dt} = L\frac{di}{dt} + i\frac{dL}{dt} . \qquad (1)$$

In this equation the positive sign corresponds to the state when the switches S are closed and the negative sign corresponds to the switches open but with current flowing through the diodes.

If the switches are closed when dL/dt is positive, power flows from the electrical supply to become mechanical output power. When the switches are opened with dL/dt still positive, part of the stored magnetic energy is returned to the supply and part is converted into mechanical form. If current flows when dL/dt is negative, a braking torque is produced.

Using the above equation, analytical expressions can be derived for the average torque and mean and rms currents in terms of the physical dimensions and electrical parameters of the motor. The calculation of average torque is based on the change of system co-energy when the rotor moves through an angle equal to the rotor pole pitch ϕ,

$$T = \frac{1}{\phi}\oint\psi.di. \qquad (2)$$

2.2 Normalised Equations

An analysis must include the stator and rotor pole arcs (β_s, β_r), air gap lengths, numbers of turns of windings, as well as the instants of switching on and off, supply voltage (V) and speed (ω). The following parameters will be used to characterise the motor:

α, ratio of max. to min. inductance, L_i/L_o;

k_d, dead-zone coefficient $(\beta_r-\beta_s)/\beta_s$; ϕ/β_s;

k_c, k_e, turn-on and turn-off coefficients; and it is helpful to normalise variables as follows:

angle, $\bar{\theta}=\theta/\beta_s$; time, $\bar{t}=t/\beta_s/\omega=\bar{\theta}$;

inductance, $\bar{L}=L/L_i$; Flux linkage, $\bar{\psi}=\psi/V\beta_s/\omega$;

current, $\bar{i}=i/V\beta_s/\omega L_i$.

Fig. 2 shows inductance variation, switch-on and off points and the current and flux linkage waveforms. When the switches are closed, the flux linkage rises linearly (dψ/dt=V) from zero to a peak value of $(k_c+1-k_e)V\beta_s/\omega$ and after switching off, falls linearly (dψ/dt=-V). Rising and falling periods are equal and their sum must be less than one cycle of the inductance or the flux will increase without limit.

The analysis to find the current, flux linkage and torque is performed by dividing the waveform into 7 Regions (R1 to R7) and applying the appropriate parameters and initial conditions. As an illustration, consider Region 1 as illustrated.

$$\bar{L}_1(\bar{t}_1) = [1 + (\alpha-1)(k_{c1}-\bar{t}_1)]/\alpha$$

and solving $d\psi/dt = V$ yields

$$\bar{\psi}_1(\bar{t}_1) = \bar{t}_1 \text{ from } 0 \text{ to } \bar{\psi}_1 = k_{c1}$$

$$\bar{i}(\bar{t}_1) = \alpha\bar{t}_1/[1 + (\alpha-1)(k_{c1}-\bar{t}_1)]$$

$$\text{from } 0 \text{ to } \bar{i}_1 = \alpha k_{c1}$$

$$\bar{\psi}(\bar{i}) = \bar{i}[1 + (\alpha-1)k_{c1}]/[\alpha + (\alpha-1)\bar{i}].$$

Substituting the values for $\bar{i}$ and $\bar{\psi}$ into Eqn. 2 gives the average torque per phase

$$T = [V^2 \beta_s^2/\omega^2 \phi L_o]F \qquad (3)$$

where $F = \frac{1}{\alpha}\oint \bar{\psi}.d\bar{i}$ is a function of the dimensionless parameters α, k_d, ϕ/β_s, k_c and k_e.

As a further example, the rms current in the switches is given by

$$I = [(V\beta_s/\omega L_o)\sqrt{\beta_s/\phi}]R \qquad (4)$$

where $R = \sqrt{\frac{1}{\alpha^2}\int_0^{(1-k_e)} \bar{i}^2 \overline{dt}}.$

This analysis yields cumbersome algebra but readily allows the study of the effects of parameters on performance. It is used below to examine means of control of motor speed.

3 SPEED CONTROL

Rearranging Eqn. 3 to give

$$\omega = \frac{\beta_s}{\sqrt{\phi L_o}}.V.\sqrt{\frac{F}{T}} \qquad (5)$$

reveals two methods of speed control - by varying (i) the supply voltage (V) and (ii) the switching angles (F), and that without control, the speed is inversely proportional to $\sqrt{T}$.

Note from Eqn. 3 that the torque is proportional to F and that for a given voltage and speed, F is proportional to peak flux linkage,
$$\hat{\psi} = (k_c+1-k_e)V\beta_s/\omega.$$

The two control strategies will now be examined in the context of a torque/speed characteristic with constant torque and constant power regions, typical of traction applications.

Consider voltage control at constant torque. From Eqn. 5 V is proportional to ω and $\bar{\psi}$ and I can be seen to be constant. In fact the current waveform remains the same, except for change in frequency. The s.r. motor is seen to be like the d.c. motor in that voltage control naturally gives speed control with constant torque, the flux and currents remaining constant.

For control by switching angles, the speed depends on k_c and k_e for a given load characteristic (Eqn. 5). There is no analytical solution for k_c and k_e but a graphical method is shown in Fig. 3.

In quadrant I, F (for given k_d, ϕ/β_s, α) is plotted against conduction period (k_c+1-k_e) for various values of k_e. In II the curve of F against ω for the particular load torque/speed characteristic is plotted. In IV the ordinate is ω and the abscissa is (k_c+1-k_e). For a given motor operating on a given voltage, $\hat{\psi}$ is proportional to $(k_c+1-k_e)/\omega$, therefore the ratio of the coordinates of a point in IV is proportional to $\hat{\psi}$.

Choosing a point in the curve in quadrant II determines possible combinations of k_c and k_e. (Move horizontally to the right). Having chosen a particular pair, a point is defined in IV (by transferring from I and II to IV) corresponding to a particular $\hat{\psi}$. In this way various control strategies can be explored and Fig. 3 illustrates a case in which k_e is held constant and k_c is varied. $\hat{\psi}$ increases as the speed falls below 'base speed' (the highest speed of the constant-torque region), but falls as the speed rises in the constant-power region. Angle control is thus seen to be

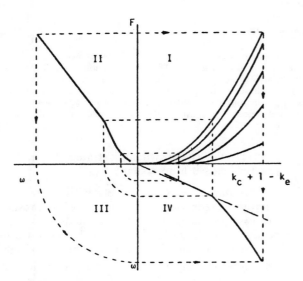

FIG 3 Control diagram

analogous to control by field weakening in a
d.c. motor and to be suitable for meeting
constant-power loads.

4   CONCLUSION

The analysis leading to equations defining the
behaviour of an ideal doubly salient switched
reluctance motor has been outlined and
expressions for torque and rms current have been
given in terms of normalised parameters.

Two methods of speed control have been identified
and an elegant graphical method for choosing the
parameters to meet any required load characteristic has been presented. It has been illustrated by application to a characteristic with
constant torque and constant power regions.

5   ACKNOWLEDGEMENTS

The authors thank their colleagues for their
support and encouragement and Dr Čorda thanks
the University of Leeds for financial support.

6   REFERENCES

1   Stephenson, J.M. and Čorda, J.:
     'Computation of torque and current in doubly
     salient reluctance motors from nonlinear
     magnetisation data', Proc. IEE, 1979,
     126(5), pp 393-396.
2   Čorda, J. and Stephenson, J.M.: 'An
     analytical estimation of the minimum and
     maximum inductances of a double-salient
     motor', Proc. of Int. Conf. on Stepping
     Motors and Systems, University of Leeds,
     1979, pp 50-59.
3   Ray, W.F. and Davis, R.M.: 'Inverter
     drive for doubly salient reluctance motor',
     IEE J. Electr. Power Appl., 1979, 2(6),
     pp 185-193.
4   Davis, R.M., Ray, W.F. and Blake, R.J.:
     'An inverter drive for a switched
     reluctance motor', Int. Conf. on Electrical
     Machines, Athens, 1980.
5   Lawrenson, P.J., Stephenson, J.M., Fulton,
     N.N. and Čorda, J.: 'Switched reluctance
     motors for traction drives', Ibid.
6   Lawrenson, P.J., Stephenson, J.M.,
     Blenkinsop, P.T., Čorda, J. and Fulton, N.N.:
     'Variable-speed switched reluctance motors'
     IEE Proc. B, Electr. Power Appl., 1980,
     127(3), pp 253-265.
7   Davis, R.M., Ray, W.F. and Blake, R.J.:
     'Inverter drive for switched reluctance
     motor: circuits and component ratings',
     Ibid 1981, 128(2), pp 126-136.
8   Blenkinsop, P.T.: 'A novel, self-commutating,
     singly-excited motor', PhD Thesis, University
     of Leeds, 1976.
9   Čorda, J.: 'Switched reluctance machine
     as a variable-speed drive', PhD Thesis,
     University of Leeds, 1979.

Microprocessor Control of a Variable Reluctance Motor

P.H. Chappell, W.F. Ray and R.J. Blake

Proceedings IEE, Vol. 131, Pt. B, No. 2, March 1984, pp. 51-60

Indexing terms: *Microprocessors, Control systems, Reluctance motors*

Abstract: A working closed-loop microprocessor control system is described for a variable reluctance motor. The system incorporates features such as the separate control of both lead angle and conduction angle. The paper addresses the problem of implementation of the control strategy rather than the best choice of control switching angles. Rotor-position measurement and motor-phase-current measurement are used for feedback purposes. Interface circuitry between the microprocessor system and power system is designed to make ~~efficient~~ use of both hardware and software. Some waveforms are presented to illustrate the relative idle time of the microprocessor.

List of abbreviations

APC   = angular position control (Figs. 3 and 10)
CCC   = chopped current control (Figs. 3 and 9)
CMOS = complementary metal oxide semiconductor
CPU   = central processing unit
CTC   = counter/timer circuit
DAC   = digital to analogue convertor
MSI   = medium-scale integration
NMOS = n-channel metal oxide semiconductor
PIO   = parallel input/output controller
PROM = programmable read-only memory
RAM   = random-access memory
SSI   = small-scale integration
TTL   = transistor-transistor logic
VR   = variable reluctance [11]
V_s = supply voltage

1 Introduction

1.1 Background

It is well known [1, 2] that for maximum torque production the switching angle of a stepping motor must vary with the switching rate and that in order to ensure the appropriate switching angle is applied for a particular speed, it is necessary to have a reference measurement of rotor position. This is a move from the traditional open-loop control, where rotor position relative to stator excitation on a phase-to-phase basis is dependent on load torque and speed, to closed-loop control where the rotor-position/stator-excitation relationship is dictated by the controller. The disadvantage of the necessity for position measurement is compensated by the improved motor performance in terms of torque, stability, higher-speed operation and guaranteed step integrity.

Publications concerning the closed-loop control of stepping motors have in general addressed either or both of two problems:

(i) What are the optimum switching angles and how may these best be calculated or found experimentally?

(ii) How may the switching strategy be implemented in a practical closed-loop system?

This paper deals exclusively with problem (ii) from a microprocessor viewpoint and therefore presupposes that the two problems may be separated. This assumption is partly justified by previous papers [3–9] concerning microprocessor control of stepping motors, although these are relatively limited in number as a proportion of stepping-motor publications.

Most attention has been paid to problem (i). Since switching or lead angle is the only control parameter (torque increases with lead angle until a maximum torque value is reached), steady-state running at a fixed speed requires a unique lead angle for a particular load torque; this may be achieved with or without a microprocessor using a simple speed feedback loop [7]. The attention therefore has been directed to the cases of acceleration and deceleration, mainly the former, with the objective of achieving minimum time position control. The criteria used therefore are either maximisation of torque at all speeds [2–4, 7] or minimisation of time for a specified angle of rotation [5, 6, 9]. Attention has also been given to the problem of damping as the rotor settles at its desired position [5, 9].

The approach has generally been to derive a mathematical relationship between torque, speed and switching angle, to compute the optimum angles in accordance with the above criteria, and to compare the predicted trajectory with that obtained experimentally for the same switching strategy. Although the agreement between theory and experiment is only approximate due to the difficulty of modelling the motor, reasonable predictions of optimal switching angles or sequences can be obtained [2]. However, it has been found that simplified switching strategies using for example only three values of lead angle for acceleration have yielded equally acceptable results [3, 6, 9]. Furthermore, the complexity of computation has not been appropriate for on-line optimal control unless simplifications are made [7, 8].

In general it has been necessary to obtain the required switching strategy either by off-line computation [2, 3, 6] or by using a microprocessor as an experimental test aid [5, 9], and then to store the table of angles (or time intervals) in the memory for closed-loop microprocessor

Paper 2980B (P1, C6), received 5th April 1983

The authors are with the Department of Electrical and Electronic Engineering, University of Nottingham, University Park, Nottingham NG7 2RD, England

implementation. This separates the problems of the determination and implementation of the switching strategy.

Previous microprocessor implementations have been directed at stepper motors with phase currents of a few amperes and with step angles of typically 15° [3, 5, 6]. Optical encoders have been used for position measurement [2, 3, 6, 7, 9, 14], in one case [3] two encoders, one for forward and one for reverse running, although current waveform detection methods have also been used [8, 10]. A higher resolution may be obtained by following the encoder with a phase-locked loop [2]; this technique is used for the system described in this paper.

The sole control parameter has been the switching or lead angle: that is the rotor angle at which driving voltage is applied to a given phase in advance of the detente position for that phase. The turn-on of a phase has been synonymous with the turn-off of the previous phase, and this is generally an inherent constraint of the power switching circuit. Since the timing of lead angle for each phase does not overlap, it is straightforward to use the microprocessor as the timing or counting element, and in many cases no additional integrated circuits apart from the microprocessor and its associated storage have been required.

1.2 Objective

In contrast, the object of this paper is to examine the microprocessor control of a motor where turn-on and turn-off angles are independently controlled and are no longer synonymous; furthermore, to examine the implications of operating a 10 kW system using thyristors rather than transistors as the main switches, thereby requiring additional switches for forced commutation. The questions that are addressed are:

(i) How may the timing be implemented for both switch-on and conduction angles considering that the conduction angles for the phases may overlap?

(ii) How accurately may the angles be set, taking into account angular delays corresponding to microprocessor interrupt and computation time?

(iii) What precautions need to be taken for operation in the electrically noisy environment of high-power switching circuits?

(iv) How can the state of the commutation circuit be controlled to prevent commutation failures?

(v) What proportion of microprocessor time is needed to fulfil its essential roles of timing and sequencing the switching of the phases?

In general, how can (a) a microprocessor be used for controlling a more flexible and higher powered motor drive circuit, assuming a moderately large step angle (15°) and assuming the switching angles appropriate for each speed and torque level have been predetermined and stored and (b) will sufficient time remain for some supervisory and perhaps health-monitoring functions [16]?

1.3 Drive system

The system selected as the most appropriate for microprocessor implementation is taken from a previous pub-

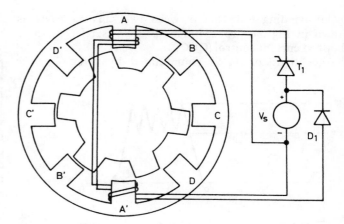

Fig. 1 *Basic operation of a VR motor showing the A phase only*

lication [11] by Davis, Ray and Blake. Rieke [14] has used a microprocessor for controlling a VR motor, but the operation was limited compared to that in Reference 11.

Fig. 1 shows the motor cross-section having a 15° step angle and utilising a secondary winding to return stored energy to the supply at the cessation of excitation for each phase. (Firing T_1 switches the supply voltage V_s across the main winding of phase A, causing clockwise rotation and on switching off T_1 the current transfers to the closely coupled bifilar winding and the return diode D_1).

The inverter circuit is shown in Fig. 2 and is fully

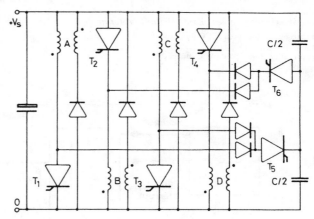

Fig. 2 *Power electronics*
Four-phase circuit using winding resonant reversal

described in Reference 11. Currents are built up and decayed sequentially in the four phase windings A, B, C and D by turning on and off the main thyristors $T_1, ..., T_4$. Commutation of the main thyristors is achieved using the commutation thyristors T_5 and T_6. This presupposes that the commutation capacitor is charged with the correct polarity; for example, the junction point P must have a voltage in excess of $+V_s$ for the commutation of T_2 or T_4.

Unlike conventional stepper-motor drives the controller must cater for two different modes of operation depending on motor speed. At low speeds the motor current in a particular winding is cycled between two current levels as shown in Fig. 3a [13–15]. This process is referred to as chopped current control (CCC).

After thyristor T_1 is commutated by firing T_5, the

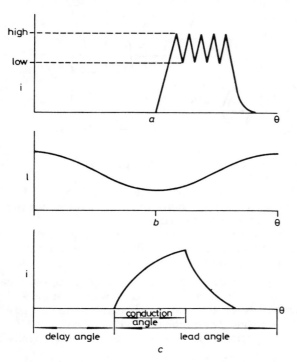

Fig. 3 *Showing for a given phase, relative to rotor angle θ*

a Phase current *i* at low speed (CCC mode)
b Inductance *l*
c Phase current *i* at high speed (APC mode)

current in winding A falls. When T_1 is refired to increase the current in A, T_6 must also be fired to reset the polarity of the commutation capacitors [11]. A minimum on time for T_1 must, therefore, be allowed for this reset process. A minimum off time for T_1 may also be necessary to avoid the danger of a progressive increase in peak current [11]. For very low values of torque it is necessary to operate with a fixed peak current and to allow the current to become discontinuous by increasing the off time. Chopped current control has therefore two submodes of operation depending on torque level; it mainly involves the measurement and comparison of the phase currents as described in Section 2.4.

The controller must also know the sequence in which the windings must be supplied with chopped current to produce torque. This information is obtained from a rotor-position transducer comprising two slotted optical units displaced by 15° and a disc with six teeth. The signals produced may be decoded to indicate which phase is torque productive besides giving a position reference pulse every 15° (Fig. 6).

At higher speeds, torque is produced by single current pulses [11] using angular position control (APC) with respect to rotor position. In APC operation two control parameters are used: the delay angle and conduction angle. Delay angle represents the rotor angle at which a given phase is turned on, which, in general stepper motor terminology, is referred to as switching angle or lead angle in so far that it is in advance of the detente position. However, in practice, it is not possible to time an advance, rather one has to time a delay from some previous position reference point; hence the term delay angle is considered to be more representative and is used in this paper. Fig. 3c

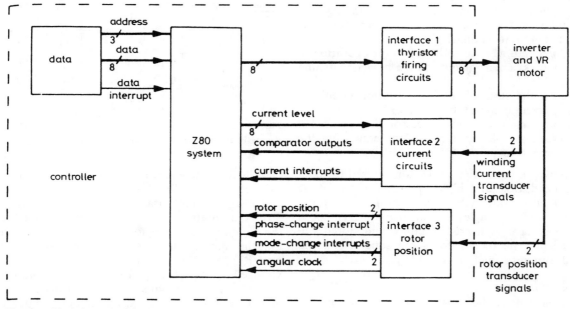

Fig. 4 *Block diagram of the system*

354

illustrates the control angles in relation to a typical single current pulse.

While the motor may be considered as similar to a stepper [12], the controller for this system must be capable of executing significantly more control functions than for conventional stepper-motor controls. In particular, it must be able to change between different operating modes, handle combinations of control parameters, undertake the timing of intervals which may overlap, and accept information from both current and position transducers. These tasks indicate the advisability of microprocessor implementation to avoid the proliferation of control electronics.

2 Hardware

2.1 General

The control strategy for the system has already been described in Reference 11. However the previous implementation used discrete integrated-circuit logic and analogue elements. The change to microprocessor control has implications both for the interfacing between the power system and the controller, as described in this Section, and in the organisation of the control functions by the microprocessor software described in Section 3.

A block diagram of the complete system is shown in Fig. 4, in which the data block represents the user controls and the inverter and motor provide the winding current and rotor-position feedback signals. Where possible, CMOS devices are used for the interface circuits and are operated at 10 V to give good noise immunity. The 10 V logic signals of the transducer interface circuits are then converted to the NMOS logic levels of the microprocessor system using simple transistor buffers which were found to give good noise immunity while performing the necessary logic level shift.

2.2 Microprocessor system

Central to the controller is the Z80 microprocessor system (Fig. 4) which comprises:

(i) central processing unit (CPU)
(ii) clock (2.5 MHz)
(iii) three parallel input/output controllers (PIO)
(iv) one counter/timer circuit (CTC)
(v) 1 kbyte of RAM

Provision is made for programmable read-only memory (PROM), although a development system's RAM is used in target system evaluation. As well as allowing for the transfer of data between the CPU and power system, the PIO devices are used to handle interrupt signals. Timer functions during CCC and counting functions during APC are performed by the four eight-bit counter/timers in the CTC.

2.3 Interface 1: thyristor firing circuits

There are six thyristors in the inverter, four main and two commutating, and so one PIO port is used to handle the firing of thyristors. A change in output of a port line from logical 0 to 1 causes a thyristor to be fired. This edge is converted into a pulse using a monostable circuit in order to avoid using microprocessor time to perform a waiting task. Between the output to the monostable and thyristor gate there is a pulse amplifier and pulse transformer. Firing of more than one thyristor is achieved by the output of a byte containing more than one set bit. During the CCC mode, main and commutating thyristors are fired together as described earlier.

2.4 Interface 2: current circuits

The current transducers produce analogue voltages approximately proportional to the sum of the main and bifilar winding currents: this is called the phase current. These transducers are used in the CCC mode to detect current levels. A sequence of events during this mode (illustrated in Fig. 3) is to fire a main thyristor causing the current in a winding to increase until it reaches a demanded level (high), then fire a commutating thyristor causing the current to decrease until it reaches a lower demanded level (low), then refire a main thyristor and so on.

One way of implementing this process using a microprocessor is to connect a current transducer output to an analogue-to-digital convertor and feed the digital data into a PIO port. Then, after firing a main thyristor, the microprocessor can sample the current and compare this value with the parameter stored in memory. When the data received from the analogue to digital convertor is the same or greater than the demanded level (high), the commutating thyristor is fired and the current subsequently decreases. The microprocessor is thus totally dedicated to this task.

A better approach was adopted which uses a digital-to-analogue convertor (DAC) (see Fig. 5). Here, the microprocessor sets the demanded level (high) on the DAC and

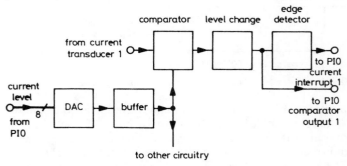

Fig. 5 Block diagram of interface 2

fires a main thyristor. The interrupt system is enabled and the microprocessor waits for an interrupt. Comparators are used to compare the demanded level set on the DAC with the output of the current transducers. When a comparator changes state the microprocessor is interrupted, commutates the main thyristor, then sets the demanded low level on the DAC and enables interrupts. Similar circuitry is provided for other phases, otherwise false interrupts may occur. For example, at a phase change, the current may be decaying in one phase while the current is increasing in the next phase. The decaying current is of no interest and so software is used to mask off this interrupt line thereby eliminating any false interrupts.

2.5 Interface 3: rotor position circuits
The rotor-position transducers are used to indicate which phase of the motor is torque productive. Fig. 6 shows the operating principle of the rotor-position transducers. The relative position of the toothed disc is adjusted to give slotted-optodevice transitions at suitable rotor reference

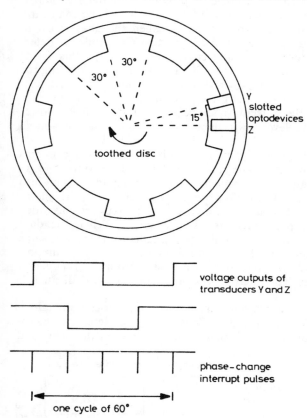

Fig. 6   *Operating principle of the rotor-position transducers*

positions. For example, the minimum inductance of phase A may coincide with the positive edge of the voltage output of the slotted optodevice Y. The negative edge then indicates the minimum inductance position for phase C. The slotted optodevice Z is positioned 15° from X to give similar outputs in relation to phases B and D.

The rise and fall of the voltage output from the phototransistor of a slotted optodevice is slow and noisy. A comparator with hysteresis is used to clean up the digital signal from a transducer before being fed into the microprocessor system via a single transistor buffer (Fig. 7).

The two Gray coded signals from the transducers are fed into an exclusive NOR gate, the output of which is fed to an edge detecting circuit. The latter circuit produces a pulse every time one of the transducer outputs changes

state (Fig. 6). This pulse is used as an interrupt signal and indicates to the microprocessor system when to change to the next phase, i.e. a phase change interrupt. The transition

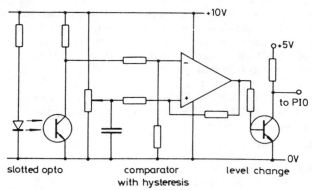

Fig. 7   *Rotor-position transducer interface*

between the CCC and APC modes is initiated from a simple timing network which uses the rotor-position transducer signals to generate an interrupt pulse at the required speed. For an increasing speed the transition occurs at a higher speed than that for decreasing speed, i.e. hysteresis is included to avoid indecision.

The microprocessor system requires an angular clock so that it can correctly position current pulses in the APC mode regardless of speed. This clock is generated from the rotor position signals multiplied by 60 using a phase-locked loop [2]. A clock edge is thus produced for every half degree. To interface the clock to the counter, which is part of the microprocessor system, a TTL gate is used.

2.6 Data
The control parameters of the system are stored as bytes of data in RAM. These are:
  (i) high and low current levels used during CCC
  (ii) on and off times used during CCC
  (iii) delay and conduction angles used during APC

Control of the system involves changing these parameter values while the system is running. An external circuit generates an interrupt pulse which is fed into the strobe input of a PIO port. The interrupt program reads the state of two ports: one for the parameter and one for the address. Sets of switches with pull up resistors provide an easy solution for changing the states at port lines in the laboratory.

3   Description of the software

3.1   General
The software is divided into five sections:
  (i) initial
  (ii) CCC mode
  (iii) mode transition
  (iv) APC mode
  (v) data

The programs within these sections are executed in sequence as indicated in Fig. 8. After executing the initial

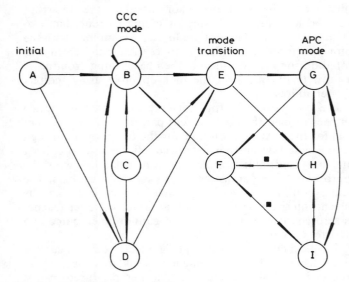

Fig. 8 *Simplified sequence diagram*

A   initial and priming
B   phase change
C   timer
D   CCC, fire main and commuting thyristors
E   mode transition (accelerating)
F   mode transition (decelerating)
G   phase change, load delay angle
H   fire main thyristor, load conduction angle
I   fire commutating thyristor
■   nested interrupts

program, the microprocessor performs all other tasks using interrupt routines. Interrupts for CCC, APC and mode transition are serviced as they occur. Coincident interrupts are processed by the Z80 interrupt structure. If two interrupts occur together, one routine is serviced while the other waits for the first to finish. Interrupt contention is negligible during CCC since events occur at a slow rate compared with the execution times of routines. Also only one phase is active at a time. However, during APC events for one phase can overlap in time with events for another phase. Errors produced as a result of this overlap are discussed in Section 4.1. The decelerating mode transition routine uses nested interrupts.

Changing control parameter values is a background task and the other routines have a higher priority so they can suspend the execution of the data routine. The Z80 maskable interrupt mode 2 is used throughout the software. The main function of the software is to recognise the position of the rotor and then to fire the appropriate thyristors. The software for this task may be written in a number of ways:

(i) use logical instructions to decode the rotor position bits

(ii) use a look-up table containing bytes which fire the correct thyristors where the rotor position bits act as an address pointer for a table

(iii) use the rotor position bits to direct program flow to separate routines for each phase; each routine being identical except for bytes which are required to fire thyristors associated with a given phase

Methods (i) and (ii) use a single general routine for all phases while method (iii) has repeated blocks of program. Therefore, in general, the amount of memory required for methods (i) and (ii) is less than that for method (iii). However, methods (i) and (ii) use microprocessor time to decode information from the transducers and sort out which thyristors to fire, while method (iii) immediately directs program flow to a specific block of program for a particular phase. If speed of execution is important in order to obtain good resolution, then method (iii) is better than either methods (i) or (ii). Where speed of execution is not critical, as in CCC, method (ii) is used, while for APC method (iii) is more attractive.

Modular interrupt programs have been used by Rieke [14] to control a reluctance motor in a single mode of operation equivalent to CCC mode where the current levels are set external to the microprocessor system.

3.2 Initial routine

This part of the software performs all the usual tasks associated with the setting up of a microprocessor system; for example, instructions are executed which load interrupt vectors into the PIO and CTC devices. Control parameter values, appropriate for starting the motor, are loaded into the target system's RAM. The commutating capacitors are primed into one of two states depending on rotor position; either ready to commutate phases A or C or to commutate phases B or D. Priming is performed by firing the commutating thyristor for phases A and C followed by the thyristor for phases B and D.

It is essential that the microprocessor system knows the state of the commutating capacitors. Whenever the commutating thyristor for phases A and C is fired, a bit in RAM is reset and if the thyristor for phases B and D is fired this bit is set. The program starts the CCC sequence by firing a main thyristor which causes the current in a main winding to increase. The microprocessor loads a parameter value, representing peak current, into the DAC and enables the interrupts. The initial routine functions for a stationary or nonstationary rotor in one direction at any speed up to the mode transition speed.

3.3 Chopped-current-control routines

The control strategy during CCC may be summarised as follows:

(i) maintain peak current within commutation limits

(ii) fire commutation reset thyristor with main thyristor and allow a minimum on time for the reset process

(iii) keep trough currents as high as possible, but constrained by (i) above; i.e. allow for a minimum off time

(iv) keep the frequency in the CCC mode as low as possible to minimise power loss.

The CCC mode uses three general routines: phase change, CCC and timing. These routines produce the current waveforms associated with the low-speed operation of the motor. Fig. 9 shows such a waveform in the lower trace

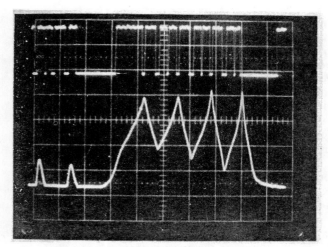

Fig. 9 *Oscilloscope photograph during motoring CCC mode*

Top trace: end of interrupt daisy chain (active low)
Bottom trace: phase current (40 A/div.)
Speed = 200 rev/min, torque = 114 Nm, time base = 2 ms/div.

which at the start shows two commutation reset pulses associated with the previous phase. The top trace shows the end of the microprocessor daisy chain. This chain shows the activity of the microprocessor when low and idling in a simple loop when high. A number of timing tasks are performed during the execution of the phase-change routine, as can be seen by the long low level on the daisy chain. At each of these peaks and troughs, microprocessor activity can be seen: here the CCC routine is being executed. Between some of the troughs and peaks there is a low on the daisy chain indicating that the timing routine has been executed. The three routines are described, bearing in mind the oscilloscope photograph (Fig. 9), the controller hardware (Figs. 4–7) and the control strategy described previously.

Execution of the phase-change routine normally results in the commutation of the thyristor energising the previous phase. However, this process is skipped if the routine has been executed immediately after a decelerating mode-transition routine (Section 3.5). A look-up table method is used to determine which thyristors to fire (method (ii), Section 3.1). The two phase bits from the rotor position transducer are stored in the low order, eight bit register of a 16-bit register pair. A firing byte is read from the look-up table using register indirect addressing. Output of this byte to the port connected to the thyristor firing circuits causes a main thyristor to be fired and the current in a winding rises. The phase-change routine is similar to the initial routine in that it fires a main thyristor and sets the high current level on the DAC so as to generate an interrupt pulse for the CCC routine.

Execution of the CCC routine results in the firing of either a commutating thyristor or a main with commutating thyristors. As it is a general routine, it has to receive information to decide which of these two sets of thyristors it should fire. This decision is achieved by reading a port connected to the output of the current comparators (Fig. 5)

which tells the microprocessor whether the interrupt occurred as a result of rising or falling current. The two bits received from this port are logically combined with the rotor-position data in the 16-bit register pair which was stored during the execution of the phase change routine. A firing byte is read from the look-up table using indirect register addressing.

The comparator output bits are also used to decide whether to load the peak-current parameters: high current level for the DAC and on time for the timer, or the trough parameters: low current level and off time. The current interrupt is disabled while a timer is loaded with the appropriate time parameter and its interrupt enabled. The timing routine has a very simple job of enabling the current interrupt on timing out. So until the timer reaches its zero count, the current routine cannot be executed and thyristors fired.

A typical current waveform (Fig. 9) has peaks of current which occur at about the same level while troughs occur at progressively lower levels. This effect is produced by the interaction between the CCC and timing routines. There is a low on the daisy chain at the first trough of the current waveform (Fig. 9). Here the timer routine and chopping routine are executing one after the other. During the transition from high to low, the current has dropped to the low level set on the DAC and has caused the current interrupt event to be stored in a PIO device. However, as the current interrupt has been disabled, the CCC routine is not executed. When the timer reaches its zero count for minimum off time it interrupts the microprocessor, and the timer routine enables the current interrupt. The CCC routine is then immediately entered on exit from the timer routine.

When the current is rising, there is microprocessor activity between a trough and peak. Here the timer has reached its zero count for minimum on time and enabled the current interrupt. When the high level is reached, a comparator changes state, generating an interrupt pulse for the CCC routine. In Fig. 9, peaks of current are thus under control from the demanded high current level while troughs are under control of the demanded off time.

3.4 Angular-position-control routines

Operation in this mode involves the positioning of single current pulses, one current pulse per phase. For each phase, a main thyristor is fired at a specified rotor position and a commutating thyristor fired at another later position. These two angular positions are measured from a reference point with an angular clock. There are four reference points (phase changes) per angular cycle of 60° given by the two rotor-position transducers and angular information is produced from the phase-locked loop which produces a clock pulse every half degree. Counting from a phase change using the clock gives the position at which to fire a main thyristor. This angular displacement defines the delay angle. Counting on from this position to another position gives the point at which to fire the commutating thyristor. The angle between the position of firing the main thyristor and firing the commutating thyristor defines the conduction angle for the thyristor. The counting is executed by auxiliary counters, as in previous paper [7, 9] in order to free the microprocessor for its other duties.

The events described in the above paragraph are summarised:

(i) detect a phase change (reference point) and load a delay-angle parameter into a counter

(ii) detect first end of count, fire a main thyristor and load a conduction-angle parameter into a counter

(iii) detect second end of count and fire a commutating thyristor

It is important to realise that the events described above are for one phase only and that they occur at rotor angles determined by the delay and conduction-angle parameters. As there are four phase circuits, then in total 12 events occur over a 60° rotor angle. The events for one phase can overlap in time with events of another phase and can also coincide causing angular errors (see Section 4.1).

Each phase uses one counter to count the delay-angle parameter down to zero and then to count the conduction-angle parameter down to zero. On reaching a zero count, the counter generates an interrupt and program execution is transferred to the interrupt address associated with the counter. As there are four counters, only four interrupt routines can be addressed at a given time. The software within one of these routines would have to work out whether the interrupt has been generated as a result of loading the delay-angle parameter or as a result of loading the conduction-angle parameter.

A better method to produce fast program execution is to change interrupt addresses within routines. During the execution of the phase-change routine the interrupt address of a counter is changed to point to the routine where a main thyristor is fired. Execution of this routine changes the interrupt address to point to the program which causes the commutating thyristors to be fired. Associated with each counter are two types of routine; execution of one fires a main thyristor while execution of the second causes commutation. In all, there are eight routines associated with the firing of thyristors (i.e. method (iii), Section 3.1).

Another set of routines are required for the reference points. Here, a phase change causes an interrupt routine to be executed which loads delay-angle parameters into counters. There are five blocks of program. The first block uses the two rotor-position bits to transfer program execution to one of four locations. At each of these locations (one for each phase) the delay-angle parameter is loaded into the appropriate counter, the interrupt address associated with the counter is changed, the counter started, interrupts enabled and program execution returned to the main program.

The oscilloscope photograph (Fig. 10) shows a single pulse of current typical of APC. There is microprocessor activity at the start of the pulse which corresponds to a position where a main thyristor is fired and a counter loaded with the conduction-angle parameter. At the peak of current, the count has reached zero and interrupted the

microprocessor (third low on the daisy chain). Here a commutating thyristor is fired and the current falls. The second low in the daisy chain corresponds to a phase-change event associated with a different phase than the one showing the pulse of current. A phase-change event for one phase can thus overlap in time with the execution of the counter routines for another phase.

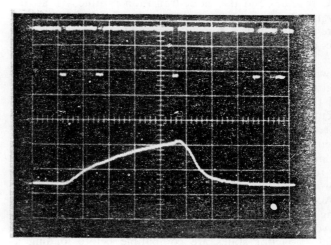

Fig. 10 *Oscilloscope photograph during motoring APC mode*
Top trace: end of interrupt daisy chain (active low)
Bottom trace: phase current (40 A/div.)
Speed = 750 rev/min, power output = 2.5 kW, time base = 0.5 ms/div.

3.5 Mode-transition routines

There are two interrupt routines for the mode transition: one for accelerating from CCC to APC and one for decelerating. Events during the CCC mode, though amplitude controlled, are subject to minimum time constraints which require the CTC to use the microprocessor clock. Single-pulse-mode events, however, are angle dependent; for example, the CTC uses an external rotor-angle clock from the phase-locked loop. Consequently mode-transition routines are concerned mainly with this change of task. During mode transition, a large part of the processing involves looking at the rotor position information, inferring that the commutation capacitors are in their correct state, waiting for routines to clear or events to take place and changing interrupt addresses.

The tasks for the decelerating routines are summarised:

(i) stop the execution of the APC phase change routine by disabling its interrupt

(ii) observe the state of the four counters and wait until their routines have finished firing the commutating thyristors (i.e. have finished task (iii) in Section 3.4)

(iii) enable the CCC phase change interrupt

The APC routines cannot be stopped simply by disabling all their interrupts since a main thyristor may have been fired and must therefore be commutated. Producing a single current pulse starts with the execution of the phase change routine and ends with the execution of a commutation routine ((i), (ii) and (iii) of Section 3.4). Stopping the execution of the phase-change routine by disabling its interrupt eventually causes all phases to be commutated and no more APC events to occur.

On disabling the phase-change interrupt, some counters may still be counting down to zero and continue to fire thyristors. The decelerating mode-transition routine has therefore to monitor their progress and to wait for APC events to cease. Information has to transfer between the APC routines and the mode-transition routine. An auxillary register is used for this transfer of information. As there are four phases, four bits are used in a register. A bit is set during the execution of the phase change routine and reset during the commutation routine. All phases are shown to be commutated by the resetting of all four bits. If a bit is not zero, then a phase has not been commutated and the microprocessor waits for an APC routine to be executed by enabling interrupts and executing a HALT instruction. After servicing an APC routine, program execution is returned to the mode-transition routine where the microprocessor disables interrupts. It again examines the register and loops round until all the bits are zero, showing that all the phases have been commutated.

Accelerating mode transition is easier than the decelerating routine. The routine consists of commutating the active phase and then checking that the phase and commutating capacitor states are compatible before loading a counter with a delay-angle parameter and enabling interrupts. Execution of this mode-transition routine starts the APC routines by executing instructions which are equivalent to those of the APC phase change routine; i.e. by carrying out task (i) in Section 3.4.

3.6 Data routine
This routine performs a background task and the port generating the interrupt has lowest priority. It is a very simple routine and two ports are used to obtain an address and a parameter. Only three bits are required to decode an address of a parameter stored in RAM. The unwanted bits are masked off and a register pair used to hold the full address of the parameter. The six parameters, listed in Section 2.6, are stored in RAM and they are used for all the phases. Should a particular operating condition be required which uses different values for different phases, then this is easily implemented in software.

The operator has control over the parameter values which are read into the microprocessor system. For example, the on and off times are required to have minimum values as described in the Introduction. Some constraints also apply to the other parameter values; for example, the high current level should not be less than the low current level. Although checking procedures could be included in the data software it is assumed that the parameter values are loaded into the system by an intelligent operator.

4   Discussion

4.1 Designed performance and practical limitations
During CCC the microprocessor spends about 50% of its time idle and during single-pulse mode the microprocessor is mostly idle. CCC events are partly position dependent in the following sense: one phase is active, a phase change occurs, the next phase is active and so on. Within a phase, the events are independent of rotor position as they depend on the current-level parameters (high and low) and on the time parameters (on and off). APC events, on the other hand, are solely position dependent and hence the microprocessor has more idle time.

The objective of the controller in the APC mode is to position accurately single pulses of current. Due to microprocessor execution time, delays can occur which may result in an error in setting delay or conduction angles, dependent on rotor speed. However, since the specified angles are stored as parameter values, the parameter values can be adjusted to allow for the angular delays and ensure that the thyristor firings do occur at the desired rotor positions.

Nevertheless, it is instructive to estimate the magnitude of these angular delays. The time taken from a phase-change event to the loading of a parameter into a counter, as explained in Section 3.4, is about 60 μs; 30 μs is required to detect a zero count and to load the counter with the conduction angle and a further 10 μs to fire the main thyristor making a total time error between the detection of a phase-change event and firing of the main thyristor of about 100 μs. At 750 rev/min the rotor moves through 0.45° in 100 μs while at 3000 rev/min the rotor moves through 1.8°. The time from detecting zero count of conduction angle to firing a commutation thyristor takes about 20 μs. Since the loading of conduction angle occurs about 10 μs before firing the main thyristor, the total time error in the conduction angle is 10 μs due to the order of events. (Counter events always occur before the firing of thyristors: if a main thyristor had been fired and then the counter loaded with a parameter, the two errors would add together.)

The delay-angle error could be reduced at the expense of extra hardware by reducing the execution time spent decoding the phase information. Improvements could be made using four separate interrupt pulses and the necessary microprocessor hardware to allow for separate interrupt vectors, one for each of the four phases. A time error of less than 30 μs would be achieved for the detection of a phase change to the loading of a parameter into a counter. However, this is not strictly necessary since, as described above, the stored parameter values can be adjusted to ensure accurate thyristor firings.

Another source of angular error occurs as a result of routine contention and since this is an irregular event it cannot be eliminated by adjustment of stored parameter values. Phase-change events can overlap in time with the firing of thyristors (Section 3.4) which results in one routine having to wait for service until another routine has finished. Any error which is produced as a result of two routines contending for service adds to either the delay or conduction angle. The maximum value of this error depends on the execution times for routines which for the experimental system is about 100 μs. However, in practice, this source of error produced very little effect on the controllability of the system. A possible explanation is that if one routine is serviced first during four phase changes it is serviced second during the next four phase changes, and a reduced effect is produced over a rotor revolution.

Methods for reducing routine contention have to be considered carefully. The use of nested interrupts would not eliminate the error since the chosen priority would either produce a larger error in the on angle and a smaller error in the conduction angle or vise versa. Elimination of routine contention can be achieved by freeing the microprocessor of the task of firing the thyristors. A way of implementing this improvement is to use eight counters to fire the thyristors directly. Execution of the phase-change routine would then cause the microprocessor to load two parameters into two counters; the on-angle parameter and the off-angle parameter. The design of such a system requires not only a significant increase in hardware but careful consideration of the operation of the other tasks of the microprocessor. For example, the microprocessor needs to have direct control over the firing of thyristors in the CCC mode. Further, the microprocessor needs to know the state of the counter for the mode-transition declerating routine, so if counters external to the microprocessor system are used they have to feed signals back to the microprocessor system.

The angular errors described above may always, of course, be reduced by reducing the execution times for the routines, i.e. by using a microprocessor system which is clocked at a higher rate. Improvements in the resolution of angles may also be achieved in the present experimental system. The required rotor angular range is 60° which in the present system requires parameters using seven bits for 1/2° resolution. As eight bits are available 1/4° resolution could be provided by doubling the angular clock frequency.

4.2 System monitoring

Using a microprocessor for the control of the system allows for flexibility in the sense that tasks may be added or modified using software. Also the software may be readily used for health-monitoring features so that functional operation of the system is maintained by the correct sequential firing of thyristors. The state of various parts of the system (microprocessor and power electronics) may be expressed in a binary form; for example, a thyristor is either on or off. Bits stored in memory showing the states of the parts of the system allow for individual routines to examine the bits and determine subsequent operation. The state of the system is then globally available to all routines. This approach is very useful; for example, in the mode transition decelerating routine where all APC events are allowed to finish by observing bits in a register which show the activity of the counters and hence thyristors. As a background task the microprocessor could check that the firing issued to the power electronics actually caused thyristors to be fired; for example, by observing the current waveforms from the current transducers. This background software could act as a health monitor for the whole system.

A microprocessor controller easily lends itself to these health-monitoring features. A dedicated SSI/MSI controller can also have these features but would significantly add to the physical size and cost. A microprocessor controller is also more reliable than a dedicated system in the sense that it can readily handle irregular events such as an interrupt pulse which results in the execution of a routine out of sequence. Such an event need not cause a system malfunction; instead the routine can examine the bits showing the state of the system and decide on the correct subsequent sequence of events. This is illustrated by considering an example which can occur at low speed or during a decelerating mode transition. Suppose that the main thyristor of phase A has been fired using the phase-change routine (B Fig. 8). Another phase-change interrupt occurs rather than the normal sequence of events which would be the timing and CCC routines (C and D Fig. 8). Execution of routine B causes the commutating thyristors for phase A to be fired and the microprocessor reads the port connected to the rotor-position transducers. The microprocessor receives data to show that the rotor is in the torque productive region for phase C and not as would normally be expected for phase B. Unfortunately the polarity of the commutating capacitors is incorrect for commutating the current of phase C. Instead of firing the main thyristor for phase C, the microprocessor waits until the rotor moves to phases B or D. Operation then continues normally with the main thyristor of phase B or D being fired. Alternatively the microprocessor could decide to swing the commutating capacitors into the polarity ready to commutate the current in phase C and then fire the main thyristor for phase C.

4.3 Interference

Power electronic systems create an electrically noisy environment for the control electronics which can result in erroneous operation. Some difficulty was experienced in making the microprocessor system respond correctly to interrupt signals. The mechanisms behind the behaviour of the controller to interference are not clear. No single technique produced a satisfactory answer to these problems, and successful operation was achieved using a number of heuristic approaches. Most of the interference is caused by electromagnetic radiation from the power system when the thyristors are fired. CMOS devices operate correctly in this environment when powered at 10 V while single transistor buffers provide the necessary logic level shift to interface to the microprocessor system. Lead lengths are kept to a minimum and critical signals brought as near to the microprocessor system as possible before level shifting.

Early work on the project where PIO ports were polled in a single program during the CCC mode produced satisfactory operation without requiring any of the above interference minimisation techniques. The chopping frequency was low since the rate of servicing tasks was slow, as would be expected from such an approach. However, the program demonstrated that, despite erroneous signals arriving at the microprocessor ports, the correct sequence of firing thyristors could be maintained. This operation

361

was achieved using software methods such as those described in Section 4.2. The microprocessor knew which thyristors it had fired and which thyristors it should fire next according to information received from the transducers. However, if the signals suggested an abnormal sequence the software could allow for a spurious but not catastrophic sequence to occur.

4.4 Hardware

Despite the use of multiple blocks of program to speed up the execution times of some tasks the amount of memory required is small. The program can easily be accommodated in a single 2 kbyte device. Further, routines require only a small amount of RAM for the storage of registers and parameters. Extra memory would be required for the storage of parameters associated with the implementation of higher levels of control. The complete microprocessor system has 26 devices (mainly 4000 series CMOS) plus the microprocessor devices, which are eight in number.

5 Conclusions

Closed-loop microprocessor control has been demonstrated in the laboratory for a VR motor incorporating additional features when compared with previously described microprocessor stepper controls. First, turning on a given phase is not synonomous with turning off the previous phase, and hence separate control of both lead and conduction angle has been necessary. Secondly, feedback information has been taken for both rotor position and winding currents. Thirdly, a high-power system has been used involving forced commutation of the drive thyristors and requiring careful synchronisation of commutation circuit state to avoid commutation failure.

Despite the increased requirements of the control functions, it has been shown that an eight-bit microprocessor of moderate speed (2.5 MHz) is capable of providing the necessary basic control, provided some auxiliary hardware is used, with microprocessor idling time of at least 50%. Switching angles may be set with acceptable accuracy. The total control package has a relatively small component count and, unlike its discrete counterpart, additional tasks to meet particular applications, health monitoring etc., may be added using additional software only.

The use of stored bits to show the binary state of individual components is easily implemented with a microprocessor and allows for diagnostic and health-monitoring features. By appropriate choice of signal levels, interference problems arising from the proximity of the power electronics have been solved.

6 Acknowledgments

The authors wish to thank the UK Science and Engineering Research Council for their financial support of this project. Thanks are also due to Prof. R.L. Beurle for the use of departmental facilities at Nottingham University and to Dr. B. Jasper (Zilog, Northern Europe) for helpful discussions.

7 References

1 FREDRIKSEN, T.R.: 'Applications of the closed loop stepping motor', *IEEE Trans.*, 1968, **AC-13**, pp. 464–474
2 ACARNLEY, P.P., and GIBBONS, P.: 'Closed-loop control of stepping motors: prediction and realisation of optimum switching angle', *Proc. IEE.*, 1982, **129**, (4), pp. 211–216
3 WELLS, B.H.: 'Microprocessor control of step motors'. Proceedings 5th annual symposium on incremental motion control system and devices, University of Illinois, 1976
4 LAFRENIERE, B.C.: 'Interactive microprocessor controlled step motor acceleration optimisation'. Proceedings 7th annual symposium on incremental motion control systems and devices, University of Illinois, 1978
5 MIYAMOTO, H., and GOELDEL, C.: 'Optimal control of a variable reluctance step motor by microprocessor'. Proceedings 8th annual symposium on incremental motion control systems and devices, University of Illinois, 1979
6 VALLON, B., and JUFER, M.: 'Microprocessor optimisation of stepping motor dynamic performance'. Proceedings International conference of stepping motors and systems. University of Leeds, 1979
7 RADULESCU, M.M., and STOIA. D.: 'Microprocessor closed-loop stepping motor control'. Proceedings international conference on stepping motors and systems, University of Leeds, 1979
8 LANGLEY, L.W., and KIDD, H.K.: Closed loop operation of a linear stepping motor under microprocessor control'. Proceedings international conference on stepping motors and systems. University of Leeds, 1979
9 KENJO, T., and TAKAHASHI, H.: 'Microprocessor controlled self optimisation drive of a step motor'. Proceedings 9th annual symposium on incremental motion control systems and devices, University of Illinois, 1980
10 KUO, B.C., and CASSAT, A.: 'On current detection in variable reluctance step motors'. Proceedings 6th annual symposium on incremental motion control systems and devices, University of Illinois, 1977
11 DAVIS, R.M., RAY, W.F., and BLAKE. R.J.: 'Inverter drive for switched reluctance motor: circuits and component ratings', *IEE Proc., B, Electr. Power Appl.*, 1981, **128**, (2), pp. 126–136
12 HARRIS, M.R.: 'Brushless motors—a selective review', Drives/Motors/Controls, University of Leeds. 1982. pp. 84–90
13 HUGHES, A.: 'Using stepping motors', Drives/Motors/Controls, University of Leeds, 1982, pp. 109–123
14 RIEKE, B.: 'The microprocessor control of a four phase star-connected multi-pole reluctance motor'. International conference on electrical machines, Athens, Sept. 1980. pp. 394–401
15 BYRNE, J.V., and LACY, J.G.: 'Characteristics of saturable stepper and reluctance motors', in 'Small electrical machines'. *IEE Conf. Publ.*, **136**, 1976, pp. 93–96
16 SPENCER, P.T.: 'Microprocessors applied to variable-speed-drive systems', *Electron. & Power*, 1983. **29**, pp. 140–143

Four-Quadrant Brushless Reluctance Motor Drive

T.J.E. Miller, P.G. Bower, R. Becerra and M. Ehsani

IEE Conference on Power Electronics and Variable Speed Drives, London, July 1988

INTRODUCTION

The switched reluctance motor drive has often been identified as a high-efficiency integral-horsepower drive, and less attention has been focussed on smaller sizes where servo-quality or near-servo-quality drives are in demand. The brushless PM motor is advancing in this area, but the advantages of the SR motor are considerable. Its attractive control characteristics are perhaps not widely recognized.

Both machines are shaft-position switched d.c. brushless motors. The PM motor is cylindrical and doubly-excited (with rotor magnets and multiple stator windings). The SR motor is singly-excited (multiple stator windings and no magnets), with different numbers of salient poles on the stator and rotor, Fig. 1. The PM motor has superior efficiency in small frame sizes, and may have less torque ripple. The SR motor is simpler in construction and as free from the operational problems associated with magnets as it is from their benefits. Less obvious is that for the same amount of control logic and only 2/3 the power devices, the control performance of the SR drive is less constrained than that of the PM motor, and it has two or three times the speed range. 'Control performance' is here intended to mean the facility for precision speed control, with operation in all four quadrants and rapid response in terms of the torque/inertia ratio and the ability to switch quickly between quadrants. The SR drive is therefore a candidate for near-servo-quality speed-controlled small drives, and it should have a cost advantage. The price paid is likely to include a higher noise level and increased torque ripple, and a less linear torque/current relationship. In respect of the noise and torque ripple, it should be pointed out that the brushless PM motor is unusually quiet. The other drive technologies competing in this area, such as small a.c. drives and a.c. and d.c. commutator motors, are by no means as quiet, and by this standard of comparison the SR drive does not have the 'noise problem' that is sometimes ascribed to it.

CONVERTER CIRCUITS

The torque is independent of the direction of the phase current, which can therefore be unidirectional. This permits the use of unipolar controller circuits with a number of advantages over the corresponding circuits for a.c. or PM brushless motors, which require alternating current.

In Fig. 2 is shown a circuit suited for transistors (bipolar, field-effect, or insulated-gate). The phases are independent. In this respect the SR controller differs from the a.c. inverter and has inbuilt protection against shoot-through failures. The upper and lower phaseleg switches are switched on together at the start of each conduction period or 'stroke', and off at the commutation point. During the stroke, one or both of them may be 'chopped' according to some control strategy, such as maintaining the current within a prescribed 'hysteresis band'. This mode is necessary at low speeds when the self-e.m.f. of the motor is much smaller than the supply voltage. At high speeds both devices may remain on throughout the conduction period and the current waveform adopts a 'natural' shape depending on the speed and torque. It is convenient in the logic design to use one transistor primarily for 'commutation' and the other for chopping. At the end of the stroke when both switches are turned off, any unconverted magnetic energy is returned to the supply via the diodes. When they become forward-biased, the diodes connect $-V$ across the winding to reduce its flux-linkage quickly to zero.

The phase inductance varies with the rotor position. Therefore if fixed-frequency chopping is used, the current ripple varies. If hysteresis-type current regulation is used, the chopping frequency varies as the poles approach alignment. Chopping frequencies above 10kHz are usually desirable, as in other types of drive, to minimize acoustic noise. In large drives (say, above 20kW) it is difficult to chop at such high frequency because of device limitations, making it difficult to achieve quiet low-speed operation in larger motors.

Small drives often use p.w.m. control over the entire speed range. In such cases the branches shown in dotted lines can be removed from the circuit of Fig. 2, saving two transistors and two diodes. One chopping transistor supplies controlled voltage which is commutated by the lower transistors to the motor phases in proper sequence, under the control of the shaft position sensor and gating logic. This circuit requires only n+1 transistors and n+1 diodes for a motor with n phases. A three-phase motor thus requires only 4 transistors and 4 diodes. There is little loss of functionality with this circuit relative to the full circuit having 2n transistors in Fig. 2, and at low speeds it runs more smoothly. Its main limitation is on the overlap between phases; but this only becomes a problem at very high speeds.

CONTROL

In motoring, the pulses of phase current must coincide with a period of increasing inductance, i.e. when a pair of rotor poles is approaching alignment with the stator poles of the excited phase. The timing and dwell of the current pulse determine the torque, the efficiency, and other parameters. In d.c. and brushless d.c. motors the torque per ampere is more or less constant, but in the SR motor no such simple relationship emerges naturally. With fixed firing angles, there is a monotonic relationship between average torque and r.m.s. phase current, but in general it is not very linear. This may present some complications in feedback-controlled systems although it does not prevent the SR motor from achieving 'near-servo-quality' dynamic performance, particularly in respect of speed range, torque/inertia, and reversing capability.

The regulating strategy employed has a marked effect on the performance and operating characteristics. Ref. [2] discusses two methods of control; current-controlled p.w.m. and voltage-controlled p.w.m. In the former, the torque is roughly proportional to the current reference. With loads such as fans and blowers, speed adjustment is possible without tachometer feedback, but in general feedback is needed to provide accurate speed control. The current regulator requires current transducers, but they can be grounded at one end if the upper transistor is always the one used for chopping.

The alternative is fixed-frequency p.w.m. of the voltage with variable duty-cycle, d. Current waveforms are shown in Fig 3. After commutation the current decays rapidly through the diodes. The reverse voltage applied is effectively 1/d times the forward voltage applied during the conduction period. With

fixed turn-on and commutation angles, the motor tends to have a more nearly constant speed characteristic, as with armature-voltage control in a d.c. motor.

In some cases the pulse train from the shaft position sensor may be used for speed feedback, but only at relatively high speeds. At low speeds a larger number of pulses per revolution is necessary, and this can be generated by an optical encoder or resolver, or alternatively by phase-locking a high-frequency oscillator to the pulses of the commutation sensor [6]. Systems with resolver-feedback or high-resolution optical encoders can work right down to zero speed.

Current feedback can be added to the circuit of Fig. 2 to provide a signal which, when subtracted from the voltage reference, modulates the duty cycle of the p.w.m. and 'compounds' the torque-speed characteristic. It is possible in this way to achieve under-compounding, over-compounding, or flat compounding just as in a d.c. motor with a wound field. For many applications the speed regulation obtained by this simple scheme is adequate. For precision speed control, normal speed feedback can be added. The current feedback can also be used for thermal overcurrent sensing.

When the p.w.m. duty cycle reaches 100% the motor speed can be increased by increasing the conduction angle or its phase advance; or both. When these increases reach maximum practical values, the torque becomes inversely proportional to speed squared, but they can typically double the speed range at constant torque. The speed range over which constant power can be maintained is also wide, and very high maximum speeds can be obtained, as in the synchronous reluctance motor and induction motor.

There is not the limitation imposed by fixed excitation as in the usual squarewave PM brushless motor. However, the interior-magnet motors described by Jahns [7] have a significant speed-range advantage over the conventional brushless PM configuration. These motors are underexcited synchronous machines requiring sinewave or six-step currents, and are not directly comparable with the electronically 'commutated' SR and brushless PM motors.

COMMUTATION CONTROL

The commutation requirement of the SR motor is very similar to that of a PM brushless motor; indeed the same shaft position sensor and decoding logic can be used with delta-connected diode-blocked phase windings.

The commutation of brushless PM motors requires decoding of preconditioned logic-level signals from the shaft sensor such as a Hall-effect sensor with a magnetized ring, or optical interrupters. Relatively simple logic provides forward and reverse operation with fixed conduction angles. Firing angle advance results in a limited increase in the speed range, and fixed angles are normally used. (However, see [7] and [9]).

In the case of the switched reluctance motor, some quite complex schemes have been used for controlling the commutation angles, including continuous variation with speed and load, and microprocessor control [6].

But small SR motors can be commutated as simply as PM brushless motors, with logic-selectable 'modes': fixed firing angle combinations derived from the shaft position sensor. Without increasing sensor complexity or cost, sufficient modes can be constructed by combinatorial logic to provide not only four quadrants of operation, but also a selection of different conduction angles in motoring. With only 2 or 3 motoring modes it is possible to double the speed range.

It can be shown, and will be published in a future paper, that the simplest possible bidirectional scheme for a 3-phase motor uses only 2 sensors (Hall or optical). In general 2 or 3 sensors can be used with any 3-phase SR motor, and either 2 or 4 with a 4-phase motor. For symmetrical forward and reverse operation there are restrictions on the number of sensors and their placement. The scheme described here is an illustration and not necessarily the best for all applications.

The sensors are fixed in relation to the stator. The torque of one phase is given approximately by

$$T = \frac{1}{2} i^2 \frac{dL}{d\theta}$$

where L is the phase inductance and i is the phase current. With 6 stator poles and 4 rotor poles, forward motoring current pulses must coincide with rising L, which usually terminates at the 'aligned' position, e.g. from 60 to 90 degrees. However, conduction from 60 to 90 deg. would be too late. There would be too little time to build up the current at the beginning of the stroke, and the diode current would continue past the aligned position at 90 deg., producing unwanted negative torque. With a simple scheme the conduction angle must be at least 30 deg. so that torque can be produced at any rotor position. The precise amount of phase advance required to optimize the performance at any speed and load is a complex function of many parameters, but with fixed angles and p.w.m. voltage control, considerable latitude is permissible, especially if the 'last drop' of efficiency is not too important.

Fixed angles of 45 and 75 deg. give efficient operation at high speed and a wide speed range under voltage p.w.m. control. For forward braking the combinatorial logic switches the conduction period to 75/105 deg. In reverse the complementary angles are used, that is, 45/15 for motoring and 15/-15 for braking. The scheme has the limitation of lack of starting torque at standstill at the 45 degree positions. This can be overcome, at the expense of symmetry, by rotating the sensors slightly, but for loads requiring maximum torque at low speed, 3 sensors must be used. Fig. 3 shows current waveforms of two phases of a 3-phase motor operating with a mode-switched commutation scheme that has several symmetrical forward and reverse motoring and braking modes, derived from 3 sensors. The chopping frequency (5kHz) can be clearly seen on these waveforms.

Fig. 4 shows the speed/torque capability of one small drive under two different modes, indicating at least a doubling of torque at fixed speed, or a doubling of speed at fixed torque. The upper or 'boost' mode is less efficient, but it can be used either intermittently or restricted to high-speed operation.

A small laboratory drive of about 100W rating has been tested over a speed range from a few rpm up to 8,000rpm with four-quadrant capability. The dynamic reversing capability is particularly impressive and is achieved simply by mode-selection logic that is the equivalent of reverse-commutation in the PM brushless motor. Its implementation requires the same amount of logic as for the PM brushless motor; only the gating is different. In the test system, a programmable logic array is used for the combinatorial logic and a single p.w.m. chip for voltage control. This results in a two-chip open-loop control that can be extended to incorporate a speed loop with only one additional chip. The 3-phase prototype machine also employs a converter circuit with only four IGT switches and four diodes [2]. Its noise level and torque ripple are far below the level previously experienced with SR drives.

CONCLUSION

Commutation of the SR motor is as simple as that of the brushless PM motor, using a similar shaft position sensor and gating logic. Speed and torque can be modulated by commutation angle control over a wide range. Combined with voltage p.w.m., sufficient flexibility is obtained by switching between small number of modes or commutation angle combinations to at least double the speed and torque control ranges compared to those possible with fixed angles. Four-quadrant symmetrical operation is easily achieved, wit extremely rapid reversing. The entire control is readily implemented using any of the norma methods (gate array, programmable logic array custom LSI, or microprocessor).

Acknowledgements

The work was performed in Glasgow partly unde the Glasgow University SPEED programme, a research consortium fully supported by industrial companies in the UK and the United States, and partly under a joint brushless drives research project between Glasgow and Texas A&M Universities, which is separately supported by General Electric Company. Acknowledgment is made to GE and to the subscribing companies of the SPEED program. Thanks are also due to Mr. A.K. Evans and Mr. A. McLaren of Glasgow University for their participation.

REFERENCES

1. Lawrenson, P.J. et al (1980). Variable-speed switched reluctance motors. Proceedings IEE, 127, Pt. B, 253-265

2. Miller, T.J.E. (1987). Brushless reluctanc motor drives. IEE Power Engineering Journal, 1, 325-331

3. Davis, R.M. et al (1981). Inverter drive for switched reluctance motor; circuits and component ratings. Proceedings IEE, 128, Pt. B, 126-136

4. Bass, J.T. et al (1987). Development of a unipolar converter for variable-reluctance motor drives. IEEE Transactions, IA-23, 545-553

5. Bass, J.T. et al (1987). Simplified electronics for torque control of sensorless switched reluctance motor. IEEE Transactions, IE-34, 234-239

6. Bose, B.K. et al (1986). Microcomputer control of switched-reluctance motor. IEEE Transactions, IA-22, 708-715

7. Jahns, T.M. et al (1986). Interior magnet synchronous motors for adjustable-speed drives. IEEE Transactions, IA-22, 738-747

8. Harms, H.B. and Erdman, D.M. (1985) Electronically commutated f.h.p. motor drives Motorcon Proceedings

9. Jahns, T.M. (1984). Torque production in P synchronous motor drives with rectangular current excitation. IEEE Transactions, IA-20, 803-813

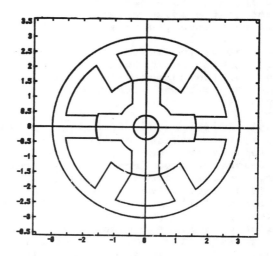

Fig 1. Cross-section of SR motor.

The rotor is aligned with phase 'a' (wound on the two horizontal poles) i.e. $\theta = 0$

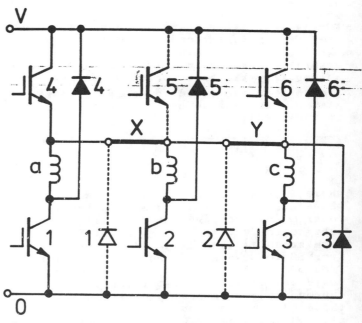

2. Converter circuit for SR motor control.

Shown connected for 4-transistor operation with links X and Y in place and the components in dotted branches removed. For 6-transistor operation links X and Y must be removed.

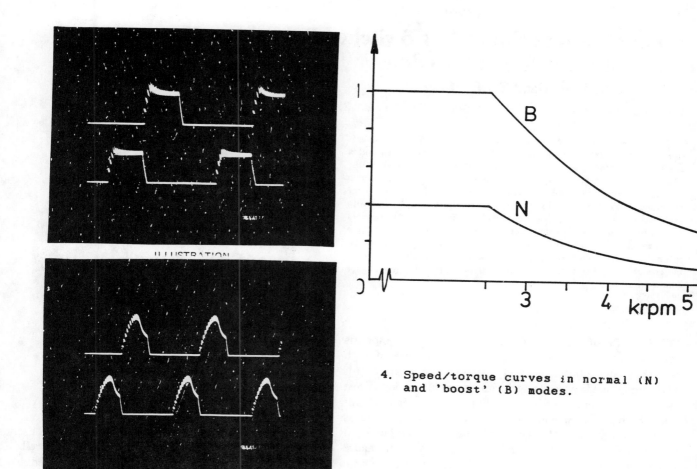

4. Speed/torque curves in normal (N) and 'boost' (B) modes.

3. Current oscillograms.

   Upper – normal mode at low speed
   Lower – 'boost' mode at high speed

Robust Torque Control of Switched-Reluctance Motors Without a Shaft Position Sensor

J.T. Bass, M. Ehsani and T.J.E. Miller

IEEE Transactions, Vol. IE-33, No. 33, August 1986, pp. 212-216

Abstract—This paper describes a technique for stabilizing the operation of variable-reluctance stepping-motor drives operating without a shaft-position sensor. In such systems there is a trade-off between the system efficiency and the torque margin (or pull-out torque) which depends on the width of the phase conduction pulse width. The scheme described in the paper permits the motor to run in the steady state with both narrow conduction pulse widths and high efficiency. Under transient or overload conditions the conduction pulse width is increased in response to a change in the dc link current, providing an increase in available torque. Tests on a small motor drive have produced a steady-state torque margin of over 300 percent and of 200 percent under step-change conditions.

I. Introduction

THE APPLICATION of small brushless dc motors is becoming more widespread as the advantages of high efficiency and adjustable-speed operation are realized [6]. This trend is made possible by the relative reduction in the cost of power electronic components and associated control hardware. Both the stepper motor and the conventional brushless dc motor are capable of adjustable speed operation, but it has been usual to apply these machines in high performance systems where their control characteristics are more important than their energy conversion properties, such as efficiency and power per unit volume [5]. Such control-type motors are typically expensive and they usually incorporate permanent-magnet materials which increase the cost and require special consideration in both design and application to avoid demagnetization. The variable-reluctance (VR) stepping motor, particularly the large step-angle motor such as the "6/4" motor (Fig. 1) is an important alternative to motors using permanent magnets and although it is inherently less efficient than the brushless dc motor it has the advantages of robustness and low cost, as well as the complete absence of permanent magnets.

This paper is concerned with VR motors designed for highly efficient power conversion. Even in small sizes where the net energy cost may be small, efficiency is still important in many applications, particularly those in which battery power is used and wherever the size of the motor is an important consideration. In the design of VR motors the pole geometry must be arranged to give high efficiency while satisfying the require-

Manuscript received July 8, 1985; revised January 16, 1986.
J. T. Bass is with Bass Engineering Company, Longview, TX 75602.
M. Ehsani is with Texas A & M University, College Station, TX 77843.
T. J. E. Miller is with General Electric Company, Corporate Research and Development, Schenectady, NY 12301.
IEEE Log Number 8609195.

368

Fig. 1. Cross section of 3-phase VR motor.

ments for self-starting and stable operation over the entire speed range. In most stepper systems the "dwell" or conduction angle (phase current pulse width) is fixed at 120 electrical degrees; [1], but in the systems described here it is modulated to optimize the system efficiency over the entire operating range. It is found that in general there is a tradeoff between efficiency and stability that depends on the strategy for controlling the dwell angle. For high efficiency it is generally desirable to maintain a narrow dwell angle. Under these conditions the pull-out torque is minimal and the stability is poor: the lightest torque disturbance can cause the motor to lose synchronism. A wide dwell angle, on the other hand, provides a large pull-out torque and stable operation, but the efficiency is poor. When designing for the very highest efficiencies it is impossible to achieve a satisfactory compromise between the efficiency and the stability by geometrical design alone. This paper describes a method for achieving this compromise by means of the external control strategy for the dwell angle.

The most direct way to implement a dwell-modulation control scheme is by means of a shaft-position sensor or encoder. This way, the controlled variables are the ON angle, the OFF angle (both referenced to the rotor position), and the current or voltage applied to the phase windings. This mode of control is naturally adapted for the control of torque; for speed control, an outer control loop may be added. In low-cost applications, particularly at low power levels, the shaft-position sensor is unacceptably costly, and represents an additional potential source of failure. Therefore, a sensorless system is preferred. In such systems the ON and OFF angles cannot be controlled directly, but their difference, the dwell angle, can be controlled, as can the current (or voltage) applied to the windings. A third variable must be controlled to determine the state of the system uniquely, and this is the frequency, which directly determines the speed provided that the motor is in synchronism with the current pulses fed from the power converter. This "open-loop" frequency-controlled system is the one normally used in stepper and brushless dc systems, except that the dwell angle is normally fixed.

Previous workers have described methods of improving the operating range of such open-loop systems without resorting to the use of shaft position sensing, [1]. These methods can be divided into "open-loop" [2] and "current-sensing" [3] controls. The open-loop controller sends a series of pulses to the motor without any indication of the rotor position. The current-sensing schemes sense the freewheeling current in the "off" phases (Fig. 2) and attempt to derive the rotor position from the waveshape [4], providing a signal from which the conduction pulses can be referenced. Neither of these methods is as robust, in a control sense, as using a position encoder. In the case of the motor described in this paper, the rated-load efficiency when using these methods is no better than 14 percent.

II. Open-Loop System

The motor can be operated "open-loop," where each phase is fired with a fixed pulse width or "dwell." With a wide dwell, the system is stable and can withstand a large load torque disturbance. But the efficiency is low. With a narrow dwell, the efficiency is high, but the motor will stall under even a light load disturbance. To characterize the tradeoff between stability and efficiency, the operating curves of the motor will be examined.

Fig. 3 is a set of computed torque/angle curves for the motor used in our experiments and shows the relationship between the turn-off angle θ_2 and the average torque, for lines of constant dwell θ_D. The angles are defined in Fig. 4 with reference to the inductance profile of one phase. In an open-loop system the turn-off angle θ_2 cannot be controlled, but is instead a result of the dwell angle and the load at that particular speed. The difference between the equilibrium point (the aligned position of the rotor an stator poles) and the actual turn-off angle is defined as the torque angle θ_T. For a given dwell, the torque increases with the torque angle until the maximum or "pull-out" torque is reached. Since there is no position feedback in the open-loop system, the rotor will lose synchronism and stall if the load exceeds the pull-out torque. The torque/angle curves in Fig. 3 show two possible operating points for a load of 1.8 oz-in. With $\theta_T \doteq 15°$, the difference between the pull-out torque and the load torque is very small. This difference is defined as the torque margin. With $\theta_D = 30°$ and the same load torque, the torque margin is much

greater. The motor can, therefore, withstand a much greater torque disturbance when it is operated with a wide dwell.

Fig. 5 shows the relationship between the torque angle and the system efficiency for lines of constant dwell. For this

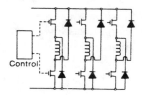

Fig. 2.   Converter circuit.

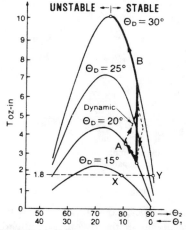

Fig. 3.   Pull-out torque versus commutation angle with conduction angle as parameter.

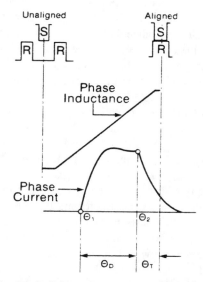

Fig. 4.   Definition of angles θ_1, θ_2, ThD, and ThT.

particular motor the efficiency peaks when the torque angle is 10° irrespective of the dwell. From the example in Fig. 3 with a load torque of 1.8 oz-in, θ_T was 10° for a dwell of 15°, and for $\theta_D = 30°$, $\theta_T = -0.5°$. The efficiency for the former operating point is 59 percent while for the latter it is only 22.5 percent. A dwell of 30° produced a comfortable torque margin but a poor efficiency, while with $\theta_D = 15°$ the efficiency was high but the torque margin was low. There is no single dwell angle that provides an acceptable compromise between torque margin and efficiency.

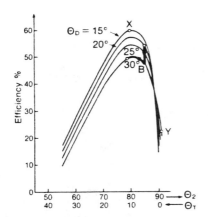

Fig. 5. Efficiency versus commutation angle with conduction angle as parameter.

III. Sensorless Control by Dwell Angle Modulation

The motor can be operated with maximum torque margin and at a high efficiency if the dwell is changed sufficiently rapidly in response to a change in the load. The dc link current is sensed as a measure of the load torque and fed back to the control system which quickly adjust the dwell.

The torque/angle curves in Fig. 3 show the dynamic trajectory A of the operating point of the open-loop system with a 20° dwell angle. The trajectory shows how the operating point changes when the load is increased from 2.5 to 3.5 oz-in. Since the dwell is fixed in the open-loop system, the torque angle changes in response to the change in load. If the load increases beyond 4.25 oz-in, the rotor loses synchronism and stalls.

If the system follows the dynamic trajectory B the effective torque margin increases dramatically. With this trajectory the torque angle is maintained at 5° and the dwell angle is increased in response to the increase in load torque. Once the dwell angle reaches 30°, it remains fixed and any further increase in torque results in an increase in the torque angle. Following the earlier example, the torque angle remains fixed at 5°, and the dwell changes from 20° to approximately 23° when the load is increased from 2.5 to 3.5 oz-in. With the open-loop system the dwell remained fixed at 20° and the torque angle advanced from 5 to 9.5°. Using trajectory B the load torque can be increased well beyond the original pull-out

torque for $\theta_D = 20°$. The pull-out torque for trajectory B is 10.2 oz-in compared with 4.25 oz-in for the open-loop system (trajectory A). Trajectory B is mapped on Fig. 5, which shows that the efficiency is close to the maximum at all operating points; it, therefore, achieves the best obtainable combination of efficiency and stability.

In Fig. 6 trajectory B can be mapped on the graph of I_{dc} versus θ_D. The dwell generated by the sensorless control system is a function of the dc link current I_{dc}. If the function shown in Fig. 6 is used, the steady-state operating point lies on that curve, and the location depends only on the load. As a result, the torque angle is indirectly fixed at 5° and the dwell is increased or decreased as the load torque increases or decreases. This allows the motor to operate efficiently at all operating points along the trajectory and maintain the maximum torque margin.

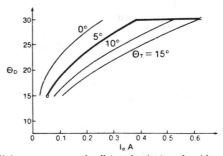

Fig. 6. DC link current versus dwell (conduction) angle with torque angle as parameter.

Ideally the operating point remains on the trajectory at all times. However, in practice it deviates from the trajectory in a transient situation. The plots in Fig. 3 show the quasistatic and transient trajectories in the torque/angle plane. The transient trajectories are for a step increase of load torque from 2.1 to 5.7 oz-in and then a step decrease back to 2.1 oz-in. When the load torque is increased, the operating point initially follows the dwell curve and the torque angle increases. This occurs because the controller does not increase the dwell until it senses an increase in the dc link current. When the torque angle increases, the link current increases, causing the controller to increase the dwell angle according to the desired trajectory. The dwell angle will continue to increase until the operating point lies on the desired trajectory. Similarly, when the step decrease occurs, the operating point initially follows a constant dwell curve. Then the controller decreases the dwell angle until the operating point is on the desired trajectory. The example indicates that the operating point will return to the desired curve when a load or other disturbance forces it away from the trajectory.

The block diagram of the controller is shown in Fig. 7. The dc link current is filtered by a four-pole butterworth filter to obtain the dc value I_{dc}. The function generator is a nonlinear amplifier used to produce a voltage representing the desired

370

dwell angle. Its transfer function is the desired trajectory and is taken directly from the dwell/current plane. This ensures that the steady-state operating point is on the desired trajectory. The gate signal generator converts the reference voltage to a three-phase gate-pulse with a pulsewidth proportional to θ_R. The implementation of the filter, the function generator, and gate signal generator circuits is the subject of a paper presently under preparation.

IV. EXPERIMENTAL RESULTS

The controller described above was built and both the steady-state and dynamic capabilities were tested. The steady-state performance improved considerably over the performance of the open-loop system, as expected. The system can produce over 4 per-unit torque while maintaining a system efficiency greater than 35 percent (which is high for motor drives in this size). The steady-state performance will be presented first, followed by the dynamic performance.

The control circuit is designed to give a constant torque angle of 5°. Table I shows the system efficiency and quasistatic torque margin along with the results for the open-loop system with dwell angles of 15 and 30°. It is clear that trajectory B almost combines the pull-out torque of the 30°

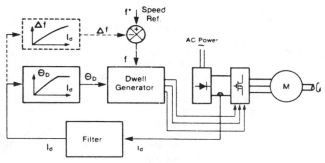

Fig. 7.   System block diagram.

TABLE I

System	Max. Torque (oz-in)	Efficiency @ 2.25 W (%)
Trajectory B	9.52	35.2
Open-loop $\theta_D = 15$ deg	3.03	42.7
Open-loop $\theta_D = 30$ deg	10.0	14.0

open-loop system with an efficiency approaching that of the 15° open-loop system.

The dwell modulation was tested dynamically by stepping the torque from no-load to some final value. The results of these measurements are shown in Table II. The dynamometer required 100 ms to step the torque from no-load to the value listed in Table II.

The oscillograms in Fig. 8 show a typical torque step and the dwell angle response. The initial dwell is 15°. It increases

very quickly to 30°, then follows an underdamped response to the new operating point. The oscillogram is Fig. 9 shows the first cycle of the response in Fig. 8. The dwell voltage and the corresponding pulses from the dwell generator are shown for one phase. The dwell changes from 15° to 30° after four pulses, which is equivalent to one revolution.

The sensorless control system using dwell modulation performs well for both steady-state and dynamic conditions. In the steady state, the maximum torque produced by dwell modulation is more than three times the maximum torque produced by the open-loop system with comparable efficiency. In addition, the modulated system can withstand a step torque more than twice that of the open-loop system. This stability actually exceeds what is normal for integral-horsepower induction motors operating at fixed frequency, and is clearly adequate for most load applications.

V. SENSORLESS CONTROL BY FREQUENCY MODULATION

The control system described so far keeps the motor running at a fixed synchronous speed. In applications where precise

TABLE II

System	Max. Step Torque (oz-in)
Open-loop (15 deg. dwell)	2.60
Dwell-modulation system	6.00

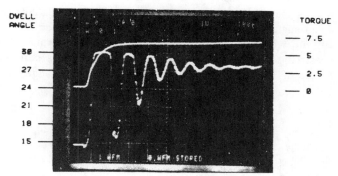

Fig. 8.   Step response showing the applied torque step (upper trace) and the dwell reference signal.

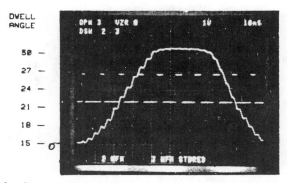

Fig. 9.   Step response—detail of first cycle of underdamped response, showing the individual transistor gating pulses and the dwell reference signal.

371

speed control is not required, a further improvement in stability can be achieved by introducing droop or regulation into the control loop. When the torque increases, the speed reference signal is reduced. In many cases, e.g., with fans and centrifugal pumps, the load torque decreases rapidly with speed so that a useful increase in torque margin can be obtained with a relatively small reduction of speed. The dotted elements in Fig. 7 show the structure of a control system that uses this principle. The dc link current is again sensed as a measure of load torque and applied to a nonlinear amplifier. The transfer function of the amplifier is designed to constrain the operating point to follow a trajectory in the speed/current plane that maximizes the system efficiency at each value of torque. Test results are shown in Table III.

The concept of pull-out torque for quasi-stable load changes is now no longer applicable: when the load torque increases the motor simply slows down. However, pull-out is still possible when the load torque changes rapidly. Table IV shows the step response tested in the same way as in Table II for the dwell-modulation system.

The dynamic torque margin is even smaller than that of the open-loop unstabilized system. It also depends on the inertia of the load, which is an undesirable feature not found in the dwell-modulation system.

VI. CONCLUSION

We have demonstrated a control system for VR stepper motors that increases the pull-out torque by modulating the dwell angle or conduction angle in response to the dc supply current, which is an indirect measure of the load torque. For quasistatic changes of load torque the system has a pull-out torque of four times the rated torque. For step changes the

pull-out torque is more than twice the pull-out torque of the open loop, with the unstabilized system having comparable efficiency. The motor can be operated with a narrow dwell and high efficiency. Without the controller the torque margin is then too small; but with the controller the torque margin is dynamically and almost instantaneously increased so that the high efficiency is obtained with no penalty in stability. The control system is completely independent of load inertia.

A modified control system has been demonstrated in which the dwell angle is kept fixed while the speed reference signal is reduced in response to an increase of dc supply current (load torque). This scheme effectively produces a drooping speed/torque characteristic while maintaining synchronism between the motor and the converter current pulses. It is applicable as an additional means of stabilization for systems where precise speed control is not required, such as fan and blower drives. The speed regulating controller is too slow to respond to rapid changes of load torque and its response depends on the load inertia. The signals representing the speed drop and the dwell modulation can be summed to give a controller that combines the advantages of both strategies.

REFERENCES

[1] J. R. Frus and B. C. Kuo, "Closed-loop control of step motors without feedback encoders," in *Proc. Fifth Annual Symp. on Incremental Motion Control Systems and Devices*, B. C. Kuo, Ed. (Urbana-Champaign, IL), May 1976, pp. CC1–CC11.

[2] B. C. Kuo and A. Cassat, "On current detection in variable-reluctance step motors," in *Proc. Sixth Annual Symp. on Incremental Motion Control Systems and Devices*, B. C. Kuo, Ed. (Urbana-Champaign, IL), May 1977, pp. 205–220.

[3] A. J. C. Bakhuizen, "On self-synchronization of stepping motors," *Proc. of the Int. Conf. on Stepping Motors and Systems.* England: University of Leeds, Sept. 19–20, 1979, pp. 77–83.

[4] P. P. Acarnley. *Stepping Motors: A Guide to Modern Theory and Practice*, IEE Control Engineering Series, no. 19. New York: Peter Peregrinus Ltd., 1982, pp. 123–126.

[5] P. M. Bartlett and H. Shankwitz, "The specification of brushless dc motors as servomotors," *Powerconversion International*, pp. 10–17, Sept. 1983.

[6] W. R. Pearson and P. C. Sen, "Brushless dc motor propulsion using synchronous motors for transit systems," *IEEE Trans. Ind. Electron.*, vol. IE-31, pp. 346–351, Nov. 1984.

TABLE III

Frequency Modulation System $\theta_D = 15$ deg.					
Torque (oz-in)	Power (W)	Shaft θ_D (deg.)	Speed (rpm)	Current (mA)	Link Eff (%)
2.20	2.45	8.70	1506	145	42.2
2.59	2.69	9.00	1400	147	45.7
2.99	2.88	9.10	1304	158	45.4
3.45	3.06	9.20	1200	171	44.9
4.69	3.47	9.10	1000	204	42.4
5.71	3.79	9.20	899	230	41.1
7.08	4.18	9.60	799	266	39.2
8.85	4.58	9.40	698	311	36.7

TABLE IV

System	Max. Step Torque (oz-in)
Frequency Modulation	2.30
Dwell Modulation	6.00
Open Loop, 15 deg dwell	2.60

Simplified Electronics for Torque Control of Sensorless Switched Reluctance Motor

J.T. Bass, M. Ehsani and T.J.E. Miller

IEEE Transactions, Vol. IE-34, No. 2, May 1987, pp. 234-239

Abstract—Sensorless control of the switched-reluctance (SR) stepping motor using pulsewidth modulation was introduced in an earlier paper by the authors. This scheme senses the dc link current and permits the motor to operate efficiently with a narrow pulsewidth under normal conditions. Changes in load are sensed by the dc link current, and the pulsewidth is adjusted appropriately. This paper describes the electric implementation of the controller.

I. Introduction

SMALL brushless dc motor applications are becoming more widespread for their high efficiency and adjustable-speed capability. The switched-reluctance (SR) stepping motor, particularly the large-step-angle motor such as the "6/4" motor is an important alternative to motors using permanent magnets. This paper describes the implementation of a sensorless control scheme for SR motors which result in a high-efficiency drive system.

The most direct method of controlling the SR motor is by using a shaft position sensor or encoder. The shaft position sensor is used to directly control the stator current ON and OFF angles referenced to the rotor position. However, in low-cost applications, particularly at low power levels, the shaft position sensor is too costly, and it also decreases the reliability of the system. Therefore a high-performance sensorless system is preferred.

Sensorless schemes presented in [1] can be divided into open-loop and current sensing controls. The current sensing controls sense the freewheeling current in one of the OFF phases of the drive converter to estimate the rotor position from the current waveshape. This estimated position is used to determine when one phase should be switched off and another one switched on. The open-loop system sends a series of stator current pulses without any indication of rotor position. The pulsewidth is defined as the difference between the ON and OFF angles. The frequency of the stator current pulses determines the motor speed. In order to maintain synchronism even when the load increases or is disturbed, these sensorless controls drive the motor with a much larger pulsewidth angle than is necessary for the nominal load. Consequently, the efficiency of these systems is quite low. For the motor used in our investigations, the efficiency at rated load is no better than 14 percent using these sensorless control methods.

Manuscript received September 30, 1985; revised April 2, 1986.
J. T. Bass and M. Ehsani are with Texas A & M University, College Station, TX 77843.
T. J. E. Miller is with General Electric Company, Corporate Research and Development, Schenectady, NY 12301.
IEEE Log Number 8613368.

A sensorless control using pulsewidth was introduced in [1]. This scheme senses the dc link current to determine the pulsewidth. This permits the motor to operate efficiently with a narrow pulsewidth under normal conditions. Any change in the load is sensed by the controller through the dc link current, and the pulsewidth is adjusted appropriately. This paper describes the electronic implementation of the above control system.

II. The Control System

The block diagram of the controller is shown in Fig. 1. The dc link current is filtered to obtain the dc value, I_d. The function generator is a nonlinear amplifier that uses I_d to produce a voltage representing the desired pulsewidth. The gate signal generator produces a three-phase gate pulse with a pulsewidth proportional to the function generator output.

Since a position encoder is not being used, the control system cannot fix the torque angle directly. The torque angle is defined as the mechanical angle between the rotor and stator poles at the instant when a stator winding is deenergized. Instead, the dc link current I_d is sensed and the pulsewidth or dwell angle is set according to a predetermined function of the dc link current. Fig. 2 shows a trajectory on the graph of I_d versus Θ_d that maintains the torque angle at 5° until the dwell reaches 30°. If the transfer function of the mentioned function generator is the trajectory shown in Fig. 2, the steady-state operating point produced by the controller lies on that curve, and the location depends only on the load. As a result, the torque angle is indirectly fixed at 5° and the stator current pulsewidth is increased or decreased as the load torque increases or decreases.

The cutoff frequency of the I_d filter affects the dynamic response of the system. For fast response of the control system, it is desirable to have the cutoff frequency as high as possible. However, if the cutoff frequency is too high, the ripple voltage on I_d becomes significant, with possible negative impact on the drive system. A four-pole butterworth filter consisting of two second-order stages was used, and the circuit was constructed so that the cutoff frequency could be changed quite easily [2]. Thus the trade-off between system dynamic response and efficiency could be measured experimentally. An instrumentation amplifier proceeded the filter to accurately amplify the low-level signal from the shunt in the dc link.

A fast Fourier transform of the dc link current indicated an 80-dB attenuation of the 300-Hz harmonic component is necessary to produce I_d with sufficiently low ripple when the

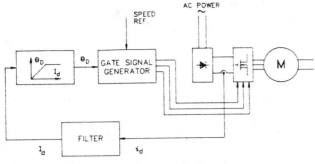

Fig. 1. System block diagram.

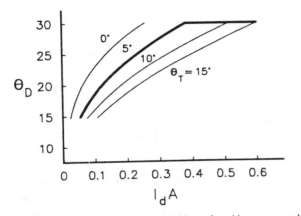

Fig. 2. DC link current versus pulsewidth angle with torque angle as parameter.

motor is operated at the rated 1500-rpm ripple. With a four-pole filter, the cutoff frequency needs to be 30 Hz to obtain this attenuation. But, as stated before, the system was tested with a 25-Hz and a 100-Hz dc link cutoff filter to evaluate the dynamic performance of the system.

The output of the dc link filter goes to the nonlinear amplifier which determines the trajectory that the operating point follows. When choosing the operating trajectory, a trade-off exists between system efficiency and stability, with a small torque angle produces a low efficiency with good stability, and a large torque angle produces a high efficiency with poor stability [3]. Therefore the nonlinear amplifier was designed so that the function can be easily changed between the above extremes. The nonlinear amplifier produces a maximum of three breakpoints, and clamps the output at 0 and 5 V. The 0-V level corresponds to a pulsewidth angle of 15° while the 5-V level corresponds to 30°. Therefore the maximum and minimum pulsewidths possible are 30 and 15°. The curves presented in Fig. 2 were computer generated by a computer model. In order to follow the desired trajectory, the nonlinear function must be obtained from experimental measurements of the test motor. The experimental curves for a 2.5-W SR motor running at 1500 rpm are shown in Fig. 3 with two trajectories shown with bold lines that approximate torque angles $\theta_t = 9$ and 5°.

374

The outputs of the operating point function generator and a 76.8-kHz square-wave generator are fed to the inverter gating signal generator. The square-wave frequency is divided by 256 to obtain the 300-Hz fundamental frequency. The 300-Hz signal is then converted to a three-phase inverter gate signal

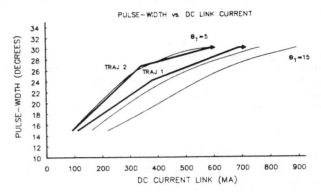

Fig. 3. Experimentally determined stable region with the two trajectories used to test the pulsewidth modulation.

such that only one phase will be on at any time. The function generator output is converted to a digital word and loaded into an 8-bit counter which is synchronized with the 300-Hz fundamental. The width of the gate pulse (dwell angle) produced by the counter is determined by the digital word. The most-significant bit of the counter is always high so the minimum pulsewidth is 15° and the maximum pulsewidth is 30°. The gate pulse produced by the counters is anded with the three phases. The resulting three-phase output has a pulsewidth proportional to the analog voltage output of the function generator.

The control circuit is designed so that all the parameters which affect the system performance significantly can be changed easily. These parameters include the mentioned desired trajectory and the cutoff frequency of the filter. Consequently, the system could be easily tested in many different configurations. The experimental results for the system are presented in the Section III.

III. EXPERIMENTAL RESULTS FROM PULSEWIDTH MODULATION

The control system described above was built and both the steady-state and dynamic capabilities were tested. As expected, the cut-off frequency of the filter and the desired trajectory had a significant effect on the dynamic performance of the system. The steady-state performance improved considerably over the performance of the open-loop system. The system can produce over 4-per-unit torque (1 pu torque = rated torque of 2.25 oz·in) while maintaining a system efficiency greater than 40 percent. The steady-state performance will be presented next followed by the results from the dynamic tests.

The system steady-state performance was obtained for both trajectories shown in Fig 3. The steady-state results for trajectory 1 are given in Table I. The results show that the system efficiency is high for a wide range of operating points. Consequently, the same motor and control system can be used to drive any load efficiently up to the maximum torque the motor can produce.

Table II gives a comparison of the sensorless systems and the open-loop systems. The results verify that the trajectory with the largest torque angle (trajectory 1) produced the highest system efficiency. Also, the sensorless control by pulsewidth modulation produced a maximum torque three

TABLE I
STEADY-STATE RESULTS FOR TRAJECTORY 1. (PULSEWIDTH MODULATION, TRAJECTORY 1, 100 HZ FILTER, Vdc = 40.0 V, SPEED = 1500 rpm)

Torque (oz-in)	Shaft Power (W)	θ_t (deg)	θ_d (deg)	Link Current (mA)	Eff (%)
.05	.05	3.0	15.4	66	1.9
2.03	2.26	7.25	16.2	135	41.9
4.02	4.46	8.00	20.0	243	45.9
6.03	6.70	7.75	23.3	369	45.4
8.05	8.94	7.75	27.1	513	43.6
10.03	11.14	9.00	29.8	680	41.0

TABLE II
COMPARISON OF STEADY-STATE SENSORLESS AND OPEN-LOOP RESULTS

System	Maximum Torque (oz-in)	Efficiency at 2.25 W (%)
Traj. 1, 100 Hz Filter	10.03	40.3
Traj. 2, 100 Hz Filter	9.52	35.2
Traj. 2, 25 Hz Filter	9.13	34.3
Open loop, 15 deg.	3.03	42.7
Open loop, 30 deg.	10.0	14.0

times that of the 15° open-loop system with system efficiencies comparable to the open-loop system. Therefore the sensorless system produced a large torque margin and high system efficiency, simultaneously. As stated earlier, this cannot be accomplished with an open-loop system.

The pulsewidth modulation was tested dynamically by stepping the torque with a Magtrol dynamometer from no load to some final value. The rise time of the step was approximately 100 ms. Two different dc link cutoff frequencies were used to determine what effect it has on the dynamic response of the system. The step torque was increased until the rotor lost synchronization with the stator and the rotor stalled. The largest step the motor could withstand without stalling is shown in Table III. The results indicate trajectory 2 can withstand a slightly larger step than trajectory 1. But, both trajectories can withstand a significantly larger step when the 100-Hz dc link filter is used.

The sensorless control system using pulsewidth modulation performs well for both steady-state and dynamic conditions. In the steady state, the maximum torque produced by pulsewidth modulation is more than three times larger than the maximum torque produced by the open-loop system. In addition, the sensorless system can withstand a step torque that is more than

twice the step capabilities of the open-loop system. In Section IV another method of implementing the sensorless control will be investigated and compared to this implementation.

IV. PULSEWIDTH MODULATION USING SIMPLIFIED ELECTRONICS

The previous implementation is very general in order to test the concept of pulsewidth modulation. The circuit design can be simplified considerably by using a different topology for the dc link filter, the gate signal generator, and the amplifier which determines the operating trajectory. The simplified design will be presented next followed by experimental results comparing the designs.

TABLE III
COMPARISON OF RESULTS FROM DYNAMIC TESTS

System	Step Torque (oz-in)
Traj. 1, 25 Hz Filter	3.70
Traj. 1, 100 Hz Filter	5.90
Traj. 2, 25 Hz Filter	4.10
Traj. 2, 100 Hz Filter	6.00
Open loop, 15 deg dwell	2.60

A. DC Link Filter

The dc link filter topology is simplified by eliminating the instrumentation amplifier. A 1-Ω resistor was used to sense the dc link current, the voltage across the resistor was nominally in the 100-mV range. Therefore, an instrumentation amplifier was no longer necessary. The filter topology has changed from a four-pole butterworth to a second-order passive filter followed by a noninverting amplifier. The poles of the filter are located at 25 Hz (see Fig. 4). The dynamic performance of this topology was slightly lower, when compared to the previous circuit, but was still acceptable. The amplifier has a gain of ten so that 10 V corresponds to a 1-A dc link current.

B. Trajectory Generator

The amplifier which determines the operating point trajectory had a nonlinear gain. The nonlinear amplifier is required if the trajectory follows a line of constant torque angle. However, a linear trajectory that approximates the desired torque angle can be used. An example of a linear trajectory for the pulsewidth modulation is shown in Fig. 5. The linear trajectory is shown along with two trajectories used to test the pulsewidth modulation. The linear trajectory approximately follows the high efficiency trajectory (trajectory 1) over the typical region of operation. For larger loads, the linear trajectory deviates from trajectory 1 resulting in a lower system efficiency. However, with a rated steady-state load, the operating point enters this region only during transient overloading conditions. The linear amplifier used to realize the trajectory of Fig. 5 has a positive gain with a negative offset to set the nominal current at 1500 rpm (see Fig. 4).

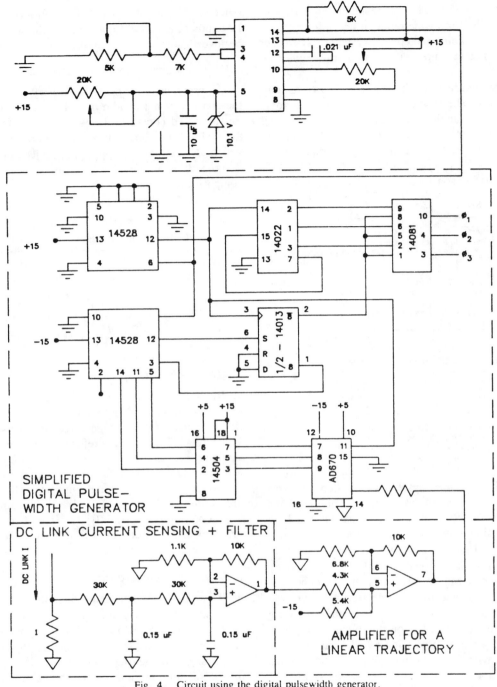

Fig. 4. Circuit using the digital pulsewidth generator.

C. Gate Signal Generator

The gate signal generator was redesigned using two different topologies. The first is a simplified version of the original pulsewidth generator, while the second uses an analog rather than a digital approach to the design of the signal generator. Each topology was evaluated with the above dc link filter and linear amplifier. The simplified version of the original gate signal generator is shown in Fig. 4. This circuit uses a 4-bit instead of an 8-bit counter to divide f_{in}. The V_{dwell}

is converted to a 3-bit digital word by the A/D converter. Therefore, the primary difference between this circuit and the original one is that the pulsewidth is quantized to eight discrete widths by the new circuit while it was quantized to 128 widths using the original circuit.

The analog approach to the gate signal generator is shown in Fig. 6. An integrator is used to generate a ramp. The output of the integrator V_{int} is clamped at zero by the FET when f_{in} is

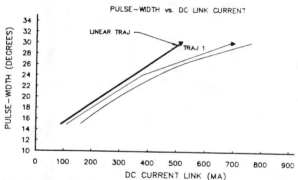

PULSE-WIDTH vs. DC LINK CURRENT

Fig. 5.   Linear trajectory for the sensorless control by pulsewidth modulation.

high. After f_{in} goes low, the integrator output begins to ramp up. When V_{int} reaches V_{dwell}, the comparator output changes from high to low. Therefore, the pulsewidth is directly proportional to V_{dwell}. In addition, the OR gate insures that the gate pulse is high when f_{in} is high. By this circuit, the minimum dwell angle is kept at 15° and the maximum at 30°.

The two-gate signal generators described above are combined with the dc link filter and linear amplifier to form two separate sensorless control systems. These systems, shown in Figs. 4 and 6, provide a simple load-independent method of controlling an SR motor without using a shaft position sensor. The experimental results from both circuits will be presented in the Section V.

V. EXPERIMENTAL RESULTS WITH THE SIMPLIFIED CONTROL CIRCUITS

The steady-state and dynamic capabilities of the simplified sensorless control circuits were tested. As with the previous control circuits, the system efficiency was high for a wide range of operating points. As shown in Table IV, the new topologies do not produce as much torque as the nonlinear trajectories, but they are within 15 percent of the nonlinear system. Fig. 5 shows the trajectory for the linear amplifier lying between trajectory 1 and trajectory 2. Therefore, the system efficiency for the linear operating trajectory should be higher than for trajectory 2 and lower than for trajectory 1. The data in Table IV confirm this intuitive expectation. It should be noted that the operating point trajectory for the linear amplifier can be shifted so that it follows trajectory 1 for

TABLE IV
COMPARISON OF STEADY-STATE RESULTS FOR THE LINEAR AND NON-LINEAR TRAJECTORIES

System	Maximum Torque (oz-in)	Efficiency @ 2.25 W (%)
Trajectory 1	10.0	42.7
Trajectory 2	9.52	34.1
Linear Trajectory Using Analog Gate Pulse Gen.	8.97	40.9
Linear Trajectory Using 3 bit Gate PulseGen.	8.68	41.0

TABLE V
DYNAMIC RESULTS FOR THE LINEAR AND NONLINEAR TRAJECTORIES

System	Step Torque (oz-in)
Trajectory 1 (nonlinear)	5.90
Trajectory 2 (nonlinear)	6.00
Linear Trajectory Using Analog Gate Pulse Gen.	4.95
Linear Trajectory Using 3 bit Gate Pulse Gen.	5.00

narrow pulsewidths dwells. In that case, the system efficiency would be approximately the same as for trajectory 1 for pulsewidths up to 24°.

The system was tested dynamically by stepping the load from no load to some final value. The largest step torque which each system can withstand is compared to the previous results in Table V. The dynamic performance of the previous systems is slightly better than the systems using a linear trajectory. The slight decrease in dynamic performance can be attributed to the different dc link filter topology. However, the linear systems can still withstand a step torque which is several times the nominal load.

VI. CONCLUSIONS

A general nonlinear circuit and a simplified linear circuit have been presented that implement the sensorless control by pulsewidth modulation for a SR motor. Although the linear approach is much simpler than the nonlinear circuits, both the steady-state and dynamic performance of the simplified topologies were comparable to the more complicated nonlinear systems. The pulsewidth modulation with a linear trajectory provides a simple practical method of stabilizing the SR motor at rated speed without using a shaft position sensor.

REFERENCES

[1] J. T. Bass. M. Ehsani, and T. J. E. Miller. "Robust torque control of switched-reluctance motors without a shaft position sensor," submitted to *IEEE Trans. Ind. Electron.*

[2] J. Millman and C. C. Halkios, *Integrated Electronics: Analog and Digital Circuits and Systems.* New York: McGraw-Hill, 1972, pp. 449–453.

[3] J. T. Bass, "Internship experience at general electric corporate research and development center," Dept. Electrical Engineering. Texas A&M University, College Station. TX. Res. Rep.. May 1985.

[4] J. V. Wait. L. P. Huelsman, and G. A. Korn, *Introduction to Operational Amplifier Theory and Applications.* New York: McGraw-Hill, 1975, pp. 176–187.

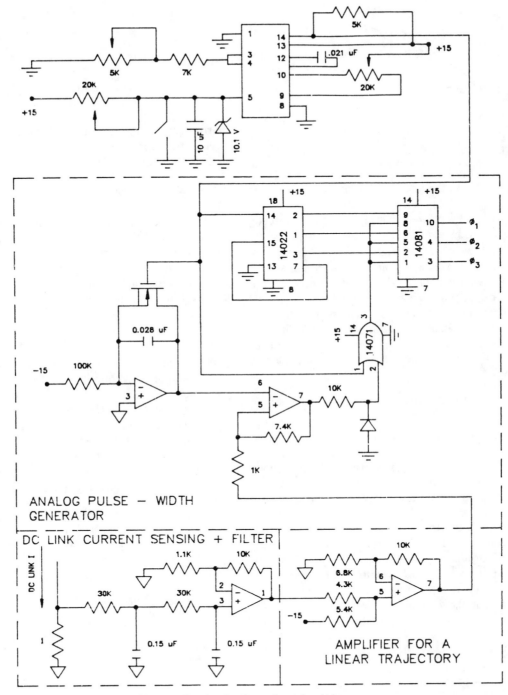

ANALOG PULSE – WIDTH
GENERATOR

DC LINK CURRENT SENSING + FILTER

DC LINK I

AMPLIFIER FOR A
LINEAR TRAJECTORY

Fig. 6. Circuit using the analog pulsewidth generator.

Detection of Rotor Position in Stepping and Switched Reluctance Motors by Monitoring of Current Waveforms

P.P. Acarnley, R.J. Hill and C.W. Hooper

IEEE Transactions, Vol. IE-32, No. 3, August 1985, pp. 215-222

Abstract—The paper describes new methods of detecting rotor position in stepping and switched motors, with chopper or series–resistance drives, by monitoring winding currents. In contrast to previous methods of waveform detection, the new techniques are reliable over the complete speed range. It is shown that the most useful indicators of rotor position are: i) current rise times arising from chopping an unexcited phase at low current and ii) the initial rate of current rise as a phase is switched on in a series–resistance drive. Implementation is via simple low-cost electronic circuits and the paper includes discussion of how the techniques can be applied to closed-loop stepping motor control, ministep drives, and optimization of step response.

NOMENCLATURE

i	Instantaneous phase current.
I	Average current level in excited phase.
l	Incremental inductance.
p	Number of rotor teeth.
R	Phase circuit resistance.
t_{decay}	Current decay time.
t_{rise}	Current rise time.
V	Supply voltage.
δI	Current excursion during chopping.
θ	Rotor position.
ψ	Flux linkage.

I. INTRODUCTION

CLOSED-LOOP stepping motor control has the attractive properties of guaranteed step integrity and optimal performance for varying load conditions [1], [2]. However, this form of control has failed to attain widespread popularity because the usual optical methods of rotor position detection [3] are expensive, unreliable, and only capable of producing quantized position information (usually one signal per motor step). For true optimal control continuous monitoring of rotor position is required, necessitating either an expensive high-resolution encoder or the synthesis of additional position information using a phase-locked-loop [2] to generate several signals per motor step. Similar problems of position detection have arisen in the recent development of switched-reluctance motors [4], where precise timing of phase excitation relative to rotor position is needed for operation over continuous speed ranges with varying loads.

In attempts to eliminate the optical encoder, several authors [5]–[7] have described so-called "waveform detection" of rotor position. With this technique the modulating effect of the

Manuscript received May 4, 1984. This work was supported by the UK Science Research Council and the British Technology Group.
The authors are with Cambridge University Engineering Department, Cambridge, UK.

motional EMF on the current waveform is observed electronically, and since the instantaneous motional EMF is dependent on rotor position, it is possible to deduce rotor position. Waveform detection has the advantage of low cost, but is unreliable in many situations, because, for example, it is unable to operate at low speeds where the motional EMF approaches zero. Furthermore, it appears that waveform detection can only be applied to the constant-voltage drives commonly used for small (<1 kW) motors.

This paper describes alternative and more versatile methods of position detection based on monitoring phase current with the aim of detecting the rate of change of phase current, rather than the motional EMF. Since the rate of current change is dictated by the incremental inductance [9] of the phase circuit, and the incremental inductance is in turn a function of rotor position and phase current, rotor position can be deduced from a knowledge of phase current and its rate of change. Such a scheme has the important advantage that it is useful even at zero speed where there is no motional EMF. In fact the motional EMF is a possible source of difficulty with the new method, since in general it influences the rate of current change, but by careful choice of monitoring conditions these difficulties are avoided.

After establishing the fundamental relationship between rate of current change and the motor's magnetizing characteristic, the direct detection of rotor position with both "chopper" and "series–resistance" drives [1] is considered. Subsequently, applications of the method to variable-speed drives, mini-step drives, and damping are discussed. Throughout the paper attention is centered on variable-reluctance stepping and switched motors, since the variation of incremental inductance with rotor position is largest in these types, but the position detection method is applicable to any motor (including the hybrid stepping motor), in which the inductance variation is significant.

II. ROTOR POSITION DETECTION BY CURRENT WAVEFORM MONITORING

A. General Principles

In a doubly salient stepping or switched motor the flux linkage with a phase winding is a function of rotor position and phase current; typical characteristics are shown in Fig. 1. When rotor teeth are aligned with the stator teeth of the relevant phase the flux linkage, for a given phase current, is maximized. However, at this aligned position the relationship between flux linkage and phase current is extremely nonlinear in a well-designed machine, because the tooth iron is magnetically saturated at the rated phase current. Therefore,

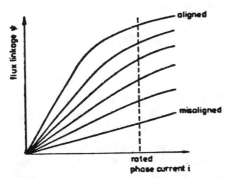

Fig. 1. Typical flux linkage versus current characteristics at various rotor positions.

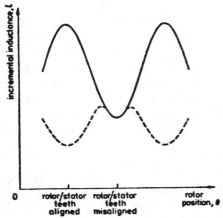

Fig. 2. Typical incremental inductance versus rotor position characteristics. —————— Low-winding current, —— ——— High (rated)-winding current.

$$V = Ri + \frac{\partial \psi}{\partial i} \cdot \frac{di}{dt} + \frac{\partial \psi}{\partial \theta} \cdot \frac{d\theta}{dt}$$

$$= Ri + l \frac{di}{dt} + \frac{\partial \psi}{\partial \theta} \cdot \frac{d\theta}{dt} . \qquad (2)$$

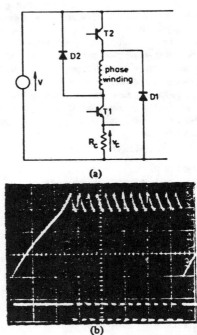

(a)

(b)

Fig. 3. (a) One phase of a chopper drive circuit. (b) Chopping characteristics. Upper trace: phase current. Lower trace: chop control signal.

the incremental phase inductance—the gradient of the flux linkage/current characteristic—at this position is high for low values of phase current, but decreases rapidly as the current is increased towards its rated value, as shown in Fig. 2.

At the rotor position where the rotor teeth are completely misaligned with the phase stator teeth, the magnetic circuit is dominated by the large airgap between stator and rotor teeth, and despite local magnetic saturation of the tooth corners, the flux linkage is linearly dependent on phase current. The phase incremental inductance at this misaligned position has a value which is independent of phase current, and which is between the values for aligned position incremental inductances for low- and high-phase currents (Fig. 2).

For most doubly salient motors with windings concentrated on stator teeth or poles the mutual coupling between phases is small, so the voltage equation for one phase can be written in terms of the current and flux linkage for that phase as

$$V = Ri + d\psi/dt \qquad (1)$$

and, since the flux linkage ψ is a function of phase current and rotor position

380

Rearranging (2) to give an expression for the rate of change of phase current with time

$$\frac{di}{dt} = \frac{V - Ri - \frac{\partial \psi}{\partial \theta} \cdot \frac{d\theta}{dt}}{l} . \qquad (3)$$

The incremental inductance l appearing in the denominator of (3) is rotor-position dependent, so the rate of change of current is also related to rotor position. However, this relationship is complicated by the dependence of incremental inductance on current and by the appearance of a "motional EMF" term $(\frac{\partial \psi}{\partial \theta} \cdot \frac{d\theta}{dt})$ in (3). Errors can arise from these effects when the current gradient is used for rotor position detection, but by careful matching of the detection technique to drive circuit and operating mode [8] the errors can be minimized or eliminated.

B. Chopper Drives

The circuit diagram for one phase of a unipolar chopper drive circuit is shown in Fig. 3(a). The drive incorporates a high-supply voltage, which if applied continuously to the winding would result in a phase current of many times rated. When the phase is first turned on, transistor switches $T1$ and

T2 are conducting and the phase current builds up at a rate dictated by the effective phase inductance. A small resistor R_c is connected in series with the collector of T1 and the voltage V_c dropped across this resistor acts as a monitor of the instantaneous phase current. When the phase current reaches a level which slightly exceeds that required, transistor T2 is switched off and the phase current freewheels around the closed path which includes the winding, T1, R_c, and D1. This current path has a high inductance and low resistance, so the decay of current is slow, as shown in Fig. 3(b). When the current has decayed to a level slightly below that required, the reduction in monitoring voltage V_c is used to turn on T2, so that the high-supply voltage is applied to the phase for a short time, causing the current to rise back to above the required level. The cycle of operation repeats, with T2 turning on and off, until the end of the phase-excitation interval is reached. Both transistors are then switched off and the current freewheels around the path which includes the winding D2, the supply voltage, and D1. As the high-supply voltage is included in this path the current is forced rapidly to zero and a large proportion of the stored inductive energy is returned to the supply.

During chopping the phase current swings around the required level at a rate dictated by the incremental inductance of the phase at that current level, and as the incremental inductance is rotor-position dependent, the instantaneous rotor position can be deduced from the chopping characteristics. For normal operation of a stepping or switched motor the phase current is either held at its rated value or reduced to zero, and, therefore, there is the possibility of detecting rotor position from variation of chopping characteristics in a phase which is either ON (with phase current at rated) or OFF (with phase current zero). In each case the current swing during chopping occurs around a constant average current level, so the variation of incremental inductance with current level is unimportant.

C. Chopping at Rated Current

In this operating condition the phase current varies between $I + \delta I/2$ and $I - \delta I/2$, so during the rise time t_{rise} the current increases δI, and during the decay time t_{decay} the current decreases δI. Assuming that the current swing is small relative to the mean current level, the current rise or decay is linear and the corresponding times can be deduced from (3)

$$t_{\text{rise}} = \frac{I \delta I}{V - RI - \frac{\partial \psi}{\partial \theta} \cdot \frac{d\theta}{dt}} \qquad (4)$$

$$t_{\text{decay}} = \frac{I \delta I}{RI + \frac{\partial \psi}{\partial \theta} \cdot \frac{d\theta}{dt}}. \qquad (5)$$

According to the particular application (see Section III), it may be necessary to detect rotor position either continuously or at a single well-defined position per step, and either over a wide speed range or with the rotor stationary. Considering first the question of continuous position sensing, the current rise, and decay times are shown to be proportional to incremental inductance. However, the most notable feature of the incremental inductance/rotor position characteristic at rated current (Fig. 2) is that around the position where stator/rotor teeth are misaligned there are four rotor positions corresponding to each value of incremental inductance. Because of this ambiguity in rotor position corresponding to a given incremental inductance, and hence rise or decay time, continuous position detection is unreliable for this operating condition. However, detection of a specific rotor position is still possible. With the machine producing maximum motoring

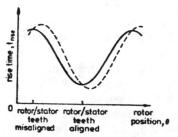

Fig. 4. Effect of motional EMF on current rise time versus rotor position characteristics. —— Low speed (motional EMF → 0), — · — High speed (motional EMF → supply voltage).

torque, for example, each phase is switched on at the misaligned teeth position and switched off at the aligned teeth position. In Fig. 2 it can be seen that the phase turn-off position corresponds to the minimum incremental inductance, so this position may be detected at low speeds by seeking the minimum rise or decay times in the chopping of the excited phase.

Consider now the problem of position detection over a wide speed range. During normal chopping operation the dc supply voltage V is much greater than the resistive voltage drop RI and the motional EMF $\frac{\partial \psi}{\partial \theta} \cdot \frac{d\theta}{dt}$. Since the dc supply voltage appears in the numerator of the rise-time expression (4), but not the decay time expression (5), the rise time corresponding to a given position is relatively immune from changes in running speed. Even the rise time's immunity breaks down at the highest operating speeds, where the motional EMF becomes a significant fraction of the supply voltage, as illustrated in Fig. 4. However, one important and useful exception occurs at the aligned and misaligned tooth positions where the rate of change of flux linkage with position ($\partial \psi/\partial \theta$) is zero, and hence the instantaneous motional EMF is zero. Therefore, the current rise and decay times are independent of speed at these particular rotor positions, which (perhaps fortuitously!) correspond to the positions where commutation between phases should occur for maximum torque production.

In summary, by monitoring the chopping characteristics for an excited phase it is possible to detect a single rotor position per step, and the most reliable indicator of rotor position is the current rise time at aligned and misaligned tooth positions. An application of this principle is described in Section III-A.

D. Chopping at Zero Current

An alternative strategy for direct position detection with a chopper drive is to observe chopping behavior in a phase which is turned off. During normal operation of the drive, of course, the current in such a phase would be zero, but by setting the required current level to be a small fraction of the rated current, chopping behavior can be observed without diminishing the torque-producing capability of the motor. For chopping at currents near zero the variation of the phase incremental inductance with rotor position is a smooth near-sinusoidal curve (Fig. 2), and, therefore, continuous monitoring of rotor position is possible.

In this situation the phase current increases from zero to δI during the rise time t_{rise}. The voltage dropped across the winding resistance by this small current is negligible in comparison to the high dc supply voltage. Furthermore, the flux linkage with the winding tends to zero as the current approaches zero, so the rate of change of flux linkage with rotor position ($\partial \psi / \partial \theta$), and hence the motional EMF, can be neglected. Substituting into (3) with these simplifications included gives

$$t_{rise} = l\delta I / V. \tag{6}$$

The natural decay of current during chopping is exponential with a long time constant, since the freewheeling path (via $T1$ and $D1$ in Fig. 3) has a high inductance and low resistance. In the case of chopping at low currents this slow current decay may be unacceptable if the rotor position is to be continuously monitored. However, rapid current decay can be obtained by switching off both transistors in the chopper drive ($T1$ and $T2$ in Fig. 3) and forcing the phase current back to zero by causing it to flow against the dc supply voltage via diodes $D1$ and $D2$. The effective supply voltage is then $-V$ for a decay of current from δI to zero in time t_{decay}, so substituting into (3) gives

$$t_{decay} = l\delta I / V \tag{7}$$

assuming that the supply voltage is again much greater than the resistive voltage drop and the motional EMF. During the decay time the phase current does not flow through the current sensing resistor connected in series with $T1$, because this transistor is now nonconducting. Consequently the end of the decay interval cannot be found by detecting zero current in the current sensing resistor and one of three alternative methods must be adopted.

a) Insert an additional current sensing resistor in series with diode $D1$ and detect the end of current decay directly.

b) From (6) and (7) the rise and decay times at a given rotor position are equal, so the drive transistors can be turned off for a time equal to the previous rise time.

c) The largest incremental inductance, and, therefore, the longest decay time, occurs at the teeth aligned position, so the transistors can be switched off for a fixed time which is sufficiently long to ensure current decay is complete even at this "worst case" position.

In summary, monitoring of chopping behavior in a phase which is turned off has the advantages that the rotor position can be deduced with no ambiguity and immunity from errors due to motional EMF. However, a slight modification to the chopper drive operation is required to ensure rapid current decay and a consequent high sampling rate.

E. Series–Resistance (L/R) Drives

Simple drive circuits incorporating a series-forcing resistance are used for a wide range of applications involving low-power stepping motors. Typical current waveforms for this type of drive are shown in Fig. 5. At low speeds (Fig. 5(a)) the current in an excited phase rises exponentially to the rated current level and subsequently the current gradient is zero, so there is no possibility of detecting rotor position from current gradient in a turned-on phase. For high-speed operation (Fig. 5(b)) the effects of motional EMF produce a very different

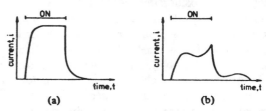

Fig. 5. Typical current waveforms for a constant-voltage drive. (a) Low speed (phase-excitation time ≫ phase-electrical time constant). (b) High speed (phase-excitation time = phase-electrical time constant).

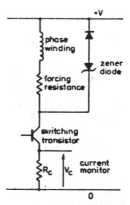

Fig. 6. A constant-voltage drive suitable for position detection by monitoring initial current rise.

current waveform, in which the changes of current gradient are a function of rotor position and speed [5]–[7].

As an alternative to current monitoring throughout the excitation period of a phase, it is profitable to concentrate on the situation when the phase is first turned on. The phase current and flux linkage are zero, so the resistive voltage drop and motional EMF are negligible, and the initial rate of current rise (from (3)) is

$$\frac{di}{dt}\bigg|_{i=0} = V/l. \tag{8}$$

In this expression the relevant incremental inductance is that associated with low-phase currents (Fig. 2), which leads to a well-defined unambiguous relationship between current gradient and rotor position, independent of operating speed.

Closed-loop control with a series–resistance drive is implemented by repeatedly turning on the next phase in the excitation sequence and examining the initial current gradient. If the gradient matches the gradient obtained by turning on at the optimum rotor position for excitation change [2] then the excitation change is allowed to proceed. However, if the optimum position has not been attained, the current gradient is incorrect and the small phase is reduced to zero, in preparation for another turn-on attempt a short time later.

Several small modifications to the simple series-resistance drive are required if this form of rotor position detection is to be implemented. A small current-sensing resistor must be included in the circuit and is most conveniently located in series with the phase transistor emitter (Fig. 6). Following an attempt to change excitation at an incorrect position, the residual phase current must be forced to zero as quickly as possible. The natural decay of current through a freewheeling diode is too slow to allow position sampling at a useful rate, so a zener diode is placed in the freewheeling path. If the diode's breakdown voltage is equal to the dc supply voltage the current is forced back to zero at a rate equal to the current gradient during turn-on. The maximum voltage appearing at the transistor collector is then twice the dc supply voltage and the transistor rating must be increased accordingly. In these circumstances the end of the current decay interval may be found by any of the three methods described in Section II-D.

For a series–resistance drive the optimum rotor position for excitation changes is a function of the stepping rate [2]; the "switching angle" is speed-dependent. With conventional optical methods of position detection the detected rotor position is inevitably fixed, and switching angle variation is awkward. However, the new method described here allows switching angle to vary with speed simply by relating the required current gradient to the instantaneous stepping rate.

The technique of monitoring initial current gradient in the next phase to be excited is also applicable to chopper drives. In that situation the technique has the advantage, over monitoring of chopping characteristics, that it is able to detect position even at the very highest speeds, where chopping operation may have ceased because the current build-up time exceeds the phase excitation period (Fig. 8).

In summary, position detection by observing initial current rise in a series–resistance drive is suitable for continuous position monitoring and is immune from motional EMF effects. A few low-cost modifications to the drive circuit are required for effective implementation. An application of the technique to step-response damping is described in Section III-C.

III. IMPLEMENTATION AND APPLICATIONS

In Section II it is shown that rotor position can be detected by monitoring one of three distinct current waveform features, namely on-phase chopping, off-phase chopping, and initial current rise. According to the operating characteristics of the motor, it may be necessary to detect the rotor's arrival at a distinct point within a step (e.g., in closed-loop stepping motor position control) or to monitor rotor position continuously (e.g., during ministepping). Three examples of the way in which the new position detection methods can be applied are described in this section.

A. Closed-Loop Control of Stepping and Switched Motors

For closed-loop control of a stepping motor or switched-reluctance motor [4] the pattern of excited phases must be changed as the rotor reaches a uniquely defined point during each step of its motion. An optical position encoder, producing one pulse per step of rotor movement, may be employed, but, because these motors are often driven with chopper-type circuits, direct detection of rotor position by monitoring chopping behavior is a reliable and cost-effective alternative.

For detection of the position from chopping in a turned-on phase maximum immunity from motional EMF effects is obtained by observing current rise time (Section II-C). A circuit suitable for the detection of a specified current rise time (T) is shown in Fig. 7(a). The system monitors the chop control signal to determine the latest rise time, which is compared to the period of a positive-edge triggered monostable using a D-type flip-flop, producing a "low" output if the latest rise time is longer than that specified.

A complete cycle of operation is illustrated in the timing diagram of Fig. 7(b), which has been made clearer by portraying operation at a much lower frequency than would be chosen in practice. At the beginning of the cycle the current rise time is short, so the positive-edge triggered monostable output is still "high" when the current rise is complete. The transition of the chop control signal "high" → "low" produces a short duration pulse from the negative-edge triggered monostable and this pulse acts as a clock signal for the D-type flip-flop, which, as its D input is "high," produces a "high" output.

The circuit remains in this state until time t_1, when a current rise time longer than T occurs. The chop control signal "high" → "low" transition, and, therefore, the flip-flop clock pulse, now occurs when the monostable output and flip-flop D input, are "low" so that the flip-flop output changes "high" → "low." After a further two-chop cycle, however, at time t_2 the current rise time falls below T and the flip-flop output returns to the "high" state.

Fig. 8(a) shows the variation of no-load running speed with monostable period, and hence detected rotor position, for a closed-loop stepping motor system. For short monostable

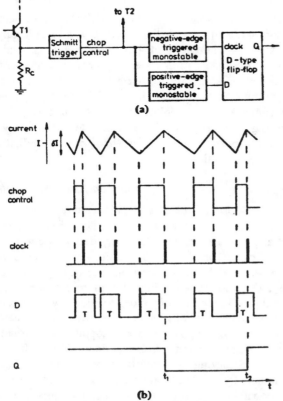

(a)

(b)

Fig. 7. Position detection by monitoring chop current rise time. (a) Control circuit. (b) Control waveforms during one step.

periods the detected position is too close to the phase detent position for the motor to produce significant torque (the effective switching angle is negative) and the useful range of stepping rates is severely restricted. As the monostable period is increased, the rotor's arrival at a progressively earlier position within each step is detected. The motor torque then increases, since the switching angle approaches its optimum value, giving a wider speed range.

Phase current waveforms for operation at maximum stepping rate with short and long monostable periods are shown in Fig. 8(b) and (c). The turn-off of the phase immediately after a rise time shorter than that specified by the monostable period can be seen in both cases. In Fig. 8(c) the useful operating speed range is limited by the finite time required for the phase current to build up to a level where chopping commences. For operation beyond this limiting speed, rotor position detection

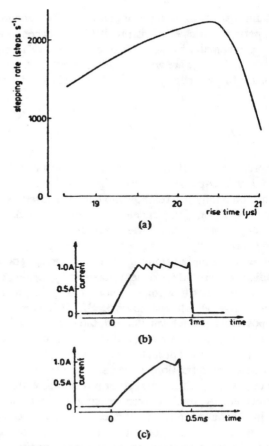

Fig. 8. (a) Stepping rate versus detected rise time characteristics for a stepping motor with closed-loop control. (b) Current waveform at 1000 steps s^{-1}. (b) Current waveform at 2000 steps s^{-1}.

based on initial current gradient (Section II-C) could be employed.

B. Ministep Drives

In a ministep system the motor's resolution is improved by partially exciting two or more phases, so that the rotor is positioned at intermediate points between the positions obtained by full-phase excitation [1]. Up to 32 ministeps per full step may be available. However, the inherent weakness of these systems is that they are open-loop, and a load torque displaces the rotor from the required position introducing an error which may even exceed one *full* step. Obviously, it would be useful to check the rotor's position, and possibly take corrective action, but the cost of an optical encoder with ministep resolution is prohibitive.

Since partial phase excitation in a ministep drive is usually implemented by varying the current reference level in a chopper drive, the technique of detecting position by monitoring chopping characteristics is applicable. As the current levels in the excited phases vary during ministepping it is appropriate to observe chopping in the unexcited phases, so that only one incremental inductance/position characteristic is in use.

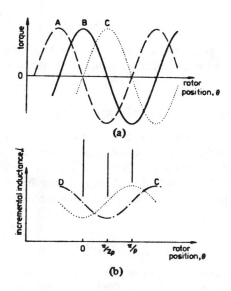

Fig. 9. Closed-loop control of a ministep drive for a 4-phase motor. (a) Static torque versus rotor position characteristics at rated phase current. (b) Incremental inductance versus rotor position characteristics at 0 phase current.

The principle can be illustrated by referring to Fig. 9 which shows, for a four-phase motor, the variations of static torques and low-current incremental inductances with position for several phases. Ministepping for positions between the detent positions for phases A ($\theta = 0$) and B ($\theta = \pi/2p$) involves partial excitation of phases A and B. For example, by exciting phases A and B equally, the rotor is positioned at $\theta = \pi/4p$, midway between the separate detent positions. However, if a load tends to move the rotor towards $\theta = 0$ a position error will appear, unless the current in phase A is reduced and in phase B is increased, enabling the motor to produce a torque to balance the load at the correct rotor position.

Considering the variations of incremental inductance for phases C and D in the range $0 \leqslant \theta \leqslant \pi/2p$ (Fig. 9(b)), we see that the characteristics increase or decrease steadily over the complete range, so that any intermediate rotor position is uniquely defined by the chopping characteristics of phases C and D. Therefore, high-accuracy ministepping, independent of load, can be obtained by adjusting the currents in phases A and B until the chopping characteristics of phases C and D correspond to the demanded ministep position. For maximum resolution, it is important to monitor both of the unexcited phases, because the slope of each inductance/position characteristic approaches zero at one end of the position range. A similar approach can be adopted for other pairs of excited phases, e.g., with B and C on, chopping is monitored in A and D.

C. Damping of Step Responses

A stepping motor system has inherently poor damping, so the response to a single excitation change (the "single-step response") is highly oscillatory; the rotor undergoes many cycles of oscillation before settling at the required position. In an application involving repeated single steps, such as

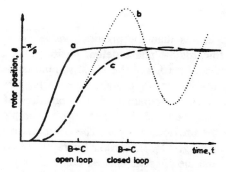

Fig. 10. Two-step responses of a 4-phase motor. (a) Optimized response for the unloaded motor. (b) Response with the same controller settings as in (a), but with load inertia = 4 × rotor inertia. (c) Response for the loaded motor with closed-loop control using continuous position monitoring.

positioning of a print head, the system's long settling time may cause an unacceptable delay in positioning.

One scheme for improving settling time [1] utilizes the poor damping to some advantage by allowing the system to overshoot an intermediate position and arrive at the required position with zero velocity. Suppose, for example, in Fig. 9 the motor is to move from the phase A detent position ($\theta = 0$) to phase C detent ($\theta = \pi/p$). Initially phase A is switched off and phase B switched on, so the motor moves a step to the phase B detent ($\theta = \pi/2p$). Because of the poor damping, however, the rotor overshoots this position by almost 100 percent and arrives in the proximity of the phase C detent with zero velocity. At this instant of maximum overshoot the excitation is changed from phase B to phase C and the rotor settles at the required position (Fig. 10(a)). Precise timing of the excitation change from B to C is needed to ensure that the rotor is at the position of maximum overshoot, giving minimum subsequent settling time.

This damping scheme is perfectly satisfactory for constant load operation, but if the load inertia increases, for example, with no adjustment of the timing, a highly oscillatory response may result (Fig. 10(b)). In systems where load changes can be significant, continuous monitoring of the rotor position is needed to detect the position of maximum overshoot, and hence match excitation timing to load conditions. Referring to Fig. 9(b), the incremental inductance of phase C increases throughout the required rotor movement, so the point of maximum overshoot can be detected by monitoring the rate of current rise following an attempt to switch on phase C and finding the slowest current rise, hence maximum phase C inductance. Using this closed-loop control scheme the excitation change from phase B to phase C always occurs at the position corresponding to minimum settling time (Fig. 10(c)).

A simple circuit to detect the optimum excitation change position is shown in Fig. 11(a). The phase transistor is turned on and off repeatedly for equal times T, which allows a small rise and decay of the phase current. At the end of the turn-on interval the phase current is sampled, by closing the analog switch $S1$. This latest value of current is compared to the previous value contained at the output of the hold circuit. If the latest value is less than the previous value the rotor is still advancing towards the phase detent position, the comparator output remains at zero, and the latest sample is transferred to the hold by briefly closing the analog switch S2. Attempts to turn on the phase continue until the position of maximum overshoot is attained and the latest current sample is slightly larger than the previous sample. The comparator output then changes state (Fig. 11(b)), further operation of S1 and S2 is suspended, and the phase is excited continuously, holding the rotor at the required position.

IV. CONCLUSIONS

This paper describes new techniques [10] for the detection of rotor position in stepping and switched motors by monitoring current waveforms. The techniques are applicable to any motor in which the winding inductances vary with rotor position, and to both chopper and series–resistance drive circuits. Continuous position monitoring, or detection of a well-defined position per step independent of operating speed, can be obtained by careful choice of waveform characteristic on which detection is based. Implementation of the technique is by simple low-cost electronic circuits, which are an attractive alternative to the optical encoders presently used for position detection in switched motors. By incorporating closed-loop control, with continuous monitoring of rotor position, the positional accuracy of a ministep drive is greatly improved and step responses can be optimized for all load conditions.

ACKNOWLEDGMENT

The authors would like to acknowledge the facilities for research provided by the Cambridge University Engineering Department, Cambridge, England, and to thank Richard Danbury, Ward Crawford, and Nigel Doran for their work on the initial investigation into the detection techniques.

REFERENCES

[1] P. P. Acarnley, *Stepping Motors: A Guide to Modern Theory and Practice*. London, England: Peter Peregrinus, 1982.

[2] P. P. Acarnley and P. Gibbons, "Closed-loop control of stepping motors: Prediction and realisation of optimum switching angle," *IEE Proc. B, Electr. Power Appl.*, vol. 129, no. 4, pp. 211–216, 1982.

[3] P. A. Lajoie, "The incremental encoder—An optoelectronic commutator," in *Proc. 2nd Ann. Symp. on Incremental Motion Control Systems and Devices*, Univ. of Illinois, Chicago, IL, 1973.

[4] P. J. Lawrenson, J. M. Stephenson, P. T. Blenkinsop, J. Corda, and N. N. Fulton, "Variable-speed switched reluctance motors," *IEE Proc. B, Electr. Power Appl.*, vol. 127, no. 3, pp. 253–265, 1980.

[5] A. J. C. Bakhuizen, "On self-synchronisation of step motors," in *Proc. Int. Conf. on Stepping Motors and Systems*, Univ. of Leeds, Leeds, UK, 1979.

[6] B. C. Kuo and A. Cassat, "On current detection in variable-reluctance step motors," in *Proc. 6th Ann. Symp. on Incremental Motion Control Systems and Devices*, Univ. of Illinois, Chicago, IL, 1977.

[7] M. Jufer, "Self-synchronization of stepping motors," in *Proc. of the Int. Conf. on Stepping Motors and Systems*, Univ. of Leeds, Leeds, UK, 1976.

[8] A. Hughes, P. J. Lawrenson, and P. P. Acarnley, "Effect of operating mode on torque/speed characteristics of variable-reluctance motors," in *Proc. International Conf. on Stepping Motors and Systems*, Univ. of Leeds, Leeds, UK, 1976.

[9] P. P. Acarnley and A. Hughes, "Predicting the pullout torque/speed curve of variable-reluctance stepping motors," *IEE Proc. B, Electr. Power Appl.*, vol. 128, no. 2, pp. 109–113, 1981.

[10] UK Patent Application 8 307 047, "Stepping motors and drive circuits there for."

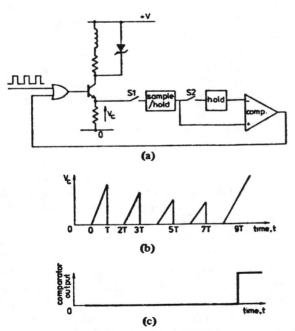

Fig. 11. Detection of maximum overshoot from initial current rise with a constant-voltage drive. (a) Control circuit. (b) Monitoring voltage waveform, showing the reduced initial current gradients. (c) Comparator output, showing detection of maximum overshoot.

A Simple Motion Estimator for VR Motors

W.D. Harris and J.H. Lang

IEEE Industry Applications Society Annual Meeting, Pittsburgh, PA, October 1988

ABSTRACT

This paper presents a simple motion estimator for inverter-driven variable-reluctance motors. The estimator probes unexcited phases with short voltage pulses from the inverter, and evaluates the resulting currents to measure the phase inductances. From these inductances, instantaneous motor position is estimated. Individual position estimates are optionally combined by a state observer to produce smoothed position and velocity estimates. Next, the secondary phenomena of eddy currents in the motor laminations, inverter switching noise, magnetic coupling between motor phases, and quantization introduced by digital implementation are all examined for their effects on estimator performance. Each phenomenon is addressed by a modification of the estimator. Finally, the estimator is evaluated experimentally using an inverter-driven three-phase motor having six stator poles and four rotor poles. The estimator is implemented digitally with an Intel 8031 microcomputer and little extra hardware. Further, it is embedded into the closed-loop control of the inverter and motor. The joint operation of the estimator and controller is quite robust. The closed-loop system maintains stability and proper control when started from rest or when subjected to abrupt load changes. The position accuracy of the estimator itself is measured as approximately ±1% of an electrical cycle over the entire velocity range of the motor.

(1)   INTRODUCTION

Traditionally, a motor is thought of as an actuator. However, it can often be a very good sensor of the motion which it actuates. This is possible when its voltages and currents possess sufficient information to determine its position and velocity. Generally speaking, to use this information requires a motion estimator which accepts measured voltages and currents as inputs, and produces position and velocity estimates as outputs. In recent years, a variety of motion estimators have been studied for a variety of motors. For variable-reluctance motors in particular, see [5,1] and the references cited therein.

This paper presents a simple motion estimator for inverter-driven variable-reluctance motors. The estimator probes unexcited phases with short voltage pulses from the inverter, and evaluates the resulting currents to measure the phase inductances. From these inductances, instantaneous motor position is estimated. Individual position estimates are optionally combined by a state observer to produce smoothed position and velocity estimates. Similar motion estimators have been suggested elsewhere, for example in [1]. A main contribution of this paper then is to examine several secondary phenomena which complicate the implementation, and limit the

accuracy and bandwidth, of the estimator. These phenomena include eddy currents in the motor laminations, inverter switching noise, magnetic coupling between motor phases, and quantization introduced by digital implementation. Each phenomenon is examined for its effects on motion estimation, and is addressed by a modification of the estimator.

This paper is organized as follows. Section 2 presents the operation of the estimator. Section 3 examines the secondary phenomena listed above for their effects on motion estimation. Modifications of the estimator are suggested to minimize these effects. Section 4 presents the experimental evaluation of the estimator. Section 5 provides a summary and conclusions.

(2)   ESTIMATOR OPERATION

Estimator operation is presented with reference to Figure 1. During normal operation, the current in a variable-reluctance motor phase is driven to zero for some nonzero period of time. During that period, the phase inductance can be simply measured to yield information from which motor position can be estimated. To do so, the inverter is used to probe the phase by alternately switching it on and off, thereby generating current pulses. The shapes of the current pulses are functions of phase inductance and hence motor position as shown in the figure. In particular, the peak of a current pulse, i_P, is

$$i_P = \frac{VT}{L(\theta_P)} \tag{1}$$

where V is the inverter voltage, T is the period for which the phase is switched on, L is the phase inductance and θ_P is the motor position at the time of the peak. The assumptions behind (1) are that current pulses start at zero, that the voltage across the phase resistance is negligable, that the phase is magnetically independent and linear, and that eddy currents in the motor laminations are negligable. Note that T can be made short enough so that the voltage across the phase resistance is negligable. Small current pulses also limit magnetic saturation and undesirable torque. Note too that if the phase is independent, its switching can be timed so that each current pulse does start at zero. From (1), θ_P can be determined from measurement of i_P, and knowledge of V, T and L. In general, V, T and L are not known exactly, and i_P is not measured exactly. Thus, the inversion of (1) does not actually yield θ_P. Rather, it yields the estimate Θ given by

$$\Theta = \theta_P + \Psi \tag{2}$$

where Ψ results from measurement error, modeling error and numerical quantization resulting from digital implementation.

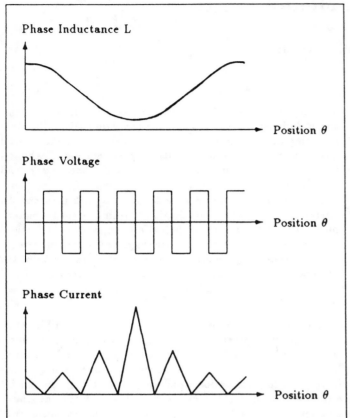

Phase Inductance L

Position θ

Phase Voltage

Position θ

Phase Current

Position θ

Figure 1: Phase inductance, voltage and current during probing of the phase.

There are many methods by which individual position estimates can be combined to yield smoothed estimates of position and velocity. Even if smoothing is not required, individual estimates must be combined in some manner because L is a double-valued function of position over one electrical cycle. Therefore, position can not be uniquely determined within an electrical cycle from a single inductance measurement. Simultaneous measurements from two different phases, however, are always sufficient to uniquely determine position. For example, by probing all phases prior to start-up, the initial position of the motor can be determined quite accurately.

Perhaps the simplest method of combining individual position estimates is through threshold evaluation of i_P. Consider, for example, motive commutation given the phase inductance of Figure 1; the treatment of regenerative commutation is similar. Independent of velocity, motive torque is produced by the phase as its inductance increases in time. Allowing for excitation advances to establish and extinguish its currents, the phase is usually switched on somewhat before its minimum inductance and switched off somewhat before its maximum inductance. Thus, prior to switching the phase on, its inductance is decreasing and successive i_P are increasing; this relation can be seen in Figure 1. To commutate the phase,

successive measurements of i_P are compared to a threshold. When i_P exceeds the threshold, the phase is switched on, and the phase previously switched on is switched off. Probing commences with the next phase in the commutation sequence. In this way, commutation is combined with the position estimator, and position itself is not explicitly estimated. Torque control can be easily included by varying the threshold or by chopping the phase current during the time the phase is switched on. Note that individual position estimates are actually combined because commutation occurs as i_P crosses the threshold from below. Knowledge of the direction of threshold crossing comes from the previous i_P.

The simple motion estimator described above is somewhat limited because it does not explicitly produce estimates of position and velocity. Further, it uses past measurements to estimate present position only in a minor way, and may therefore react seriously to the noise in Ψ; there is no smoothing. A state observer is a relatively simple means of addressing these objections [4]. Moreover, state observers provide other benefits as described below.

To develop a state observer compatible with digital implementation, model the mechanical dynamics of the motor as

$$\frac{d}{dt}\begin{bmatrix} \theta(t) \\ \omega(t) \end{bmatrix} = \begin{bmatrix} 0 & 1 \\ 0 & -\frac{B}{J} \end{bmatrix}\begin{bmatrix} \theta(t) \\ \omega(t) \end{bmatrix} + \begin{bmatrix} 0 \\ \frac{1}{J}(\tau_M(t) + \tau_U(t)) \end{bmatrix} \quad (3)$$

where t is time, ω is motor velocity, τ_M is modeled torque, τ_U is unmodeled torque, and B and J are the combined motor and load viscous damping and inertia, respectively. Motor torque and some load torques might, for example, be included in τ_M while other load torques might be included in τ_U. Next, temporally discretize (3) to yield

$$\begin{bmatrix} \theta(t_{n+1}) \\ \omega(t_{n+1}) \end{bmatrix} = \Phi(\Delta_n)\begin{bmatrix} \theta(t_n) \\ \omega(t_n) \end{bmatrix} \\ + \Gamma_M(t_{n+1}, t_n) + \Gamma_U(t_{n+1}, t_n) \quad (4)$$

$$\Phi(\Delta_n) \equiv \begin{bmatrix} 1 & \frac{J}{B}(1 - e^{-B\Delta_n/J}) \\ 0 & e^{-B\Delta_n/J} \end{bmatrix} \qquad \Delta_n \equiv t_{n+1} - t_n$$

$$\Gamma_{M,U} \equiv \int_{t_n}^{t_{n+1}} \begin{bmatrix} \frac{1}{B}(1 - e^{-B(t_{n+1}-\sigma)/J}) \\ \frac{1}{J}e^{-B(t_{n+1}-\sigma)/J} \end{bmatrix} \tau_{M,U}(\sigma)\, d\sigma$$

where t_n is the nth sampling time. Given (4), an appropriate state observer is

$$\begin{bmatrix} \tilde{\theta}(t_{n+1}) \\ \tilde{\omega}(t_{n+1}) \end{bmatrix} = \Phi(\Delta_n)\begin{bmatrix} \tilde{\theta}(t_n) \\ \tilde{\omega}(t_n) \end{bmatrix} + \Gamma_M(t_{n+1}, t_n) \\ + \begin{bmatrix} G_\theta(\Delta_n) \\ G_\omega(\Delta_n) \end{bmatrix}(\tilde{\theta}(t_n) - \Theta_n) \quad (5)$$

where a tilde indicates an estimated state, G_θ and G_ω are gains which might depend on Δ_n, and Θ_n is from (1) and (2) at t_n with $\tilde{\theta}(t_n)$ used to uniquely invert (1). The implementation of (5) should be simple when lookup tables are used to compute Φ and Γ_M, which is practical at least when the number of allowable Δ_n is limited, and when τ_M is reasonably constant over a sampling interval. A lookup table can also be used to invert (1). Finally, note that (5) is a predictor because Θ_n is used to produce an estimate of $\theta(t_{n+1})$. This need not be the case, but it is convenient because that part of the interval $t_n < t < t_{n+1}$ not used for estimation can then be used to calculate a control for t_{n+1} based on $\tilde{\theta}(t_{n+1})$, which is the best available estimate of $\theta(t_{n+1})$.

In addition to performing its primary function of smoothing sequential position and velocity estimates, Equation (5) can be rearranged and solved for the interval Δ_n after which a particular position is expected to occur. This interval can be timed using a microcomputer interrupt to trigger a control action such as commutation. In this way, interpolation between sampling times is possible using (5).

Observer performance is examined through its estimation errors. By combining (5), (4) and (2), with the understanding that θ_P at t_n is $\theta(t_n)$, the estimation errors are found to propogate according to

$$\begin{bmatrix} \bar{\theta}(t_{n+1}) \\ \bar{\omega}(t_{n+1}) \end{bmatrix} = \begin{bmatrix} \Phi(\Delta_n) + \begin{bmatrix} G_\theta(\Delta_n) & 0 \\ G_\omega(\Delta_n) & 0 \end{bmatrix} \end{bmatrix} \begin{bmatrix} \bar{\theta}(t_n) \\ \bar{\omega}(t_n) \end{bmatrix}$$
$$+ \Gamma_U(t_{n+1}, t_n) - G_\theta(\Delta_n)\Psi_n \qquad (6)$$

where an overbar indicates an estimation error, defined as the estimated state minus the actual state, and Ψ_n is from (2) at t_n. The gains G_θ and G_ω should be chosen so that (6) is stable. Beyond that, they could be chosen to minimize the estimation errors in some sense [3], or to establish a specified observer bandwidth [3,4].

Finally, it should be noted that phase probing need not be restricted to periods during which phases are otherwise unexcited. However, even in the absence of phase resistance, eddy currents, magnetic dependence on other phases and magnetic nonlinearities, (1) becomes

$$L(\theta_P)i_P - L(\theta_I)i_I = VT \qquad (7)$$

where the subscript I denotes an initial value at the time the phase is switched on. From (7) alone, it is no longer possible to determine position without an assumption like $\theta_P \approx \theta_I$, which introduces estimation error at high motor speeds. As the current increases to the point where the phase magnetically saturates and the voltage across its resistance is important, (7) becomes even more complex. Nonetheless, it is important to realize that the phase currents and voltages contain information concerning motion at all times, not just at a limited number of probing instants. In this spirit, observers which include the electrical dynamics of the motor offer higher-accuracy position estimation [5], but at the cost of increased microcomputer computing requirements.

(3) SECONDARY PHENOMENA

While the estimator described above works well in principle, its successful implementation requires additional effort. In particular, the secondary phenomena listed above can degrade its performance. These phenomena must often be addressed by modifications of the estimator as discussed below.

(A) Eddy Currents In The Motor Laminations

Eddy currents in the motor laminations can distort the current pulses depicted in Figure 1, thereby complicating inductance measurement. To examine this distortion, consider the model inductor shown in Figure 2. The inductor comprises a magnetically permeabale core with an air gap which is excited through a winding with N turns. The core is constructed with M laminations each having width W, thickness δ, permeability μ and conductivity σ. The mean length around the core laminations in the direction of the magnetic field is X and the air gap separation is G. In a simple way, this inductor

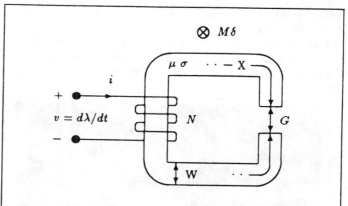

Figure 2: Inductor used for modeling the effects of eddy currents on the current pulses.

represents the motor laminations and air gaps as seen by the phase being probed.

Let the magnetic field outside and between the laminations be H. Given H, the magnetic field inside a lamination is calculated from the magnetoquasistatic Maxwell's Equations [6,2]. This magnetic field is then multiplied by μ, integrated over the cross section of the lamination, summed over all M laminations and summed over all N winding turns to yield the magnetic flux λ linked by the winding. The magnetic field outside the laminations is neglected when determining λ. Assuming that the magnetic field inside all laminations is initially zero, this results in

$$\lambda(s) = \mu M N \delta W \, F(s) \, H(s) \qquad (8)$$

$$F(s) \equiv \left[1 - \sum_{Odd\ n} \frac{8}{n^2\pi^2} \frac{\tau_n s}{\tau_n s + 1} \right] \qquad \tau_n \equiv \frac{\mu\sigma\delta^2}{n^2\pi^2}$$

389

where (8) is expressed in the frequency domain using the single-sided Laplace Transformation with s as the complex frequency. From (8), the voltage v at the terminals of the winding is

$$v(s) = \mu M N \delta W \; s F(s) \; H(s) \qquad (9)$$

where the winding resistance has been neglected. Next, the current through the winding is determined by integrating Ampere's Law from the magnetoquasistatic Maxwell's Equations using a closed contour that follows the magnetic field between two laminations and then across the air gap. Only the winding current passes through the surface spanning such a contour. This results in

$$N i(s) = X H(s) + \frac{G}{\mu_o M N \delta W} \lambda(s) \qquad (10)$$

which, when combined with (8) and (9), yields

$$\frac{i(s)}{v(s)} = \frac{1}{M N^2 \delta W s} \left[\frac{G}{\mu_o} + \frac{X}{\mu F(s)} \right] \qquad (11)$$

as the transfer function of the inductor from its applied voltage to the resulting current.

Equation (11) can now be used to examine the effects of eddy currents on the current pulses observed during probing. To do so, consider the low-frequency approximation to (11)

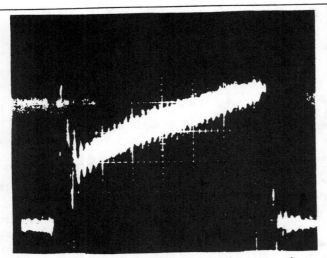

Figure 3: The current pulse response of a motor phase to an applied voltage step. The vertical and horizontal scales are 0.04 A and 10 μs per division, respectively.

in which only the first term in the summation within $F(s)$ is retained. Further, let v be a step to the inverter voltage V. The inverse Laplace Transformation of (11) then yields

$$i(t) = \frac{Vt}{M N^2 \delta W} \left(\frac{G}{\mu_o} + \frac{X}{\mu} \right)$$
$$+ \frac{8 X \tau_1 V}{M N^2 \mu \delta W} \left(1 - e^{-t/((1 - 8/\pi^2)\tau_1)} \right) \qquad (12)$$

as the resulting current. This current contains two terms. The first term is the current-ramp response of the inductor to an applied voltage step in the absence of eddy currents. The second term corresponds to the effects of eddy currents and decays exponentially with the time constant $(1 - 8/\pi^2)\tau_1$ to a current-step response.

An experimental motor, described further in Section 4, is used to verify (12). In the terms of Figure 2, it is modeled roughly by $X = 0.35$ m, $G = 5.1 \cdot 10^{-4}$ m, $W = 0.03$ m, $M\delta = 0.089$ m, $\delta = 4.6 \cdot 10^{-4}$ m, $N = 100$, $\mu = 1000 \, \mu_o$ and $\sigma = 2 \cdot 10^6$ S. Note that the low value of μ is consistent with the low magnetic flux densities in the laminations during probing; at higher magnetic flux densities, $\mu = 6500 \, \mu_o$ is more appropriate. With $V = 35$ V, and the rotor poles aligned with the corresponding stator poles, the phase current pulse is as shown in Figure 3. Note that the inverter is such that when a phase is switched off, current measurements for that phase are disabled, hence the phase current appears to fall immediately to zero; in reality it does not. Ignoring, the inverter switching noise, the current pulse in Figure 3 is well predicted by (12). Using the parameters listed above, the exponential decay time is calculated to be 10 μs, the current step is calculated to be 0.16 A, and the zero-frequency inductance is calculated to be 0.039 H. These calculations certainly agree with Figure 3 to the extent that Figure 2 actually models the experimental motor. This is true for all motor positions and velocities.

As demonstrated by Figure 3, the application of (1) can result in a significant inductance measurement error in the presence of eddy currents. One way to address the eddy currents is to wait for their decay, and then make a two-point measurement of inductance as in (7). Unfortunately, waiting limits the maximum sampling rate of motor position. Additionally, inverting the two-point measurement requires an assumption like $\Theta_P \approx \Theta_I$, which introduces estimation error at high motor speeds. A more accurate but more complex inversion would involve knowledge of motor velocity. In any case, waiting for the eddy currents to decay is practical in many situations and is done here. For the case of the present experimental motor, probing is effected by switching the phase on for 62 μs and off for 63 μs. The sampling rate is thus 8 kHz. Phase current measurements are taken 15 μs after switching the phase on and just prior to switching it off.

(B) Magnetic Coupling Between Motor Phases

While the magnetic coupling between phases in a variable-reluctance motor rarely affects its operation as a motor, it can significantly affect its operation as a sensor. In particular, the small current pulses in Figure 1 can be distorted by such coupling even if the coupling is also small. There are two common mechanisms by which phases can become magnetically coupled. The first is the traditional mutual inductive coupling which occurs primarily because the windings of neighboring phases occupy the same slots. The second occurs when one phase saturates some region of the motor laminations which is shared by the magnetic flux path of another phase, such as the stator and rotor shells. Present experience indicates that the first mechanism is most significant. This is plausible because a phase usually experiences probing at positions corresponding to near minimum inductance, that is, when its magnetic circuit is dominated by its air gaps.

To examine the effects of mutual inductive coupling on the probing of a phase, consider the evolution of the resulting current pulses. In the absence of phase resistance, eddy currents and magnetic nonlinearities,

$$v_1 = \frac{d}{dt}(L_1(\theta)i_1 + L_{12}(\theta)i_2) \tag{13}$$

where i_1, v_1 and L_1 are the current, voltage and inductance of the probed phase, L_{12} is the mutual inductance between the probed phase and a second phase, and i_2 is the current in the second phase. During probing, (13) yields

$$L_1(\theta_P)i_{1_P} - L_1(\theta_I)i_{1_I} = \\ VT - L_{12}(\theta_P)i_{2_P} + L_{12}(\theta_I)i_{2_I} \tag{14}$$

which reduces to (7) in the absence of mutual inductance, and further to (1) when $i_{1_I} = 0$. From (14) it becomes apparent that even if L_{12} is much smaller than L_1, current pulse evolution can be significantly affected by mutual inductive coupling because i_2 is typically much greater than i_1.

One way to address to the effects of mutual inductive coupling between phases is to determine the positions for which the coupling is tolerable and restrict probing to those positions. This is done here. To determine the allowable probing positions, $L_{12}(\theta)$ is measured, worst-case i_{2_P} and i_{2_I} are assumed as functions of θ, and (14) is examined for the error in determining θ_P. An allowable probing position corresponds to acceptable error. Whenever probing is suspended, estimated position and velocity are used to predict when probing once again becomes allowable. For the experimental motor described below, during motoring operation, the allowable probing positions for a given phase are in the vicinity of those corresponding to minimum inductance. Fortunately, these are precisely the positions for which probing is desired.

(C)   Inverter Switching Noise

Figure 3 shows considerable inverter switching noise at the beginning of the current pulse. If current measurement were not disabled by the inverter when the phase is switched off, similar noise would appear after the peak of the current pulse.

Similar noise also appears in the current pulse when other phases are switched on and off. Thus, to maintain measurement accuracy, such inverter switching noise must be avoided at times when the phase currents are to be measured for probing. The same is true when excited phase currents are to be measured during chopping. Thus, probing of unexcited phases and chopping of excited phases must be synchronized to guarantee that no phase is switched on or off when the current in any phase is measured. If inverter control and current measurement are supervised by the same microcomputer, the required synchronization is a simple matter. This is done here.

(D)   Quantization

When the motion estimator is implemented with a microcomputer, the microcomputer and its peripherals introduce numerical quantization errors during the inversion of (1) or (7), the computation of (5) if implemented and the analog-to-digital conversion of a current pulse measurement. The effects of these errors can be included in (2) and (6) if desired. Of these errors, the current measurement conversion error appears to exercise the greatest influence over the complexity and cost of the implementation. Presumably, only one fixed-range converter is desired for conversion during both phase probing and chopping. This converter must therefore match the large dynamic range of fully-excited phase currents and the fine-scale accuracy required during probing. Present experience indicates that an 8-bit analog-to-digital converter is sufficient. If the probing current pulses are limited in amplitude to 10% of the fully-excited phase currents, then approximately 25 levels are available for quantizing the probing current pulses and hence inductance measurements. If the minimum phase inductance is a small fraction of the maximum phase inductance, then the variation in measured inductance, from which position is estimated, is quantized over approximately 4% levels. This results approximately in a 2% position quantization of each electrical cycle, or a ±1% position estimation error, since the inductance variation is traversed twice during an electrical cycle. The actual estimation accuracy is worse in the vicinity of minimum and maximum inductance, and better in between. Further, this accuracy is degraded as the minimum inductance becomes a significant fraction of the the maximum inductance. Nonetheless, the preceding calculation indicates that respectable accuracy is achievable from (1) even with simple hardware.

(4)   EXPERIMENTAL EVALUATION

The motion estimator is evaluated experimentally using a three-phase variable-reluctance motor with four rotor poles and six stator poles. Each pole subtends a 40° arc. Each phase is wound series-aiding with 50 turns per pole on two opposite stator poles. The motor outer diameter, rotor diameter, stack length and air gap separation are 0.15 m, 0.084 m, 0.088 m and $2.5 \cdot 10^{-4}$ m, respectively. The dependence of flux linkage on current and position is shown in Figure 4 for

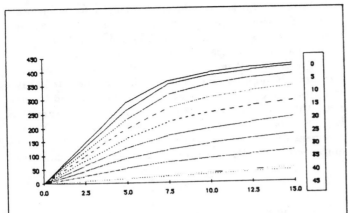

Figure 4: Dependence of flux linkage on current and position for one phase of the experimental motor. The vertical axis is flux linkage (mWb), the horizontal axis is current (A) and the family parameter is position (mechanical degrees).

a phase which has its poles aligned with those on the rotor at $\theta = 0°$. During motoring, the maximum phase current is typically 15 A, which produces a torque of approximately 6.8 Nm. All probing currents are less than 1 A, which results in torques well below 1 Nm.

The simple motion estimator is implemented with an Intel 8031 8-bit microcomputer and an Analog Devices 7828 8-bit analog-to-digital converter. The observer is not implemented. Rather, the motion estimator is combined with threshold detection commutation as described above. Further, motor velocity is controlled closed-loop through the chopping level of the excited phase currents. Probing and chopping are synchronized to avoid the effects of inverter switching noise. Probing occurs at an 8kHz rate, which allows sufficient time for eddy currents in the motor laminations to decay. Probing is limited roughly to the 15° prior to commutation near minimum phase inductance to avoid the effects of mutual inductive coupling between phases. This distance is half that between commutation positions, so, when commutation occurs, half the time since the last commutation position is used to predict when probing next becomes allowable.

Figure 5 shows the operation of the simple motion estimator combined with threshold detection commutation as described above. For this experiment, V is 35 V and the motor speed is 1500 rpm. Note again that phase current measurement is disabled by the inverter when a phase is switched off. Initially, the bottom phase current is excited and chopped. Approximately 15° before commutation to the top phase, probing of the top phase commences. The current pulse peaks rise as the minimum inductance position is approached. Suitably near this position, as determined by the commutation threshold, the top phase is switched on and the bottom phase is switched off as described above. This process repeats with the next phase.

392

The operation of the motion estimator with commutation is quite robust. For all speeds from standstill to 6000 rpm, commutation consistently occurs within ±1°, or ±1.1% of an

Figure 5: Currents for two phases during motor operation by a controller based on the simplest motion estimator. The vertical and horizontal scales are 1 A and 1 ms per division, respectively.

electrical cycle, of the desired position independent of load or load changes. This is consistent with the discussion of quantization presented above. On very rare occasions, the commutation position is missed. For this reason a timer is set to force commutation when it is predicted that the commutation position is significantly passed. Finally, preliminary experience with an observer to smooth the position estimates indicates that nearly an order of magnitude improvement in commutation accuracy is possible with the use of an observer.

(5)   SUMMARY AND CONCLUSIONS

This paper presented a simple motion estimator for inverter-driven variable-reluctance motors. The estimator probed unexcited phases with short voltage pulses from the inverter, and evaluated the resulting currents to measure the phase inductances. From these inductances, instantaneous motor position was estimated. An optional companion state observer was also developed to produce smoothed position and velocity estimates if desired.

Several secondary phenomena were identified as having an important influence on the implementation and performance of the motion estimator. The phenomena were eddy currents in the motor laminations, inverter switching noise, magnetic coupling between motor phases, and quantization introduced by digital implementation. Each phenomenon was examined for its effects on estimator performance, and each was addressed by a modification of the estimator or its implementation. Eddy currents were addressed by allowing them to decay and then making two-point current measurements. This limited estimator bandwidth. Inverter switching noise was addressed by preventing the switching of any phase when the current in any phase was to be measured. The easiest implementation of this prevention required the use of one microcomputer for both estimation and inverter control. The magnetic coupling between phases was addressed by suspending phase probing at positions where the effects of this coupling were significant. Motion prediction was used to determine when probing was again allowable. Quantization was addressed by using sufficiently accurate digital hardware.

The motion estimator, without the state observer, was evaluated experimentally using an inverter-driven three-phase motor having six stator poles and four rotor poles. It was implemented with an 8-bit Intel 8031 microcomputer and little extra hardware. Further, it was embedded in the closed-loop control of the inverter and motor. The joint operation of the estimator and controller was quite robust. The closed-loop system maintained stability and proper control when started from rest or when subjected to abrupt load changes. The position accuracy of the estimator itself was measured as approximately ±1% of an electrical cycle over the entire velocity range of the motor. In summary, the estimator worked well.

Finally, preliminary experimental investigation of the companion state observer indicates that it can improve the accuracy of the simple motion estimator by nearly an order of magnitude. Further, the use of the observer permits the simple motion estimator to run successfully at much higher speeds where there is time for only one or two current pulses per phase per electrical cycle. In this case, the additional use of (5) for interpolation between position sampling times becomes necessary.

ACKNOWLEDGEMENTS

This paper is based primarily on the S. M. Thesis of the first author [2]. That thesis was supported under research agreements with the Copeland Corporation of Sidney, OH and the Emerson Electric Company of Hazelwood, MO.

REFERENCES

[1] P. P. Acarnley, R. J. Hill and C. W. Hooper, "Detection of rotor position in stepping and switched motors by monitoring of current waveforms", *I.E.E.E. Transactions on Industrial Electronics*, 32, 215–222, August 1985.

[2] W. D. Harris, *Practical Indirect Position Sensing for a Variable Reluctance Motor*, S. M. Thesis, Massachusetts Institute of Technology, Cambridge, MA, May 1987.

[3] H. Kwakernaak and R. Sivan, *Linear Optimal Control Systems*, John Wiley and Sons, 1972.

[4] D. G. Luenberger, *Introduction to Dynamic Systems*, John Wiley and Sons, 1979.

[5] A. Lumsdaine, J. H. Lang and M. J. Balas, "A state observer for variable reluctance motors", *Proceedings of the Incremental Motion and Control Systems Symposium*, 267–273, Urbana-Champaign, IL, June 1986.

[6] H. H. Woodson and J. R. Melcher, *Electromechanical Dynamics*, John Wiley and Sons, 1968.

A State Observer for Variable Reluctance Motors

A.H. Lumsdaine, J.H. Lang and M.J. Balas

15th Incremental Motion Control Systems Symposium, University of Illinois, Urbana-Champaign, Illinois, June 1986, pp. 267-273

ABSTRACT

A state observer driven by measurements of phase voltages and currents is developed for variable reluctance motors. The exponential stability of the observer error in a neighborhood of the origin is proven for the case of constant rotor velocity. For this case, experiments are provided to demonstrate the global operation of the observer. In these experiments, the observer often estimates rotor position to better than 1 part in 50,000 of a revolution.

INTRODUCTION

The variable reluctance motor (VRM) is found in applications which range from incremental position actuation to traction. Figure 1 shows the cross section of a typical three phase VRM which illustrates the essential characteristics of any VRM. The motor consists of steel stator and rotor laminations, and copper phase windings. Each phase is wound with alternating magnetic polarity on symmetrically located stator poles. Due to the symmetry of the phases, there is negligible mutual inductance between them. The excitation of a phase magnetizes both the stator and the rotor. This produces a torque causing the rotor to align its poles with those excited on the stator. Since the torque is attractive, it is independent of the sign of the phase current. Sequential phase excitation causes rotor motion which synchronously aligns the rotor poles with those excited on the stator.

To obtain high performance from a VRM, its phases must be excited in synchronism with the position of its rotor. A shaft encoder or resolver typically provides the rotor position information necessary for this closed loop excitation. In some applications these sensors are undesireable for reasons of cost, size, weight or reliability. In contrast, this paper examines an alternative to direct rotor position measurement, namely the estimation of rotor position using an observer driven by measurements of phase voltages and currents.

Rotor position estimators driven by measurements of phase voltages and currents were previously reported for the VRM [1-4] and various permanent magnet motors [5-8]. These estimators primarily used zero crossing and peak detection to confirm coarse rotor motion rather than provide continuous position estimation. In contrast, this paper develops and demonstrates a continuous rotor position estimator for the VRM from a state observation viewpoint.

OBSERVER THEORY

Observer theory is well developed for linear systems [9,10] and partially developed for nonlinear systems [11-14]. However, the use of observers with electrical machines is rare despite the advantages they offer. Notable exceptions are concerned with the estimation of rotor flux in induction machines [15-17].

Before developing an observer for the variable reluctance motor, it is worthwhile to review the theory of observers for linear systems. An observer for a linear system takes the form of a simulation which is dynamically corrected according to errors between the simulated and measured outputs of the system. The states of the observer become the estimates of the states of the system. To illustrate, consider the linear system of

$$x' = A x + B u \qquad (1)$$

$$y = C x \qquad (2)$$

where, u is the known input vector, x is the state vector, y is the measured output vector, A, B, and C are coefficient matrices, and ' denotes temporal differentiation. An observer for this system is

$$\tilde{x}' = A \tilde{x} + B u + F (C \tilde{x} - y) \qquad (3)$$

where $\tilde{}$ denotes an estimate. The last term in (3) is the difference between the simulated and measured outputs of the observed system multiplied by the observer gain matrix F.

To evaluate the performance of the observer, consider its error dynamics. Define the observer error by

$$\bar{x} = \tilde{x} - x . \qquad (4)$$

Substitution of (1-3) into the temporal derivative of (4) yields

$$\bar{x}' = (A + FC) \bar{x} . \qquad (5)$$

The inital condition for (5) is the initial estimation error. If F is chosen so that (A+FC) is a stable matrix, then $\bar{x}$ decays to zero as time proceeds, and $\tilde{x}$ converges to x. The rate of convergence is governed by the stability of (A+FC). If (A+FC) is time invariant, then it is stable if and only if its eigenvalues all have negative real parts, and these eigenvalues determine the rate of convergence of the observer. This suggests that arbitrarily fast convergence can be obtained by the proper selection of F. While this is true in theory if certain observability criteria are met [9,10], strong feedback through F also amplifies the measurement disturbances which will certainly be present in y. Therefore, a compromise is usually made during the selection of F. One such compromise results in the Kalman Filter which selects F to yield the minimum mean square state estimation error in the presence of a specific noise model [18].

VRM MODEL

The essential electromechanical operation of a magnetically linear VRM can be modelled by

$$\lambda_n' = - R_n H_n(\theta) \lambda_n + v_n \tag{6}$$

$$\omega' = - \frac{D}{J} \omega - \frac{1}{2J} \sum_n \frac{dH_n(\theta)}{d\theta} \lambda_n^2 \tag{7}$$

$$\theta' = \omega \tag{8}$$

$$i_n = H_n(\theta) \lambda_n . \tag{9}$$

Here, λ_n, v_n, i_n, R_n, and H_n are the flux linkage, voltage, current, resistance and reciprocal inductance of phase n, respectively, and ω, θ, D and J are the velocity, position, damping and inertia of the rotor, respectively [19]. Magnetic saturation and hysteresis are omitted in (6,7,9).

Let the VRM have N phases and M rotor poles. Then, H_n is periodic in rotations of $\theta = 2\pi/M$ and the individual reciprocal inductances are related to each other according to

$$H_n(\theta) = H_1(\theta - \frac{2\pi(n-1)}{N}) \tag{10}$$

where the numbering of the phases varies according to the details of VRM construction. For convenience, two rotational cycles are defined. The angular rotation of $\theta = 2\pi$ is termed a mechanical cycle while the angular rotation of $\theta = 2\pi/M$ is termed an electrical cycle.

The model of (6-9) is condensed using

$$\lambda = [\lambda_1 . . \lambda_N]^* \tag{11}$$

$$i = [i_1 . . i_N]^* \tag{12}$$

$$v = [v_1 . . v_N]^* \tag{13}$$

$$R = \text{Diagonal} [R_1 . . R_N] \tag{14}$$

$$H(\theta) = \text{Diagonal} [H_1(\theta) . . H_N(\theta)] \tag{15}$$

where * denotes algebraic transposition. Further, for simplicity, the case of infinite rotor inertia, $J \to \infty$, is considered. Using (11-15), (6-9) become

$$\lambda' = - R H(\theta) \lambda + v \tag{16}$$

$$\omega' = 0 \tag{17}$$

$$\theta' = \omega \tag{18}$$

$$i = H(\theta) \lambda \tag{19}$$

as $J \to \infty$. The assumption of an infinite rotor inertia reduces the VRM to constant velocity operation, as shown in (17). This assumption is of practical importance since an observer can often be designed to converge much faster than VRM velocity can respond to torques.

VRM STATE OBSERVATION

In the preceding VRM model, v is the known input, i is the measured ouput, and λ, ω and θ are the states. In order to estimate the states, an observer of the form of

$$\tilde{\lambda}' = - R H(\tilde{\theta}) \tilde{\lambda} + v + F_\lambda(\tilde{\theta},\tilde{\lambda}) (\tilde{i} - i) \tag{20}$$

$$\tilde{\omega}' = F_\omega(\tilde{\theta},\tilde{\lambda}) (\tilde{i} - i) \tag{21}$$

$$\tilde{\theta}' = \tilde{\omega} + F_\theta(\tilde{\theta},\tilde{\lambda}) (\tilde{i} - i) \tag{22}$$

$$\tilde{i} = H(\tilde{\theta}) \tilde{\lambda} \tag{23}$$

is developed. Following the theme of (5), this observer combines the VRM model of (16-19) with the additional dynamic correction terms present in (20-22). The correction terms are comparisons of the estimated and measured outputs multiplied by gains which depend on $\tilde{\theta}$ and $\tilde{\lambda}$ to reflect the principal nonlinearities of the VRM. This observer is similar in form to several of those considered in [11-12].

Consider the observer error dynamics. Define the observer errors according to

$$\bar{\lambda} = \tilde{\lambda} - \lambda \tag{24}$$

$$\bar{\omega} = \tilde{\omega} - \omega \tag{25}$$

$$\bar{\theta} = \tilde{\theta} - \theta \tag{26}$$

$$\bar{i} = \tilde{i} - i . \tag{27}$$

It is important to realize that an observer, as well as a shaft encoder or resolver, can not distinguish one basic cycle from another without the aid of an external marker. Consequently, the preceding errors are defined modulo an electrical cycle. Substitution of (16-23) and (27) into the temporal derivatives of (24-26) yields

$$\bar{\lambda}' = R (H(\theta) \lambda - H(\tilde{\theta}) \tilde{\lambda}) + F_\lambda(\tilde{\theta},\tilde{\lambda}) \bar{i} \tag{28}$$

$$\bar{\omega}' = F_\omega(\tilde{\theta},\tilde{\lambda}) \bar{i} \tag{29}$$

$$\bar{\theta}' = \bar{\omega} + F_\theta(\tilde{\theta},\tilde{\lambda}) \bar{i} \tag{30}$$

as the observer error dynamics.

To complete the observer, it is necessary to select F_λ, F_ω and F_θ so that (28-30) are stable. To do so, a simple observer is considered in which

$$F_\lambda(\tilde{\theta},\tilde{\lambda}) = R . \tag{31}$$

In this case, (20) and (28) become

$$\tilde{\lambda}' = - R i + v \tag{32}$$

$$\bar{\lambda}' = 0 , \tag{33}$$

respectively. From (33), it is apparent that the observer of (20-23) with (31) is only a simulator of flux linkage. That is, (16) is simulated without dynamic correction. This is also evident by comparing (16) with (32). This form of observer is often justified by the operation of the VRM and its inverter. Since the torque produced by a VRM is independent of the sign of its currents, a simple unipolar inverter is often used to excite a VRM. A typical inverter for the VRM of Figure 1 is shown in Figure 2. When a switch pair closes, the power supply is connected to the corresponding phase so as to increase the current in that phase. When the switch pair opens, the power supply is reconnected to the phase through a diode pair so as to decrease the current. If the current then reaches zero, the diode pair opens and the current remains zero until the switch pair closes again. Consequently, VRM currents are always nonnegative, and driven to zero for a nonzero duration once per electrical cycle to avoid counter productive torque. In this case, the inverter maintains the current and hence flux at exactly zero until the next electrical cycle. Consequently, (32) can be integrated from the exact initial condition of $\tilde{\lambda} = 0$ when the inverter switches are closed. Thus, in the absence of disturbances and modelling errors, $\bar{\lambda} = 0$.

The gains F_ω and F_θ must still be determined so that (29-30) are stable. To facilitate this, (29-30) are linearized about zero observer error with $\bar{\lambda} = 0$ to yield

$$
\begin{bmatrix} \bar{\theta} \\ \bar{\omega} \end{bmatrix}' = \begin{bmatrix} F_\theta(\theta,\lambda) \frac{dH(\theta)}{d\theta} \lambda & 1 \\ F_\omega(\theta,\lambda) \frac{dH(\theta)}{d\theta} \lambda & 0 \end{bmatrix} \begin{bmatrix} \bar{\theta} \\ \bar{\omega} \end{bmatrix}
$$

$$
= A(\theta,\lambda) [\bar{\theta} \ \bar{\omega}]^* . \tag{34}
$$

The stability of (34) guarantees the stability of the observer error in a neighborhood of the origin. The dynamics of (34) are periodic due to (17-18), the periodicity of H, and the further assumption that the VRM is excited by a periodic control law. Therefore, a stable (34) is exponentially stable because a Floquet transformation can be found which makes it time invariant [9,20].

Since (34) is exponentially stable in a neighborhood of the origin when it is stable, the omission in (21) of the torque terms from (7) will lead to bounded $\bar{\omega}$ and $\bar{\theta}$ for sufficiently large J or small torques. Similarly, sufficiently small errors encountered in the simulation of λ using (32) will lead only to bounded $\bar{\omega}$ and $\bar{\theta}$.

One set of stabilizing F_ω and F_θ is

$$
F_\omega(\theta,\lambda) = - G_\omega \frac{(\frac{dH(\theta)}{d\theta} \lambda)^*}{|\frac{dH(\theta)}{d\theta} \lambda|^2} \tag{35}
$$

$$
F_\theta(\theta,\lambda) = - G_\theta \frac{(\frac{dH(\theta)}{d\theta} \lambda)^*}{|\frac{dH(\theta)}{d\theta} \lambda|^2} \tag{36}
$$

where G_θ and G_ω are positive constants. For these gains, the state matrix of (34) is constant and exponentially stable provided that $(dH/d\theta)\lambda$ never vanishes. That $(dH/d\theta)\lambda$ must never vanish places an operating constraint on λ which is almost certainly met.

Unfortunately, the gains of (35-36) can exhibit very small denominators in practice which amplifies disturbances and modelling errors. While they do work, their experimental performance is poor, and an alternative is sought. A suitable alternative replaces the denominators in (35-36) by unity, and, for additional simplicity, λ^* by [1 . . 1]. Then,

$$
F_\omega(\theta,\lambda) = - G_\omega [1 . . 1] \frac{dH(\theta)^*}{d\theta} \tag{37}
$$

$$
F_\theta(\theta,\lambda) = - G_\theta [1 . . 1] \frac{dH(\theta)^*}{d\theta} . \tag{38}
$$

To guarantee the stability of (34) with (37-38), consider the Lyapunov function V where

$$
V = [\bar{\theta} \ \bar{\omega}] Q(\theta,\lambda) [\bar{\theta} \ \bar{\omega}]^* \tag{39}
$$

$$
Q(\theta,\lambda) A(\theta,\lambda) + A(\theta,\lambda)^* Q(\theta,\lambda) = - I \tag{40}
$$

and I is the identity matrix. With positive G_ω and G_θ in (37-38), A in (34) has eigenvalues with negative real parts. Therefore, the time varying solution Q to (40) is always symmetric and positive definite. Finally, substitution of (40) into the temporal derivative of (39) yields

$$
V' = - (\bar{\theta}^2 + \bar{\omega}^2) + [\bar{\theta}^2 \ \bar{\omega}^2] Q'(\theta,\lambda) [\bar{\theta}^2 \ \bar{\omega}^2]^* \tag{41}
$$

which is negative definite if G_ω and G_θ are sufficiently large so that Q' has less than unity magnitude. In this case, (34) is stable [20]. To determine sufficient conditions for stability, solve (40) with (34) and (37-38) for Q, take the magnitude of its temporal derivative, and bound this magnitude by unity. This results in

$$
G_\omega > |\frac{2 G}{(1-G) S(\theta,\lambda)}| \tag{42}
$$

$$
G_\theta > |\frac{S'(\theta,\lambda)}{2 G S^2(\theta,\lambda)}| \tag{43}
$$

$$
S(\theta,\lambda) = [1 . . 1] \frac{dH(\theta)^*}{d\theta} \frac{dH(\theta)}{d\theta} \lambda > 0 \tag{44}
$$

$$
0 < G < 1 . \tag{45}
$$

Here, G is an arbitrary constant subject to (45) which trades off the inequalities which bound G_ω and G_θ. From (42-43), the lower bound on G_θ is proportional to ω while the lower bound on G_ω is independent of ω. Thus, the lower bounds can be satisfied even at $\omega = 0$. Both lower bounds are inversely proportional to the amplitude of λ indicating that the VRM must be excited for the observer to function.

EXPERIMENTS

Experiments are provided to illustrate the operation of the observer of (20-23) which uses (31) and (37-38). For these experiments, a VRM with N = 3, M = 141, R = 2.40 Ω, B = 0.126 kg-m^2/s and J = 9.78 kg-m^2 is used. The large J is obtained by adding extra inertia to the rotor. The reciprocal inductance H_1 is well modelled by

$$
H_1(\theta) = \sum_{m=0}^{3} H^{(m)} \cos(m M \theta) \tag{46}
$$

where the reciprocal inductance coefficients $H^{(0)}$ through $H^{(3)}$ are 7.855, -2.980, 0.396 and 0.042 Henries^{-1}, respectively. The inverter power supply is 12.4 V. The control law for all experiments switches a phase on/off when the VRM reaches a position of minimum/maximum inductance for that phase. In response, the VRM rotates with $\omega = 1.62$ rad/s. Finally, the VRM is fitted with a highly accurate shaft resolver from which θ and $\bar{\theta}$ can be measured.

Figures 3-10 present the results of various

experiments. In each figure, all positions and velocities are presented in electrical cycle units as opposed to the mechanical cycle units used elsewhere. Thus, these figures present positions and velocities multiplied by M, which is 141.

The control law initially obtains VRM position from the resolver. Voltages and currents are recorded and processed offline by the observer. Figure 3 shows one set of experimental voltages and currents. Figures 4 and 5 respectively show $\hat{\theta}$ and $\bar{\omega}$ for $G_\theta = 0.00880$ Henry-rad^2/A-s and $G_\omega = 0.402$ Henry-rad^2/A-s^2. Figures 6 and 7 respectively show $\hat{\theta}$ and $\bar{\omega}$ for $G_\theta = 0.0166$ Henry-rad^2/A-s and $G_\omega = 0.830$ Henry-rad^2/A-s^2. In each experiment, the initial $\hat{\theta}$ is a parameter, and the initial $\bar{\omega}$ is zero. To illustrate the effects of modelling error, and to illustrate the operation of a computationally simple observer, (46) is simplified in the observer to include only a constant and a first harmonic. With the gains corresponding to Figures 4 and 5, the results of Figures 8 and 9 are achieved. Small but noticable estimation errors are present in the figures, and the periodicity of these errors indicates that they are driven by modelling errors.

Finally, the observer of Figures 4 and 5 is operated in real time providing the position information for the control law. In these experiments, $H^{(0)}$ through $H^{(3)}$ are inadvertently set to 14.7, -9.81, 3.76 and -2.09 Henries^{-1}, thereby introducing modelling errors. Figure 10 shows $\hat{\theta}$ and $\bar{\omega}$ after a transient in which $\bar{\omega}$ is cleared. The stability of the closed loop control system is maintained with the use of the observer, and the behavior of the observer in this situation is consistent with Figures 4 and 5 given the presence of large VRM modelling errors.

In all experiments, the rms $\hat{\theta}$ is less than 0.01 $2\pi/M$, which is more than sufficient for the high performance commutation of a VRM. In those experiments with little VRM modelling error, the rms $\hat{\theta}$ is less that 0.0025 $2\pi/M$. Since M = 141 in the experimental VRM, the observer can determine rotor position to within 1 part in 50,000 of a mechanical cycle in the absence of VRM modelling errors.

SUMMARY AND CONCLUSIONS

The development of a simple state observer for a VRM was presented. The underlying theme was the use of observer theory to advance electrical machine systems, and conversely, the use of electrical machine systems to provide novel and interesting applications for observer theory. The observer was based upon the simplifying yet practical assumptions of linear VRM magnetics and constant VRM velocity. The observer was driven only by measurements of phase voltages and currents. The exponential stability of the observer error was proven and demonstrated experimentally. The observer was able to determine rotor position to within 1 part in 50,000 of a revolution in the absence of VRM modelling errors, and was successfully used as the basis for closed loop control. Further, the observer operated successfully in the presence of large VRM modelling errors.

It should be emphasized that the present

observer operates successfully when $\omega = 0$. The only requirement for this is that at least two phases be excited. Two phases are required since the self inductances through which the observer views rotor position are double valued functions of θ per electrical cycle. Consequently, measurements from at least two phases are necessary to resolve position modulo an electrical cycle if the VRM is stationary. In contrast, an observer developed for a permanent magnet motor, for example, would not operate successfully at $\omega = 0$ since it would view position through the mutual inductances between the magnet and the phases. These mutual inductances require time variation to induce a voltage at the terminals of a phase. However, the phases need not be excited for this voltage to be observed.

Several extensions and improvements of the present observer would be worthwhile. First, the observer could be studied in the presence of nonconstant ω and nonlinear VRM magnetics. Second, (31) could be modified so that dynamic correction is present in (20). Third, the observer could be made adaptive so as to follow variations in R, B and J. Fourth, the observer could be modified to work with other motors. In particular, the modification of (31) is an important extension. It is observed experimentally that if (20) with (31) is not reset by the combined action of the VRM control law and inverter at a rate of approximately $RH^{(0)}$, then the present observer performs poorly, primarily because of modelling errors in H and R. This extension has been successfully demonstrated in the laboratory, and presently awaits a formal stability proof and criteria for gain determination.

ACKNOWLEDGEMENTS

The work reported here is based on [21] and was supported by the United States Department of Energy and the Aerospace Corporation of Washington, DC under Aerospace Corporation contract W-0399NV. The Superior Electric Company of Bristol, CT provided the experimental VRM. The Packard Electric Division of the General Motors Corporation of Warren, OH, provided the computer system which supported the offline and real time observer. The first author was supported by a General Motors Fellowship during the course of the work. The third author is on sabbatical leave from the Rensselear Polytechnic Institute of Troy, NY.

REFERENCES

[1] J.R. Frus and B.C. Kuo, 'Closed-loop control of step motors without feedback control', Proceedings of the 5th Incremental Motion Control Systems And Devices Symposium, CC/1-CC/11, Urbana, Illinois, 1976.

[2] B.C. Kuo and A. Cassat, 'On current detection in variable reluctance step motors', Proceedings of the 6th Incremental Motion Control Systems And Devices Symposium, 205-220, Urbana, Illinois, 1977.

[3] A. Pittet and M. Jufer, 'Closed-loop control without encoder of electromagnetic step motors', Proceedings of the 7th Incremental Motion Control Systems And Devices Symposium, 37-44, Urbana, Illinois, 1978.

[4] P.P. Acarnley, R.J. Hill and C.W. Hooper, 'Detection of rotor position in stepping and switched motors by monitoring of current waveforms', IEEE Transactions on Industrial Electronics, 32, 215-222, 1985.

[5] W.C. Lin, B.C. Kuo and U. Goerke, 'Waveform detection of permanent magnet motors: parts I and II', Proceedings of the 8th Incremental Motion Control Systems And Devices Symposium, 227-256, Urbana, Illinois, 1979.

[6] B.C. Kuo and K. Butts, 'Closed loop control of a 3.6° floppy disk drive pm motor by back emf sensing', Proceedings of the 11th Incremental Motion Control Systems And Devices Symposium, Urbana, Illinois, 1983.

[7] T. Higuchi, 'Closed loop control of pm step motors by back emf sensing', Proceedings of the 11th Incremental Motion Control Systems And Devices Symposium, Urbana, Illinois, 1982.

[8] V.D. Hair, 'Direct detection of back emf in permanent-magnet step motors', Proceedings of the 12th Incremental Motion Control Systems and Devices Symposium, 211-219, Urbana, Illinois, 1983.

[9] D.G. Luenberger, 'An introduction to observers', IEEE Transactions on Automatic Control, 16, 596-602, 1971.

[10] D.G. Luenberger, Introduction to Dynamic Systems, John Wiley, 1979.

[11] F.E. Thau, 'Observing the state of nonlinear dynamic systems', International Journal of Control, 17, 471-479, 1973.

[12] S.R. Kou, D.L. Elliot and T.J. Tarn, 'Exponential observers for nonlinear dynamic systems', Information and Control, 29, 204-216, 1975.

[13] D. Bestle and M. Zeitz, 'Canonical form observer for nonlinear time-variable systems', International Journal of Control, 38, 419-431, 1983.

[14] A.J. Krener and W. Respondek, 'Nonlinear observers with linearizable error dynamics', SIAM Journal of Control and Optimization, 23, 197-216, 1985.

[15] A. Bellini, G. Figalli and G. Ulivi, 'Realization of a bilinear observer of the induction machine', Proceedings of the Second International Conference on Electrical Variable Speed Drives, 175-178, London, 1979.

[16] Y. Dote, 'Stabilization of controlled current induction motor drive system via new nonlinear state observer', IEEE Transactions on Industrial Electronics and Control Instrumentation, 27, 77-81, 1980.

[17] G.C. Verghese and S.R. Sanders, 'Observers for faster flux estimation in induction machines', Proceedings of the IEEE Power Electronics Specialists Conference, Toulouse, 1985.

[18] H. Kwakernaak and R. Sivan, Linear Optimal Control Systems, John Wiley, 1972.

[19] H.H. Woodson and J.R. Melcher, Electromechanical Dynamics, Volume 1, John Wiley and Sons, 1968.

[20] R.W. Brocket, Finite Dimensional Linear Systems, John Wiley and Sons, 1970.

[21] A. Lumsdaine, 'Control of a Variable Reluctance Motor Based on State Observation', S.M. Thesis, Massachusetts Institute of Technology, Cambridge, MA, November 1985.

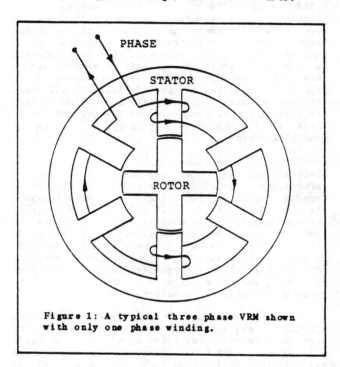

Figure 1: A typical three phase VRM shown with only one phase winding.

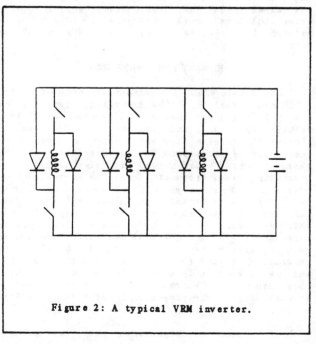

Figure 2: A typical VRM inverter.

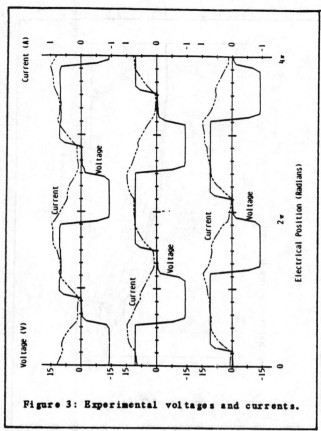

Figure 3: Experimental voltages and currents.

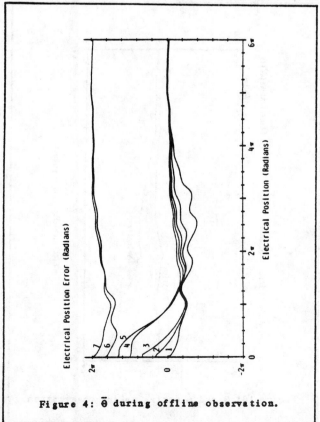

Figure 4: $\bar{\theta}$ during offline observation.

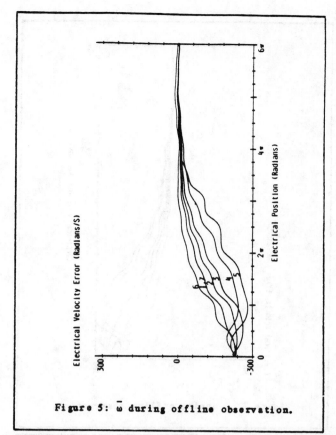

Figure 5: $\bar{\omega}$ during offline observation.

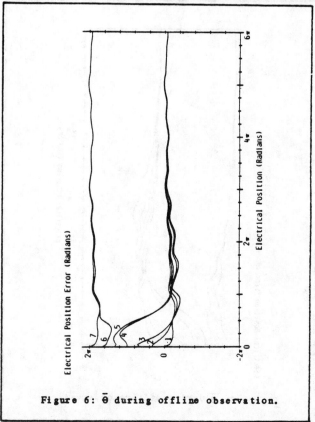

Figure 6: $\bar{\theta}$ during offline observation.

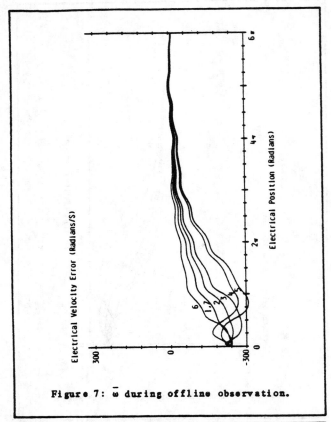

Figure 7: $\bar{\omega}$ during offline observation.

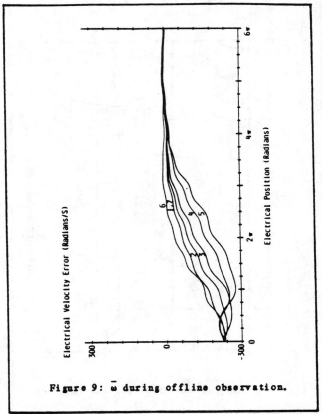

Figure 9: $\bar{\omega}$ during offline observation.

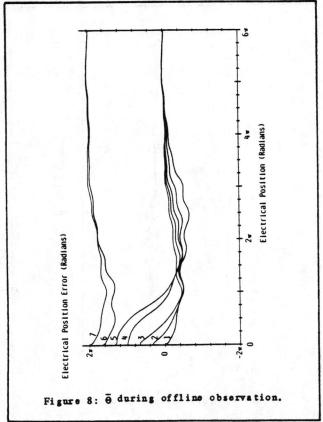

Figure 8: $\bar{\theta}$ during offline observation.

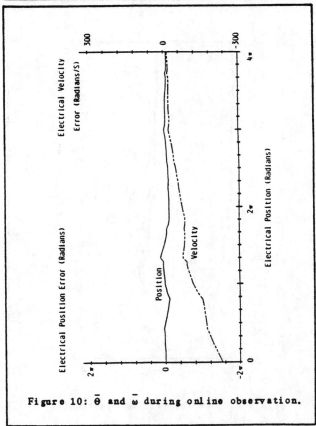

Figure 10: $\bar{\theta}$ and $\bar{\omega}$ during online observation.

Section 3
Applications

3.1 - Applications

A High Performance Variable Reluctance Drive: A New Brushless Servo

J.V. Byrne, J.B. O'Dwyer and M.F. McMullin

PowerConversion International magazine, February 1986, pp. 60-66

Synchronously commutated variable - reluctance machines have a history stretching back 150 years to the first "Electromagnetic Engines". In this decade, because of their inherent robustness and simple unipolar drive requirements, they have again been applied, this time as brushless industrial drives in the Kw range. The system described is distinguished from other "stepper" motors by its power and speed range, and from other high power reluctance drives by its quiet operation and low torque ripple.

This article describes the special adaptations and control strategies that make possible high-performance servo application. A special magnetic design is employed to control torque sensitivity as a function of angle. Current is shaped to vary with angle in a complementary manner. Torque contributions from two phases sum to a constant. The system does not require costly resolvers or delicate encoders.

Long before DC servos were dreamed of, in the period 1835 to 1860, between the discovery of the electromagnet by Oersted and the development of the ring-wound armature by Pacinotti and Gramme, there ruled supreme that dinosaur of electrical machines, the "Electromagnetic Engine". *(Appendix 1)*.

Electromagnetic engines used the pull of sequentially-excited DC electromagnets to achieve continuous torque. A contemporary reference[1] to the first American Patent (Taylor, 1838) states: "A series of electromagnets... are alternately and almost instantaneously magnetized and demagnetized, without any change of polarity whatever taking place, and certain other masses of iron are brought successively under the influence of the (electromagnets), which later are demagnetized as soon as their attractive power ceases to operate with advantage . . . they always act attractively only, or with such a preponderance of attractive forces, as to exercise a uniform moving force".

These "engines" were the earliest variable-reluctance machines, and they ran self-synchronized, like brushless DC motors today. Cam-driven switches doubled as angle sensors and power controllers, doing the familiar work of valves in steam engines (the electromagnet was the cylinder). Some electromagnetic engines had connecting rods and flywheels and looked like steam engines.

The characteristics of these early machines were clearly described by Noad[2] in 1859: he noted that structural problems arose from the large pulsating forces, that output was reduced when mechanical clearances had to be increased, and that contact erosion was severe. Clearly the unlaminated cores were lossy; excessive amounts of magnetic field energy were stored during each working stroke, this energy then being wastefully dissipated in arcing at the commutating switches. With hindsight we can see many reasons for the demise of the electromagnetic engine.

After the advent of the DC machine, 100 years were to elapse before two important developments made the reluctance motor viable again:
(1) The advent of high-power semiconductor switches in the 1960's.

(2) An understanding of the improvements to be had in energy conversion efficiency through exploitation of magnetic saturation. Although it had been shown as early as 1851 by Lord Kelvin that linear, singly-excited devices can at best convert only half the electrical input to mechanical work, the other half being stored in the field (Karapetoff[3] called this 50-50 split which occurs at constant current, "KELVIN'S LAW"). The exploitation of magnetic non-linearity to increase[4] and, ideally, double (5,6) the specific output of reluctance machines is quite recent.

The majority of high-power reluctance motor drives developed in the last 10 years were, like their predecessors, "engine-like" in the fundamentally pulsating nature of the total torque. [7-10]. *Figure 1* shows a typical asymmetric torque-angle relation for one phase of a 10kW motor designed for a variable-speed drive application[10], where, apart from the ability to start against load, tight speed control was not needed near zero speed. At high speeds, such machines are usually voltage-fed and the torque is controlled by phase-advancing the start of conduction. Then, the torque profile over the working stroke becomes progressively more asymmetric and peaky, resembling that of an I.C. engine, and requiring the flywheel effect of the rotating masses to smooth it out.

Thus at first sight, the variable reluctance machine with its engine-like ancestry may seem an unlikely candidate for servo applications, despite the appeal of its mechanical simplicity and unipolar drive requirements. At the start of the research program described in this article, the primary design objective was identified to be: finding a means for controlling torque contributions of individual phases so that they would be free from abrupt transitions and sum to a constant. In controlling the torque-producing forces within these machines it was hoped to also control the source of electromechanical noise and

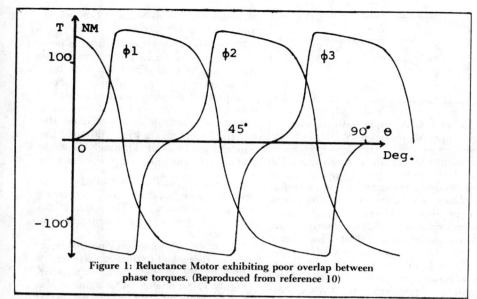

Figure 1: Reluctance Motor exhibiting poor overlap between phase torques. (Reproduced from reference 10)

vibration which has also become associated with them. Initially, the range of available strategies was constrained by the availability of coarse digital shaft position signals only. An associated article[11] describes one initial approach, based on "rectangular torque" target waveforms, with current ramping to smooth the transitions from phase to phase. Another initial approach, with "trapezoidal torque" target waveforms, is briefly referred to in Section 3. For the final approach described here, continuous shaft position information was available and an analog control system was used for current profiling and speed control. An associated article[12] describes the digital implementation and detailed control strategies which were carried forward into production.

Note on interpretation of results: "Static torque-angle (or T-θ) curves" imply those measured with constant winding currents. To a first approximation, their magnitude will be proportional to current and the output torque for a system can be obtained as the sum, for all phases, of the product of phase currents and static characteristics.

THE RELUCTANCE MOTOR AS A SERVO DRIVE

Servodrives are characterized by (a) low inertia (b) high peak torques for acceleration (c) high ratio of torque/volume (d) low torque ripple (e) quiet running.

Traditionally, servomotors have been permanent magnet DC machines with phase controlled SCR, or PWM type transistor drives. Ferrite magnet motors are generally used for less demanding applications, and Samarium Cobalt magnet motors where maximum performance is required. In the past few years there has been a general move to brushless drives, with a consequent improvement in the above performance criteria by a factor of about two.[13]

With the move from brush type to brushless DC machines there has also been an upsurge of interest in finding an alternative to systems using permanent magnets. This is because in many industrial applications the full performance of modern brushless DC servodrives is not used to the full and there is a natural desire to trade unused performance for cost. The material costs of even high energy permanent magnets are typically quite a small fraction of the overall system cost, but their elimination, coupled with the consequent savings in material and labour for securing them to a shaft rotating at high speeds, will always be very attractive to the design engineer who has been given the task of reducing his system cost by 10%.

The induction motor has been studied as a servodrive for several years now. It holds a facination for the servo designer with the above task. However while the "Standard" induction motor can be purchased extraordinarily cheaply in comparison to servomotors of the same rating, it is now becoming generally accepted that the number of modifications necessary to the "standard" motor, and the complexity of the controller required, result in a system which has very little overall cost advantage.

The reluctance motor was chosen by Inland for development as a servodrive because some previous experience in the area suggested that the problem of electromechanical noise could be coped with, and the system offered some unparalleled advantages:
1. Extreme motor simplicity.

2. Good torque/weight ratio.
3. Synchronous type operation is very suitable for precision speed regulation.
4. Simplified drive circuits result from unipolar current requirements.

The development program began in 1981 and has yielded a system exhibiting full servo performance. The fact that this most difficult of applications has been filled by the reluctance motor has far reaching significance for its future use in many other areas.

CHRONICLE OF DEVELOPMENT PROGRAM

From an early stage the control of torque ripple, and consequently electromagnetic noise generation, was identified as being a key factor in the development program. In order to place a controlled torque on the motor shaft it is necessary to carefully program the contributions of several phase windings excited either sequentially or with overlap.

One of the first schemes which was implemented in an effort to obtain this controlled torque is fully described by Byrne and Devitt.[11] In this scheme the magnetic structure was designed to give rectangular, overlapping phase torque contributions and when excited with trapezoidal currents constant shaft torque would result. (Figure 2) The results of this work were very encouraging, particularly the Torque/Inertia and Torque/Weight ratios obtained. The scheme was also successful in obtaining low torque ripple but only at low speeds. At higher speeds there were difficulties in obtaining the desired current waveforms due to effective back emf's within the phase windings.

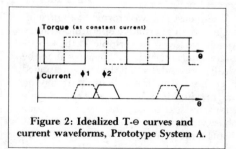

Figure 2: Idealized T-θ curves and current waveforms, Prototype System A.

In another scheme, which will be described fully at a later date, a three phase machine was used in which the magnetic structure was designed to have a trapezoidal torque-angle characteristic. When excited with a rectangular current waveform a constant shaft torque would result. Figure 3a The experimental results verified that the torque waveforms were reasonably good

403

(Figure 3b) although more suggestive of a sinewave than a trapezoid. The resultant

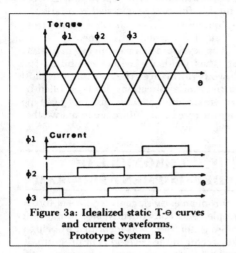

Figure 3a: Idealized static T-⊖ curves and current waveforms, Prototype System B.

shaft torque had a low ripple content with no abrupt changes. The results for this machine complement the results for the machine previously described. The torque/inertia and torque/weight ratios were lower but there was no difficulty in obtaining the desired current waveforms over the full speed range of the machine.

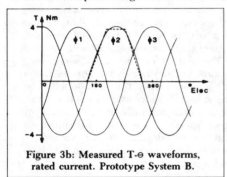

Figure 3b: Measured T-⊖ waveforms, rated current. Prototype System B.

The measured waveforms of *Figure 3b* suggested an alternative operating strategy to achieve low torque ripple. If the torque angle characteristic is considered to be ideally sinusoidal and if the current applied to each winding is a half sinusoid, having its maximum where the torque-angle characteristic peaks, the torque for one phase varies as the square of the sine of the displacement angle. The torque for an adjacent phase, situated 90 electrical degrees away, similarly varies as the square of the cosine of the angle. In virtue of

$$\sin^2 + \cos^2 = 1$$

the resultant torque is angle independent, *(Figure 4)*.

As can be seen from Figure 3b for the "trapezoidal - torque" prototype, the actual torque-angle curves did approach a sinusoidal waveshape. The magnetic design

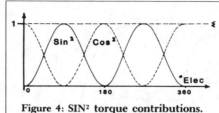

Figure 4: SIN² torque contributions. Operating strategy 'C'.

philosophy of this machine was applied to the four phase prototype machine described by Byrne and Devitt in an effort to obtain a machine with a sine-cosine torque-angle characteristic, high torque/inertia and high torque/weight ratios.

An essential feature of the Sine-Cosine strategy was the availability of absolute position information, to a reasonable degree of accuracy, to allow synthesis of the required current waveforms. While there are such transducers on the market it was considered that none of them fulfilled the requirements of the application: low cost combined with tolerance of harsh environments. A special device was developed for the application and will be the subject of future publications.

DESCRIPTION OF THE SYSTEM

The machine configuration is an 8-pole, 4-phase stator and 6-pole rotor. The measured torque-angle characteristic for the overlapping rotor stator poles is, as intended, sinusoidal. This is achieved by carefully controlling the magnetic

permeance of the phases, as a function of rotor position *(Figure 5)*.

Since only unidirectional currents flow in each phase a simple and robust power circuit can be used *(Figure 6)*. With this circuit there is no possibility of a shoot through fault occuring. Each transistor is operated as a switch in a P.W.M. controller. When the transistor is switched off the current continues to flow through the winding and is diverted through the diode and onto the auxiliary rail. The voltage on this auxiliary rail is controlled by the switching of a linear inductor transferring energy to the main rail. The action of this circuit is described in more detail in reference 12.

The shaft angle sensor provides two linear position dependant triangular wave signals offset by 90 degrees from each other. *(Figure 7)* An analog tacho signal is also available from the transducer, generated by differentiating the appropriate position signal. This gives a low ripple, wide bandwidth velocity signal comparable to the best quality brush type tachos used with servo drives. This device will be described in detail at a later date. The linear signals from the transducer are converted to sinusoids using shaping networks. A conventional speed regulator is used and the output can be regarded as a torque demand signal. In this system the two position information signals, SIN ⊖ elec and COS ⊖ elec, are then multiplied by the torque demand signal using analog multipliers. The outputs from the multipliers and their inverse are rectified to provide the reference

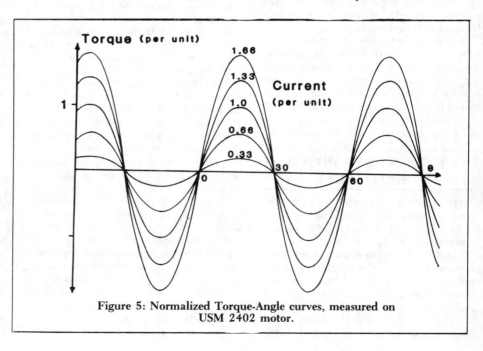

Figure 5: Normalized Torque-Angle curves, measured on USM 2402 motor.

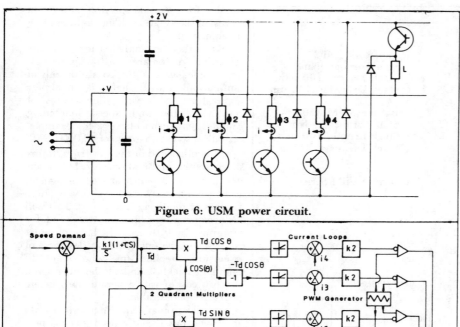

Figure 6: USM power circuit.

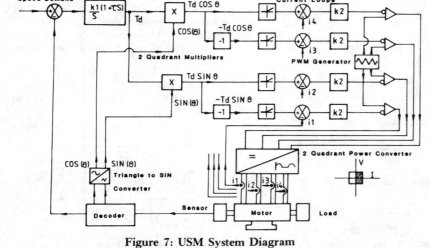

Figure 7: USM System Diagram

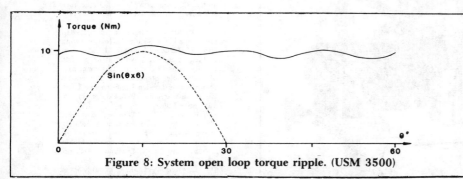

Figure 8: System open loop torque ripple. (USM 3500)

TABLE 1			
	TT2952	USM 2402	BHT 2203
Continuous Rating (Stall) Nm	4.1	4.0	8.5
Rotor Inertia Kg. M²	0.001	0.00065	0.00076
Active Motor Length mm.	104*	100	95
Stator Diameter mm	110	112	112

* Including commutator. 74mm without.

current waveforms for the four phases. Four conventional P.W.M. currents loops, one for each phase, are used to independently control the phase currents. In this case the current is always controlled in two phases to maintain constant torque.

The drive unit carried into production is essentially similar to that described, but is implemented in a digital manner. This permits the use of more sophisticated control techniques than are possible with analog circuits. A full description is to be found in the companion article by Murphy et al[12].

SYSTEM PERFORMANCE

The shaft torque/angle characteristic of the system, operating under current control alone, is shown in *Figure 8*. The ripple torque has a magnitude of 5% (average to peak) and, very significantly, contains no abrupt changes. This means that the effects of any torque ripple can be virtually eliminated under closed velocity loop conditions. A current sinewave is shown for reference

Systems up to 3kW continuous rating have been tested, and run very quietly. They are considered to be comparable with any normal servo from this point of view. This is the first time that such performance has been achieved with variable reluctance machines.

Table 1 gives comparative data for three Inland servodrives. These are:
(1) TT2950, a high performance Samarium Cobalt DC brush motor.
(2) USM2404, a high performance reluctance motor (or Unipolar sysncronous machine).
(3) BHT 2203, a high performance brushless DC motor.

It can be seen that the stack lengths, or active motor volumes, are approximately equal for all three. The Samarium Cobalt brush motor has been regarded as the industry standard for high performance for many years. The brushless DC motor, however, delivers more than twice the continuous torque from the same frame size, with a 24% power inertia rotor. Clearly, this is the drive for applications requiring the very highest performance. The reluctance motor, however, equals the DC motor on torque/volume, and betters it on inertia. It has very clear advantages over the other two systems in terms of simplicity and ruggedness. (For fairness, the commutator has been included in the active length of the brush motor, as this is an overhead not carried by the brushless systems).

Figure 10 shows a system performance curve for the USM 2402 A. Clearly it is free from the typical commutation limits of brush type servomotors and exhibits a good

Top: USM 2402
Stall Torque: 4Nm
Stack Length: 100 mm
Stack Diameter: 112 mm

Bottom: USM 3500
Stall Torque: 10 Nm
Stack Length: 100 mm
Stack Diameter: 165 mm

Figure 9: USM Series.

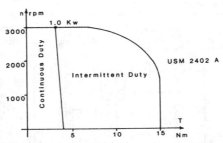

Figure 10: System Performance curve.

APPENDIX 1

The Electromagnetic Engine as described in reference 2

steam engines.

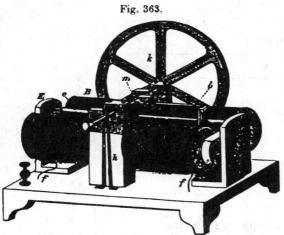

Fig. 363 represents a small working model of an electro-magneto-motive engine constructed by Mr. Bain, with some few improvements by the publishers of this work. On to a stout mahogany board are fixed the brass uprights *E E*; to these are attached the electro-magnets *A B*, covered with stout wire; through the upper part of these uprights, and above the

Fig. 363.

ratio of acceleration torque to continuous rating at every speed.

Figure 9 shows two motors from the Inland USM range.

CONCLUSIONS

The electromagnetic engine has come of age. In fulfilling the performance requirements of a modern servo system it has demonstrated itself worthy of consideration for many applications in the modern world.

Its most attractive features are its simplicity and robustness, making it a possible drive solution for even mass market products like domestic appliances. Beyond that, its syncronous nature makes it very suitable for precision speed control (as in servos) or speed ratio holding (as in printing machinery).

Two other special features of the reluctance motor are worth singling out. The simple passive rotor, consisting only of laminations on a shaft, is ideal for applications with very high speeds of rotation. Some high speed machining spindles and turbo blowers, for example, operate at speeds in excess of 40,000 RPM.

At the other end of the spectrum, rotor-stator tooth combinations can be chosen which effectively realize an electromagnetic gear within the machine, giving high torques at relatively low speeds. This will reduce, or even eliminate, the need for transmission systems, which is of great interest, on the one hand, to designers of precision mechanisms who want to eliminate backlash and, on the other hand, in very cost sensitive applications where gears are undesirable.

The Electromagnetic Engine is an old man who has recently been given a new lease on life. We will hear more of him in the coming years.

Note: Patents are pending on the systems described in this article.

REFERENCES

1. Mechanics Magazine, 1840, No. 109, p. 694.
2. H.M. Noad, "A Manual of Electricity", (Lockwood, London, 1859), pp. 674-684.
3. V. Karapetoff, "Mechanical Forces between Electric Currents and Saturated Magnetic Fields", Trans. AIEE, 46, (1927), pp. 563-569.
4. J. Jarret, "Machines Electriques a Reluctance Variable et a dents Saturees", Tech. Mod., 1976, 2, pp. 78-80.
5. J.V. Byrne & J.G. Lacy, "Electrodynamic System Comprising a Variable Reluctance Machine", British Patent No. 1321110, 1973.
6. J.V. Byrne, "Tangential Forces in Overlapped Pole Geometries Incorporating Ideally Saturable Materials", IEEE Trans. on Magnetics, Mag-8, No. 1, 1972, pp. 2-9.
7. J.V. Byrne and J.G. Lacy, "Characteristics of Saturable Stepper and Reluctance Motors". IEE (London) Conference on Small Electrical Machines, 1976, Conf. Publ. No. 136, pp. 93-96.
8. J.V. Byrne and J.B. O'Dwyer, "Saturable Variable Reluctance Machine Simulation Using Exponential Functions", Proc. of the International Conference on Stepping Motors and Systems, Leeds, England, 1976, pp. 11-16.
9. P.J. Lawrenson et al., "Variable-Speed Switched Reluctance Motors", Proc. IEE, (1980), 127, Pt.B, pp. 253-265.
10. J.V. Byrne and M.F. McMullin, "Design of a Reluctance Motor as a 10kW Spindle Drive", Proc. PCI/Motorcon Conference, Geneva, Switzerland, 1982.
11. J.V. Byrne & F. Devitt, "Design and Performance of a Saturable Variable Reluctance Servo Motor", Motorcon 1985.
12. J.M.D. Murphy, M.G. Egan, P.F. Kenneally, J.V. Lawton, M.F. McMullin, "A High-Performance Variable Reluctance Drive: Achieving Servomotor Control", Motorcon 1985.
13. A.C. Stone and M.G. Buckley, "Ultra High Performance Brushless DC Drive", Proc. Drives/Motors/Controls, Brighton, England, 1984, pp. 86-91.

Design of a Reluctance Motor As a 10kW Spindle Drive

J.V. Byrne and M.F. McMullin

MOTOR-CON Proceedings, Geneva, Switzerland, September 1982, pp. 10-24

ABSTRACT

The design of a saturable reluctance motor to meet a specification for a 10kW spindle is described. Emphasis is placed on achieving a wide constant power range. Trajectories in the ϕ-i plane, derived from the magnetization characteristics, are used to illustrate motor operation. Experimental results are presented showing how shaping of the rotor poles can improve torque smooth ess. An inverter drive is described and good rorelation is shown to exist between experimentally measured and predicted trajectories. A numerical simulation of the system is used to investigate the motor's response to changes in the control variables and likely problem areas for the control system are outlined.

1. INTRODUCTION

In a book published in 1859 Noad[1] made a review of what were then called "Electromagnetic Engines". He attributed the construction of what seems to have been the first variable reluctance motor to Davidson, in 1837. In the following half century reluctance motors underwent considerable development and in his book Noad records that an attempt was even made to propel a railway carriage with such an "engine". However just one year after this was written Pacinotti, in 1860, invented the ring wound d.c. machine which in time was to completely superceed the reluctance motor as a source of mechanical power. One hundred years was to elapse before advances in power semiconductor switching devices made the reluctance motor a viable electric drive again. Initial interest was in their use as positioning devices and much research has gone into the development of "Stepping Motors", these being generally in the fractional HP range. The work described here, which was carried out in University College Dublin between 1975 and 1979 was amoung the first attempts to apply a reluctance motor as a conventional industrial drive in the kW range*.

In recent years there has been a clear trend in the machine tool industry towards replacing conventional commutator type d.c.

*The work was carried out under the sponsorship of Lucas Industries.

spindle and feed drives by "brushless" motors of one type or another, in the interest of reduced maintainance and increased reliability. As in the rest of the electric drives industry, most research has concentrated on the development of inverters for induction and syncronous motors, generally using conventional polyphase A.C. machines. The design problem is then to synthesise sinusoidal current waveforms of adequate smoothness to ensure ripple free low speed operation. Work on less conventional variable speed drives, such as the reluctance motor, has had the advantage of designing the motor and drive as a matched pair, with minimum overall complexity.

At this point it is worthwhile to review the general characteristics of reluctance motors:

(1) The motor is very simple, and potentially cheap to construct.

(2) The rotor contains no current carrying conductors and hence the motor is "brushless".

(3) The simple rotor construction makes very high speed operation, and hence high power output, possible.

(4) Since all electrical windings are in the stator, their cooling is very efficient.

(5) The windings need only carry unidirectional currents, and hence the number of power semiconductor switches in the drive in relativly low.

(6) The specific output of modern machines designed to operate in a magnetically saturable mode is high.

(7) It will be shown that a wide speed range at constant power can be covered at constant voltage through the mechanism of advancing the winding firing angle.

(8) Rotation is inherently incremental, the number of steps per revolution being determined by the number of stator and rotor poles.

The last three characteristics listed carry behind them what could be considered the disadvantages of reluctance motor systems:

(1) Highly saturated iron paths imply working with relativly small airgaps, and hense tight mechanical tolerances. Also the very non linear nature of the motor makes them difficult to analise.

(2) The sensitivity of the motor to change in firing angle is
very non linear and the control stratagies required are complex,
if this mechanism is to be used.

(3) Incremental motion can be an advantage if it is exploited to
give a simple open loop positioning system, but a problem when
smooth rotation down to crawl speeds, quietness, and lack of
vibration are requirments.

However, in spite of these problems, the other major advantages
of reluctance motor systems made, and continue to make, their
development as high power variable speed drives very attractive.

The principal target specifications for the system in question
where: 5kW continuous rating and 10kW for 10 minutes in any hour.
Constant torque range 80 - 8000 RPM reversible. Constant power
range 800 - 7000 RPM reversible. Class F insulation, maximum
ambient temperature 50°C.

2. A MODEL OF A SATURABLE RELUCTANCE MOTOR

The most complete representation of the working stroke of a V.R.
motor is given by its trajectory in the flux linkage-current
plane. The principal electric circuit and magnetic variables
are simultaneously displayed and simple graphical interpretations
for work output, torque and back e.m.f. possible. The trajectories are a particularly powerful aid to the understanding of
systems working in magnetically non-linear regions of their
characteristics.

A V.R. motor can be characterized statically by a family of magnetisation curves, $\psi(i,\theta)$, flux linkage as a function of current
and rotor angle. Curves from a typical family are shown in fig.1
for $\theta = \theta min$, $\theta = \theta max$ and $\theta = \theta s$, these being the minimum and
maximum reluctance rotor positions, and the angle where suppres-
sion of the winding current commences. A typical flux-current
trajectory, T, is drawn for the winding excited with a constant
positive voltage between θmax and θs, and an equal negative
voltage after θs.

It can be shown that the total mechanical energy output per
working stroke is $\oint i.d\psi$, or the area enclosed by the trajectory.[2]

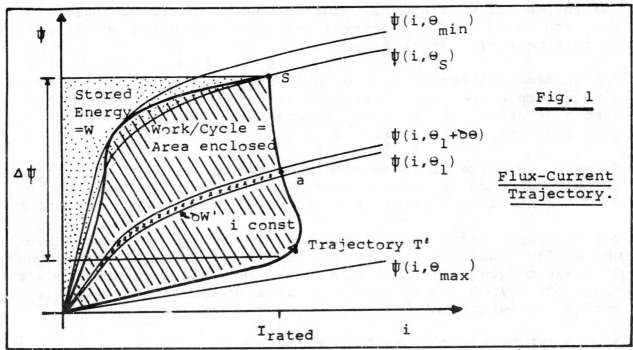

Fig. 1

Flux-Current
Trajectory.

The instantaneous torque produced at a point 'a' on the trajectory is:

$$Ta = \left.\frac{\delta W'}{\delta \Theta}\right| i \text{ constant}$$

Where "W'" is the co-energy of the winding[2]. This is illustrated in fig. 1 as the change in co-energy, $\delta W'$ for a small angular increment $\delta \Theta$, devided by the increment.

A simple, though quite accurate electrical model of the reluctance rotor is shown in fig. 2. The quantity $\frac{d\psi}{d\Theta}$ $(i, \Theta,$ is derived from the magnetisation characteristics surrounding the trajectory. L is the average value of incremental inductance in the saturation region, and R is the winding resistance.

The energy stored in the winding at any point is given by:

$$W = \int_0^\psi i \, (\psi', \Theta). \, d\psi$$

and is represented by the area to the left of the magnetisation curve for the point in question. An example is shown in fig.1 for the point S, where de-enerisation commences.

Perhaps the most striking fact in studying Fig.1 is that the reluctance motor, operating in a saturated mode, is a very non linear device. It cannot be analysed in any depth with simple models.

The advantages of saturation have been described at length[3] but
they can be quickly seen in fig. 1. The work output per cycle
is given approximatly by the area enclosed between the curves
ψ (i,θ max), ψ (i,θ min) and a vertical line through I rated.
For the case shown where ψ (i,θ min) exhibits heavy saturation the
area enclosed approaches that of a rectangle, $\Delta\psi$. I rated. If,
however, ψ (i,θ min) were a straight line passing through the
origin and point S, the area enclosed could approach a maximum of
only half this value. Furthermore, the energy stored in the
winding would approximatly double for the linear case. This
would cause many problems since this energy must be either
dissipated, or returned regenerativly to the supply at the end of
the working stroke.

Maintaining a relativly small airgap is obviously essential to
achieving the type of saturation characteristics shown.

The methods of constructing trajectories and determining work
output and torque are well known[4].

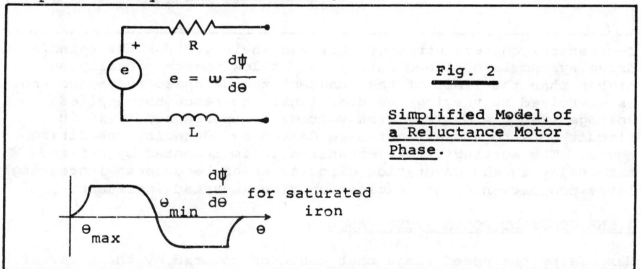

Fig. 2

Simplified Model, of
a Reluctance Motor
Phase.

3. CONTROL STRATEGY

The reluctance motor has 3 mechanisms by which its trajectories
can be controlled. These are the applied voltage, V, firing
angle advance, $\propto$, and working stroke extension, γ (these are
defined assumed that there are base values for normal conditions).
The first, applied voltage, operates directly to increase the volt
- seconds applied to the winding at a constant speed, as does the
stroke extension. Advancing the firing angle allows higher currents to be achieved, by applying voltage before the back e.m.f.
component shown in fig.2 has reached a signifigant magnitude.

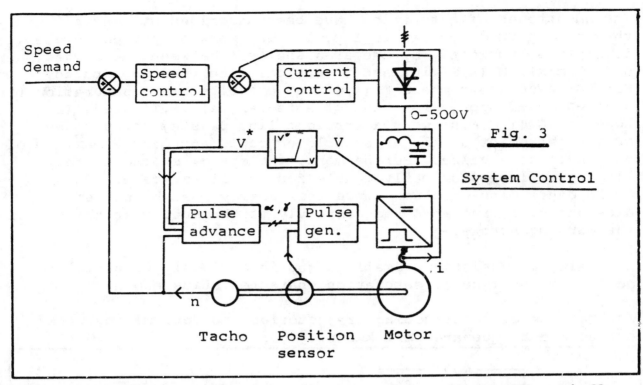

Speed demand

Speed control

Current control

0-500V

Fig. 3

System Control

V*

V

Pulse advance

Pulse gen.

n

Tacho   Position   Motor
          sensor

i

The general control strategy that was envisaged for the spindle
drive system is outlined in fig. 3. At low speeds, ideally no
higher than the limit of the constant torque speed range, control
is exercised by varying the d.c. link, and hence the applied
voltage. As the maximum link voltage is approached, a limit
circuit acts to answer increased demand by advancing the firing
angle. The working stroke extension is implemented by a fixed
time delay in the comutation circuit, which becomes an increasingly
large proportion of the excitation time as speed increases.

4 THE CONSTANT POWER SPEED RANGE

Maximising the speed range that could be covered by the motor at
full power and full rail voltage, i.e. through the mechanisms of
firing advance and stroke extension, was crucial to the ecomony of
the controller. If this was short of the specified constant power
speed range of the system it would be necessary to use a controller
which would operate at base speed with both full current and a
reserve of voltage, i.e. have a higher power rating than the motor.
It will be shown that, firstly, a wide constant power speed range
implies designing a machine in which the ratio of the minimum and
maximum reluctance characteristics is high, and secondly, that
this implies a machine with relatively low reluctance flux paths,
and hence low mmf requirements. If we assume that the constant
power speed range, n, is covered at the maximum controller
voltage V, we can write for the trajectory peak flux linkages

$$\hat{\psi}m = V \frac{\Theta m}{\omega m}$$

$$\hat{\psi}b = V \frac{\Theta b}{\omega b}$$

Where b and m imply maximum and base speeds, ω 'is angular velocity, and Θ is the angular excitation period in radians. These equations yield,

$$\frac{\omega m}{\omega b} = \frac{\hat{\psi}b}{\hat{\psi}m} = \frac{\Theta m}{\Theta b} = n \qquad (1)$$

To obtain quantative results it is necessary to idealize the characteristics and the trajectories as shown in fig. 4.

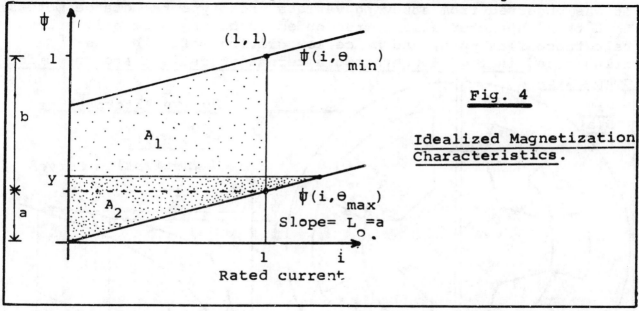

Fig. 4

Idealized Magnetization Characteristics.

The minimum reluctance curve shows abrupt saturation and has an incremental inducance equal to that of the maximum reluctance curve. Trajectories are shown for base and top speed conditions with areas A_1 and A_2. The former is assumed to fill all the area bounded by the magnetization curves and the rated current. The latter is assumed to fill an area limited by a horizantal line corresponding to the peak flux linkage of the top speed trajectory. Experience has shown that these idealizations are reasonable. It is assumed that both these trajectories correspond to rated power output, i.e. $A_1 = nA_2$. What we require is an expression for n as a ratio of the magnetisation curves. We can define this ratio as:

$$K = \frac{b + a}{a} = \frac{1}{a} \qquad (2)$$

413

From Geometry $A_1 = b$

$A_2 = \dfrac{y^2}{2a}$ (3)

Therefore $b = n \dfrac{y^2}{2a}$ (4)

Re writing equation 1 gives:

$$y = \frac{1}{n}\frac{\Theta m}{\Theta b} \qquad (5)$$

Substituting gives:

$$n = \tfrac{1}{2}\left(\frac{\Theta m}{\Theta b}\right)^2 \frac{1}{ab} \qquad (6)$$

$$= \tfrac{1}{2}\left(\frac{\Theta m}{\Theta b}\right)^2 \frac{K^2}{K-1} \qquad (7)$$

Which expression is directly proportional to K when its value is large. Thus, a wide constant power speed range implies a high ratio of minimum and maximum reluctance magnetisation curves. It was observed from studying various prototype reluctance motors that this characteristic corresponded to having a relativly low reluctance flux path, and hence low driving mmf. This is illustrated in fig. 5 where two alternative designs are shown for comparison.

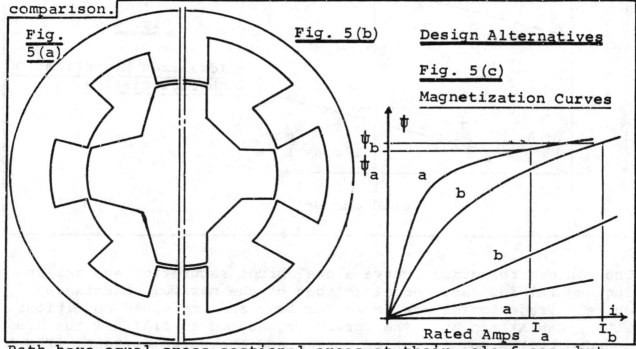

Fig. 5(a)

Fig. 5(b)

Design Alternatives

Fig. 5(c)

Magnetization Curves

Rated Amps

Both have equal cross sectional areas at their pole faces, but (a) uses extensive taper of the poles to ensure that areas of saturated iron and high mmf drop are always concentrated in short lengths of the magnetic circuit . At rated current and full over-lap of the poles the flux linkage achieved would be close for both designs, (b) perhaps having a lower pole flux, but more turns. At the crossed rotor position, however, (b) would have a much higher inductance, since this will be determined principally by the number of turns per coil, the surrounding iron not having a great influence. Typical characteristics are shown in (c).

414

From another point of view, 5 (a) is a motor with relatively large iron, and small mmf, area. This implies a relativly high ratio of ironborne to air-bourne flux, where the air-bourne flux can be linked to a leakage   type incremental inductance.

5. MOTOR GEOMETRY

A 6 pole stator, 4 pole rotor geometry was selected, with proportions generally as outlined in fig 5(a). It is generally desirable to use the minimum possible number of stator poles, since controller complexity and operating frequency increase with the number of phases. It is not possible to operate a 2 phase machine reversibly, since a symmetrical rotor gives indeterminate initial direction of rotation, and hence this possibility was not considered. An 8/6 pole geometry (4 phases) seemed likely to give increased specific output, but also reduced constant power speed range, since the reduced ratio of rotor pole - interpole region implies reduced scope for firing angle advance. Due to pressures of time a computer aided design was not attempted. Results from previous prototypes were analyzed and extrapolated, model geometries were cut in mild steel, and small scale iterative calculations of proposed changes were made. The final stack dimensions selected were 305mm diameter and 190mm length, the airgap diameter being 173mm.

6. TORQUE SMOOTHNESS

Fig. 6 shows the static torque-angle characteristics for the three phases of the final motor geometry, each excited with a constant current, and later superimposed. The characteristics can be divided into 4 distinct regions. Around the maximum reluctance position (0° for phase 1) the motor and stator poles are a long way from overlapping. At about 10° position, however, the rotor and stator poles are coming into close proximity and a rapidly rising torque region commences. At approximatly 12° actual overlap of the poles has begun and torque is largly constant with angle, showing some roll off due to saturation effects.
At approximatly 37° full overlap of the poles is being approached and torque rolls of rapidly with angle, due to extensive saturation in the flux paths.

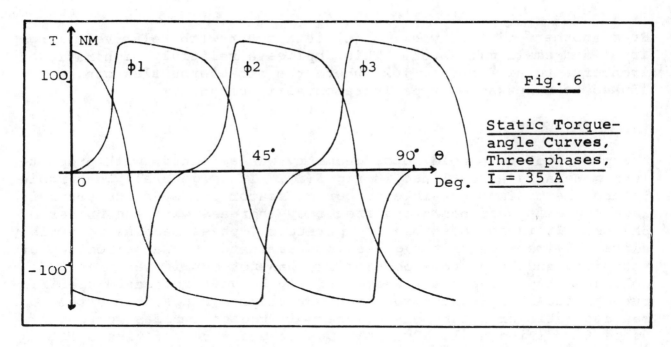

Fig. 6

Static Torque-
angle Curves,
Three phases,
I = 35 A

The torque at the intersections between phases was approximatly
54% of the peak value. If switching between phases were chosen
to occur at this point the torque ripple would be considerable,
although its effects could be reduced with a tight velocity loop.
The most serious problem to emerge, however, was the very rapidly
rising nature of the torque where pole overlap commences. Its
effects was similar to that of an impulsive blow to the poles
setting up noise and vibration in the motor.

Methods of increasing torque smoothness were examined. The rotor
of an early mild steel model geometry was cut to have wedge shaped
corners taking up some 20% of its leading edge. Rate of rise of
torque was reduced, although not as much as was desired. Total
work per stroke, however, was also reduced due to the lower cross
sectional area of pole face, and flux path. It was concluded that
more extensive pole shaping was needed, with an increased rotor
pole arc to compensate for lost pole face area.

7. THE INVERTER

A variable voltage d.c. link and purely commutating 'block' inverter
was chosen to drive the motor. PWM type inverters are essential
when driving rotating field type machines at low speeds, where the
purity of the sinusoidal current waveform is critical. It was not
felt that such current shaping would be necessary for the V.R.
motor. Furthermore, there is an inherent stability problem limiting high speed operation of PWM drives, as the fundamental
frequency approaches the modulation frequency. For the present
V.R. motor, requiring 12 switchings per revolution and 7000 RPM

top speed, the block inverter seemed a more natural choice. Also, the controlled rectifier limits the voltage of the d.c. link to its design value, thus avoiding line voltage surges reaching other parts of the circuit, and the use of overated devices. Switching conditions in the bridge were not severe and it was possible to avoid the use of inverter grade thyristors altogether.

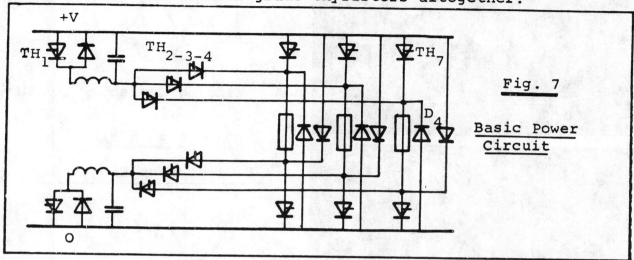

Fig. 7

Basic Power Circuit

The basic power circuit is shown in fig. 7. The windings are excited by 2 main thyristors (e.g. TH7 upper and lower) which are fired simultaneously, and apply the rail voltage V to the windings. At the end of the working stroke the main thyristors are commutated and winding currents are allowed to run down through diodes (say D4, upper and lower) into V, in the reverse sence. The resonance components of the commutation circuits are common to the 3 phases, commutation being directed by 3 selection thyristors, TH$_{2-3-4}$.

8. SYSTEM OPERATION.

The motor-drive system was run under open loop conditions to investigate its operation at the principal points defined in the specifications. Fig. 8(a) shows winding voltage and current waveforms for 10kw, 800 RPM operation. Fig. 8(b) shows the corresponding flux linkage-current trajectory. This was displayed on a CRO by deriving a flux signal from the winding voltage, with compensation for resistive drops. Fig. 9 shows corresponding waveforms for the top speed condition. This had to be simulated at a reduced speed, due to power limitations in the dynamometer, and rotor imbalance. Applied voltage was reduced in proportion to speed, all other paramaters being unchanged. The continuous motor rating with air-through cooling, was found to be 19.7 kw at 1200 RPM. The constant power speed range for the system was 5.9:1 on the basis of equation 1 and the trajectories achieved.

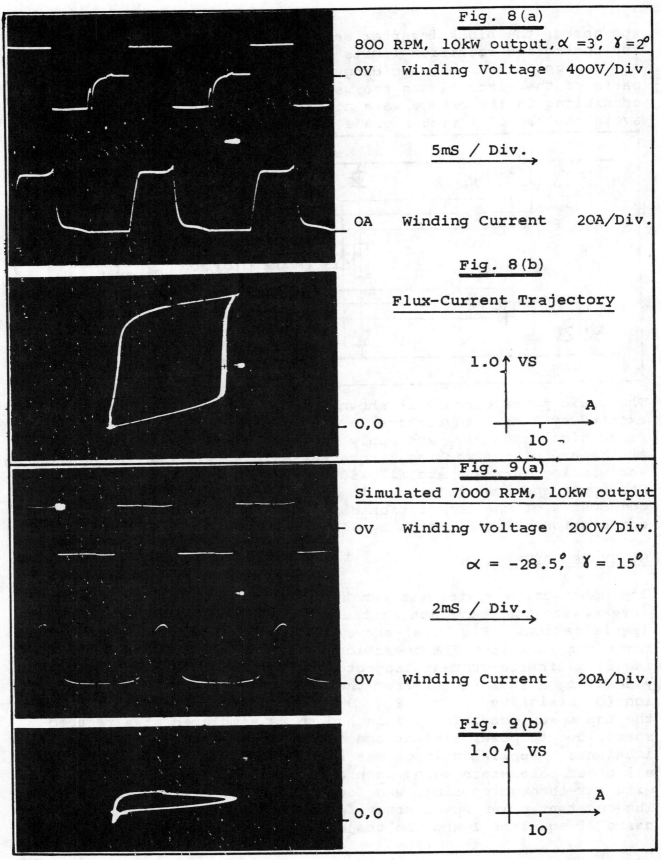

Fig. 8(a)

800 RPM, 10kW output, $\alpha = 3°$, $\gamma = 2°$

0V   Winding Voltage   400V/Div.

5mS / Div.

0A   Winding Current   20A/Div.

Fig. 8(b)

Flux-Current Trajectory

1.0 VS

A

0,0

10

Fig. 9(a)

Simulated 7000 RPM, 10kW output

0V   Winding Voltage   200V/Div.

$\alpha = -28.5°$, $\gamma = 15°$

2mS / Div.

0V   Winding Current   20A/Div.

Fig. 9(b)

1.0 VS

A

0,0

10

9. COMPUTER SIMULATION

A numerical simulation of the system was made, using a program previously developed for a similar reluctance motor-drive. The program stored magnetization data for the machine, ψ (θ, i), as a set of 23 exponential curves, and calculated trajectories by a numerical integration technique. Torque was calculated at points along the trajectory, and an average value derived. Predicted motor performance was compared with measured results, and agreement found to be very close. A calculated trajectory is shown in fig. 10.

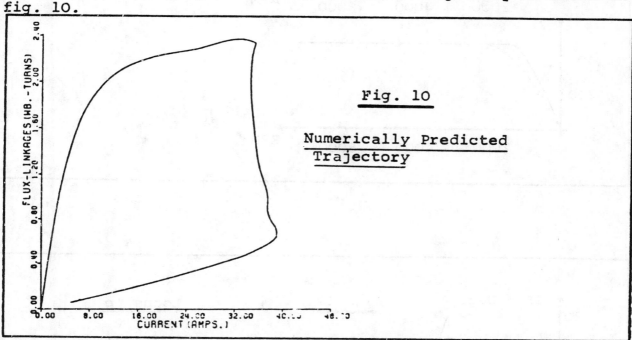

Fig. 10

Numerically Predicted
Trajectory

The program allowed a considerable amount of data to be generated with relative ease and it was used to investigate motor performance over a wide range of operating conditions. Fig. 11 shows the range of α and V over two operating loci,γ being implemented as a speed dependent extension, varying from 0° to 15° at top speed. Locus 'a' corresponds to full output over the entire speed range, i.e. 120 Nm from 80 to 800 RPM, and 10kW from 800 to 700 RPM, locus 'b' corresponds to half this output. The control stratagy is generally as described in section 3. It can be seen that the required response of the control arms appear smooth, if not actually linear, however, a more detailed examination along locus 'a' showed a considerable range in the firing angle, see fig. 12. Such variations would make it difficult to optimise an automatic control strategy for the motor.
Investigations did show that the sensitivity could be reduced by commencing firing advance before maximum voltage was reached, as shown in locus e, fig. 12.

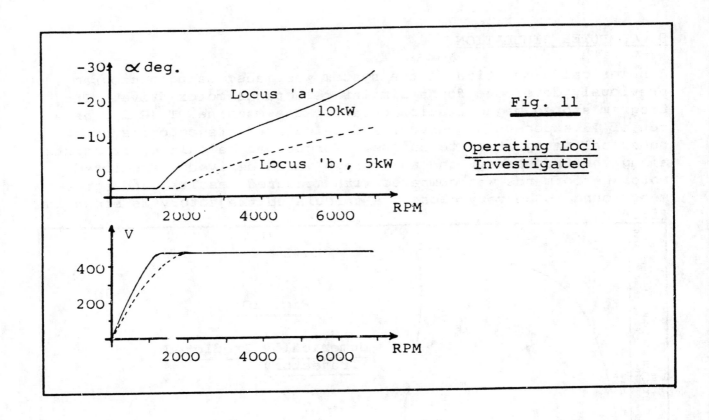

Fig. 11

Operating Loci
Investigated

Locus 'a'
10kw

Locus 'b', 5kW

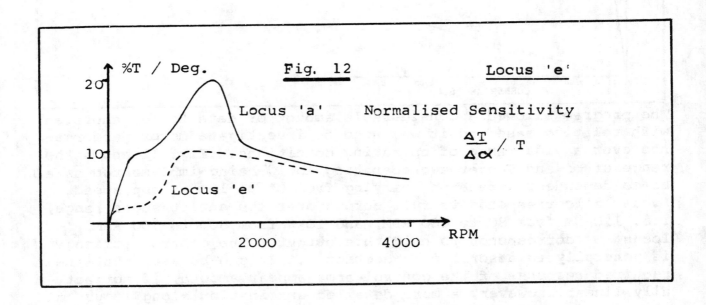

Fig. 12

Locus 'e'

Normalised Sensitivity

$$\frac{\Delta T}{\Delta \alpha} \Big/ T$$

Locus 'a'

Locus 'e'

10. CONCLUSIONS

It was shown that the variable reluctance motor is capable of operating as a high output drive with wide constant power speed range. Motor and controller are simple, and potentially very economical. Due to its very non linear nature, accurate design of the magnetic circuit requires the use of computer aided flux plotting techniques. Prediction of system operation requires the calculation of flux linkage - current trajetories by numerical integration. Torque smoothness, vibration and general noise problems in variable reluctance motors are areas requiring much study. The control requirments of these motors are complex.

REFERENCES.

(1) Noad, H.M.: "A Manual of Electricity"
     Lockwood, London, 1859. pp. 674-684.

(2) Woodson, M.M. and Melcher, J.R.:
     "Electromechanical Dynamics". Wiley,
     1969. pp. 60-84.

(3) Byrne. J.V.: "Tangential Forces on overlapped pole geometries incorporating ideally saturated materials". IEEE Transations on magnetics, Vol. Mag-8, No. 1. March 1972.

(4) Byrne. J.V. and O'Dwyer, J.B.:
     "Saturable variable reluctance machine simulation using exponential Functions". Proc. of the international conference of stepping motors and systems, 1976, University of Leeds.

ACKNOWLEDGEMENT

We are very grateful for the support given by Lucas Industries (UK) to University College Dublin, for this research work.

A Low-Cost, Efficient 1kW Motor Driver

Bruce Powell, Motornetics Corp., Santa Rosa, California

MOTOR-CON Proceedings, Atlantic City, April 1984, pp. 229-237

A compact, efficient, reliable and low cost 1.2 KW (continuous) amplifier is designed for a new variable reluctance motor intended for direct drive robotics and industrial automation. The amplifier design benefits from some of the unique characteristics of the motor.

INTRODUCTION

Robot manufacturers are eager to eliminate expensive and troublesome reduction geartrains in their robot joints. The advent of precision servo motors capable of supplying hundreds of foot-lbs at low speeds meets this need, and powerful, reliable, and low cost amplifiers are required to drive these motors.

This amplifier design achieves its size, weight, efficiency, and cost goals by taking advantage of the drive requirements of a new variable reluctance AC synchronous servo motor.

The result is a compact, inductorless half-bridge amplifier with a power to weight ratio of more than 1000 watts/lb., and a combined amplifier and motor efficiency of over 70 percent.

MOTOR CONSIDERATIONS

The amplifier is intended to power a precision ultra high-torque motor for direct drive applications in industrial automation and robotics. Performance and reliability are paramount in these applications, and cost becomes increasingly important as these industries become more competitive.

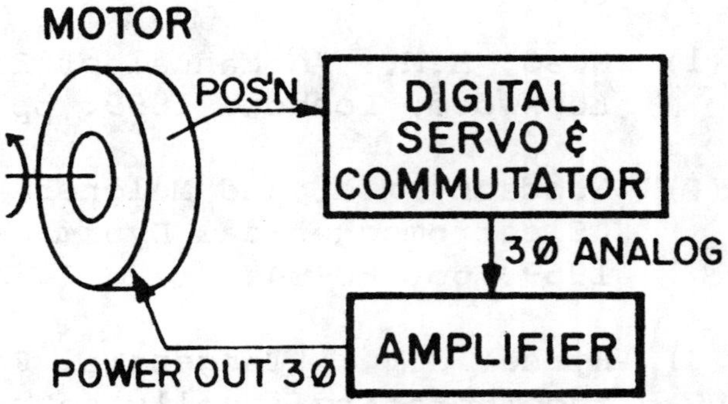

FIG. 1: SYSTEM BLOCK DIAGRAM

The motor is a three phase variable reluctance type, with 300 poles so that 150 electrical cycles are required for one revolution. The motor is designed to operate in a servo with a position accuracy of 30 arc-seconds while providing 250 ft-lbs of torque. The amplifier must supply up to 9 amps per phase based on the motor's torque constant.

The motor includes an integral position transducer (resolver) which is decoded by a digital servo circuit. The servo compares the desired position, velocity, and acceleration instructions with its computed values and modulates current in each motor phase.

A typical application may be a direct joint motor in a robot arm. In addition to high torque and high reliability, the motor must have excellent low speed operating characteristics, and "stiffness" for operation in a servo loop.

Stiffness is obtained by rapid motor response and high servo loop gain. Rapid motor response requires a high motor voltage, and high servo gain requires minimum amplifier delay.

High efficiency is an important intermediate goal. Poor efficiency results in high power dissipation which affects reliability and performance. Derated performance affects the size, weight, and cost. The motor is very efficient by virtue of its brushless design, low winding resistance, and low iron losses.

From the amplifier's viewpoint the motor may be considered an efficient 1KW transformer with small iron losses and no brush losses. The "secondary" of this hypothetical transformer is the mechanical energy output imparted to rotor. The motor requires a three-phase unipolar drive at high controlled current levels. High voltages are desirable for operation at high motor speeds, and for rapid response.

The interdependence of these system attributes is illustrated in Figure 2.

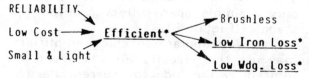

"*": considerations for Amplifier Design

FIGURE 2

The Unipolar Approach--An Attractive Motor

In conventional servo motor systems, permanent magnets are used to exert a force on a current carrying conductor. This force is given by $\overline{F} = \overline{I} \times \overline{B}$ The geometry of this type of motor is illustrated in Figure 3.

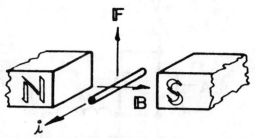

FIG. 3: PERMANENT MAGNET MOTOR FORCE GEOMETRY

Conventional permanent magnet motors require a bipolar drive. The most common figuration is the "H" bridge (Figure 4). Only a unipolar supply is needed, and by conducting transistors A or B, currents of either sign will be realized. The disadvantage of this technique are numerous expensive power transistors, complex drive circuits, difficult motor current measurement, and a tendency toward catastrophic failure due to cross-conduction.

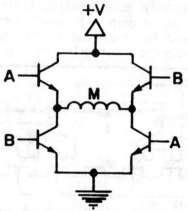

FIG. 4: "H" BRIDGE CONFIGURATION

Another method is the "T" bridge. (Figure 5) Its major disadvantage is the expense and complexity of adding a second, negative power supply. Power supply "pumping" may occur by charging the load with one supply, and discharging into the other. With the "T" configuration one must also make sure that cross-conduction does not occur.

A variable reluctance motor exerts a force on the rotor by magnetic attraction only. The magnetic field generated in the stator windings interacts with a passive ferromagnetic rotor: the magnetic orientation of the rotor material will be aligned with the stator

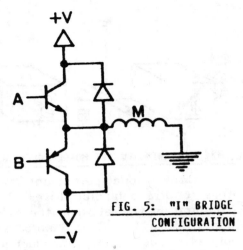

FIG. 5: "T" BRIDGE
CONFIGURATION

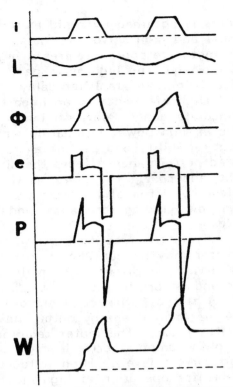

FIG. 7: V-R MOTOR OPERATING PARAMETERS

magnetic field regardless of its direction. The force is a function of the flux density and the permeance of the flux path. The change in the flux path permeance causes the winding inductance to vary sinusoidally with the rotor angle. If current is applied to the winding as the inductance is increasing (rotor approaching alignment) then mechanical energy will be imparted to the rotor. This process is illustrated in Figure 6.

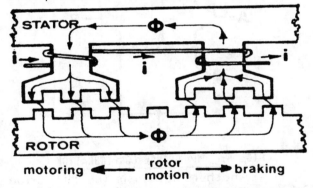

FIG. 6: SIMPLIFIED V-R MOTOR GEOMETRY

For simplicity, only one motor phase is shown. When the rotor moves to the left in the illustration, positive work is imparted to it. When the rotor moves to the right, dynamic braking occurs. (Note that the direction of the current never changes.) Figure 7 illustrates the relationship between work output and current input. This simplified analysis assumes there are no motor losses, no saturation effects, and no secondary effects on the inductance.

Also, the current waveform is a simple trapezoid. The positive trend in the energy of the motor represents the mechanical energy imparted to the rotor.

Since the direction of the magnetic flux (Φ) does not need to change, the currents may be unipolar, thus providing significant simplification of the amplifier.

A linear unipolar amplifier needs only a simple open-collector output stage. (Figure 8)

A unipolar switching amplifier may use this circuit, but the speed is limited by the inductor current decay time. A simple half-bridge circuit (Figure 9) ensures the current fall time will be fast since no circulating currents are needed.

In both cases the complexity and expense of the output stage is sharply reduced. The expense of two high-power supplies as in the "T" configuration is avoided. Reliability is improved by the elimination of cross-conduction paths, especially at higher powers where the power transistors are slower.

424

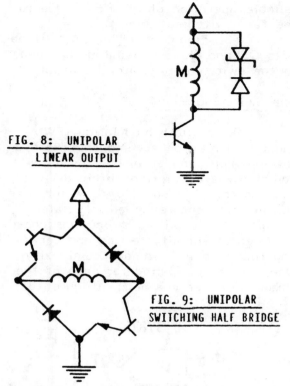

FIG. 8: UNIPOLAR
LINEAR OUTPUT

FIG. 9: UNIPOLAR
SWITCHING HALF BRIDGE

The Switching Amplifier

Switched-mode operation is the obvious choice for an efficient amplifier in the 1 KW power range.

The advantages of a switched-mode output to a linear output is significantly less power dissipation resulting in smaller size, weight and cost. Using one half-bridge per phase requires only six transistors and one unipolar supply for a three-phase amplifier.

Motor Switching Losses

Amplifier switching losses may be minimized by keeping the switching frequency as low as possible without entering the audio frequency range. The amplifier's three channels are synchronized to switch at a 20 khz rate. Motor losses are minimized by constructing the motor core with thin, insulated ferrous laminations. The chosen material has low hysteresis losses and will support a high flux density. The motor is therefore very efficient and operates much like a "good transformer."

Therefore, no additional switching regulator inductor is required since the motor windings themselves are good inductors.

The winding inductance must be large enough to prevent large ripple currents but small enough to keep the time constant L/R within performance requirements.

This compromise value is 100 to 200 mH.

The Fet Output

The power MOSFET has many advantages in this circuit design. The higher speed, simpler drive circuitry, and lower on-state power dissipation for devices in the 200 volt, 10 amp range makes the design simple, efficient, and economical.

Due to the device geometry and positive temperature coefficient of R_{DS} (on), the MOSFET shows no second breakdown effect within the range of its maximum ratings. The gate can be modeled by a capacitor, making the drive for these devices very simple and direct.

An IRF-642 FET was chosen, which has a 200V breakdown, 0.2 ohm R_{DS} (on), and 12.5 amp continuous drain current capability. The TO-220 package is compact and simple to mount. The half bridge output circuit is shown in Figure 10. The lower transistor in this illustration, Q1, is driven by a self

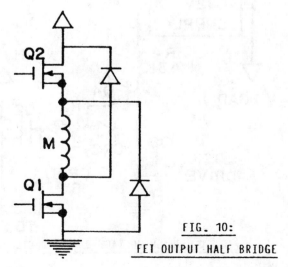

FIG. 10:
FET OUTPUT HALF BRIDGE

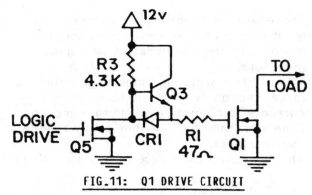

FIG. 11: Q1 DRIVE CIRCUIT

biased emitter follower and open drain FET push-pull stage, shown in Figure 11. This configuration provides a low impedance source and sink so that the gate capacitance in Q1 may be charged and discharged quickly.

R1 limits the gate time constant to 47 ns and dissipates high frequency ringing in the gate circuit.

Figure 12 shows the drive circuit for the upper transistor, Q2. This circuit is essentially the same, except the transistors Q4 and Q6 must be rated for the H.V.D.C. level plus a 12 volt bias. This bias is provided by a 12 volt floating supply referenced to the source voltage of Q2. C1 is placed directly at the drain lead of Q2, to bypass high frequencies on the H.V.D.C. bus. All traces are kept to a minimum length and a ground plane is employed. These measures are crucial for the

reliable operation of the FET's due to the fast switching times.

The "logic drive" of small signal Fets Q5 and Q6 are tied together and to the switching regulator logic output.

The Switching Regulator

Pulse width modulation (PWM) regulation may be accomplished with a few op-amps and a comparator or with commonly available monolithic devices such as the Unitrode 1524A, which offer the needed features in an integrated circuit. A monolithic device was chosen in order to simplify testing and service, and reduce board size and complexity. One device is required for each of the three amplifier channels. (Figure 13)

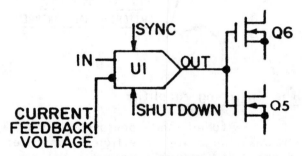

FIG. 13: SWITCHING REGULATOR BLOCK DIAGRAM

The IC accepts as inputs the control voltage, a feedback voltage, and a sync input to synchronize the three amplifier channels at 20 khz as well as a "Shutdown" input which may be used by error detection circuits to inhibit operation. The output of the switching regulator IC drives small signal FET's Q5 and Q6. (Figure 11 and 12)

The Control Feedback

The control feedback is implemented as a voltage proportional to motor current. (Figure 14) The voltage V_{R6} across .05 ohm resister R_6 is proportional to the load current during the transistor-conduction half of the cycle. Similarly, the voltage V_{R5} across R_5 is proportional to the load current during the reverse conduction half of the cycle. (This voltage is negative going.)

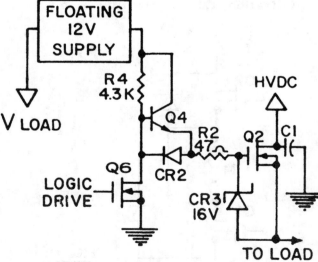

FIG. 12: HIGH VOLTAGE Q2 DRIVE CIRCUIT

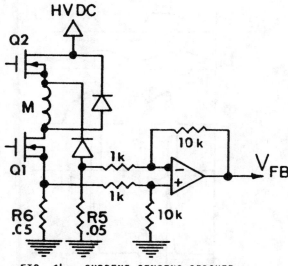

FIG. 14: CURRENT SENSING CIRCUIT

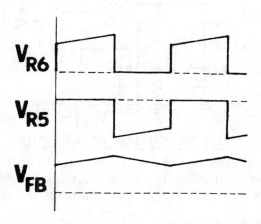

FIG. 15: CURRENT SENSING WAVEFORMS

An op-amp amplifies the difference $V_{r6} - V_{r5}$, providing a useful current feedback signal V_{FB}. (Figure 15)

The Motor Voltage

The traditional motor DC power supply is typically large, expensive, and wasteful of power. A 1 KW transformer and power conversion devices may represent a substantial investment. By designing a high voltage supply which eliminates the transformer, large savings in cost, dissipation, size, and weight result.

The amplifier obtains a nominal 160 volts DC by full wave rectification of the 115 VAC line, allowing the use of common 200 V MOSFET's. (Figure 16)

If the amplifier is constructed with 400 volt devices, the amplifier may be operated from a 230 VAC line.

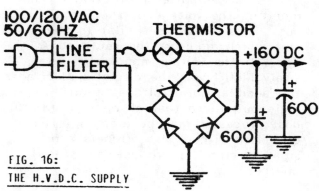

FIG. 16:
THE H.V.D.C. SUPPLY

The disadvantage of direct line rectification is that line isolation is not provided, so optocouplers are used to isolate the power amplifier from the control circuits.

If a system designer desires more isolation, a common 1:1 ISO transformer may be placed in the line. The transformer may power several circuits. This flexibility benefits the system designer.

Amplifier Protection Circuits

The amplifier-motor system is not sensitive to fluctuations in the AC line voltage or frequency within the range of 90 - 130 VAC, 50-60HZ, due to closed current, position and velocity loops.

Below 90 VAC, the 12 and 5 volt DC supplies may not be regulated, causing unpredictable operation.

Above 130 VAC, the H.V.D.C. supply may reach a voltage level close to the breakdown voltage of the MOSFET's.

The AC line voltage is checked by the line voltage Monitor circuit. (Figure 17) The rectified AC line is filtered and divided down for comparison with a voltage reference. Hysteresis currents ensure jitter-free response to line voltages outside the specified range.

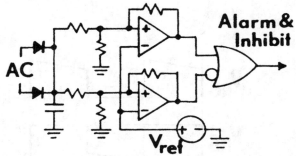

FIG. 17:   THE LINE VOLTAGE MONITOR

The circuit output is used to inhibit the switching regulators and to sound an external alarm and logic output.

Controlling Regenerative EMF

The spinning motor and load has a kinetic energy given by $K = \frac{1}{2} I \, w^2$ (J; kg-m², rad/sec). During regenerative braking the displaced kinetic energy is transformd into electrical energy, less losses and friction. For an efficient motor and amplifier such as this one, the losses are low and, therefore, much of this energy must be absorbed by the amplifier.

The regeneration mechanism of a variable reluctance motor is quite different from a permanent magnet DC motor. In Figure 6, the stator and rotor section is shown undergoing dynamic braking when the rotor moves to the right.

The amplifier applies current "i" for a short time at the point where the stator-to-rotor permeance is at its maximum.

The winding inductance decreases as the rotor moves out of alignment with the stator.

When the current reaches its control value, kinetic energy is taken out of the motor as a function of the decrease in inductance. Ideally, the energy $U = \frac{1}{2} L \, i^2$ will be reduced in proportion to the change in the inductance L provided the current "i" is held constant.

A simplified analysis shown in Figure 18 makes this process readily apparent. (The assumptions made for Figure 7 apply to Figure 18 as well.)

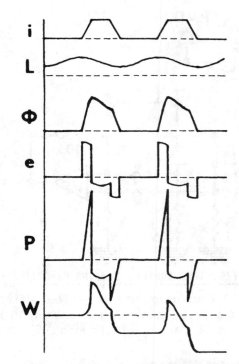

FIG. 18:   V-R MOTOR REGENERATING PARAMETERS

Since the net work "delivered" to the rotor is negative, the rotor undergoes deceleration. The energy returned from the motor is converted to a stored charge in the filter capacitors on the H.V.D.C. bus:

$$U_K = \tfrac{1}{2} I w^2 \longrightarrow U_B = \tfrac{1}{2} L i^2 \longrightarrow U_E = \tfrac{1}{2} \, q^2 / C$$

The H.V.D.C. bus voltage increases by

$$\Delta V_c = \Delta q_c / C$$

If the D.C. bus voltage approaches the operating limits of the amplifier then some means must be taken to discharge the capacitor energy.

An amplifier protection circuit monitors the voltage of the H.V.D.C. bus. This circuit is similar to the circuit which monitors the rectified A.C. line. If the voltage approaches excessively high levels the circuit will trigger the audible alarm/voltage "fault" warning logic output, so that an operator or the host processor may be alerted. The detection circuit also triggers a dissipation circuit to reduce the stored charge.

428

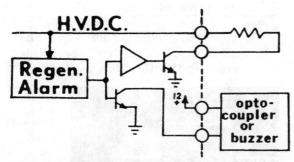

FIG. 19: DISSIPATIVE SHUNT REGULATOR

The amplifier may dispose of this energy two ways: a dissipative shunt regulator or an AC line regenerative regulator. For most applications not requiring rapid deceleration with high inertia loads, an external resistor (mounted within the amplifier case) may be connected as shown in Figure 19. The H.V.D.C. bus is brought out to a connector as well as the open collector of a "dump" transistor. The resistor may be sized according to the application, and in most cases a 90 ohm 20 watt resistor is adequate.

For more demanding applications, an AC line regeneration circuit may be added externally.

An opto-coupler connected to the logic output may drive the trigger of the logic circuit which synchronously connects the H.V.D.C. bus to the AC line. The circuit must deliver the excess charge without producing excessively high peak currents, which requires passive or active current limiting. This more expensive approach is only justified for applications with large amounts of continuous regenerative braking, such as a motor used to unreel cable against a tension.

Input Signal Conditioning

The amplifier requires three analog inputs representing motor current in each phase. Optical isolation between the amplifier and external circuits is imperative for maximum reliability.

Three opto-couplers receive the signals, which are pulse-width modulated at 70 khz. Since the analog signals are unipolar, 0% duty cycle represents zero current.

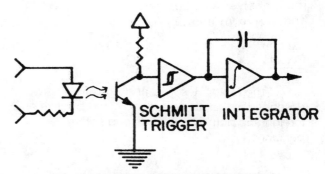

FIG. 20: INPUT SIGNAL CONDITIONING

The phototransistor outputs are squared by Schmitt triggers, and three integrators convert the signals back to their analog form. (Figure 20)

The output is used as the command voltage for the switching regulators.

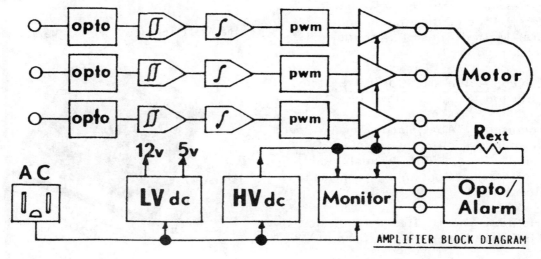

AMPLIFIER BLOCK DIAGRAM

Conclusion

This amplifier is an example of how economical and reliable motor current amplification in the 1 KW range can be obtained by using up-to-date power control technology and by considering the unique characteristics of the modern brushless, non permanent-magnet variable reluctance motor.

Motor features such as brushless unipolar operation and low losses are taken advantage of to make the amplifier much simpler, and the use of direct line operation, power FET's and switching regulator IC's helps keep the amplifier efficient, compact, and inexpensive without sacrificing reliability and performance.

Patents

The Motor Resolver and Amplifier described in this article are protected by several foreign and domestic patents and patent pendings.

References:

D.C. Motors, Speed Controls, Servo Systems, Engineering Handbook, Electro-Craft Corporation, 5th Edition, 1980

B.C. Kuo, "Permeance Model of Low-Resolution Variable Reluctance Motors," Proc. 20th Annual Incremental Motion Control Systems and Devices, 1983

B.C. Kuo, Automatic Control Systems, 4th Edition, Prentice-Hall, Inc. 1982

J. Tomasek, "Analysis of Torque-Speed Performance Limits in Brushless DC Motors," Proc. 20th Annual Incremental Motion Control Systems and Devices, 1983

D. Halliday and R. Resnick, Physics, 3rd Edition, John Wiley & Sons, 1978

Operating Waveforms

These photographs were taken at 6 amps motor current per phase; horizontal deflection is 10us/cm

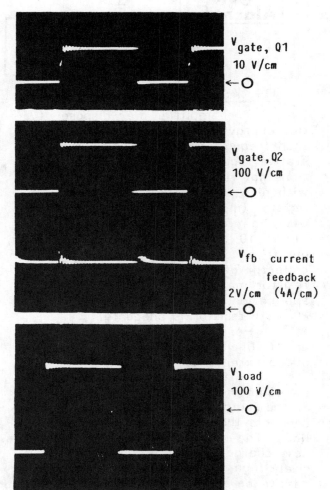

V_{gate}, Q1
10 V/cm
←O

V_{gate}, Q2
100 V/cm
←O

V_{fb} current feedback
2V/cm (4A/cm)
←O

V_{load}
100 V/cm
←O

The Amplifier

Weight 1 lb. 1 oz.
(0.74 kg)

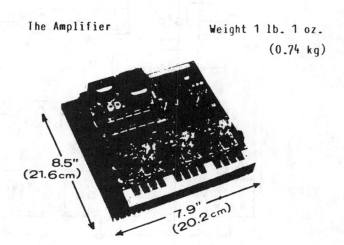

8.5"
(21.6cm)

7.9"
(20.2 cm)

Ultra-High Torque Motor System for Direct Drive Robotics

Ross Welburn, Motornetics Corp., Santa Rosa, California

MOTOR-CON Proceedings, Atlantic City, April, 1984, pp. 17-24

ABSTRACT

This paper describes a unique motor system incorporating a recently developed dual stator annular rotor reluctance motor operated as a brushless DC motor. This new motor is combined with a simple unipolar switching amplifier and an integral high resolution synchro/resolver positioning device yielding an ultra high torque positioning and velocity servo system. Specifically designed for direct drive robot applications, it eliminates moving parts and their backlash for a more accurate, reliable and cost effective robot "muscle."

INTRODUCTION

Most motors produce their maximum power at several thousand RPM. By increasing the number of poles, the speed can be reduced while the torque is increased so the maximum power is developed at a few hundred RPM. Increasing the number of poles in a motor has a practical limit depending on the size of the motor and its cost. Historically, when a lower speed or more torque is required, a mechanical reduction is usually installed between the motor and load. These mechanical reduction devices are commonly used in constant speed machines where backlash, wear, lubrication and accuracy are of secondary importance. In a positioning servo, however, these limitations are severe, and precision gears with anti-backlash devices must be used. At lower speeds and higher torques, the gear-trains become more complex and the advantages of direct drive motors are multiplied.

Robots are the extreme example; requiring speeds of less than one half a revolution per second, requiring very high torque to lift a load at the far end of an arm, and requiring a high angular position accuracy and resolution. A robot with several axes of motion is more sensitive to the disadvantages of gear reduction since multiple points of backlash will interact, resulting in an unstable and inaccurate positioning servo.

The cost of the computer and control electronics is falling while the costs of precision gear and mechanical assemblies, as well as qualified production and service technicans are rising rapidly. From a cost standpoint, the balance between electronics and mechanics is shifting in favor of intelligent, flexible electronics with simple, durable mechanics. The goal is fewer moving parts and the limit is one moving part per point. This ultra high torque motor, when incorporated as an integral part of the robot arm assembly, achieves that goal.

THE OVERALL SYSTEM

The system consists of four complementary devices: the "Megatorque"™ motor, the power amplifier, the "Reactasyn"™ resolver and the servo processor unit "SPU." Each block was designed while considering the others to yield a higher performance at lower cost.

THE MOTOR

The motor is a three phase variable reluctance device with no permanent magnets. It is a synchronous motor operated as a brushless DC motor with electronic commutation. The motor has a thin, annular rotor mounted between two concentric stators. (Fig. 1) Magnetic teeth on the rotor and the stators react to produce torque by sequentially energizing the poles. The torque per pound of iron is multiplied several times because both inner and outer stators react with the rotor.

Figure 1 – Motor Laminations

Since the rotor has a thin cross section, the rotating weight and inertia is minimized. Just as important, however, is the short flux path circuit. Instead of a long 180° flux path through the rotor, the flux travels radially through the thin rotor cross section from outer stator to inner stator, and back. (Fig. 2) This short rotor flux path means that a lower resistance flux path exists so that the motor has a higher flux per ampere-turn; therefore a higher torque per watt invested.

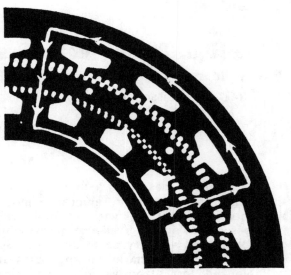

Figure 2 – Motor Flux Path Circuit

The rotor is electrically inactive; that is, there are no windings and therefore no heating due to current flow as is the case of DC and induction motors. The flux passes straight through the rotor providing torque without heat, which is important for a motor that must produce full torque at standstill or low speeds. In most conventional servo motors the rotor heat must be dissipated by convection and radiation at low speed, which is not very effective. Electrically active rotors must either force air cool or derate the torque of the motor at low speeds. This is especially true of PC moving coil motors which have no rotor iron and a very short thermal time constant. The small amount of heat developed in the Megatorque motor originates in the stators and is then conducted out of the motor through the motor case. Heat removal by conduction is orders of magnitude more efficient than heat removal by convection and radiation.

Increasing the number of poles in a motor increases the torque and reduces the speed while the output power remains the same, analogous to an ideal gear reduction. Adding teeth to a synchronous motor is a way of increasing the number of poles without adding more windings. The teeth on the motor and stators' face multiply the torque.

Since this motor is designed for applications requiring high torque at low speeds, reducing speed is part of the design criteria (eliminating mechanical speed reductions).

This motor requires 150 cycles per revolution. A conventional motor would require 300 poles per phase or 900 windings which is, of course, not practical. The "Megatorque" motor has 18 windings per stator.

It might be interesting to note that although this motor has stator and rotor teeth similar to a stepping motor, the design considerations are different because this motor is designed to operate in a position-based brushless electronic commutation loop. Stepping motors must not pass the position of maximum torque from step to step (excess position lag) or they will lose a step. This motor would be a poor stepper because the first step would violate the stiffness curve (torque vs. angular displacement) which would result in a low useable torque and unstable dynamic operation.

Since the Megatorque motor has low internal losses and therefore low inherent damping, the dynamic response is rapid. Damping and settling time is managed by the servo system. In order to get the most torque per pound of iron, the motor laminations are driven up to hard saturation. Since there are no permanent magnets, demagnetization due to the high flux density is not a problem. The resulting distortion is corrected by modulating the winding current vs. position by the pattern stored in ROM's in the electronics section.

Most motors are not designed to be operated continuously at heavy load without a limited duty cycle to prevent overheating. The Megatorque motor, however, is designed to be operated continuously at saturation. Robotics applications frequently require continuous operation at heavy load; for instance continuous acceleration and deceleration may require maximum

torque 100% of the time. By design, this motor does not overheat because a disproportionate amount of the stators' area is reserved for copper winding space so that winding IR losses can be minimized, through the use of heavier guage wire. The lack of permanent magnets in this motor means that frameless motors can be assembled easily by the user, the motor freewheels with power off and most importantly the motor requires only unipolar currents for operation. The amplifier complexity may be reduced by one-half since it can be unipolar.

The mechanical mounting of the motor in some applications is more complicated. If the user requires a large pass through hole in the motor or high bearing loads, large diameter bearings and an unusual rotor mounting are required. Adding two stators only slightly increases the complexity. Figure 3 shows possible mounting configurations: "A" using a cantilevered rotor ring and two identical small diameter bearings. Currently if the motor is supplied in a case, it will be of type "B" design as it is more tolerant of a wide variety of bearing loads.

THE POWER AMPLIFIER

The power amplifier is a direct line operated half bridge switching regulator. (Fig 4) The AC line is first rectified then filtered and the resulting 150 volts DC is pulse-width-modulated by the amplifier. The amplifier drives the motor windings directly requiring no switching inductor. Line isolation for the control inputs is provided by opto-couplers which protect against ground loops as well.

MOSFETS switches are used to simplify the input drive and to eliminate load line shaping components. Because the "on resistance" is low enough and the switching times are fast, the power transistors dissipate less heat than the return diodes.

This reluctance motor simplifies the power amplifier design in several ways:

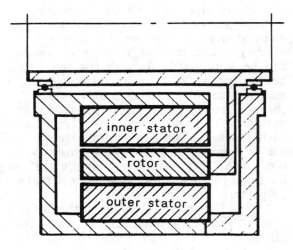

Figure 3a:
Cantilevered Rotor Design

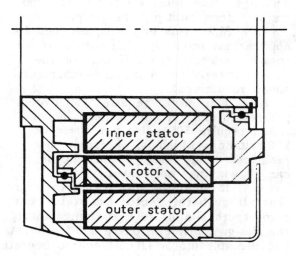

Figure 3b:
Cantilevered Stator Design

First, since only unipolar currents are required, the full "H" bridge is eliminated, and second, the reluctance motor winding inductance is very low in losses and can tolerate high switching ripple currents with minimal heating. The elimination of the full "H" bridge means that half the number of switch transistors are required and no "dead time" protection circuits are required to protect against failures caused by "cross conduction."

At high amplifier power levels, these advantages become acute because switching transistors in larger sizes have lower gain and slower switching speeds, thus compounding the risk of "cross-conduction".

The amplifier is operated in a current loop with two ground-referenced current sensing resistors, while the PWM switches at a constant 20 kHz.

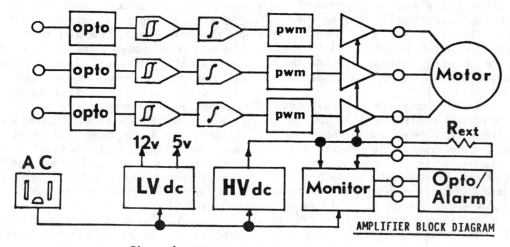

Figure 4 - Power Amplifier Block Diagram

THE "REACTASYN" tm RESOLVER

Conventional large diameter thin ring type position and velocity sensors are uncommon and expensive. This new reactive resolver, however, uses the same lamination design as the motor's, simplifying the mounting by matching the shape and reducing the costs because the same punch die processing and inventory is used.

Figure 5 shows that only one stator is required, in this case the outside stator. When the teeth are aligned, the inductive reactance is higher than when the teeth are mis-aligned. The reactance of each of the three phases is compared and the difference is converted by a simple circuit to the standard synchro/resolver format to be digitized by a resolver-to-digital converter (RDC).

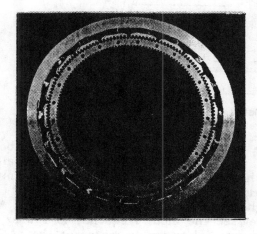

Figure 5 – Reactasyn^{TM} Resolver

Synchro/resolvers have many advantages in motor applications: they are insensitive to the motor's temperature rise, insensitive to the motor's bearing run-out or axial end play, and insensitive to electrical noise from the motor as well as the industrial environment. The "Reactasyn" resolver has these advantages plus the advantages of a reactance sensing device. There are no primary windings to energize in this resolver, hence slip rings or rotary transformers are eliminated.

The primary and secondary windings are combined in the Reactasyn so that all the active magnetic area is utilized which increases the accuracy and the output signal level, while requiring only three signal wires and a common in the interconnect cable.

The overwhelming advantage of reactance sensing is that it allows the use of ferro-magnetic teeth to multiply the number of poles without having to wind many windings. The Reactasyn resolver output is analogous to a 150-speed resolver; that is, for one mechanical revolution the electrical output rotates 150 times.

The Reactasyn resolver has a higher inherent resolution than a conventional resolver so that a low-cost 10 bit RDC can achieve a 153,600 ppr (150 cycles $\dot{x}$ 2^{10}). If a 12 bit RDC is used, 614,600 ppr resolution is available, which is about equivalent to a 19 bit incremental system. The speed is limited to about 100 rpm which is sufficient in robotic applications, due to the speed limitations of currently available 12 bit RDC's.

The tooth pitch of this resolver is the same as the motor's, so it can also be used for commutation information, eliminating optical or "hall" effect sensors. Unlike a totally incremental encoder, however, the position data is absolute over one tooth pitch, ensuring proper phasing and reliable motor starting when power has been reapplied. The positional accuracy of the Reactasyn resolver is within one half of an arc-minute, including the RDC (Figure 6).

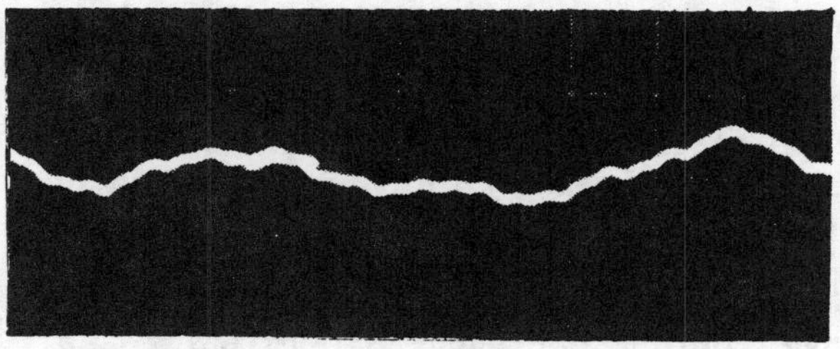

Figure 6 – Reactasyn accuracy
Vert: 1 arc-min/cm
Horiz: 1 Reactasyn cycle(144 arc-min)

A digital velocity loop is practical because of the high resolution of the Reactasyn, thus no tachometer is required for a servo loop. Although the Reactasyn resolver is actually a three phase device and perhaps should be called a "synchro" it is converted in the reactance detector electronics to the resolver format to be compatible with the more common resolver to digital convertor (RDC).

THE SERVO AND COMMUTATOR

The servo loop incorporates a commercially available microprocessor and digital servo positioning system.[3] This servo system was originally designed to be used with a DC motor with an incremental optical encoder. An interface called the commutator board was developed to convert to the brushless DC motor and to read a RDC. The digital servo system requires no tachometer since it derives velocity from the position sensor vs. time in the control algorithms. The loop parameters are software programmable as well as move commands through an RS-232 data link. The commutation board (Figure 7) contains the electronics for commutation of the motor based on the Reactasyn resolver position and combines it with the analog error signal from the servo loop's A/D input. The commutation board also supplies the 3 Khz AC for exciting the Reactasyn resolver as well as the reactance detectors for the RDC.

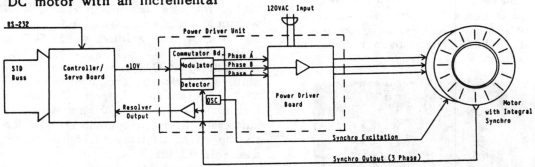

Figure 7 - Megatorque^TM Motor System

SYSTEM PERFORMANCE

The speed/torque curve in Figure 8 illustrates clearly the high output torque at low speed of the 14 inch diameter motor. At speeds below .2 rev/sec (12 rpm) the torque is over 200 ft-lbs. At speeds over .4 rev/sec the torque is less than half the static torque. This rapid reduction in torque is primarily a limitation of the power amplifier's output voltage. If the voltage is increased at higher speed the torque also increases because the slope of the DI/DT would increase similar to the "back EMF" limiting the speed of DC motors.

The maximum power output is .58 hp at .33 rev/sec (20 rpm). The power output decreases below that speed because the motors torque doesn't increase as rapidly as the speed is dropping due to magnetic saturation effects. Above .33 rps, the power output drops gradually with speed as the motor is voltage limited. Most of the drop in output power at higher speeds is losses in the motor. At 1 rev/sec the input frequency is 150 hz, a conventional two pole motor at that AC frequency would be spinning at 9000 rpm. This motor will operate at higher speeds (several hundred rpm) but the torque is falling rapidly and might only be useful in applications requiring a high slewing speed.

The maximum efficiency of this motor and power amplifier combination is 70% under full load at .5 rps (30rpm) which is quite high compared to a system with gear reduction. In robotic arms much energy is consumed at slow or stopped conditions where the efficiency of any system is close to zero, so little if any increase of efficiency may be realized.

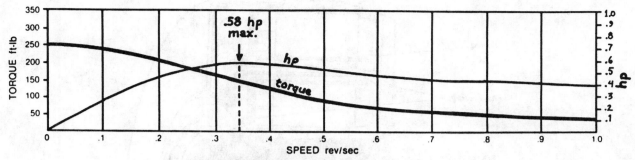

Figure 8 – Speed/Torque Curve

APPLICATIONS

Robotics and factory automation are the most obvious applications. The high output torque of this motor allows direct drive operation for the first time in robotics. The elimination of gears in positioning servos means backlash errors, wear and lubrication are eliminated while simplicity and reliability are increased. The response time is also increased by the elimination of the gear reducer, because the effective inertia of a motor increases as the square of the gear ratio.

Many systems for example, using harmonic drives, typically use low inertia, high performance ironless rotor motors to minimize the motor's contribution to the system inertia. Other applications also require high torque at a low speed; indexing tables, rotating antennas, large material take-up reels, etc. What they all have in common is the need to control position accurately and are currently using a higher speed motor with preloaded gears thus degrading that position.

CONCLUSION

A digital direct drive motor system has been demonstrated that is a departure from the conventional philosophy of using higher speed motors and anti-backlash gears. Designing a mechanical positioning system around a motor that develops its maximum power at 20 rpm significantly simplifies the product yielding higher performance at lower cost. A positioning system was described which stressed that it is necessary to consider the power amplifier as well as the position sensor when designing a brushless DC motor.

The motor design also easily lends itself to different sizes, from a 6" diameter motor with 35 ft-lbs of torque, up to a 22" motor that provides over 1000 ft-lbs.

Patents

The motor, resolver, and amplifier described in this article are protected by several foreign and domestic patents and patents pending.

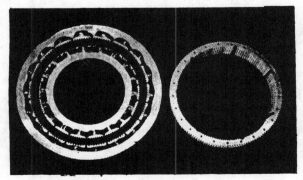

Figure 9a – Frameless Motor

Figure 9b – Motor with case

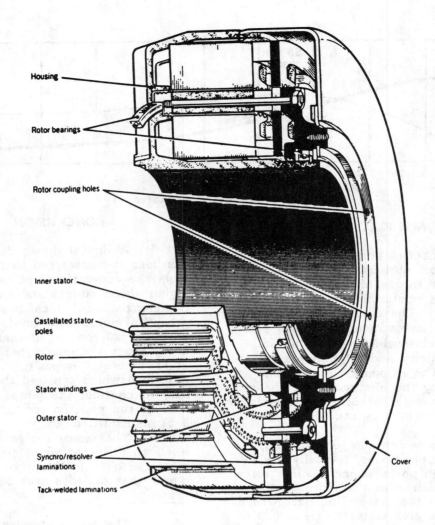

Housing
Rotor bearings
Rotor coupling holes
Inner stator
Castellated stator poles
Rotor
Stator windings
Outer stator
Synchro/resolver laminations
Tack-welded laminations
Cover

Figure 10 – Cutaway drawing of Megatorque motor with integral Reactasyn position transducer

References:

B. Powell, "A Low Cost, Efficient 1KW Motor Driver," PCI/MOTOR-CON, Atlantic City, N.J., 1984

I. Cushing, "A New High Accuracy Angular Position Transducer, " PCI/ MOTOR-CON, Atlantic City, N.J., 1984

J. Tal, "Improved Design with General-Purpose Digital Motion Controller," 12th Annual Symposium Incremental Motion Control Systems and Devices, Champaign, IL.

H. Asada, T. Kanade and R. Reddy, "Design Concept of Direct-Drive Manipulators Using Rare-Earth Torque Motors," Robotics Institute, Carnegie-Mellon University, Pittsburgh, PA 15213

High Performance Switched Reluctance Brushless Drives

W.F. Ray, P.J. Lawrenson, R.M. Davis, J.M. Stephenson, N.N. Fulton and R.J. Blake

IEEE Transactions, Vol. IA-22, No. 4, July/August 1986, pp. 722-730

Abstract—Switched reluctance (SR) drives offer the advantages of simple and robust motor construction, high speeds, high overall efficiencies over a wide operating range of torque and speed, simple power converter circuits with a reduced number of switches, and excellent controllability. The basis of these claims is explained. The history of the SR system, in particular the extensive research at Nottingham and Leeds Universities, and the basic operating principles and design considerations for motors and power converters are reviewed. Alternative configurations are discussed. The difficulties of establishing a simple mathematical model for the motor and of calculating torque and inverter VA requirements are examined. A comparison of the VA requirements for SR and pulsewidth modulation (PWM) is given for a 7.85-kW system. Measured drive performance is discussed in terms of efficiency over an operating envelope, specific output, controllability, and power converter ratings. Earlier traction drives, incorporating a constant power range and including regeneration, are reported together with the constant torque industrial drives and more recent traction extensions. The long-term potential of the drive is discussed for a wide range of applications. A comprehensive list of references is provided.

I. INTRODUCTION

THE SWITCHED reluctance (SR) variable-speed brushless drive is attracting increasing attention from analysts and research organizations and now has its own place in the program of recognized machines and drives conferences. However, as well as (and despite) the theoretical arguments which are developing, and no doubt will continue, as to whether or not the SR system should be expected to have intrinsic fundamental advantages over existing variable-speed systems, SR motors and controllers have been built, tested, and marketed and now constitute a serious alternative to dc and ac inverter drives.

It is now two years since Tasc Drives Ltd., UK, launched the first commercially available industrial SR "Oulton" drive, using a transistor-commutated thyristor controller. Recently, an improved version, using gate turn-off (GTO)

Paper IPCSD 85-54, approved by the Fractional and Integral Horsepower Subcommittee of the Industrial Drives Committee of the IEEE Industry Applications Society for publication in this TRANSACTIONS. Manuscript released for publication December 21, 1985. This work was supported in part by Chloride Technical Ltd., and by Lucas Electrical Vehicle Systems Ltd.
W. Ray and R. Davis are with the Electrical and Electronic Engineering Department, Nottingham University, and with Switched Reluctance Drives, Ltd., Springfield House, Hyde Terrace, Leeds LS2 9LN, England.
P. Lawrenson and J. Stephenson are with the Electrical and Electronic Engineering Department, Leeds University, and with Switched Reluctance Drives, Ltd., Springfield House, Hyde Terrace, Leeds LS2 9LN, England.
N. Fulton and R. Blake are with Switched Reluctance Drives, Ltd., Springfield House, Hyde Terrace, Leeds LS2 9LN, England,
IEEE Log Number 8608168.

Fig. 1. 7.5-kW "Oulton" GTO power converter (courtesy of Tasc Drives Ltd.).

thyristors, shown in Fig. 1, has appeared. This 7.5-kW SR drive has been the subject of an independent in-depth study [1] by the Electrical Research Association at Leatherhead, UK, who conclude that it offers "significant potential advantages over induction motor drives in efficiency, reliability and cost."

The prognosis of potential cost and reliability is due to the simplicity and ruggedness of the motor construction, exceeding even induction machines, and the simplicity of the power converter circuit which requires only one semiconductor switch per phase and which has no "shoot through" fault current path. The performance measurements indicate that the drive maintains a very high overall efficiency over a large proportion of its torque–speed envelope and has excellent controllability with a high torque-to-inertia ratio and a fast response to change in demanded torque. These features are fundamental and apply for power ratings ranging from a few watts up to megawatts and from low to high speeds.

However, although the SR system is gaining attention and recognition, it is still not as widely known or understood as the authors feel it should be. The object of this paper is to explain the basic operating principles, to review the progress that has been made to date, and to report on the measured performance and future potential of the SR drive.

II. Brief History

The SR motor is not a new concept. Early inventors of "electromagnetic engines" (e.g., Davis [2]) understood the switched reluctance principle but were unsuccessful in their attempts to build a motor due to poor electromagnetic and mechanical design and to the unavailability of suitable switching devices. An interest in switched field machines [3] was revived in the 1960's with the advent of the thyristor, and in the early 1970's patents were filed by Amato, Unnewehr, and Bedford [4]–[8], and Byrne and Lacey [9], among others. Published papers [10] at this time were rare. The work was mainly directed to finding alternative motors for vehicle propulsion [11]. However, serious development of these ideas did not appear to take place, probably due to the apparent (and unnecessary) complexity of either motors or power converters, or to an insufficiently competitive performance stemming from an insufficient understanding of the design principles.

The major work [12] which laid the foundation for the practical design of SR motors of competitive size and performance was published in 1980[1] and led to the commercial development of the Oulton drive. An interest in the SR system arose partly, but not entirely, from a desire to minimize the number of switching devices for the power converter, and particular attention was paid to the minimization of the VA ratings required. Using, in the first instance, a simplified linear model for the motor, important properties for the most beneficial current waveforms were identified [13], and various patents were granted [14]–[16]. The development of a representative nonlinear model [12] enabled these properties to be confirmed and accurate predictions to be made of drive performance. The prototype models confirmed the high efficiencies, and a further paper [17] described various power-switching circuits and control philosophies.

The work on the battery vehicle drive was extremely encouraging and Switched Reluctance Drives Ltd. was formed in 1980 by the originators of the program to carry out research, development, and design work and to license SR technology to manufacturers for industrial and other applications. Further papers (e.g., [18], [19]) outlined the progress made. Publications from other authors are now appearing; in particular, Harris and Finch [20] have been examining the benefits of multiple teeth per pole, and Miller [21] has been re-examining the converter VA-rating requirements based on a quasi-linear motor model.

III. Principles of Operation

Fig. 2 illustrates the principal features of an SR drive. The motor has salient poles on both the stator and the rotor and must not be confused with a conventional reluctance machine which has the well-known distributed three-phase winding in

[1] This paper followed an extensive program of research at Leeds and Nottingham Universities sponsored from 1974 by Chloride Technical Ltd. for battery-vehicle applications.

the stator slots and which is energized from sine wave supplies. The SR motor is similar to a variable reluctance (VR) stepping motor and, indeed, may appear identical except that

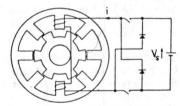

Fig. 2. Four-phase SR motor showing switching circuit for one phase.

the latter is, generally, designed as a low-power positioning device which runs in synchronism with a square wave supply, whereas the former differs in its design proportions to give an efficient and smooth variable-speed power drive for which switching of the supply is dictated by the rotor position. In its overall behavior the SR motor has more affinity with the dc brushless drive.

The motor shown in Fig. 2 has eight stator poles with simple concentrated windings and six rotor poles. The windings on the diametrically opposite poles are connected in series to form phases—four in this case. No windings of any kind exist on the rotor. Different numbers of poles may be used—for example, a 6–4 combination will give a three-phase machine with a 30° step angle, or a 4–2 combination, such as that described by Byrne and Lacey [9], will give a two-phase motor with a 90° step angle which, although unidirectional, can be designed to be self-starting from any position. The 8–6 motor shown has a step angle of 15° which, for various reasons, was found to be the most suitable for the battery vehicle and industrial applications. A more detailed discussion of pole combinations and phase numbers is given in [12].

Fig. 2 also shows, for one phase, one of the basic electrical circuits which may be used. When the switches are closed, the current builds up in the winding under the excitation of the direct voltage supply. When the switches are opened, the current transfers to the diodes, and the winding sees the reverse supply voltage, which causes the current to collapse. Pulses of current are thereby supplied to each phase in sequence, and for motoring operation each pulse causes the most adjacent rotor pole to move toward alignment with the energized stator pole. It will be seen that the rotor steps around in the opposite direction to the sequence of the stator pole excitations according to the well-known stepper motor principle. However, to think in terms of steps is only helpful from the viewpoint of understanding the rotation—in practice, the current pulses are controlled to occur at specific rotor angles to give a smooth transition, which, generally, means that the phase winding is substantially de-energized before the poles align.

A given phase undergoes a cyclic variation of inductance as rotation occurs. By making the simplistic assumption that the inductance is independent of the current, this variation is shown in Fig. 3. A motoring torque is produced if current exists during the interval when the inductance is increasing. Fig. 4(a) shows typical current pulses for operation at high speed. The energy is supplied during the period up to the commutation point; some is converted to mechanical output, some is stored in the magnetic field, and some is lost in the copper or iron. During the period after commutation, the field

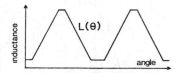

Fig. 3.   Linearized inductance–rotor angle variation for one phase.

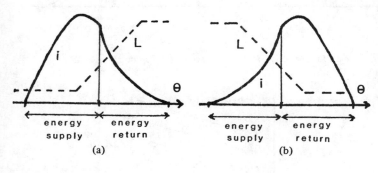

(a)   (b)

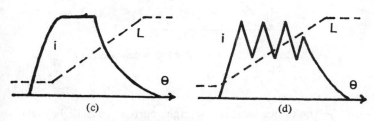

(c)   (d)

Fig. 4.   Typical phase current pulse shapes related to phase inductance. (a) High-speed motoring. (b) High-speed regenerating. (c) Flat topped. (d) Low-speed chopping.

energy is partly returned to the supply and partly converted to further mechanical output and losses. The mean current or power drawn from the supply depends on the difference between the current time areas of the energy supply and the energy return periods.

If the current pulse is appropriately positioned to occur while the rotor pole is departing from the stator pole (i.e., during decreasing inductance), then it takes the shape shown in Fig. 4(b), which is the mirror image of the motoring pulse. In this case a net mean current is returned to the supply, and the machine acts as a brake or generator. In the energy supply period, energy is taken both from the supply and (to a lesser extent) from the shaft and stored in the field; during the energy return period, further energy is taken from the shaft and returned, with the field energy, to the supply.

It will be seen that the winding current is unidirectional and independent of the direction of torque, which is an important

and advantageous feature of the drive. It is thus possible to control the drive for both motoring and braking/regeneration with a single switching device per phase. The electrical circuit of Fig. 2 can be modified to achieve this as shown later.

A rotor position transducer is an essential element of the system. To achieve the best performance for a particular operating speed and torque, the current must be initiated and commutated at prescribed rotor angles.

At high speeds the switches must be closed in advance of the rising inductance region to allow time for the current to build up. During the rising inductance region, a "back electromotive force (EMF)" exists which can cause the current to fall before commutation (Fig. 4(a)) or remain substantially constant, if it equals the supply voltage as shown by the flat-topped waveform of Fig. 4(c). This mode of behavior is called "single pulse."

At low speeds the current is maintained between the upper and lower chopping levels by repeatedly opening and closing the switches during the rising inductance period as shown in Fig. 4(d). This mode of behavior is called "chopping."

IV. CALCULATION OF TORQUE

The flux linkage $\psi(\theta, i)$ for a phase winding varies cyclically with the rotor position θ and the phase current i; typical variations are shown in normalized terms in Fig. 5(a). The voltage v applied to the winding is related to ψ at any instant by

$$v = Ri + \frac{d\psi(\theta, i)}{dt} = Ri + \frac{\partial \psi}{\partial i}\frac{di}{dt} + \frac{\partial \psi}{\partial \theta}\frac{d\theta}{dt}. \quad (1)$$

The SR motor phase can, therefore, be represented by the equivalent circuit of Fig. 6, where the incremental inductance $L_i(\theta, i) = \partial \psi / \partial i$ and the speed dependent back EMF $(\partial \psi / \partial \theta)(d\theta / dt)$ may be subdivided into two sources e_f and e_m. $i \cdot e_f$ is the rate of energy storage in the magnetic field due to changing inductance, and $i \cdot e_m$ is the gross mechanical output power. The iron loss effect is neglected.

Unlike most machines, this equivalent circuit is not very helpful, since the "constants" L_i and $\partial \psi / \partial \theta$ vary with θ and i, as does the ratio e_f / e_m. It is, therefore, not possible to calculate torque from simple equivalent circuit considerations, which is frustrating to those who wish to predict performance.

Instantaneous torque $\tau(\theta, i)$ must be calculated from the winding coenergy $W'(\theta, i)$ by the relationship

$$\tau(\theta, i) = \frac{\partial W'}{\partial \theta} \quad (2)$$

where

$$W'(\theta, i) = \int_0^i \psi(\theta, i) \, di. \quad (3)$$

441

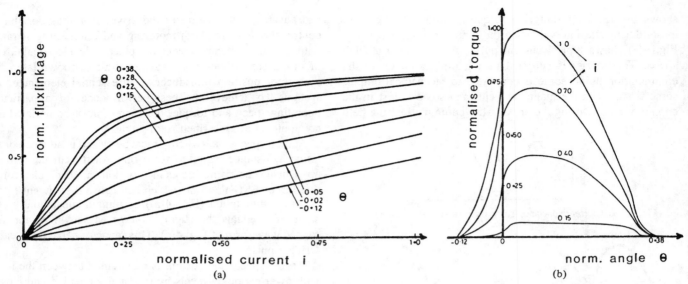

Fig. 5. Typical variations with rotor angle θ and current i for one phase. (a) Flux linkage ψ. (b) Torque τ.

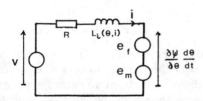

Fig. 6. Equivalent circuit for SR motor phase. e_f represents energy storage due to inductance change. e_m represents energy converted to mechanical output.

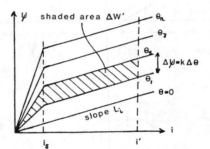

Fig. 7. Quasi-linear flux linkage characteristics.

Typical variations of $\tau(\theta, i)$ are shown in normalized terms in Fig. 5(b). Since $i \cdot e_m = \tau \cdot d\theta/dt$, the ratio

$$\frac{e_f}{e_m} = i \left(\frac{\partial \psi}{\partial \theta}\right) \Big/ \left(\frac{\partial W'}{\partial \theta}\right) - 1. \qquad (4)$$

Various attempts may be made to approximate the ψ–θ–i model in order to calculate torque. Fig. 7 shows a typical case, similar to that used by Miller [21], for which

$$\psi = L_i i + k\theta, \qquad (i \geqslant i_s, \ 0 \leqslant \theta \leqslant \theta_n)$$

$$\psi = \left\{L_i + \frac{k\theta}{i_s}\right\} i, \qquad (i \leqslant i_s, \ 0 \leqslant \theta \leqslant \theta_n) \qquad (5)$$

where L_i, k, and i_s are constants, and where θ is measured over the period of increasing inductance.

It will be seen from Table I that for a given i, the torque is constant over the period $0 < \theta < \theta_n$ as shown in Fig. 8(a), which should be compared with Fig. 5(b). The deficiency of the quasi-linear model is apparent. Nevertheless, this model does make two useful predictions that are confirmed in practice.

442

1) The torque increases with the current, essentially in the manner shown by Fig. 8(b).
2) The ratio of energy converted to energy stored is greater for a saturated motor. This has been shown by Byrne [22] and Miller [21]. However, accurate calculation of the mean torque can only be achieved by a) computation of the $\psi(\theta, i)$ characteristics, b) definition of a driving voltage waveform $v(\theta)$, c) integration of (1) to yield a current waveform $i(\theta)$, d) transposition using the $\tau(\theta, i)$ characteristics to give a torque waveform $\tau(\theta)$, and e) integration of the torque waveform to determine the mean torque.

Computer algorithms, developed by the team and refined over a period of years, enable this process to be executed and sufficiently accurate predictions of performance to be made [12]. A more detailed examination of torque calculation is given by Stephenson and Corda [23], [24].

V. CIRCUITS AND DEVICES

Fig. 9 shows three basic power circuits. The first, used in Fig. 2, requires two switches per phase, each rated at the supply voltage V_s.

TABLE I

INSTANTANEOUS TORQUE AND BACK EMF RATIO FOR QUASILINEAR
MODEL OF FIG. 7

	$i \, \partial\psi/\partial\theta$	$\tau = \partial W'/\partial\theta$	e_m/e_f
$i \leqslant i_s$	ki^2/i_s	$ki^2/2i_s$	1
$i \geqslant i_s$	ki	$k(i - (i_s/2))$	$(2(i/i_s) - 1)$

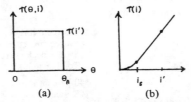

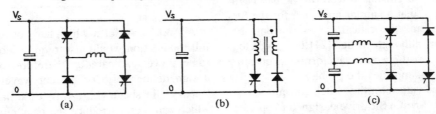

Fig. 8. Torque–angle–current relationship for quasi-linear model of Fig. 7.

Fig. 9. Basic power converter circuit configurations. (a) Requiring two switches per phase. (b) Requiring one switch per phase. (c) For higher voltages, with dc supply center tapped.

The second circuit (Fig. 9(b)) requires only one switch per phase but requires a secondary bifilar winding on each coil. The switch voltage rating is now $2V_s + \Delta V$ (ΔV is due to imperfect coupling [17]). It has the disadvantage that more connections are necessary between the motor and the power converter, and the winding utilization within the motor is reduced. However, it has attractions for applications where the supply voltage is sufficiently low such that switches of $2V_s + \Delta V$ rating are of little penalty. This circuit was used for the battery vehicle application.

The third circuit (Fig. 9(c)), which shows two phase windings, again requires only one switch per phase but is more attractive at higher voltages because the dc supply is center tapped. The phase windings, which must be even in number, are designed for $V_s/2$ (and, therefore, will draw twice the current compared with the former circuits), and the switches must be rated for V_s. This circuit is used for the Oulton drive.

The center tap of circuit 9(c) is formed by capacitors and its potential is maintained by the symmetry of the mean-phase currents. This requires adjustment of the chopping levels at low speed [19].

It will be seen that the total VA requirements for the switches are essentially the same for the three circuits of Fig. 9. However, the single device per phase has advantages from the viewpoint of assembly and cost.

Due to its enhanced peak-to-mean current rating and the current waveforms for the SR motor, the thyristor is the natural choice of switching device rather than the transistor. Earlier power converter designs were, therefore, based on

forced commutated thyristors, and, due to the absence of antiparallel diodes, savings in commutation circuitry compared with conventional inverters could be made by employing group commutation [17], [25]. However, GTO thyristors have recently become available and offer a more attractive alternative. Bipolar transistors and metal–oxide–semiconductor (MOS) devices are possibilities, although probably at lower powers. The technology is changing fast, and the most economic solution one year may be overturned the next.

As for PWM inverter drives, the switching circuit is buffered from the ac supply by a dc link capacitor. Therefore, no difference exists from the viewpoint of the supply power factor, which depends entirely on the impedance of the supply and any added reactance on either the ac or dc side of the diode rectifier.

VI. VOLT–AMPERE RATINGS

Various attempts have been made to estimate converter volt–ampere ratings from linear [17] and quasi-linear [21] models of the SR motor for comparison with the three-phase PWM induction-motor circuit. The PWM case is relatively easy to estimate. Assuming that the devices are rated for the full power condition corresponding to a fully modulated (quasi-square wave) motor-line voltage and that the output power P depends entirely on the fundamental component, then

$$P = \frac{3}{\pi} V_s \hat{I}_1 \eta_1 \cos \phi_1 \qquad (6)$$

where

V_s dc link supply voltage,
$\hat{I}_1$ peak line current for fundamental component,
ϕ_1 current phase lag for fundamental component,
η_1 efficiency operating from sine wave supplies.

The motor line current at full power is shared largely between the two switches, and the contribution to the rms current by the diodes is negligible. The switches, however, must be rated for the peak motor current. The rms and peak ratings required are, therefore,

$$I_{sw\ rms} > k_r \hat{I}_1 / 2$$

$$I_{sw\ pk} > k_p \hat{I}_1 \qquad (7)$$

where k_r and k_p are multiplying factors (>1) to account for the harmonic content of the current waveform. It has been demonstrated in Section V that a similar approach for the SR case is not really viable, and theoretical values can only be obtained by computation, although Miller [21] has predicted ratings based on a flat-topped current waveform and quasilinear characteristics which simplifies the problem.

Table II gives measured values for the switch currents at full power for a 7.85-kW 1500-r/min SR drive designed to operate using the circuit of Fig. 9(c) from a minimum dc link voltage of 475 V. These values are compared with those for a typical induction motor operating at the same power and link voltage. The current multipliers k_r and k_p of (7) are taken to be 1.1 and 1.25, respectively, with $\eta_1 \cos \phi_1 = 0.85^2$. Miller takes k_r as 1.15 but neglects the effect of η_1.

However, switch ratings are not necessarily based on the full power condition if additional slow-speed torque is required. Table II also gives the current values for slow speed rotation at 140-percent torque for the SR drive and compares these with the corresponding PWM case. It is assumed that I_1 will be 140 percent of the 100-percent torque value (in practice, I_1 will be slightly greater) and that $k_p = 1.04$ and $k_r = 1.0$. However, the switch rms current is also reduced by a factor of $\sqrt{2}$ since the motor current is now shared between two switches and two diodes.

Taking into account the four switches used by the SR converter, compared to six for PWM, Table II also compares the overall kVA requirement. A small but significant overall advantage exists in favor of the SR system, but this advantage is enhanced by the lower number of switches required. The comparison will vary slightly, depending on the type of induction motor used. In general, as the drive rating becomes larger, k_r will be worse—or some modulation at full power will be necessary to reduce harmonics. Some evidence exists that the SR advantage increases with the power rating.

The slightly greater kVA/kW requirement of 10.9, estimated by Miller [21] from theoretical considerations of the SR system, deserves comment. It may appear logical that a flat-topped current waveform is best for minimizing the peak current required. However, it results from an incompletely

modulated voltage supply; the maximum available volt seconds of the supply have not been utilized and, with a greater flux linkage excursion, the current excursion can be reduced.

TABLE II
COMPARISON OF SR AND PWM CONVERTER RATINGS FOR A 7.85-kW 1500-r/min 475-V dc LINK DRIVE

	SR Motor (a)	Induction Motor (b)	Ratio (4 × (a))/ (6 × (b))
100-percent torque, 100-percent speed			
Switch rms current, A	17.7	13.2	0.90
Switch peak current, A	40	30	0.89
Total switch kVA/kW	9.7	10.9	0.89[a]
140-percent torque, slow speed			
Switch rms current, A	17.5	11.9	0.98
Switch peak current, A	45	35	0.86

[a] Ratio (a)/(b).

A PWM waveform which had not reached the quasi-square full modulation would also require increased currents for the same power. A paradox therefore exists—it is possible to design deliberately for current waveforms which are more "peaky" in shape (Fig. 4(a)) by utilizing early turn-on, but which can thereby reduce the peak current requirement [17], [26].

The appropriate design for an SR drive involves a careful balance between the motor and the power converter constraints in conjunction with a balance between top-speed, base-speed, and low-speed starting conditions for a particular drive specification. The SR system has a variety of parameters, such as the number of phases, poles, pole arcs, diameters, turns, converter circuits, and switching angles, which offer great flexibility in the design and enable, albeit by complex calculation, the optimum combination to be utilized. .

VII. CONTROL

The SR system requires, for the 8–6 motor, a four-phase position reference signal which is simply obtained from two optical or magnetic heads. Once rotating, a finer resolution of position is obtained using a phase-locked loop, and speed is measured using an F–V converter. The four-phase system also requires a current transducer for each pair of phases.

The control system requires a memory for the correct switching angles for each torque–speed condition within the single-pulse mode and for the appropriate current levels in the chopping mode, together with logic to fire and commutate sequentially the power switching devices. No requirement exists for complex modulation strategies as with the PWM. At present the memory is analog, and the controller uses discrete IC's. With due development this will be replaced by a microprocessor and/or uncommitted logic array (ULA) controller.

VIII. PERFORMANCE

Fig. 10 shows the power–speed characteristic and efficiency contours for the 50-kW prototype battery-vehicle drive with a base speed of 750 r/min. The integrated assembly of the forced/through ventilated power converter and motor is shown in Fig. 11. Despite its early design, the range of high efficiency is noteworthy.

The Oulton industrial drives produced by Tasc Drives Ltd. have a constant torque characteristic, rather than the constant

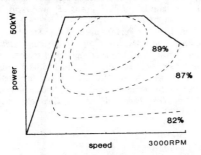

Fig. 10.   Power/speed envelope and efficiency contours for 50-kW prototype SR traction drive.

Fig. 11.   50-kW prototype SR battery-vehicle drive.

power characteristic of Fig. 10 used for traction. Table III gives the efficiencies at half and full speed for the four drives presently in the family. A typical 22-kW four-pole induction motor on a sine-wave supply would have an efficiency of about 89 percent at full power but significantly less when supplied from an inverter.

Although a drive may be designed, say, for a constant torque characteristic to 1500 r/min, modification to the control strategy can give increased power and torque and operation to higher speeds; i.e., the drive can meet other more demanding applications. As an example, Fig. 12 shows the torque/power/speed characteristics and efficiencies obtained by a prototype system developed by SR Drives for a special application which requires high torques at low speeds and a series-type characteristic up to 3000 r/min. A standard Tasc 180 motor, normally rated (100-percent torque and power) at 22 kW at 1500 r/min, is supplied by a GTO converter. The figure clearly shows the high low-speed torque (210 percent) and the maximum power (132 percent) obtainable (limited here by the GTO ratings), and the contours of constant efficiency empha-

size the fact that the high overall efficiencies are maintained over a wide range of loads and speeds. The discontinuity of the efficiency contours at 625 r/min is caused by the change from the chopping mode to the single-pulse mode of control.

At the present state of development, no known upper limit exists to the prospective range of torques and powers for SR drives. Some years ago the authors carried out a feasibility study for a traction drive, developing 220 kW at 1390 r/min,

TABLE III
OVERALL EFFICIENCIES OF OULTON SR DRIVES

| Size | 750 r/min | | 1500 r/min | | Peak Torque Inertia (rad/s^2) |
	kW	Efficiency (percent)	kW	Efficiency (percent)	
D112	2.0	80	4.0	82	7640
D132	3.75	83	7.5	85	5600
D160	7.5	85	15.0	87	3580
D180	11.0	88	22.0	91	2680

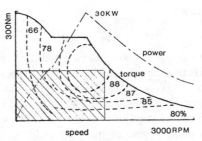

Fig. 12.   Torque/speed envelope and efficiency contours for high-performance 180-frame GTO drive based on standard Tasc 180 motor (industrial constant torque shown shaded).

the conclusion of which was that the thyristor-switched SR drive should have a considerable cost advantage over the inverter-fed induction motor for this application [19].

However, performance is not limited only to the overall efficiencies that can be achieved. Since for the 8–6 system switching angles are set every 15° of rotor rotation, the developed torque can be changed from full torque to zero or to maximum braking in a very short time. Virtually no time delays exist between the controller demanding torque and the motor developing the same. This, coupled with the high torque-to-inertia ratio (see Table III for general industrial drives) enables fast speeds of response to be achieved and has important implications in the machine tool and servo-motor application areas. A prototype 3-kW machine-tool drive was capable of braking from 4000 r/min to rest (admittedly on a free shaft) in 0.3 s. It is surprising that the limiting factor can be the ability of the control electronics to keep abreast of the rapidly changing situation.

A performance feature which often gives rise to question, stemming from stepper-motor experience, is stability. The SR system is not synchronous in the sense that the rotor takes up a load angle with respect to an alternating or pulsed excitation.

445

With reference to the rotor position measurement, the switching angles are set to produce a demanded torque which will accelerate or decelerate the motor. No inherent instabilities exist in the system; the only way instability can arise is due to improper design of a speed-control feedback loop, as is true for any speed-control system.

IX. POTENTIAL

The application potential for the SR drive is considerable; indeed, few applications exist for which it will not be a future contender. The high efficiencies, essential for the original battery vehicle work, and high speeds lead toward fan and pump drives, and, increasingly, industrial drives will become more energy-efficient. The robust brushless construction and good thermal features (heat loss is largely confined to the stator) make the drive attractive for mining and flameproof applications in addition to traction. The simplicity and low cost have implications for domestic appliances and the controllability for machine tools and robotics. The developments and designs are in hand, covering powers from tens of watts to megawatts.

X. CONCLUSION

SR drives have been shown to be capable of providing high levels of performance over a wide range of specifications in a robust form. This is achieved by sophisticated design of motors, power electronics, and control strategies, including closely integrated design of the motor and the converter. To achieve this many variables must be included in the design optimization.

The SR drive is new compared with its competitors, and many ideas are yet to be explored and much refinement of the detailed technology to be undertaken before it can be said to have reached its full potential. In addition, the development of new and improved electronic devices is constantly improving its capabilities.

ACKNOWLEDGMENT

The authors wish to thank Dr. M. F. Mangan for his support and assistance, the Departments of Electrical and Electronic Engineering at Leeds and Nottingham Universities for the use of their facilities, and Tasc Drives Ltd., whose importance to the work will be clear from the text. Appreciation and thanks are also due to the more recent members of the team of SR Drives Ltd., including P. D. Webster, S. P. Randall, D. M. Sugden, and H. C. Lovatt.

REFERENCES

[1] A. W. Woods and R. G. Cann, "A comparison of induction and switched reluctance motor drives," *Elec. Res. Ass. Rep.* 84-0220, Mar. 1985.
[2] A. F. Anderson, "Discussion on variable-speed switched reluctance motor systems," *Proc. Inst. Elec. Eng., Pt. B*, vol. 128, p. 265, Sept. 1981.
[3] ——, "The thyristor control of a reluctance motor," Ph.D. dissertation, Univ. of St. Andrews, England, 1965.
[4] US Patent 3560817.
[5] US Patent 3560818.
[6] US Patent 3560819.
[7] US Patent 3560820.
[8] US Patent 3679953.
[9] J. V. Byrne and T. G. Lacy, "Electrodynamic system comprising a variable reluctance machine," British Patent 1321110, 1970.
[10] L. E. Unnewehr and W. H. Koch, "An axial gap reluctance motor for variable speed applications," *IEEE Trans. Power App. Syst.*, vol. PAS-92, pp. 367–376, 1974.
[11] H. Bausch and B. Rieke, "Performance of thyristor-fed electric car reluctance machines," in *Proc. Int. Conf. Electrical Machines*, 1978.
[12] P. J. Lawrenson et al., "Variable-speed switched reluctance motors," *Proc. Inst. Elec. Eng., Pt. B*, vol. 127, pp. 253–265, July 1980.
[13] W. F. Ray and R. M. Davis, "Inverter drive for doubly salient reluctance motor: Its fundamental behaviour, linear analysis and cost implications," *Proc. Inst. Elec. Eng., Pt. B, Elec. Power Appl.*, vol. 2, pp. 185–193, 1979.
[14] British Patent 1597486.
[15] British Patent 1604284.
[16] British Patent 2037103B.
[17] R. M. Davis et al., "Inverter drive for switched reluctance motor: Circuits and component ratings," *Proc. Inst. Elec. Eng., Pt. B*, vol. 128, pp. 126–136, Mar. 1981.
[18] R. M. Davis, "The switched reluctance drive," in *Proc. Conf. Drive/Motors/Controls*, 1983, pp. 188–191.
[19] W. F. Ray et al., "Industrial switched reluctance drives—Concepts and performance," in *Proc. Inst. Elec. Eng. Conf. Power Electronics and Variable Speed Drives*, 1984, pp. 357–360.
[20] J. W. Finch et al., "Variable speed drives using multi-tooth per pole switched reluctance motors," in *Proc. 13th Symp. Incremental Motion Control Systems and Devices*, 1984, pp. 293–301.
[21] T. J. E. Miller, "Converter volt-ampere requirements of the switched reluctance motor drive," presented at *IEEE Ind. Appl. Soc. Annu. Meeting*, Chicago, IL, Oct. 1984.
[22] J. V. Byrne and J. B. O'Dwyer, "Saturable variable reluctance machine simulation using exponential functions," *Proc. Int. Conf. Stepping Motors and Systems*, 1976, pp. 11–16.
[23] J. Corda and J. M. Stephenson, "Analytical estimation of the minimum and maximum inductances of a doubly salient motor," in *Proc. Int. Conf. Stepping Motors and Systems*, Sept. 1979, pp. 50–59.
[24] J. M. Stephenson and J. Corda, "Computation of torque and current in doubly salient reluctance motors from non-linear magnetisation data," *Proc. Inst. Elec. Eng.*, vol. 126, pp. 393–396, May 1979.
[25] W. F. Ray et al., "Switched reluctance motor drives for rail traction—A second view," *Proc. Inst. Elec. Eng., Pt. B*, vol. 131, pp. 220–225, Sept. 1984.
[26] W. F. Ray and R. M. Davis, "Reluctance electric motor drive systems," British Patent 1591346.

William F. Ray graduated from Cambridge University, England, in 1963.

After spending a year with Elliott Bros. at Rochester, he joined the Electrical Engineering Department at Nottingham University where he is now a Senior Lecturer teaching automatic control and power electronics. He has acted as a consultant for a number of companies including GEC, British Rail, Brush and Chloride Technical, and is a founder director of Switched Reluctance Drives Ltd.

Peter J. Lawrenson (SM'67–F'76) received the M.Sc. and D.Sc. degrees from Manchester University, England, in 1956 and 1971, respectively.

He spent five years with the Metropolitan Vickers Company before joining the Electrical Engineering Department at Leeds University. He was promoted to a Chair in 1966 and to Head of Department in 1974. He was elected to the Fellows of the Royal Society (FRS) in 1981. He has acted as Chairman of several boards at the university and as a Consultant to more than 30 companies. He is a founder and Chairman of Switched Reluctance Drives Ltd.

Dr. Lawrenson has acted as chairman of several boards for the Science and Engineering Research Council and the Institution of Electrical Engineers.

Rex M. Davis graduated from Queen Mary College, London, England, in 1954.

He joined the Brush Electrical Engineering Company Ltd. at Loughborough where he became Head of Electrical Research. He joined Nottingham University in 1966 and was promoted to Senior Lecturer in 1968. In 1969 he published a widely used book on power electronics. He has acted as a Consultant to several companies including GEC, Rolls Royce, British Rail, and Chloride Technical. He is a Founder and Director of Switched Reluctance Drives Ltd.

Mr. Davis has served on several committees and boards at the SERC and the Institution of Electrical Engineers.

J. Michael Stephenson graduated from Leeds University and received the Ph.D. degree in 1958 and 1974, respectively.

He spent three years with AEI, Rugby, working on the design of large dc machines before joining the Electrical Engineering Department at Leeds University. He was promoted to Senior Lecturer in 1978. His book, *Per Unit Systems,* has been widely used by machine designers. He has acted as a consultant to ten companies, including British Rail, S.U. Fuel Systems, and Chloride Technical, and is a Founder and Director of Switched Reluctance Drives Ltd.

Dr. Stephenson has served on several Institution of Electrical Engineers committees and the organizing committees for four international conferences in electrical machines.

Norman N. Fulton graduated from Strathclyde University and received the Ph.D. degree, in 1969 and 1973, respectively.

He spent eight years with Hoover Ltd. where he became Superintendent of the Motor Laboratory. In 1977 he joined Leeds University as a Research Fellow and was promoted to Senior Research Fellow in 1979. There he worked on battery-vehicle SR motor development. He joined Switched Reluctance Drives Ltd. in 1983 as General Manager and Chief Motor Designer.

Roy J. Blake graduated from Surrey University, received the M.Sc. degree at Imperial College, London, and the Ph.D. degree from Nottingham University, in 1974, 1975, and 1986, respectively.

He then spent a year with UKAEA, Culham, and a year with Eurotherm Ltd., Worthing, before joining Nottingham University in 1977 as an Industrial Research Fellow where he worked on the development of power-electronic SR drives for battery vehicles. He joined Switched Reluctance Drives Ltd. in 1983 as Chief Electronics Engineer.

High Performance MOSFET Switched Reluctance Drives

D.M. Sugden, R.J. Blake, S.P. Randall, J.M. Stephenson and P.J. Lawrenson

IEEE Industry Applications Society Annual Meeting, Atlanta, GA, October 1987

ABSTRACT

This paper describes the application of MOSFET switches to high-performance, low-power, doubly-salient, switched reluctance drives in the power range 100W-3000W. The basic features of SR motors are reviewed together with the range of possible power electronic circuits used to switch the motor currents. Design principles are discussed in the context of technical and economic factors and practical results are presented. The low power MOSFET-switched SR drive is shown to provide excellent performance and show great commercial potential.

(1) INTRODUCTION

The doubly salient switched reluctance (SR) drive is now accepted as a serious competitor to inverter-fed induction motors and controlled dc motors for a very wide variety of variable-speed applications. Advances in the design of SR motors and their control electronics together with the steady progress in power switching technology, have combined to produce a drive of outstanding performance in terms of specific output, efficiency and controllability using a cheap, simple, rugged and reliable variable speed motor. For perhaps the first time, the complete drive is engineered as a unified system - the machine and converter design being interactive and interdependent, producing an optimum total drive package.

This is particularly demonstrated by a recent extensive programme of development of small drives with particular reference to the power range 100W to 3000W and torques 0.5Nm to 10Nm. Results of this programme form the basis for this paper.

The majority of the substantial body of previously published work on SR systems [1 to 12] has focused on physical principles of the system and the performance of medium and high power SR drives for general industrial and traction duty. This paper redresses the balance by concentratring on low power drives which use the rapidly advancing MOSFET technology in the power converter.

(2) REVIEW OF SR DRIVE FEATURES

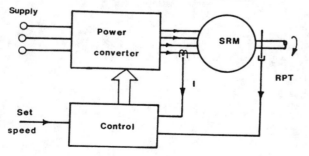

Fig. 1. Basic SR Drive.

The principles of operation of SR drives are now widely understood and will not be explained here. It is appropriate, however, to briefly review the principal features which characterise the system. Fig 1 shows the basic components of the SR drive system.

The stator of the machine consists of a series of wound poles, the coils of which may be energised in sequence (from the power converter). Fig 2 illustrates a 3-phase motor (with the coils of only one phase shown) although work is in progress on 1,2,3 and 4-phase drives. The rotor is a simple toothed structure with no magnets, coils or brushes. The motor is therefore extremely robust and simple to manufacture.

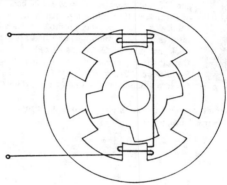

Fig. 2. Basic SR Motor.

The rotor position transducer, together with the electronic control system, ensures that the coils are excited at the desired instant to develop the required torque in the most efficient way.

Although the principle is very simple, the design and control of these drives are sophisticated. Some of the many design parameters will emerge below, but it should be emphasised that the design of the system for a given application is a complex matter, involving electromagnetic, power electronics and control electronic sub-systems in the total drive optimisation. Design and operating experience is increasing* rapidly and development is directed to increased performance and range of powers and applications.

It is the flexibility of the control of the motor excitation which is largely responsible for the drive's excellent performance and high efficiency over a wide speed range. At low speeds the converter acts as a chopper supplying a controlled magnitude of current to the motor windings. The higher the level of the current supplied, the higher is the developed drive torque. At higher speeds, however, the drive runs in a 'single-pulse mode' whereby the converter supplies single pulses of current to the motor phases in sequence at carefully-timed instants. In this way

---

* S P Randall and Dr J M Stephenson are also with the University of Leeds

the drive can be run at any load point with the control parameters set by the electronics to give the best drive efficiency.

The characteristics of a given 'standard' drive can be programmed to meet a particular application (eg a constant torque drive, a fan drive, a constant power drive) in the optimum way. However the choice at the design stage of control strategy is very important for a new system with particular regard to speed, torque and power ranges, system efficiency profile with load and speed, specific output, starting torque, need for reversibility, braking etc.

The SR drive therefore lends itself to the design of standard drive 'packages' which offer extremely good control and application flexibility.

An important feature of the SR drive is that the direction of current in the motor windings is immaterial, leading to simple power converter circuits. A number of circuit configurations are available to the designer, each with its own merits and the one chosen will depend on the application for the drive. For low power drives, the four circuits shown in Fig 3 are worthy of serious consideration. It should be noted that none of these circuits includes the shoot-though fault path unavoidable in most converters supplying bidirectional current including inverters.

(3) POWER SWITCHING DEVICE REQUIREMENTS

The power converter circuit can be implemented in a number of ways as described previously but to a first order the total switching VA requirement is the same for any of the circuit options. The individual device ratings are, however, influenced by circuit topology. The two-switch-per-phase converter gives a switch voltage rating of the dc supply voltage plus allowances for overshoots and a safety margin. The one-switch-per-phase converters give switch voltage ratings of 2-3 times the dc supply voltage. Current ratings are of course dependent on motor power output but experience has shown that in a design where the motor and power electronics are optimised the switch peak current rating is approximately twice the rms requirement. However, the peak rating chosen is usually somewhat higher to provide for overcurrent protection with an adequate safety margin. The switching frequency of the power devices is dependent on the mode of control. In the single pulse mode, frequencies are very modest with for example, a value of 800Hz for a 3-phase motor running at 12,000 rev/min. In chopping the current in the motor winding is kept constant by closed-loop feedback and switching frequencies of 20kHz are used so that acoustic noise is minimised.

In the context of the selection of switching devices it is helpful to make a comparison with the corresponding needs of inverters for supplying synchronous reluctance motors or, more commonly, induction motors. The synchronous reluctance motor requires an inverter of larger VA rating than the corresponding induction motor. The SR motor on the other hand requires a total switch VA similar to or less than the induction motor. (This important point has been made for integral horsepower drives in a previous paper by the Authors [5] which itself followed, and was accepted by, the authors of an earlier report [6].

In comparing the switching device requirements of SR drives and inverter fed induction motors it is useful to consider a specific example. The power converters for both systems are required to operate from the single phase 220V/240V ac mains supply. The power converter configuration chosen for the SR motor was the two switch per phase and this together with the induction motor inverter is shown in Fig 4.

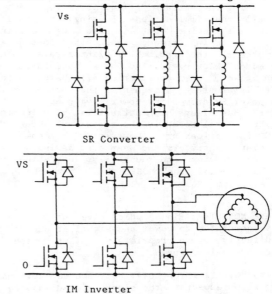

SR Converter

IM Inverter

Fig. 4. SR Converter compared with a Variable Frequency Inverter.

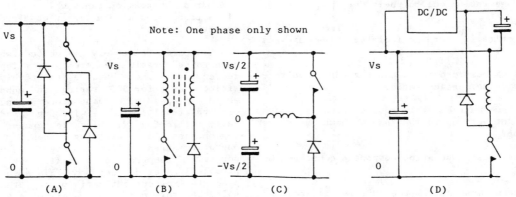

Note: One phase only shown

(A)   (B)   (C)   (D)

Fig. 3. SR Power Converter Circuit Configurations.

449

The voltage rating of the switches for both systems is identical with 450-500V being adequate for the nominal 320V link. When considering the current ratings of the switches a useful picture may be obtained by comparing the peak and rms currents for full speed full torque and low speed full torque.

Drive requirements and safe operating areas are similar to MOSFETs although the peak to rms current ratio is not as good. It seems likely that IGTs will replace bipolar transistors in the higher power ranges because of their simple low-cost drive requirements, but clearly the speed at which this happens will depend on the pricing policy of the manufacturers. However, they are unlikely to compete with MOSFETs at lower powers even though they use the silicon area more efficiently.

Low power GTOs are available which are suitable for drive power levels of 500W-3kW but their gate driving, snubbing and low switching speed make them unsuitable unless high voltages are being used. They are however widely accepted as the best switch for inverters above the 100kVA level and as a serious contender for inverters in the 10kVA to 100kVA region.

When assessing the various devices it is important to embrace all the implications of using the device. These include device cost, ease of driving, snubbing, switching speed, conduction losses and most important, their suitability/inter-relationship with the motor (keeping in mind the various forms that this can take). When considering all these factors the MOSFET is at present frequently the optimum choice for low power SR drives for power levels up to 2-3kW. The MOSFET has always been attractive on technical grounds but until recently the costs have been too high. MOSFET prices are, however, now competing with bipolar transistors and when cost savings due to simple gate driving and snubberless operation are taken into account the MOSFET is often the most economic solution.

The lower MOSFETs (Fig 4) which have their source terminals connected to power and signal ground may be driven directly by a proprietary buffer which is controlled by logic signals.

The upper MOSFETs which have their source terminals floating with respect to the signal ground require an isolated drive system. (P- channel MOSFETs could have been used which would require a simpler level shifting drive arrangement but at present they are not economic at this voltage rating due to their very poor use of silicon.)

The design criteria for the isolated drive are as follows:

1. The system must provide a constant drive voltage over a wide mark space ratio.

2. The system must be able to withstand a common-mode dv/dt of up to 15kV/µs without spurious switching.

3. The system must have a low component count and be low cost.

4. The system must switch the MOSFETs as fast as practicable.

450

Many methods of achieving these requirements are possible and three commonly used solutions are as follows:

1. Transformer isolation.

2. Optical isolation.

3. Capacitative isolation.

The peak or rms current ratings for the switches in the SR converter are generally somewhat lower than those for a variable frequency inverter of an equivalent power rating by a factor of between 5 to 10%. Moreover, it should be noted that these ratings apply to drives which have motor volumes much smaller than the equivalent induction motor and which operate over considerably wider speed ranges. The advantage of the SR drive, that there is no shoot-through path in the power converter, is emphasised by the Figs 3 & 4. This means that fault conditions can easily be monitored and corrective action taken before device currents exceed their normal operating levels. Thus, unlike the induction motor inverter, the switch current rating does not have to be increased by a large amount to cope with fault conditions.

(4) SWITCHING DEVICE SELECTION AND PERFORMANCE

The power switching requirements for low-power SR drives may be met by a number of semiconductor technologies as follows:

    *   Bipolar Transistors
    *   MOSFETs
    *   Insulated Gate Transistors (IGT)
    *   Gate turn-off Thyristors (GTO)

Bipolar transistors are an established technology. Devices provide good use of silicon area for a particular current rating and thus have the potential for low cost. Conduction losses are dependent on a constant voltage drop characteristic and are relatively low. However, switching speeds can be slow and safe operating areas limited such that snubbing is often required. Gains are also relatively low meaning that devices are usually configured as Darlingtons giving poorer use of silicon area and slower switching speeds. In common with inverters the SR converter requires switching devices with a fairly high peak to rms current ratio which most bipolar transistors or Darlingtons cannot meet without excessive base drive. Hence the device selection is dominated by peak currents which necessitates the use of a transistor with a larger continuous rating than is actually needed.

MOSFETs are relatively new devices which are still undergoing rapid technical development and cost reductions. They have a relatively poor use of silicon area for 500V ratings and conduction losses, which are based on a temperature dependent resistance, are relatively high. Switching speeds however are fast, allowing switching frequencies up to the 1MHz region and low switching losses. Moreover, safe operating areas allow snubberless operation and a positive temperature coefficient allows easy paralleling of devices. MOSFETs also exhibit a high peak to rms current ratio making them particularly suitable for motor drive circuits. Perhaps their greatest advantage over bipolar transistors and GTOs is that drive power is very low due to a very high gate input impedance and this enables very simple, efficient driving circuits to be used.

IGT's are devices which aim to combine the best features of bipolar and MOSFET transistors. Development of the devices has been aimed at voltages in excess of 500V and current ratings in excess of 10A where MOSFET conduction losses become excessive. There is no fundamental reason why devices with lower voltage and current ratings cannot be produced but that market is adequately met by MOSFETs at present. Devices introduced so far have shown conduction losses and switching times similar to bipolar Darlingtons. Each method has advantages and disadvantages which need to be carefully considered for each particular application.

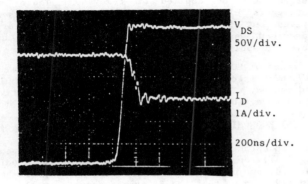

Fig. 5. MOSFET Switching Perfomance.

Fig 5 illustrates the performance of one circuit which has been developed using pulse transformer isolation. This achieves all the criteria listed earlier and requires only 3 components per MOSFET, giving a very economic gate drive system. The design is based on a widely used idea [13] in which the transformer is only energised when the gate capacitance needs to be charged or discharged. At other times the gate voltage is held reasonably constant due to the very high impedance of the gate terminal causing the gate capacitance to remain charged. This method allows the use of a transformer with a small voltage-time product and ensures that the gate voltage remains near constant over a wide mark space range. The transformer also has a very small inter-winding capacitance and this ensures that the system has a high dv/dt immunity. MOSFET switching times are only marginally slower than those achieved for the directly driven devices.

(5) OVERALL DRIVE PERFORMANCE

Fig 6 shows an example of the construction of a small SR drive comprising motor with the rotor position transducer, control electronics PCB and power switching PCB. The performance of this type of low-power SR drive is outstanding in every respect. The output torque/speed envelope is shown in Fig 7. The maximum output power is 700W continuous (Class B rise) with a short-term rating of 1.2kW. The drive can maintain a constant torque (continuous rating) up to a speed of 4000 rev/min and provides a high peak output particularly at low speeds. It should be noted that the peak torque below 4000 rev/min was limited by the rating of the dynamometer not by the SR drive.

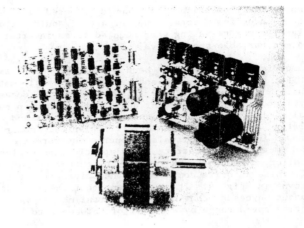

Fig. 6. Complete SR drive system

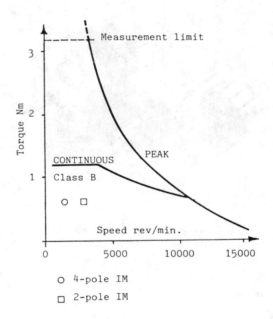

Fig. 7. Small SR Drive Torque Output

The overall efficiency for the SR drive is very high and remains high over a very wide speed and load range. The drive described above exhibits an overall efficiency of between 73% and 77% and other MOSFET drives developed by the authors return efficiency overall figures up to 90%.

For comparison outputs from two separate standard 3-phase, sinusoidally fed induction motors with identical outside diameters and airgaps to the single SR motor but built in the same frame, though some 18% longer than the SR motor, are shown on the graph. It is evident that at both 4-pole and 2-pole speeds the SR motor output torque is approximately twice the induction motor torque for the same temperature rise. This is remarkable bearing in mind that the single SR drive is designed to operate over a very wide speed range, whereas the two induction motors are separately optimised to their particular operating speeds. An SR drive designed to provide full power at 1500 rev/min may be expected to exhibit an even more

451

impressive comparison. It should be noted also that the induction motor performance would suffer significant de-rating when coupled to an inverter to form a variable-speed drive.

Comparisons have also been made between small MOSFET-based SR drives and series universal motors which are so commonly used in many domestic appliances. In particular an SR motor has been built using identical lamination stack dimensions and end plates as a 'state-of-the-art' universal motor as shown in Fig 8. This also compares the rotors of the two motors to illustrate the simplicity and relative smaller size of the SR rotor. The output performance of this SR drive compared to the universal motor is shown in Fig 9. It is apparent that the SR drive output exceeds that of the universal motor over the entire speed range by a factor of between 2 to 3.5.

Fig. 8. SR Rotor and Universal Motor Armature.

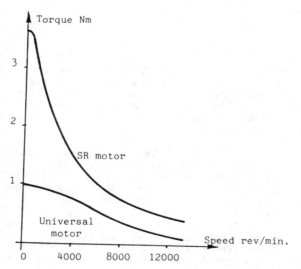

Fig. 9. Comparison between SR and Universal Motors (peak outputs)

The SR drive is extremely controllable, it can be considered to be similar to a dc machine with separate control of armature current and field current. This aspect gives the drive programmable torque-speed characteristics with a degree of flexibility that cannot be matched by inverter-fed induction motors or easily be matched by brushless PM motors. The SR drive is also capable of 4-quadrant operation with no additional power electronics apart from recovery or dumping of regenerated energy.

Moreover, the dynamic performance parameters of SR drives are also very impressive. For example the motor of Fig 7 exhibits a peak torque/inertia ratio of 80000 rads/sec^2 which provides very rapid acceleration rates. The open loop torque bandwidth for the lower speed ranges approaches 1kHz indicating how quickly torque can be applied or removed. These two parameters exceed those of many rare earth PM servo drives but it should be noted that the SR drive described was in fact designed for a very low cost domestic appliance application.

A range of MOSFET-based drives up to a power level of 3kW and a speed range of 20000 rev/min have been built and tested. The excellent performance of these drives leads to a wide variety of important applications.

(6) ECONOMIC FACTORS AND APPLICATION POTENTIAL

The acceptability of many small SR drives depends critically on cost competitiveness. Extensive studies carried out by the authors and their colleagues have clearly shown that the SR system competes well with all common forms of low power variable-speed drives. The importance of this major activity in proving the viability of the new drives cannot be over-emphasised. Not only does it require a thorough knowledge of the various possible circuit configurations and control philosophies, it also demands well informed discussions of the various integration technologies with manufacturers against a background of realistic production volumes for different and overlapping applications.

The motor is particularly simple to manufacture having no commutator, windings or magnets on the rotor and also using significantly less active material than, for example, a cage induction motor or a series universal motor. It also avoids the cost, temperature sensitivity, demagnetisation limit, brittleness and other manufacturing problems of permanent magnet motors. Hence the manufactured cost and the operational features of the SR motor are very attractive compared to conventional motors particularly in high-volume applications.

The electronic control system has a similar though somewhat reduced complexity level to a variable frequency inverter but such electronics can now be sourced at very low cost for high volume applications. Low cost electronics depends heavily on a high level of integration and the authors will report on their developments to this end in due course. The major components of a complete integrated system are illustrated in Fig 10. The power switching elements (MOSFETS and diodes) are contained in a hybrid module with an electrically isolated base plate. The main control system is contained in two custom ICs. The dc link capacitor and supply rectifier are standard components.

Fig. 10. Integrated SR Control System
(Major Components)

Using this type of integrated control system the SR drive is, of course, highly attractive in cost terms compared to inverter-fed induction motors or brushless PM motors. It is cost-effective even in the most cost-conscious of all markets, that of the triac controlled universal motor since, although the electronics are still somewhat more expensive than a triac controller, the motor is much cheaper. This fact is worthy of note since the triac-controlled universal motor has previously been the 'benchmark' for low cost variable-speed drives.

Although this paper has taken the example of mains-supplied drives, another very important application area is automotive auxiliary equipment supplied from 12 or 24V. Here again the combination of MOSFETs and SR motor provide a low-cost drive of unsurpassed performance and ruggedness. Similarly, general purpose variable-speed drives are also very attractive because of the high performance to cost ratio of SR technology .

Clearly the range of attractive application areas for small SR drives is vast. Some typical examples of variable-speed applications where the low-cost, high-performance and ruggedness of the SR system are important factors are listed below.

* Domestic appliances
* Hand-held tools
* Air conditioning plant
* Refrigeration equipment
* Office machines
* Automotive auxiliaries
* General-purpose drives
* Servo and robotic drives
* Machine tool drives

The authors and their colleagues expect to report on these and other applications in due course.

(7) CONCLUSIONS

This paper has described the operation and performance of a range of low power switched reluctance drives which has been developed as part of an extensive applications programme. MOSFET switches have been shown to be particularly appropriate for low powers.

These drives not only have outstanding performance characteristics compared to most conventional drives but are also very cost competitive.

The commercial exploitation of low-power SR technology is just beginning, but provides drive systems of outstanding technical and commercial potential for a very wide range of applications and is certain to have a major impact in many markets previously supplied by induction motors and commutator motors. The authors and their colleagues will be reporting on some of these in the near future.

(8) ACKNOWLEDGEMENTS

The authors wish to acknowledge the involvement and support of their colleagues at SR Drives Ltd.

REFERENCES

[1] Webster, P D and R J Blake (1986). Switched reluctance drives for light rail traction. Coloquium on 'Electrical Vehicle Electronics & Control' 1986/123 21 Nov 86, London.

[2] Blake, R J, P D Webster and D M Sugden (1986). The application of gtos to switched reluctance drives. PEVD Proc, IEE 264.

[3] Ray, W F, R M Davis and R J Blake (1986). The control of SR motors. CAMC 86 Proc, Minnesota.

[4] Stephenson, J M, P J Lawrenson and N N Fulton (1986). High-power switched reluctance drives. ICEM Proc, Munich.

[5] Ray, W F, P J Lawrenson, R M Davis, J M Stephenson, N N Fulton and R J Blake (1985). High-performance switched reluctance drives. IEEE-IAS-1985 Proc.

[6] Miller, T J E (1984). Converter volt-ampere requirements of the switched reluctance motor. IEEE-IAS-1984 Proc.

[7] Harris, M R, J W Finch, J A Mallick and T J E Miller (1986). A review of the integral horsepower switched reluctance drive. IEEE Trans on Circuits and Systems.

[8] Byrne, J V, M F McMullin and J B O'Dwyer (1985). A high-performance variable reluctance drive: a new brushless servo. MOTOR-CON Proc.

[9] Bass, J T, M Ehsani, T J E Miller and R L Steigerwald (1985). Development of a unipolar converter and variable reluctance motor drives. IEEE-IAS-1985 Proc.

[10] Blake, R J, R M Davis, W F Ray, N N Fulton, P J Lawrenson and J M Stephenson (1984). The control of switched reluctance motors for battery electric road vehicles. PEVD Proc.

[11] Ray, W F, R M Davis, J M Stephenson, P J Lawrenson, R J Blake and N N Fulton (1984). Industrial switched reluctance drives - concepts & performance. PEVD Proc.

[12] Davis, R M, W F Ray and R J Blake (1981). Inverter drive for switched reluctance motors : circuits and component ratings. IEE Proc, vol 128.

[13] Wood, P (1985). Transformer isolated hexfet driver provides very large duty cycle rating. I R Hexfet Databook, application note 950A.

A Current-Controlled Switched-Reluctance Drive For FHP Applications

T.J.E. Miller and T.M. Jahns

Conference on Applied Motion Control, Minneapolis, June 1986, pp. 109-117

1. INTRODUCTION

The use of adjustable-speed brushless dc drives is becoming rapidly more widespread in almost all motion-control applications. There are several reasons for this. The technology of motion control is itself developing rapidly because of the trend towards automation in machine tools and complete factories. This trend is partially fueled by the possibilities opened up by advanced control techniques based on computers, microprocessors, and digital signal processing (DSP) chips; but an increasingly important influence is the rapid advances being made in power electronics, especially in switching devices such as power MOSFETs and switching power modules with integrated control functions. When solid-state power electronics began to be applied to motion control, the bipolar junction transistor and the dc commutator motor were the main players. Today the availability of intelligent MOS-gated power switches makes it more desirable than ever before to exploit the advantages of the brushless dc motor.

The brushless dc motor is generally understood to be an inside-out dc motor in which the excitation system (usually a permanent magnet) is mounted on the rotor and the armature winding is mounted on the stator. By far the commonest version has the magnets arranged cylindrically on the rotor surface, held in place by adhesive in motors with low peripheral speeds, or by some form of retaining ring or binding in motors with high peripheral speeds. This construction, known as the surface permanent-magnet (SPM) motor, is well established and has been shown to have a high efficiency over its speed range. It is relatively simple to control by PWM of the phase currents, and it can be operated successfully without a shaft position sensor [1]. Recently the introduction of Neodymium-Iron-Boron magnet material has extended the capabilities of this motor by a significant degree. SPM motors built with Ferrite, Rare Earth/Cobalt and Neodymium-Iron-Boron magnets have been developed for innumerable applications and new ones are appearing almost daily. It is interesting to note in the proliferation of this technology that many applications are now being served by brushless drives where constant-speed or switched-resistor control was formerly the rule; for example, air-moving applications in vehicles and aircraft.

Figure 1 shows a comparison of four alternative forms of brushless motor, in which the SPM motor is No. 2. Motors 1 and 3 are variants of the Interior Permanent Magnet (IPM) motor [9,10] in which the magnets are mounted internally within a laminated rotor structure. No. 3 was developed specially for aerospace applications and since its construction requires the welding of magnetic and nonmagnetic

materials, its use is restricted to special applications and it is not discussed further. No. 1, the IPM motor, has been described in Ref. [9]. It is a hybrid between a synchronous reluctance motor and a pure PM motor, and was derived from earlier versions developed for line-start integral-horsepower industrial energy-saver motors [11,12]. Its characteristics are quite different from those of the SPM motor, and although they use the same inverter circuit the IPM motor needs sinusoidal currents (or a PWM approximation) with accurate phase angle control, so that it is more logical to consider this system as an ac drive rather than a brushless dc drive. Among its advantages are very low torque ripple, low magnet weight, high efficiency, and the ability to produce torque even when the magnets are demagnetized (partially or completely).

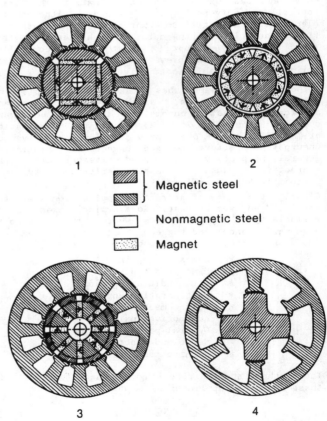

▨ ▨	} Magnetic steel
☐	Nonmagnetic steel
▨	Magnet

Fig. 1 Configuration of four classes of brushless or ac motor (1) Interior Permanent Magnet ac (IPM); (2) Surface Permanent Magnet (SPM); (3) Interior Permanent Magnet with circumferentially oriented magnets; (4) Switched Reluctance

Dr. Timothy J.E. Miller and Dr. Thomas M. Jahns, General Electric Company, Corporate Research and Development, Schenectady, NY (518) 387-5058

The fourth motor in Fig. 1 is the switched reluctance (SR) motor, [2-8]. It belongs to the class of single-stack variable-reluctance (VR) stepping motors, but it is usually operated with controlled gating angles. Control of the gating angles (with respect to rotor position) permits this motor to be designed for higher torque/volume ratio and higher efficiency than would normally be found in conventional stepping designs, and recent studies have shown performance comparable with the best ac induction motor drives and dc commutator motor drives [5,13]. An especially attractive feature of the SR motor is its physical simplicity, which makes it robust, reliable, and able to operate over a wide range of temperature in any kind of environment.

This paper describes an exercise in which a SR motor was designed, built, tested, and demonstrated as an adjustable-speed blower drive of the type that might be used in an automotive or aircraft air-conditioning application. The drive was powered from a 12V dc battery and one of the objectives was to evaluate the C-Dump converter described in [3] as a means for minimizing the losses associated with forward voltage drops in the power switches. The C-Dump converter exploits the unipolar nature of the SR motor by using only one switch per phase (instead of the two per phase that are needed with the other motors in Fig. 1). The phase currents are commutated into a "lossless" suppression circuit. Some of the limitations of this technique are discussed. A further objective was to evaluate the size and efficiency of the SR motor relative to a dc commutator motor.

2. PRINCIPAL FEATURES OF BLOWER DRIVE MOTOR

Fig. 2 shows the rotor of the SR motor. A three-phase motor was chosen to minimize the number of switches and also to minimize the magnetic losses in the iron. The iron losses turned out to be small anyway, but it is worth noting that with 12 steps per revolution the switching frequency in each phase at a given speed is the same as that of an 8-pole ac motor or SPM motor, i.e., (rev/min)/15. At the top speed of 3300 rev/min the switching frequency in each phase is thus 3300/15 = 220 Hz. A four-phase motor would have 8 stator poles and 6 rotor poles and the frequency at this speed would be 330 Hz in each phase. Higher frequencies are met in chopping the current at low speeds, as will be seen. The rotor is constructed from silicon-steel laminations compressed between two washers. The stack length is approximately 1.75 inches for a motor designed to deliver about 180 Watts at 3300 rev/min with free convection cooling. The stator is shown in Fig. 3. The endwindings are relatively short, which makes the stack length long in relation to the overall length of the motor, and provides a robust construction. There are no crossovers between different phase windings in the end-turn region, and this is considered to be desirable for reliability. It also further reduces the weight of copper in the end turns and may improve their cooling by leaving spaces open for cooling air. The fill factor of copper in the slots was 29% in the motor as built, and although higher figures are possible the design was biased towards very low material cost. There is a clear passage for cooling air between the two coil sides sharing each slot, but no special effort was

made to optimize the heat transfer to the air flowing through it.

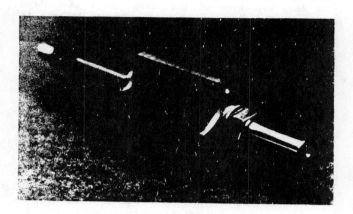

Fig. 2   Rotor of switched reluctance motor

Fig. 3   Stator of switched reluctance motor

3. CONVERTER DETAILS

Fig. 4 shows the essential topology of the converter circuit, which is a C-Dump circuit of the type described in Ref. [3]. Each phase has only one power switch, and the current in each phase winding is unidirectional. Since the sign and magnitude of the torque are both independent of the direction of the current, there is no loss of material utilization resulting from the unipolar operation, as there would be in an ac or PM motor. Iron losses in the stator teeth are reduced because the flux in them is unidirectional.

The phase current is commutated into the dump capacitor (which is common to all three phases) when the phase transistor switches off. Energy continues to be converted into mechanical work during the freewheeling phase, but a proportion of the magnetic energy stored at the point of commutation is inevitably transferred to the dump capacitor. The voltage of the dump capacitor is pumped up by the repeating commutations of the phases, and a step-down chopper is included in the circuit to relieve the dump capacitor of its charge in a controlled manner. The chopper maintains a power balance in which energy is returned to the dc battery at the same average rate at which the phases dump it into the capacitor. By this means the dump capacitor voltage is maintained constant, typically at a value around twice the dc battery voltage.

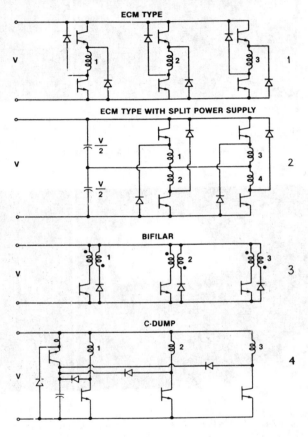

SWITCHED RELUCTANCE INVERTER CIRCUITS

Fig. 4 Converter circuits for SR motor drive

4. Basic Efficiency Limitations

The following zeroth-order scaling analysis illustrates some of the basic problems in achieving high efficiency in small reluctance motors.

4.1 Excitation Penalty and Effect of Physical Size

The excitation power for a reluctance motor is supplied through a component of the phase current. In SPM motors all the excitation is supplied by the magnet. The reluctance motor thus has an inherent disadvantage with regard to its efficiency, since the excitation component in the phase current adds to the losses incurred in the winding resistances and the power switches. This "excitation penalty" is the price paid for eliminating the magnets. A

similar difference exists between synchronous and induction motors.

The relative excitation penalty of the reluctance motor is a function of physical size. In small motors the effect of iron saturation is, to a first approximation, negligible, and the mechanical output power P_m is determined by the product of the square of some current parameter i and an inductance difference parameter ΔL

$$P_m \propto i^2 \Delta L$$

where ΔL has dimensions $\mu_0 N^2 A/g \propto s^2/g$. Here s is a generalized linear dimension representing the size of the motor and g is a generalized parameter representing the mechanical airgap. Note that s and g do not necessarily scale together in small motors, since g may reach some minimum value as s is decreased, to allow for mechanical clearance and manufacturing tolerances. The resistive loss P_r will similarly be given by an equation of the form

$$P_r = i^2 R$$

where R has dimensions $\rho s/A \propto s^{-1}$. Assume that as the motor size s is changed, the current level is adjusted such that the resistive losses are proportional to the area of surfaces available for cooling. Then $P_r \propto s^2$ and

$$i^2 R \propto \frac{i^2}{s} \propto s^2$$

so that

$$i \propto s^{3/2}$$

Under this constraint the scaling of the mechanical power is given by

$$P_m \propto i^2 \Delta L \propto s^3 (s^2/g) = s^5/g$$

The per-unit resistive loss P_r/P_m is therefore given by

$$\frac{P_r}{P_m} \propto \frac{s^2}{s^5/g} = \frac{g}{s^3}$$

If g scales with s, then P_r/P_m scales with s^{-2}, but if the gap is fixed at some minimum-clearance value then P_r/P_m scales with s^{-3}.

This formulation is now repeated for the SPM motor. If the resistive losses are again constrained to be proportional to the area of surfaces available for cooling, then the current scales with $s^{3/2}$ and $P_r \propto s^2$ as in the reluctance motor. However, the mechanical output power is now given by an equation of the form

$$P_m \propto i\phi$$

where ϕ is a flux parameter of the form BA. The

456

flux density B is typically close to the remanent flux-density of the magnet and is not sensitive to the mechanical airgap except in extremely small motors. Consequently $\phi \propto s^2$ and since $A \propto s^2$

$$P_m \propto is^2 \propto s^{3/2} s^2 = s^{5/2}$$

The per-unit resistive loss is therefore given by

$$\frac{P_r}{P_m} \propto \frac{s^2}{s^{5/2}} = s^{-3/2}$$

Comparing the scaling of P_r/P_m for reluctance and SPM motors, it is clear that in the reluctance motor the per-unit resistive losses increase at a faster rate than in the SPM motor as the size is reduced. For applications in which efficiency and/or power density is the prime consideration, the SPM motor is liable to have the advantage over the reluctance motor. The price paid for this includes the material and assembly costs of the magnets and their attendant temperature limitations. For applications where the first cost of the motor is a prime consideration, or where the temperature range is too great for PM motors, the reluctance motor is liable to have the advantage.

This analysis is of course too narrow to provide a complete basis for comparing different types of motor. In a given application the tradeoff is generally much more complicated. The analysis does, however, bring out the expected sensitivity of small reluctance motor efficiency to the circuit resistances, and in the following sections it will be seen that these resistances had to be systematically reduced to achieve acceptable efficiency in the 12V system described.

4.2 Analysis of Resistive Losses

The SR drive described in this paper is typical of small reluctance motor drives in that resistive losses are predominant. The following analysis identifies some of the determinants of these losses and their variation with speed.

Fig. 5a shows the inductance profile of the motor, i.e., the variation of phase self-inductance with rotor position. The ideal profile is calculated from the pole geometry assuming no fringing flux, no leakage, and no iron saturation. Measured profiles are shown for phase currents of 10, 20 and 30 A. The effect of saturation is to flatten the profile and reduce the available torque per ampere. In Fig. 5b is shown an idealized current waveform with a flat top, synchronized with the inductance profile so that positive motoring torque is produced during periods of increasing inductance according to the equation

$$T = \frac{1}{2} i_0^2 \frac{dL}{d\theta}$$

The rate of change of inductance with rotor position can be approximated as $\Delta L/\beta$ where $\Delta L = L_a - L_u$ is the difference between the "aligned" and "unaligned" inductances and β is the stator pole arc. (See Ref. [2]). With N_r rotor poles and q phases the average torque is then

$$T_a = \frac{1}{2} i_0^2 \Delta L \frac{qN_r}{2\pi}$$

and the average mechanical power is $P_m = \omega T_a$.

The resistive losses are given by

$$P_r = i_0^2 R$$

in each phase, where R is the total series resistance of the phase winding and the on-state resistance of the MOSFET power switch. The duty cycle of current in each phase is assumed to be $\beta/(2\pi/N_r)$ from Fig. 5b, so that with q phases the mean resistive power loss is given by

$$P_r = \beta R i_0^2 \frac{qN_r}{2\pi}$$

The resistive losses are now normalized to the mechanical power P_m, giving per-unit resistive losses:

$$\frac{P_r}{P_m} = \frac{2\beta R}{\omega \Delta L}$$

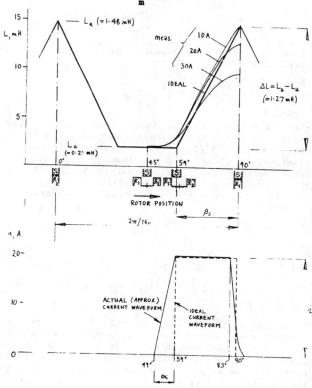

Fig. 5(a) Idealized and Measured Inductance Profiles
(b) Idealized Current Waveform

This parameter can be used to gauge the performance of different designs with regard to their resistive losses. Before proceeding to do this, however, we should make a slight correction for the initial ramp section of the phase current (i.e., through angle α in Fig. 5b). Assuming that the power switch is turned on α radians before the approaching pole corners of rotor and stator reach alignment, the current rises according to the equation

$$i = \frac{V}{R}\,[1-\exp(-Rt/L_u)]$$

At the point where the pole corners are aligned, $t = \alpha/\omega$ and

$$i = \frac{V}{R}\,[1-\exp(-R\alpha/\omega L_u)]$$

The condition for flat-topped current is $di/dt = 0$, so that

$$V = Ri + i\omega\frac{dL}{d\theta} \approx (R + \omega\frac{\Delta L}{\beta})\,i$$

and

$$i = \frac{V}{(R + \omega\Delta L/\beta)}$$

is the flat-topped current. Equating this current with the exponential, the solution for α is

$$\alpha = \frac{\omega L_u}{R}\,\ln[1 - 1/(1 + \omega\Delta L/\beta R)]$$

If R is sufficiently small the current ramp is linear and

$$\alpha = \frac{L_u}{\Delta L}\,\beta = \frac{\beta}{\lambda - 1}$$

where $\lambda = L_u/L_u$ is the inductance ratio. With a linear current ramp the mean squared value during the ramp is $i^2/3$. The resistive losses are augmented by $i^2 R/3$ flowing through an angle α of rotor motion, which is equivalent to $i^2 R$ flowing through an angle $\alpha/3$. The effect on the per-unit resistive losses can be incorporated by writing

$$\frac{P_r}{P_m} = \frac{2\beta' R}{\omega\Delta L} \qquad (A)$$

where

$$\beta' = \beta + \frac{\alpha}{3} = \beta\left[1 + \frac{1}{3(\lambda - 1)}\right]$$

A similar correction can be made for the negative current ramp during the freewheeling period, but this is not included here. The losses during the freewheeling period are controlled by different circuit parameters, and in the C-Dump converter the sizing and efficiency of the chopper play a significant role in defining their value.

5. Performance of 180W SR Drive

The motor as built for 12V operation had an inductance ratio of 7.0 with $\Delta L = 1.27$mH and a phase resistance of 0.08Ω at temperature. Fig. 6a shows the original design of the converter phaseleg using one MOSFET (IRF540) per phase and a current-sensing

resistor of 0.1Ω. The total series resistance per phase is $R = 0.265\Omega$, and using Equation (A) the per-unit resistive losses are plotted as a function of speed in Fig. 7, labelled SR1. For comparison the load power is plotted on the same graph, along with the per-unit resistive losses of a dc commutator motor that was designed to fit in the same frame as the SR motor, with the same stack length. From this curve it appears that the general variation of resistive losses with speed is similar in the two motors. However, it must be pointed out that the SR motor is operating with fixed gating angles to permit the use of a very simple control (see below). With gating angle control the losses in the SR motor could be significantly reduced at part load. Since the treatment in this paper is restricted to constant gating angles, the results must be treated as a worst case that includes the efficiency penalty due to fixed gating angles.

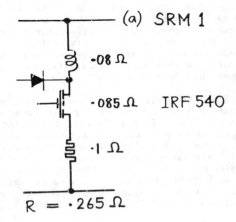

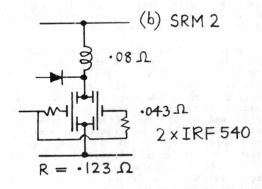

Fig. 6   Phaseleg circuits for SR motor drive

For many purposes the losses achieved with the SRM1 design would be acceptable, since the overall efficiency exceeds that of the DC drive with series resistance control, and the SR motor is about 20% shorter than the dc motor. However, the development exercise was pursued further to see what efficiency gains could be achieved with reasonable design changes.

From Fig. 6a the resistive losses are
apportioned between the phase winding resistance,
the MOSFET, and the current shunt in proportion to
their resistances, and the motor losses are the
smallest of the three. Two design changes were
therefore made in the converter circuit. First, the
MOSFET was replaced by two parallel MOSFETs
providing half the resistance of each one. The
original MOSFETs were possibly rated at too high a
voltage for this application and therefore this
substitution is roughly equivalent to replacing the
original transistor with one of lower $R_{DS(ON)}$ but
with a lower voltage rating. Gate lead resistors
were added to suppress oscillations between the
parallel transistors.

The second design change was to eliminate the
current sensing resistor and to use the voltage drop
across the MOSFETs as the current sensing signal.
An exclusive-OR logic with diodes was used to
control the switching of the phases separately, and
this scheme worked perfectly well. The combined
result of the changes is seen in Fig. 6 in which the
total series resistance is reduced to 0.123Ω and
labelled SRM2. The per-unit resistive losses are
plotted as a function of speed on Fig. 7. This
represents a major improvement in efficiency, the
full-load losses having been reduced to 57W from the
117W obtained with SRM1.

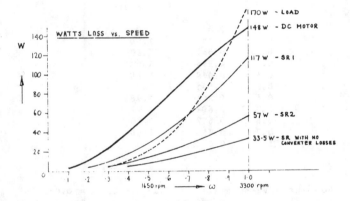

Fig. 7 Resistive loss vs. Speed

Also plotted on Fig. 7 is the per-unit
resistive loss curve for the motor by itself, i.e.,
assuming zero external resistance in the power
switches. This represents a lower limit to the
resistive losses that can be reduced only by
reducing the motor resistance or by improving its
inductance ratio.

To qualify the comparison between the SR and
the dc commutator motors, the weights of active
material should be included. Both motors have
approximately the same weight of iron. The dc motor
has 0.28 lb of copper (excluding the commutator) and
0.594 lb of ferrite magnet, which costs roughly
about the same as the copper. The SR motor has 0.49
lb of copper and no magnet. The SR motor thus uses
56% of the active material (copper plus magnet)
compared with the dc motor. It also has no
commutator and no brushgear. Of course it needs an
electronic converter, but to achieve continuously

variable speed the dc motor would also require an
electronic converter.

6. Further Means for Loss Reduction

The C-Dump converter was developed to exploit
the unipolar operation of the SR motor to reduce the
number of power switches per phaseleg from two to
one. This objective is met at the price of the
additional components in the step-down chopper and
the dump-capacitor, and the losses associated with
them. The additional components include the chopper
transistor, inductor, and freewheel diode, together
with some means for sensing the current in the
chopper transistor. The voltage rating of the
chopper transistor is the same as that of the
phaseleg transistors, but its current rating may be
smaller. The losses associated with the chopper are
not negligible and are subject to some fundamental
constraints which are discussed next.

The average power passed through the chopper
and returned to the dc supply is not necessarily a
small proportion of the mechanical output power of
the motor. This can be seen from Fig. 8, which
shows the trajectory of phase winding flux-linkage ψ
and current i during a single step. The diagram is
drawn for single-pulse operation; that is, the
phaseleg transistors are not chopping and the
current is limited by the back-emf of the motor. At
the commutation point C there is a certain amount of
stored magnetic energy in the phase winding. After
commutation the current freewheels into the dump
capacitor and the magnetic energy is divided between
mechanical work and stored energy on the capacitor.
The ratio of dumped energy W_d to converted energy W_m
can be estimated from the areas in Fig. 8.

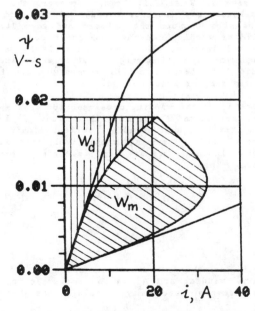

Fig. 8 Flux-Linkage/Current Trajectory showing the
relation between converted and dumped energy

It can also be estimated from the theory of Ref.
[2]. From Equation 20 of Ref. [2] it can be shown

459

that if W_m is the energy converted into mechanical work at each step, then with a flat-topped current waveform

$$\frac{W_d}{W_m} = \frac{\lambda}{\lambda - 1}$$

where λ is the inductance ratio discussed earlier. Since the motor is usually designed for the largest possible inductance ratio (to maximize the torque per unit volume), the dumped energy is comparable in magnitude to the converted energy. The ratio W_d/W_m may be lower than indicated by this linear theory if the motor is running below the speed at which the current waveform is flat-topped. Such is the case in Fig. 8. A further reduction in W_d/W_m is obtained if the iron is saturated in the aligned position [2], but in small motors this effect is weak. In any case the power converted by the chopper is comparable to the motor output power, so that the efficiency of the chopper is significant in determining the efficiency of the entire drive.

6.1 Bifilar-Wound Motor

An alternative inverter circuit that does not need the C-dump capacitor or the step-down chopper is obtained by making the motor phase windings bifilar and connecting the secondary in series with the freewheel diode across the dc supply, as shown in Fig. 4. (See also Ref. [2]) If the bifilar turn ratio is unity then W_d has the same value as in the C-dump circuit when the capacitor voltage is maintained at twice the battery voltage. Conversion to a bifilar configuration would eliminate the chopper losses and in the present drive system this would save 15 W at full load. However, the motor copper losses may increase because of the partition of slot area between the primary and secondary of the bifilar winding and the consequent reduction in copper utilization. The following analysis derives the optimum apportionment of slot area and is included here because it is general enough to be useful in stepper motor design.

Let A_o be the winding cross-section area and R_o be the resistance of the monofilar winding. Let i_o be the total r.m.s. winding current (main plus freewheel) and i_m and i_f respectively the r.m.s. currents in the primary and secondary bifilar windings. Then

$$i_o^2 = i_m^2 + i_f^2$$

and

$$A_o = A_m + A_f$$

where A_m is the cross-section area of the main (primary) winding and A_f is the cross-section area of the freewheel (secondary) winding. Also let

$$x = i_f^2/i_o^2$$

and

$$a = A_f/A_o$$

Then the resistive winding loss is given by

$$P_r = R_m i_m^2 + R_f i_f^2$$

where $R_m = R_o/(1 - a)$ and $R_f = R_o/a$. Hence

$$P_r = R_o i_o^2 \left[\frac{1 - x}{1 - a} + \frac{x}{a} \right]$$

Now for a given value of x we can minimize the copper loss with respect to the area ratio a. By differentiation the optimum value of a is

$$a = -x + \frac{[x(1 - x)]^{1/2}}{1 - 2x}$$

From this the accompanying table can be constructed showing the ratio of winding resistive loss relative to the monofilar winding.

Bifilar/Monofilar Loss Ratio and Optimal Apportionment of Winding Area

x	a_o	Loss Ratio
0.0	0.0	1.0
0.1	0.250	1.60
0.2	0.333	1.80
0.3	0.396	1.92
0.4	0.449	1.98
0.5	0.495	2.0
0.6	0.551	1.98
0.7	0.604	1.92
0.8	0.667	1.80
0.9	0.750	1.60
1.0	0.0	1.0

For the present motor at full load, $x = 0.081$ so that the optimum area ratio is $a = 0.229$ and the loss ratio is 1.55, i.e., conversion to a bifilar configuration would increase the motor copper losses by 55%. This is more than enough to outweigh the 15 W saved in the converter, so it was concluded that the bifilar arrangement would be less efficient than the C-dump system.

The present comparison is restricted to circuits with one power switch per phaseleg. If two are used [5,7] the overall system efficiency may be higher, but circuits of this type were not evaluated here. The first circuit in Fig. 4 is of this type.

7. Control Aspects

The control scheme adopted was a simple current-chopping arrangement with fixed gating angles. The gating angles were determined by a slotted disc rotating between three optical interrupters, one for each phase. This "optical commutator" was mounted at the non-drive end of the motor. The use of a physical shaft position sensor is convenient for a demonstration system such as the

460

present one. However, for future systems which do not require precision torque control, the shaft sensor can be eliminated by means of several alternative methods (see for example, Ref. [4]).

The speed is regulated by raising or lowering the current reference so that the motor produces more or less torque. In the system as built there is no speed loop, although one could be added without difficulty [6].

Fig. 9 shows a typical current waveform during operation at an intermediate speed. Chopping continues through the step, and the variation of the current chopping band reflects the change in phase inductance as the rotor rotates.

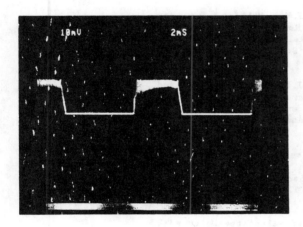

Fig. 9 Typical current waveform at part load

8. Conclusions

This paper presents some of the results of an evaluation of a 12V 180W adjustable-speed drive based on a switched reluctance motor with a C-Dump converter. Particular attention is given to the resistive losses and it is shown that the SR motor is a considerable improvement over the dc brush motor drive, with significantly less copper and no magnets. The absence of the commutator permits the SR motor to be about 20% shorter than the dc commutator motor, although some of this space saving is used up by a shaft-mounted position sensor used for commutation. In other development drives [4] this sensor has been eliminated and it is not regarded as an essential element.

The control scheme for the SR drive is based on current regulation by means of a simple hysteresis-band PWM algorithm with fixed gating angles. This leads to a very simple control implementation in which the torque is controlled by adjusting the current reference, - an adequate procedure for blower and pump applications where precision and rapid dynamic response are not the most important requirements. This simple scheme does not exploit the possibility for loss reduction through gating angle control at part load. A worthwhile future development would be to provide this additional degree of control, preferably without a great

increase in control complexity. One of the attractive features of the SR system as described here is the simplicity of the control algorithms.

The acoustic noise level of the SR motor was comparable to that of the dc commutator motor and the chopper noise was comparable to that of a somewhat larger surface-magnet motor drive. The general EMI levels produced by the SR drive were comparable to those produced by the surface magnet motor drive, although these comparisons were not precisely quantified. An earlier version of the SR motor was very noisy with bronze sleeve bearings, but quiet operation was restored with anti-friction ball bearings. This corroborates earlier experience and suggests that in general the noise level from SR machines is sensitive to mechanical alignment and manufacturing tolerances.

The paper includes a scaling analysis to show that reluctance motors suffer from an "excitation penalty" that reduces their efficiency or power/volume ratio as a result of losses incurred by the excitation component of phase current. Such results are widely known to motor design engineers, though not necessarily very accessible in the literature. It is shown that the excitation penalty increases as the motor size decreases, and if the mechanical airgap reaches a minimum value (to allow for clearance and manufacturing tolerances) the excitation penalty increases at an even faster rate. In small motors the provision of magnets for excitation rapidly alleviates this problem by eliminating the excitation penalty. PM motors are still subject to a general reduction in efficiency as their size decreases, but it is not as severe as in the reluctance motor. At a power level of 200W, even at 12V the reluctance motor is a very viable technology, particularly in view of the low materials cost, the simplicity of construction, the wide range of operating temperature, and the general robustness of motor and converter.

The paper describes the elimination of the current-sensing resistor from the phaselegs to reduce the series resistance and eliminate a component of resistive losses. The forward voltage drop across the power switches (MOSFETs) is used to measure the instantaneous phase current. The scheme works successfully with a diode-based exclusive-OR logic that permits independent control of the phase currents.

9. Acknowledgments

Several colleagues contributed to the work reported in this paper, and our thanks are due to all of them. Particular thanks are due to W.R. Oney for his work in the motor design, and to Frank Forbes of the Motor Technology Department in Ft. Wayne for his guidance.

10. References

1. Harms, H.B. and Erdman, D.M.: 1985 Motorcon paper

2.   Miller, T.J.E.: "Converter Volt-Ampere Requirements of the Switched Reluctance Motor Drive", IEEE Transactions, Vol. IA-21, No. 5, Sep/Oct 1985, pp 1136-1144

3.   Bass, J.T., Miller, T.J.E., and Ehsani, M.: "Development of a Unipolar Converter for Variable Reluctance Motor Drives", IEEE Industry Applications Society Annual Meeting, Toronto, October 1985

4.   Miller, T.J.E. and Bass, J.T.: "Stabilization of Variable-Reluctance Motor Drives Operating Without Shaft Position Feedback", Incremental Motion Control Systems Symposium, University of Illinois (B.C. Kuo, ed.) Urbana-Champaign, Illinois, June 1985

5.   Harris, M.R., Finch, J.W., Mallick, J.A. and Miller, T.J.E.: "A Review of the Integral Horsepower Switched Reluctance Drive", IEEE Industry Applications Society Annual Meeting, Toronto, October 1985

6.   Bose, B.K., Miller, T.J.E., Szczesny, P.M. and Bicknell, W.H.: "Microcomputer Control of Switched Reluctance Motor", IEEE Industry Applications Society Annual Meeting, Toronto, October 1985

7.   Lawrenson, P.J., Stephenson, J.M., Blenkinsop, P.T., Corda, J., and Fulton, N.N.: "Variable-Speed Switched Reluctance Motors", IEE Proceedings, Vol. 127, Pt. B, No. 4, July 1980, pp. 253-265

8.   Lang, J.H.: "A Variable-Reluctance Motor with Maximized Ratio of Torque to Mass", Conference on Applied Motion Control, Minneapolis, MN, Jun 10-12, 1986

9.   Jahns, T.M., Kliman, G.B. and Neumann, T.W.: "Interior Permanent Magnet Synchronous Motors for Adjustable-Speed Drives", ibid.

10. Jahns, T.M., "Flux-Weakening Regime Operation of an Interior Permanent Magnet Synchronous Motor", IEEE Industry Applications Society, Annual Meeting, Denver, Colorado, October 1986

11. Richter, E., Miller, T.J.E., Neumann, T.W., and Hudson, T.L.: The Ferrite Permanent-Magnet AC Motor - A Technical and Economical Assessment", IEEE Transactions, Vol. IA-21, No. 4, May/June 1985, pp. 644-650

12. Miller, T.J.E., Richter, E., and Neumann, T.W.: "A Permanent-Magnet Excited High-Efficiency Synchronous Motor with Line-Start Capability", IEEE Industry Applications Society, Annual Meeting, Mexico City, October 1983

13. Byrne, J.V., McMullin, M.F., and O'Dwyer, J.B.: "A High Performance Variable Reluctance Drive: A New Brushless Servo", Motorcon 1985 Proceedings, pg. 147-160.

A Review of the Integral-Horsepower Switched Reluctance Motor Drive

M.R. Harris, J.W. Finch, J.A. Mallick and T.J.E. Miller

IEEE Transactions, Vol. IA-22, No. 4, July/August 1986, pp. 716-721

Abstract—An evaluation of the capabilities of the switched reluctance (SR) motor drive, particularly in small integral-horsepower sizes, is presented, and some of its special features are discussed. The simplicity of the construction of the rotor together with certain advantages in the power circuit, such as unipolar operation and the independence of the phases, are described along with some of the important performance parameters, which are compared with those of typical induction motor drives. It is shown that the ruggedness and simplicity of the SR drive are accompanied by a performance profile that matches that of modern induction motor drives in torque per unit volume, efficiency, converter voltampere requirements, and other parameters. A comparison of three SR motors, including one low-inertia design and one with two stator teeth per pole, shows torque/inertia ratios several times greater than those for induction motor drives.

INTRODUCTION

THE switched reluctance (SR) motor drive has been investigated and developed in the past decade by several organizations with results that are much more promising than those obtained in previous work (see [1]–[4]). The definitive work by Lawrenson *et al.* at the Universities of Leeds and Nottingham is particularly cited for bringing out and developing many of the special features of these drives, although excellent work has been done elsewhere in Europe and the United States since the early 1960's. Modern power semiconductor switching devices, such as the power transistor and the gate turn-off thyristor, are now capable of bringing out the optimum performance from the motor, which was not possible even a decade ago. At the same time the electromagnetic analysis of doubly salient motors has progressed to the point where the geometry of laminations can be optimized for particular applications [6], [7].

The most striking feature of the SR motor is the complete absence of rotor windings or a commutator, although a shaft

Paper IPCSD 85-53, approved by the Fractional and Integral Horse Power Subcommittee of the Industrial Drives Committee of the IEEE Industry Applications Society for presentation at the 1985 Industry Applications Society Annual Meeting, Toronto, ON, October 6–11. Manuscript released for publication December 21, 1985. This work was supported in part by a Visiting Research Fellowship completed by one of the authors at the Corporate Research and Development Center, General Electric Company.

M. R. Harris and J. W. Finch are with the Department of Electrical and Electronic Engineering at The University of Newcastle-upon-Tyne, England, NE 1 7RU.

J. A. Mallick is with the General Electric Company, Corporate Research and Development Center, P.O. Box 8, KWC-425, Schenectady, NY 12301.

T. J. E. Miller is with the General Electric Company, Corporate Research and Development Center, P.O. Box 43, Building 37, Room 380, Schenectady, NY 12345.

IEEE Log Number 8608163.

position sensor is usually necessary to control the switching of the power semiconductor devices. The rotor is mechanically and thermally robust, with no cast-aluminum cage or permanent magnets (see Fig. 1). The stator is almost equally robust with a small number of coils similar to the field coils of a dc motor and with very short endwindings, a factor that leads to high efficiency and robustness as well as keeping the frame length short in relation to the stack length; see Figs. 2 and 3.

The simplicity of the motor is also reflected in the power electronic converter. Several circuit topologies are possible (Fig. 4), and all of them share the advantage that no "shoot-through" path exists through the power devices, since a phase winding always exists in series with each device. This advantage was pointed out in [4], but several other features are noteworthy. Converters can be built with only one series switch per phase, and this paper reviews some of the most common of these along with a completely new one [8]. The phases operate independently to a degree not possible in ac drives, so that the loss of one phase permits the drive to continue at reduced power without the problems of negative-sequence fields and high short-circuit currents.

This paper represents an attempt to define more closely than hitherto the capabilities of the SR drive relative to today's induction motor drive. The parameters evaluated include efficiency, torque per unit volume, copper utilization, the torque/inertia ratio, and the kVA requirement and are based on an extensive series of tests and computations. Further improvements will, no doubt, continue to be made in SR drive systems as the basic principles and characteristics become better appreciated and understood.

BASIC CHARACTERISTICS OF THE SR DRIVE

The basic characteristics of the SR motor are well-known, since it is a member of the class of single-stack variable-reluctance (VR) stepper motors. It is, however, designed for efficient power conversion rather than as a torque motor or control-type motor, and the pole geometry and control strategies differ accordingly. The tooth numbers are typically small, giving a large step angle, and the conduction angle is, generally, modulated as a function of both speed and torque to optimize operation as an adjustable-speed drive. Figs. 1–3 show components of an experimental 5-hp motor designed for a high torque/inertia ratio.

In this paper an attempt is made to compare some of the most important performance parameters of three SR motor drives and to illustrate some of the similarities and differences

Fig. 1. Rotor of low-inertia SR motor.

Fig. 2. 5-hp SR motor on test stand.

Fig. 3. 5-hp motor components.

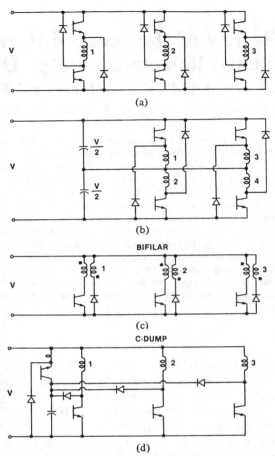

(a)

(b)

BIFILAR

(c)

C-DUMP

(d)

Fig. 4. SR power circuits. (a) Circuit 1. (b) Circuit 2. (c) Circuit 3. (d) Circuit 4.

between them and typical standard- and high-efficiency induction motor drives in the low integral-horsepower range. The three SR drives are 1) the 5-hp low-inertia motor already mentioned and shown in Figs. 1–3, 2) a three-phase motor designed and built at Newcastle-upon-Tyne University with two teeth per stator pole, and 3) an experimental 10-hp design calculated to fall between the standard- and high-efficiency induction motor drives in its overall performance. The lamination geometry of the Newcastle motor is shown in [5].

POWER CIRCUIT AND CONTROL CHARACTERISTICS

Fig. 4(a)–(d) show several alternative power circuits that have been built for SR drives; see also [10]. Circuit 1 is similar to that used for ac PWM motor drives except that the motor windings are inserted in the phase legs in series with the power devices instead of taking off a connection from each midpoint

464

to each terminal of a three-phase motor. Both power devices (shown as transistors in Fig. 4) are switched on and off together, instead of in pairs, one from each phase leg as in the ac PWM drive. The devices are protected from shoot-through faults by the motor windings, and the phases (phase legs) are completely independent. Circuit 2 is a variant of circuit 1 in which two phaselegs are "horizontally" split, and the upper and lower halves are used independently to control separate phases, making a total of four phases. A split capacitor is required in the dc link, and it is important to keep the upper and lower phases balanced to maintain correct voltage levels. This makes it difficult to design with this circuit for an odd number of phases.

Circuit 3 is the simple bifilar-winding circuit which has, apparently, the greatest simplicity in the converter. However, an imperfect coupling between the primary and secondary phase windings causes voltage transients to appear across the phase transistors at switchoff, and heavy snubbering may be required to protect them. The bifilar winding has a poor space factor in the motor and requires double the number of connections compared with the other circuits.

Both circuits 2 and 3 use only one power device per phase instead of the two that are normally used in ac motor drives. This can lead to a space saving in the converter and a reduction in losses, since now only one forward voltage drop occurs in series with each motor winding instead of two. The kVA of these converters is, however, unaffected by the circuit configuration, being a fundamental requirement of the motor. The voltage rating of the power devices in circuits 2 and 3 is, therefore, double the value in circuit 1 and is basically twice the dc link voltage (plus an allowance for transients and abnormal operating conditions).

Circuit 4 is a new converter described in [8]. It is intended to achieve the advantages of a circuit with only one switch per phase without the need for either a bifilar motor winding or a split dc link. Unconverted energy from all three phases is dumped into a common capacitor during the freewheeling period of each "working stroke" or step, and the accumulated energy in the capacitor is returned to the dc supply via a step-down chopper. The dump-capacitor voltage is maintained at a level above the dc supply voltage and can be controlled in such a way as to optimize the motor phase current waveform.

Typical current waveforms are shown in Fig. 5 for the low-inertia 5-hp motor, and Fig. 6 shows the trajectory of the operating point on the plane of the phase flux-linkage and the phase current. The area enclosed by this trajectory is proportional to the developed torque, being equal to the energy converted in each step. The trajectory is necessarily bounded by the extreme magnetization curves, which correspond to the aligned and unaligned positions of stator and rotor poles. A "good" motor design is one with a large separation between these extreme curves. For maximum torque per unit volume the curves should have as little saturation as possible, but for minimum converter kVA the effect of saturation is beneficial, even though it increases the motor size needed for a given torque [1].

It is an inherent feature of the SR motor that if commutation is delayed too long, then at sufficiently low speed a tendency exists for the current to increase as the operating point moves up into a region of high magnetic saturation where the magnetization curves are not only flatter but closer together, reducing the effective back EMF and leading to a potential runaway condition that can only be prevented by commutation or by chopping the phase current. Uncontrolled, this tendency towards a spiky current would raise the converter kVA per kilowatt of output power at a rapid rate and would put the SR motor at a disadvantage relative to ac drives, where the effects of saturation are more benign. In the studies that follow, comparisons between the SR and the ac induction motor drive are drawn for flat-topped current waveforms in the SR motor, i.e., at a level of output where the current spike has not yet set in. The flat-topped current waveform represents a natural condition associated with the base speed of the motor and is discussed at some length in [1]. Fig. 7 shows a torque/speed capability diagram for the low-inertia 5-hp motor.

PARAMETER COMPARISON OF SR MOTORS

The three motors to be evaluated have already been described. The low-inertia motor is the only one with a bifilar

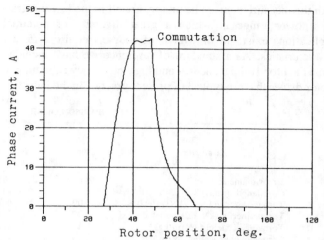
Fig. 5.   Current waveform of 5-hp SR motor.

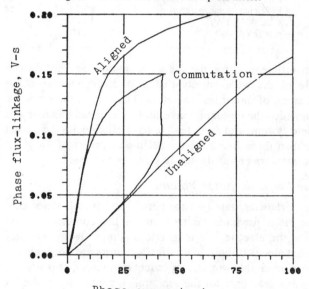
Fig. 6.   Flux-linkage/current trajectory for 5-hp motor.

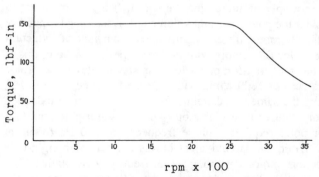
Fig. 7.   Torque–speed capability of 5-hp motor.

winding, and the results are corrected to correspond to a monofilar winding of the same copper weight. Table I is a summary of the main parameters evaluated. It contains a mixture of calculated and test results. The methods of calculation have been carefully cross checked and agree well with the measurements. The motors are all in the low integral-horsepower range in which a great diversity of potential applications exist, all of which, of course, require different characteristics. As a standard of reference we have included some data for 10-hp standard- and high-efficiency ac induction motor drives that are typical of present practice in the industry.

power density, although the torque density may be quite high (particularly in motors of instrument size).

Table I shows that the Newcastle motor and the 10-hp motor have about the same rated torque per unit stator volume, their temperature rises being comparable. (Both motors are totally enclosed and fancooled.) The 5-hp motor, however, has a

TABLE I
COMPARISON OF SR AND INDUCTION MOTOR DRIVES

Parameter	5-hp	Newcastle	10-hp	Standard Induction Motor	High-Efficiency Induction Motor
Stator diameter (mm)	200	165	205	221	221
Core length (mm)	83	108	179	95	140
Torque/stator volume (kN·m/m³)	6.09	8.70	8.68	11.2	7.56
Torque/inertia (kN·m/kg·m²)	8.97	5.32	3.74	1.59	1.07
Torque/EM weight (N·m/kg)	1.24	1.56	1.43	1.52	1.02
Torque/copper weight (N·m/kg)	3.10	7.15	6.93	7.72	5.93
Efficiency (motor, percent)	87.5	85.6	88.3	85.0	89.8
Efficiency (total, percent)	85.3	82.2	85.7	—	—
Peak (kVA/kW)	19.2	22.4	11.2	11.4	10.4
rms (kVA/kW)	6.60	7.84	5.50	4.74	4.26

System efficiencies for the induction motors are not quoted because of the wide variety that is met, depending on the combination of motor and inverter. (The best of these are the ones in which the motor and inverter are designed together as a complete system, and these tend to have efficiencies somewhat higher than those listed for the SR motor drives in Table I.) The parameters are evaluated in turn, as follows.

Torque Per Unit Stator Volume

This parameter, rated torque per unit of overall stator core volume, is a fundamental measure of the combined useful effect of the electric and magnetic loadings. It is obviously important since, if the figures for SR motors are not high and at least comparable to those of induction motors, it is hard to sustain any argument for the economic manufacture of the SR motor. The torque per unit volume is more meaningful than the power per unit volume since it is possible to raise the power simply by increasing the speed. It is not important to consider the power per unit volume unless the speed capability of the SR motor differs significantly from that of induction motors; for SR motors with one tooth per stator pole, this is not the case. An SR motor with eight stator poles and six rotor poles runs at 1800 r/min with a phase frequency of 180 Hz, having the same speed/frequency ratio as a 12-pole induction motor. A three-phase SR motor with six stator poles and four rotor poles requires a phase frequency of 120 Hz to run at 1800 r/min and is equivalent in this respect to an eight-pole induction motor. Although the frequency is higher, the magnetic loading is lower in the SR motor than in the induction motor, and, therefore, the core losses do not become prohibitive until very high speeds are reached or unless the motor has multiple stator teeth per pole. (The Newcastle motor has two stator teeth per pole.) Conventional stepper motors often have many more teeth per pole, and this limits their

significantly lower torque/volume, resulting from its small rotor diameter and short stack length. The other two compare favorably with the induction motors, falling between the standard-efficiency and the high-efficiency versions. Note that the Newcastle motor is somewhat at a disadvantage because of its smaller stator diameter compared with the other motors.

Torque/Inertia Ratio

This parameter is important as a measure of the speed with which the motor can respond to changes in speed reference and as a measure in all motion control applications. Table I shows that the SR motors are all much better than the induction motor in the torque/inertia (T/J) ratio, the worst one being the 10-hp motor, which is still $3741/1592 = 2.35$ times better than the best induction motor. The induction motors are, of course, designed for general-purpose applications and do not have specially small rotor diameters, but even so, the 10-hp SR motor achieves this large advantage in the torque/inertia ratio with no penalty in any other performance parameter.

The low-inertia 5-hp motor has, as expected, the highest value of T/J at 8972 N·m/kg·m², which is $8972/1592 = 5.6$ times better than the standard-efficiency induction motor and 8.35 times better than the high-efficiency induction motor. This high value is attributed to the special design of the 5-hp machine, which has a rotor/stator diameter ratio of only 0.45, significantly lower than in any of the other motors. The high T/J of the 5-hp motor is achieved with 81 percent of the rated torque per unit stator volume of the high-efficiency induction motor and 54 percent of that of the standard-efficiency induction motor; its efficiency at rated load is midway between the efficiencies of the two induction motors. To achieve this, it is necessary to drive the 5-hp motor into the current spike regime discussed earlier, so that the peak kVA per kW of shaft output is $19.2/11.4 = 1.68$ times that of the standard-

466

efficiency induction motor and 1.85 times that of the high-efficiency induction motor. This parameter is defined as the product of the peak device current times the peak device voltage times the number of main power switching devices divided by the shaft output power. (Peak device current usually occurs at the commutation point; see Fig. 5.) If the rms phase current is substituted for the peak current, the rms kVA/kW of the 5-hp motor is 6.60/4.74 = 1.39 times that of the standard-efficiency induction motor and 1.55 times that of the high-efficiency induction motor. The factors leading to a high value for these parameters are discussed in [1].

Torque Per Unit Electromagnetic Weight

In many applications the total electromagnetic (EM) weight is important as a measure of the cost and the suitability of the motor for applications where light weight is at a premium. All of the SR motors in Table I have a higher torque per unit EM weight than the high-efficiency induction motor, but only the Newcastle motor with two teeth per stator pole exceeds the figure for the standard-efficiency induction motor. This parameter does not bring out any inherently great advantage of either type, both being subject to the same trade-off between weight and efficiency.

Torque Per Unit Copper Weight

This parameter, rated torque per unit of stator copper weight, is obviously important in view of the relatively high cost of copper. Again, no inherent advantage exists for either type of motor, and both types are subject to the same trade-off between efficiency and copper weight. The SR motor is, perhaps, more sensitive than the induction motor because of its higher electric loading and lower magnetic loading, and because it is singly excited with a relatively poor effective power factor compared with a high-efficiency induction motor. One striking result in Table I is the poor copper utilization in the 5-hp motor, which results from the small rotor diameter. A small rotor diameter is inconsistent with a large aligned/unaligned inductance ratio, and this leads to a large MMF requirement.

Converter Voltamperes

We have already defined the peak and rms kVA per kW of the shaft output as a measure of the converter rating. A fundamental theory of this parameter, which is akin to a theory of the power factor in the SR motor, is given in [1], and a further development of these ideas will be published shortly. It is important to note from Table I that the SR motor is not at a serious disadvantage relative to the induction motor in the converter voltampere requirement, when designed for comparable overall performance. The 10-hp design has an rms kVA/kW of 5.5/4.74 = 1.16 times that of the standard-efficiency induction motor and 5.5/4.26 = 1.29 times that of the high-efficiency induction motor. The induction motor figures are based on six-step voltage-source inverters; with PWM inverters their kVA requirements would be significantly higher and,

therefore, closer to the figures for the SR motor. For the 5-hp low-inertia motor the kVA/kW is markedly higher than for any of the other motors for reasons already discussed. This motor has a somewhat spiky current at the 15.8 N·m quoted in Table I. At 10.9 N·m, however, the current is more nearly flat topped, giving a peak kVA/kW of 12.4 and an rms kVA/kW of 6.4, values which are much closer to those observed for the 10-hp SR motor. The high figure of 22.4 peak kVA/kW for the Newcastle motor is also associated with a tendency to a spiky current waveform. In the 10-hp SR motor the current waveform is not spiky at all but has a rounded top and a crest factor (peak/rms) that is better than that of an induction motor operating with six-step excitation.

Special Features of Motor with Two Stator Teeth per Pole

With two or more stator teeth per pole the angular rate of change of inductance is increased, so that for a given exciting MMF, the average torque can be expected to be greater. This expectation is justified by detailed analysis [5]. At the same time the frequency/speed ratio is approximately doubled compared to that of a machine with the same number of phases and stator poles. Consequently, the iron losses are higher, and this limits the high-speed capability. Table I shows that the Newcastle motor does, indeed, have a high torque per unit of stator volume—the highest of all of the SR motors tested. It still falls between the values for the two induction motors, however. The torque per unit of electromagnetic weight is also the highest of all of the SR motors and is about equal to that of the standard-efficiency induction motor. The torque/inertia ratio is 5325/1592 = 3.34 times that of the standard-efficiency induction motor and 4.95 times that of the high-efficiency induction motor, but this is achieved with a rather peaky current that leads to a high value for the converter voltamperes. It is not yet known whether this is an inherent feature of motors with two teeth per stator pole, or whether other particular features of this motor lead to a spiky current. Work on this investigation is still in progress.

CONCLUSION

The SR drive has obvious appeal as a rugged low-cost adjustable-speed drive. The motor is physically simpler than almost any other motor in the range of powers and applications considered, and the converter can be designed in several different ways to meet application requirements. All SR converters have inherent protection against shoot through, and several topologies exist that use only one switching device per phase, leading to a reduction in size and forward conduction losses. The independence of the phases is a further feature that improves the reliability and the ability to continue to operate after certain inverter or motor faults. Even when motor faults do occur, the simplicity of the construction is such that repair—for example, the replacement of a stator coil—is a simple matter, so repair costs for this type of motor should be extremely low.

The parameter comparison reported in this paper shows that SR motors can be designed to equal the overall performance of the induction motor drive and, in some respects, to exceed it. Comparing the performance profile of the 10-hp design, it clearly matches the best features of both the standard-efficiency and the high-efficiency induction motors by combining these in one drive. The somewhat sceptical response to the initial development of this class of motor drives is not justified in view of these figures. In particular, the often-expressed belief that the reluctance motor requires high peak currents and a large converter rating is simply not true, this impression probably being a hangover from the sinewave or synchronous reluctance machine, which does have a significantly lower power factor than comparable ac induction or permanent-magnet motors [9]. In terms of the converter kVA, whether based on peak or rms currents, the SR motor has about the same requirement as a PWM inverter-fed induction motor. At the same time, it requires only three or four main power switches (transistors or GTO's) compared with the six required for ac motors. One parameter in which the SR motor significantly outperforms the comparable induction motor is the torque/inertia ratio, which is on the order of two to three times that of induction motors having the same efficiency and converter rating. Still greater values are possible, as in the low-inertia 5-hp design, but at the expense of converter rating and efficiency. Among the reasons for the high T/J ratio are the salient-pole structure of the rotor, which removes a great deal of weight from the outermost rotor diameter, and the tendency to design SR motors with smaller rotor/stator diameter ratios (though not as small as those of classical 15° instrument-size stepper motors).

Acknowledgment

The authors would like to thank several colleagues, especially F. G. Turnbull and W. R. Oney for their work in both design and testing, T. W. Neumann for a great deal of helpful discussion and analysis, and J. A. Rulison for much of the experimental work. Also acknowledged is the support of J. C. Bunner at the DC Motor and Generator Department.

References

[1] T. J. E. Miller, "Converter volt-ampere requirements of the switched reluctance motor drive," in *Proc. IEEE Ind. Appl. Soc. Annu. Meeting*, Oct. 1984.

[2] B. D. Bedford, US Patent 3 678 352, 1972.

[3] B. D. Bedford, US Patent 3 679 953, 1972.

[4] P. J. Lawrenson *et al.*, "Variable-speed switched reluctance motors," *Proc. Inst. Elec. Eng.*, vol. 127, Pt. B, pp. 253–265, July 1980.

[5] J. W. Finch *et al.*, "Variable speed drives using multi-tooth per pole switched reluctance motors," in *Proc. 13th Annu. Sym. Incremental Motion Control Systems and Devices*, May 1984.

[6] J. W. Finch and M. R. Harris, "Linear stepping motors: An assessment of performance," in *Proc. Int. Conf. Stepping Motors and Systems*, Sept. 1979.

[7] J. Corda and J. M. Stephenson, "Analytical estimation of the minimum and maximum inductances of a double-salient motor," in *Proc. Int. Conf. Stepping Motors and Systems*, Sept. 1979.

[8] J. T. Bass *et al.*, "Development of a unipolar converter for variable reluctance motor drives," in *Proc. IEEE Ind. Appl. Soc. Annu. Meeting*, Oct. 1985.

[9] T. J. E. Miller *et al.*, "A permanent-magnet excited high-efficiency synchronous motor with line start capability," in *Proc. IEEE Ind. Appl. Soc. Annu. Meeting*, Oct. 1983.

[10] R. M. Davis *et al.*, "Inverter drive for switched reluctance motor: Circuits and component ratings," *Proc. Inst. Elec. Eng.*, vol. 128, Pt. B, pp. 126–136, Mar. 1981.

Martyn R. Harris received the degree from University College, London (UCL), England.

He moved to a research post on electrical machines at AEI Ltd., Trafford Park, Manchester. He subsequently became a Design and Development Engineer in the Large Industrial Machines Division, concentrating on induction motor development. In 1961 he was appointed Lecturer at UCL, responsible for teaching and research in electrical machines and electromagnetic theory. He subsequently became a Senior Lecturer and then a Reader in 1976. From 1974 to 1977 he was an Honorary Research Associate at the University of Leeds, pursuing collaborative work on stepping motor development. He became a Professor at Newcastle University in 1978 and now works with a group of staff covering research in large and small electrical machines and power systems closely linked to industry. He has published papers in electromechanics and has acted as a consultant with many national and international companies. In 1984 he was a Visiting Research Fellow at General Electric, Schenectady, NY, working on switched reluctance motors for the adjustable speed drives program.

Mr. Harris is a past Chairman of the Electrical Machines Professional Group Committee of the Institution of Electrical Engineers, (IEE), a past member of the Power Divisional Board and is presently a member of the IEE Council.

John W. Finch was born in Durham, England, in 1946. He received the degree with first-class honors in electrical engineering from University College, London, England, in 1967, and the Ph.D. degree from Leeds University.

After a period of employment with International Computers Ltd., he lectured in control engineering at University College North Wales, Bangor, North Wales. Since 1978 he has been employed as a Lecturer at Newcastle University where his research interests have centered on the control and design of electrical machines and drives. He has acted as a consultant to several national and international firms.

Dr. Finch is a chartered engineer. He is a past recipient of the Institution of Electrical Engineers Control and Instrumentation Division Heavyside Premium. He received the Faculty Engineering Prize at University College, London, and his Ph.D. dissertation was awarded the F. W. Carter Memorial Prize.

John A. Mallick (M'79) received the S.B. degree in 1973, the S.M. and E.E. degrees in 1976, and the Sc.D. degree in 1979, all from the Massachusetts Institute of Technology, Cambridge.

In 1979 he joined the General Electric Corporate Research and Development Center in Schenectady, NY, as an Electrical Engineer. Since 1983 he has also been an Adjunct Professor of electrical engineering at Union College, Schenectady. His research interests are electromechanics, electromagnetics, numerical simulation, and VLSI systems architectures.

Dr. Mallick is a member of Eta Kappa Nu and Sigma Xi.

Timothy J. E. Miller (M'74–SM'82), for a photograph and biography please see page 715 of this TRANSACTIONS.

Step Motors That Perform Like Servos

K.A. Regas and S.D. Kendig

Machine Design, December 10, 1987, pp. 116-120

Though variable-reluctance motors are normally open-loop steppers, one new type performs as a servomotor in closed-loop systems.

KENNETH A. REGAS
Development Engineer
Hewlett-Packard Co.
San Diego, CA

STEPHEN D. KENDIG
Kernco Inc.
Danvers, MA

Variable-reluctance (VR) motors are widely used as step motors. Until recently, their torque-vs-rotor position characteristics were unsuitable for closed-loop systems. A new type of VR motor, however, exhibits torque characteristics suitable for this application. When commutated electronically and powered by current-sourcing drivers, the new motors outperform comparable dc motors at speeds up to 4,000 rpm and more.

VR basics

Conventional VR motors contain a multipole stator set up for two, three, or four-phase operation, and a multipole rotor. The stator typically contains more poles than the rotor. When the phases are energized in sequence, the rotor steps through angles equal to the pitch of the rotor poles minus the pitch of the stator poles.

A variable-reluctance motor, here disassembled, contains an eight-pole rotor and a six-pole stator. The motor, manufactured by Warner Electric Brake & Clutch Co., is simpler, more reliable, less costly, and develops more torque than comparable dc motors.

A three-phase VR step motor, for example, typically contains 12 stator poles on a 30° pitch and 8 rotor poles on a 45° pitch. The step angle is 15°.

Motor designers select the number of teeth on the stator and the spacing between them to optimize operation in step mode. The teeth normally have a uniform width to facilitate the winding of stator coils. The width of the teeth at the tooth face usually is the same as or slightly greater than the width of the space between the teeth. That is, the ratio of tooth-width to space-width, called tooth ratio, is typically ≥ 1. Windings generally fill the space between teeth.

Rotor teeth are the same width as stator teeth. Thus since there are fewer rotor teeth than stator teeth, the tooth ratio is generally about 0.5.

If the motor shaft is positioned so its rotor teeth are misaligned with the active stator teeth, flux density is higher near the tip of the teeth than in the body when the motor is energized. The teeth saturate at a relatively low value of current but do so only near the tip.

When the teeth are aligned, however, flux density is essentially uniform along their length. The entire length of the teeth can saturate, but only with a relatively high current.

Torque is proportional to the rate

TOOTH-SHAPE MAKES THE DIFFERENCE

When operated in step mode, the motors diagrammed here both rotate 15° per step. But their torque-vs-rotor position qualities differ appreciably, a result of differing reluctance characteristics.

In conventional VR motors, reluctance of the magnetic circuit varies with rotor position. When the rotor and stator teeth are misaligned, the teeth saturate at a low level of current. But only corners of the teeth saturate. Here, tooth reluctance is a small percentage of the total for the entire magnetic circuit.

When the rotor and stator teeth on an ordinary motor are aligned, the teeth saturate only at high current. But the entire length of each tooth saturates. Tooth reluctance is large with respect to total circuit reluctance. Thus, reluctance for conventional VR motors varies with shaft position.

Teeth in the new VR servomotor, on the other hand, saturate at a low current regardless of rotor-stator alignment. But in all positions, the tooth shape limits saturation to a region near the tip. Reluctance of the teeth is constant regardless of rotor position and is a small part of total circuit reluctance.

Also, when a phase of the new motor is energized, the associated rotor and stator teeth already partially overlap. The constant tooth reluctance and initial overlap together tend to smooth out torque ripple.

The new motors are designed to be driven with the tips of the stator teeth in saturation. In this regime, torque varies linearly with current.

A two-axis controller developed by Semifusion Corp. powers variable-reluctance motors at acceleration rates up to 100,000 rad/s². The controller interfaces with HP71B and PC-compatible computers.

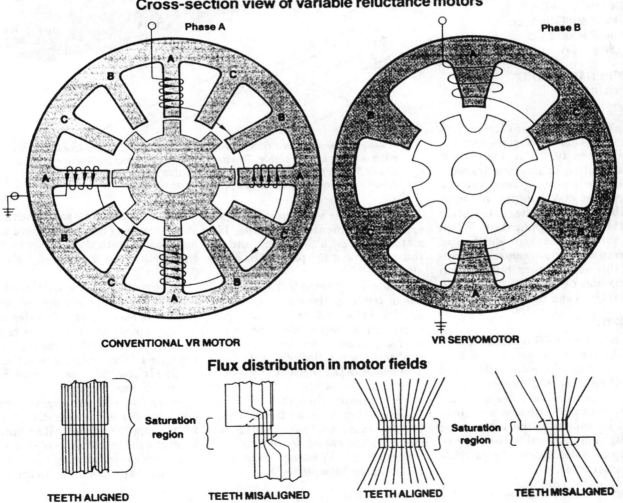

Cross-section view of variable reluctance motors

CONVENTIONAL VR MOTOR

VR SERVOMOTOR

Flux distribution in motor fields

TEETH ALIGNED

TEETH MISALIGNED

TEETH ALIGNED

TEETH MISALIGNED

470

of increase of flux carried by rotor and stator teeth as they rotate into alignment. Both air-gap reluctance and tooth reluctance simultaneously decrease as conventional VR step motors rotate into detent position. The sharp peaks in torque that result are ideal for step motors but produce variations in torque versus rotor position. The so-called torque ripple is undesirable in servo applications.

A new approach

VR servomotors, unlike conventional VR motors, contain a rotor with more teeth than the stator. The tooth ratio on the rotor typically runs from about 0.75 to 0.9 and the stator tooth ratio is about 0.5. Stator teeth are tapered and typically are twice as wide at their base as at the face.

One three-phase version of the VR servomotor, for example, contains an eight-pole rotor and six-pole stator. The step angle is 15°. Tooth ratio is 0.78 on the rotor and 0.5 on the stator.

The tapered shape would make it difficult to wind coils on the teeth

TORQUE FOR PRECISE POSITIONING

In positioning servos, it is important to minimize torque variations, or ripple, with rotor position because such variations affect closed-loop poles of a servo. The sharp peaks in torque-vs-rotor position of conventional VR steppers are inimical to this end. Because of their different tooth geometry, VR servomotors achieve smooth enough torque for closed-loop operation.

VR servomotors are commutated electronically from encoder position data, much as are brushless dc motors. The selection of which winding to energize at any particular rotor position determines the direction of torque. The amount of current supplied to the motor determines the magnitude of the torque. Position accuracy depends on the resolution of the encoder, not on step size.

were it not for two nonmagnetic spacers. The spacers, mounted on each end of the stator stack, contain the same number of teeth as the stator. But spacer teeth have a uniform width equal to the base dimension of the stator teeth.

When a motor is energized, flux density is relatively low at the base of the stator teeth and progressively higher towards the tip. Unlike conventional VR steppers, VR servomotors are designed to be driven beyond saturation. Here, the ta-

pered shape confines saturation to a region near the tip. This keeps tooth reluctance small and relatively constant versus rotor position. And because of the high tooth ratio on the rotor, when opposing teeth are energized they already overlap somewhat. Thus, the increase in flux as they rotate into alignment is dominated by the increase in overlap rather than by changes in fringing fields. Together, constant tooth reluctance and tooth-to-tooth overlap greatly re-

THE PRIMARY ADVANTAGE

Two VR servomotors power a new plotter recently developed by HewlettPackard. The new motors, which are manufactured by Warner Electric Brake & Clutch Co., replaced more costly samarium-magnet dc motors. Despite the lower cost of the new motors, the VR servosystems cost about 30% more than the dc systems, a result of more costly drivers. The higher cost is justified by improved performance.

Torque rating for the VR servomotors exceeds that for the dc motors by a factor of almost three. The higher torque is available at speeds up to 4,000 rpm or more. And the VR system provides 70% greater acceleration than dc drives in this range, despite an inertia that is 1.7 times higher.

Torque rating for VR motors exceeds that for dc motors because the VR motors can more easily dissipate the heat resulting from energy losses. VR losses are primarily caused by hysteresis and eddy currents in the

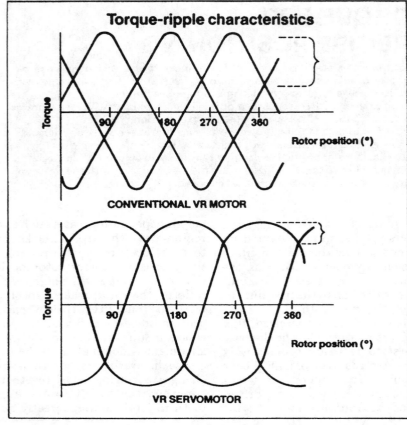

Torque-ripple characteristics

Torque

Rotor position (°)

CONVENTIONAL VR MOTOR

Torque

Rotor position (°)

VR SERVOMOTOR

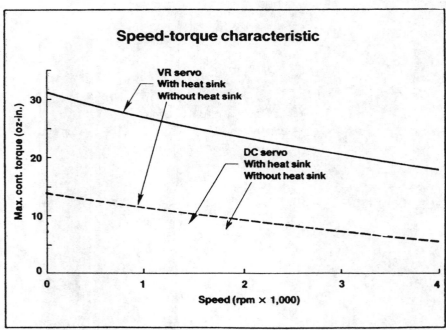

Speed-torque characteristic

VR servo
With heat sink
Without heat sink

DC servo
With heat sink
Without heat sink

Max. cont. torque (oz-in.)

Speed (rpm × 1,000)

duce torque ripple. And because they are driven beyond saturation, these motors develop more torque than do conventional types.

Rotors for these motors differ from those of conventional VR types in that they are formed with silicon steel laminations. The structure minimizes hysteresis and eddy-current losses.

Because they contain no commutator or magnets, the new motors are simpler, more reliable, and less costly than dc motors. As in dc brushless types, electrical losses arise primarily in the stator rather than in the rotor making them relatively easy to cool compared to conventional dc motors.

Control modes

VR servomotors do not operate in either step or microstep mode. Instead, they are commutated electronically like dc brushless motors. Windings in VR servomotors, however, exhibit much higher inductance than those in dc motors. The inductance varies widely with both current and rotor position. Thus, precise control calls for current-source drivers rather than the voltage-source drivers that normally power dc motors.

The inductance of VR motors introduces a highly variable, low-frequency (as low as 12 Hz) electrical pole in the servo loop when power comes from voltage-source drivers. The pole degrades servo-system performance. But using a current-source driver, the pole shifts to a high frequency — generally above 1 kHz — where it has negligible effect.

Current-source drivers generally do not fit in easily with the digital control systems normally used for motion control systems. The problem is that the combination normally calls for a costly A-to-D converter to change the analog current feedback voltage to a digital signal. A specially developed controller for the motor, however, avoids the cost of the converter.

In this controller, an application-specific IC (asic) converts a digital motor command to a pulse-width modulated (PWM) digital signal. A low-pass filter converts the PWM signal to a dc signal having a sawtooth component. The average value of the sawtooth varies linearly

stator. Losses in dc motors arise primarily from resistance of the rotor windings.

The accompanying graph displays the continuous torque available from comparable VR and dc motors. The

graph was plotted from experimental data where motor temperature rise was held constant at a rated value. Thus, the graph is not a typical speed-torque curve where voltage is held constant.

with the duty cycle of the PWM signal.

Sense resistors in the motor circuit supply a current feedback signal. A comparator processes both the saw-tooth wave and the current signal to control switches in an H-bridge motor driver. The comparator turns the switches on when the value of the current signal is below the sawtooth. The action applies full forward voltage to the motor, forcing motor current toward maximum. When the value of the current signal is greater than the saw-tooth, the comparator turns the switches off. Motor current is forced to pass through diodes against full reverse voltage. Current is driven quickly to zero.

When the current signal lies within the sawtooth portion of the dc signal, the switches turn on and off at the same frequency as the PWM signal. The ratio of on-time to total time, or duty cycle, automatically adjusts to keep current constant. This duty cycle generally differs from that for the PWM signal itself.

Driver gain, in a small-signal sense, is proportional to the power-supply voltage and inversely proportional to the amplitude of the sawtooth wave. The sawtooth amplitude is proportional to $N(N-1)$, where $N =$ the PWM duty cycle. Current-driver gain, and, thus, the frequency of the pole associated with the current loop, varies from infinity, when $N =$ either 0 or 1, to a minimum value when $N = 0.5$.

Disadvantages of the variable gain are avoided by adjusting the current gain of the controller. Current gain at $N = 0.5$ is set to keep the associated pole at a sufficiently high frequency so that its effects in the servo loop are negligible. Pole positions for other values of N are at higher frequencies where they are even more negligible. This arrangement can make current gain high enough to permit operation with unregulated power supplies.

Switches in the drivers reside in three H-bridges, one bridge for each phase. The bridges each consist of two transistors and two diodes, a low cost configuration. The approach suits VR motors because the direction of current flow is the same for either CW or CCW. In VR motors, the sequence in which the phases are energized determines the direction of motor rotation. ∎

ASIC MATES CURRENT DRIVER WITH DIGITAL CONTROLLER

Analog feedback from a current-source driver must be converted to digital form for use by a digital controller. The accompanying diagram discloses one low-cost technique for handling this chore.

An application-specific IC converts a digital motor command into a PWM signal. A low pass filter then converts the PWM signal to dc with a sawtooth component.

A comparator compares the dc signal with an analog current signal that is supplied by a differential amplifier. Comparator output controls switches in bridge circuits that power the motor. When on, the switches apply full voltage to the motor, driving motor current towards a maximum value. When off, they force motor current to flow against full reverse voltage, driving motor current towards zero.

When the current signal exceeds the sawtooth wave, the switches go off. When the current signal is less than the sawtooth, the switches go on.

When the current signal lies between the sawtooth valleys and peaks, the switches turn on and off at the same frequency as the PWM signal. The duty cycle for the switches automatically regulates the average level of current.

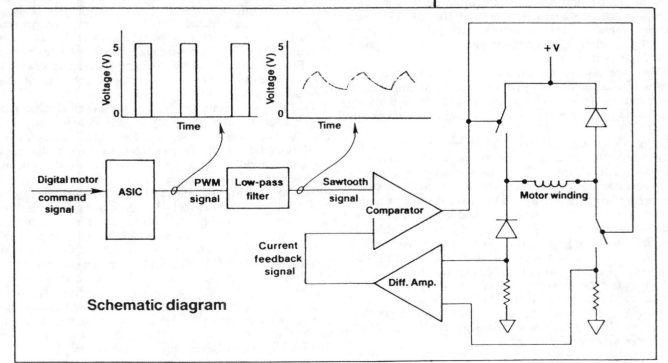

Schematic diagram

Switched Reluctance Motor Drives for Rail Traction: Relative Assessment

P.S.R. French

Proceedings IEE, Vol. 131, Pt. B, No. 5, September 1984

Abstract: The paper reports an assessment of the technical and economic benefits of switched reluctance motor systems applied to railway propulsion in motor sizes of 100–200 kW. Several forms of power electronic circuit have been considered for providing the pulsed DC supply required. Comparison is made with inverter-fed induction motor drives using conventional McMurray three-phase PWM thyristor inverters. Further comparisons are drawn against chopper-fed DC motor systems such as are currently entering service on various railways. It is concluded that the motor itself should offer lower maintenance requirements, higher reliability and a greater power/weight ratio than either the induction motor or the DC motor. The total system, however, appears to show no clear benefits over the induction motor system, and both types are at present more expensive than DC motor systems. (A second view on switched reluctance motor drives by Ray *et al.* is also published in the same issue.)

1 Introduction

The reluctance motor has evolved in various forms over several decades and has found commercial applications in general industry. Recently it has emerged as a serious competitor for both the DC and induction motors in traction applications, and has been applied, with success, to road traction at a power level of around 20 kW.

Much development work is required before the reluctance motor can be applied to rail traction, but it has reached the stage where it is appropriate to conduct a thorough appraisal of its technical features and possible economic benefits in comparison with both existing DC systems and projected induction motor and inverter systems.

Such an appraisal has recently been conducted at a power level applicable to rapid-transit rail applications, in which a high proportion of axles is motored. The individual motor rating is comparatively low (100–200 kW), and thus represents a realistic step in the extrapolation of existing designs, enabling a comparison to be made against known standards.

There are always difficulties in comparing projected designs of equipment with existing equipment, and careful interpretation of the results and consideration of the variable factors must be employed in drawing the final conclusions.

Designs of reluctance motor have been commissioned, based upon the internal space envelope of a standard 140 kW traction motor. The power conditioning equipment necessary for the reluctance motor was compared with a standard McMurray, 12-thyristor inverter, capable of operating in both sine-wave modulated and quasi-square-wave modes. The McMurray circuit was chosen as the yardstick, being the most commercially widespread competitor. Regulated-source-type inverters, and transistor inverters at present, tend to be more expensive.

This paper presents results of the quantitative and qualitative assessments made between the two tried systems. Also included to keep the costings in perspective are relative through-life costs, estimated for DC and induction motor systems.

2 Benefits offered

DC series motors, traditionally employed for railway propulsion, are troublesome machines. The arduous electrical and mechanical conditions under which these machines operate make commutation very difficult, and the presence of large quantities of dirt and dust in the cooling air can damage the armature insulation. Consequently, rapid brush wear and deterioration of the commutator surface are common, as are insulation failures necessitating a full rewind after less than five years service.

The leading objective of all proposals to use AC motors in traction is to eliminate the commutation problem and so to achieve a higher availability. Confining insulated coils to the stationary member by using a squirrel-cage induction motor will give improved reliability to the insulation system. In the switched reluctance motor the stator winding comprises a few (typically eight) coils of large section and simple shape. This feature is expected to lead to further improvements in the reliability of the insulation system and to simplify rewinding when it becomes necessary. Furthermore, the absence of any winding at all on the rotor will make for a greatly enhanced integrity of that member.

One of the attractive features of the switched reluctance motor system is the possibility of maintaining full power over a wide speed range, whereas a DC series motor exhibits a falling power characteristic unless the commutation conditions allow field weakening to be employed by divert resistors or tapping.

However, using AC motors brings a heavy penalty on first cost, because, although AC motors are cheaper than equivalent DC motors, the control equipment is always more expensive, often by a gross margin. The switched reluctance motor proposal is an attempt to minimise this penalty. The cost of the electrical components of the motor is very much less than for an equivalent induction motor, and by using pulsed DC rather than bi-directional currents it is possible to halve the number of switching elements in the control equipment in certain applications. It is shown in Section 4.3, however, that the conditions in railway applications are unfortunately not favourable.

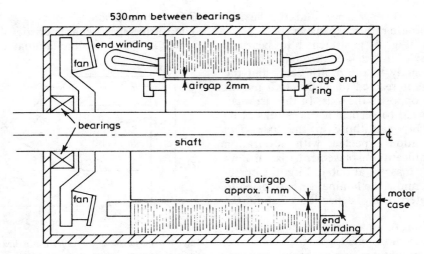

Fig. 1 *Cross-sections of switched reluctance and induction motors of identical frame sizes*

3 Motor design

3.1 General description

The principles of the operation and design of switched reluctance motors with salient-pole rotor and stator have been set out by Lawrenson *et al.* [1]. The motor described in this study was designed for British Rail Research by Leeds University. The particular arrangement found most suitable for the present application has eight stator teeth and six rotor teeth, although other combinations are available. Around each stator tooth is placed a coil; pairs of coils on opposite sides of the stator are connected in series to form the four independent phases. Each coil has a rectangular shape and approximately square cross-section.

3.3 Mechanical design considerations

Two types of motor mounting are in common use: bogie mounted with flexible drive, and axle hung. In the bogie-
The conductor material used is of 25 mm² cross-section, and made up from a number of smaller strands so that eddy-current losses are restricted. The rotor consists of the shaft with its laminations and a disc for supplying angular-position information to the drive electronics. During one revolution of the rotor each stator coil experiences the passage of six identical rotor teeth, and six pulses must be applied; that is the phase switching frequency has to be six times the rotational frequency. The base speed of the motor is 1390 rev/min and the top speed is 3475 rev/min. The corresponding stator switching frequencies are 139 Hz and 347.5 Hz.

In view of the high frequencies involved it is worth-while using a low-loss grade of magnetic material for the cores. Transil 315-35 was specified; it is a low hysteresis material with a thickness of 0.35 mm.

3.2 Dimensions

The overall dimensions of the motor case are a 480 mm diameter and a 710 mm length. The mass is approximately 900 kg. The lamination dimensions were specified to enable the motor to fit within the frame of a known induction motor built for trials in an experimental railway vehicle. Fig. 1 presents the longitudinal section of the motor in outline form, along with the induction motor for comparison. The switched reluctance motor has a greater specific volume in terms of stator-core volume to produce an equivalent output. The low winding overhang enables this to be achieved within the same space envelope as the induction motor.

mounted case, Fig. 2, the motor is protected from much of the shock loading by the primary suspension and by soft

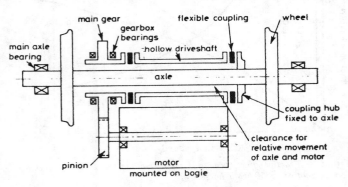

Fig. 2 *Bogie mounted motor*

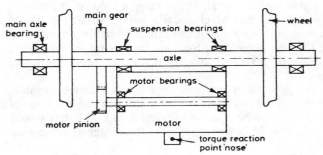

Fig. 3 *Axle-hung motor*

mountings for the motor. The gearbox and the flexible drive for this arrangement are rather expensive.

An axle-hung motor, Fig. 3, is attached to bearings fitted directly on the axle and drives through a simple pinion and gear fixed rigidly to the motor shaft and the axle, respectively; torque is reacted through a soft mount at the side of the motor opposite the axle. In this arrangement the motor is subjected to vertical accelerations at its bearings that are roughly half those at the axle. The maximum rotor acceleration expected, with a duration sufficient to give significant displacement, is around 300 ms^{-2}. For the rotor mass of about 150 kg, the corresponding force is 45 kN, and if the laminations are fitted to a shaft of 75 mm diameter, then a deflection of about 0.4 mm would result.

To reduce the deflections to an acceptable level, some or all of the following features must be adopted:

(i) Place the rotor laminations under heavy precompression so that the stack contributes to the overall stiffness of the rotor.

(ii) Place the bearings as closely as possible to the rotor lamination stack. The short length of the stator winding overhang helps in attaining this. However, it is necessary in an axle-hung motor to keep the drive-end bearing close to the pinion, and some compromise will be necessary.

(iii) Keep construction and assembly tolerances, and bearing clearances to the minimum practicable values.

(iv) Use the largest possible shaft diameter permitted by the electromagnetic design

If these measures are adopted, a gap of 1 mm should be maintained in an axle-hung motor. In a bogie-mounted motor a smaller gap would be possible but at the cost of closer machining and assembly tolerances.

3.4 Airgap

The airgap used in the 140 kW induction motor whose frame has been adopted for the present application is 2 mm. The penalty for using a larger gap would be increased magnetising current. In traction motors it is considered that airgaps of less than about 2.5 mm should be avoided. This experience is derived from DC motors, however, in which a large gap has several electromagnetic benefits apart from the obvious mechanical one.

To obtain the best performance from a reluctance motor, it is necessary to make the inductance of the stator coils vary to the largest possible extent with rotor position. It is essential, therefore, to keep the airgap between rotor and stator teeth as small as possible; a value of 0.5 mm or less would be ideal. Mechanically, such a small gap would be unacceptable owing to the motor shaft deflections predicted in the preceeding Section. Increasing the airgap to 1 mm incurs a penalty of only 6% in the rated output of the motor.

A compromise is required between the electromagnetic and mechanical requirements outlined. For the purposes of the study it has been assumed that a gap of 1 mm would be acceptable in both respects for the switched reluctance motor.

3.5 Performance

The torque/speed characteristic of the switched reluctance motor in Fig. 4 has the familiar traction-curve knee point

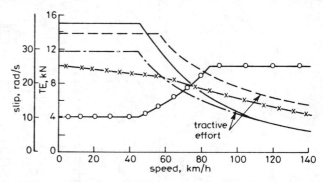

Fig. 4 *Performance curves for a switched reluctance motor and a typical variable frequency induction motor drive*

- - - - switched reluctance motor
———— induction motor
—·—·- continuous rating of induction motor
-×-×-× switched reluctance motor continuous at 5 A/mm²
-o-o-o- slip frequency

at the end of the constant-torque region. Extension of the conduction period above this speed increases the effective voltage applied to the motor, and enables a constant-power region to be obtained. Almost any electromagnetic machine will produce constant power if the applied voltage is increased in proportion to the square root of the speed. This occurs naturally in present-day diesel/electric locomotives whose generators display high regulation. The Class 253 high-speed diesel locomotive characteristic in Fig. 5 is an example. In the switched reluctance motor pulse extension from 120° at the knee point speed to 180° at maximum speed is equivalent to a voltage increase of 1.5:1 and gives the motor a constant-power speed range of 2:1. The induction motor can produce a constant-power region either by increasing the rotor slip speed over a range limited to about 2:1 or by increasing the stator voltage. The constant-power range of either type of drive

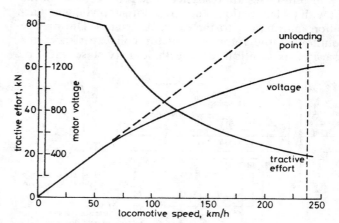

Fig. 5 *Tractive-effort/speed and motor-voltage/speed curves for Class 253 locomotive (2250 brake horsepower)*

may be extended by oversizing either the machine or the power-conditioning equipment.

If the gears are chosen to give a top speed of 145 km/h (90 mile/h), which is a usual figure for outer suburban stock, then the tractive effort to be expected at the rail at base speed is 12.8 kN (allowing 2% loss in the gears). This is not a particularly high figure compared with DC motors; however, the overall train performance benefits substantially from the constant-power sections of the characteristics.

A typical performance curve for a variable-frequency induction motor is also shown in Fig. 4. Gear ratios of 69:14 have been taken in both cases. Considerable difference will be noted in the performance characteristics of the two machines. The acceleration characteristics are based upon machine electromagnetic limitations up to base speed, and upon constant line voltage thereafter. The switched reluctance motor has the better acceleration characteristic.

The continuous rating curves are based upon equal current density in each machine of 5 A/mm². This is a low figure for the induction motor; typical figures for slotted windings in traction applications lie between 5 and 7 A/mm². For the reluctance motor this basis may be optimistic; its windings are very similar to the field coils of a DC motor, where 3–4 A/mm² is typical. For continuous rated current, the induction motor gives the higher torque at low speeds. However, above about halfspeed, the switched reluctance motor can benefit from the improving form factor, whereas the induction motor is slip limited. This enables the switched reluctance motor output to exceed that from the induction motor at the higher speeds.

In general, it may be said that, for accelerating and sustaining high speed, the switched reluctance motor is the more suitable machine. However, for hauling heavy loads at low speeds, or for a stop/start duty cycle with a midrange average speed, the induction motor is the most suitable machine. The conditions for the latter case are the ones which are most often required for a motor of this rating in a rail-traction application.

The torque generated by the switched reluctance motor, particularly at high speed, is uneven. At very low speeds, in the chopped-voltage-drive mode, the torque ripple has a peak-peak value of a few percent of the mean, and near the top speed the torque can drop, momentarily, almost to zero. The frequency of the ripple is 24 times the rotational frequency and therefore extends from 0 to 1390 Hz. Although the ripple is not seen as a potentially serious problem area, it will be necessary to design the motor mounting and the transmission without sharp resonances in this frequency range. In addition, the stresses in the pinion will be adversely affected to a small extent. It is desirable that a resilient transmission member be employed to alleviate torque pulsation effects.

4 Power conditioning

4.1 Requirements
It has been shown [2] that the switched reluctance motor will accept a pulsed DC waveform, and it is this which

gives it a unique flexibility in drive circuit options. It is important to note, however, that the voltage must be allowed to reverse in order to satisfy the demands of rapid negative flux changes. For this reason a halfbridge inverter must be used to drive the motor and not simply a number of phase displaced choppers. This distinction has important repercussions upon overall system costs.

For the greater part of the speed range a single drive pulse per phase suffices. The applied average winding voltage can be regulated by pulse-width control, and the output torque optimised by relative angular timing of the pulse. At low speeds a chopping mode of control must be used to obtain a smooth torque output. Fig. 6 illustrates

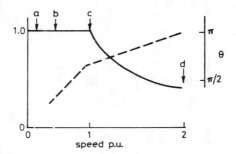

Fig. 6 *Thyristor conduction angle and operating points related to the traction characteristics*

– – – – thyristor conduction angle θ
—— p.u. tractive effort

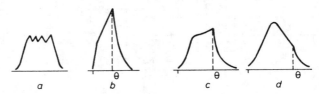

Fig. 7 *Typical motor waveforms for the operating points shown in the traction characteristics of Fig. 6*

the drive conditions for the motor in relation to speed. Fig. 7 shows the amplitude and form of the current pulses which can be expected over the speed range. Wide variation of the form factor is evident.

4.2 Circuit configuration
Four of the very many options for driving the switched reluctance motor are shown in Fig. 8. For certain components it is necessary to use series or parallel configurations to achieve the required rating. Minor components such as snubbers have been omitted for clarity. The original circuit shown in Fig. 8a is described in Reference 2; high voltages are generated within this circuit, and therefore the alternatives shown in Figures 8b–d were considered by British Rail Research instead.

The simplest circuit which can be used to drive a fourphase switched reluctance motor is shown in Fig. 8c. This circuit has a minimum number of active switching devices

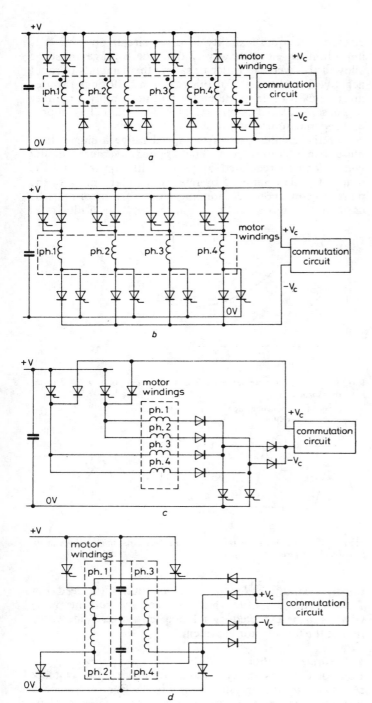

Fig. 8 *Four-phase switched reluctance motor drive circuits*

a Original circuit for a bifilar wound motor
b Eight-limb double ended drive circuit
c Four-limb double ended drive circuit
d Four-limb single ended drive with split supply

and good utilisation of power semiconductor device ratings below the base speed. This is because each thyristor conducts twice per cycle in the single pulse mode, giving a total conduction period of up to 180° per thyristor. Unfortunately this circuit does not allow load pulse width extension beyond 90°, and it therefore has limited application.

478

The circuit in Fig. 8*d* has half the output voltage of the circuit in Fig. 8*b* for the same input voltage but requires parallel thyristors to produce the same output power. Circuit-complexity is reduced, but commutation currents are increased. Of the four options considered, the circuit in Fig. 8*b* is the most suitable for the line-fed applications of rail traction.

The switched reluctance motor drive has one of its most important advantages in its commutation circuit. It is possible to use one commutation circuit to commutate all phases of the drive. This is termed group commutation, and it permits an overall economy of components. It is possible to use the same technique in three phase inverter drives but it does not always produce a net benefit.

In the group commutation configuration, the electrodes of the thyristors of the same supply polarity can all be connected to one commutation circuit through a diode gate arrangement. Separate positive and negative commutation circuits may be required. In the simple basic drive circuit of Fig. 8*a* one circuit can switch both negative and positive polarity thyristors, but the voltages generated in the process are up to three times line voltage, and many series devices would be required in a line-fed traction application.

Circuits to commutate the inverters in Figs. 8*b* and 8*c* have a rather less stringent specification as peak voltages are better controlled. An example, illustrated in Fig. 9, makes much better use of component voltage capability.

The circuit in Fig. 9 provides sequential commutation of

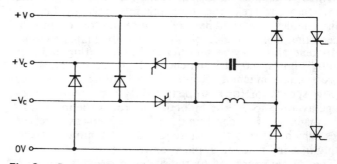

Fig. 9 *Commutation circuit for a switched reluctance motor*

both positive and negative thyristors. One capacitor bank and four thyristors are required for the complete circuit. Two thyristors are required to initiate commutation and conduct the resonant discharge current. Two more are required between positive and negative commutation pulse injection points, $+V_c$ and $-V_c$, to prevent restriction of the reverse voltage available to the motor windings.

4.3 Switching devices

The high peak-to-average values of the load waveforms, especially in the lower speed range, suggest that the thyristor is the obvious switching device for use with the switched reluctance motor. However, developments reported in References 3 and 4 imply that the transistor has a peak current capability which may be exploited in this type of application in the future.

This assessment has been based on the thyristor as the main switching element, and to appreciate the final conclusions it is necessary to consider the pertaining price structures of thyristors in relation to rating. Fig. 10 shows thyristor price trends of suitable inverter grade devices from two manufacturers. An increasing slope is apparent beyond 600 V in these graphs. This is attributable to many factors, among which are a high basic packaging cost and best device yields in the middle voltage range.

The basic switched reluctance motor drive circuit in Fig. 8a requires a device voltage capability of twice peak supply voltage plus a contingency to allow for the effect of winding leakage inductance. For a drive fed from a battery of, say, 200 V, it makes little difference to the cost whether the required device voltage rating required is 200 V or 500 V. However, when the supply voltage is approaching

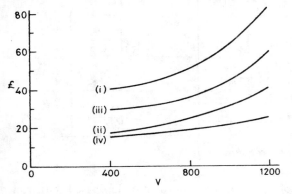

Fig. 10 *Thyristor price variation with rated forward voltage (mid-1981 prices)*

(i)   350 A device, first manufacturer
(ii)   80 A device, first manufacturer
(iii) 350 A device, second manufacturer
(iv) 220 A device, second manufacturer

1000 V the overvoltage requirement leads to many series devices and such an unacceptable cost penalty that other solutions must be sought.

4.4 Passive components

In inverter drives, and traction drives in particular, filter and commutation capacitor costs share prominence with those of the semiconductors. It is therefore of prime importance, in cost minimisation, to choose circuit configurations which enable best use to be made of capacitor ratings. In heavy-duty traction applications the input filter capacitor must be a high-quality type. Performance demands made upon this expensive component by the power output stage have a big effect upon overall material cost. Hence both the drive system choice and its mode of operation can affect the filter capacitor cost contribution. The repercussions of system choice are more serious if the specification for the filter input inductor is affected. This is generally the most expensive single item of equipment in the power-conditioning equipment. The filters for the alternative drives considered have all been designed to reduce DC link voltage ripple to the same level.

The commutation capacitor is another major cost inescapable in an inverter drive. It can be minimised by the choice of a circuit which has the fewest elements, the best RMS rating of components in that circuit, and makes least demand upon the capacitor in terms of such parameters as inductance.

The remaining significant cost factors, which must be taken into account in a relative cost assessment, are the general complexity of the circuits in terms of component numbers and the internal heat dissipation of the unit.

The overall component count, coupled with a knowledge of the mechanical design, provides a measure of the assembly labour in manufacture. Similarly, a knowledge of the component heat dissipation provides an indication of the bulk of the equipment, through surface area and heat sink weight.

5 System cost comparisons

5.1 Preliminaries
Costs have been investigated for the double ended four-phase switched reluctance motor drive as in Fig. 8b. The widely used McMurray inverter drive in Fig. 11 has been taken as the comparative standard. The costs presented in the Tables have been obtained both from component supply quotations and discussions with traction-equipment manufacturers.

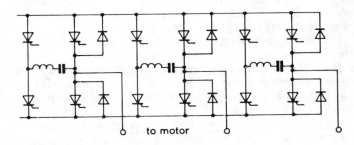

Fig. 11 *McMurray inverter circuit*

The costs listed here must be taken as guide figures in which market factors may lead to fluctuations in any of the elements. Because of this, and the sensitivity of costs in general, manufacturers do not wish to be quoted directly in a paper such as this. Specific component references are therefore avoided and brief technical specifications given *in lieu.*

5.2 Basis of comparison
The induction motor and switched reluctance motor drives are fundamentally different in their operating principles and in their utilisation of similar basic components. The machines may conveniently be compared on the basis of equivalent frame size, but for the power-conditioning equipment a different approach is required. For the latter, the most satisfactory basis of comparison is produced from a calculation of the equipment rating for each type of

motor, when the same principal components are used in each case. This yields specific cost (£/kW) figures which are not influenced by the anomalies in the pricing structures of components of different types and ratings.

The components which principally determine the rating of the power-conditioning equipment are the main load-current-carrying thyristors, though these may not be the most expensive group of components. The dissipation in the main load-current-carrying thyristors forms the ultimate factor in equipment steady-state rating. The comparison method was, therefore, to select an inverter-grade thyristor which was as near as possible to the equipment rating required for the study, and then to calculate the performance of the device in the circuits to drive the induction and switched reluctance motors.

For the purpose of the assessment an inverter-grade thyristor was chosen with an average current rating of 480 A, a turn off time of 35 μs and a voltage rating of 1800 V. The calculated equivalent power ratings using this device were then comparable with the acceleration ratings of traction motors for outer-suburban multiple-unit stock.

5.3 Rating points

In suburban applications, traction motors are usually operated at about twice their continuous rating when accelerating up to base speed. The power conditioning must be rated to sustain current at the maximum accelerating tractive effort for several minutes. Worst-case rating points must be considered for each equipment, but these are not necessarily at the maximum-power points.

The worst-case design point for the induction motor drive is at the base speed where the carrier frequency must be at its highest, before the transition to six-step operation begins. At this point, the 480 A thyristor has a power dissipation of 480 W, and the inverter has an output of 350 kW.

In the reluctance motor drive, the worst case for commutation is at the lowest speed in the single pulse/phase/cycle region, where the current to be commutated is 1670 A. However, the worst-case dissipation for the main load-current thyristors in this drive occurs at maximum speed. The load conditions which give the 480 A thyristor the same power dissipation as in the induction motor drive occur at 3475 rev/min, when the inverter is delivering a power of 450 kW.

The filter capacitor design criterion for the switched reluctance motor has been taken at the lowest speed in the single pulse per cycle control regime, where maximum ripple voltage occurs. The equivalent worst-case condition for the induction motor occurs just above base speed in the six-step mode. For both equipments the design figure is 300 V peak to peak at full inverter rating. The DC line voltage ripple frequency is almost equal for these two design conditions. At high speeds the supply ripple currents of the switched reluctance motor drive improve considerably and are of a similar order to those of the induction motor drive when it is operated in the PWM mode.

5.4 Motor costs

The comparative cost breakdowns for the induction and reluctance motors are shown in Fig. 12. Copper costs in

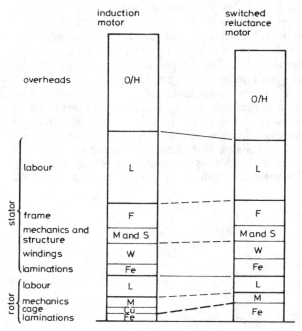

Fig. 12 *Motor cost breakdowns*

the rotor of the reluctance motor are eliminated, but iron costs are increased by the greater active volume and use of a more expensive material. The total rotor-material costs are therefore similar for the two machines. Labour cost in the rotor is reduced, even though thinner laminations are required in a greater number; this is the biggest benefit.

The stator copper volume of the reluctance motor is lower, but only by about 5%. The iron cost is increased by over 50%. The result, after considering all other factors, brings the reluctance motor material cost to 5% more than that of the induction motor. The stator labour costs of the reluctance motor are low because of the electromagnetic simplicity. The simplifications effected apply to a low fraction of the construction cost of a rail traction motor and the savings are not great. The total labour saving in both rotor and stator is about 5% with respect to the induction motor.

Overheads must be applied to the combined costs to cover the administration, testing, selling costs and return on capital investment. The overheads are taken as 50% of the net factory cost in each case. Overall, the cost of the switched reluctance motor, including an allowance for uncertainties in estimation, is predicted to be between 0 and 10% lower than the induction motor.

5.5 Power equipment costs

A typical cost breakdown for an induction motor drive is shown in Fig. 13. This costing may be used as a basis for extrapolation to obtain equivalent costs for a switched re-

Table 1: Induction motor drive (rated power 350 kW) major power circuit component costs

Component	Rating	Configuration	Quantity	Unit cost £	Total £
Load switching thyristors	$I_{F(av)} = 480$ A $V_{RRM} = 1.8$ kV $tq = 36$ μS		6	165	990
Commutation thyristors	$I_{F(av)} = 480$ A $V_{RRM} = 1.8$ kV $tq = 35$ μS		6	165	990
Flywheel diodes	$I_{F(av)} = 380$ A $V_{RRM} = 1.5$ kV		6	29	174
Thyristor snubber circuits		series capacitor/ resistor network	12	8	96
Current rise rate limiting reactors	4 μH		6	13	78
Commutating capacitors	15 μF 850 V 61 A RMS	3 banks of 2 in series 9 in parallel	54	29	1566
Commutating inductors	8.6 μH 560 A RMS		3	120	360
Filter capacitor	850 μF 125 A RMS 1700 V	2 in parallel	2	290	580
Commutation energy control: diodes			6	18	108
resistors			3	80	240
					5182

Table 2: Switched reluctance motor drive (rated power 450 kW) major power circuit component costs

Component	Rating	Configuration	Quantity	Unit cost £	Total £
Load switching thyristors	$I_{F(av)} = 480$ A, $V_{RRM} = 1.8$ kV, $tq = 35$ μS		8	165	1320
Commutation thyristors	$I_{F(av)} = 480$ A, $V_{RRM} = 1.8$ kV	2 in parallel	4	165	660
Auxiliary commutation thyristors	$I_{F(av)} = 350$ A, $V_{RRM} = 1$ kV	2 in series	4	48	192
Snubber circuits		series capacitor/ resistor networks	16	8	128
Current rise rate limiting reactors	4 μH		2	13	26
Commutation pulse injection diodes	$I_F = 320$ A, $V_{RRM} = 1.8$ kV		8	36	288
Commutation diodes (1 and 2)	$I_F = 800$ A, $V_{RRM} = 1.8$ kV		4	60	240
Commutation capacitors	15 μF, 850 V, 61 A	2 in series, 26 in parallel	52	29	1508
Commutation inductors	11 μH		1	43	43
Filter capacitors	1000 μF 100 A RMS 1700 V		6	274	1640
					6045

luctance motor drive. The principal difference between the two drives lies in the active switching device circuits, but these, and their associated electronic controls, account for a fairly small proportion of the total cost. The major share of the total is taken up by the factory assembly and general overhead costs. Also the design and development contribution will be pronounced in low-quantity orders. A portion of the material including input filtering, protection, switchgear etc. will be common to both drives.

Tables 1 and 2 list the components and costs for the two systems. Because the drive systems have different powers they must be compared on a cost/kW, i.e. specific cost, basis as in Table 3, columns (a) and (b). In column (c)

account has been taken of the development potential of the switched reluctance motor drive. There is possible scope for thyristor and capacitor cost reduction in the commutation circuit by the use of alternative circuit designs. It is also possible that economy could be made in the cost of the filter capacitor by the use of alternative switching strategies and by matching the ripple current rating to specific vehicle duty cycles. These factors have been used to obtain the minimum projected cost for the switched reluctance motor drive. Table 4 compares the total semiconductor dissipation in both drives. This Table provides a measure of the total cooling capability to be provided in each drive. The peak power dissipation does not necessarily occur in

Table 3: Induction motor and switched reluctance motor drives: Comparison of the specific costs of power circuit components

Component group	(a) Induction motor drive (350 kW)		Switched reluctance motor drive (450 kW) (b) As costed		(c) Minimum projected cost	
	Cost	Cost/kW	Cost	Cost/kW	Cost	Cost/kW
	£	£/kW	£	£/kW	£	£/kW
Load current switching devices and components	1290	3.7	1672	3.7	1672	3.8
Active commutation devices	1146	3.3	1182	2.6	1044	2.3
Passive commutation components	2166	6.2	1551	3.5	1243	2.8
Filter capacitor	580	1.7	1640	3.6	1120	2.5
Total cost	5182		6045		5079	
Total cost/kW		14.9		13.4		11.3

Table 4: Main power circuit component dissipation

Drive system	Circuit, element	Quantity	Device worst-case, dissipation	Total dissipation	Dissipation/kW of OP
			W	W	W/kW
Switched reluctance motor drive (450 kW)	Load thyristors	8	480	3840	8.5
	Commutation thyristors	4	790	3160	7.0
	Auxiliary commutation thyristors	2	426	852	1.9
	Commutation pulse injection diodes	8	400	3200	7.1
	Commutation diodes (1)	2	720	1440	3.2
	Commutation diodes (2)	2	330	660	1.5
	Total				29.2
Induction motor Drive (350 kW)	Load thyristors	6	480	2880	8.2
	Commutation thyristors	6	640	3840	8.5
	Flywheel diodes	6	370	2220	6.3
	Total				23.0

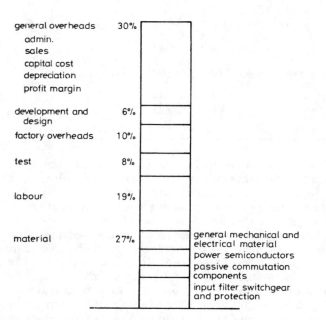

Fig. 13 *Typical cost breakdown for a thyristor PWM induction motor drive*

each device simultaneously, so the figures quoted may possibly be reduced by sharing heatsinks between certain devices, but it is a useful measure.

Application of the tabulated results coupled with a knowledge of material and work content enables the relative power equipment costs to be obtained as in Fig. 14. This Figure also features the minimum projected cost of the switched reluctance motor drive and the costs of a chopper-fed DC motor drive of equivalent rating. This figure has been scaled to the current market price for induction motor drives of 20 kW continuous rating. The material costs which are common translate without change. The active material costs change in proportion to the tabulated results. The extra complexity of the reluctance motor drive circuit will increase the wiring and assembly cost of manufacture by with respect to the inverter. The need for greater heat dissipation will increase the factored material and mechanical assembly costs by 22%. Test time for the two drives is not likely to be significantly different in the two cases; also works overheads should be very similar.

The general overhead contribution, which covers sales costs, administration, development, plant and return on capital investment, is usually allocated *pro rata* with factory cost. It is adjusted according to such factors as specialised application engineering content, production quantity etc.

6 System cost discussion

The costs in Fig. 14 present a realistic assessment situation which applies to traction equipment at this power level at mid-1981 prices. The cost magnitudes are different from those encountered in the industrial field, and the proportions are very different from those which apply in the low-power or battery-road-vehicle applications.

At present the nearest commutatorless-motor-equipment competitor to the chopper-fed DC motor drive costs approximately 50% more. However, system costs have shown a continuously converging trend, and will continue to do so as new semiconductor devices and inverter techniques become available.

To each of the drive system costs must be added the equivalent capital sum derived from the maintenance costs in the following Section. These costs go a little way

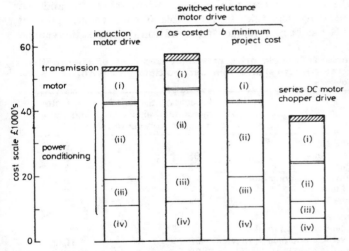

Fig. 14 *200 kW drive system cost comparisons (January 1981 prices)*

(i) motor
(ii) overheads design and cost
(iii) labour
(iv) material

towards bridging the gap, and the maintenance factor is labour intensive and likely to rise in real terms while power-equipment costs continue to fall.

Of the three drives it is apparent that the switched reluctance motor drive makes particularly heavy demands upon the input filter capacitor because of the way in which it is operated. It would therefore benefit from further development to reduce the high peak currents which occur in the lower part of the speed range. It must be noted that the switched reluctance motor is a form of synchronous machine and therefore separate power equipments are required for each motor, because it is not practicable to synchronise the four axles of a power car. Quantity dis-

count will be negotiable as a result, but this could be outweighed by the loss of economy of scale when the use of a single large equipment is precluded. The current practice for induction motor drives is to feed many motors in parallel from one large equipment, but wheel diameters must be matched closely. The induction motor drive thus benefits from the possible economy of scale.

7 Maintenance and reliability

Semiconductor controlled railway stock has a proven high reliability. BR experience with the Class 314 convertor-controlled fleet for the Glasgow suburban area, where each unit covers 80–90 thousand miles per year, has shown an absence of failures after two years running with 16 units. Nevertheless vehicle power electronic equipment will incur annual costs through preventive maintenance and repair following in-service failures.

Japanese experience [5] has shown that out of a fleet of 1500 cars, the majority of which are resistance controlled, approximately 1 in every 109 fail annually. Chopper-controlled vehicles have been shown to be considerably better than this. Such vehicles tend to show similar failure ratios during their initial settling in period, but subsequently reliability is notably improved. In the chopper-controlled vehicle fleet described in Reference 5, of a class of 210 vehicles, no failures occurred in the last two years, and of another class of 95 cars no failures occurred in four years. The probability of failure may therefore be taken as 1 in at least 400 with an upper limit which is unknown. The cost of in-service failures can therefore be apportioned to the vehicles of a fleet in this ratio.

The annual depreciation cost of a four-car multiple-unit train consisting of one power car and three trailers costing £820 k, with a service life of 25 years, is £97.4 k with an average interest rate of 11%. If 1 unit in 100 fails per annum, is out of service for a week and costs £1.5 k to repair, then the total cost of this rate of failure, including both capital and repair charges, is only £34 per vehicle per annum, which is acceptable. (However, as there will be at least one spare unit in the fleet, it is doubtful whether either capital or loss of revenue charges need be considered here.)

Routine maintenance required to equipment contributes more to the annual cost. If this is carried out on an annual basis, and the chopper-controlled power car of each four-car multiple unit required one man for six hours, then the annual cost will be £76. This work will include inspection and attention to mechanical components, contactors etc., cleaning of semiconductor heat sinks and functional checks of the control system. There are no figures available to indicate the levels of preventive maintenance required for either induction motor or reluctance motor drives. As the work to be done is approximately 50% mechanical it will not increase in direct proportion to the complexity of the power electronic circuits of the other types of drive. By increasing the electrical part of the cost in proportion to the complexity of the power electronics we derive an annual cost of £114 for the inductor motor control and £152 for the switched reluctance motor controller.

DC traction motors require periodic attention to the brushes and commutator, and regular inspection to check for wear in that area. Both induction motor and reluctance motor systems would eliminate this requirement. In an EMU application this would represent an annual saving in labour and material of £100–200 per motor. Such scheduled maintenance is generally carried out when the equipment is otherwise not in traffic, so that there is no effective reduction in the availability of the vehicle involved.

Rewinding of motors is, for some types, carried out on a schedule basis to reduce the risk of in-service failure, but for others it is done as necessary. A ten year life is a typical figure for many types of motor. The stator windings of induction motors are similar to DC machine armature windings so far as the insulation system is concerned, and would be subjected to similar levels of vibration and contamination by dirt. There seems to be no reason to assume any significant difference in winding life. The cost of rewinding a DC motor of about 200 kW capacity is in the region of £4000 including retaping the field and interpole coils, or closer to £6000 if the commutator has to be replaced. On an annual basis the cost of repairs is in the region of £500 per motor. Induction motors would cost a little less to rewind and, of course, would not call for commutator replacement, so that an annual average cost of £350 could reasonably be anticipated for rewinding. To this figure should be added a contribution associated with failures of cage bars: a fairly common occurrence with industrial motors. It is impossible to quantify this figure but it should not be large because any motor type which shows frequent bar failure would probably be unacceptable to the railway operations.

The annual cost of repairs to reluctance motors should be much smaller than for either DC or induction motors. The windings of the reluctance motors are similar in some ways to DC motor field coils which generally give few in-service failures. It would be wasteful to rewind on a schedule basis, and failures should be infrequent. It should be remembered that the experience of DC machine field winding, working at typically 50 V, is likely to give an optimistic forecast for reluctance motors working at typically 750 V. The cost of carrying out a rewind would be considerably less for the reluctance motor than for either DC or induction motors. (£1000–2000 is to be expected.) The expected annual cost is therefore likely to be less than £100.

The total annual costs for both motors and power electronics are given in Table 5. The costs for the electronic

Table 5: Maintenance and repair costs per motor per year

System	DC motor and chopper	Induction motor and inverter	Switched reluctance motor and drive
Motor	650	510	100
Power control	19	29	38
Total annual	669	541	138
Equivalent capital	5630	4560	1160

power equipment has been split between the four motors of the power car, referred to in the previous Section. Also included is the equivalent capital sum for a 25 year depreciation period and an average interest rate of 11%. The equivalent capital sums in Table 5 may be added to the costs charted in Fig. 14, where they will tend to reduce the cost differentials between the drives.

8 Energy

The annual energy consumed by an electric multiple unit depends upon the given route and the schedule operated. For suburban vehicles, such as the Class 507 Liverpool/Southport stock, the energy consumed is likely to lie between 410 and 640 MWh per annum. An average figure of 530 MWh has been taken for use in the following analysis. A portion of the energy is dissipated in the vehicle traction equipment, and this may be estimated from the system efficiencies and the time the vehicle spends at maximum power. This yields a contribution to the equipment operating costs set out in Table 6 together with the efficiencies of the three drives considered. All three drives are capable of regenerative braking, resulting in a saving of between 10 and 20% of the total energy used. The exact figure again depends upon the route and the operating schedule, so for Table 6 an average figure of 15% has been assumed. Regeneration decreases the vehicle

Table 6: Vehicle drive efficiency and cost of energy dissipated per drive of a four-car multiple unit with four axles powered

	Induction motor drive	Switched reluctance motor drive	DC motor and chopper drive
Efficiency, %			
Gear train	98	98	98
Motor	93	94	93
Power conditioning and filter	92	94	96
Overall	84	87	87
Energy			
Total dissipated in vehicle equipment, MWh	100	81	81
Annual cost, £ per four-car unit	2500	2030	2030
Cost per motor			
Annual cost, £	625	507	507
Equivalent capital, £	5260	4270	4270

energy used to 450 MWh, but increases the energy through the traction equipment to 620 MWh. The energy costs in Table 6 have been calculated taking a figure of 2.5 pence/kWh, consistent with 1981 price levels. The values obtained have been converted to equivalent capital sums which may be added to the equipment costs in Fig. 14.

9 Conclusions

Chopper drive systems are being introduced continually on the railways of the world, and total numbers in all countries have already reached several thousands. At present a differential exists between costs of a DC drive system and any of the brushless equivalents. Continuing development of the inverter has narrowed the differential over the last decade, and in the foreseeable future the inverter drive system will be a commercially advantageous option. This has already occurred in the industrial market, at low powers, due to the introduction of transistor inverters.

If the choice of a brushless motor can be justified on special mechanical, performance or environmental grounds, then either the induction or switched reluctance motors are equally viable alternatives. This study has shown that it is feasible to design a switched reluctance traction motor in the same space envelope as an induction motor. If the current density of 5 A/mm^2 can be realised or exceeded, then the motor will be competitive. On grounds of overall cost for the complete drive systems, there is little to choose between induction or switched reluctance motor drives. It is therefore likely that the choice will depend more on the different characteristics of the drives, and the overall system implications, than on individual drive cost. The switched reluctance motor still has development potential which, as Fig. 14 demonstrates, could bring its costs a little below those of an induction motor drive, for a rail traction application at 200 kW continuous rating.

The study has been carried out at the high voltage levels of rail traction, where the extra complexity caused by the need for high-voltage capability in the drive weighs marginally against the switched reluctance motor at present. At lower voltages and lower powers the relative proportions in Fig. 14 become radically altered, and there is a level at which the switched reluctance motor has a substantial advantage over the induction motor. There are also circumstances where it competes favourably with a DC motor. There is no doubt that it will become increasingly important in the industrial scene in the next few years.

9   Acknowledgments

The author is obliged to the British Railway's Board for permission to publish this paper. The author wishes to thank the teams at the Universities of Leeds and Nottingham who have carried out design studies on reluctance motor drive systems on behalf of the Director of Research. The author is also grateful for information supplied by the Director of Mechanical & Electrical Engineer's staff, and for the assistance of their professional colleagues in the Traction Industry with the preparation of equipment costings.

10   References

1 LAWRENSON, P.J., STEPHENSON, J.M., BLENKINSOP, P.T., CORDA, J., and FOULTON, N.N.: 'Variable-speed switched reluctance motors', *IEE Proc. B, Electr. Power Appl.*, 1980, **127**, (4), pp. 253–265

5 MIZUNO, Y., ARAI, M., JIMBO, Y., SHIRASHAJI, A., and KANEDA, J.: 'History of development, operation, result and new techniques of thyristor chopper controlled train at Teito Rapid Transit Authority'. Proceedings of the MITRE Corporation International Conference on advanced propulsion systems for urban rail vehicles, Washington DC, February 1980

6 'Les nouveaux moteurs electrique d'automobiles', *L'Electricien*, 1969, **97**, pp. 87–91

2 RAY, W.F., and DAVIS, R.M.: 'Inverter drive for doubly salient reluctance motor: its fundamental behaviour, linear analysis and cost implications', *IEE J. Electr. Power Appl.*, 1979, **2**, (6), pp. 185–193

3 FISHER, G.A.: 'High power transistor inverters—potential for single device operation at 1000 A and 800 V'. Proceedings of 16th Universities Power Engineering Conference, April 1981

4 FRENCH, P.St-J.R., MORRIS, R.J., and WHITTING, G.: 'The development of a high current 1000 Vcex transistor and its application to inverter circuits'. Proceedings of the AEI conference on semiconductori di alta potenza ed applicazione, Genoa, May 1981

Switched Reluctance Motor Drives for Rail Traction: A Second View

**W.F. Ray, R.M. Davis, P.J. Lawrenson,
J.M. Stephenson, N.N. Fulton and R.J. Blake**

Proceedings IEE, Vol. 131, Pt. B, No. 5, September 1984, pp. 220-264

1 Introduction

This short paper has been written in response to an invitation from one of the Honorary Editors and P.St.J.R. French. It gives the authors' view of the capability of switched reluctance (SR) systems for rail traction and, in particular, of the conclusions which can be properly drawn, in the light of the feasibility study carried out for British Rail in 1980 and of subsequent related publications.

In 1980 the authors were engaged in a research programme, at the Universities of Leeds and Nottingham, into the development of a new kind of electrical drive system, the 'switched reluctance' (SR) drive, for a 50 kW battery-powered electric vehicle. This work was funded by Chloride · Technical Ltd. At the invitation of British Rail Research, they undertook a feasibility study of an SR drive applied to rail traction.

The results of this study, in particular the dimensions and characteristics of the SR motor and its operating values of current and torque, have formed the basis of a paper by French ([1], see pp. 209–219), giving an assessment of the SR traction drive compared to an induction motor version.

In interpreting French's paper, it is important that the reader should be aware, both of the criteria set for the original feasibility study, and also of relevant developments which have been published in the intervening four years. In this paper, therefore, the authors provide comments on both of these matters. In particular, they establish a sound basis for comparison between an SR drive and an inverter-fed induction motor drive for the application considered by French, and they outline a recast SR drive design, drawing on published industrial drive experience, which is shown to have considerable advantages over the induction motor drive, especially in cost. It is not the authors' intention to criticise French, who has attempted, on the basis of an early feasibility study, to draw cost and performance comparisons between an SR system and other types. The intention is rather to illuminate the subject further in the light of our developing knowledge and experience of SR drives, and to avoid any misevaluation of SR drives which might otherwise occur.

Paper 3322B (P2, P6), received 20th June 1984

Mr. Ray and Mr. Davis are with the Department of Electrical & Electronic Engineering, The University, Nottingham NG7 2RD; Prof. Lawrenson and Dr. Stephenson are with the Department of Electrical Engineering, University of Leeds, Leeds LS2 9JT; and Dr. Fulton and Mr. Blake are with Switched Reluctance Drives Ltd., Leeds, England

486

2 Original (bifilar) SR design

The feasibility rail study conducted for British Rail by the universities was very directly influenced by the authors' experience with the road-going battery-vehicle project, particularly in the following two aspects:

(*a*) The traction characteristics proposed by BR were similar to those for the battery vehicle, and the authors' work had shown that the latter could be achieved very effectively by an S.R. system. The characteristic agreed as the target had constant torque, corresponding to a tractive effort of 13.8 kN, up to 0.4 p.u. speed, and constant power of 225 kW above this speed (curve (i) in Fig. 1). This characteristic gave somewhat less torque at lower speed than

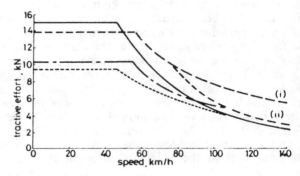

Fig. 1 *Tractive effort/speed envelopes*

—— induction motor
- - - switched reluctance motor
   (i) original specification
   (ii) revised specification
— · — switched reluctance motor rating at 5 A/mm²
········· estimated rating of induction motor at 5 A/mm²

more conventional systems, but considerably more at higher speeds. Nevertheless, at the time, this was agreed to be sufficiently similar in overall performance to enable the broad capabilities of the SR system to be assessed

(*b*) Up to that time, the only practical experience of SR technology in a traction drive had been obtained at the relatively low voltage of 160 V, derived from a lead-acid battery. At this voltage, the power convertor circuit considered to be the most cost-effective utilised bifilar motor windings, as shown in Fig. 2a and this circuit was subsequently developed and described in various publications [2, 3].

In the interest of building the new feasibility study as directly and as economically as possible on the existing experience, it was decided that the motor should be designed for a bifilar winding, so allowing the use of the same basic power convertor circuit. It was explained and

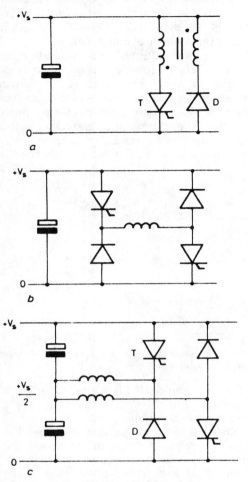

Fig. 2 *Main power convertor circuit alternatives*

a Bifilar configuration used for battery-vehicle drives and for the original SR feasibility study, showing one phase

b Configuration, proposed by French, with two main thyristors per phase, showing one phase

c Split supply configuration used for the revised design, showing two phases

understood that this would not be the appropriate circuit for the much higher voltages of the railway application, but it was agreed, nevertheless, to be acceptable for the initial assessment of the broad capabilities of SR drives. Based on this circuit, anticipated values of motor winding currents and power electronic component voltages and currents were provided to BR for the various operating conditions defined by the agreed torque/speed envelope.

These decisions were taken at the outset; the study served its purpose of exploring, for the first time, the general capabilities of SR drives for traction characteristics and ratings in excess of 100 kW. However, the study was not intended to provide the basis for a firm cost comparison with alternative systems.

This is recognised by French, in proposing his alternative power convertor for use with a nonbifilar motor. His main circuit, with two main thyristors per phase, is shown in Fig. 2*b*, and is the same as that proposed by Amato [4] in 1969. However, his commutation circuit is different and, although it has certain merits, the authors consider it to be unnecessarily expensive, particularly in the amount of commutation capacitance used. His circuit, therefore, like the bifilar arrangement, does not provide for meaningful comparison with alternative systems. It is shown by the authors in Section 3 that a more realistic system can be based on an alternative power convertor circuit already published [2, 5], currently marketed for industrial applications, and on a modified version of the commutation circuit proposed by French.

Ideally, the comparison should be made on the basis of identical performance characteristics for both induction motor and SR drives. However, since the machines are inherently different in nature, identical characteristics are not easily obtained, and it is therefore necessary to use characteristics which result in broadly equivalent train performance. The characteristic previously used for the SR drive, i.e. curve (i) in Fig. 1, has therefore been modified to that shown in curve (ii), to match the induction motor characteristic more closely.

On the matter of the induction motor characteristic referred to by French, the authors were suprised at the high output claimed at a stator current density of 5 A/mm^2. Discussion with a leading manufacturer of traction machines has shown that the tractive effort to be expected at 5 A/mm^2 for a motor having the dimensions given by French and the same maximum output is as shown in the dotted Curve in Fig. 1. The tractive effort below 46 km/h is lower than that claimed by French by some 20%, and lies below the corresponding curve for the SR drive throughout the speed range.

3 Revised system design

3.1 Scope of revision

In comparing the two systems, the authors have not undertaken further calculations to find an 'optimum' design for the SR system. In suggesting revisions to the original SR system (as modified by French), the magnetic circuitry of the motor, i.e. the lamination geometry and overall dimensions, has not been altered. Moreover, no attempt has been made to define the best power convertor circuit; calculations have been made using circuits already published or suggested.

This is not to imply, however, that with developing expertise the design of both motor and power convertor could not now be improved upon. It is simply that time has not been available to undertake a full investigation, and that a new stance would not be appropriate in the present context.

3.2 Motor

The only change to the motor is to revise the winding

design to produce the revised tractive effort/speed characteristic shown in Fig. 1.

The low-speed torque for the SR system is unchanged at 8% below the induction motor up to 0.4 pu speed, and a constant power (225 kW) characteristic is maintained to 0.52 pu speed. Above 0.52 pu speed, the power is allowed to fall with speed, giving a 'series' characteristic. The excess power above the constant-torque region has thus been reduced, but still gives at least 15% more than the induction motor at all speeds. This brings the two systems much more into line, although it still leaves the SR system with a superior overall performance. Advantage has been taken of the reduction in high-speed power to greatly increase the number of turns of the motor winding, with consequential reductions in peak currents at lower speeds. The benefits to the system are summarised in Section 4.

3.3 Power convertor

It had always been understood that a nonbifilar system would be appropriate for rail-traction application. A nonbifilar power convertor circuit, with which the authors have had much practical experience, is that shown in Fig. 2c, as used in the range of industrial SR drives marketed by TASC Drives Ltd., to the design of the authors, and as described in Reference 5. By using a split supply, derived from the filter capacitors, only one main thyristor is required per phase winding. In order to maintain the balance of the split supply at low speeds in the chopping mode, two phases are excited simultaneously at all times. This gives enhanced torque for a given level of current, since two phases are torque productive rather than one, as is well-known practice with low-power stepper motors. The original bifilar system did not have the capability of simultaneous phase excitation in the chopping mode, and this limitation has been incorrectly assumed by French to apply to the circuit of Fig. 2b.

It should be noted that for the split-supply circuit in Fig. 2c, the winding and main thyristor currents are double that for Fig. 2b, as the voltage is half, and hence the combined current rating for the 4 main thyristors is the same as for the 8 main thyristors of Fig. 2b. The authors therefore do not claim any advantage in combined ratings for this circuit, but would expect some advantage to result from fewer devices and a reduced number of motor connections.

The commutation circuit proposed by French, which is one of several alternatives, has some significant merits. A single commutation capacitor is used to service all main thyristors, and no resonant reversal circuit is required. Having commutated one main device for a given phase, the capacitor polarity is correct for the commutation of the companion device for that phase. The reason for the inclusion of the 11 μH commutation inductor and its associated diode network is unclear to the authors. Presumably its function is to make the commutation time less dependent on load current. However, it effectively operates in parallel with the motor winding, and takes a peak current of the order of 2000 A, compared to a maximum commutation current for the original design of 1670 A. It therefore

approximately doubles the capacitor current, whereas without it the capacitance could be reduced by 50% to satisfy the same thyristor reverse voltage time. Its elimination would thus result in significant cost saving (£917 or 15% for the 450 kW SR drive using French's values), in addition to reducing the number of components.

There are, of course, other alternative commutation circuits to that used by French. However, so as not to cloud the comparison and to avoid the necessity for further investigation to assess the relative merits of alternatives, the authors have used the commutation circuit proposed by French, but without the commutation inductor, for the relative assessment of the revised design.

The complete circuit is given in Fig. 3 and comprises, in

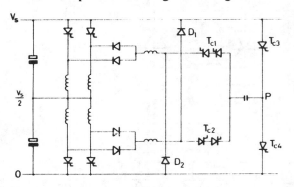

Fig. 3 *Revised 337.5 kW power convertor, using modified version of commutation circuit proposed by French*

addition to the filter capacitor, 10 thyristors, 6 diodes, a commutation capacitor and 2 (di/dt) inductors. There would be, in addition, various snubbing networks.

It should be noted that the commutation thyristors T_{C1} and T_{C2} will see half the supply voltage, in addition to the commutation capacitor voltage, assuming the snubber networks to hold point P at half supply potential with the commutation circuit in the 'off' state. It is therefore assumed that a voltage rating of 2.7 kV is appropriate, and hence two devices in series are used for T_{C1} and T_{C2}. Furthermore, at commutation the main diodes D_1 and D_2 see the capacitor voltage in addition to the supply voltage, and hence a voltage rating of 3.6 kV is necessary for these devices, although only 1.8 kV has been quoted by French. All other devices can be rated at 1.8 kV.

It should also be realised that with simultaneous chopping in two phases and a single commutation circuit, an extra sophistication of control would be required over that necessary for the system in French's paper, where chopping only occurs in one phase at a time. Although the authors have not worked with the proposed commutation circuit, they are nevertheless confident that this control can be achieved.

The proposed circuit of Fig. 3 looks very attractive alongside the McMurray inverter for the PWM induction motor system: in particular, there are fewer thyristors and no commutation inductors. The complexity penalty and low efficiency attributed by French to his SR circuit do not apply in this case.

488

4 Comparison

4.1 Motor

Despite the increased number of turns in the motor winding and the energisation of phases in the chopping mode for 50% duty rather than 25%, it is possible to reduce the motor winding current density. This implies a significant reduction in the winding currents. For the revised design, with the new torque/speed envelope shown in Fig. 1, the current density is 14.0 A/mm^2 at stall, and not greater than 7.6 A/mm^2 while rotating. This compares with 15.6 A/mm^2 at stall and a maximum of 11.7 A/mm^2 while rotating for the original feasibility study.

4.2 Main thyristors and basis of comparison

As a result of the reduced winding currents, the worst-case operating values originally given to BR for the various power convertor components can be relaxed, or, alternatively, the rated power for the drive using the same components can be reassessed.

The comparison undertaken by French is based on rating the drive to utilise a particular inverter-grade main thyristor, rated at 480 A mean, 1800 V and 35 μs. The resulting drive ratings were 350 kW for a PWM drive using 6 of these thyristors, and 450 kW for an SR drive using 8 (based on the feasibility study current values and the inverter circuit proposed by French). While the authors do not necessarily feel that this is the best criterion for comparison, it is nevertheless helpful to adopt the same approach, so as to assess the relative improvement resulting from the revised design.

The worst-case thyristor current for the original SR design was 278 A RMS at stall for a 225 kW drive, or 556 A RMS for 450 kW, and this value applies to either main circuit of Figs. 2a or b. For Fig. 2c, with a split supply, the corresponding value is 556 A RMS for a 225 kW drive, since the winding voltage is halved. However, with the revised motor design, the maximum main thyristor current for a 225 kW drive using the circuit in Fig. 2c is 370 A RMS at stall, a reduction of 33%. Alternatively, the same thyristors can be used to give a 50% increase in power.

4.3 Revised drive rating and components

The results for the revised design are therefore expressed as a set of components for a 337.5 kW SR drive, using four main thyristors of the same type as used by French. These components are listed in Table 1, which may be compared with Tables 1 and 2 in French [1].

In undertaking this comparison, the components previously used by French are used for the revised design wherever possible, to avoid confusion arising from the relative cost advantages of various manufacturers components. The authors, therefore, have made no attempt to find alternative devices, except for the 3.6 kV main diodes, which are not the main cost elements. It has, however, been necessary to scale some of the prices quoted by French, to account for slightly different ratings.

All the semiconductors are assumed to require the same ratio of nominal mean current rating to worst-case oper-

ating RMS current as for the main thyristors, with some additional allowance being made for the commutation devices, as these are operating at relatively high peak values. Judging from the quoted prices, it is anticipated that the nominal current ratings given refer to single-side cooling.

Whereas the authors have no reason to doubt that the component costs quoted by French are representative, they would not necessarily make the same selection. The objective is to establish whether using the same or similar components with a more enlightened design, relative improvements for the overall SR drive cost can be made.

4.4 Commutation components

The worst case for commutation for the revised 337.5 kW design is at stall, with a commutation of 1150 A, compared with the 450 kW value given by French of 1670 A, at the lowest speed in the single-pulse operating mode. This is a very significant reduction, since for the same circuit (Fig. 2b) and power for the two cases, the comparison would be between 770 A and 1670 A. However, as stated earlier, the circuit of Fig. 2c operates with double the current values of that for Fig. 2b. The commutation/chopping frequency for the revised design is 15% higher than for the original design.

As a result of the reduced commutation current, the commutation capacitance (60 μF or 0.178 μF/kW) is significantly less than that specified for the circuit used by French (195 μF or 0.433 μF/kW), although, as previously stated, this could be reduced by approximately 50%. The 8 parallel $\times$ 2 series array of 15 μF 850 V units has ample RMS current rating and adequate kVAr rating, and is estimated to give a 40 μs reverse voltage for a 1400 A commutation current and 750 V supply. There is, therefore, additional commutation capability for transient situations.

The commutation thyristors need to be inverter grade, but the turn-off time is not critical; 100 μs would suffice. The cost of the 250 A 1.8 kV device is estimated as approximately half that of the main thyristors, and the cost of the corresponding 1.4 kV device is reduced in proportion.

4.5 Filter capacitor

The major criterion used by French was that the peak-to-peak ripple voltage on the link under the worst condition should be limited to 300 V. This criterion demands a certain minimum capacitance. For the original BR design, the worst-case capacitor current was approximately 500 A RMS at 240 Hz at the lowest speed (600 rpm), in the single-pulse operating mode, for a 225 kW drive. This necessitated a link capacitance of 6000 μF for 450 kW. For the revised design, the peak-to-peak voltage across each half of the link must be restricted to $(300/\sqrt{2})$ volts, to meet the same criterion. The worst-case is at the lowest speed in the single-pulse mode, as previously, but the estimated current is now approximately 315 A RMS at 200 Hz, for the 337.5 kW drive. The appropriate capacitance is therefore 3300 μF.

In addition to capacitance value, the link capacitor must also have sufficient RMS current rating. This need not equal the RMS current rating while accelerating at full

Table 1: Major power circuit component costs

Component	Rating	Configuration	Quantity	Unit cost	Total cost
				£	£
Load switching thyristors	I_F(av) = 480 A, V_{RRM} = 1.8 kV tq = 35 μs		4	165	660
Commutation thyristors	I_F(av) = 250 A, V_{RRM} = 1.8 kV		2	90	180
Auxiliary commutation thyristors	I_F(av) = 250 A, V_{RRM} = 1.4 kV	2 in series	4	70	280
Snubber circuits		Series capacitor/ resistor networks	10	8	80
Current rise rate limiting reactors	4 μH		2	13	26
Distribution diodes	I_F = 530 A, V_{RRM} = 1.8 kV		4	47	188
Main diodes 1 and 2	I_F = 530 A, V_{RRM} = 3.6 kV		2	80	160
Commutation capacitors	15 μF, 850 V, 61 A	2 in series, 8 in parallel	16	29	464
Filter capacitors	1100 μF, 113 A RMS, 850 V	2 in series, 3 in parallel	6	170	1020
					3058

Switched reluctance motor drive, rated power 337.5 kW

Table 2: Comparison of the specific costs of power circuit components for induction motor and switched reluctance motor drives

Component group	Induction motor drive (a) 350 kW		Switched reluctance motor drive (b) 450 kW		(c) 337 kW	
	(From paper by French)				Revised design	
	Cost	Cost/kW	Cost	Cost/kW	Cost	Cost/kW
	£	£/kW	£	£/kW	£	£/kW
Load current switching devices and components	1290	3.7	1672	3.7	880	2.6
Active commutation devices	1146	3.3	1182	2.6	694	2.1
Passive commutation components	2166	6.2	1551	3.5	464	1.4
Filter capacitor	580	1.7	1640	3.6	1020	3.0
Total cost, £	5182		6045		3058	
Total cost/kW, £/kW		14.9		13.4		9.1

torque (due to the relatively short time duration of this operation and the thermal intertia of the capacitors), but should exceed the RMS current for the full-power characteristic. Using the same ratio of capacitor RMS current rating to the RMS current for low speed chopping, the appropriate current rating required for the 337.5 kW drive is 337.5 A RMS.

4.6 Relative assessment

Table 2 shows the relative component costs for the 350 kW induction motor drive (a), the 450 kW SR motor drive (b), both taken from the paper by French, and the revised 337.5 kW drive described in this paper (c), following the same format used by French. The estimated component cost per kW for the revised design (£9.1/kW) is significantly lower than the 'minimum projected cost' of £11.3/ kW quoted by French, and very much less than the £14.9/ kW for the PWM drive.

5 Discussion

The two main features of the revised design are as follows:

(a) The relatively lower motor currents, arising from the more accurately aligned torque/speed envelope and from simultaneous chopping, result in a significant reduction in component cost

(b) Use of the split-supply circuit of Fig. 2c results in a significant reduction in the number of components and motor winding connections.

The proposed commutation circuit (a modified version of that proposed by French) uses a single commutation capacitor to service all main thyristors and is very cost effective, resulting in an estimated overall component cost of £9.1/kW, compared with £14.9/kW for the PWM induction motor drive. However, to the authors' knowledge, neither this commutation circuit nor that proposed by French has been experimentally proven, and therefore some caution should be exercised on the projected cost.

It will be appreciated that several alternative commutation circuits exist and, with thought, other possibilities may well emerge. For example, the cost of the power convertor using a modified form of the well-known and tried circuit by Amato [4] is estimated at £11.2/kW. However, the important factor to emerge is that with estimated component costs in the range £9–£11/kW, a significant saving can be made on the estimated £14.9/kW for the PWM drive.

The comparison of specific costs given in Table 2 indicates that, whereas some saving ensues from the semiconductors, the major saving results from the significant reduction in passive communication components. On the other hand, the SR drive, as French points out, is more demanding on its filter capacitor requirement when plastic film capacitors are used. For drives using electrolytic capacitors, there is sufficient capacitance for a given RMS current requirement to considerably reduce, if not eliminate, the cost differential for the filter capacitor between SR and PWM. Nevertheless, despite this disadvantage for the rail traction application considered in this paper, the overall component costs are significantly better for the SR system. The authors have concentrated their attention on component count and cost, although it will be seen from French's paper that component or material cost only represents a very small proportion of the overall power-conditioning cost which, including overheads, development, test and labour is given as £210/kW for a PWM inverter. The authors have no experience of these cost breakdowns for a BR traction drive, and are not therefore in a position to comment on their absolute values.

However, the authors do not agree with French, that the extra complexity of the SR drive will increase the wiring and assembly costs of manufacture over that for the PWM inverter. On the contrary, they consider that, with fewer components overall, these costs should be reduced. It is probably true that, being a much younger drive, the future development cost for the SR drive for a rail traction application will be higher than for PWM, but then, if comparing the merits of the two drives, the past development costs of PWM inverters should be considered. The authors would not accept that the SR drive is more difficult to develop, indeed it avoids many of the commutation problems, such as trapped commutation energy, associated with the PWM McMurray inverter.

Furthermore, the authors do not agree with French, that the SR system has greater heat dissipation or poorer efficiency, and that it therefore entails increased material and assembly costs. Whereas it has not been possible to undertake theoretical calculations of power convertor loss for this paper, previous experience indicates that the efficiency of the power convertor is very much better than the 94% estimated by French and, indeed, the overall efficiency for an SR drive of this power rating should be in excess of 90%, for a considerable proportion of the torque/speed operating area. It is not clear whether French has included the dissipation of trapped commutation energy in his estimation of McMurray invertor power-loss; components for this dissipation are listed in his Table 1, but are not shown in the circuit or included in the component dissipation breakdown in his Table 4.

6 Conclusion

The revised switched reluctance drive design for rail traction applications of several hundred kilowatts indicates that on estimated material and component costs, the SR drive has a significant advantage, of between 25% and 30%, over the PWM induction motor, for a similar torque/speed envelope. The SR drive has fewer major components but more motor connections (5 compared to 3). From the viewpoint of complexity it is no worse, and from the viewpoint of efficiency it is expected to be better.

The background information for a valid comparison of the overall costs, including development, labour, assembly, testing, maintenance and other overheads, is not available, and therefore no quantitative conclusion on these other cost categories can be drawn by the authors.

The studies undertaken by the authors and by P.St.J.R. French of British Rail are, however, very preliminary, with no experimental verification. Much further investigation and development is required before a precise comparison of the merits of the two drives can be made for rail traction applications. Nonetheless, it does appear fair to conclude that the advantages which SR drives are proving in industrial applications, and which French points to, should also hold in traction applications. Moreover, the SR drive has very considerable scope for future development, much more so than the induction motor system, which has already largely approached its full potential performance.

7 References

1 FRENCH, P.St.J.R.: 'Switched reluctance motor drives for rail traction: a relative assessment', IEE Proc. B, Electr. Power Appl., 1984, 131, (5), pp. 209–219
2 DAVIS, R.M., RAY, W.F., and BLAKE, R.J.: 'Inverter drive for switched reluctance motor: circuits and component ratings', ibid., 1981, 128, (2), pp. 126–136
3 DAVIS, R.M., RAY, W.F., and BLAKE, R.J.: 'An inverter drive for a switched reluctance motor'. International Conference on Electrical Machines, Athens, Sept. 1980, pp. 411–417
4 AMATO, C.J.: 'Reluctance motor power circuit'. US Patent 3 560 817, filed Jan. 1969
5 LAWRENSON, P.J., RAY, W.F., DAVIS, R.M., STEPHENSON, J.M., FULTON, N.N., and BLAKE, R.J.: 'Controlled-speed switched-reluctance motors: present status and future potential'. Drives/Motors/Control '82, Leeds, 1982, pp. 23–31

SWITCHED RELUCTANCE MOTOR DRIVES FOR RAIL TRACTION: A SECOND VIEW

This correspondence is in reply to Paper 3322B [*IEE Proc. B, Electr. Power Appl.*, 1984, **131**, (5), pp. 220–225], itself a response to my original paper on switched reluctance motor drives for rail traction [A]. There is unfortunately a discrepancy in the data between Fig. 4 of my paper and the text. The curve showing the continuous rating of the induction motor is correctly labelled, but the current density in this case is 6.4 A/mm^2, not 5 A/mm^2 as stated in the text. Based on data from test results, the induction motor has an output of 10 kN at 5 A/mm^2 and full flux. An oversight in the commutation diode voltage rating is acknowledged. The rating of commutation diodes 1 and 2 should be 3.6 kV.

I note with interest the revision to the motor design which reduces the motor currents. This is the kind of system optimisation which is needed to obtain the best from the drive. It should be noted that the amount by which the motor turns can be increased to reduce the stall current is determined by the amount which the pulse width can be extended at base speed to maintain full flux in the machine. Changing the winding turns will not change the average current requirements at maximum power but will improve the form factor.

Where I must remain in disagreement with the authors is on the subject of thyristor rating. Table 2 in their paper implies that it is possible to achieve nearly 50% more from a load current switching device in a switched reluctance motor (SRM) drive than in an induction motor drive, four devices being credited with 337 kW in the former case, while six devices produce only 350 kW in the latter.

If this were possible, then we should all be developing SRM drives, now. Unfortunately, there are fundamental laws of net power flow and conduction period which do not permit such gains.

The 'fallacy' is not difficult to spot. The authors have based their revised rating on the reduction in the thyristor dissipation in the stall case, where one motor phase may be drawing current continuously. I did not use the stall case for rating the SRM drive, intending that this special case, which is severe as a comparison point, should be catered for in transient thermal ratings and with protection devices. It is clearly stated in the paper that the comparison point for the SRM drive is taken in the upper speed range. This is a more favourable way of rating the SRM drive, and probably a more realistic one. The reduction in stall current is very helpful, but it does not alter the fact that the same average current must flow at base speed for a given power, albeit with increased conduction angle. The revised power rating of the drive with the new design of motor should be recalculated for the base speed case. The rating is dependent on the waveform at this speed, but no details of this are given by the authors. It is not a valid comparison to label the new drive 337 kW, on the basis of the extrapolations made in their paper from the stall case.

The commutation circuit given for the SRM drive in my paper is perhaps a little heavy. This is partly due to the parallel inductor, which it is possible to omit, but is also due to the need to maintain commutation at a low line voltage of 550 V, with a suitable margin. To be fair, this criterion should have been stated in the paper. If the same criterion is applied to the design by the authors, then their commutation capacitor value would have to be somewhat larger.

Since the paper was written there have been developments not only in motor and power circuits but also in devices. The most important of these is the advent of the high power GTO thyristor. We should now consider the relative economics of the induction and SRM drives when the commutation components disappear and are replaced by the GTO thyristor snubber circuits. This needs to be the subject of a further study, because of the various circuit factors and new operating modes possible. However, early indications are that, again, the output ratings obtainable from a given power semiconductor switch will be of a very similar order in both drives.

25th July 1984   P.ST.-J.R. FRENCH

Railway Technical Centre
London Road
Derby DE2 8UP
England

We welcome very much P.St.-J.R. French's letter in response to our paper, in which we discussed and developed his comparison [A] between a switched reluctance (SR) motor drive and an induction motor scheme for traction. What he has to say, including his correction concerning his assumed current densities, considerably undermines the conclusions in his original paper. Moreover, it enables us to point more forcefully to the advantages of the SRM system.

Firstly, French acknowledges that his continuous rating curve for the induction motor was drawn for a current density of 6.4 A/mm^2 and not, as he stated in his paper, for the same density (5 A/mm^2) as the curve for the SR motor. He fails to point out, however, that the comparison of machine performance in Section 3.5 of his paper was based on the assumption of equal densities, and that this comparison is thereby radically changed. On the basis of equal current densities, our studies show the output of the non-optimised SR motor to be greater than for the induction motor over the whole speed range (Fig. 1 of our paper).

The other main point arising from French's letter concerns the claim which he makes there on the basis of his understanding of the drive rating and its dependence on the main thyristors. French argues that we have wrongly assessed this rating by extrapolating only the starting condition, and that we have overlooked the more limiting running condition. He is wrong. We have not extrapolated or overlooked anything; we computed the currents over the whole operating range to find the limiting condition. Our paper presented a revised SR drive with a more closely comparable performance to that of the induction

492

motor system. This revision allowed more turns to be used for the motor windings, which led to a higher drive rating for the given thyristor.

By way of elaboration, the relative improvement in drive rating for a given thyristor is virtually the same, regardless of whether the ratio of the stall currents or the ratio of the worst-case running currents is used as the basis for the uprating. While it is true that for a given power the average supply current is the same, irrespective of motor turns, this is not even approximately true of the thyristor current. The main thyristor takes current from the supply and the main diode returns current to the supply for each phase. Thus, for the H-configuration of the four phases, we have

$$I_{av.\ supply} = 2(I_{av.\ thyristor} - I_{av.\ diode})$$

The relative magnitudes of thyristor and diode average currents are profoundly influenced by the number of turns—which, incidentally, is not, as French asserts, set by base speed conditions—and an increase in the number of turns at a given speed and input power (current) can result in thyristor and diode average currents being equally reduced. We have not violated any fundamental laws; the relative improvement in drive rating is a complex function of reduced power at high speed, increased turns and revised control strategy. SR systems are rather more complex and challenging than French seems to realise.

In conclusion on this aspect, while we would hesitate to go along with the argument which French suggests for us that 'it is possible to achieve nearly 50% more from a load current switching device in an SR drive than in an induction motor drive'—even if only because we are not party to his calculations for the induction motor scheme—it is now clear from this discussion that SR drives should offer real advantages for rail traction, as well as other applications.

In his final paragraph, French refers to the future influence of GTO thyristors. We are in no doubt about the importance of these, and confidently expect them to increase, not decrease, the overall competitiveness of SR drives.

4th September 1984

W.F. RAY
R.M. DAVIS

Department of Electrical & Electronic Engineering
The University
Nottingham NG7 2RD, England

PROF. P.J. LAWRENSON
J.M. STEPHENSON

Department of Electrical Engineering
University of Leeds
Leeds LS2 9JT, England

N.N. FULTON
R.J. BLAKE

Switched Reluctance Drives Ltd.
Springfield House
Hyde Terrace
Leeds LS2 9LN, England

Reference

A FRENCH, P.ST.-J.R.: 'Switched reluctance motor drives for rail traction: relative assessment', *IEE Proc. B. Electr. Power Appl.*, 1984, **131**, (5), pp. 209–219

3492B

Additional References

Bass, J.T., Ehsani, M., and Miller, T.J.E. (1987). Simplified electronics for torque control of sensorless switched reluctance motor. IEEE Transactions, IE-34, 234-239

Bausch, H. and Rieke, B. (1978). Performance of thyristor-fed electric car reluctance machines. Proceedings of the International Conference on Electrical Machines, Brussels, E4/2.1-2.10

Bausch, H. and Rieke, B. (1976). Speed and torque control of thyristor-fed reluctance motors. Proceedings of the International Conference on Electrical Machines, Vienna, Part I, 128.1-128.10

Bose, B.K., Miller, T.J.E., Szczesny, P.M. and Bicknell, W.H. (1986). Microcomputer control of switched reluctance motor. IEEE Transactions, IA-22, 708-715

French, P. and Williams, A.H. (1967). A new electric propulsion motor. Proceedings AIAA Third Propulsion Joint Specialist Conference, Washington, D.C., July 1967

Harris, M.R. (1975). Static torque production in saturated doubly-salient machines. Proceedings IEE, 122, 1121-1127

Harris, M.R., Andjargholi, V., Lawrenson, P.J., Hughes, A., and Ertan, B. (1977). Unifying approach to the static torque of stepping motor structures. Proceedings IEE, 124, 1215-1224

Ilic-Spong, M., Miller, T.J.E., MacMinn, S.R. and Thorp, J.S. (1987). Instantaneous torque control of electric motor drives. IEEE Transactions, PE-2, 55-61

Jarret, J. (1976). Machines electriques a reluctance variable et a dents saturees. Tech. Mod. 2, 78-80

Koch, W.H. (1977). Thyristor controlled pulsating field reluctance motor system. Electric Machines and Electromechanics, 1, 201-215

Lang, J.H. and Vallese, F.J. (1985). Variable reluctance motor drives for electric vehicle propulsion. US D.o.E. Report DOE/CS-54209-26, May 1, 1985

Nasar, S.A. (1969). DC switched reluctance motor. Proceedings IEE, Vol. 116, No. 6, 1048-1049

Pollock, C. and Williams, B.W. (1987). An integrated approach to switched reluctance motor design. European Power Electronics Conference, Grenoble

Sugden, D.M. et al. (1987). Switched reluctance drives using MOSFETs. European Power Electronics Conference, Grenoble, 935-940